AF577356

A FIRST CIRCUITS COURSE FOR ENGINEERING TECHNOLOGY

Charles Belove
New York Institute of Technology

Holt, Rinehart and Winston

New York Chicago San Francisco
Philadelphia Montreal Toronto London
Sydney Tokyo Mexico City
Rio de Janeiro Madrid

To Golda, the world's greatest grandmother;
our children, Eric and Laura;
Jeannie and Michael;
and grandchildren, Colin and Daniel

Library of Congress Cataloging in Publication Data

Belove, Charles.
A first circuits course for engineering technology.

Includes index.
1. Electric circuits. I. Title
TK454.B3993 621.319'2 81-6350
ISBN 0-03-057851-5 AACR2

Address correspondence to:
383 Madison Avenue
New York, N. Y. 10017

Printed in the United States of America

1 2 3 4 144 9 8 7 6 5 4 3 2 1

CBS COLLEGE PUBLISHING
Holt, Rinehart and Winston
The Dryden Press
Saunders College Publishing

Contents

Preface

During the past two decades, the field of electrical technology has undergone tremendous changes. The advent of solid-state integrated-circuit devices culminating in the microcomputer has revolutionized the practice of technology. However, the fundamentals of electric circuit theory have not changed. In order for the technologist to be able to shake hands with the new technology, it is absolutely necessary that he or she be thoroughly familiar with basic principles. This thought represents the motivation for this book. It is intended to provide the student with a thorough grounding in basic electric circuit theory with emphasis on topics which will be most useful in the later study of the new technologies.

The title of the book, "A *First* Circuits Course for Engineering Technology," expresses the philosophy of the text. It is specifically designed as an introduction to electric circuit theory for beginning technology students. The approach is based on over 10 years of experience teaching the course. At New York Institute of Technology, it is given in the first two semesters of the 2-year AAS program in electrical technology and the 4-year BT program in electro-mechanical-computer technology. The required mathematics background includes high school algebra and trigonometry. Concurrently, a course in college algebra and trigonometry should be taken. The mathematical level throughout the text has been carefully planned to take account of these facts. Calculus is not used at all. Instead of the derivative d/dt, when necessary, developments are carried through using the average slope $\Delta/\Delta t$.

Since the handheld electronic calculator is a fact of life today, the text and examples assume that the student has one available. For the first half of the course, the standard arithmetic operations—addition, subtraction, multiplication, and division—are all that are required. Because most electrical components have tolerances of 10 or 20 percent, all calculator results are rounded to two or three significant figures. However, for the remainder of the course, the student will find that a scientific calculator with exponential and trigonometric functions along with rectangular-polar and polar-rectangular conversion is a necessity. At the time of this writing, such calculators are available for under $30.

The book begins with a chapter that introduces the student to engineering notation, units, accuracy and rounding, a review of algebra of the straight line, and other topics.

Subsequent chapters cover the concepts of current, voltage, power, and resistance, after which classical circuit theory is presented using the time-proven approach of introducing resistive circuits first for simplicity. A chapter on dc measurements provides practical applications of the theory.

Magnetism and magnetic circuits are covered next, followed by an introduction to inductance and capacitance. A chapter on charge and discharge in RL and RC circuits then introduces the student to the exponential and the concept of transient response. Following this, a number of chapters present ac circuit analysis using phasor methods. Three-phase systems and power calculations are covered in detail.

A chapter on frequency response presents the concepts of harmonics, ac transfer functions, and resonant circuits. This is followed by an introduction to two-port networks and a chapter on ac measurements.

At appropriate points in the text, circuit analysis using the ECAP program is presented. These sections are designed to stand alone, and may be omitted without loss of continuity if the instructor so desires.

Units, symbols, and notation carefully follow USA standards USAS Y10.5-1968 and USAS Y10.19-1967 (IEEE Nos. 280 and 260). A departure from past textbook practice appears in the notation used for phasors, in that boldface type is no longer used. Instead, both phasor quantities and dc quantities are represented by lightface capital italic letter symbols. The magnitudes of phasors are represented using vertical bars. Therefore, an equation written with lightface capital italic letter symbols may require either simple algebraic operations or vector operations. This will always be clear from the context.

The text contains more than enough material for a two-semester course meeting 4 hours per week. Various suggested semester assignments will be found in the instructors manual.

Each chapter begins with a set of objectives which detail what the student is expected to learn from the chapter. In the text that follows, technical material is presented and immediately followed by one or more examples. The author's experience in teaching this course has convinced him that technology students learn best when taught by example. The text strongly reflects this philosophy. Examples are phrased in practical terms, using readily available standard component values. In addition, examples of trouble-shooting are included where appropriate. Every section is followed by learning exercises designed to emphasize the main points of the section. These exercises are provided with numerical answers in scrambled order and without units. The student must unscramble the answers and provide proper units. Following the text material, a summary of the major points of the chapter is presented. A set of review questions which require word answers is followed by a set of problems which require calculations.

In summary, this text provides the student with a comprehensive, integrated treatment of the course material, and it supplies the instructor with all the pedagogical tools required to effectively present the subject.

The author would like to take this opportunity to thank all of those who helped in the preparation of this book. Special thanks go to Mrs. Carolyn Richards, whose excellent typing and editing made proofreading a pleasure, and to Dr. Edward Nelson for his painstaking review of the manuscript and careful preparation of the solutions manual. Also thanks are due the many anonymous reviewers whose suggestions were carefully considered and often used. Considerable help was cheerfully provided by Mrs. Eileen Jackson and students Gary Wieboldt, Sandra Lewis, Kathy Sholander, Cathy Eoviero, and Tillie Hsia. Finally, eternal gratitude is due my wife for her limitless patience, continuous encouragement, and time spent transcribing many of my notes.

Charles Belove

1

Introductory Topics and Algebra Fundamentals

OBJECTIVES

Upon completion of this chapter, the student should be able to

Section

1.1 1. State the most commonly used decimal prefixes.
1.2 2. Utilize scientific notation to effectively organize and complete numerical calculations with arbitrary numbers using a calculator.
3. State all numerical values in engineering notation.
1.3 4. Define variable and parameter.
1.4 5. Describe the International System of units and state the four basic quantities of the mksa system.
6. Perform calculations and conversions involving one system of units or several systems of units.
1.5 7. Define accuracy.
8. Round an arbitrary number to a specified number of significant figures.
1.6 9. Perform calculations as required for problems involving linear equations and graph the equations.

INTRODUCTION

In order to establish a firm foundation for our study of electrical technology, we first consider several aspects of the mathematical language used by workers in the field. Experience has indicated that there are a few topics in basic mathematics which consistently require review before more technical topics are considered. In this chapter, we consider these topics and provide examples and problems so that the student can acquire facility in these areas. Additional topics in mathematics are taken up in later chapters as they are required.

ORGANIZATION OF THE TEXT

Each chapter of the text is divided into sections. For ease of reference, all equations, examples, and learning exercises are numbered to correspond with the chapter and section to which they belong. For example, Eq. (2.4-3) is the third numbered equation in Sec. 4 of Chap. 2. In addition, the objectives at the beginning and the summary at the end of each chapter are keyed to the sections.

1.1 SCIENTIFIC AND ENGINEERING NOTATION

In all areas of technology, numerical calculations are required, and it is important that the student learn to carry them out quickly and accurately. Accordingly, we spend some time in this section providing hints on how to ensure correct calculations, as well as other appropriate topics. We note at the outset that most routine calculations today are made using pocket electronic calculators, and we assume that the student has one available.

Scientific Notation

In scientific work, the magnitudes of the quantities involved vary over wide ranges. In order to facilitate calculation when very small and very large numbers are involved, a scientific notation has been developed making use of powers of 10, so that numbers can all be expressed as a number of convenient magnitude (generally between 1 and 10) multiplied by a power of 10. The number which multiplies the power of 10 is called the *mantissa,** and the power to which 10 is raised is called the *exponent.*

For numbers which are multiples of 10, the power is the number of 10s which must be multiplied together to obtain the number. The zero power is defined such that anything to the zero power is 1. In addition, $10^{-n} = 1/10^n$. Some examples are

$$\begin{aligned} 1000 &= 10^3 \\ 100 &= 10^2 \\ 10 &= 10^1 \\ 1 &= 10^0 \\ 0.1 &= 10^{-1} \\ 0.01 &= 10^{-2} \\ 0.001 &= 10^{-3} \end{aligned}$$

Rules for Finding the Power of 10

In order to determine the proper power of 10, the following rules may be applied:

1. Decimals. Shift the decimal point to the right as desired and count the

*This name is also used to refer to the fractional part of the logarithm of a number. *Mantissa* is not being used in this way here.

number of places to the original point. The count is the proper negative power of 10.
Example:

0.000 01

$0.000\,01. = 1 \times 10^{-5}$ since the decimal point is shifted five places.

2. Large numbers. Shift the decimal point to the left and count the number of places to the original point. The count is the proper positive power of 10.
Example:

10 000

$1.0000. = 1 \times 10^{4}$ since the decimal point is shifted four places.

3. For numbers which are not simply multiples of 10, the same rules apply. In all cases we make the mantissa a number between 1 and 10.
Some examples:

$$0.004\,32 = 4.32 \times 10^{-3}$$
$$0.137 = 1.37 \times 10^{-1}$$
$$1260 = 1.26 \times 10^{3}$$
$$336\,000 = 3.36 \times 10^{5}$$

Laws of Exponents

The laws of exponents as applied to powers of 10 are particularly simple:
Multiplication:

$$10^{m} \times 10^{n} = 10^{m+n} \qquad (1.1\text{-}1a)$$

Division:

$$\frac{10^{m}}{10^{n}} = 10^{m-n} \qquad (1.1\text{-}1b)$$

Raising to a power:

$$(10^{m})^{n} = 10^{m \times n} \qquad (1.1\text{-}1c)$$

In all of these m and n may be any number. For cases of interest to us at this point, m and n will be integers.

EXAMPLES

$$(100)(10\ 000) = (10^2)(10^4) = 10^{(2+4)} = 10^6$$
$$(100)(0.000\ 01) = (10^2)(10^{-5}) = 10^{(2-5)} = 10^{-3}$$
$$\frac{100}{10\ 000} = \frac{10^2}{10^4} = 10^{(2-4)} = 10^{-2}$$
$$\frac{100}{0.000\ 01} = \frac{10^2}{10^{-5}} = 10^{[2-(-5)]} = 10^7$$
$$(100)^3 = (10^2)^3 = 10^{(2\times 3)} = 10^6$$
$$(0.001)^4 = (10^{-3})^4 = 10^{(-3\times 4)} = 10^{-12}$$

Decimal Prefixes

Certain powers of 10 arise so often that a system of prefixes has been devised to identify them so that they need not be written each time they are required. Those which we use most in electrical technology are listed in Table 1.1.

EXAMPLES

$$1\ 500\ 000 \text{ ohms} = 1.5 \text{ megohm} = 1.5\ \text{M}\Omega$$
$$68\ 000 \text{ ohms} = 68 \text{ kilohms} = 68\ \text{k}\Omega$$
$$0.000\ 033 \text{ farads} = 33 \text{ microfarads} = 33\ \mu\text{F}$$
$$0.002 \text{ henries} = 2 \text{ millihenries} = 2\ \text{mH}$$
$$4 \times 10^{-9} \text{ seconds} = 4 \text{ nanoseconds} = 4\ \text{ns}$$
$$6 \times 10^{-12} \text{ seconds} = 6 \text{ picoseconds} = 6\ \text{ps}$$

The first four examples are electrical quantities which will be studied in later chapters.

TABLE 1.1
Decimal prefixes

Power of 10	Prefix	Abbreviation
10^{-12}	pico	p
10^{-9}	nano	n
10^{-6}	micro	μ
10^{-3}	milli	m
10^{3}	kilo	k
10^{6}	mega	M
10^{9}	giga	G
10^{12}	tera	T

Engineering Notation*

This system differs from scientific notation in that it uses only the prefixes in Table 1.1, all of which are powers of 1000. The quantity is specified by a numerical value (mantissa) and a unit preceded by the appropriate prefix. The numerical value should lie between 0.1 and 1000. Some examples of scientific notation and the corresponding engineering notation are given below. The student should develop the habit of using engineering notation, as we do throughout this text.

Quantity (V)	Scientific Notation (V)	Engineering Notation
0.000 000 012	1.2×10^{-8}	12 nV
0.000 000 12	1.2×10^{-7}	0.12 μV or 120 nV
0.000 001 2	1.2×10^{-6}	1.2 μV
0.000 012	1.2×10^{-5}	12 μV
0.000 12	1.2×10^{-4}	0.12 mV or 120 μV
0.001 2	1.2×10^{-3}	1.2 mV
0.012	1.2×10^{-2}	12 mV
0.12	1.2×10^{-1}	0.12 V or 120 mV
1.2	1.2	1.2 V
12	1.2×10^{1}	12 V
120	1.2×10^{2}	0.12 kV or 120 V
1200	1.2×10^{3}	1.2 kV
12 000	1.2×10^{4}	12 kV
120 000	1.2×10^{5}	0.12 MV or 120 kV
1 200 000	1.2×10^{6}	1.2 MV

In those cases in which there is a choice, the prefix to be used is chosen to minimize the number of different prefixes in the problem solution or on the circuit diagram.

• • •

LEARNING EXERCISES FOR SEC. 1.1

These will be found at the end of most sections. They are designed to provide the student with the opportunity to test his or her understanding of the important points of the section. Following each learning exercise, the numerical values of the answers are given in scrambled order. The student should unscramble the answers and provide the proper units.

1. Express the following numbers in scientific notation.
   a. 0.000 679 b. 679 000 c. 0.679

Ans. 6.79×10^{5}; 6.79×10^{-1}; 6.79×10^{-4}

*"IEEE Standard Metric Practice," Institute of Electrical and Electronic Engineers, Inc., pp. 9–10, 1979.

2. Repeat Exercise 1 using engineering notation.

Ans. 0.679×10^{-3}; 0.679; 679×10^{3}

3. The numbers in Exercise 1 are in units of meters. Express them using the standard decimal prefixes.

Ans. 0.679 m; 0.679 mm; 679 km

• • •

1.2 ARITHMETIC OPERATIONS WITH ARBITRARY NUMBERS

Now we are ready to consider arithmetic operations with very large and very small numbers. Using scientific notation, the operations with the mantissas can be done separately from those involving the powers of 10. This makes the whole process easy to check and provides a means for adequate "bookkeeping."

EXAMPLE 1.2-1 Arithmetic Operations

Do the indicated operations in the following problems. Express all answers in engineering notation.

1. Addition:

$$\begin{aligned} 720 + 1340 &= 7.2 \times 10^2 + 13.4 \times 10^2 \\ &= (7.2 + 13.4) \times 10^2 \\ &= 20.6 \times 10^2 \\ &= 2.06 \times 10^3 \end{aligned}$$

Note that both numbers were converted so as to yield the same power of 10 and mantissas greater than 1. This must be done for addition and subtraction, since it is the only way we can factor out the power of 10.

$$\begin{aligned} 0.0036 + 0.000\,72 &= 36 \times 10^{-4} + 7.2 \times 10^{-4} \\ &= (36 + 7.2) \times 10^{-4} \\ &= 43.2 \times 10^{-4} \\ &= 4.32 \times 10^{-3} \end{aligned}$$

2. Subtraction:

$$\begin{aligned} 6700 - 72\,000 &= 6.7 \times 10^3 - 72 \times 10^3 \\ &= -65.3 \times 10^3 \end{aligned}$$

3. Multiplication:

$$
\begin{aligned}
6700 \times 800 &= (6.7 \times 10^3)(8 \times 10^2) \\
&= (6.7 \times 8)(10^3 \times 10^2) \\
&= 53.6 \times 10^5 \\
&= 5.36 \times 10^6
\end{aligned}
$$

Note that we do not have to use the same power of 10 for both numbers for multiplication. The power is chosen so that the mantissa will be convenient. These comments also apply to division.

$$
\begin{aligned}
(0.0013)(0.04) &= (1.3 \times 10^{-3})(4 \times 10^{-2}) \\
&= (1.3 \times 4)(10^{-3} \times 10^{-2}) \\
&= 5.2 \times 10^{-5} = 52 \times 10^{-6}
\end{aligned}
$$

4. Division:

$$
\begin{aligned}
\frac{6400}{80\,000} &= \frac{6.4 \times 10^3}{8 \times 10^4} \\
&= \frac{6.4}{8} \times \frac{10^3}{10^4} \\
&= 0.8 \times 10^{-1} \\
&= 80 \times 10^{-3}
\end{aligned}
$$

$$
\begin{aligned}
\frac{0.000\,36}{0.03} &= \frac{3.6 \times 10^{-4}}{3 \times 10^{-2}} \\
&= \frac{3.6}{3} \times \frac{10^{-4}}{10^{-2}} \\
&= 1.2 \times 10^{-2} = 12 \times 10^{-3}
\end{aligned}
$$

5. Powers:

$$(0.0002)^3 = (2 \times 10^{-4})^3$$

Since $(ab)^n = a^n b^n$, this becomes

$$
\begin{aligned}
(2 \times 10^{-4})^3 &= (2)^3 \times (10^{-4})^3 \\
&= 8 \times 10^{-12}
\end{aligned}
$$

• • •

Estimating the Answer

In the preceding example, the arithmetic was quite simple and no special means were required to arrive at the correct answer. In more complicated cases, it is a good idea to estimate the answer before carrying through the detailed calculation. For example, consider multiplying 45 by 24. This is almost the same as multiplying 40 by 25 which is 1000 ($40 = 4 \times 10$, so $40 \times$

$25 = 4 \times 25 \times 10 = 100 \times 10$). Thus the answer should be very close to 1000. The actual product is 1080. If the answer arrived at had been 10.8, we would have immediately suspected that a factor of 100 had been lost in the process and would have gone back to check.

In a more complicated case involving both multiplication and division, a good procedure to follow is:

1. Convert all numbers to scientific notation, with rounded mantissas.
2. Estimate the answer.
3. Use a pocket or desk calculator to find the exact answer and check it against the estimate.
4. Express the answer in engineering notation.

This procedure is illustrated in the example which follows.

EXAMPLE 1.2-2 Detailed Calculation

An investigation leads to the formula for x in meters.

$$x = \frac{(0.0012)(370)}{(534)(0.07)} \text{ m}$$

Find x.

Solution

1. Scientific notation:

$$x = \frac{(1.2)(10^{-3})(3.7)(10^{2})}{(5.34)(10^{2})(7)(10^{-2})}$$

$$= \frac{(1.2)(3.7)}{(5.34)(7)} \times 10^{[(-3+2)-(2-2)]}$$

2. Estimate answer:

$$x \approx \frac{(1)(4)}{(40)} \times 10^{-1} = 10^{-2} = 0.01$$

3. Exact answer (pocket calculator):

$$x = 0.011\ 878\ 009\ 63 \text{ m}$$

The number of digits in the exact answer from the calculator represents accuracy far beyond any we might normally require. As explained in Sec. 1.5, we "round off" the number to $x = 0.011\ 9$ m.

4. In engineering notation, the answer is

$$x = 11.9 \text{ mm}$$

• • •

(a)

(b)

(c)

FIGURE 1.1

Photographs of pocket calculators. (a) Shirt pocket scientific calculator (courtesy of Casio, Inc.). (b) Pocket scientific calculator (courtesy of Hewlett-Packard Company, Inc.). (c) Programmable pocket calculators with optional printer (courtesy of Texas Instruments Incorporated).

Electronic Pocket and Desk Calculators

As noted previously, portable electronic calculators are available in a wide range of prices, styles, and capabilities. Photographs of some of these are shown in Fig. 1.1. The less expensive models perform the operations of addition, subtraction, multiplication, and division. Most of them automatically keep track of the decimal point. The number of significant figures varies, in some models running as high as 12. In almost all cases the number of significant figures is adequate for our purposes and the user will round off the final answer as desired.

More expensive models provide readouts in scientific notation (sometimes this is called floating-decimal-point representation). Some models offer the user a choice of fixed or floating-point output. In addition to the four basic operations, some models perform other types of scientific calculations, such as square roots, squares, reciprocals, trigonometric, exponential, and logarithmic functions, and can store numbers for later recall and use.

The most expensive models contain memory banks for retaining several intermediate values in a calculation and have other advanced features.

An appropriate type for the technologist would be a model utilizing floating-decimal-point notation and providing trigonometric, exponential, and logarithmic functions in addition to the basic arithmetic operations. All of the calculations in the examples in this text are done on this type of calculator, and the results are rounded off as discussed in Sec. 1.5.

• • •

LEARNING EXERCISE FOR SEC. 1.2

1. Evaluate the following, giving the answers in engineering notation.

a. $\dfrac{(6.3)(40)}{(0.0072)}$ b. $\dfrac{(630)(0.004)}{(72)}$ c. $\dfrac{(0.006\,3)(0.04)}{(0.000\,072)}$

Ans. 3.5; 35×10^3; 35×10^{-3}

• • •

1.3 NOTATION

Throughout this book, we refer to physical quantities such as distance, voltage, and current. These quantities are *variables*, that is, quantities which can have more than one value, or may change with time. In addition to variables, we deal with *parameters*, which are constant, or fixed, numbers. In electrical technology, a typical parameter is *resistance*, which appears in circuit equations, along with the variables voltage and current.

For convenience, each of these quantities is represented by a single letter symbol when we write equations. For example, distance is often represented by the letter *x*, voltage by *v*, and current by *i*. As we see in subsequent chapters, variables can be constant in the sense that they do not vary with time. When this is the case we use *CAPITAL ITALIC* (slanted) letter symbols to represent the variables. Thus, *V* and *I* represent constant values of voltage and current. When the variables can change with time, we use *lower-case italic* letter symbols such as *v* and *i* to represent instantaneous values. Parameters are almost always represented by *CAPITAL ITALIC* letters.

In Sec. 1.4 we consider the units which must accompany each variable or parameter. In order to distinguish the letter symbol for the unit from the letter symbol for the variable or parameter, we use ROMAN (upright) letters for the unit symbols. Thus, V is the letter symbol abbreviation for volt, which is the unit of voltage, while A is the letter symbol for the ampere, the unit of current. Letter symbols and units are introduced as needed.

1.4 UNITS

The importance of a thorough understanding of units cannot be stressed enough. Most of the problems with which technologists are concerned result in an answer which is a *number*. In almost every case the number has associated with it a *unit*. In all cases the information concerning the unit is as important as the number which gives the size of the quantity involved. For example, if you were told that an athlete ran 1 in 4, you would not have the slightest idea what he had accomplished. On the other hand, if you were told that he ran 1 mile (mi) in 4 minutes (min) his accomplishment would be clear. A moment's thought will indicate that the frequent use of units in our daily lives is accomplished almost automatically. For example, if we purchase something at the store for 40 cents and give the storekeeper a $1 bill, we make use of our knowledge of the units of our money system in order to check the change. We know that the $1 bill is equivalent to 100 cents; thus our change should be 60 cents. Further, the change will undoubtedly be given to us in different coins. We must know the value of each of these in cents in order to check that we have received the correct amount, and we do all this with very little effort. Our aim is to develop the same facility with technical systems of units.

Systems of Units

In technical work, we must be able to describe whatever we do in such a way that anyone working in the same field will have no difficulty understanding exactly what we mean or in duplicating our results. In order to do this, we need a *system of units* that can be reproduced anywhere.

Today there are two principal systems of units in use. One of these is the English system used in our country. This is slowly being replaced by the

TABLE 1.2
Basic quantities

Quantity	Symbol	Unit	Abbreviation
Length	l	Meter	m
Mass	m	Kilogram	kg
Time	t	Second	s
Current	i	Ampere	A

International System of Units (abbreviated SI units), which includes the metric, or mksa, system used in electrical technology. The basic units of the mksa system are the meter (length), kilogram (mass), second (time), and ampere (current). The standards for these quantities have been established internationally and are the same all over the world.

The basic quantities, along with their SI units, usual mathematical symbols, and abbreviations are shown in Table 1.2. Other physical quantities, such as power, energy, and so on, are derived from the basic set. These derived quantities are introduced as needed throughout the text.

Before concluding this section, let us consider the relation between the magnitudes of some of the quantities involved in the English system and SI units. Since most of us are familiar with the magnitude of the various quantities in the English system, this may provide a means for obtaining some feeling for the relative sizes of some of the quantities used in the International System. A comparison of some often used quantities is made in Table 1.3.*

Conversion of Units

Because of the fact that different systems are in use, we often have to convert from one system to another. This can often be a confusing process prone to many errors unless careful "bookkeeping" methods are used. We illustrate the process in the following examples.

TABLE 1.3
Comparison of some English and SI units

	English	SI
Length	1 inch (in) = 0.0254 meter (m)	1 meter = 39.4 inches
Mass	1 slug = 14.6 kilograms (kg)	1 kilogram = 0.0685 slug
Force and weight	1 pound (lb) = 4.45 newtons (N)	1 newton = 0.224 pound
Energy	1 foot-pound (ft·lb) = 1.36 joules (J)	1 joule = 0.738 foot-pound

*A unit conversion table is in Appendix A.

EXAMPLE 1.4-1 Unit Conversion

a. Convert 1 foot (ft) to centimeters (cm).
b. Convert 9 ounces (oz) to newtons (N).
c. Convert 6 joules (J) to inch-pounds (in·lb).
d. Convert 60 miles per hour (mi/h) to kilometers per hour (km/h).

Solution

a. The basic idea to recognize here is that we can multiply any number by 1 without changing its value, so for this operation we use a ratio of units that reduces to 1. For this example, 12 in = 1 ft and 2.54 cm = 1 in, so that

$$\frac{12 \text{ in}}{1 \text{ ft}} = 1 \quad \text{and} \quad \frac{2.54 \text{ cm}}{1 \text{ in}} = 1$$

The actual conversion is then written out as follows:

$$(1 \cancel{\text{ft}})\left(\frac{12 \cancel{\text{in}}}{1 \cancel{\text{ft}}}\right)\left(\frac{2.54 \text{ cm}}{1 \cancel{\text{in}}}\right) = 30.5 \text{ cm}$$

Note how the dimensions ft and in cancel out as indicated by the lines through them. Also note that any of our ratios of units can be inverted without changing its value. This is illustrated in the next example.

b. For this conversion, we use

$$\frac{16 \text{ oz}}{1 \text{ lb}} = 1 \quad \text{and} \quad \frac{1 \text{ lb}}{4.45 \text{ N}} = 1$$

Then

$$(9 \cancel{\text{oz}})\left(\frac{1 \cancel{\text{lb}}}{16 \cancel{\text{oz}}}\right)\left(\frac{4.45 \text{ N}}{1 \cancel{\text{lb}}}\right) = 2.5 \text{ N}$$

Again, observe the manner in which this is written out so that all factors are accounted for and all unit cancellations are clearly indicated. This kind of careful bookkeeping will help to avoid all kinds of unnecessary errors.

c. $$(6 \text{ J})\left(\frac{1 \text{ ft}\cdot\text{lb}}{1.356 \text{ J}}\right)\left(\frac{12 \text{ in}}{1 \text{ ft}}\right) = 53.1 \text{ in}\cdot\text{lb}$$

d. $$\left(60 \frac{\text{mi}}{\text{h}}\right)\left(\frac{5280 \text{ ft}}{1 \text{ mi}}\right)\left(\frac{12 \text{ in}}{1 \text{ ft}}\right)\left(\frac{0.0254 \text{ m}}{1 \text{ in}}\right)\left(\frac{1 \text{ km}}{1000 \text{ m}}\right) = 96.6 \frac{\text{km}}{\text{h}}$$

The student should practice this type of conversion until he or she is quite familiar with it.

• • •

The next example illustrates conversions within one system of units.

EXAMPLE 1.4-2 More Unit Conversion

a. Convert 1200 grams (g) to kilograms.
b. Convert 62×10^8 cm to meters.
c. Convert 0.076 h to microseconds (μs).
d. Convert 0.016 microfarads (μF) to picofarads (pF).

Solutions

As in the previous example, we multiply by 1 using the form of a ratio of units.

a. $$(1200\text{ g})\left(\frac{1\text{ kg}}{1000\text{ g}}\right) = 1.2\text{ kg}$$

b. $$(62 \times 10^8\text{ cm})\left(\frac{1\text{ m}}{100\text{ cm}}\right) = 62 \times 10^6\text{ m}$$

c. $$(0.076\text{ h})\left(\frac{60\text{ min}}{1\text{ h}}\right)\left(\frac{60\text{ s}}{1\text{ min}}\right)\left(\frac{10^6\ \mu\text{s}}{\text{s}}\right)$$

$$= (7.6)(10^{-2})(6)(10^1)(6)(10^1)(10^6) = 274 \times 10^6\ \mu\text{s}$$

Note the use of scientific notation in this example to avoid decimal point errors.

d. $$(0.016\ \mu\text{F})\left(\frac{10^6\text{ pF}}{1\ \mu\text{F}}\right) = 0.016 \times 10^6\text{ pF} = 16\,000\text{ pF}$$

For conversions within the metric system, powers of 10 often have to be juggled around. The type of bookkeeping illustrated in this example will ensure that they are handled properly.

• • •

Checking with Units

Every equation we write must have the property of dimensional homogeneity. In simpler language, this means that all terms which are added in the equation must have exactly the same units. This fact is used in two ways. First, to check calculations, and, second, to find the units of newly defined quantities. The student should get into the habit of checking units every time an equation is written.

For example, suppose an equation works out to be

$$R_T = \frac{R_1}{R_1 + R_2}$$

where R represents electrical resistance in ohms, for which the unit symbol is the Greek capital letter omega (Ω). Upon checking the dimensions of each side we immediately realize that an error has been made, since the left side has dimensions of ohms while the right is dimensionless (Ω/Ω). What probably occurred is that a resistance factor multiplying the right side was lost somewhere along the line. A dimensionally correct formula would be

$$R_T = \frac{R_1 R_2}{R_1 + R_2}$$

Here the right side has dimensions of $(\Omega)^2/\Omega = \Omega$ as required.

It is important to note that correct dimensions *do not* guarantee that the equation is correct! In mathematical terms, correct dimensionality is a necessary, but not sufficient, condition that an equation be correct.

Dimensionality can often be used to jog the memory in order to recall an equation. For example, consider remembering the formula for velocity in terms of time and distance. We might initially remember it as $v = d \times t$. In terms of units, this says that velocity has the units distance × time, which is incorrect. The only way that distance and time can be combined to yield the correct units is as the ratio d/t (for example, miles/hour) so that $v = d/t$ is the correct formula.

A further point to note again is that both sides must have the same units. If v is desired in meters per second then d must be in meters (or if in another unit it must be converted to meters) and t must be in seconds. Although this may appear obvious and extremely simplistic, experience has indicated that students often make such mistakes. Correct application of the dimensionality rule will almost always detect such errors.

• • •

LEARNING EXERCISES FOR SEC. 1.4

1. Convert the following to meters.
   a. 1 ft b. 1 yard (yd) c. 1 mi

Ans. 1610 m; 0.305 m; 0.915 m

2. Convert 0.000 032 h to
   a. seconds b. milliseconds c. microseconds

Ans. 115.21; 0.115 2; 115 200

• • •

1.5 ACCURACY AND ROUNDING

In this section, we consider certain aspects of the representation of data. Accuracy and the accepted conventions for conveying significant figures are discussed, along with techniques for the proper rounding of numbers.

Accuracy

When dealing with measurements, *accuracy* is defined as closeness to the truth. We are concerned with accuracy whenever we make a measurement. For example, suppose we order an item of furniture that appears as though it may just fit through our front door. If the fit is close enough we will have to be concerned with the accuracy of the ruler we use to measure the width of our door and the accuracy of the dimensions of the furniture as supplied by the manufacturer.

It is easy to envision a situation which would result in disaster. For example, if the quoted furniture dimension of interest is 36 inches and our ruler indicates that the door measures 36.5 inches we may feel that there is sufficient clearance and that the delivery will proceed smoothly. However, if our ruler is in error by 1 inch, so that the true door measurement is 35.5 inches while the furniture dimension of 36 inches is accurate, the delivery will not proceed smoothly, and in fact the furniture will not go through the door. From this example it is clear that we must know the accuracy of our measuring equipment.

In the preceding example, if we knew in advance that our ruler had *accuracy limits* of 36.5 inches ± 1 inch we would be certain that the true value of the door measurement of 36.5 inches actually lies somewhere between 35.5 and 37.5 inches. We could then anticipate possible trouble. The only way that we can come any closer to the true value is to use a more accurate ruler.

These ideas are considered further in Chap. 6, where we consider electrical measurements.

Another important factor is the mathematical accuracy of the calculations we make using the results of measurements. In such cases we often make computations in which the mathematical operations are carried to several unnecessary decimal places. This extra work gives a false impression of the precision or degree of fineness of the measurement. For most of our work, instruments are generally not correct to more than three significant figures, and the values marked on components are frequently correct to only two or even one significant figure.

Significant Figures

In technical work, the accuracy of a number representing a measurement is expressed by the number of *significant figures* used. The number of significant figures is the number of digits in the number, excluding leading zeros. Leading zeros are required in order to show the proper location of the decimal point. Some examples are

378 is correct to *three* significant figures
0.02 is correct to *one* significant figure

2500 is correct to *four* significant figures
0.002 50 is correct to *three* significant figures

The last digit in each of these numbers is understood to have an uncertainty of ±1/2 unit. Thus in the first example the actual number is between 377.5 and 378.5. In the second example the actual number is between 0.015 and 0.025, in the third between 2499.5 and 2500.5, and in the fourth between 0.002 495 and 0.002 505.

Rounding

Since the precision of a calculated result cannot be greater than the precision of any of the numbers entering into the calculation, the result of all computations should be *rounded off* accordingly. The following rules should be applied:

1. If the number *dropped* is *greater* than 5, the last figure *kept* is increased by 1. Thus 43.52 and 43.7 rounded to two significant figures are both 44.
2. If the number dropped is *less* than 5, the last figure kept is not changed. Thus 43.42 and 43.2 rounded to two significant figures are both 43.
3. If the number dropped is *exactly* 5, add 1 to the last figure kept if it will make it even, otherwise do not change the last figure kept. Thus 43.5 rounded to two significant figures is 44, while 42.5 becomes 42.

In most applications, powers of 10 are used so that no zeros will appear beyond the uncertain digit. For example, if a resistor has been measured as 375 000 Ω, correct to three significant figures, it can be written several ways, that is,

375 kΩ
0.375 MΩ
375×10^3 Ω

When the resistance is specified this way, there is no ambiguity in interpreting its value and the precision with which it is known. In this case it is understood to lie between 374.5 and 375.5 kΩ. Excluding values of precision components and precision instrument readings, most of the numbers with which we deal are not known to this accuracy.

Calculations done with pocket electronic calculators lead to answers containing from 5 to 12 significant digits. However, we must remember that the numbers which go into these calculations are seldom known to more than two or three significant digits. Thus if we write down the answer exactly as read from the calculator, the reader will have a false impression of the accuracy of our result. In general, the answer on the calculator should be rounded off so that the number of significant figures in the result is no more than the number of significant figures in the numbers entering into the calculation. We will make it a rule that all results are to be rounded to, at most, three significant digits since this is sufficient for our purposes.

• • •

LEARNING EXERCISE FOR SEC. 1.5

1. Round the following numbers to three significant digits and express the results in engineering notation.
   a. 486 372 b. 0.004 863 c. 0.485 5

Ans. 4.86×10^{-3}; 486×10^{3}; 0.486

• • •

1.6 ALGEBRA REVIEW

The only mathematics required for the first half of this text is algebra. In the second half, trigonometry is used along with various other topics, which are introduced when needed. In this section we review proportion and linear equations, topics from algebra which are extremely important in what is to follow.

Proportion

The most used equation in electrical technology is Ohm's law, which relates current and voltage in a resistance. At this point we are not concerned with the law's physical interpretation, but rather with the mathematics. The law involves two variables, V and I, and one parameter, R. Expressed as an equation, it is

$$V = RI \tag{1.6-1}$$

where

V = dependent variable, voltage in volts, V
I = independent variable, current in amperes, A
R = parameter, resistance in ohms, Ω

This equation represents a *proportion*, in which the variable I is multiplied by the parameter R to give the value of the variable V. We describe this in words by saying that "V is proportional to I, with R the constant of proportionality."

We use the form of the equation above when R and I are known and we are required to find the unknown value of V. Very often I or R is unknown. When this is the case, it is convenient to rearrange Eq. (1.6-1) in order to isolate the unknown on the left side of the equation. For example, suppose V and R are given, and we are required to find the unknown value of I. We write the equation in the form

$$RI = V$$

and then, noting that we wish to isolate the variable I, we divide both sides of the equation by R so that I will stand alone on the left side. This yields

$$\frac{\cancel{R}I}{\cancel{R}} = \frac{V}{R}$$

$$I = \frac{V}{R} \qquad (1.6\text{-}2)$$

If, instead of dividing by R, we divide by I, the result is

$$\frac{R\cancel{I}}{\cancel{I}} = \frac{V}{I}$$

$$R = \frac{V}{I} \qquad (1.6\text{-}3)$$

In each of the equations above, we have isolated the unknown quantity on the left-hand side, and arranged all the known quantities on the right-hand side. Thus, known values are easily substituted in the right-hand side to yield a direct answer.

This technique of isolating the unknown on the left-hand side should be used whenever similar situations arise.

EXAMPLE 1.6-1 *Working with a Proportion*

A variable y is proportional to a variable x, with a constant of proportionality m. It is known that $y = 12$ when $x = 3$. Find the value of y when $x = 16$.

Solution

The statement of the problem tells us that

$$y = mx \qquad (1.6\text{-}4)$$

The known data can be used to find m so we manipulate Eq. (1.6-4) in order to isolate m. This yields

$$m = \frac{y}{x} \qquad (1.6\text{-}5)$$

Substituting the given values

$$m = \frac{12}{3} = 4$$

Thus the proportion is

$$y = 4x \tag{1.6-6}$$

and when $x = 16$

$$y = (4)(16) = 64 \tag{1.6-7}$$

• • •

Graphical Interpretation

Let us return to the form

$$I = \frac{V}{R} \tag{1.6-2}$$

and plot it on I, V coordinates using the parameter value $R = 5$. Since the proportion is a linear (straight line) relationship we only need two points, or one point and the slope of the line, in order to plot the graph. We will find two points. The easiest point to find is where $V = 0$; at this point $I = 0$. Any convenient value of V will serve for the second point; we choose $V = 10$, at which point $I = 10/5 = 2$. These are summarized in this data table:

Point	V	I
P_1	0	0
P_2	10	2

In the resulting graph shown in Fig. 1.2, voltage V is the independent variable (plotted on the x axis) and current I (plotted on the y axis) is the dependent variable. Graphs such as these are used frequently when we begin our study of circuits in Chap. 4.

Linear Equations

A large percentage of the mathematics with which we are concerned involves linear equations (like the proportion), which graph as straight lines. Equations that graph as curves are called nonlinear equations. The slope-

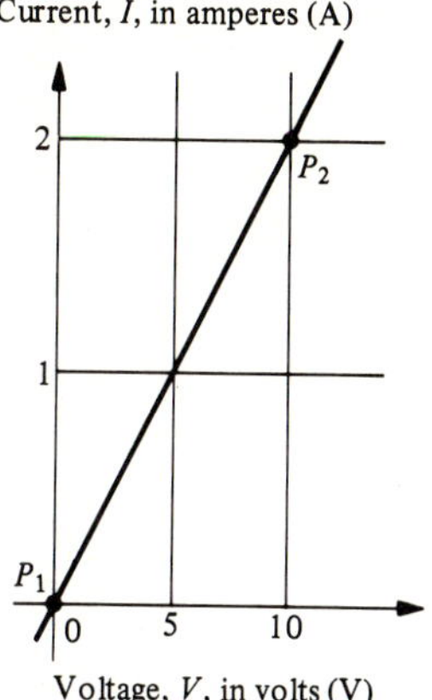

FIGURE 1.2
Graph of $I = V/R$ with $R = 5$.

intercept form of the straight line equation is

$$\boxed{y = mx + b} \tag{1.6-8}$$

where

y = dependent variable
x = independent variable
m = slope of the straight line
b = value of the y intercept

In order to illustrate the significance of the various terms, we consider a numerical example in which $m = 2$ and $b = 3$:

$$y = 2x + 3 \tag{1.6-9}$$

The graph of this equation, shown in Fig. 1.3, is easily constructed from the slope-intercept form by following four simple steps.

1. Set $x = 0$. Then $y = b = 3$. This gives the coordinate on the y axis.
2. Set $y = 0$. Then $x = -b/m = -3/2$. This gives the coordinate on the x axis.
3. Draw a straight line through the two points.

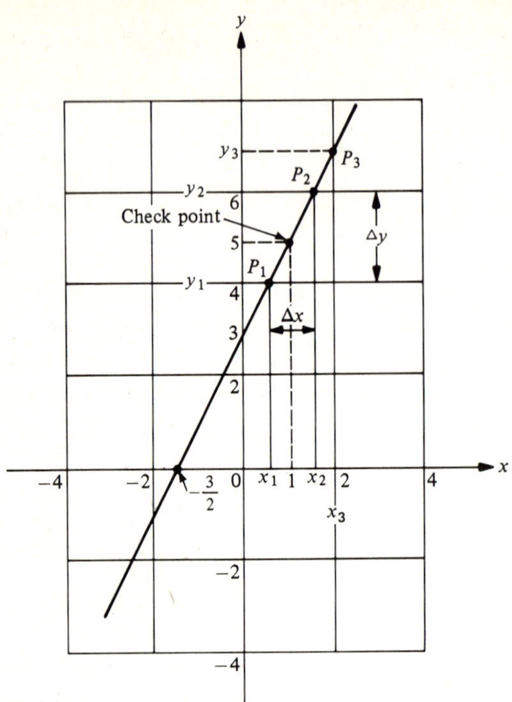

FIGURE 1.3
Graph of $y = 2x + 3$.

4. As a check, choose a convenient value for x (say x = 1), find y (y = 5), and see that this third point lies on the same straight line (it does).

Let us first note that in the graph the point where the line cuts the y axis is at x = 0, y = b. This point is determined by inspection when the equation is in slope-intercept form.

Second, note the slope of the line. In order to get at the significance of the slope, consider a change (increment) in x between points P_1 and P_2. There is a corresponding increment in y in the positive y direction. The x increment is called Δx where Δ (delta) means "the change in." From the diagram

$$\Delta x = x_2 - x_1 \tag{1.6-10}$$

The y increment corresponding to this Δx is

$$\Delta y = y_2 - y_1 \tag{1.6-11}$$

By definition, the slope of the line is the ratio of the change in y between points P_1 and P_2 to the corresponding change in x between these same two points. Thus

$$\text{Slope} = \frac{\Delta y}{\Delta x} = \frac{y_2 - y_1}{x_2 - x_1} = m \tag{1.6-12}$$

The change in y is often called the "rise," since it is the distance that y *rises* vertically when we move from point 1 to point 2. Also, the change in x is called the "run," since it is the distance that x *runs* horizontally when we move from point 1 to point 2. Thus, a convenient way to remember the formula for slope is that it is *the rise over the run*. For our example,

$$m = \frac{y_2 - y_1}{x_2 - x_1} = \frac{6 - 4}{1.5 - 0.5} = 2 \tag{1.6-13}$$

This, of course, checks the value of the number which multiplies x in Eq. (1.6-9). (This number is called the *coefficient* of x.) Since we are considering a straight line, the slope will be the same anywhere along the line. The slope will also be the same for any size Δx, as long as we remember to take the corresponding value of Δy. Thus, for a straight line, the slope is constant. Also note that the magnitude of the slope is a measure of the *steepness* of the line. For example, a line with a slope of 4 units would be twice as steep.

It is important to recognize the physical significance of the slope. It represents the *rate of change* of y with respect to x, that is, *the number of units change in y for a one-unit change in x*. If we know the slope of a straight line, we can predict, for any point on the line, the change in y for *any* change in x.

Let us return to our example given in Eq. (1.6-9). Consider that we are starting at point P_1 in Fig. 1.3 and wish to determine the change in y if x changes by 1.5 units. Since the *rate of change* of y with respect to x is +2, all we need do is multiply the change in x (1.5 units) by the *rate of change* (+2) to find the corresponding change in y (+3 units). Note that we are not finding y, but we are finding the *change* in y, Δy. Thus the actual value of y at our *new point* P_3 is the value at P_1 ($y_1 = 4$) plus the change ($\Delta y = 3$) which gives the final value $y_3 = 7$.

In order to express this mathematically, we proceed as follows:

At point P_3:

$$y_3 = mx_3 + b \tag{1.6-14}$$

At point P_1:

$$y_1 = mx_1 + b \tag{1.6-15}$$

Subtracting,

$$y_3 - y_1 = m(x_3 - x_1)$$

Add y_1 to both sides:

$$y_3 = y_1 + m(x_3 - x_1) \tag{1.6-16}$$

$$= y_1 + m\,\Delta x$$

Substituting the values $y_1 = 4$, $m = 2$, $\Delta x = 1.5$, we get

$$y_3 = 4 + (2)(1.5) = 7$$

The concept of slope as rate of change is an extremely important one, and we will be making considerable use of it as we proceed. In the preceding example, the slope is positive, causing the straight line to rise vertically as one moves along it from left to right. The calculations involved are very simple. In the next example, we consider a negative slope, which sometimes causes difficulty.

EXAMPLE 1.6-2 *Operations with a Linear Equation*

Consider the equation

$$16x + 5y = 39$$

a. Plot the equation.
b. Use the slope concept to find the value of y when $x = -2$.

Solution

a. The first step is to write the equation in slope-intercept form.

$$5y = -16x + 39$$

$$y = -\frac{16}{5}x + \frac{39}{5}$$

$$= -3.2x + 7.8$$

Following the steps set down previously we have

1. With $x = 0$, $y = 7.8$.
2. With $y = 0$, $x = 2.44$.

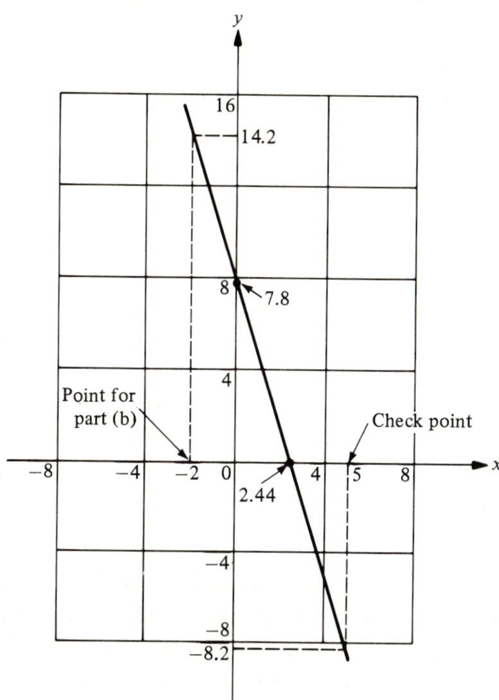

FIGURE 1.4
Graph for Example 1.6-2.

3. The graph is shown in Fig. 1.4.
4. As a check take $x = 5$, then $y = (-3.2)(5) + 7.8 = -8.2$

b. Since the slope is -3.2, y *decreases* 3.2 units for each unit *increase* in x. If we use $x = 0$, $y = 7.8$ as our starting point, then x *decreases* by 2 units to the specified point. Thus the change in y is an *increase* of $2 \times 3.2 = 6.4$ units and the new value of y is $7.8 + 6.4 = 14.2$. To check this mathematically we use Eq. (1.6-16).

$$\begin{aligned} y_2 &= y_1 + m\,\Delta x \\ &= 7.8 + (-3.2)(-2) \\ &= 7.8 + 6.4 \\ &= 14.2 \end{aligned}$$

• • •

LEARNING EXERCISES FOR SEC. 1.6

1. Find the slope of the following straight lines.
   a. $y = 6x + 3$ b. $3y - 0.18x = 0.9$ c. $120x = 0.6 - 0.02y$

Ans. −6000; 6; 0.06

2. Find the *y* intercepts for each of the straight lines of Exercise 1.

Ans. 3; 0.3; 30

• • •

Summary

In this chapter we have considered certain topics which arise frequently in electrical technology. It is essential that they be understood before we proceed to Chap. 2, where we begin our study of electricity. The important points concerning each topic are summarized below.

Section

1.1 1. The most important decimal prefixes are: mega (M) = 10^6, kilo (k) = 10^3, milli (m) = 10^{-3}, micro (μ) = 10^{-6}, nano (n) = 10^{-9}, pico (p) = 10^{-12}.

1.2 2. A typical number, such as 120 000, is represented in scientific notation as 1.2×10^5; this number in engineering notation is 120 k.

3. The laws of exponents are

$$10^m \times 10^n = 10^{m+n}$$

$$\frac{10^m}{10^n} = 10^{m-n}$$

$$(10^m)^n = 10^{m \times n}$$

1.3 4. A variable is readily changed, while a parameter remains constant.

1.4 5. The basic units of the mksa system are the meter (m), kilogram (kg), second (s), and ampere (A).

6. When performing unit conversions, multiply by 1 in a form such as (12 in)/(1 ft) or (60 s)/(1 min).

7. When dealing with equations describing any physical phenomena, the dimensions of each term must be the same.

1.5 8. Accuracy is closeness to the truth. It is often expressed in terms of *accuracy limits.* As an example, 12 in ±1 in is known to lie between 11 and 13 in.

9. In a number expressed in terms of significant figures, the last digit has an uncertainty of ±½ unit.

10. To round off a number increase (decrease) by 1 the last digit kept if the number dropped is greater than (less than) 5. If the number dropped is exactly 5, add 1 to the last figure kept if it will make it even, otherwise leave it alone.

11. In all calculations, the result should be rounded off to no more than three significant digits.

1.6 12. The equation for a straight line is

$$y = mx + b$$

where m = slope and b = y intercept.

13. The increment Δx is used to specify a *change* in x, that is,

$$\Delta x = x_2 - x_1$$

QUESTIONS FOR REVIEW

Sec. 1.1

1. Describe scientific notation as used in expressing numerical quantities.
2. State the laws of exponents for multiplication, division, and raising to a power.
3. List the important decimal prefixes.

Sec. 1.2

4. Describe the use of scientific notation in performing calculations involving very large and very small numbers.
5. Describe engineering notation.
6. How does engineering notation differ from scientific notation?

Sec. 1.3

7. What is a variable? What is a parameter?
8. To what do the letters SI refer?
9. A quantity is symbolized as capital italic *V*. What does this signify?
10. What is the significance of a lower-case italic *i*?
11. What is the significance of a roman capital A?

Sec. 1.4

12. Why must the unit always be specified with a number?
13. Name the basic quantities and their units in the mksa system.
14. How does one go about converting units from one system to another?
15. An equation consists of several terms added together. What must be true of the units of each term?

Sec. 1.5

16. What are "accuracy limits"?
17. What is meant by the phrase "the number of significant digits"?
18. Describe how you would "round off" a calculator reading to three significant digits.

Sec. 1.6

19. Write a typical equation which represents proportion. In the equation, distinguish among the variables and the parameters.
20. Describe a convenient way to graph a proportion.

21. What is the slope-intercept form of the straight-line equation? Define all of the variables and parameters in the equation.
22. What is the significance of the slope of a straight line?

PROBLEMS (Answers to selected odd-numbered problems are found at the back of this book.)

Sec. 1.1

1. Express the following as powers of 10.
   a. 1 000 000 d. 0.000 000 1
   b. 0.000 01 e. 10 000 000
   c. 10 000 f. 0.01
2. Express the following as a number between 1 and 10 times a power of 10.
   a. 1 365 000 d. 0.03
   b. 0.000 012 73 e. 24 300 000
   c. 14 000 f. 0.016 400
3. Reduce all of the following to a single power of 10.
   a. $(100)(1000)$
   b. $(0.01)(0.0001)$
   c. $(0.001)(10\,000)$
   d. $(0.001)(10^6)$
   e. $\dfrac{100}{10\,000}$
   f. $\dfrac{100}{0.0001}$
   g. $\dfrac{0.01}{10^6}$
   h. $\dfrac{100}{10^8}$
4. Do the following calculations. All results are to be in terms of powers of 10.
   a. $(10\,000)^3$
   b. $(0.0001)^{1/2}$
   c. $(1\,000\,000)^{1/3}$
   d. $\dfrac{(0.01)^2(1000)^3}{0.001}$
   e. $\dfrac{[(100)(0.001)]^4}{(10\,000)^{-2}}$
   f. $\dfrac{(0.0001)^{-3}}{10^6}$
5. Do the following calculations. All results are to be a number between 1 and 10 times a power of 10.
   a. $10 + 100$ d. $0.000\,10 - 0.000\,001$
   b. $0.01 + 0.1$ e. $1\,000\,000 + 1000$
   c. $10\,000 - 1000$ f. $0.01 - 0.1$
6. Repeat Prob. 5 for the following:
   a. $10^2 + 10^3$ d. $10^{-1} - 10^{-3}$
   b. $10^{-2} + 10^{-5}$ e. $10^5 - 10^6$
   c. $10^4 + 10^6$ f. $10^{-4} - 10^{-2}$
7. In Probs. 5 and 6 express all results in engineering notation. Assume that all numbers represent amperes (A).
8. Express the following in engineering notation.
   a. 0.0001 F c. 10 000 Ω e. 0.004 27 N
   b. 0.01 H d. 637 000 g f. 1 362 000 in

Sec. 1.2

9. Do the following calculations. All results are to be a number between 1 and 10 times a power of 10. Estimate all answers in advance.

a. $\dfrac{(62)(3.47)}{(0.012)(23\,000)}$ b. $\dfrac{(1.6)^2(24)^{1/2}}{12}$

10. Repeat Prob. 9 for the following:

a. $\dfrac{(3700)(0.046)}{(82)(6\,720\,000)}$ c. $\dfrac{(7.3)(630)}{(42)}$

b. $\dfrac{(7.92)(0.73)^2}{(42)^3}$ d. $\dfrac{(1.3)(0.062)}{(70)(8.3)}$

Sec. 1.4

11. Perform the following conversions.
    - a. 3 s to microseconds
    - b. 0.01 F to microfarads
    - c. 1.7 min to seconds
    - d. 3.2 h to seconds
    - e. 0.000 006 3 s to nanoseconds
    - f. 0.000 006 3 s to microseconds
12. Convert the following:
    - a. 732 000 Ω to megohms
    - b. 0.016 μF to nanofarads
    - c. 78 mH to henrys
    - d. 1 m^2 to square centimeters
    - e. 1.320 cm^2 to square meters
13. Convert the following:
    - a. 30 mi/h to kilometers per hour
    - b. 20 ft to meters
    - c. 1 mi to meters
    - d. 7.3 lb to newtons
    - e. 6 in to centimeters
    - f. 12 ft·lb to joules
    - g. 18 slugs to kilograms
14. Check each of the terms below to determine if the final dimension is length.
    - a. 6 m/20 mi
    - b. (3 mi/h) (20 s)
    - c. (67 cm) (20 mi/h)
    - d. ($80\ m^2$) ($6\ cm^{-1}$)
    - e. (7 cm/s)/20 s
    - f. (20 mi) (2 μs)
15. Check each of the equations below for dimensional homogeneity. Units are t in seconds, v in centimeters per second, x in centimeters.

a. $t_4 = \dfrac{t_1 t_2}{t_1 + t_2}$ c. $v_3 = \dfrac{v_1}{v_0 - v_5}$

b. $x_4 = v_3 t_3 + x_0$ d. $t_7 = \dfrac{v_7}{x_7}$

Sec. 1.5

16. Given the following numbers, to how many significant figures is each number correct? Give the range of the actual number for each case.
    - a. 62
    - b. 62.0
    - c. 62.00
    - d. 0.0132
    - e. 0.013 20
    - f 0.013 200
17. Round off the following to the indicated number of significant figures.
    - a. 376 000 to two figures
    - b. 0.004 54 to two figures
    - c. 1235 to three figures
    - d. 1245 to three figures
    - e. 0.007 65 to one figure
    - f. 0.17 to one figure

Sec. 1.6

18. When an automobile travels at constant velocity V, the distance traversed S is directly proportional to the time t. The constant of proportionality is V. Write this as an equation in three different forms, isolating a different variable or parameter on the left side of each one.
19. In Prob. 18 the automobile has been traveling for 45 min at constant velocity and has covered 30 mi.
    a. If it continues at the same velocity, how far will it go in 3.6 h?
    b. Plot a graph of distance traveled vs. time.
20. Plot the following equations on the same set of axes.
    a. $x + 2y = 0$ b. $x + 2y = 5$ c. $x + 2y = 10$
    Comment on the relationship among the three graphs.
21. Graph the following equations. y and i are dependent variables.
    a. $y = 6x + 5$
    b. $y = -20x + 11$
    c. $i = 100v + 10^3$
    d. $v + 2i = 7$
    e. $y - 6x = 24$
22. For each of the equations of Prob. 21 find m, b, and the x or v intercept.
23. A straight line passes through (3, 4) and y increases 3 units for each 6-unit increase in x. Find the equation of the line.
24. A straight line passes through (10, 20) and x decreases 5 units for each 10-unit increase in y. Find the equation of the line.
25. Given the following measured values of V and I, plot the points, draw a straight line for best fit through them, and determine the equation relating V to I by finding m and b graphically.

V	−4	−2	2	6	10
I	17	11	−5	−22	−38

2 Current, Voltage, and Power

OBJECTIVES

Upon completion of this chapter, the student should be able to

Section

Section		
2.1	1.	Describe the function of the electric circuit and name the fundamental electrical quantities.
2.2	2.	Describe the process that sets charges in motion.
	3.	Describe the property of atomic structure that makes a given material a conductor or an insulator.
2.3	4.	State the definition of current in terms of charge flow.
	5.	Define the positive direction for current flow in terms of charge polarity and direction.
	6.	Determine current from a specified charge flow and vice versa.
2.4	7.	Describe a basic electric system in terms of energy input, energy transfer, and energy output.
2.5	8.	State the law which relates voltage to energy per unit charge and use it, given appropriate data, to find the various quantities.
2.6	9.	Describe the various types of batteries and other sources of electric energy.
2.7	10.	State that power is the time rate of change of energy and can be found for an electric element from the product of current and voltage.
	11.	Use the formula to calculate power in typical situations.
2.8	12.	Describe a typical system in terms of information input and output.
2.9	13.	Describe some typical signals.

INTRODUCTION

Without electricity, modern technology could not exist. It has advantages offered by no other form of energy. It is readily converted into other energy forms, such as mechanical energy, light, sound, and heat. In addition, it is the most easily controlled form of energy, being used to control systems ranging in size from a pocket calculator, weighing a few grams, to a huge space vehicle.

In order to effectively deal with such systems, the technologist must be able to analyze and design electric circuits. In this chapter, we consider the fundamental quantities that determine the nature of electricity. These must be thoroughly understood before we begin our study of actual circuits.

2.1 FUNCTION OF THE ELECTRIC CIRCUIT

Every electric circuit either transfers or transforms electric energy. This is accomplished by applying electric pressure to tiny particles of electric charge, causing them to move around the circuit. The motion of charge causes energy to be transferred from one element of the circuit to another or to be transformed from one form of energy to another. For example, in a flashlight, chemical energy is transformed into electric energy in the battery. The electric energy is transferred through the conductors to the bulb, where it is transformed into light energy and some heat energy. A block diagram of this simple electric system is shown in Fig. 2.1. This same block diagram describes many practical electric systems. In most systems, the circuit block is much more complicated than the conductors in the flashlight. In a radio receiver, for example, the energy source is the antenna, which converts the energy transmitted through the air by the radio station to a form of electric energy which can be processed by the circuit. The circuit consists of a number of electronic devices that process this energy so that it is in a form suitable for the load, usually a loudspeaker.

From this discussion we see that energy is the fundamental quantity or end product involved in electric circuits. Charge and electric pressure are the fundamental quantities which provide the means needed to transfer or transform the energy. The transfer of charge from one point in a circuit to another is called *electric current*, and the electric pressure which causes the charge transfer is called *voltage*. Circuits are analyzed and designed in

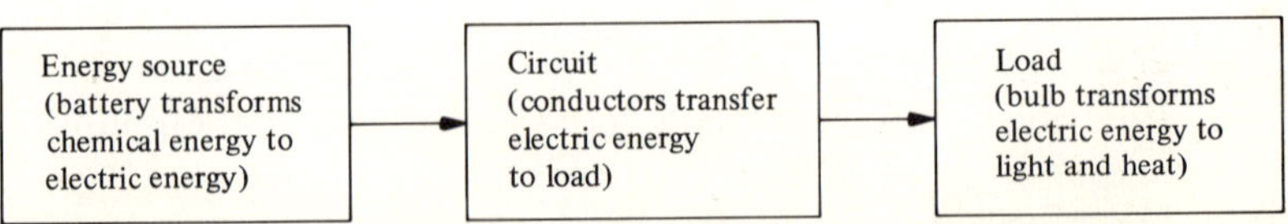

FIGURE 2.1
Block diagram of flashlight.

terms of current and voltage, and after this power and energy calculations are made.

2.2 ELECTRIC CHARGE

We have stated above that electric charge is a fundamental quantity. The smallest charged particles are the *electron* and the *proton*. Charge has *polarity;* the proton positive and the electron negative. This polarity determines whether a given charge will attract or repel other charges. The SI unit for charge is the *coulomb* and the abbreviation for this unit is C. Compared to the electron, the coulomb is a very large quantity of charge, for it takes 6.25×10^{18} electron charges to make one coulomb. Thus we have

$$1 \text{ coulomb (C)} = 6.25 \times 10^{18} \text{ electrons} \qquad (2.2\text{-}1)$$

The charge on a single electron can be found by taking the reciprocal of the number of electrons per coulomb. This will give us the number of coulombs per electron. Thus

$$\frac{\text{Coulombs}}{\text{Electron}} = \frac{1}{6.25 \times 10^{18}} = 0.16 \times 10^{-18}$$

and the charge on one electron, Q_e, is

$$Q_e = -0.16 \times 10^{-18} \text{ C} \qquad (2.2\text{-}2)$$

where Q is the letter used as a symbol for charge and the subscript e indicates an *electron* charge. We have also included the negative sign to explicitly show the polarity of the charge.

Force on Electric Charge

It has been found experimentally that when two charged bodies are relatively close to each other, a force of attraction will exist between them if the charges are opposite in polarity and a force of repulsion will exist if they are the same. The characteristic that leads to this phenomenon is the *electric force* between charged particles. This force obeys the inverse square law and can be calculated from the formula

$$F = \frac{kQ_1Q_2}{R^2} \qquad (2.2\text{-}3)$$

where

F = force in newtons, N

Q_1 and Q_2 = charge in coulombs, C
R = distance between charges in meters, m
k = constant, $9.0 \times 10^9 \dfrac{N \times m^2}{C^2}$

Note that if both charges have the same polarity, the resulting force will be positive, while if the polarities are opposite, the force will be negative. Thus a repulsive force is considered positive, while an attractive force is negative.

The electric force provides the mechanism for setting charges in motion so that the resulting current can transfer energy.

EXAMPLE 2.2-1 Force on Charged Bodies

A pair of positively charged ping-pong balls is suspended from fine silk threads as shown in Fig. 2.2. Each carries a charge of 3 μC and they are initially 0.2 m apart. Find the force of repulsion between them.

Solution

Using Eq. (2.2-3) we have

$$F = \frac{kQ_1Q_2}{R^2} = \frac{(9.0 \times 10^9)(3 \times 10^{-6})(3 \times 10^{-6})}{(0.2)^2} = 2.025 \text{ N}$$

• • •

Atomic Theory

Electrons and protons are combined with other particles in the atom, the basic building block for all of the materials that we know. The electrical

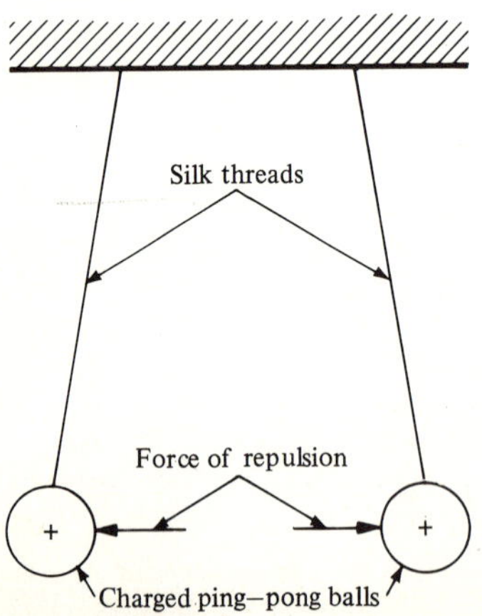

FIGURE 2.2
Example 2.2-1.

characteristics of substances containing a particular atom are determined by the arrangement of charged electrons and protons within the atom. Normally, equal amounts of positive and negative charge are present in every atom, so that any electrical effect is cancelled. In order to make use of the substance electrically, we must do work (equivalent to applying pressure) in order to separate the electrons and protons, so that the atom will no longer be electrically neutral. When this is done, the electrons move about in the material and we have a flow of charge that forms the electric current.

Certain materials have an atomic structure that allows for relatively free movement of some of the electrons. Most metals fall in this category, copper being especially useful for this purpose. A diagram of the atomic structure of copper is shown in Fig. 2.3. The copper atom is characterized electrically by a nucleus that contains 29 positively charged protons. There are 29 negatively charged electrons arranged in *shells* which revolve around the nucleus, much like the earth and other planets revolve around the sun. According to atomic theory, each shell can hold a fixed number of electrons; 2 in the innermost shell, 8 in the second shell, and 18 in the third shell. The outermost shell can hold 32 electrons, but in the copper atom there is only one. This electron is not as strongly attracted to the nucleus as the electrons in the fully occupied shells and is called a *free electron.* Copper has many of these free electrons, so that when voltage is applied, relatively large currents will flow. Thus, copper is classed as a *conductor,* a material which allows electric current to flow freely. Most metals and some liquids are good conductors. On the other hand, there are many materials

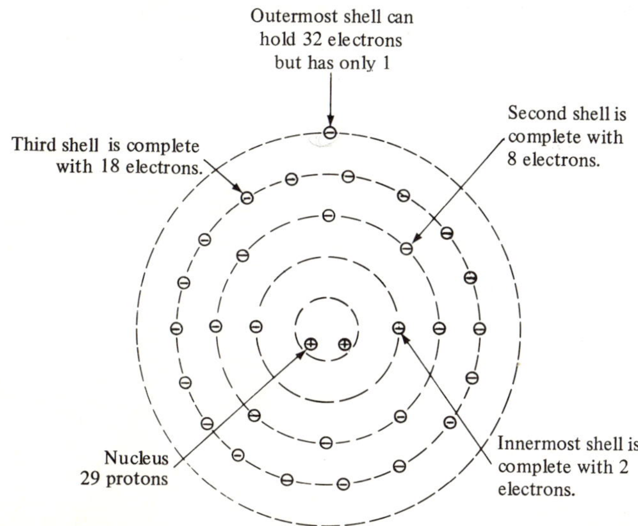

FIGURE 2.3
Atomic structure of copper.

which have relatively few free electrons and these are known as *insulators.* Typical insulating materials are air, glass, and mica, and it is very difficult to achieve any current flow in these materials. Materials such as silicon and germanium, which fall between conductors and insulators, are called *semiconductors.* These materials form the basis for the transistor, the integrated circuit, and other modern electronic devices.

• • •

LEARNING EXERCISES FOR SEC. 2.2

1. How many electrons are contained in 0.53 C?

2. A charged body contains the equivalent of 3 billion electrons. What is its charge in coulombs?

3. Two electrons are placed 0.01 mm apart. What is the electric force between them?

Ans. 2.3×10^{-18}; 3.31×10^{18}; 0.48×10^{-9}

• • •

2.3 ELECTRIC CURRENT

When charges such as electrons are free to move about in a material, application of voltage will cause a net motion of the free charges from one point to another. Such a movement of charge constitutes an electric *current.* Consider the cylindrical copper wire shown in Fig. 2.4. Assume that a voltage is applied so that the free electrons from the copper atoms effectively move from left to right as shown. Then, in SI units, if 1 C of charge passes through a cross section of the wire in 1 s, the magnitude of the current is 1 *ampere* (A). Thus

$$1\ \text{A} = 1\,\frac{\text{C}}{\text{s}} = 6.25 \times 10^{18}\,\frac{\text{electrons}}{\text{s}} \tag{2.3-1}$$

The diagram in Fig. 2.4 can also represent the flow of water through a pipe. Simply visualize the electrons shown as particles of water. If we want

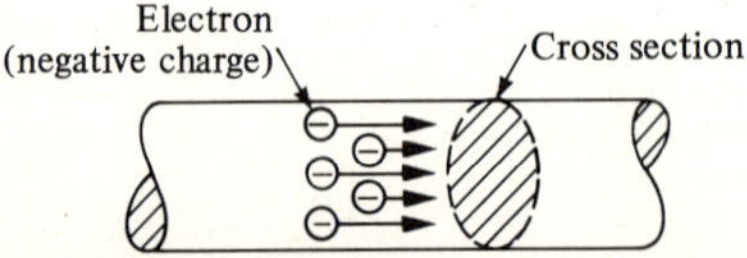

FIGURE 2.4
Electron motion in a conductor, analogous to fluid flow in a pipe.

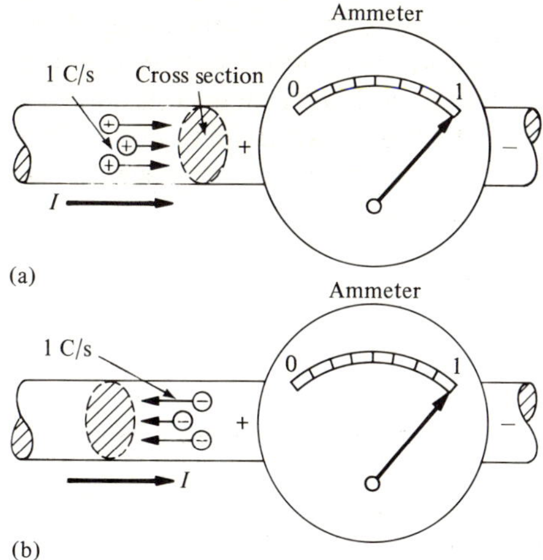

FIGURE 2.5
Illustrating the positive direction for current. (a) Positive charge. (b) Negative charge.

to measure the flow we would count the number of gallons passing the cross section in a convenient interval of time, such as 1 s. This is directly analogous to the flow of electricity in terms of coulombs per second. In the hydraulic case the flow is the gallons per second and in the electrical case the current is the coulombs per second passing through a given cross section.

We immediately recognize that the flow of water in a pipe can be in one of two directions. Clearly the same must be true of the flow of charge. However, there is an additional complication in that charge can have either positive or negative polarity. In order to be consistent, we define a current as being positive in a given direction if positive charge flows in that direction.* This is shown in Fig. 2.5a. In the diagram, 1 C/s of *positive* charge passes through the cross section in a direction from left to right. According to our definition, the current is 1 A in that direction. We indicate this current schematically by the line with the directional arrowhead and the symbol I, the usual letter symbol for current. Electrically, the same effect is achieved by 1 C/s of negative charge passing through the cross section in the opposite direction, as shown in Fig. 2.5b. In both cases an instrument which measures current† would register a reading of 1 A as shown on the figure.

*This definition is used by most people today. In the past, current was defined by some as being positive in the direction of electron flow (negative charge).

†Such instruments are called *ammeters*. They have two terminals marked + and − and when current flows *in* to the + terminal, the meter needle moves upscale, indicating positive current. Some ammeters are capable of measuring both positive and negative currents.

In metallic conductors, current flow is due to motion of free electrons. However, in semiconductor materials used in transistors and integrated circuits, current flow can also be due to *holes,* which act much like positively charged electrons. Thus positive current in a semiconductor can be due to either type of charge separately or a combination of the two. Our definition of current is designed so that the circuit theory to be developed in this book will be uniform and consistent regardless of whether the current is due to positive or negative charge carriers or any combination of both of them.

EXAMPLE 2.3-1 Charge Flow

In a certain material, the flow of charge through a cross section is monitored. A zero-center ammeter capable of measuring both positive and negative currents is connected to the wire. What will it read for the different conditions shown in Fig. 2.6a and Fig. 2.6b?

Solution

Fig. 2-6a: The effect of the 10 C/s of negative charge flowing from right to left is the same as that of 10 C/s of positive charge from left to right. This effectively adds to the 20 C/s of positive charge already flowing from left to right. Thus we have 10 + 20 = 30 C/s flowing from left to right into the + terminal of the ammeter and the reading will be 30 A.

Fig. 2-6b: In this case the net flow of charge is 20 C/s of negative charge flowing into the positive terminal of the ammeter. Thus, the meter will read −20 A. The significance of the negative sign is simply that current as it is defined in terms of positive charge is actually flowing from right to left in the wire. We consider this point in more detail in Chap. 4.

• • •

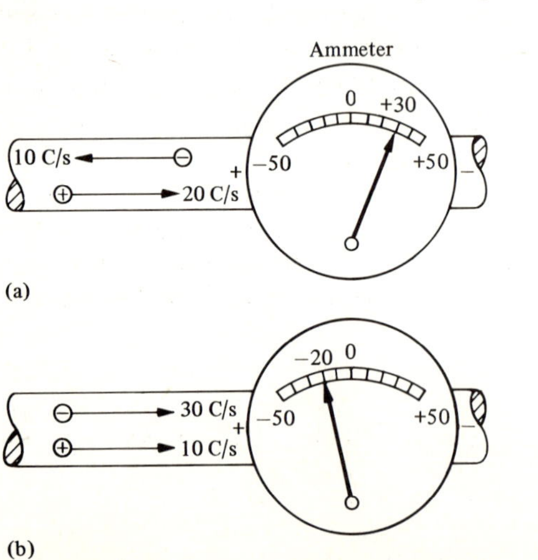

FIGURE 2.6
Example 2.3-1.

Current in Terms of Charge

From the preceding discussion and example we see that when the charge flow in C/s (the rate-of-change of charge) is constant, the current can be expressed mathematically in the form

$$\boxed{I = \frac{Q}{t}} \tag{2.3-2}$$

where

I = current in amperes
t = time in seconds
Q = charge through cross section in time t, in coulombs

The box around the equation indicates that this is an important basic formula, and that it should be memorized.

If the charge flow is not constant, we must use a more general equation:

$$I = \frac{\Delta Q}{\Delta t} \tag{2.3-3}$$

where ΔQ* is the *change* in charge through the cross section in time Δt. The basic formula $I = Q/t$ is a special case of this more general form. For both formulas Q will be positive or negative according to the polarity of the charge. The formula automatically takes care of the current direction for either polarity. We illustrate typical calculations in the examples which follow.

EXAMPLE 2.3-2 Finding a Current in Terms of Charge Flow

The charge through a cross section of a certain conductor is monitored and it is found that at time $t_1 = 2$ s a total of 4 C of positive charge has passed. The charge flow continues linearly and at $t_2 = 6$ s, 12 C of charge has passed. Find the current in the wire during this interval.

Solution

The data are illustrated graphically in Fig. 2.7a. From the graph we see that in the interval $\Delta Q = Q_2 - Q_1 = 12 - 4 = 8$ C. Then

$$I = \frac{\Delta Q}{\Delta t} = \frac{8\text{ C}}{4\text{ s}} = 2\text{ A}$$

*Refer to Sec. 1.6 for the definition of Δ as *change* or *increment*.

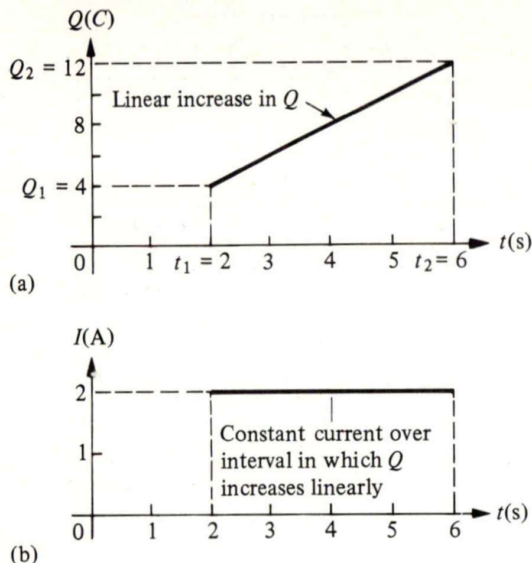

FIGURE 2.7
Example 2.3-2.

Observe that since $I = \Delta Q/\Delta t$, we can state in words (see Sec. 1.6) that the current is equal to the *slope* of the graph of charge vs. time. Since the slope is constant from 2 to 6 s, the current is constant throughout this interval and is in the direction of charge flow because the charge is positive. This is shown graphically in Fig. 2.7b.

It is important to note that the current would also be 2 A if the charge had changed from 0 to 8 C or from 8 to 16 C. It is the 8-C *change* in Q that determines the current.

• • •

EXAMPLE 2.3-3 *Finding the Amount of Charge in a Given Current*

A wire carries a constant current of 8 A. How much charge flows along the wire in 10 min?

Solution

Starting with $I = \Delta Q/\Delta t$ we find that

$$\Delta Q = I\Delta t$$

For our problem $I = 8$ A and $\Delta t = 10$ min $= 600$ s. Then

$$\Delta Q = (8 \text{ A})(600 \text{ s}) = 4800 \text{ C}$$

and 4800 C of charge pass through a cross section of the wire in 10 min.

• • •

EXAMPLE 2.3-4 *Finding the Current When the Charge Varies with Time*

It is found that the charge through a conductor cross section varies with time as shown in Fig. 2.8a. Find the current.

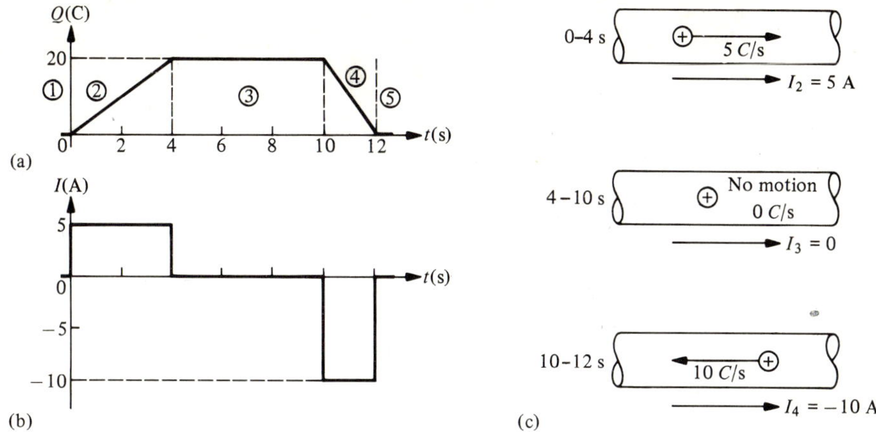

FIGURE 2.8
Example 2.3-4. (a) Graph of charge versus time. (b) Current waveform. (c) Charge flow direction during intervals 2, 3, 4.

Solution
This problem is best solved by considering each line segment of the charge curve separately. These are indicated on the graph as intervals 1 through 5.

Interval 1: Here the charge is zero and is not changing, so

$$\Delta Q = 0 \qquad \text{for all } \Delta t \text{ and therefore } I_1 = 0$$

Interval 2: Here the charge increases linearly from 0 to 20 C in a time of 4 s. Thus

$$I_2 = \frac{\Delta Q}{\Delta t} = \frac{20 - 0}{4 - 0} = 5 \text{ A}$$

Interval 3: The charge remains at 20 C from 4 to 10 s. Thus

$$\Delta Q = 20 - 20 = 0 \qquad \text{so that } I_3 = 0$$

Note the fact that Q = 20 C in this interval does not mean that charge is flowing; it simply means that 20 C have passed the point of measurement. There is current in the conductor only when the charge is changing.

Interval 4: Here the charge decreases from 20 C to 0 in the 10 to 12 s interval. Thus

$$I_4 = \frac{\Delta Q}{\Delta t} = \frac{0 - 20}{12 - 10} = \frac{-20}{2} = -10 \text{ A}$$

Interval 5: This is the same as intervals 1 and 3 where $\Delta Q = 0$, so that $I_5 = 0$.

The graph of current vs. time is shown in Fig. 2.8b. We can interpret the graph in the following way: during the 0 to 4 s interval, positive charge flows in a certain direction in the wire at the rate of 5 C/s, so the current is 5 A in that direction, as shown in Fig. 2.8c. At t = 4 s the flow stops abruptly and no additional charge passes the measurement point until $t = 10$ s. From $t = 10$ to 12 s positive charge flows in the *opposite* direction, so the rate of change is -10 C/s, corresponding to a current of 10 A in the opposite direction. This is also illustrated in Fig. 2.8c.

• • •

LEARNING EXERCISE FOR SEC. 2.3

1. a. 12 C/s of negative charge flows from left to right in a wire. Find the magnitude and direction of the current.
   b. It takes 6 s for 0.3 C of positive charge to flow from left to right along a wire. Find the magnitude and direction of the current.

Ans. R-L; L-R; 50 mA; 12 A

• • •

2.4 WORK AND ENERGY

If we lift a weight against the force of gravity, we are doing *work*. Energy is defined as the ability to do work. Thus whenever energy is expended, work is done. Some of the different forms of energy are mechanical energy, electric energy, acoustical energy (sound), and thermal energy (heat). Electric energy is obtained by converting some other form of energy. The most well known of these conversions are (1) electrochemical conversion, in which chemical energy is converted to electric energy, as in a battery, (2) electromechanical conversion, in which a rotating generator produces electric energy from mechanical energy of rotation, and (3) direct conversion of light into electricity, such as in *solar cells*.

We have stated at the beginning of this chapter that in electric systems energy is either transferred or transformed. A simple example which will illustrate these important concepts is shown in Fig. 2.9. In the diagram we show a battery, to which is connected a pair of wires which in turn are connected to a light bulb. This simple system contains the three elements that are common to virtually all practical electric systems. These include: (1) an element which introduces energy into the system, in this case, the battery; (2) an element called the *load* which makes use of, or converts the energy, in this case, the light bulb; and (3) means by which the energy is transferred from the battery to the load, in this case, the wire conductors.

Now let us consider the mechanism by which this energy transfer is effected. In the battery, chemical energy is converted to electric energy by the chemical action, which causes electrons to be removed from the atoms of one electrode and transferred through the *electrolyte* (material or liquid surrounding the electrodes) to the other electrode. Thus the two electrodes have become charged. The electrode from which the electrons have been removed is called the positive electrode since it now has a net positive charge, and the other is the negative electrode since it has a net negative charge equal in magnitude to the charge on the positive electrode. The chemical energy has been converted to an *electric potential* energy *differ-*

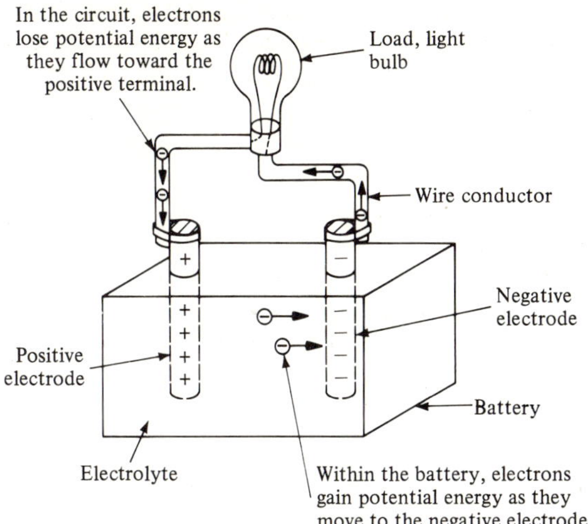

FIGURE 2.9
Simple electric system illustrating energy conversion.

ence between the free electrons on the negative electrode and those on the positive electrode. This potential difference (usually called *voltage*) provides the means for establishing the flow of charge (in this case, the free electrons) which will transfer the energy to the light bulb. The electrons experience a potential rise within the battery so that those at the negative terminal are at a higher potential energy level than those at the positive terminal. They have moved against the electric forces acting on them because they have been forced to move away from the positive terminal (where the electric force is attraction) toward the negative terminal (where the force is repulsion). This potential energy difference is now available for use in the circuit. When we connect the wires to the battery, the electrons at the negative terminal will be attracted toward the positive terminal. They move toward the positive terminal through the circuit external to the battery, which consists of the conducting wires and light bulb. During the time that they spend moving through the external circuit between the two terminals, they lose the potential energy that they gained within the battery. This energy is converted into light and heat energy in the bulb. Thus we have an energy balance, where the chemical energy generated in the battery is exactly equal to the energy converted to light and heat in the lamp.

At this point we must distinguish between two kinds of energy change. The first involves the energy change which takes place inside the battery. Here, the electrons experience a potential rise and are thus available to form a current which can transfer energy from the battery. The battery is thus a *source* of potential energy. It is sometimes described as a source of *electromotive force*. On the other hand, in the external circuit, the potential energy change is a decrease usually called a *voltage drop*. Associated with the voltage drop is the energy conversion in which the electric energy is converted to some other form, such as heat. When analyzing or designing circuits we do not need to distinguish between these two, and we use the letter symbol V (voltage) for both as described in the next section.

2.5 VOLTAGE

In Sec. 2.4 we learned that the conversion of any form of energy to electric energy or the conversion of electric energy to another form is accompanied by the movement of charge, which is brought about by application of an appropriate potential energy or voltage. When the voltage is constant, the quantities of interest, that is, energy, charge, and voltage, are related by the equation*

$$\boxed{V_{ab} = \frac{W}{Q}} \tag{2.5-1}$$

where

V_{ab} = potential difference in volts between points a and b
Q = amount of charge in coulombs moving from a to b
W = energy level difference of the charge as it moves from a to b. Energy is measured in units of *joules*. The voltage between points a and b is 1 V if the energy involved in moving 1 C of charge from a to b is 1 joule (J).

Voltage always has a polarity associated with it. A positive sign indicates at which of the two points the charge is at a higher energy level. This is illustrated in Fig. 2.10 where point a is at a higher energy level than point b, as indicated by the positive sign at a and negative sign at b. We call V_{ab} the *voltage drop* from a to b and we always specify voltage as being *between two points;* it is incorrect to refer to voltage *at a point*. Similarly, with current we specify current *through* a wire or an element; it is incorrect to refer to *current across* an element. These concepts are pursued further in Chap. 4.

*A more general formula that applies when the energy per unit charge is not constant is $V_{ab} = \Delta W/\Delta Q$.

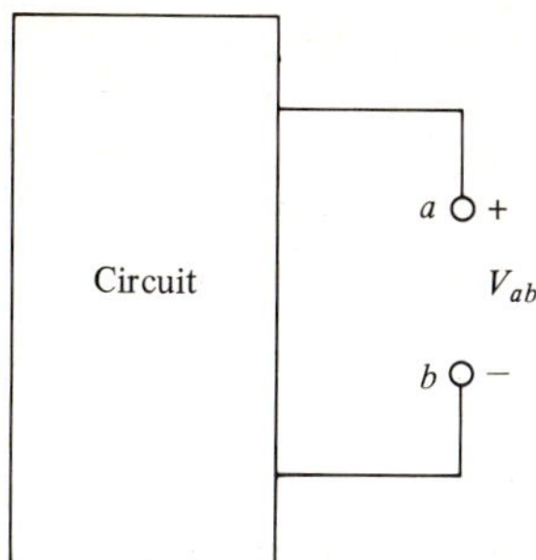

FIGURE 2.10
Illustrating voltage polarity.

EXAMPLE 2.5-1 *Finding a Voltage*

A charge of +12 C loses 9 J of energy as it moves from point *a* to point *b* in a particular circuit. Find the voltage drop V_{ab}.

Solution

$$V_{ab} = \frac{W}{Q} = \frac{9\text{ J}}{12\text{ C}} = 0.75\text{ V}$$

so that point *b* is 0.75 V *lower* in voltage than point *a*, because the positive charge *lost* energy in moving from *a* to *b*.

• • •

EXAMPLE 2.5-2 *Finding Charge and Voltage*

A current of 2 A flows through a heater and converts 60 J of electric energy to heat energy in 8 s. Find the voltage drop across the heater.

Solution

We first find the charge transported through the heater during the 8-s interval. This is

$$Q = I \times t = 2\text{ A} \times 8\text{ s} = 16\text{ C}$$

Then the voltage drop across the heater is

$$V = \frac{W}{Q} = \frac{60\text{ J}}{16\text{ C}} = 3.75\text{ V}$$

• • •

LEARNING EXERCISES FOR SEC. 2.5

1. A +5-mC charge gains 3 μJ of energy in moving from point *a* to point *b*. Find the voltage drop V_{ab}.

2. A single electron gains 0.3 pJ of energy as it moves from point *a* to point *b*. Find the voltage drop V_{ab}.

Ans. 1.88; 0.6 (scrambled units of millivolts and megavolts)

• • •

2.6 VOLTAGE SOURCES

In Sec. 2.4 we stated that all practical electric systems included an element which introduces energy into the system. These elements are called *sources* and they come in many varieties. The voltage sources to be discussed in this section provide a voltage that is constant with time as the energy input to the system. They are usually called dc supplies where the initials dc are an abbreviation for *direct current.* This terminology is used to describe any current in which the charge flow is in one direction or any voltage for which the polarity does not change. In order to avoid confusion, a single symbol, shown in Fig. 2.11a, is used in electric circuit diagrams to represent all dc voltage sources. The longer line in the symbol represents the more positive terminal. For example, if the source is rated at 6 V, then $V_1 = 6$ V and the upper terminal is said to be 6 V positive with respect to the lower terminal. Alternately, we can describe the situation by stating that there is a voltage *drop* of 6 V from the upper terminal to the lower terminal.

Batteries

The most familiar dc voltage source is the chemical battery, in which electrochemical energy is converted to electric potential energy as was described in connection with Fig. 2.9. Photographs of some typical batteries

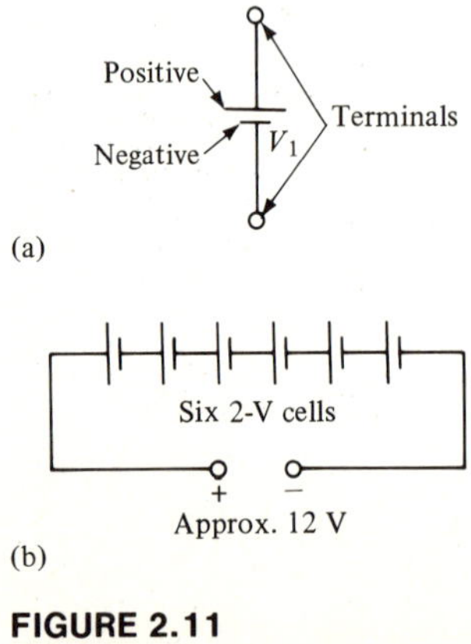

Positive
Negative
V_1
Terminals
(a)

Six 2-V cells
+ −
Approx. 12 V
(b)

FIGURE 2.11
Circuit symbols. (a) dc voltage source or battery. (b) Series connection of individual cells.

are shown in Fig. 2.12. All batteries are made up of combinations of basic cells, the cell being the fundamental unit for each particular combination of materials that causes the chemical-energy to electric-energy conversion. These basic cells fall into two categories: *primary* and *secondary*. The secondary cell can be recharged (as in the typical automobile battery), while

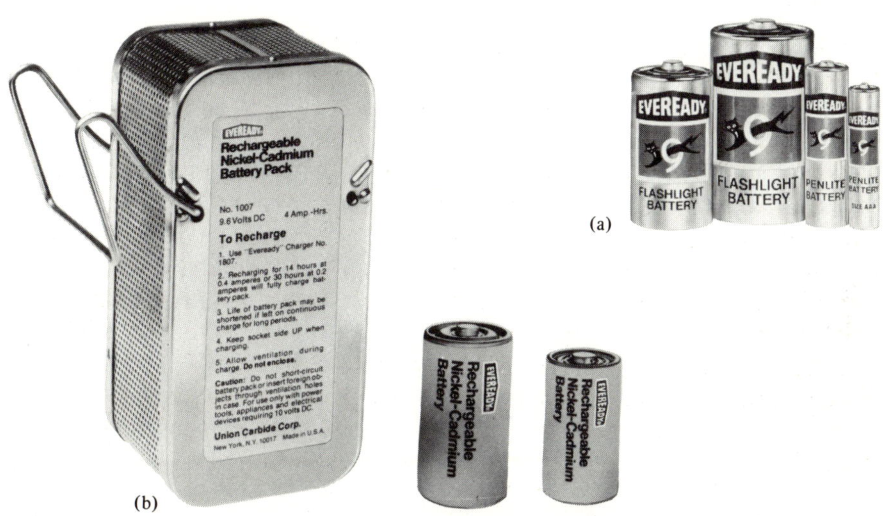

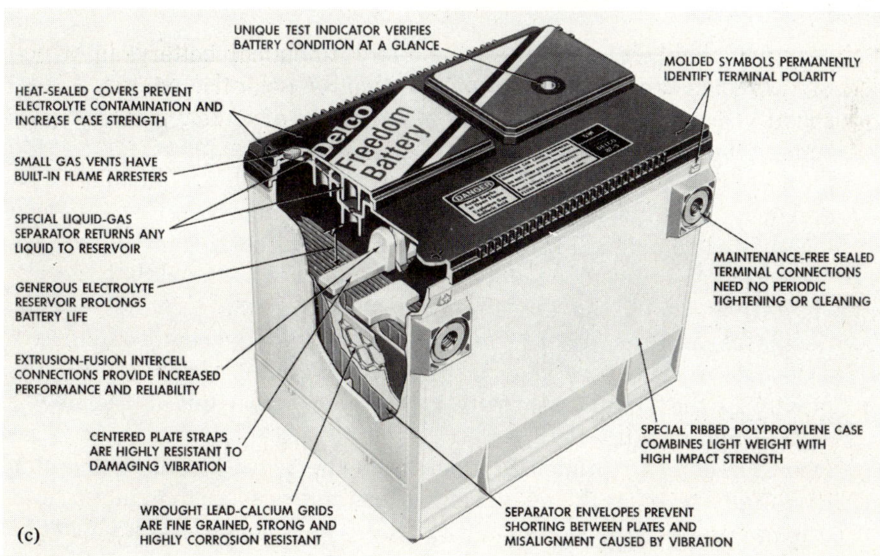

FIGURE 2.12

Photographs of typical batteries. (a) General-purpose carbon-zinc dry cells (courtesy of Union Carbide Corporation). (b) Rechargeable nickel-cadmium batteries (courtesy of Union Carbide Corporation). (c) Automotive lead-acid battery (courtesy of Delco-Remy Division, General Motors Corp.).

the primary cell cannot. Some of the currently popular types along with their important properties are:

1. The carbon–zinc dry cell. This is a primary cell which comes in many sizes, as shown in Fig. 2.12. The basic cell voltage is 1.5 V and by combining individual cells in a single case, voltages of any multiple of 1.5 V are obtained. An example is the 9-V battery used in many portable transistor radios. An advantage of this type is its low cost. The principal disadvantage is that its *shelf life* is not very long because of local chemical action and drying of the paste electrolyte. Thus manufacturers often place an expiration date on their batteries beyond which the battery will not perform according to specifications. The shelf life for medium size dry cells is about a year, while for the very small sizes it may be only a few months. Other types of dry primary cell are the alkaline–manganese–zinc cell (1.5 V), and the mercury cell (1.3 V).

   In order to compare relative capacities of different batteries, a rating called the ampere-hour (A·h) is used. For a typical 1.5-V flashlight battery (D-cell size) the rating is 4500 mA · h at 100 mA. (Recall that 1 mA = 10^{-3}A). This means that if we use a flashlight bulb that requires 100 mA, then the flashlight can be operated for a time interval

$$t = \frac{4500 \text{ mA}\cdot\text{h}}{100 \text{ mA}} = 45 \text{ h}$$

   before the light will start to become dim. If the bulb draws 200 mA, then the expected life is 22.5 h. A typical rating for a 9-V transistor battery is 400 mA·h at 8 mA.

2. Among the secondary cells, the lead-acid cell used in automobiles is the most well known. The basic cell has a terminal voltage of about 2.2 V, and a typical automobile battery is made up of six cells connected together as shown in Fig. 2.11b. This is a *series* connection, which we study in detail in Chap. 4. A typical rating for an automobile battery is 70 A·h at 3.5 A. As the energy is used up, the terminal voltage decreases, and the battery can no longer supply the required current. The state of discharge can be determined by measuring the specific gravity of the liquid electrolyte in the cell. When this falls below a certain figure, it is time to recharge the battery. This process involves forcing a current through the battery in the reverse direction, which essentially reverses the chemical process and restores the battery to its original condition. The advantage of this type of battery is that it can be used over and over again. A disadvantage is that the acid used in the electrolyte can cause serious damage if spilled. Thus this type of cell is unsuitable for portable equipment. In addition, if long periods of storage are required, the electrolyte must be drained from the cell and stored separately.

The nickel–cadmium battery is a completely sealed secondary cell which is popular for calculators and other small portable equipment. It is available in a wide range of capacities, from 50 mA·h to 50 A·h. When fully charged, the individual cells have a terminal voltage of 1.2 V. Recharging is necessary when this drops to 1.1 V. Compared to the primary dry cell, the nickel–cadmium type is quite expensive; however, it can be recharged several hundred times before requiring replacement. In addition, it can be stored for several months as long as it has been fully charged.

Rotary Generators

For many applications, batteries are not suitable because they must be replaced or recharged periodically. In many of these applications a rotating generator is used. These devices can be designed for virtually any terminal voltage, 120 and 240 V being standard for power applications. In this type of energy conversion, mechanical energy is used to rotate conductors in a magnetic field, causing electric energy to be generated. The mechanical energy can be obtained from the potential energy in falling water, as in a hydroelectric power station, or from the conversion of the chemical energy in coal or oil into heat energy. This is then converted into mechanical energy by steam or gas turbines which drive the shaft of the rotating generator. Atomic energy is also used for this purpose.

Rotating generators are used to generate both direct current and alternating current. *Alternating current* is abbreviated ac, and refers to any voltage or current which changes polarity periodically. Most of the world's energy supply is generated by this type of machine.

Electronic Energy-Conversion Devices

The voltage sources described previously operated by converting chemical, mechanical, or atomic energy to electric energy. Electronic devices are available which convert from one type of electric energy to another. For example, battery eliminators are available which can be used in place of batteries to power portable electronic equipment such as transistor radios or portable cassette recorders. These power supplies plug into an ordinary electric outlet and convert the alternating voltage (ac) supplied by the power company to the constant voltage (dc) required to operate the equipment. The conversion is effected by a process called *rectification,* which will be studied in electronics courses. Thus in this device, electric power is converted from ac form to dc form. The same rectification process is used in laboratory power supplies, power supplies for plating operations, and for any application where dc is required. Conversion from dc to ac is also possible. This is done in electronic systems called *inverters.*

All of the sources discussed up to this point have been *constant voltage* sources. They are characterized by a terminal voltage which remains essentially constant for any load current within the operating range of the supply. By using transistors in electronic circuits it is possible to implement a *constant current* supply in which the current is constant regardless of the voltage across the terminals. Such devices are useful in many applications.

Other Sources

Many other sources of energy exist. One type currently under development that holds promise for the future is the *fuel cell.* In this device, chemical energy is converted to electric energy as in a battery, but the chemical action does not affect the electrodes; only the chemical fuel is consumed in the process. Thus a fuel cell can supply energy as long as proper fuel is supplied. In the cell, hydrogen and oxygen are combined to form water. The energy released by this process is used to separate electric charges so that a positive charge appears at one electrode and a negative charge at the other. Currently, fuel cells are used in space exploration vehicles.

Many energy-conversion devices are quite specialized and operate at very low efficiency so that they are not practical for household or industrial use. An interesting example is the *solar cell,* in which semiconductor devices convert light energy from the sun directly to electric energy by providing sufficient energy to cause conduction electrons in the semiconductor material to move, thus forming a current. These *solar cells* are connected in banks to form *solar batteries,* which are used on communications satellites to operate radio receivers and transmitters. Although inefficient, they have extremely long life and among other applications, have helped to make communications satellites possible.

• • •

LEARNING EXERCISE FOR SEC. 2.6

1. a. A 12 A·h battery is used in a pocket calculator. If the average current is 28 mA, how long will the battery last?
   b. A portable radio requires 32 mA. If it must operate for 100 h between battery changes, what must the battery rating be?

Ans. 3.2; 429

• • •

2.7 POWER

When we work with energy conversion systems we are concerned with the *amount* of energy converted and the *rate* at which the conversion takes place. In fact, the rate of conversion of energy, which is the same as the rate at which work is done, is often more important than the amount of work done. For example, consider that we have to raise the temperature of a room from 0 to 25°C. It takes approximately the same *amount* of heat energy

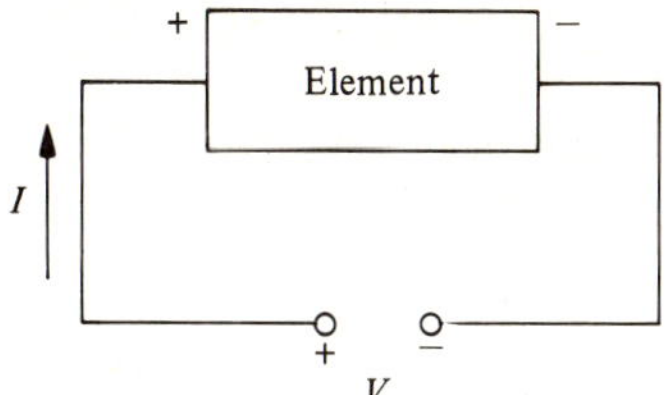

FIGURE 2.13
Element absorbing power.

to accomplish this in 15 min as it does in 30 min. However, the *rate* at which we must supply heat is twice as great for the shorter time interval.

The rate at which energy is used or supplied is called *power* and is measured in joules per second (J/s). The SI unit for power is the watt (W), which is 1 J/s. The general formula for power is*

$$\boxed{P = \frac{W}{t}} \tag{2.7-1}$$

where

P = power in watts
W = energy in joules used or supplied
t = time in seconds

Making use of previous results, we can derive a general formula for the power delivered to any element in an electric circuit as follows: Suppose a current I through an element causes a voltage drop V across the element in the direction of the current as shown in Fig. 2.13. The charge transferred through the element in time t is, from Sec. 2.3,

$$Q = It \tag{2.7-2}$$

The energy delivered to the element is, from Sec. 2.5,

$$W = QV \tag{2.7-3}$$

This energy is *delivered* to the element because the charge is going from a point of higher potential energy to one of lower potential energy. The power

*As with Eqs. (2.3-2) and (2.5-1) this equation can also be written in a more general form in terms of changes, that is: $P = \Delta W/\Delta t$.

is found by using Eqs. (2.7-2) and (2.7-3) in Eq. (2.7-1) in the following way: substitute Eq. (2.7-3) into Eq. (2.7-1) to get

$$P = \frac{W}{t} = \frac{QV}{t} = V\left(\frac{Q}{t}\right) \tag{2.7-4}$$

From Eq. (2.7-2) we recognize that $(Q/t) = I$ so that

$$\boxed{P = VI} \tag{2.7-5}$$

In the formula, P is power delivered to or absorbed by the element in watts (W), V is the voltage across the element in volts (V), and I is the current in amperes (A) through the element in the direction of the voltage drop. If the current is in the other direction, the power is negative and is actually being delivered by, rather than to, the element. These concepts are illustrated in the examples which follow.

EXAMPLE 2.7-1 Power and Energy

Find the power absorbed by the element in Fig. 2.13 if it is a flashlight bulb and if $V = 3$ V and $I = 25$ mA. How much energy is delivered to the bulb in 5 min?

Solution

From Eq. (2.7-5)

$$\begin{aligned} P &= VI \\ &= (3\text{ V})(25\text{ mA}) = 75\text{ mW} \end{aligned}$$

From Eq. (2.7-1), the energy delivered is

$$\begin{aligned} W &= Pt \\ &= (75 \times 10^{-3}\text{W})(5\text{ min})(60\text{ s/min}) = 22.5\text{ W}\cdot\text{s} = 22.5\text{ J} \end{aligned}$$

• • •

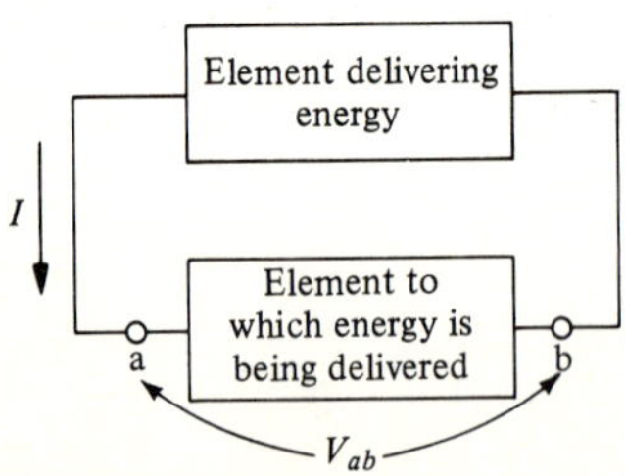

FIGURE 2.14
Example 2.7-2.

EXAMPLE 2.7-2 *Voltage and Energy*

In the circuit of Fig. 2.14, the current I is 325 mA and energy is delivered by the element at the rate of 80 J/s. Find the power and the voltage V_{ab}.

Solution

Since the element is *delivering* energy at the rate of 80 J/s, the power is 80 W. In a device that is *delivering* energy, current *leaves* the device at the higher voltage terminal. Thus, noting the direction of I shown on the diagram we see that terminal a is at a higher voltage than terminal b and, from Eq. (2.7-5)

$$V_{ab} = \frac{P}{I} = \frac{80\ \text{W}}{0.325\ \text{A}} = 246\ \text{V}$$

• • •

EXAMPLE 2.7-3 *Power Required to Perform a Specific Task*

A winch is being designed to lift weights from railroad cars. It is found that the mechanical work required to lift a typical weight a distance of 10 m is 10^6 J.

a. Assuming that the electric motor operating the winch converts electric energy to the mechanical energy required to lift the weight without loss, find the power rating of the electric motor required if the weight is to be lifted in 5 min.

b. Repeat the calculation for 2.5 min.

c. If 220-V motors are to be used, find the required currents for (a) and (b).

Solution

a. Assuming ideal conversion, we require 10^6 J of electric energy in 5 min. Thus

$$P = \frac{W}{t} = \frac{10^6\ \text{J}}{300\ \text{s}} = 3330\ \text{W} = 3.33\ \text{kW}$$

The minimum motor rating would be 3.33 kW. Most motors are rated in *horsepower* (hp) where

$$1\ \text{hp} = 746\ \text{W} = 0.746\ \text{kW}$$

Even though this is not an SI unit, we should be familiar with it because it is still used extensively. Our 3.33-kW motor rating can be converted to horsepower as follows:

$$\frac{3.33\ \text{kW}}{0.746\ \text{kW/hp}} = 4.46\ \text{hp}$$

Thus a motor rated at least 5 hp is required.

b. For a 2.5-min interval

$$P = \frac{W}{t} = \frac{10^6\ \text{J}}{150\ \text{s}} = 6670\ \text{W} = 6.67\ \text{kW}$$

Converting to horsepower

$$P = \frac{6.67 \text{ kW}}{0.746 \text{ kW/hp}} = 8.94 \text{ hp}$$

Here a standard 10-hp motor would be used. Note that the power required to do the work in 2.6 min is exactly twice the power required to do it in 5 min, and a much more expensive motor is required because of the higher *rate* at which the work must be done.

c. For the motor of part (a) $P = 3.33$ kW and, using $P = VI$,

$$I = \frac{P}{V} = \frac{3330 \text{ W}}{220 \text{ V}} = 15 \text{ A}$$

For the motor of part (b) $P = 6.67$ kW and

$$I = \frac{P}{V} = \frac{6670 \text{ W}}{220 \text{ V}} = 30 \text{ A}$$

• • •

LEARNING EXERCISE FOR SEC. 2.7

1. a. A motor operates an electric car for 30 min. During this time 860 J of energy is expended. What is the power dissipated during this interval?
   b. A plating bath requires 32 V at 120 A in order to silver-plate tableware. How much power is involved?
   c. Find the current required to operate a 150-W lamp connected to a 117-V circuit.

Ans. 3.84; 0.48; 1.28

• • •

2.8 SIGNALS

We have stated that all electric systems contain at least one element which introduces energy into the system, and the function of the system is to transfer or transform this energy. According to modern theory, *information* can be thought of as a form of energy, so that in many systems *information transfer* or *processing* is the main function. The typical system has one point at which energy is introduced in the form of an *input signal* and a point at which the energy or information is to be utilized in the form of an *output signal*. This is shown in block diagram form in Fig. 2.15a. A little reflection will show that many of the electric and electronic devices which we take for granted in our daily lives are complete systems which can be categorized in this way. For example, the radio receiver (Fig. 2.15b) that we all use can be thought of as a system in which the input signal is the electromagnetic wave picked up by the antenna and the output signal is the sound wave generated by the loudspeaker. The most important characteristic of these signals is

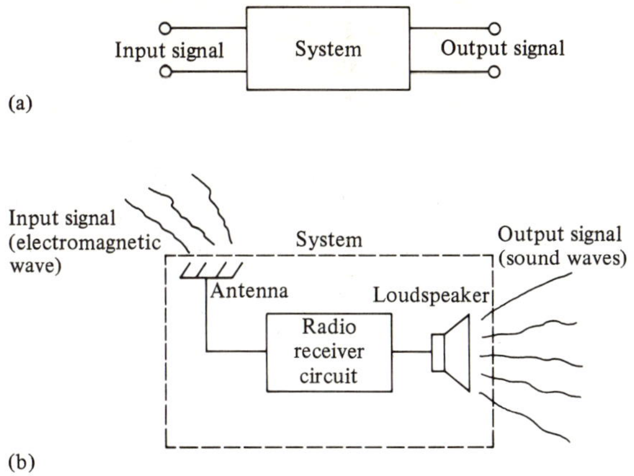

FIGURE 2.15
Systems. (a) Block diagram. (b) Radio system.

that they are carriers of *information*. The system can be thought of as *processing* the input signal so as to perform any required modifications of the information which it contains.

Signals come in a variety of shapes and sizes. Some typical examples are shown in Fig. 2.16. Figure 2.16a illustrates what a speech wave might look like when displayed on an oscilloscope after being converted to a voltage by means of a microphone. This is classified as an *analog* signal; the information is contained in the time variation of the signal. Analog signals can be represented by a series of sine waves (sinusoids) one of which is shown in Fig. 2.16b. The sine function is extremely important in electric circuit theory (in most power generation systems, the supply voltage is generated in the form of a sine wave), and we discuss it in more detail in a later chapter.

Figure 2.16c shows an amplitude-modulated signal as used in conventional AM radio. This is the actual waveform transmitted as an electromagnetic wave.

Finally, in Fig. 2.16d, we show a pulse signal. In digital systems, such as the digital computer, information is transmitted throughout the system by series of pulses such as those shown.

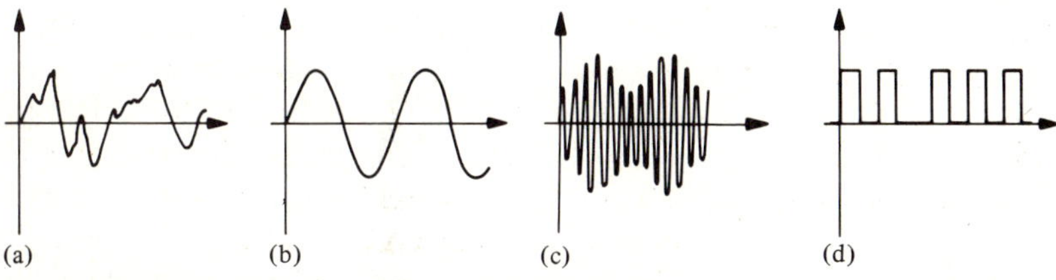

FIGURE 2.16
Typical signals. (a) Speech signal. (b) Sine wave. (c) AM wave. (d) Digital signal.

DC Signals

In this text we are beginning our study of electric circuits. Typically, circuit problems can be classified as either analysis or design. In analysis, we are given a particular circuit and an input signal; the problem is to determine the resulting output signal. In design, specifications are given on the input signal and the output signal; the problem is to design a circuit which will process the given input signal to yield the desired output signal.

One of the most important of the signals from the point of view of learning circuit theory is that called *dc*. As we noted previously, the letters actually stand for *direct current,* but are often used to indicate any signal which is constant, that is, does not vary with time. We begin our study of circuits in Chap. 4 using direct currents and voltages as the basic signals, so that we do not have to learn about time variation at the same time. When considered as a carrier of information, however, dc is about the least useful signal that one can imagine. Recall that a dc current or voltage is constant. For example, a battery is a source of dc voltage. A typical battery, such as one used in a transistor radio, has a voltage of 9 V. Thus the only information available in this dc voltage is the value +9 V. This signal is extremely uninformative; the battery is used as a source of power rather than a carrier of information. As we progress in our studies, different forms of signals will be introduced.

Circuit Symbols

Sources to supply power and almost any waveform of current or voltage are available today. In order to avoid confusion, all voltage sources are represented on circuit diagrams by one of two standard symbols. The symbol for the dc voltage source is shown in Fig. 2.17a. As noted previously, the longer line is the positive terminal, so it is not necessary to explicitly show the polarity of the voltage. Sometimes, more than one pair of lines is used, the longer line still indicating the positive terminal. The standard symbol for a general voltage source is shown in Fig. 2.17b. Here the polarity signs are

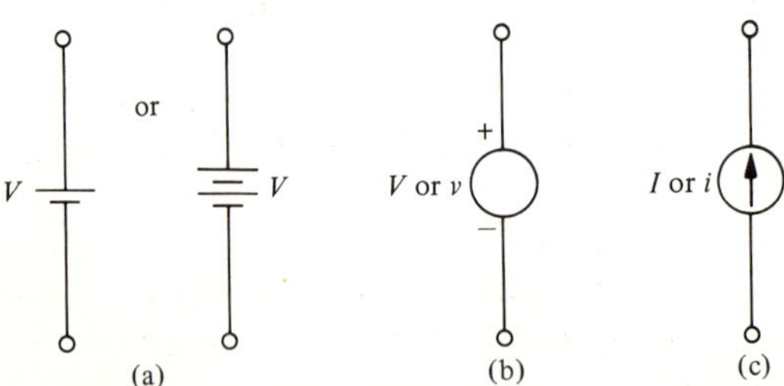

FIGURE 2.17
Circuit symbols for sources. (a) dc voltage source. (b) General voltage source. (c) General current source.

explicitly indicated and the voltage can be either dc as indicated by the capital italic letter V or any ac waveform, as indicated by the lower-case italic letter v. The standard current source symbol is shown in Fig. 2.17c. Here current from the source flows in the direction of the arrow inside the circle. The source can be either dc as indicated by the capital I or ac as indicated by the lower-case i.

SUMMARY

In this chapter we have introduced those quantities which are fundamental to the operation of electric systems. With a thorough understanding of these quantities, the student will be able to proceed to subsequent chapters in which we study such systems. The essential points of the chapter are summarized below.

Section

2.1

1. Electric circuits transfer or transform energy by virtue of the motion of electric charge (current) around the circuit caused by the application of voltage.
2. Fundamental quantities are energy and charge, and current and voltage.

2.2

3. Charge is measured in coulombs (C) and 1 C carries as much charge as 6.25×10^{18} electrons.
4. The polarity of the electron charge is negative.
5. There is an electric force between charged particles that acts on free charges in conductors so that they move around the circuit to form currents.
6. In the copper atom, there is a single free electron in the outermost shell, which is easily set in motion by the application of voltage.

2.3

7. The unit of current is the ampere (A) and if 1 C/s passes a given cross section, the current is 1 A.
8. Current is defined as being positive in the direction of flow of positive charge.
9. When the charge flows at a constant rate, the current formula states that $I = Q/t$

2.4

10. A typical electric system includes a source that introduces energy, a circuit which transfers the energy, and a load which uses or converts the energy.

2.5

11. The formula for voltage is $V = W/Q$.
12. Voltage is always specified or measured between two points. The voltage from the higher energy level point to the lower energy level point is called a *voltage drop*.

2.6

13. Voltage sources that provide constant terminal voltages are called dc supplies.
14. Typical dc supplies are batteries, rotary generators, and electronic generators.

2.7 15. Power is the time rate of change of energy; $P = W/t$.
16. In terms of voltage and current $P = VI$.

2.8 17. In many systems, information is the quantity of interest, and circuits transfer or process information.
18. Information is carried by signals in the form of currents and voltages, which vary with time.

Important Quantities

Quantity	Letter Symbol	SI Unit and Abbreviation
Energy	W	joule (J)
Charge	Q	coulomb (C)
Current	I	ampere (A)
Voltage	V	volt (V)
Power	P	watt (W)
Force	F	newton (N)

QUESTIONS FOR REVIEW

Sec. 2.1

1. Name several electric systems that transfer energy. Name several that transform energy.
2. Draw a block diagram of several electric systems with which you are familiar.

Sec. 2.2

3. Describe the electron and the proton.
4. What is the charge on one electron?
5. How does the electric force between charged particles depend on the magnitude of the charges? The distance between them? How does the polarity of the charges affect the force?
6. Describe atomic structure in terms of electron shells.
7. What is a conductor? An insulator? A semiconductor?

Sec. 2.3

8. Describe the mechanism of charge flow in a conductor.
9. A stream of negative charge flows from east to west. What is the direction of the resulting current?
10. Why is it sometimes necessary to use the formula $I = \Delta Q/\Delta t$ to find current?

Sec. 2.4

11. Explain the terms work and energy and give examples of each.
12. Describe some energy-conversion devices with which you are familiar. State the forms of energy which are converted.
13. What are the basic components of most practical electric systems?
14. Describe the energy transfer mechanism in a battery.

Sec. 2.5

15. What is the definition of voltage in terms of energy and charge?
16. State the meaning of the term "voltage polarity."
17. Compare voltage and current with regard to the terms polarity and direction.

Sec. 2.6

18. Describe several sources of voltage.

Sec. 2.7

19. What is the definition of power in terms of energy and time? In terms of current and voltage?
20. Positive charge moves from a point of low potential energy to a point of higher potential energy in an element. Is energy being delivered to or from the element?

Sec. 2.8

21. Describe a typical electric system in terms of signals.
22. Describe some signals with which you are familiar.

PROBLEMS

Sec. 2.1

1. Use a block diagram similar to Fig. 2.1 to describe several electric systems with which you are familiar. Specify the forms of energy involved at each step, and indicate whether the energy is transferred or transformed.

Sec. 2.2

2. How many electrons are required to provide 0.02 C of charge? What will the polarity of this charge be?
3. Compute the magnitude and direction of the force between electrons which are 10^{-10} m apart. What will the force be if the distance is doubled?
4. Find the distance between two charges of 5 and -4 C if the force acting on one due to the other is 15 N.
5. A charge of 4 C attracts a second charge, which is 3 m from the first charge with a force of 8 N. Find the magnitude of the second charge.

Sec. 2.3

6. Electrons travel along a wire in a computer at the rate of 9×10^{20} electrons per minute from left to right. What is the current and what is its direction?
7. A current of 25 mA flows through a wire leading to a TV picture tube. How long does it take for 50 C to pass a given cross-sectional area of the wire?
8. A direct current of 15 mA flows for 5 s in a solid-state photographic timer. How much charge is transported as a result?
9. 30 C of charge pass through a cross section in 12 s. What is the current?
10. A current in a telephone wire is composed of 3.6 mC/min. Express this in amperes.
11. How many coulombs of charge pass through a light bulb in 1 min if the current is constant at 400 mA?

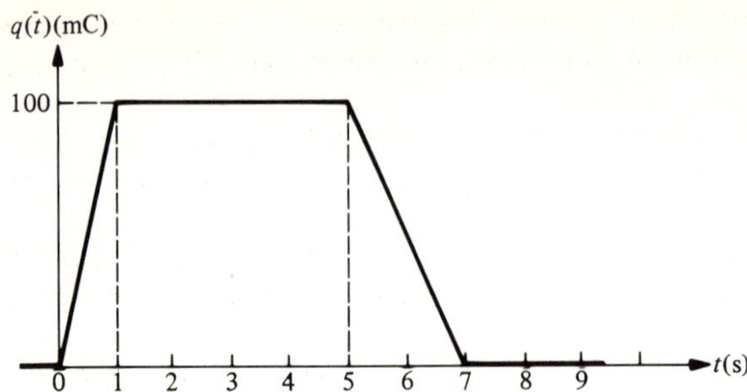

FIGURE 2.18

12. How long will it take for 12 C of charge to pass through a TV game power supply if the current is constant at 250 mA?
13. In a certain solid-state ignition system, a positive charge of 36 C/s flows from left to right and negative charge of 17 C/s flows from right to left. Find the current and its direction.
14. A linearly increasing charge is clocked at 12 mC after 3 s and at 62 mC after 7 s. Plot the charge and the resulting current.
15. A graph of the charge that has flowed through a cross-sectional area of a wire is shown in Fig. 2.18. Draw a graph of the current flow in the same direction.
16. Current in a computer terminal is measured and found to be a constant value of 250 mA from $t = 2$ to $t = 10$ s. Plot two charge vs. time curves which would produce this current.

Sec. 2.5

17. Find the difference in the energy level of a 3-C positive charge when it is moved from one point to a second point with a potential difference of 2 V between them, the second point being at a higher potential.
18. Determine the change in potential energy of a −8-C charge when it is moved from point *a* to point *b* in Fig. 2.19.
19. Find the potential difference v_{ab} if a −5-C charge gains 18 J of energy in being moved from point *a* to point *b*.
20. A charge gains 40 J of energy when moving through a potential drop of 18 V. What is the value of the charge? Is it positive or negative?
21. A current of 4 mA flows through a pocket calculator having a potential drop of 10 V in the direction of current flow. How much energy is transferred in a 1-s interval? Is the energy delivered to or from the element?
22. A flasher requires a current of 2.5 A over a 5-ms interval. This current produces a voltage drop of 50 V across the flash tube. How much energy is transformed in the tube?

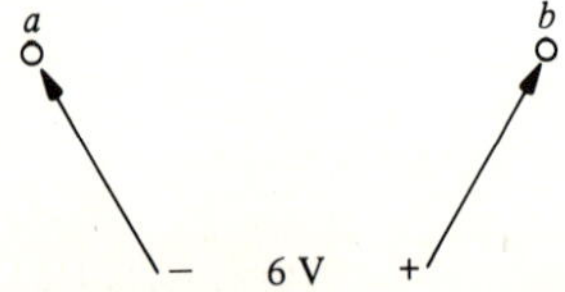

FIGURE 2.19

23. In a battery, 15 C of charge moves between the positive and the negative terminals. Five joules of chemical energy is expended in the process. Find the terminal voltage of the battery.
24. In an electric heater, 530 J of energy is converted when the voltage drop across the terminals is 440 V. How much electric charge moves from one terminal to the other?
25. The current from a battery is maintained at 0.7 A for 2 min. During this interval the terminal voltage of the battery is 1.2 V. How much chemical energy is transformed?
26. How long will a constant current of 150 mA have to flow through a light bulb rated at 6 V if 4 J of energy is to be transferred to the bulb?

Sec. 2.6

27. An automotive battery has a rating of 220 A·h. What is the maximum current it can supply for 50 h?
28. A test shows that a battery designed for portable tape recorders can supply 800 mA for 52 h before the terminal voltage starts to decrease. What is the A·h rating of the battery?
29. A flashlight battery is rated at 750 mA·h. How long can it supply a current of 125 mA before the terminal voltage begins to decrease?
30. An automobile battery is rated at 100 A·h. Plot a graph of maximum steady current vs. discharge time.

Sec. 2.7

31. How many joules of energy are converted to light and heat in a 60-W bulb in a year?
32. An electric heater is rated at 1000 W. How long will it take to convert 2×10^5 J to heat?
33. The energy converted by a bank of solar cells is monitored and a graph of the result is shown in Fig. 2.20. Plot a graph of the power dissipated during the time interval shown.
34. A 9-V battery stores 3600 J of energy. How long will the battery last if 10 mA of current is drawn? If the current drain is 5 mA? What is the maximum average current drain if the battery is to last for at least 50 h?
35. Find the power in the circuit element of Fig. 2.21. Is this power delivered to or from the element?
36. An electric generator delivers 5 kW of electric power at a voltage of 100 V. What is the current flowing through the generator? How much energy is generated in an hour?

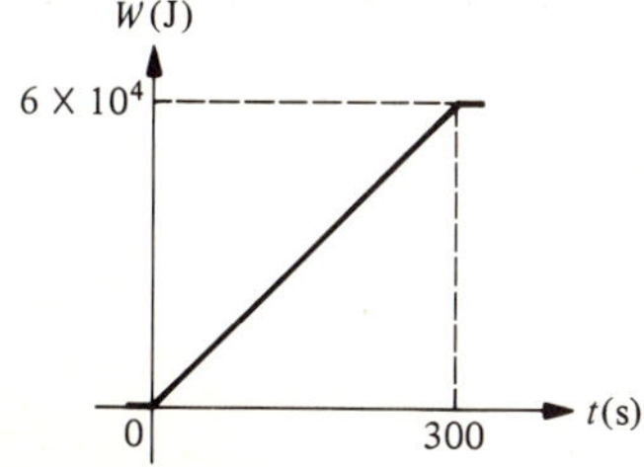

FIGURE 2.20

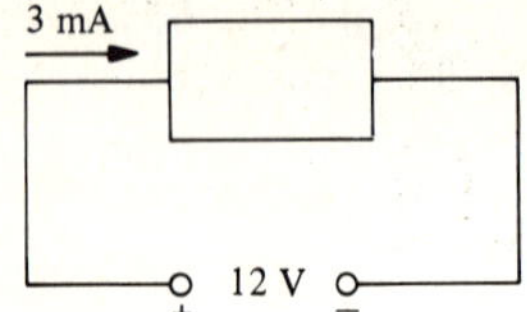

FIGURE 2.21

37. A circuit element dissipates 10 W and has a voltage of 30 V across it. Find the current through the element and its direction.
38. A current of 25 mA flows through a model railroad car motor that absorbs 2 W of power. Find the voltage across the motor and its polarity (relative to the current flow direction).
39. A 50-W bulb is switched on at $t = 0$ and off at $t = 10$ s. Plot a graph of energy converted vs. time.
40. Refer to Example 2.7-3. We are redesigning the winch in that example to operate on a model railroad. We find that the mechanical work required to lift a typical weight a distance of 10 cm is 1 J.

    a. Find the power rating in watts and horsepower of the motor required if the weight is to be lifted in 5 s.

    b. Our power supply is rated at 6 V. How much current is required by the motor?
41. A battery in a portable radio uses 15 J of chemical energy in 1 min. During this time the battery current is constant at 25 mA. Find the battery voltage.

3
Resistance

OBJECTIVES

Upon completion of this chapter, the student should be able to

Section		
3.1	1.	State the letter symbols, units, and unit symbols for resistance and conductance.
3.2	2.	Determine the resistance of a conductor when given its dimensions and resistivity.
	3.	Determine the conductance of a resistor.
3.3	4.	Use wire tables to determine the resistance of a given length of wire of a given gauge.
	5.	Determine the wire size required to achieve a specific resistance when the length of run is specified.
3.4	6.	Determine the dimensions of a specified deposited resistor, given the sheet resistance of the material.
3.5	7.	Determine the change in resistance of a conductor when the temperature changes.
3.6	8.	Describe some of the different resistor types.
3.7	9.	Identify the value and tolerance of a resistor from the color-coded bands.
	10.	Describe the color bands for specific resistance values.
3.8	11.	State the six numbers in the 20% preferred value series.
	12.	Evaluate the range of any resistance in the 20% series.
	13.	Describe some nonlinear resistors.

INTRODUCTION

In Chap. 2, we learned that there will be current in an electric circuit when energy is supplied from a source. The amount of current for a given voltage is dependent on the electrical resistance of the circuit, and according to Ohm's law, *resistance* is the constant of proportionality that relates the voltage to the current. In this chapter we study the resistance of materials of different geometries and learn how to calculate this important parameter. Then in Chap. 4 we consider Ohm's law and its applications.

3.1 RESISTANCE AND CONDUCTANCE

All materials have electrical *resistance*. The word resistance is used because this parameter determines the extent to which current is *resisted* in a circuit. In some materials it is difficult to get charge to flow; these are characterized by high resistance and are called *insulators*. The opposite is true in low-resistance materials called *conductors*.

A *resistor* is a two-terminal circuit element which is described electrically by its resistance. Resistors come in all shapes and sizes, and to avoid confusion a single circuit symbol, as shown in Fig. 3.1, is used in circuit diagrams to represent any resistor. The letter symbol *R* is universally used for the resistance parameter. The unit for resistance is the ohm,* for which the abbreviation is Ω (capital Greek omega). Commercial resistors are available with values of resistance ranging from less than 1 Ω up to 100 MΩ. Power ratings of such resistors range from fractions of a watt to kilowatts. Typical values of resistance as used in a transistor radio might range from 100 Ω to 1 MΩ at power ratings from 1/8 W to 2 W.

The *conductance* of a resistor is the reciprocal of its resistance. The letter symbol for conductance is G and its unit is the siemens,† abbreviated S. Since resistance and conductance are reciprocal, they are related by the equation

$$G = \frac{1}{R} \qquad (3.1\text{-}1)$$

• • •

LEARNING EXERCISE FOR SEC. 3.1

1. A resistor is found to have a resistance of 330 000 Ω. What is its resistance in (a) kilohms, (b) megohms? What is its conductance in (c) millisiemens, (d) microsiemens?

Ans. 3.03; 0.33; 330; 0.003 03

3.2 RESISTANCE OF CONDUCTING WIRES

The hydraulic analog for current is useful in developing a physical feeling for resistance. Recall that fluid flow is the analog of current, while pressure

*After Georg Simon Ohm, a German physicist. His famous formula was published in 1827.
†After Ernst von Siemens.

FIGURE 3.1
Circuit symbol for resistance.

is the analog of voltage. The amount of fluid flow through a pipe is proportional to the pressure, just as current through a resistance is proportional to voltage. Consider that we have two pipes of the same length but of different diameters, as shown in Fig. 3.2a. If we apply the same pressure to both pipes, the resulting flow will be less in the one with the smaller diameter. Thus smaller cross-sectional area leads to larger resistance to flow and we find that hydraulic resistance is *inversely* proportional to the cross-sectional area. A different situation is shown in Fig. 3.2b where we have two pipes of the same diameter but different length. In this case, for the same pressure applied, there will be less flow through the longer pipe because of the work that must be done to overcome the friction between the fluid and the walls of the pipe. As a result, the hydraulic resistance is *directly* proportional to length.

Just as the hydraulic resistance of a pipe depends on its length and its cross-sectional area, so the electrical resistance of a wire depends on its length and cross-sectional area. In the case of the pipe there is one additional factor which affects the resistance to flow; that is the coefficient of friction between the pipe wall and the fluid. In the electrical case we find a completely analogous situation. The resistance is proportional to the length of the wire, and inversely proportional to the cross-sectional area of the wire. There is a constant of proportionality, analogous to the coefficient of friction for the pipe, which depends on the material. This constant of proportionality is called the *resistivity* of the material; it is symbolized by the lower-case Greek letter rho (ρ) and has a very high value for insulators and a very low value for good conductors.

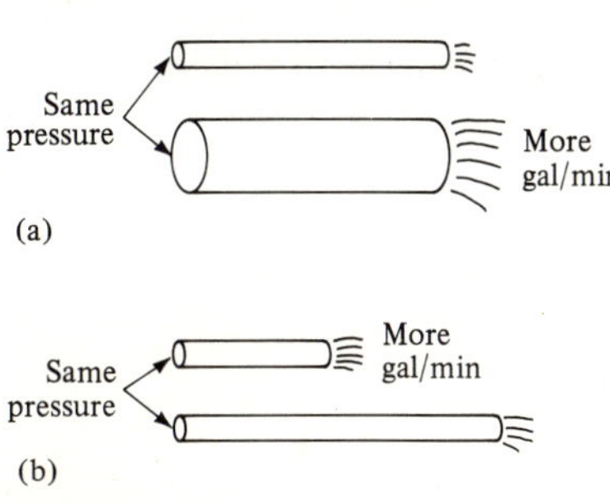

FIGURE 3.2
Fluid flow analog. (a) Same length, different diameter. (b) Same diameter, different length.

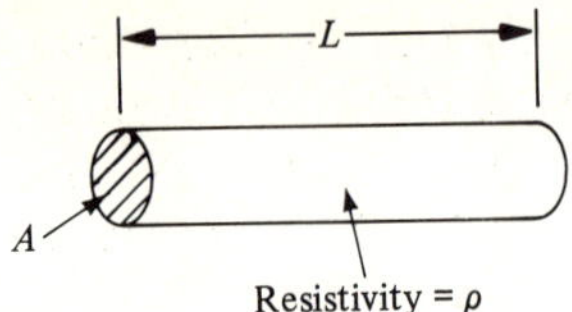

FIGURE 3.3
Typical conductor of length *L* and cross-sectional area *A*.

From the foregoing discussion we see that the resistance of a conductor depends on length, area, and resistivity. A typical cylindrical wire conductor is shown in Fig. 3.3. The formula for its resistance is

$$R = \rho \frac{L}{A} \tag{3.2-1}$$

where

R = resistance between ends in ohms (Ω)
L = length in meters (m)
A = cross-sectional area in square meters (m^2)
ρ = resistivity in ohm-meters (Ω·m)

Resistivities of various materials in use today are listed in Table 3.1. Of these, copper is the most popular material for use in wire conductors, followed by aluminum. Silver is used for electrical contacts and gold for fine wire interconnections on integrated and hybrid circuits. The other

TABLE 3.1
Resistivities of various materials

Material	Resistivity ρ at 20°C (Ω·m)
Silver	16×10^{-9}
Copper	17.2×10^{-9}
Gold	24.4×10^{-9}
Aluminum	28.3×10^{-9}
Tungsten	55.1×10^{-9}
Nickel	78×10^{-9}
Iron	120×10^{-9}
Constantan	490×10^{-9}
Nichrome	1000×10^{-9}
Carbon	$210\,000 \times 10^{-9}$

materials in the table are used for various special applications. For example, iron and constantan can be formed into devices called *thermocouples*, which are used in the measurement of high temperatures. Nichrome is used in the manufacture of heating coils and carbon is used to form resistance elements.

Before proceeding to some examples, let us check the units in the basic formula. We have, listing units in parentheses, units of

$$R = \rho \frac{L}{A} = \frac{(\Omega \cdot \text{m})(\text{m})}{(\text{m}^2)} = \Omega$$

EXAMPLE 3.2-1 Resistance of a Conductor of Square Cross Section

Find the resistance and conductance of the square conductor shown in Fig. 3.4.

Solution

From Table 3.1, ρ for copper is $17.2 \times 10^{-9}\ \Omega \cdot \text{m}$. Then

$$R = \rho \frac{L}{A} = \frac{(17.2 \times 10^{-9}\ \Omega \cdot \text{m})(10^3\ \text{m})}{(25\ \text{mm}^2)} \times \left(\frac{10^6\ \text{mm}^2}{1\ \text{m}^2}\right)$$
$$= 0.688\ \Omega$$

The conductance is

$$G = \frac{1}{R} = \frac{1}{0.688\ \Omega} = 1.45\ \text{S}$$

• • •

EXAMPLE 3.2-2 Resistance of a Cylindrical Conductor

Find the resistance and conductance of 5000 m of aluminum wire. The cross section is circular with a diameter of 2 mm. What will the resistance be if the diameter is doubled?

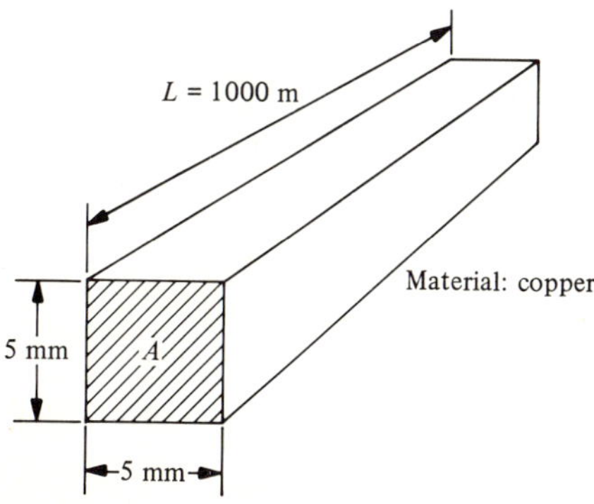

FIGURE 3.4
Conductor for Example 3.2-1.

Solution

From Table 3.1, ρ for aluminum is $28.3 \times 10^{-9}\ \Omega\cdot\text{m}$. The area of the circular cross section is $\pi D^2/4$, where $D = 2$ mm $= 0.002$ m. Then

$$R_1 = \frac{\rho L}{A} = \frac{(28.3 \times 10^{-9}\ \Omega\cdot\text{m})(5 \times 10^3\ \text{m})}{\pi[(2 \times 10^{-3})^2\ \text{m}^2]/4}$$

$$= \frac{(28.3 \times 10^{-9})(5 \times 10^3)\ 4}{\pi(4 \times 10^{-6})}\ \Omega$$

$$= 45\ \Omega$$

The conductance is

$$G_1 = \frac{1}{R_1} = \frac{1}{45\ \Omega} = 0.0222\ \text{S} = 22.2\ \text{mS}$$

If the diameter is doubled, the area is multiplied by 4 and the resistance is divided by 4. Thus for a diameter of 4 mm

$$R_2 = \frac{R_1}{4} = \frac{45}{4} = 11.2\ \Omega$$

and the conductance is

$$G_2 = 4G_1 = 88.8\ \text{mS}$$

• • •

EXAMPLE 3.2-3 *Finding Conductor Length for a Specific Resistance*

How long must a 1 mm diameter round silver wire be in order for the resistance to be 2 Ω?

Solution

From Table 3.1, ρ for silver is $16 \times 10^{-9}\ \Omega\cdot\text{m}$. In this problem, the length L is unknown, so we rearrange the basic equation $R = \rho L/A$ in order to isolate L on the left-hand side. Thus

$$L = \frac{RA}{\rho} = \frac{\pi R D^2}{4\rho} = \frac{\pi(2\Omega)(10^{-3}\ \text{m})^2}{4(16 \times 10^{-9}\ \Omega\cdot\text{m})}$$

$$= 98\ \text{m}$$

• • •

LEARNING EXERCISE FOR SEC. 3.2

1. A wire is stretched from one end to the other of a metric football field (100 m). Find the resistance if the wire has a diameter of 0.3 cm and is fabricated from (a) copper, (b) aluminum, (c) nichrome.

Ans. 0.4; 14.1; 0.243

• • •

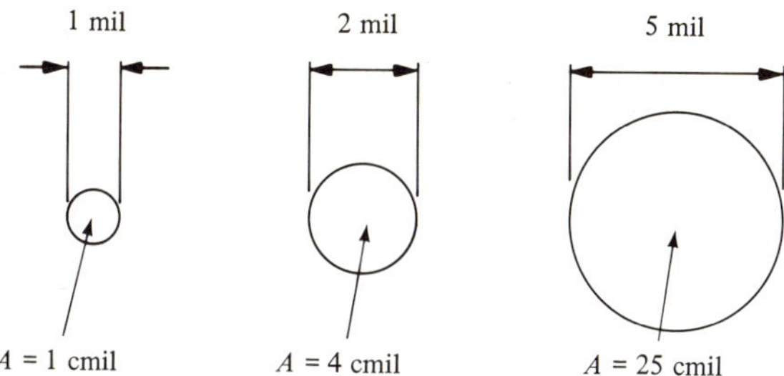

FIGURE 3.5
Illustration of circular mil unit of area.

3.3 WIRE TABLES

Up to this point, we have used SI units exclusively. In certain areas of technology, however, conversion to this universal system is taking place slowly and undoubtedly will not be complete for some time to come. The manufacture of round wire is one such area. In the United States, the American Wire Gauge (AWG) system is commonly used for specifying wire sizes. A table of AWG wire sizes, listed according to gauge number, is included in Appendix B. For each gauge number, the diameter is given in *mils* (1 mil = 0.001 in), and the area is given in circular mils (cmil), a unit of area used specifically for circular cross sections. The area of a circle in circular mils is simply the square of the diameter in mils, as shown in Fig. 3.5.

In addition, the table lists, for each wire size, factors such as the resistance per 1000 ft at a standard temperature, the weight per 1000 ft, and the maximum current that can safely be carried under various conditions.

The gauge numbers are arranged so that the wire diameter *decreases* as the gauge numbers *increase*. Thus wire resistance increases with increasing gauge number. The numbers are arranged systematically so that each succeeding number represents approximately a 25% change in cross-sectional area. If we consider a change of 3 gauge numbers, the area change is then approximately $(1.25)^3 \approx 2$ times. An increase of 3 gauge numbers then corresponds to a doubling of the resistance. For a change of 10 gauge numbers the area and resistance change by a factor of $(1.25)^{10} \approx 10$.

Use of the wire tables is illustrated in the examples that follow.

EXAMPLE 3.3-1 Resistance of a Specified Length of Wire

Use the wire table in Appendix B to find the resistance of 300 ft of AWG #20 solid copper wire at 25°C.

Solution

From the wire table, we find that AWG #20 solid copper wire has a resistance of 10.4 Ω/1000 ft at 25°C. Since resistance is directly proportional to length, we have

$$\frac{R_{300}}{R_{1000}} = \frac{300\text{ ft}}{1000\text{ ft}}$$

Therefore,

$$R_{300} = \frac{300}{1000} \times R_{1000} = 0.3 \times 10.4 = 3.12\ \Omega$$

• • •

EXAMPLE 3.3-2 Wire Size to Achieve a Specific Resistance

A loudspeaker that has a nominal resistance of 8 Ω is to be located 50 ft from the amplifier, as shown in Fig. 3.6. In order to minimize power loss in the connecting wire, it should have as small a resistance as possible.

a. Find the wire size required in order that the line resistance be less than or equal to 10% of the speaker resistance.

b. Repeat for 5%.

Solution

a. For a run of 50 ft the current must pass through 100 ft of wire. Thus we must find the wire size for which the resistance of 100 ft of wire is less than or equal to 10% of 8 Ω. Since the wire table lists wire resistance per 1000 ft we convert as in the previous example:

$$\frac{R_{1000}}{R_{100}} = \frac{1000}{100}$$

$$R_{1000} = 10R_{100} = (10)(0.8) = 8\ \Omega$$

The wire we choose must have a resistance less than or equal to 8 Ω/1000 ft. From the wire table in Appendix B, AWG #19 has a resistance of 8.21 Ω/1000 ft while #18 has a resistance of 6.51 Ω/1000 ft, so that our choice is #18 AWG, a readily available size.

b. For this case we have $R_{100} = 0.4\ \Omega$ so that $R_{1000} = 4\ \Omega$. From the wire table, #16 wire has a resistance of 4.09 Ω/1000 ft while #15 has 3.25 Ω/1000 ft. At this point, some practical judgment must be exercised. Not all wire sizes are stocked by distributors and #16 would much more likely be available, so that it would be our choice even though its resistance is slightly higher than the amount specified.

• • •

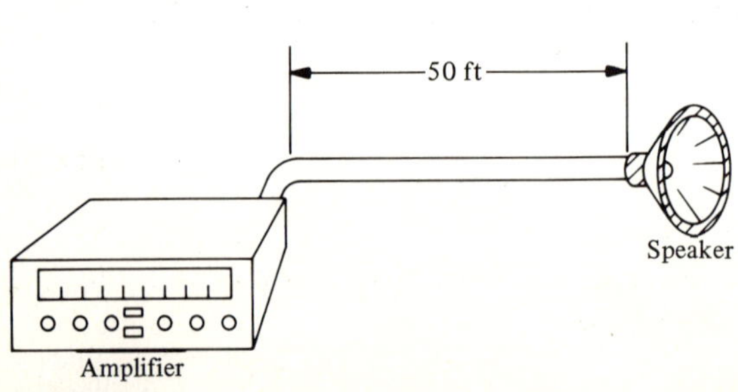

FIGURE 3.6
Example 3.3-2.

EXAMPLE 3.3-3 Finding Wire Diameter

In a certain power application it is found that a wire must have a cross-sectional area of at least 100 000 cmil. Find the wire diameter. What AWG satisfies this requirement?

Solution

The wire diameter can be found from the formula

$$A_{cmil} = (D_{mil})^2$$

Then

$$D_{mil} = \sqrt{A_{cmil}} = \sqrt{100\,000} = 316 \text{ mil}$$

From the wire table, AWG #0 meets the requirement.

• • •

LEARNING EXERCISES FOR SEC. 3.3

1. Find the resistance of 1 mi of #8 copper wire.
2. What wire size will have a resistance less than 2 Ω in 250 ft?
3. Specifications for copper wire to be used in a motor include the following: (a) minimum area 1500 cmil; (b) maximum resistance 4.5 Ω/1000 ft. What AWG wire size should be used?

Ans. 16; 18; 3.32

• • •

3.4 SHEET RESISTANCE

In modern electronic circuit construction, elements are formed by depositing layers of material on a supporting base, or substrate. Sophisticated manufacturing techniques are used in this process and the resulting devices are classified as thick-film or thin-film integrated circuits. For both types, resistors take the form shown in Fig. 3.7. The cross-sectional area is $A = TW$, so the basic formula $R = \rho L/A$ becomes

$$R = \frac{\rho L}{TW} \tag{3.4-1}$$

In most integrated circuits a standard thickness is used, so Eq. (3.4-1) can conveniently be rewritten in the form

$$R = \left(\frac{\rho}{T}\right)\frac{L}{W} \tag{3.4-2}$$

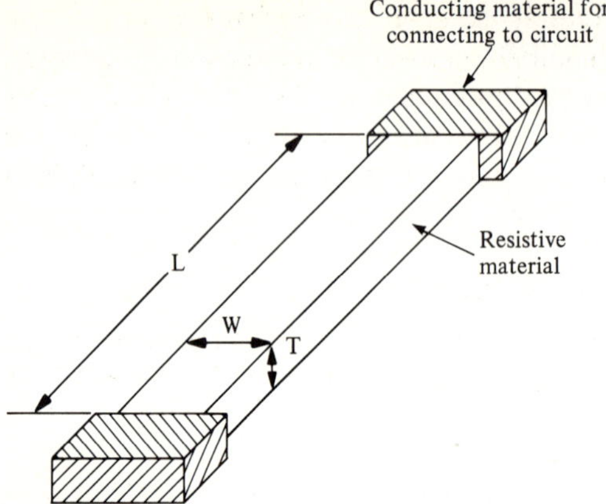

FIGURE 3.7
Deposited resistor of length *L*, width *W*, and thickness *T*.

Because the thickness T is a standard value, the factor ρ/T will be a constant for any given material. For this reason it is convenient to define a new resistivity

$$R_s = \frac{\rho}{T}\frac{(\Omega\cdot\text{m})}{(\text{m})} = \frac{\rho}{T}(\Omega) \tag{3.4-3}$$

Using this in Eq. (3.4-2),

$$R = R_s\frac{L}{W} \tag{3.4-4}$$

where R and R_s are in ohms, and L and W are usually in meters, but can be in any convenient unit of length, so long as it is the same for both. This formula states that we can find the resistance between conducting ends by multiplying R_s by the ratio of length to width.

The new resistivity R_s has an interesting physical interpretation. In order to explain this, let us first find the resistance of a 1-m square of material. In this case, $L = W = 1$ m and

$$R = R_s\left(\frac{1\text{ m}}{1\text{ m}}\right) = R_s\,\Omega$$

Thus the resistance between ends of a 1-m square is $R_s\,\Omega$. Next, we find the resistance of a 1-cm square of the same material and thickness. Then $L = W = 1\text{ cm} = 0.01$ m and

$$R = R_s\left(\frac{0.01\text{ m}}{0.01\text{ m}}\right) = R_s\,\Omega$$

Clearly, for *any size* square of this material, the resistance will be the same, so that R_s is the resistance of a *unit square*. For this reason, R_s is called the *sheet resistance* and its units are usually specified as *ohms per square* ($\Omega/\square$). This is equivalent to Ω but serves to remind us that it is the resistance per square of a layer of resistive material of constant thickness. This is illustrated in the examples which follow.

EXAMPLE 3.4-1 Design of Deposited Resistors

In a certain integrated circuit the resistive material has a sheet resistance of 120 $\Omega/\square$. Assuming a maximum width of 3 mm, design resistors of 500 Ω, 1 kΩ, and 5 kΩ.

Solution

We will assume a width of 3 mm for all three resistors. Then the only thing to be determined in each case is the length. We rearrange the basic equation $R = R_s L/W$ so that L is isolated. This yields

$$L = \frac{RW}{R_s} \tag{3.4-5}$$

where R is in Ω, R_s is in $\Omega/\square$, and, for convenience, we will use millimeters for both L and W. The resistor lengths are then

$$500\ \Omega: \quad L_1 = \frac{(500)(3)}{(120)} = 12.5 \text{ mm}$$

$$1\ \text{k}\Omega: \quad L_2 = \frac{(10^3)(3)}{(120)} = 25 \text{ mm}$$

$$5\ \text{k}\Omega: \quad L_3 = \frac{(5 \times 10^3)(3)}{(120)} = 125 \text{ mm}$$

For some applications L_1 and L_2 are reasonable lengths, but L_3 is far too long. This difficulty can be circumvented by using a zigzag type of construction, as shown in Fig. 3.8.

• • •

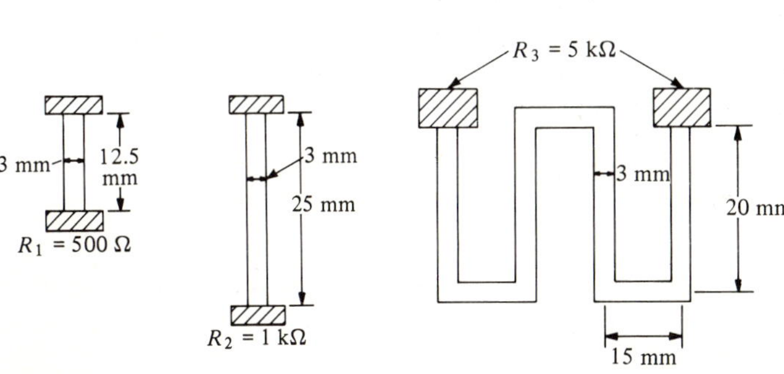

FIGURE 3.8
Resistors designed in Example 3.4-1.

LEARNING EXERCISE FOR SEC. 3.4

1. A resistive material has a sheet resistance of 55.6 Ω/□. If the maximum width is to be 5 mm, what would the length of 50-Ω, 150-Ω, and 500-Ω resistors be?

Ans. 13.5; 45; 4.5

• • •

3.5 RESISTANCE CHANGES WITH TEMPERATURE

Up to this point, we have discussed the way in which resistance varies with the geometry of the material. In addition to the dependence on geometry, the resistance of most materials increases with temperature. The variation is usually approximated by a straight line relationship, as shown in Fig. 3.9, which is accurate enough for most practical purposes over a wide range.

The Normalized Graph

Before proceeding, let us consider the graph of Fig. 3.9. An interesting feature of this graph is that the resistance axis (y axis) is *normalized,* a technique that is often used with such graphs because it makes them universal. For this case, the resistance is normalized to the value at 20°C by plotting R_T/R_{20} rather than simply R_T. The resistance at 20°C [68°F] is used because this temperature is a standard *ambient* temperature, that is, the normal temperature of the air or other medium surrounding the resistor. The graph can then be used for any value of resistance as illustrated in the following example.

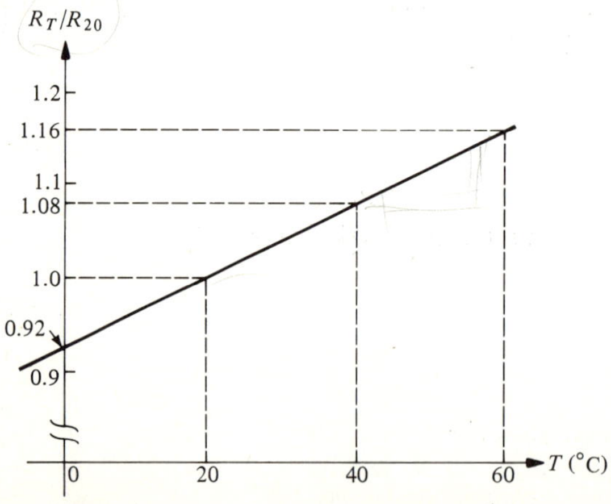

FIGURE 3.9
Normalized resistance variation with temperature for copper.

EXAMPLE 3.5-1 Using the Graph to Find the Change Due to Temperature

A copper wire has 25-Ω resistance at 20°C.

a. Use the graph to find its resistance at 60°C.
b. Repeat if the original resistance is 50 Ω at 20°C.

Solution

From the graph we see that $R_T/R_{20} = 1.16$ at 60°C. Then

a. $$\frac{R_T}{R_{20}} = \frac{R_T}{25\ \Omega} = 1.16$$

$$R_T = (25)(1.16) = 29\ \Omega$$

b. For any value of resistance at 60°C we use the same figure from the graph. Then

$$\frac{R_T}{50\ \Omega} = 1.16$$

$$R_T = (50)(1.16) = 58\ \Omega$$

• • •

The Resistance–Temperature Relation

In order to completely specify the straight line, we require two pieces of information. For this application, the slope of the line is used because it is a constant for any given material. It is called the *temperature coefficient of resistance*, and the letter symbol usually used is the Greek letter α (alpha). The second item used to specify the line is the resistance of the conductor at one specific temperature, usually normal room temperature of 20°C. The most useful form of the straight line is the point-slope form [see Eq. (1.6-13)], in which we use the normalized resistance

$$\alpha = \frac{(R_T/R_{20}) - 1}{T - 20} \qquad (3.5\text{-}1)$$

This is rearranged to isolate the unknown R_T. The result is

$$R_T = R_{20}[1 + \alpha(T - 20)] \qquad (3.5\text{-}2)$$

where

R_T = resistance in ohms at temperature T
R_{20} = resistance in ohms at 20°C
T = temperature in degrees Celsius
α = temperature coefficient of resistance in $(°\text{C})^{-1}$

Values of α for various materials are given in Table 3.2.

TABLE 3.2
Temperature coefficients of resistance at 20°C

Material	α (°C^{-1})
Silver	0.0038
Copper	0.0039
Gold	0.0034
Aluminum	0.0039
Tungsten	0.0051
Nickel	0.006
Iron	0.0055
Constantan	0.000 008
Nichrome	0.000 44
Carbon	−0.0005

Meaning of the Temperature Coefficient

The temperature coefficient of resistance has a useful physical interpretation. Let us consider that the temperature increases by 1°C. Then Eq. (3.5-2) becomes

$$R_{21} = R_{20}[1 + \alpha(21 - 20)]$$

$$R_{21} = R_{20}(1 + \alpha) \tag{3.5-3}$$

Assume that $\alpha = 0.05$. Then

$$R_{21} = 1.05R_{20} \tag{3.5-4}$$

and we see that at 21°C the resistance is 1.05 times that at 20°C. Thus the increase is 5%/°C and we can interpret 100α as the *percent increase in resistance per degree Celsius.* Referring to the table of temperature coefficients, we can now interpret the values of α in the following way. If the material is silver, its resistance increases $100\alpha = 0.38\%/°C$; aluminum increases 0.39%/°C, and so on. Carbon is the only material in the table with a negative temperature coefficient, and its resistance *decreases* 0.05%/°C.

EXAMPLE 3.5-2 Wire Resistance at Extremes of Temperature

Find the resistance of 1 mi of AWG #0 copper wire at 0°F and at 90°F. These temperatures represent typical extremes in some parts of the country.

Solution

To convert the Fahrenheit temperature T_F to the Celsius temperatures T_C we use the formula

$$T_C = \frac{5}{9}(T_F - 32)$$

when

$$T_F = 0, \qquad T_C = \frac{5}{9}(-32) = -17.8°C$$

$$T_F = 90, \qquad T_C = \frac{5}{9}(90 - 32) = 32.2°C$$

From the wire table in Appendix B, #0 wire has a resistance of 0.1 Ω/1000 ft at 25°C. Then, for 1 mi of wire

$$R_{25} = \left(\frac{0.1\ \Omega}{1000\ \text{ft}}\right)(5280\ \text{ft}) = 0.528\ \Omega$$

Now we use Eq. (3.5-2), modified to take account of the fact that the reference temperature is 25°C, and with $\alpha = 0.0039/°C$

$$R_T = R_{25}[1 + \alpha(T - 25)]$$

At $T_F = 0°F$ ($T_C = -17.8°C$)

$$\begin{aligned} R_0 &= 0.528[1 + 0.0039(-17.8 - 25)] \\ &= 0.44\ \Omega \end{aligned}$$

At $T_F = 90°F$ ($T_C = 32.2°C$)

$$\begin{aligned} R_{90} &= 0.528[1 + 0.0039(32.2 - 25)] \\ &= 0.54\ \Omega \end{aligned}$$

The change, expressed as a percentage of the low temperature value, is

$$\frac{R_{90} - R_0}{R_0} \times 100 = \frac{0.54 - 0.44}{0.44} \times 100 = 23\%$$

In succeeding chapters we consider the effect of this resistance change on the rest of the system.

• • •

LEARNING EXERCISES FOR SEC. 3.5

1. A copper bus bar has 1.2-Ω resistance at 20°C. Use the graph of Fig. 3.9 to find its resistance at 127°F.
2. An aluminum conductor has 2.3-Ω resistance at 20°C. What is its resistance at 130°C?
3. A nickel heater wire has 1.1-Ω resistance at 20°C. What is its resistance at 200° C?

Ans. 2.29; 1.34; 3.29

• • •

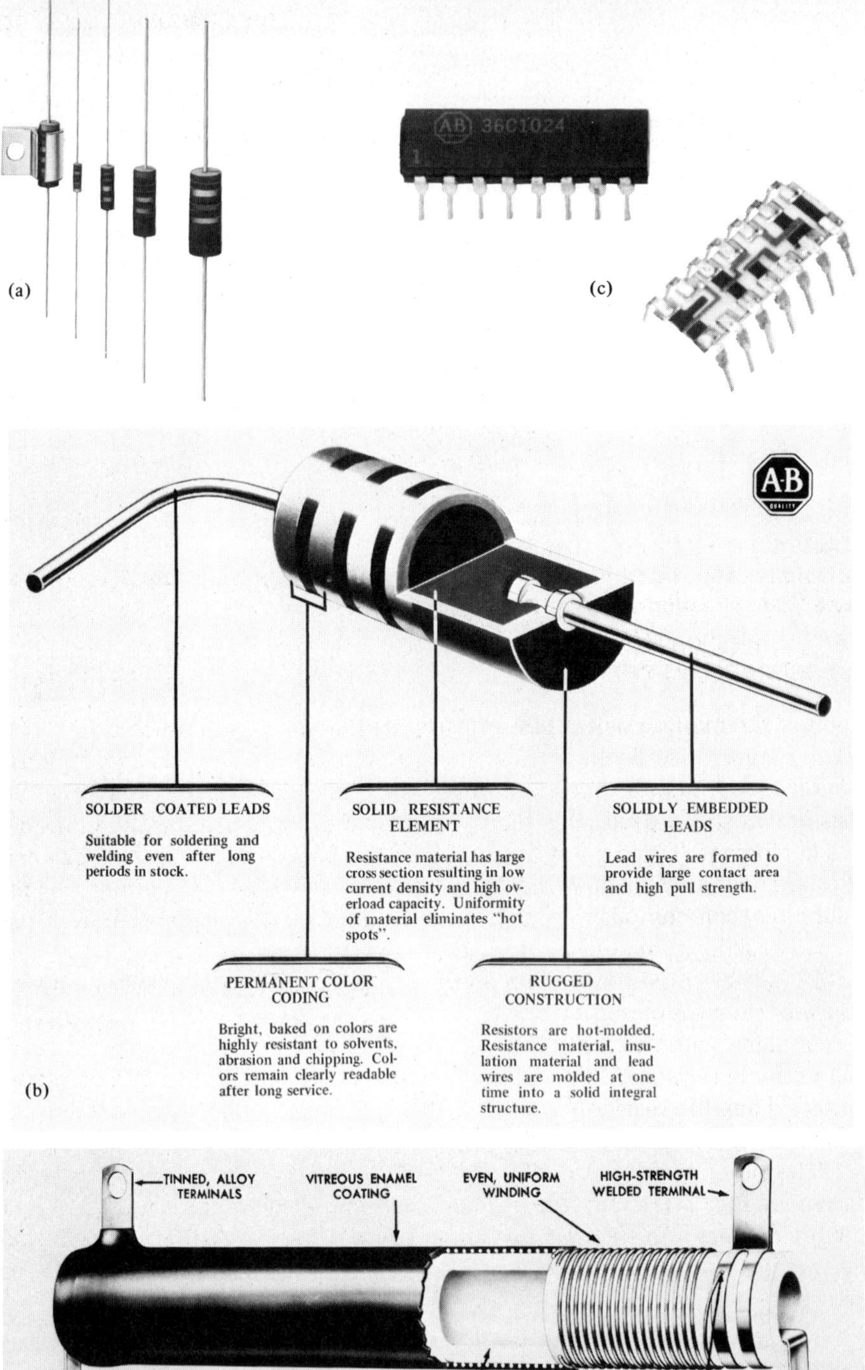

FIGURE 3.10
Commercially available resistors. (a) Carbon resistors, 1/4 W to 2 W. (b) Cutaway view of carbon resistor showing construction details. (c) Dual-in-line package resistor network for use in hybrid circuits (courtesy of Allen-Bradley Co.). (d) Wirewound resistor cutaway view (courtesy of Ohmite Manufacturing Co.).

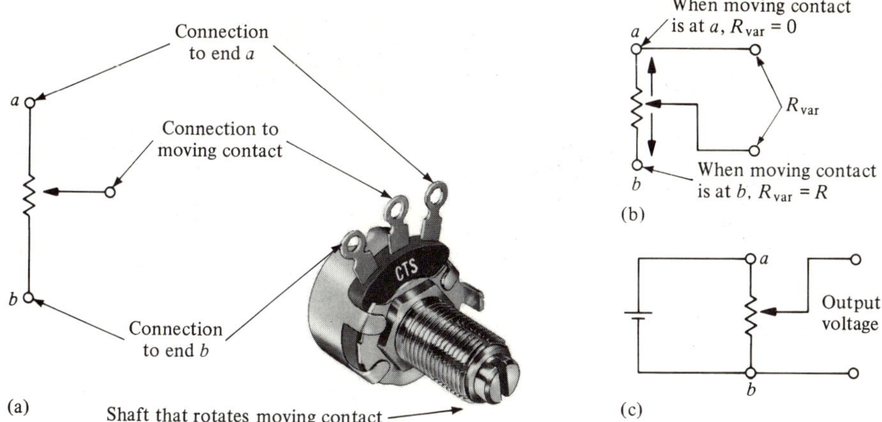

FIGURE 3.11
Variable resistor. (a) Photo (courtesy of CTS Corporation) and contact arrangement. (b) Application as a rheostat. (c) Application as a potentiometer.

3.6 RESISTOR TYPES

Some of the many types of resistors available today are shown in Fig. 3.10. The primary specification is, of course, the value of resistance. Other specifications include power rating, that is, the maximum power that the resistor can safely dissipate; tolerance (discussed in Sec. 3.8); and temperature coefficient. Typical applications and specifications for some types are given in Table 3.3. In terms of numbers, the type most used is the molded-carbon composition resistor, for which a cutaway view illustrating the construction is shown in Fig. 3.10b.

In general, resistors fall into two categories; fixed or variable. Up to now we have discussed only fixed resistors. A typical variable resistor is shown along with its circuit diagram in Fig. 3.11a. This device can be used as a variable resistance by making connections to one end and the moving contact. Then the resistance can be varied from zero Ω to the rated value, as shown in Fig. 3.11b. In this application it is called a *rheostat*.

The other way to use this device is as a *potentiometer* or "pot" as shown in Fig. 3.11c. In this application the output voltage between the moving contact and the end of the resistor is controlled by the position of the moving contact. The theory of this type of circuit is discussed in Sec. 4.5.

Nonlinear Resistors

The resistors discussed previously have all been linear, which means that the relationship between the current through and voltage across them is a straight line. The *thermistor* is a nonlinear resistor made from semiconductor material that is extremely sensitive to changes in temperature. Figure 3.12 shows how the resistance of a typical negative temperature coefficient

TABLE 3.3
Some resistor types

Application	Type	Resistance Range	Power Range	Temperature Coefficient (/°C)	Tolerances Available
General purpose	Molded-carbon composition	1 Ω to 100 MΩ	1/8 W, 1/4 W, 1/2 W 1 W, 2 W	0.0005	5%, 10%, 20%
Power	Wire-wound	0.1 Ω to 50 kΩ	2 W to 250 W	0.005 to 0.002	
Precision	Metal film Thin film Wire-wound	0.1 Ω to 100 MΩ	1/20 W to 5 W	Less than 0.001	1% or less
Variable	Carbon Wire-wound	10 Ω to 2 MΩ	Up to 5 W	0.000 02 to 0.002	
Resistance network	Thick film Thin film	10Ω to 10 MΩ	Up to 2 W/package	0.0001 to 0.0002	
High-voltage	Carbon, vacuum-sealed in glass	Up to 10^{12} Ω	Up to 40 000 V		

thermistor varies with temperature. For comparison, we have shown on the same graph the variation of resistance of copper. From the graph we see that the resistance of the thermistor decreases rapidly as the temperature increases. This makes the device useful in the measurement and control of temperature. It is also useful as a protective device in circuits where its high cold resistance limits current flow until the circuit has had a chance to warm up.

Another nonlinear resistor is the *varistor,* a device made of silicon carbide, in which the current varies as a power of the voltage. A typical characteristic curve is shown in Fig. 3.13 along with the circuit symbol for the device. Because of the shape of the characteristic, the varistor is useful for voltage surge protection for voltages up to 10 000 V.

The *photoconductive cell* is a semiconductor device in which the resistance depends on the amount of light hitting the exposed surface of the semiconductor material. As the amount of light increases, the resistance decreases, as shown in Fig. 3.14a. The considerable change in resistance makes this device useful in control applications. For example, in Fig. 3.14c bottles on a conveyor belt interrupt a light beam pointed at a photoconductive cell. Each interruption of the light beam causes a resistance change, which can be used to operate a counter. The output of the counter can be used to turn the process off after a certain number of bottles have passed, or to perform any other desired control function.

3.7 STANDARD RESISTOR COLOR CODE

Composition resistors, widely used in electronics applications, have their resistance value and tolerance indicated by a color code printed in three or

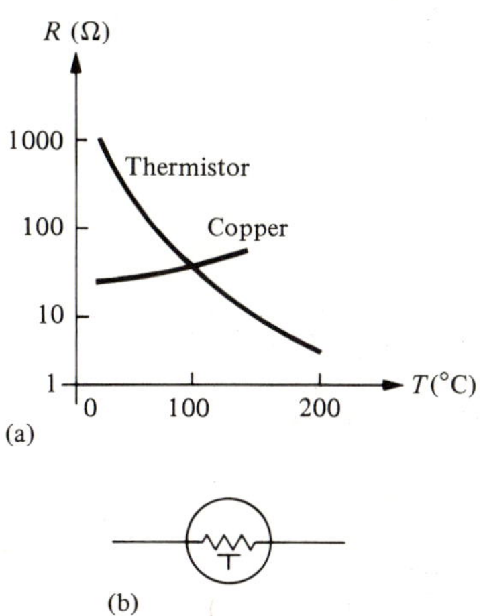

FIGURE 3.12
Thermistors. (a) Resistance-temperature curve. (b) Circuit symbol.

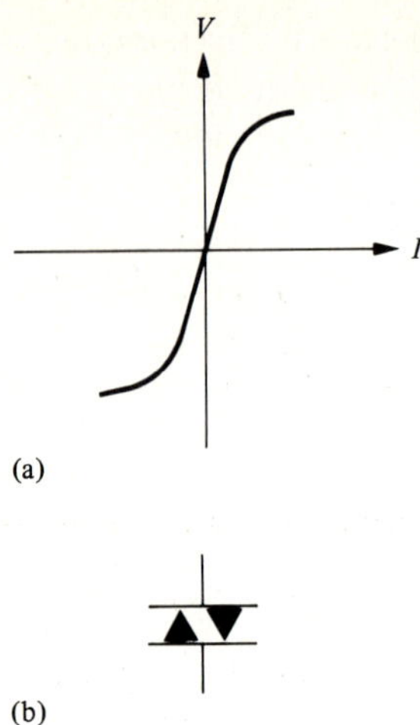

FIGURE 3.13
Varistor. (a) Characteristic curve. (b) Circuit symbol.

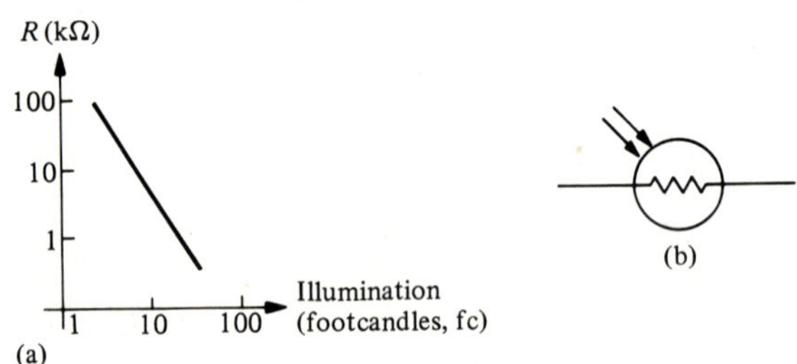

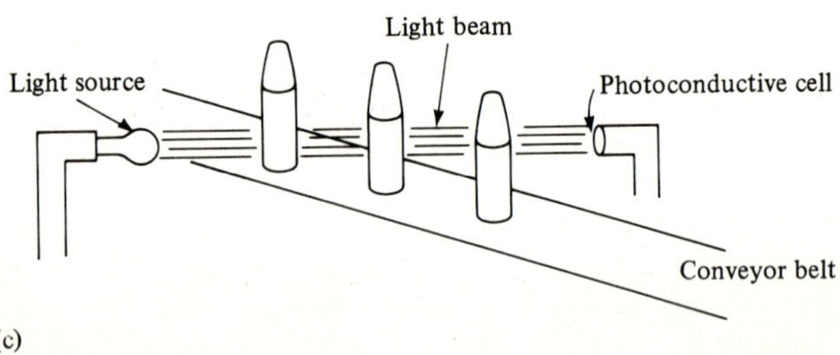

FIGURE 3.14
Photoconductive cell. (a) Characteristic curve. (b) Circuit symbol. (c) Counting application.

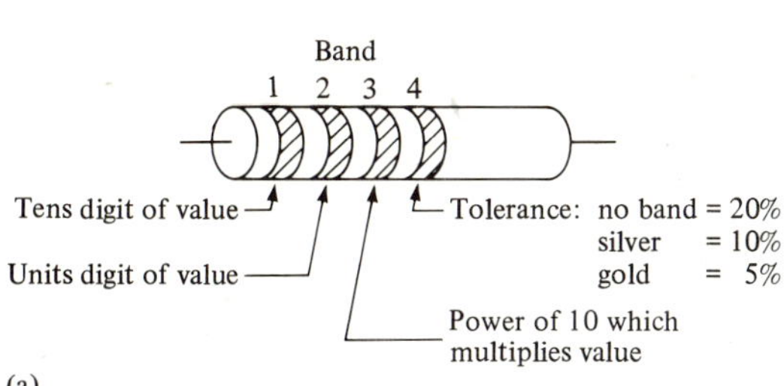

Color	Value
Black	0
Brown	1
Red	2
Orange	3
Yellow	4
Green	5
Blue	6
Violet	7
Gray	8
White	9

(b)

FIGURE 3.15
Composition resistor color code. (a) Interpretation of bands. (b) Table of color values.

four bands on the cylindrical body of the resistor. The key to the color code is shown in Fig. 3.15 along with the digit values of the various colors. Use of the code is illustrated in the following examples.

EXAMPLE 3.7-1 Finding the Value of Resistance

Find the resistance and tolerance of a resistor having the following four colored bands, starting with the one nearest the end: red, violet, orange, gold.

Solution

From Fig. 3.15 we find the following values for each color:

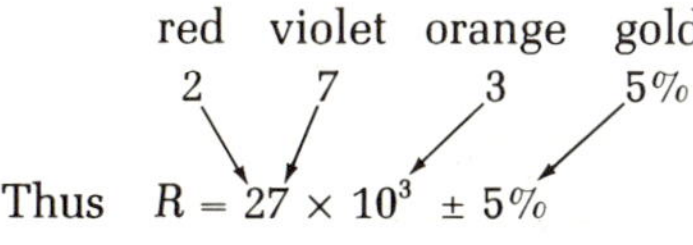

• • •

EXAMPLE 3.7-2 Finding the Colors for a Specific Resistance Value

A certain circuit requires a 10 000 Ω, 10% tolerance resistor. What color bands should we look for?

Solution

We write the resistance value and tolerance so as to isolate the tens digit, units digit, and power of 10 for the value, listing the appropriate color for each from the table in Fig. 3.15.

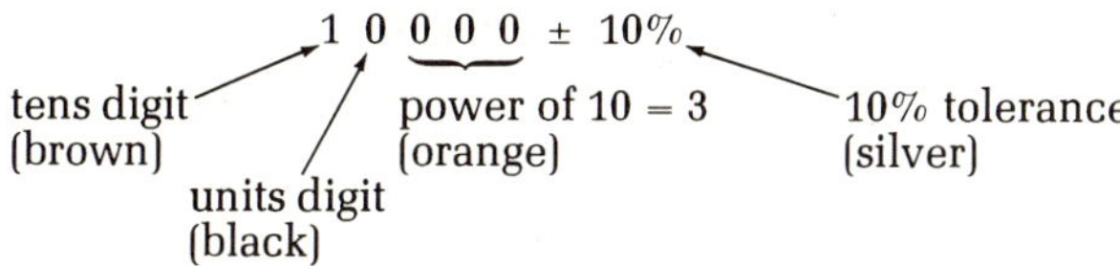

The color band sequence for this resistor, beginning with the band nearest the end is: brown, black, orange, silver.

• • •

LEARNING EXERCISES FOR SEC. 3.7

1. What is the resistance corresponding to the following color band combinations? (a) red, red, yellow; (b) yellow, violet, red; (c) brown, black, brown

Ans. 4.7; 220; 0.1

2. What are the color band combinations for the following resistance values? (a) 6.8 kΩ; (b) 330 kΩ; (c) 1.5 MΩ

Ans. blue, gray, red; brown, green, green; orange, orange, yellow

• • •

3.8 TOLERANCE AND PREFERRED VALUES

It is impossible for component manufacturers to produce every possible value which circuit designers might require. Instead they produce sets of *preferred values* with standard tolerances. A typical tolerance for carbon resistors as used in electronic circuits is ±10%. This means that if we purchase a resistor which is labeled 10 kΩ, 10%, its nominal value is 10 kΩ and the tolerance is ±10% of 10 kΩ = ±0.1 × 10 = ±1 kΩ, and its resistance must lie between 9 and 11 kΩ. Other standard tolerances are ±1, ±5, and ±20%. The 1% components are the most expensive of the four, while the 20% group is the cheapest. For economic reasons, the 20% type will be found in many circuits, because it often turns out that the component values are not critical. We will have more to say about this later in the text when circuit design is discussed.

The 20% series requires only six numbers, each of which may be multiplied by any power of 10. The numbers are

10, 15, 22, 33, 47, 68

These numbers represent the *nominal* values. The *range* of values for each is ±20%. The range of each number is given in Table 3.4. This is illustrated graphically in Fig. 3.16 and we see that all possible values are covered. In the example which follows, we consider the effect of this tolerance on a circuit design.

EXAMPLE 3.8-1

A particular circuit design calls for a 16-kΩ resistor. Find the maximum possible departure from the design value if 20% resistors are to be used in the production of the circuit.

Solution

We choose 15 kΩ as the closest value in the 20% series. Then, if we purchase a quantity of 15 kΩ, 20% resistors in order to produce the circuit, they will range in value from 12 to 18 kΩ. The worst departure from the design value of 16 kΩ will occur in circuits constructed with those at the ends of the range. For example, if the

TABLE 3.4
Ranges of 20% tolerance preferred value series

Nominal	±20% Tolerance	Range
6.8	±1.36	—
10	±2	$8 < R < 12$
15	±3	$12 < R < 18$
22	±4.4	$17.6 < R < 26.4$
33	±6.6	$26.4 < R < 39.6$
47	±9.4	$37.6 < R < 56.4$
68	±13.6	$54.4 < R < 81.6$
100	±20	—

actual resistance of the resistor taken from the box is 12 kΩ, then the departure is

$$\begin{aligned}\% \text{ error} &= \frac{\text{actual value} - \text{desired value}}{\text{desired value}} \times 100 \\ &= \frac{12\text{ k}\Omega - 16\text{ k}\Omega}{16\text{ k}\Omega} \times 100 \\ &= -25\%\end{aligned}$$

If the actual resistance is 18 kΩ, the departure is

$$\% \text{ error} = \frac{18\text{ k}\Omega - 16\text{ k}\Omega}{16\text{ k}\Omega} \times 100 = +12.5\%$$

At this point, we must return to the circuit design in order to determine the effect of these departures on the overall performance. We may find it necessary to use 10 or 5% resistors. This will be considered in Chap. 4, when we begin to design circuits.

• • •

The 10 and 5% Series

The 10% series includes all six values in the 20% series, in addition to six intermediate values, while the 5% series includes all 12 values in both 10

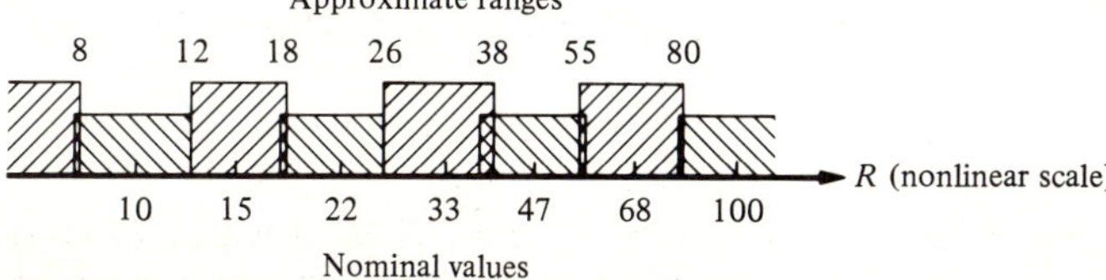

FIGURE 3.16
Ranges of 20% tolerance preferred value series.

and 20% series as well as 12 intermediate values. Since the value of any resistor is known only to two significant digits, the results of calculations involving circuits containing these resistors may be rounded to two significant digits. The numbers in the three series are tabulated below for comparison.

5% Series	10% Series	20% Series
10	10	10
11		
12	12	
13		
15	15	15
16		
18	18	
20		
22	22	22
24		
27	27	
30		
33	33	33
36		
39	39	
43		
47	47	47
51		
56	56	
62		
68	68	68
75		
82	82	
91		

• • •

LEARNING EXERCISE FOR SEC. 3.8

1. What is the range of values of a resistor which has a nominal value of 2.2 kΩ and a tolerance of (a) 20% (b) 10% (c) 5%?

Ans. 2.09–2.31; 1.98–2.42; 1.76–2.64

• • •

SUMMARY

Resistors are used in most applications of electrical technology. In this chapter we have considered those properties of materials which make them useful as resistance elements. We also discussed typical resistance element configurations. The important points concerning each topic are listed below. In Chap. 4 we will connect resistors and sources to form circuits and learn how to calculate the important variables.

Section

3.1 1. The symbol for resistance is R and its dimension is the ohm (Ω).

2. Conductance G is the reciprocal of resistance; its unit is the siemens (S).

3.2 3. Resistivity is symbolized by ρ and its unit is ohm-meters ($\Omega \cdot m$).

4. The formula for the resistance of a conductor is $R = (\rho\, L/A)\ \Omega$.

3.3 5. The area of a circle in circular mils (cmil) is the square of the diameter in mils.

3.4 6. The sheet resistance of a square of material of thickness T and resistivity ρ is $R_S = (\rho/T)\ \Omega/\Box$.

7. The resistance of a nonsquare flat resistor of length L and width W is $R = R_S(L/W)$.

3.5 8. The formula for the resistance of wire at temperature T (degrees Celsius) in terms of its resistance at 20°C is

$$R_T = R_{20}[1 + \alpha(T - 20)]$$

where α is the temperature coefficient of resistance.

3.7 9. The color bands on a resistor give the tens digit value, the units digit value, the power of 10 multiplier, and the tolerance.

3.8 10. The 20% preferred value series consists of the six numbers 10, 15, 22, 33, 47, and 68, each of which may be multiplied by any power of 10.

11. The range of any resistor R with 20% tolerance extends from $0.8R$ to $1.2R$.

QUESTIONS FOR REVIEW

Sec. 3.1

1. What differentiates insulating materials from conducting materials?
2. What differentiates semiconductors from insulating materials and conducting materials?
3. How is resistance related to conductance?
4. What are the units for resistance and conductance?

Sec. 3.2

5. Why is the resistance of a conductor proportional to cross-sectional area and inversely proportional to length?
6. Define *resistivity* and give its units.
7. What happens to the resistance of a conductor if it is stretched enough so that its cross-sectional area decreases and its length increases?
8. How does the resistance of a circular conductor vary with the diameter of the cross section?

9. Is constantan as good a conductor as copper?
10. What is the difference between resistance and resistivity?

Sec. 3.3

11. What is a circular mil?
12. A conductor has a diameter of 0.02 in. What is its area, in centimeters?
13. What is the resistance per 1000 ft of #12 wire?

Sec. 3.4

14. What is meant by the term *sheet resistance?*
15. In terms of sheet resistance, how does the resistance of a 1-cm square of a particular material compare to the resistance of a 10-m square if both have the same thickness?

Sec. 3.5

16. How does the resistance of most materials vary with temperature?
17. What does the term *ambient temperature* mean?
18. Draw a graph similar to Fig. 3.9 for carbon.
19. What is the significance of a negative temperature coefficient?
20. What is the percent increase in resistance per degree Celsius for tungsten?

Sec. 3.6

21. What is a variable resistor?
22. What is a thermistor?
23. Describe several possible uses for a photoconductive cell.

Sec. 3.7

24. Name the colors used in the standard resistor color code and give the digit value of each color.
25. Give the colors and values of the various tolerance bands.

Sec. 3.8

26. What are *preferred values?*
27. What are the numbers in the 20% series?
28. What is meant by nominal value?
29. How is the *range of values* related to the nominal value for a tolerance of ±10%?

PROBLEMS

Sec. 3.1

1. For the following resistance values find the conductance in siemens, millisiemens, and microsiemens.
   a. 150 Ω b. 22 kΩ c. 3.3 MΩ
2. For the following conductance values, find the resistance in ohms, kilohms, and megohms.
   a. 0.021 S b. 0.303 mS c. 14.7 μS

Sec. 3.2

3. Determine the resistance between the two ends of the bar shown in Fig. 3.17 if the resistivity of the material from which it is made is 5 Ω·cm.

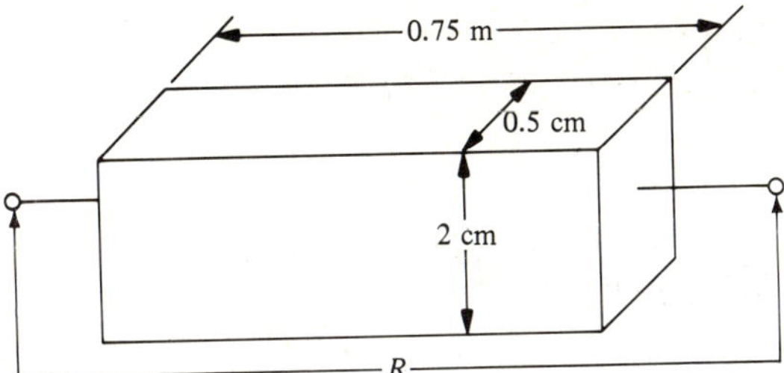

FIGURE 3.17

4. Find the conductance of the bar in Prob. 3.
5. A bar is identical to the one in Prob. 3 except for its length, which is unknown. If the resistance between the two ends of this bar is found to be 100 Ω, what is its length?
6. Find the resistance of 500 m of tungsten wire. The cross section is circular, with a radius of 2 mm.
7. Find the resistance of 750 ft of aluminum wire. The cross section is circular, with a diameter of 1/16 in.
8. Find the resistance of a cylindrical carbon rod 8 in long, with a diameter of 1/2 in.
9. A nichrome heater is constructed using circular wire of diameter 0.01 in. How long must the wire be if its resistance is to be 11 Ω?
10. A carbon bar has a resistance of 0.1 mΩ and a cross section which is 1/4 in square. How long is the bar?
11. A rectangular silver bar has a length of 0.5 m and its resistance is 0.1 Ω. Find its area in square centimeters and square inches.
12. A 1000-m length of round copper wire has a resistance of 2 Ω. Find its diameter in centimeters and inches.
13. A wire has a cross-sectional area of 3 mm^2, a length of 60 m, and a resistance of 1.8 Ω. The same material is used to manufacture wire with a 1-mm^2 cross-sectional area. How much of this wire is required for a resistance of 5 Ω?
14. A circular conducting wire is 90 m long and has a diameter of 3.0 mm and a resistance of 0.6 Ω. The same material is used to manufacture wire with a 1 mm diameter. Find the resistance of 200 m of the 1 mm wire.

Sec. 3.3

15. Work out the conversion factor from circular mils to square inches (cmil to in^2).
16. Repeat Prob. 15 for circular mils to square meters.
17. A copper wire has a diameter of 0.008 in.
    a. Find its area in circular mils.
    b. What is the closest standard gauge?
    c. Find the resistance of 400 yd of this gauge wire at 20°C.
18. Determine the gauge of the thinnest copper wire which will have a resistance no greater than 20 Ω for 50 ft.
19. Find the resistance of 50 ft of #36 wire.
20. Find the resistance of 1400 ft of # 18 wire.
21. A residence is located 100 ft from the power source. The resistance of the transmission line between them must be less than 0.05 Ω. What wire size meets

this requirement? Will this be satisfactory if the load may be as much as 100 A?

22. A 4-Ω loudspeaker is located 150 ft from the amplifier.
    a. Find the wire size required in order that the line resistance be less than 20% of the speaker resistance.
    b. Repeat for 10%.
23. What is the smallest wire size that is safe for a 15-A house circuit?

Sec. 3.4

24. Find the sheet resistance of a 1 mm thick layer of material whose resistivity is 5 kΩ·m.
25. A rectangular layer of resistive material in a hybrid circuit amplifier has a sheet resistance of 100 Ω/□. Its dimensions are 2 × 5 cm. Find the resistance between conductive connectors at each end of the 5-cm dimension.
26. Find the dimensions of a rectangular layer of resistive material whose area is to be 1 cm^2 and whose resistance, from one end to the other, is to be 1 kΩ if R_s = 50 Ω/□.
27. A thin-film resistor in a telephone amplifier is to be made from nitrided tantalum, which has a sheet resistance of 50 Ω/□. Design a 5000-Ω resistor using this material with a width of 0.2 mm.
28. Repeat Prob. 27 for tin oxide, which has a sheet resistance of 1000 Ω/□.

Sec. 3.5

29. Use the graph in Fig. 3.9 to find the resistance at 50°C of a copper bus bar that has a resistance of 0.12 Ω at 20°C.
30. Plot a graph of normalized resistance vs. temperature in degrees Celsius for nickel wire. Use it to find the change in resistance of a length of nickel wire having a resistance of 50 Ω at 20°C, when the temperature rises to 50°C.
31. 1000 ft of #24 AWG round copper wire is used in a telephone system. Find its resistance when the temperature rises to 34°C.
32. A silver conductor in a satellite receiver has a resistance of 1.5 Ω at 25°C. What is its resistance at −10°C?
33. Find the temperature at which the resistance of copper would become zero if the resistance-temperature graph (Fig. 3.9) were extended as a straight line.
34. Repeat Prob. 33 for the nickel wire of Prob. 30.
35. A nickel rod has a resistance of 1 Ω at normal room temperature (20°C). By how many degrees must the temperature change for the resistance to increase to
    a. 1.1 Ω   b. 2 Ω
36. The resistance of a copper conductor is 1.2 Ω at 25°C. At what temperature will the resistance be
    a. 1.3 Ω   b. 1.1 Ω

Sec. 3.7

37. Starting at the band nearest the edge of a composition resistor, what colors will the bands be for a 15 MΩ, 10% resistance value?
38. Starting at the end band of a resistor, the colors are brown, black, and blue. What is the value of the resistance? What is its tolerance?
39. What color bands specify the following 10% resistors?
    1 kΩ, 1.5 kΩ, 2.2 kΩ, 3.3 kΩ, 4.7 kΩ, 6.8 kΩ

40. Repeat Prob. 39 if each value is
    a. divided by 100   b. multiplied by 100

Sec. 3.8

41. Find the range of each number in the 10% series.
42. A circuit design specifies a 40-kΩ resistor. Choose a suitable nominal value and determine the maximum possible departure from the design value if 20% resistors are to be used in production.
43. Repeat Prob. 42 using 10% resistors.
44. Repeat Prob. 42 for a design value of 12 kΩ.

4
Ohm's Law-Series and Parallel Circuits

OBJECTIVES

Upon completion of this chapter, the student should be able to

Section		
4.1	1.	State Ohm's law in its three different forms.
4.2 4.3	2.	State Kirchhoff's voltage law and use it to determine unknown voltages in a series circuit.
4.5	3.	Apply the voltage divider relation to an appropriate circuit in order to determine unknown voltages.
	4.	Design a voltage divider network to meet given specifications.
4.6 4.7	5.	State Kirchhoff's current law, and use it to find unknown currents in a parallel circuit.
4.8	6.	Apply the current divider relation to an appropriate network in order to determine unknown currents.
	7.	Design a current divider network to meet given specifications.
4.9	8.	Determine the effect of an accidental open or short circuit.
4.10	9.	Analyze a loaded voltage divider or current divider network.
	10.	Design a loaded voltage divider circuit containing a bleeder resistor to meet given specifications.
4.11	11.	Determine the power supplied to a two-terminal element in terms of the voltage across it and the current through it.
4.12	12.	Determine the cost of operating an electric device for a given amount of time.
4.13	13.	Determine the efficiency of an electric system given appropriate data.

INTRODUCTION

In previous chapters we studied current, voltage, power, and resistance. In this chapter we interconnect voltage or current sources and resistors to form

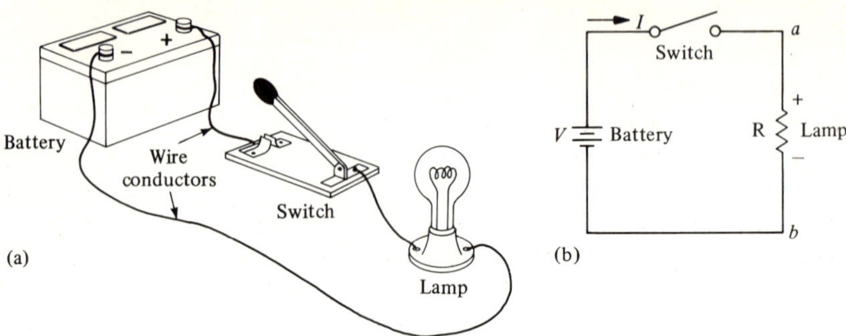

FIGURE 4.1
Electric circuit. (a) Pictorial view. (b) Schematic circuit.

circuits. We then state the basic circuit laws, and use them to determine the electrical behavior of these circuits. Our aim here is to develop understanding and facility in making routine calculations and to provide a basis from which the student can begin to obtain an intuitive understanding of the behavior of resistive circuits.

4.1 OHM'S LAW

The simplest electric circuit consists of two elements: a source, which supplies energy, and an element that absorbs energy. An example is shown in Fig. 4.1.a. In the circuit the source of energy is a battery and the element that absorbs the energy is a light bulb. Wire conductors carry the energy to and from the elements, and a switch is included so that the flow of energy can be started and stopped as desired. In practice we do not draw circuits in this manner. Instead, we use the circuit symbols introduced in Chaps. 2 and 3. Using these, the *schematic diagram* of the circuit takes the form shown in Fig. 4.1b. The light bulb is characterized by its resistance.

When the switch is open, no charge can flow. When the switch is closed so that a path for current exists, the battery voltage, current, and resistance are related by Ohm's law.* In words, Ohm's law states that the voltage is proportional to the current, with the resistance being the constant of proportionality. In equation form, this is†

$$\boxed{V = RI} \tag{4.1-1}$$

*After George Ohm, who first published this relation in 1826.
†This formula should be thoroughly understood and memorized immediately! The phrase, "Vermont equals Rhode Island" is one possible memory aid.

where

V = voltage, in volts
R = resistance, in ohms
I = current, in amperes

As an equation, this can be written in two alternate forms in order to isolate current and resistance (this was done in Sec. 1.5).

$$\boxed{I = \frac{V}{R}} \tag{4.1-2}$$

and

$$\boxed{R = \frac{V}{I}} \tag{4.1-3}$$

The units used with these equations must be consistent. The basic units are, for $V = RI$

$$\text{Volts (V)} = \text{ohms } (\Omega) \times \text{amperes (A)} \tag{4.1-4}$$

Direction and Polarity

Consider the hydraulic analog shown in Fig. 2.4. The number of gallons per minute of fluid passing through any point of the pipe in this hydraulic circuit must be the same at all points. Returning to the electric circuit in Fig. 4.1b, the motion of charge is such that the current must be the same through all points when the switch is closed. The current in the electric circuit is directly analogous to the fluid flow in the hydraulic circuit.

When the battery is supplying energy as in the circuit of Fig. 4.1b, the current comes out of the positive terminal of the battery, then through the resistor from a to b and on into the negative terminal of the battery. In accordance with Ohm's law, the current produces a voltage $V = RI$ across the resistor. The polarity is such that V is a *voltage drop* from a to b as shown by the plus sign at a and the minus sign at b. The general rule governing polarity and Ohm's law is:

Current through a resistor produces a voltage that is positive at the terminal at which the current enters.

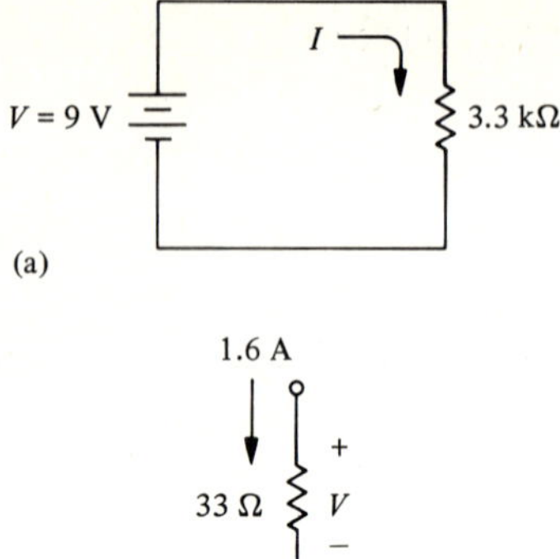

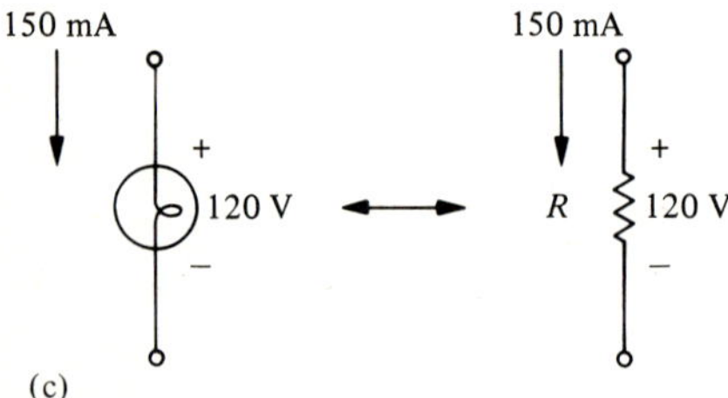

FIGURE 4.2
(a) Example 4.1-1. (b) Example 4.1-2.
(c) Example 4.1-3.

The following examples illustrate some of the ways in which Ohm's law is used.

EXAMPLE 4.1-1 Finding an Unknown Current

A 3.3-kΩ resistor is connected to a 9-V battery as in Fig. 4.2a. Find the resulting current.

Solution

Using Ohm's law in the form $I = V/R$ with $V = 9$ V and $R = 3.3$ kΩ $= 3 \times 10^3$ Ω

$$I = \frac{V}{R} = \frac{9 \text{ V}}{3.3 \times 10^3 \ \Omega} = 2.73 \times 10^{-3} \text{ A} = 2.73 \text{ mA}$$

In electronic circuits resistances are typically in the kilohm range, and currents in the milliampere range, as in this example. It is then convenient to use kilohms and milliamperes directly in Ohm's law. From Example 4.1-1 we see that current in milliamperes can be found directly by dividing volts by kilohms. Thus volts, kilohms, and milliamperes are a consistent set of units for Ohm's law. This can be proved by recalling that 1 kΩ = 10^3 Ω and 1 mA = 10^{-3} A, so that

$$V = R\,(\text{k}\Omega) \times I\,(\text{mA}) = R\,(10^3\ \Omega) \times I\,(10^{-3}\ \text{A}) = R\,(\Omega) \times I\,(\text{A})$$

in agreement with Eq. (4.1-4).

• • •

EXAMPLE 4.1-2 Finding an Unknown Voltage

In a certain circuit a 33-Ω resistor carries a current of 1.6 A. What is the voltage across this resistor?

Solution

Using Ohm's law in the form $V = RI$

$$V = RI = (33\ \Omega)(1.6\ \text{A}) = 52.8\ \text{V}$$

with polarity as shown in Fig. 4.2b.

• • •

EXAMPLE 4.1-3 Finding an Unknown Resistance

A light bulb draws 150 mA when the voltage across it is 120 V, as shown in Fig. 4.2c. What is the resistance of the light bulb?

Solution

Using Ohm's law in the form $R = V/I$

$$R = \frac{V}{I} = \frac{120\ \text{V}}{150\ \text{mA}} = 0.8\ \text{k}\Omega = 800\ \Omega$$

• • •

EXAMPLE 4.1-4 Plotting a Current–Voltage Characteristic

A 100-Ω resistor is connected to a variable dc power supply that is capable of supplying −50 to +50 V. Draw a graph of current vs. voltage as the supply voltage varies over its full range.

Solution

The circuit is shown in Fig. 4.3a, and we shall plot I in amperes vs. V in volts. On the graph the vertical axis must extend from $I = -50\ \text{V}/100\ \Omega = -0.5\ \text{A}$ to $I = +50\ \text{V}/100\ \Omega = +0.5\ \text{A}$. For the current we have the equation

$$I = \frac{V}{100}\ \text{A}$$

The graph of this equation is a straight line through the origin, as shown in Fig. 4.3b and the slope of the line is [see Eq. (1.5-2)]

$$m = \frac{1}{R} = \frac{1}{100\ \Omega} = 0.01\ \text{A/V}$$

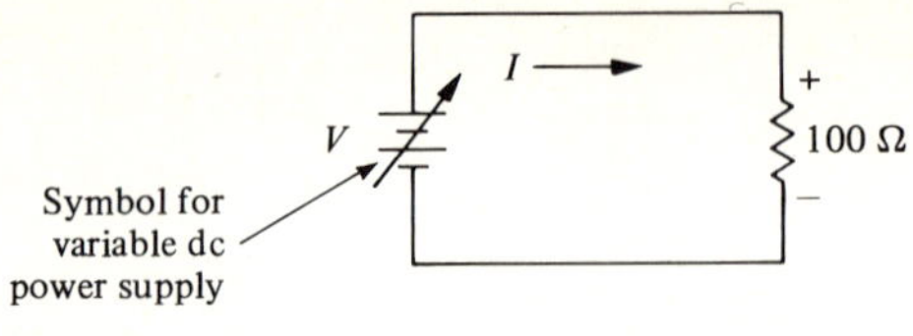

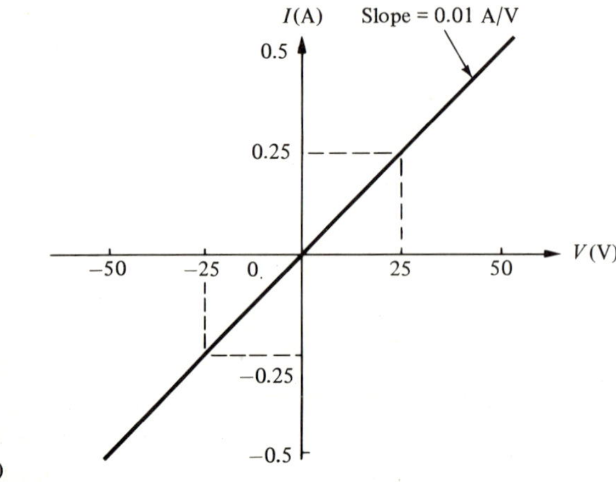

FIGURE 4.3
Example 4.1-4. (a) Circuit. (b) Graph.

Graphs such as this are called *current-voltage characteristics* (abbreviated *vi* characteristics). They are used when performing graphical analysis and design of circuits, as we see in Sec. 5.4, and in the design of transistor amplifiers.

• • •

LEARNING EXERCISES FOR SEC. 4.1

1. A dc motor used on a model railroad has a resistance of 130 Ω. How much current does it require from a 75-V supply?

2. When the motor in Exercise 4.1-1 stalls, the current rises to 850 mA. What is its resistance under this condition?

3. In order to slow the motor in Exercise 4.1-1 to 25% of full speed, the current must be reduced to 100 mA. What must the supply voltage be reduced to for this condition?

Ans. 88.2; 0.577; 13

• • •

4.2 THE SERIES CIRCUIT—KIRCHHOFF'S VOLTAGE LAW

Let us add an additional resistor to the circuit of Fig. 4.1b as shown in Fig. 4.4a. Since the current through each point of this circuit is the same, even

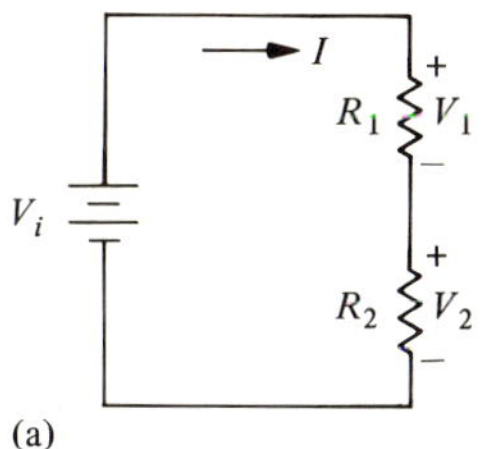

(a)

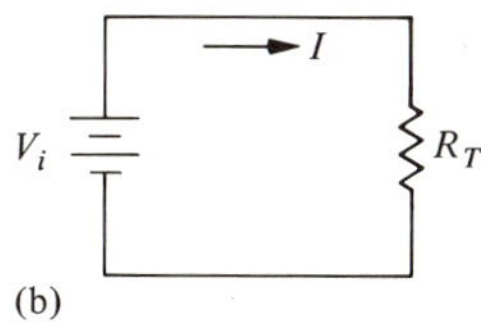

(b)

FIGURE 4.4
The series circuit. (a) Two-resistor circuit. (b) Equivalent circuit.

with the additional resistor, we use the name *series circuit* to describe such an arrangement and we say that any two or more components which carry the same current are *in series*. In the circuit of Fig. 4.4, this current is symbolized as I.

Now let us consider the effect of this current on the voltages in the circuit.

The current I through R_1 causes a voltage drop

$$V_1 = R_1 I \tag{4.2-1}$$

across R_1 and a voltage drop

$$V_2 = R_2 I \tag{4.2-2}$$

across R_2. The direction indicated for the current I produces the polarities shown on the diagram, that is, both are positive at the terminal at which the current enters. The three voltages in the circuit are related by one of the fundamental laws of electric circuits, Kirchhoff's voltage law, hereafter abbreviated KVL. For series circuits this law can be stated as follows:

In a series circuit the sum of the voltage drops across the resistors must equal the net applied voltage.

When applied to the circuit of Fig. 4.4, KVL yields the equation

$$\boxed{V_i = V_1 + V_2} \tag{4.2-3}$$

Now we substitute $V_1 = R_1I$ from Eq. (4.2-1) and $V_2 = R_2I$ from Eq. (4.2-2) into Eq. (4.2-3) to get

$$\begin{aligned} V_i &= R_1I + R_2I \\ &= (R_1 + R_2)I \end{aligned} \qquad (4.2\text{-}4)$$

Equation (4.2-4) shows that the circuit of Fig. 4.4a is *equivalent* to the circuit of Fig. 4.4b, in which $V_i = R_TI$ and where the resistance

$$\boxed{R_T = R_1 + R_2} \qquad (4.2\text{-}5)$$

is the *total* resistance of the series circuit. From this example we see that the equivalent resistance (total resistance) of two resistors in series is the *sum* of the individual resistances. This can be generalized to the following rule:

> *The equivalent resistance of any number of resistors in series is equal to the sum of the individual resistances.*

At this point, we present several examples which will illustrate typical series circuit calculations.

EXAMPLE 4.2-1 Finding an Unknown Current

The circuit of Fig. 4.5 arises often in electronic systems. In the circuit, two resistors are connected in series across a voltage source. Find the current I and the individual resistor voltage drops.

Solution

The equivalent resistance is, from Eq. (4.2-5)

$$R_T = R_1 + R_2 = 3.3\ \text{k}\Omega + 6.8\ \text{k}\Omega = 10.1\ \text{k}\Omega$$

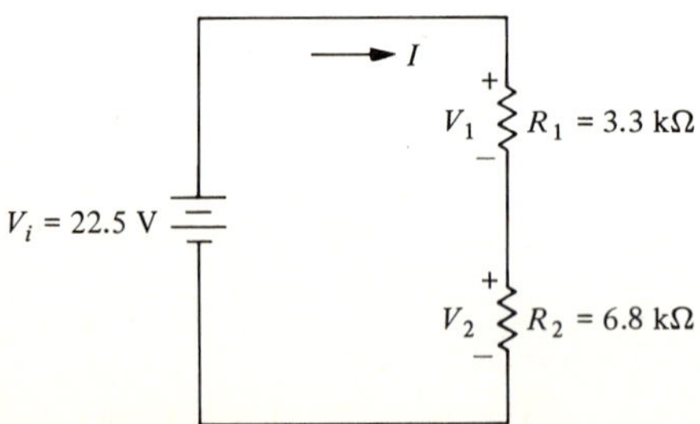

FIGURE 4.5
Circuit for Example 4.2-1.

The current is then, using Ohm's law,

$$I = \frac{V_i}{R_T} = \frac{22.5 \text{ V}}{10.1 \text{ k}\Omega} = 2.23 \text{ mA}$$

The voltage drops across the individual resistors are

$$V_1 = R_1 I = (3.3 \text{ k}\Omega)(2.23 \text{ mA}) = 7.36 \text{ V}$$

$$V_2 = R_2 I = (6.8 \text{ k}\Omega)(2.23 \text{ mA}) = 15.2 \text{ V}$$

Checking

$$V_1 + V_2 = 7.36 \text{ V} + 15.2 \text{ V} = 22.6 \text{ V}$$

This checks the value $V_i = 22.5$ V as required by KVL to sufficient accuracy.

• • •

EXAMPLE 4.2-2 Finding Unknown Voltages

In the circuit of Fig. 4.6, find the voltage drops V_1, V_2, and V_3.

Solution

Since the current is the same in both resistances,

$$V_1 = R_1 I = (22 \text{ k}\Omega)(1.8 \text{ mA}) = 39.6 \text{ V}$$

$$V_2 = R_2 I = (4.7 \text{ k}\Omega)(1.8 \text{ mA}) = 8.46 \text{ V}$$

The total voltage V_3 is found using KVL as in Eq. (4.2-3)

$$V_3 = V_1 + V_2 = 39.6 + 8.46 = 48.1 \text{ V}$$

• • •

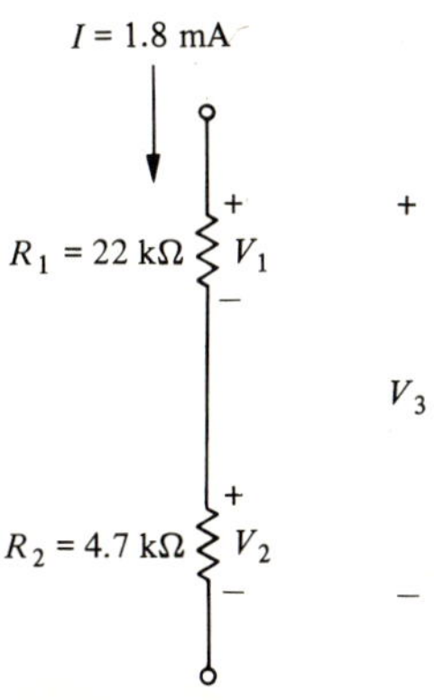

FIGURE 4.6
Circuit for Example 4.2-2.

EXAMPLE 4.2-3 Solving a Partially Specified Circuit

In the circuit of Fig. 4.7 use the information given on the figure to find the unknown voltage V_2 and unknown resistance, R_2.

Solution

Since the voltage across the 1.5-kΩ resistor is 4.2 V, the current through it must be

$$I = \frac{V}{R} = \frac{4.2\text{ V}}{1.5\text{ k}\Omega} = 2.8\text{ mA}$$

Since this is a series circuit, the current is the same through all points, thus the current through the unknown resistance is 2.8 mA. We can find the voltage across the unknown by using KVL, that is [see Eq. (4.2-3)],

$$V_i = V_1 + V_2$$

$$9\text{ V} = 4.2\text{ V} + V_2$$

from which

$$V_2 = 9\text{ V} - 4.2\text{ V} = 4.8\text{ V}$$

Finally, since we know both the voltage across and the current through the unknown resistance, we can find its resistance value using Ohm's law.

$$R_2 = \frac{V_2}{I} = \frac{4.8\text{ V}}{2.8\text{ mA}} = 1.7\text{ k}\Omega$$

• • •

LEARNING EXERCISE FOR SEC. 4.2

1. Two lamps are connected in series across 110 V. One lamp has a resistance of 80 Ω and the other has 150 Ω. Find the current through and the voltage across each lamp.

Ans. 71.7; 0.478; 38.3

• • •

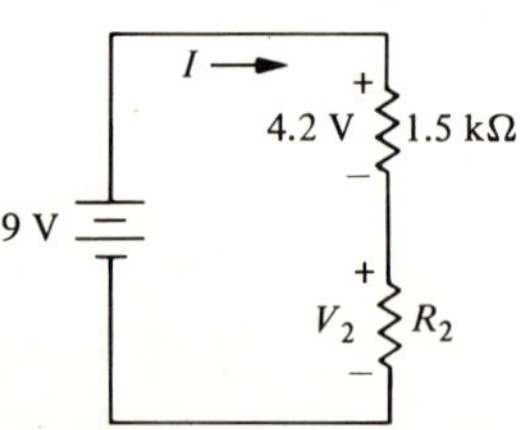

FIGURE 4.7
Circuit for Example 4.2-3.

4.3 KIRCHHOFF'S VOLTAGE LAW IN A MORE GENERAL FORM

Consider the general circuit shown in Fig. 4.8. This circuit is identical to that of Fig. 4.4 except that the three elements are unspecified, only the polarities of their voltage drops being shown. These voltage drops are related by KVL, which in general refers to voltages around a closed loop such as the circuit of Fig. 4.8. In applying KVL to such a circuit, we trace the loop beginning at any convenient point, such as a in the diagram. We then trace completely around the loop, returning to our original starting point. As we trace the loop the voltages are added together with their proper algebraic sign. In Sec. 2.5 we stated that a voltage drop would be indicated by a + sign at the point of higher energy level. In order to conform to this convention when we trace through a voltage from + to − (a voltage drop) we use a + sign for the voltage. If we trace through a voltage from − to + we call this a voltage rise and use a − sign for the voltage. An easy way to remember this process is to use the first sign encountered in the tracing direction for each voltage in the loop. In the circuit of Fig. 4.8 we begin at point a and trace the loop clockwise as shown by the arrow. The first voltage is V_i, and since the first sign encountered for V_i in the tracing direction is −, we write the term as $-V_i$. The next voltage is V_1 and the first sign encountered in the tracing direction for V_1 is +, so we write the term as $+V_1$. Similarly, for V_2 we write $+V_2$. According to KVL in its general form, these must add up to zero. In equation form

$$-V_i + V_1 + V_2 = 0 \tag{4.3-1}$$

Solving for V_i we obtain

$$V_i = V_1 + V_2 \tag{4.3-2}$$

Comparing, we find that this is identical to Eq. (4.2-3).

A mathematical statement of KVL is

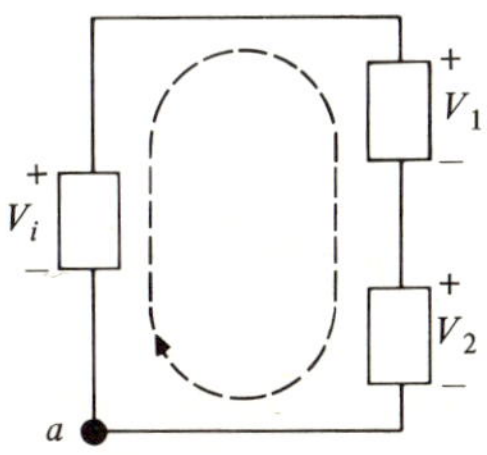

FIGURE 4.8
Circuit for illustrating KVL.

$$\sum_{\text{Loop}} V_n = 0 \tag{4.3-3}$$

In the equation the symbol Σ_{Loop} is the mathematical symbol representing the sum around the *closed* loop, and the V_n are the individual voltages with their proper signs (+ for drops, − for rises). It is important to note that the closed loop can be traced in either a clockwise or counterclockwise direction when applying KVL. The choice will usually be made for convenience. We will use the clockwise direction unless otherwise noted.

Another form for KVL can be obtained by observing Eq. (4.3-2). The voltage rises appear on the left side and the voltage drops on the right. A little thought will show that this will be true for any number of voltages, so that we can write

$$\Sigma V_{\text{rises}} = \Sigma V_{\text{drops}} \tag{4.3-4}$$

where ΣV_{rises} represents the sum of all of the voltage rises around the closed loop and ΣV_{drops} represents the sum of all of the voltage drops.

It is important to note that the loop to which we apply KVL does not have to be an isolated circuit such as that shown in Fig. 4.8. KVL is much more general than that, and applies to any closed loop in any electric circuit and is valid even if there is no current. This will be illustrated in the examples which follow.

EXAMPLE 4.3-1 Loop Tracing

A TV amplifier circuit containing five unknown elements is shown in Fig. 4.9. The voltages across four of the elements are measured and found to be as shown. Find the unknown voltage V_5.

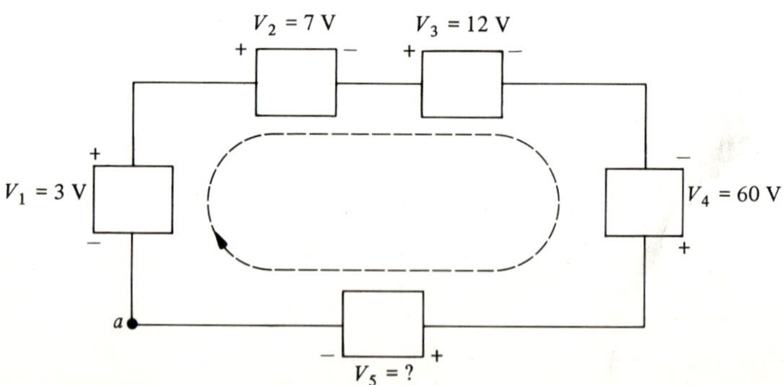

FIGURE 4.9
Circuit for Example 4.3-1.

Solution

For this problem we use KVL in the form $\Sigma V_n = 0$. The closed loop is traced clockwise beginning at point *a*. This yields

$$-V_1 + V_2 + V_3 - V_4 + V_5 = 0$$

Solving for the unknown V_5 yields

$$\begin{aligned} V_5 &= V_1 - V_2 - V_3 + V_4 \\ &= 3 - 7 - 12 + 60 \\ &= 44 \text{ V} \end{aligned}$$

Let us check this by applying KVL in the form $\Sigma V_{drops} = \Sigma V_{rises}$. Tracing clockwise we find voltage rises V_1 and V_4 and voltage drops V_2, V_3, and V_5. Substituting

$$\Sigma V_{rises} = \Sigma V_{drops}$$
$$V_1 + V_4 = V_2 + V_3 + V_5$$

Solving for V_5

$$\begin{aligned} V_5 &= V_1 + V_4 - V_2 - V_3 \\ &= 3 + 60 - 7 - 12 \\ &= 44 \text{ V} \end{aligned}$$

• • •

EXAMPLE 4.3-2 A More Complicated Circuit

In the telephone distribution circuit of Fig. 4.10, V_2 and V_3 are measured and found to have the values shown. Find the unknown voltages V_4 and V_5.

Solution

Since only voltages are specified, we use KVL to solve for the unknowns. Three different loops are shown on the diagram. Since there are two unknown voltages, only two KVL equations around two of the loops will be required to solve the

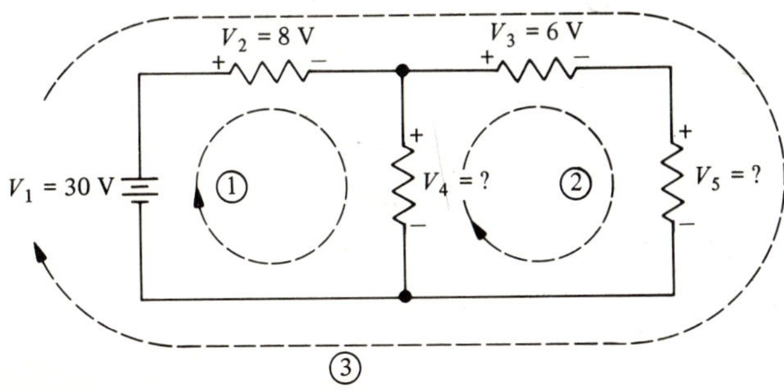

FIGURE 4.10
Circuit for Example 4.3-2.

problem. The third loop can be used for checking. We begin by writing KVL around loop 1 in order to find V_4.

Loop 1:

$$-V_1 + V_2 + V_4 = 0$$
$$V_4 = V_1 - V_2$$
$$= 30 - 8$$
$$= 22 \text{ V}$$

We could use loop 2 to find V_5. However, the solution would depend on V_4 which we found above. Thus, we use loop 3 to find V_5, since it does not include V_4.

Loop 3:

$$-V_1 + V_2 + V_3 + V_5 = 0$$
$$V_5 = V_1 - V_2 - V_3$$
$$= 30 - 8 - 6$$
$$= 16 \text{ V}$$

Now we use loop 2 as a check.

Loop 2:

$$-V_4 + V_3 + V_5 = 0$$
$$-22 + 6 + 16 = 0$$

which checks, since the voltages around the loop must add up to zero.

• • •

EXAMPLE 4.3-3 Voltage Across an Open Circuit

In the circuits of Fig. 4.11 find all unknown voltages.

Solution

Circuit *a*. Unknown voltage is V_3.
Applying KVL around the closed loop indicated by the dashed line we get

$$-V_2 - V_1 + V_3 = 0$$
$$V_3 = V_1 + V_2$$
$$= 9 + 22$$
$$= 31 \text{ V}$$

Note that the closed loop in this example includes the open circuit between points *a* and *b*, and KVL indicates that the voltage across this open circuit is 31 V. The open circuit causes the current to be zero, but has no effect on the voltage since KVL is valid with or without current.

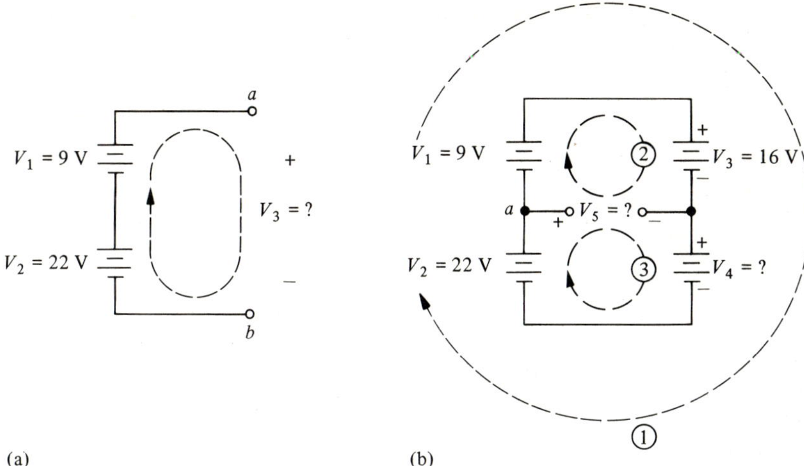

FIGURE 4.11
Circuit for Example 4.3-3.

Circuit *b*. Unknown voltages are V_4 and V_5.
First we apply KVL to loop 1 in order to find V_4.

$$\begin{aligned} -V_2 - V_1 + V_3 + V_4 &= 0 \\ V_4 &= V_1 + V_2 - V_3 \\ &= 9 + 22 - 16 \\ &= 15 \text{ V} \end{aligned}$$

In order to find voltage V_5 across the open circuit we trace loop 2 clockwise beginning at point *a*. This yields

$$\begin{aligned} -V_1 + V_3 - V_5 &= 0 \\ V_5 &= V_3 - V_1 \\ &= 16 - 9 \\ &= 7 \text{ V} \end{aligned}$$

As a check we trace loop 3 beginning at point *a*.

$$\begin{aligned} V_5 + V_4 - V_2 &= 0 \\ V_5 &= V_2 - V_4 \\ &= 22 - 15 \\ &= 7 \text{ V} \qquad \text{Check} \end{aligned}$$

• • •

EXAMPLE 4.3-4 Circuit with Three Resistors

In the circuit of Fig. 4.12 find

a. R_T
b. I

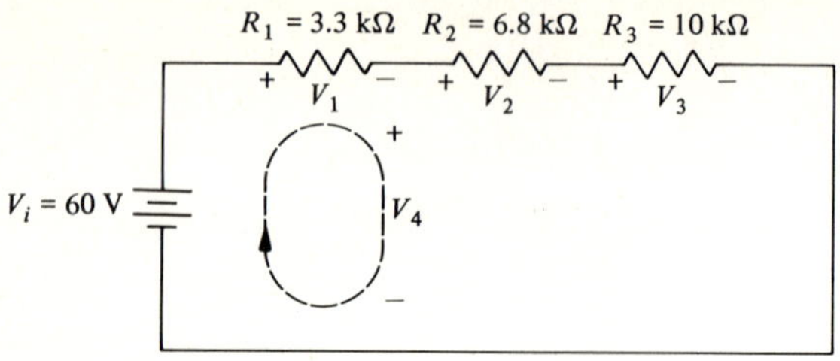

FIGURE 4.12
Circuit for Example 4.3-4.

c. V_1 through V_4
d. Check using KVL

Solution

a. Using the rule for series resistors

$$R_T = R_1 + R_2 + R_3$$
$$R_T = 3.3 \text{ k}\Omega + 6.8 \text{ k}\Omega + 10 \text{ k}\Omega = 20.1 \text{ k}\Omega$$

b. Using Ohm's law

$$I = \frac{V}{R_T} = \frac{60 \text{ V}}{20.1 \text{ k}\Omega} = 2.98 \text{ mA}$$

c. Again using Ohm's law

$$V_1 = R_1 I = 3.3 \text{ k}\Omega \times 2.98 \text{ mA} = 9.83 \text{ V}$$
$$V_2 = R_2 I = 6.8 \text{ k}\Omega \times 2.98 \text{ mA} = 20.3 \text{ V}$$
$$V_3 = R_3 I = 10 \text{ k}\Omega \times 2.98 \text{ mA} = 29.8 \text{ V}$$

To find V_4 we write KVL around the loop shown.

$$-60 \text{ V} + V_1 + V_4 = 0$$
$$V_4 = 60 - V_1$$
$$= 60 - 9.83$$
$$= 50.2 \text{ V}$$

d. According to KVL, we should have $V_i = 60$ V

$$V_i = V_1 + V_2 + V_3$$
$$= 9.83 \text{ V} + 20.3 \text{ V} + 29.8 \text{ V} = 59.9 \text{ V}$$

which checks closely.

• • •

EXAMPLE 4.3-5 Negative Voltage Drop

Find V_3 in the circuit of Fig. 4.13a.

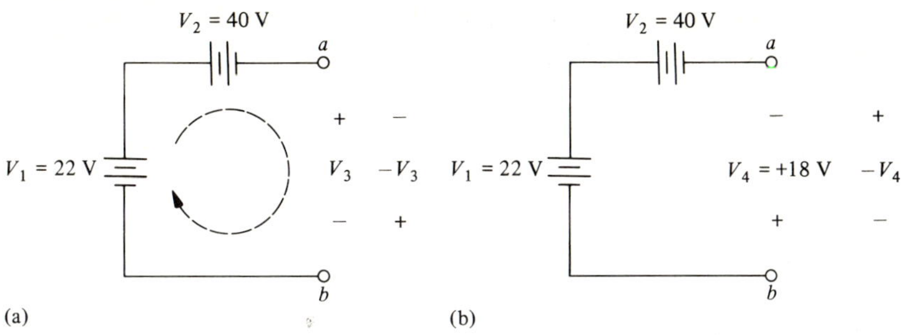

FIGURE 4.13
Circuits for Example 4.3-5.

Solution

KVL around the loop yields

$$
\begin{aligned}
-V_1 + V_2 + V_3 &= 0 \\
V_3 &= V_1 - V_2 \\
&= 22 - 40 \\
&= -18 \text{ V}
\end{aligned}
$$

The significance of the negative answer is that V_3 is actually a voltage rise from *a* to *b*. In the circuit shown in Fig. 4.13b, the voltage rise has been replaced by a voltage drop V_4 from *b* to *a* and $V_4 = -V_3 = -(-18) = 18$ V. *The two circuits in the figure are equivalent electrically.*

• • •

Combination of Voltage Sources

Any number of voltage sources connected directly in series can be replaced by a single equivalent voltage source. When this is done, care must be exercised to ensure that the polarities of the individual series sources are properly accounted for. This is illustrated in the following example.

EXAMPLE 4.3-6

In the circuits of Fig. 4.14, find the single equivalent voltage source which will replace the series combinations of sources shown.

Solution

Circuit *a*. We apply KVL around the loop in order to find V_{ab}. This is the open-circuit voltage of the series combination of sources. Starting at terminal *a* and tracing clockwise, we have

$$
\begin{aligned}
V_{ab} - 6 \text{ V} - 6 \text{ V} &= 0 \\
V_{ab} &= 12 \text{ V}
\end{aligned}
$$

If the source in the right-hand circuit has $V_{eq} = 12$ V, then it will be exactly

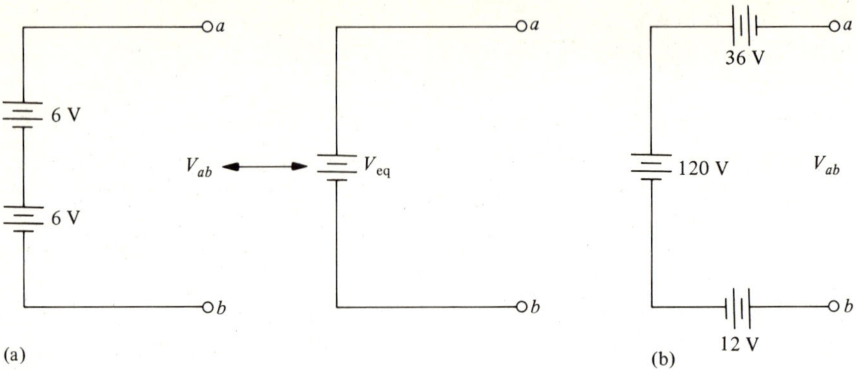

FIGURE 4.14
Circuits for Example 4.3-6.

equivalent to the left-hand circuit. Thus if a 12-V automobile battery breaks down, it can be replaced by two 6-V batteries connected in series as in Fig. 4.14a.

Circuit *b*. Proceeding as above we have

$$V_{ab} + 12\text{ V} - 120\text{ V} + 36\text{ V} = 0$$
$$V_{ab} = -12\text{ V} + 120\text{ V} - 36\text{ V} = 72\text{ V}$$

Thus a source which has $V_{eq} = 72$ V can replace the three series sources. Observe that the polarities of the individual sources are automatically accounted for correctly when KVL is used, as in this example.

• • •

It is important to note that voltage sources *should not* be placed in parallel unless it is only for a very short time. The reason for this can be seen by considering the circuit of Fig. 4.15 in which a 12-V battery is placed in parallel with a 6-V battery. Assuming that no load is connected for the moment, the no-load current *I* is

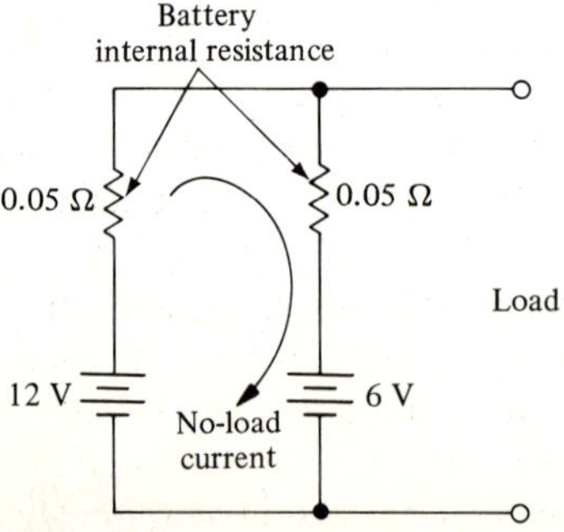

FIGURE 4.15
Batteries connected in parallel.

$$I = \frac{12\text{ V} - 6\text{ V}}{0.05\ \Omega + 0.05\ \Omega} = \frac{6\text{ V}}{0.1\ \Omega} = 60\text{ A}$$

This very large current is leaving the 12-V battery from the positive terminal and therefore discharging it, and entering the positive terminal of the 6-V battery, which is being charged. If this situation continues for any length of time, the 12-V battery will discharge to the point where it can no longer supply current, and if a load is connected, it will essentially be supplied by only the 6-V battery.

An exception to the rule concerning parallel connection of dc sources occurs when a 6-V automotive or marine battery needs a boost in order to start the engine. If a 12-V battery is available it is connected in parallel with the 6-V battery only long enough to get the engine started.

• • •

LEARNING EXERCISES FOR SEC. 4.3

1. In Fig. 4.8 $V_i = 60$ V and $V_2 = 27$ V. Find V_1.
2. In Fig. 4.9 $V_1 = 8$ V, $V_2 = -3$ V, $V_3 = -12$ V, and $V_4 = 30$ V. Find V_5.
3. You have available two batteries: 6 and 18 V. What are the different voltages that can be obtained if series combinations are allowed?

Ans. 6; 12; 53; 18; 33; 24

• • •

4.4 REFERENCE POLARITIES AND THE ZERO-REFERENCE GROUND

Reference Polarities Compared to Actual Polarities

It is important to understand the difference between reference polarities and actual polarities. A reference polarity can be arbitrarily indicated on a circuit diagram before calculations or measurements are made. For example, in the circuit of Fig. 4.16, a single resistor is drawn, and the voltage across it is labeled V_1 with polarity as shown. This is the reference polarity, since we have no advance knowledge of the actual voltage or its polarity. Now, suppose we connect a voltmeter to *a* and *b* with the positive voltmeter terminal at *a*. When the positive terminal of a voltmeter is connected to the more positive side of the voltage being measured, the meter will read upscale. If the meter reads +6 V, then the actual voltage at point *a* is 6 V positive compared to the voltage at *b*, and we write $V_1 = 6$ V. If, on the other hand, the voltmeter reads −6 V, then the voltage at *a* is 6 V negative compared to the voltage at *b*, and we write $V_1 = -6$ V. Thus we see that the reference polarity may be set down arbitrarily before any knowledge of the

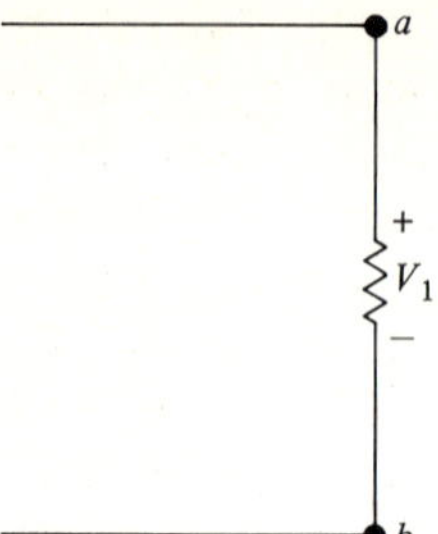

FIGURE 4.16
Illustrating reference polarity.

actual voltage is obtained. The actual polarity can be determined only through calculation or measurement.

The Use of a Zero-Reference Voltage

In many circuits, most of the elements have one terminal connected to a common point and it is convenient to consider this as a zero-voltage point. It is called *ground,* or *common* and it becomes the reference or zero-voltage point. The voltage at all other points is then specified with respect to this level. As an example, consider the circuit shown in Fig. 4.17a. In this representation, all voltages must have polarity signs so that we may write KVL equations in order to evaluate the performance of the circuit. These are reference polarities in general, as discussed in the previous paragraph, since we do not have advance knowledge of the actual polarities. Circuit diagrams of any complexity can become very cluttered using this representation. In order to avoid this we take advantage of the fact that in many circuits, most of the elements have one terminal connected to a common point. The common point is arbitrarily considered to be at zero voltage, or

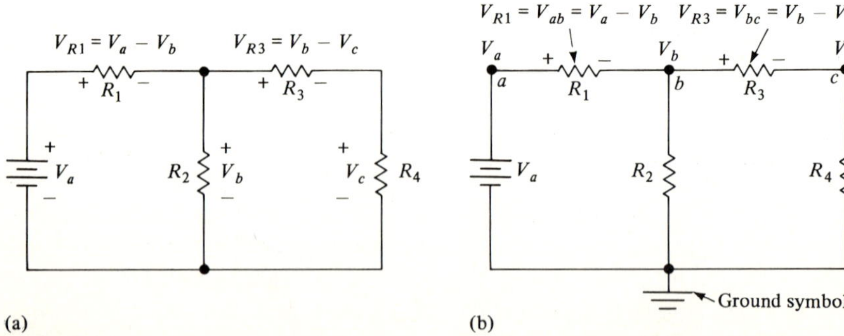

FIGURE 4.17
Illustrating the ground symbol.

ground. Using this scheme, the circuit is redrawn in Fig. 4.17b with the bottom junction considered to be at zero voltage as indicated by the ground symbol. Often this point is actually connected to a long copper rod driven several feet into the earth, or "ground." In electronic circuits, the metal base or chassis on which the elements are mounted is used as the ground or common terminal. In printed circuits a conducting path around the complete circuit is often provided as a common terminal. The voltages at the other junctions are V_a, V_b, and V_c; no polarity signs are required. With the ground symbol present we *understand* that the reference polarities of all voltages are specified with respect to ground. For example, the reference polarity for V_a is positive at point a, the reference polarity for V_b is positive at point b, and so on. The actual polarities can only be determined by calculation or measurement.

When one terminal of an element is not grounded, we must indicate the polarity. For example, in the circuit of Fig. 4.17b we may wish to specify the voltage across resistor R_1, which does not have one terminal connected to ground. This can be done in two ways. We can specify it as V_{R1} with polarity marks arbitrarily chosen as shown on the diagram. Or, we can use *double-subscript* notation where the voltage across R_1 (from point a to point b) has reference polarity V_{ab} and it is *understood* that the first subscript a is the more positive side (+) while the second subscript b carries the (−) polarity mark.

For either case we have, using KVL,

$$V_{R1} = V_{ab} = V_a - V_b \tag{4.4-1}$$

EXAMPLE 4.4-1 Voltmeter Determination of Polarity

Terminals for dc voltmeters are provided with polarity marks. In order for the voltmeter to read upscale, the terminal with the positive polarity mark must be connected to the more positive side of the voltage being measured. In Fig. 4.18a the voltmeter, when connected as shown, reads 16 V. Find V_{ab} and V_{ba}.

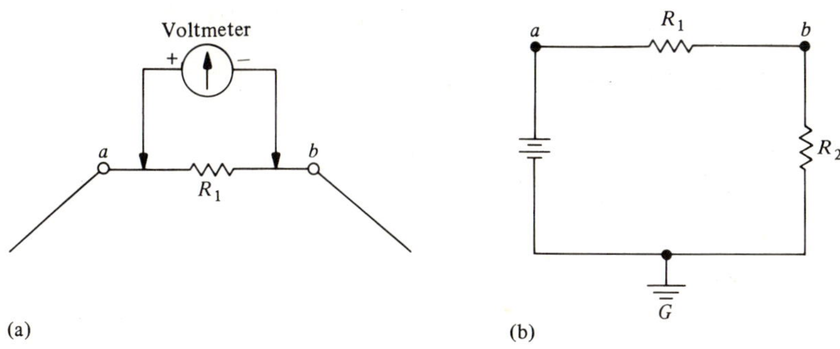

FIGURE 4.18
(a) Voltmeter circuit for Example 4.4-1. (b) Circuit for Example 4.4-2.

Solution

The voltmeter reading indicates that point a is 16 V more positive than point b. Thus, according to the definition of double-subscript notation, $V_{ab} = 16$ V. Since V_{ab} is a voltage drop of 16 V, V_{ba} must have the same magnitude. But it is a voltage rise, and we see that

$$V_{ba} = -V_{ab} = -16 \text{ V}$$

• • •

EXAMPLE 4.4-2 *Illustrating Double-Subscript Notation*

In the circuit of Fig. 4.18b measurements yield the voltages $V_a = 12$ V and $V_b = 4$ V. Find the voltage across R_1.

Solution

Let us apply KVL to the loop, denoting the ground level as G. Then, tracing clockwise starting at G, we get

$$V_{Ga} + V_{ab} + V_{bG} = 0$$

From the circuit we see that $V_{Ga} = -V_a$ and $V_{bG} = V_b$ because $V_G = 0$ so that

$$\begin{aligned} -V_a + V_{ab} + V_b &= 0 \\ V_{ab} &= V_a - V_b \\ &= 12 \text{ V} - 4 \text{ V} \\ &= 8 \text{ V} \end{aligned}$$

• • •

EXAMPLE 4.4-3 *Effect of Changing the Location of the Common Point*

a. Find all unknown voltages in the circuit of Fig. 4.19a.
b. Repeat part a if the ground is moved to point c.

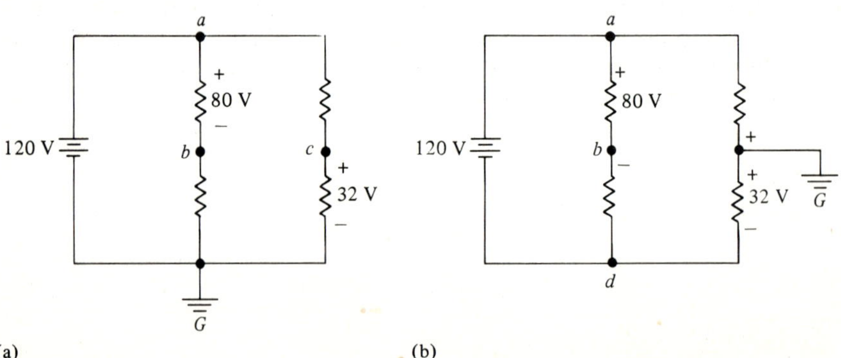

FIGURE 4.19
Example 4.4-3. (a) Original circuit. (b) Circuit with new reference location.

Solution

a. From the diagram, the given voltages are

$$V_a = 120 \text{ V} \qquad V_{ab} = 80 \text{ V} \qquad V_c = 32 \text{ V}$$

The unknown voltages are V_b, V_{ac}, and V_{bc}. Applying KVL in the form $\Sigma V_{\text{drops}} = \Sigma V_{\text{rises}}$

$$\begin{aligned} V_a &= V_{ab} + V_b \\ 120 &= 80 + V_b \\ V_b &= 40 \text{ V} \end{aligned}$$

and

$$\begin{aligned} V_a &= V_{ac} + V_c \\ 120 &= V_{ac} + 32 \\ V_{ac} &= 88 \text{ V} \end{aligned}$$

Now we trace loop *abc* in a counterclockwise direction using KVL in the form $\Sigma V = 0$ to get

$$\begin{aligned} V_{ab} + V_{bc} + V_{ca} &= 0 \\ 80 + V_{bc} - 88 &= 0 \\ V_{bc} &= 8 \text{ V} \end{aligned}$$

We can check this around the loop which includes ground

$$\begin{aligned} V_{Gb} + V_{bc} + V_{cG} &= 0 \\ -V_b + V_{bc} + V_c &= 0 \\ -40 + V_{bc} + 32 &= 0 \\ V_{bc} &= 8 \text{ V} \end{aligned}$$

b. The circuit with the new ground location is shown in Fig. 4.19b. Since the ground point in the original circuit is now point *d*, we have $V_{ad} = 120$ V. Points *a* and *b* are unaffected by the new ground location, so V_{ab} remains 80 V. The voltage from point *d* to the new ground is $V_d = -32$ V. The unknown voltages are V_{bd}, V_b, and V_a. As in the solution to part *a* we use KVL.

$$\begin{aligned} V_{ad} &= V_{ab} + V_{bd} \\ 120 &= 80 + V_{bd} \\ V_{bd} &= 40 \text{ V} \end{aligned}$$

and

$$\begin{aligned} V_{ad} &= V_{aG} + V_{Gd} \\ &= V_a - V_d \\ 120 &= V_a - (-32) \\ V_a &= 88 \text{ V} \end{aligned}$$

Finally

$$V_b = V_{ba} + V_a$$
$$= -80 + 88$$
$$= 8 \text{ V}$$

The student should check the solutions to parts a and b to be certain that they are consistent.

This example illustrates the fact that voltages *across elements* in a circuit are not affected if the ground or common point is changed! In electronic circuits the common point is usually that point which is common to the largest number of elements and is typically connected to the chassis on which the circuit is mounted. Voltages in the circuit can be both positive and negative with respect to ground. In power systems an earth ground is used to provide a measure of safety.

• • •

LEARNING EXERCISE FOR SEC. 4.4

1. You have two batteries available: 6 and 24 V. They may be connected singly or in combination any way you wish, providing one terminal is always grounded. List and draw circuit diagrams for all of the possible voltages with respect to ground.

Ans. ±24; ±6, ±30; ±18

• • •

4.5 THE VOLTAGE DIVIDER

A problem that arises frequently is the following: Given a source voltage V_i, can we arrange a circuit as in Fig. 4.20a which will have a different voltage, say V_2 as output? The circuit shown in Fig. 4.20b, known as a voltage divider, does just this. In the circuit, the input voltage V_i and the values of R_1 and R_2 are known and the output voltage V_2 is to be found in terms of

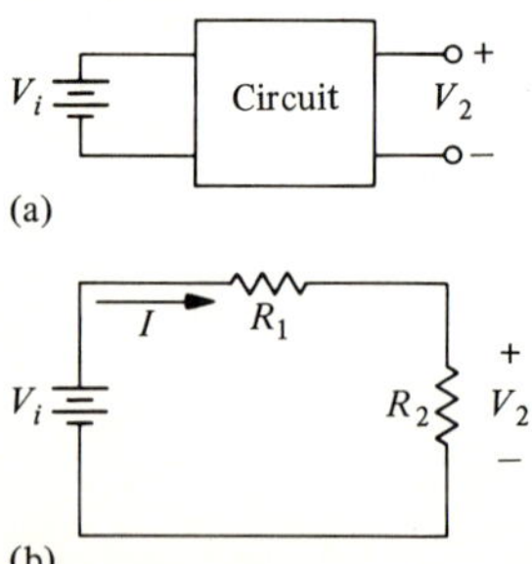

FIGURE 4.20
Voltage divider circuit. (a) Block diagram. (b) Resistive voltage divider.

these known quantities. A convenient formula can be found in the following way:

The current in the circuit is

$$I = \frac{V_i}{R_1 + R_2} \tag{4.5-1}$$

The desired voltage across R_2 is

$$V_2 = R_2 I \tag{4.5-2}$$

Now we substitute Eq. (4.5-1) into Eq. (4.5-2). This yields

$$V_2 = \frac{R_2}{R_1 + R_2} V_i \tag{4.5-3}$$

If we note that $R_1 + R_2 = R_T$, the total series resistance, we can write this as

$$\boxed{V_2 = \frac{R_2}{R_T} V_i} \tag{4.5-4}$$

A useful memory aid for this important short-form solution, called the *voltage divider formula*, is that the numerator of the resistance ratio contains the *resistance across which the desired voltage appears*, while the denominator contains the total resistance. Equation (4.5-4) is used to find V_2 without the intermediate step of finding the current, as shown in the following example.

EXAMPLE 4.5-1 Voltage Divider Calculation

The circuit of Fig. 4.21 is used to provide operating bias for the power transistors in a stereo amplifier. Find V_1 and V_2.

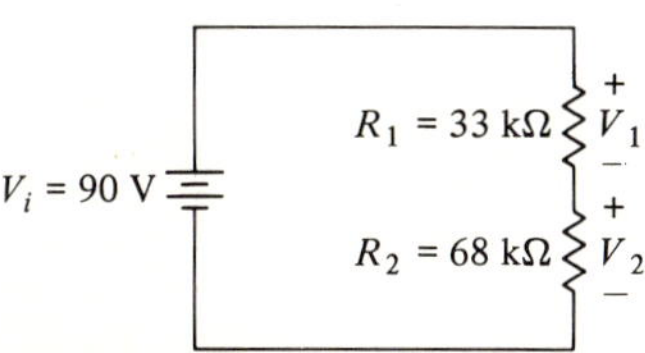

FIGURE 4.21
Circuit for Example 4.5-1.

Solution

From the diagram

$$R_T = R_1 + R_2 = 33\ \text{k}\Omega + 68\ \text{k}\Omega = 101\ \text{k}\Omega$$

Using the voltage divider formula

$$V_2 = \frac{R_2}{R_T} V_i = \frac{68\ \text{k}\Omega}{101\ \text{k}\Omega}(90\ \text{V}) = 60.6\ \text{V}$$

The voltage divider formula can be used to find V_1 by noting that the derivation would be identical in terms of R_1 and would yield the formula

$$V_1 = \frac{R_1}{R_T} V_i = \frac{33\ \text{k}\Omega}{101\ \text{k}\Omega}(90\ \text{V}) = 29.4\ \text{V}$$

A check on these calculations can be made by observing that we must have $V_1 + V_2 = V_i$ according to KVL. For the results above we find 60.6 V + 29.4 V = 90 V.

It is important to observe that this circuit has taken the 90-V voltage source and transformed it into two different voltages: 60 and 30 V.

• • •

The results of the previous example along with observation of Eqs. (4.5-3) and (4.5-4) lead to the following important facts concerning voltage dividers:

1. In Eq. (4.5-3) consider the resistance ratio $R_2/(R_1 + R_2)$. Since resistance values are always positive, the denominator is always greater than the numerator so that this ratio is always less than one. Because of this fact, the output voltage V_2 will always be less than the input voltage V_i.
2. In Example 4.5-1 we found the voltage across both of the resistors in the voltage divider. Since the current is the same through each resistance, the voltage drops (RI) across each resistance must be proportional to the value of the resistance. In the example we could make use of this fact to find V_1 after having found V_2 by simply writing

$$\frac{V_1}{V_2} = \frac{R_1}{R_2}$$

Substituting the numbers we have

$$\frac{V_1}{60.6\ \text{V}} = \frac{33\ \text{k}\Omega}{68\ \text{k}\Omega}$$

from which $V_1 = 29.4$ V. This agrees with the value found previously.

The Transfer Function Concept

In Eq. (4.5-4) let us divide both sides by V_i. The result is

$$\boxed{\frac{V_2}{V_i} = \frac{R_2}{R_T}} \tag{4.5-5}$$

If we use the component values in Fig. 4.21 this becomes

$$\frac{V_2}{V_i} = \frac{68\ \text{k}\Omega}{101\ \text{k}\Omega} = 0.673$$

Thus the *ratio* of the output voltage V_2 to the input voltage V_i is fixed by the circuit at the value 0.673. Making use of this fact we can easily find the output voltage corresponding to any intput voltage. For example, suppose the input voltage is changed to 72 V. The new output voltage is then simply

$$\begin{aligned} V_2 &= 0.673V_i \\ &= (0.673)(72\ \text{V}) = 48.5\ \text{V} \end{aligned}$$

The ratio V_2/V_i is called the *voltage transfer ratio* of the circuit, since it provides a direct measure of the transmission of voltage from the input through the circuit to the output. This is an important concept in electronics and we will refer to it quite often.

The voltage transfer ratio is symbolized by A_v, in which A indicates amplification ratio and the subscript v indicates voltage. Thus Eq. (4.5-5) becomes

$$\boxed{A_v = \frac{V_2}{V_i} = \frac{R_2}{R_T}} \tag{4.5-6}$$

The output voltage can now be expressed in the basic form

$$V_2 = A_v V_i \tag{4.5-7}$$

where $A_v = R_2/R_T$ depends only on the resistance values in the circuit.

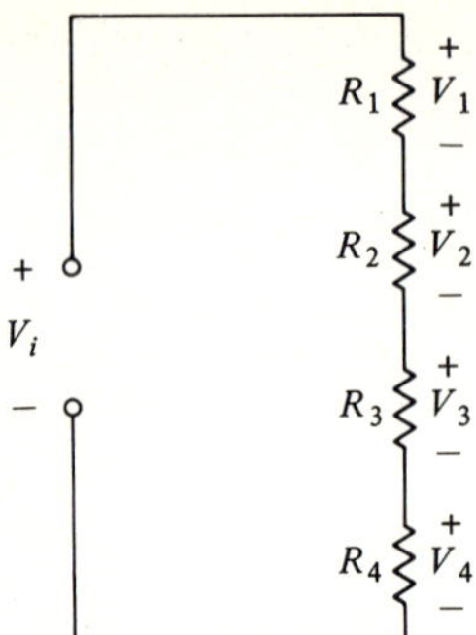

FIGURE 4.22
Voltage divider with four resistors.

Extension to More Than Two Resistances

Figure 4.22 shows a circuit with four resistors connected in series. By following the orginal derivation in Eqs. (4.5-1) through (4.5-3), it is easy to show that

$$V_n = \frac{R_n}{R_T} V_i \tag{4.5-8}$$

where $R_T = R_1 + R_2 + R_3 + R_4$ and $n = 1, 2, 3,$ or 4 depending on which voltage is desired. Clearly this formula can be generalized to a circuit with any number of resistances in series.

Design of Voltage Dividers

In the examples which follow, we illustrate several voltage divider designs.

EXAMPLE 4.5-2 Voltage Divider Design

In a certain transistor amplifier a 9-V battery is used as the main power supply and voltages of 6.3 and 4.5 V are required at various points in the amplifier. Design voltage dividers to supply these voltages from the 9-V battery. For both circuits the input resistance is to be 10 kΩ.

Solution

a. Desired voltage = 6.3 V
   For this case

$$A_v = \frac{6.3}{9} = 0.7$$

We use the standard voltage divider shown in Fig. 4.20. The input resistance of this circuit is $R_i = R_T = R_1 + R_2$. From the specifications this is to be 10 kΩ. Next we use the formula

$$A_v = \frac{R_2}{R_T}$$

Substituting given values, this becomes

$$0.7 = \frac{R_2}{10\ \text{k}\Omega}$$

from which $R_2 = 7$ kΩ. To find R_1 we note that $R_1 + R_2 = 10$kΩ, so that $R_1 = 10$ kΩ − 7 kΩ = 3 kΩ. The student should check these results to see that they do lead to a voltage divider ratio of 0.7 and an input resistance of 10 kΩ.

b. Desired voltage = 4.5 V
   For this case,

$$A_v = \frac{4.5}{9} = 0.5$$

Proceeding as before,

$$0.5 = \frac{R_2}{10\ \text{k}\Omega}$$

$$R_2 = 5\ \text{k}\Omega$$

and since $R_1 + R_2 = 10$ kΩ, we must have $R_1 = 5$ kΩ. Observe that for this case (A_v = 0.5), the two divider resistors are equal. It can be shown (Prob. 4.32) that it is true in general that a voltage divider with equal resistors yields an output voltage which is exactly one-half the input voltage.

• • •

LEARNING EXERCISES FOR SEC. 4.5

1. The resistors in a voltage divider are 22 and 47 kΩ. If the source is 28 V, what voltages are available from the divider?

2. Find A_v for both outputs in Learning Exercise 4.5-1.

Ans. 0.32; 19; 8.9; 0.68

• • •

4.6 THE PARALLEL CIRCUIT—KIRCHHOFF'S CURRENT LAW

A basic circuit closely related to the series circuit we have just studied is shown in Fig. 4.23a. The circuit is redrawn in Fig. 4.23b so as to emphasize the fact that all of the connections within the dashed line in Fig. 4.23a

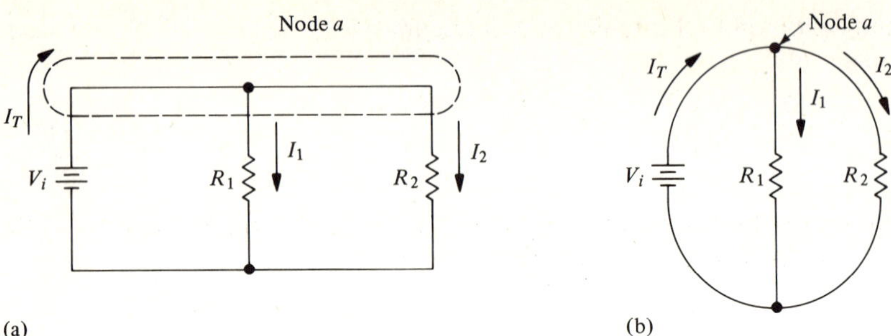

FIGURE 4.23
Parallel circuit. (a) Original circuit. (b) Circuit redrawn for clarity.

reduce to a single common point, called a junction or node. The two circuits in the figure are completely equivalent electrically. This arrangement is called a parallel circuit, and in the circuit we see that the voltage across each element is the same as the battery voltage V_i. Thus a definition of parallel elements can be stated as follows:

Two or more elements are in parallel if a common voltage appears across each of them.

An equivalent statement is that two or more elements are in parallel if they share common terminals.

The unknowns in the circuit are the currents I_1 and I_2 through the resistors and the current I_T drawn from the battery. The actual directions of these currents are shown and they are related by the second fundamental law of electric circuits, Kirchhoff's Current Law (hereafter abbreviated KCL). For a parallel circuit such as that shown in Fig. 4.23 KCL can be stated as follows:

In a parallel circuit the current drawn from the battery is equal to the sum of the currents through the resistors.

In symbols

$$\boxed{I_T = I_1 + I_2} \tag{4.6-1}$$

The individual resistor currents are found by noting that the voltage across each resistor is the battery voltage V_i. Thus

$$I_1 = \frac{V_i}{R_1} \qquad I_2 = \frac{V_i}{R_2} \tag{4.6-2}$$

Substituting Eq. (4.6-2) into Eq. (4.6-1) we find that

$$I_T = V_i\left(\frac{1}{R_1} + \frac{1}{R_2}\right) = V_i\left(\frac{R_2 + R_1}{R_1R_2}\right) \tag{4.6-3}$$

The equivalent resistance of the circuit as seen from the battery is the ratio $V_i/I_T = R_T$. From Eq. (4.6-3) this is

$$\boxed{R_T = \frac{R_1R_2}{R_1 + R_2}} \tag{4.6-4}$$

This important formula gives the equivalent resistance of two parallel resistors. When using the formula, R_1 and R_2 must be in the same units.

For resistors connected in parallel, it may be more convenient to use conductance (Sec. 3.1) to calculate the equivalent resistance. In order to show this we proceed by returning to Eq. (4.6-3) and writing it in the form

$$I_T = \frac{V_i}{R_T} = \frac{V_i}{R_1} + \frac{V_i}{R_2} \tag{4.6-5}$$

The V_i's cancel out so

$$\frac{1}{R_T} = \frac{1}{R_1} + \frac{1}{R_2} \tag{4.6-6}$$

and since $1/R_T = G_T$, $1/R_1 = G_1$, and $1/R_2 = G_2$, we have

$$\boxed{G_T = G_1 + G_2} \tag{4.6-7a}$$

If a third resistor is connected in parallel with R_1 and R_2 the total conductance will be

$$G_T = G_1 + G_2 + G_3 \tag{4.6-7b}$$

From Eq. (4.6-7) we see that when any number of resistors are in parallel, their equivalent conductance is simply the sum of the individual conductances. Thus conductance in a parallel circuit is treated exactly like resistance in a series circuit. Converting Eq. (4.6-7b) to resistance, we have

$$\frac{1}{R_T} = \frac{1}{R_1} + \frac{1}{R_2} + \frac{1}{R_3} \tag{4.6-7c}$$

The three terms on the right-hand side can be combined into a single fraction, which is inverted to give R_T. However, it is much easier to use the equation as written, making the calculation using the reciprocal function on a pocket electronic calculator. This is illustrated in the next example.

EXAMPLE 4.6-1 *Parallel Resistance*

Find the equivalent resistance of two parallel resistors if $R_1 = 10$ kΩ and $R_2 = 1$ Ω, 100 Ω, 10 kΩ, 1 MΩ, and 100 MΩ.

Solution

The calculation can be carried out by using Eq. (4.6-4) or Eq. (4.6-7a) depending on the calculator being used. On scientific calculators it is easy to add reciprocals, so Eq. (4.6-7a) was used in the form

$$R_T = \frac{1}{1/R_2 + 1/R_1}$$

The results are tabulated below, and shown graphically in Fig. 4.24.

R_1 (kΩ)	R_2 (kΩ)	R_T (kΩ)
10	0.001	0.0009999 ≈ .001
10	0.1	0.099 ≈ 0.1
10	10	5
10	10^3	9.9 ≈ 10
10	10^5	9.999 ≈ 10

From the results (and the formula) we can draw the following conclusions concerning the equivalent resistance of two parallel resistors:

1. The equivalent resistance of two parallel resistors is *always less than* either of the individual resistances. This is also true for more than two parallel resistors.
2. When two parallel resistors differ widely, the equivalent resistance is approximately equal to the *smaller* of the resistances.
3. When two parallel resistances are equal, the equivalent resistance is one-half the individual resistance.

• • •

EXAMPLE 4.6-2 *Parallel Circuit Calculations*

The circuit of Fig. 4.23 represents a stage in a transistor amplifier in which $V_i = 9$ V, $R_1 = 3.3$ kΩ, and $R_2 = 6.8$ kΩ. Find the current through, and voltage across, each resistor, the total current, and the equivalent resistance.

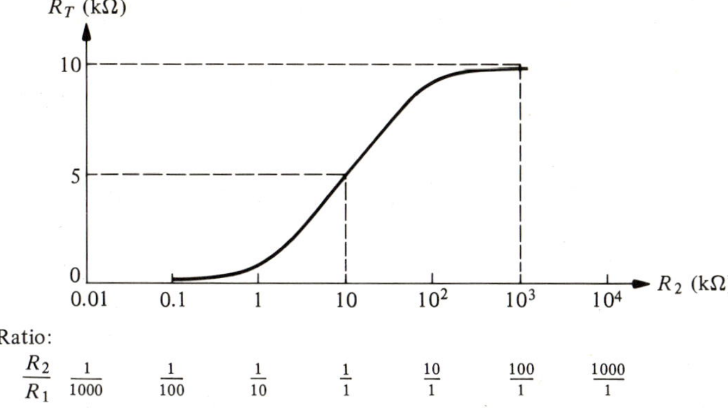

FIGURE 4.24
Equivalent resistance of R_1 = 10 kΩ in parallel with R_2.

Solution

We proceed by finding the individual resistor currents, which add up to give the total current. Since this is a parallel circuit the voltage across each of the resistors is the battery voltage. Then

$$I_1 = \frac{V_i}{R_1} = \frac{9\text{ V}}{3.3\text{ k}\Omega} = 2.7\text{ mA}$$

$$I_2 = \frac{V_i}{R_2} = \frac{9\text{ V}}{6.8\text{ k}\Omega} = 1.3\text{ mA}$$

The total current is

$$I_T = I_1 + I_2 = 2.7\text{ mA} + 1.3\text{ mA} = 4\text{ mA}$$

The equivalent resistance is

$$R_T = \frac{R_1 R_2}{R_1 + R_2} = \frac{(3.3\text{ k}\Omega)(6.8\text{ k}\Omega)}{3.3\text{ k}\Omega + 6.8\text{ k}\Omega} = 2.2\text{ k}\Omega$$

As a check, we calculate $R_T I_T$ = (2.2 kΩ) (4 mA) = 8.8 V. The expected answer is 9 V.

• • •

EXAMPLE 4.6-3 A Parallel Circuit with a Current Source

In Fig. 4.25a a current source is connected to two parallel resistors in a computer memory circuit. Find all unknown voltages and currents.

Solution

The hydraulic analog shown in Fig. 4.25b illutrates pictorially how current divides between the resistors in a parallel circuit. In the analog the large-diameter pipe would have similar hydraulic resistance and would allow greater fluid flow than the

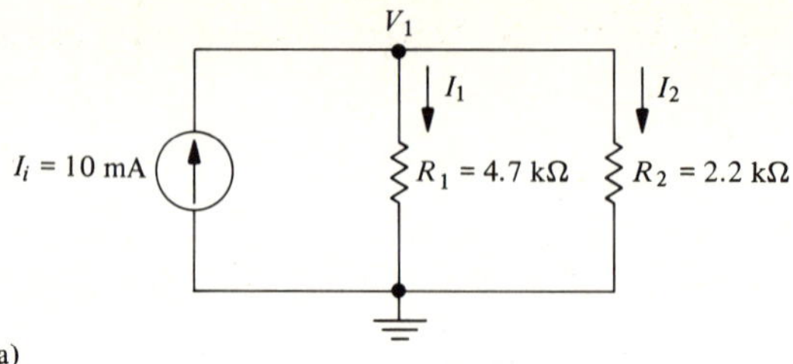

(a)

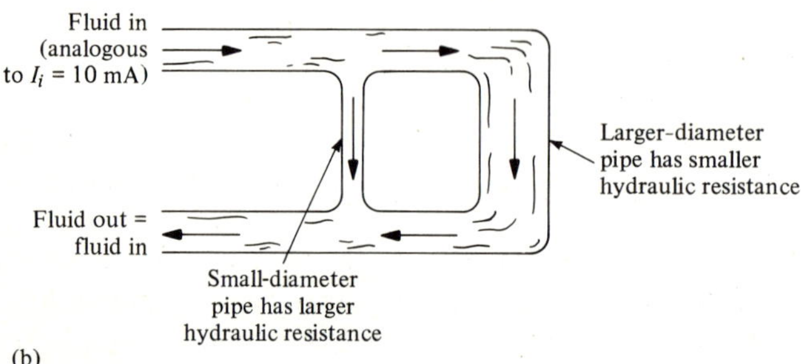

(b)

FIGURE 4.25
Parallel circuits. (a) Circuit with current source. (b) Hydraulic analog.

smaller diameter, higher-resistance pipe. The currents in the electric circuit behave in exactly the same way. In order to find these currents, we first find the equivalent resistance R_T and then, using Ohm's law, the voltage V_1

$$R_T = \frac{R_1 R_2}{R_1 + R_2} = \frac{(4.7\ \text{k}\Omega)(2.2\ \text{k}\Omega)}{4.7\ \text{k}\Omega + 2.2\ \text{k}\Omega} = 1.5\ \text{k}\Omega$$

Using Ohm's law,

$$V_1 = R_T I_i = (1.5\ \text{k}\Omega)(10\ \text{mA}) = 15\ \text{V}$$

To find I_1 and I_2, we again use Ohm's law

$$I_1 = \frac{V_1}{R_1} = \frac{15\ \text{V}}{4.7\ \text{k}\Omega} = 3.2\ \text{mA}$$

$$I_2 = \frac{V_1}{R_2} = \frac{15\ \text{V}}{2.2\ \text{k}\Omega} = 6.8\ \text{mA}$$

As a check we apply KCL

$$I_1 + I_2 = 3.2\ \text{mA} + 6.8\ \text{mA} = 10\ \text{mA} = I_i$$

The student should note carefully from this example that the current into a parallel circuit splits inversely as the resistance values. This means that the *smaller* resistance draws the *larger* current.

• • •

LEARNING EXERCISES FOR SEC. 4.6

1. In the circuit of Fig. 4.23, $R_1 = 47\ \text{k}\Omega$, $R_2 = 22\ \text{k}\Omega$, and $V_i = 45$ V. Find I_1, I_2, I_T, and R_T. Check your result.
2. In the circuit of Fig. 4.25a, $R_1 = 68\ \text{k}\Omega$, $R_2 = 22\ \text{k}\Omega$, and $I_i = 10$ mA. Find I_1, I_2, and V_1. Check using KCL.

Ans. 15; 3; 2; 0.96; 2.4; 166; 7.6

• • •

4.7 KIRCHHOFF'S CURRENT LAW IN A MORE GENERAL FORM

Consider a portion of a circuit in which a number of wires are connected to a junction (node) as shown in Fig. 4.26. In general each of the wires connected to the node will carry a different current. KCL relates all of these currents. Some of the currents will be toward the node, and some will be away. Since electric charge cannot be stored at the node,

> *the sum of the currents entering the node must be the same as the sum of the currents leaving the node.*

This can be expressed mathematically as

$$\sum_{\text{node}} I_{\text{in}} = \sum_{\text{node}} I_{\text{out}} \qquad (4.7\text{-}1)$$

All of the currents must be included.

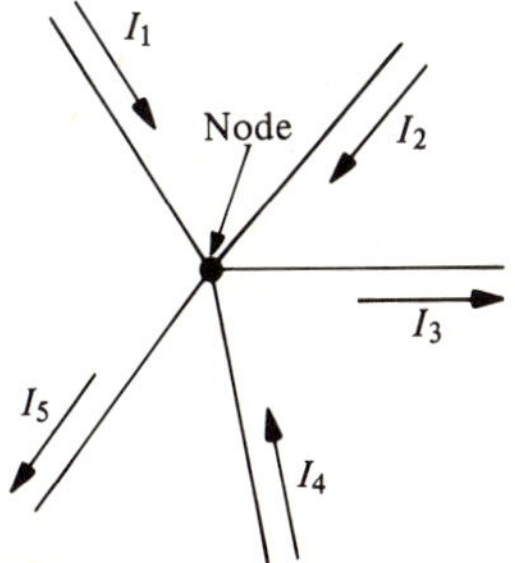

FIGURE 4.26
Circuit node.

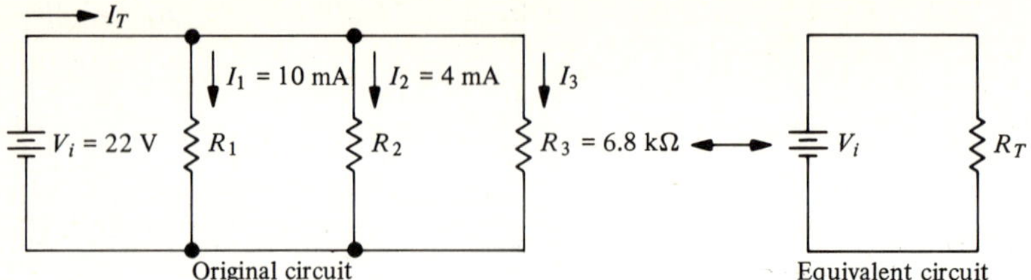

FIGURE 4.27
Circuit for Example 4.7-1.

In order to apply this to the junction shown in Fig. 4.26 we observe that I_1, I_2, and I_4 are flowing in while I_3 and I_5 are flowing out. Then, according to Eq. (4.7-1),

$$I_1 + I_2 + I_4 = I_3 + I_5 \tag{4.7-2}$$

The basic formula, Eq. (4.7-1), can be written in a more concise form by adopting the convention that currents flowing away from the node are positive while those flowing toward the node are negative. Then KCL takes the form

$$\boxed{\sum_{\text{node}} I = 0} \tag{4.7-3}$$

where all currents must be included, each with its proper sign. Applying this form to Fig. 4.26 we note that I_1, I_2, and I_4 are negative so that

$$-I_1 - I_2 + I_3 - I_4 + I_5 = 0 \tag{4.7-4}$$

If we transpose all negative currents to the right side of the equation we see that it is identical to Eq. (4.7-2). Some examples will illustrate the use of KCL.

EXAMPLE 4.7-1 Calculations with KCL

The circuit of Fig. 4.27 represents the motor circuit of a cassette recorder. Currents I_1 and I_2 are known and we are to find I_3, I_T, R_1, R_2, and R_T.

Solution

Using Ohm's law

$$I_3 = \frac{V_i}{R_3} = \frac{22\text{ V}}{6.8\text{ k}\Omega} = 3.24\text{ mA}$$

Using KCL

$$\begin{aligned} I_T &= I_1 + I_2 + I_3 \\ &= 10 \text{ mA} + 4 \text{ mA} + 3.24 \text{ mA} \\ &= 17.2 \text{ mA} \end{aligned}$$

Using Ohm's law

$$R_1 = \frac{V_i}{I_1} = \frac{22 \text{ V}}{10 \text{ mA}} = 2.2 \text{ k}\Omega$$

$$R_2 = \frac{V_i}{I_2} = \frac{22 \text{ V}}{4 \text{ mA}} = 5.5 \text{ k}\Omega$$

Using conductance

$$G_T = G_1 + G_2 + G_3 = \frac{1}{2.2 \text{ k}\Omega} + \frac{1}{5.5 \text{ k}\Omega} + \frac{1}{6.8 \text{ k}\Omega} = 0.783 \text{ mS}$$

$$R_T = \frac{1}{G_T} = 1.28 \text{ k}\Omega$$

As a check, we calculate I_T using Ohm's law

$$I_T = \frac{V_i}{R_T} = \frac{22 \text{ V}}{1.28 \text{ k}\Omega} = 17.2 \text{ mA}$$

• • •

EXAMPLE 4.7-2 Application of KCL and KVL

The circuit of Fig. 4.28 represents a typical interconnection in a telephone system. Find I_1, I_6, and V_{ad}.

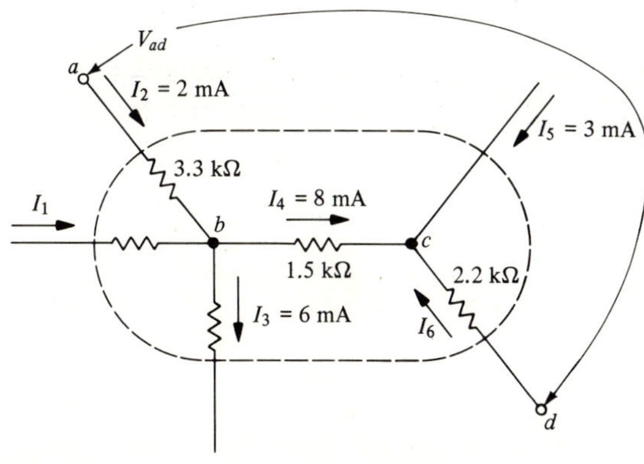

FIGURE 4.28
Circuit for Example 4.7-2.

Solution

Using KCL at node b,

$$\begin{aligned} -I_1 - I_2 + I_3 + I_4 &= 0 \\ I_1 &= I_3 + I_4 - I_2 \\ &= 6\text{ mA} + 8\text{ mA} - 2\text{ mA} \\ &= 12\text{ mA} \end{aligned}$$

Using KCL at node c,

$$\begin{aligned} -I_4 - I_5 - I_6 &= 0 \\ I_6 &= -I_4 - I_5 \\ &= -8\text{ mA} - 3\text{ mA} \\ &= -11\text{ mA} \end{aligned}$$

the negative sign indicates that I_6 is actually 11 mA flowing *away* from the junction. In order to find V_{ad} we apply KVL in the form

$$\begin{aligned} V_{ad} &= V_{ab} + V_{bc} + V_{cd} \\ &= (2\text{ mA})(3.3\text{ k}\Omega) + (8\text{ mA})(1.5\text{ k}\Omega) - (-11\text{ mA})(2.2\text{ k}\Omega) \\ &= 6.6\text{ V} + 12\text{ V} + 24.2\text{ V} \\ &= 42.8\text{ V} \end{aligned}$$

• • •

It is important to note that the entire circuit enclosed by the dashed line in Fig. 4.28 can be considered as a surface junction or node to which KCL may be applied. This is a perfectly valid application of the basic law. Considering only the conductors appearing outside of the dashed line, KCL yields

$$\begin{aligned} -I_1 - I_2 + I_3 - I_5 - I_6 &= 0 \\ -I_1 - 2\text{ mA} + 6\text{ mA} - 3\text{ mA} - I_6 &= 0 \\ I_1 + I_6 &= 1\text{ mA} \end{aligned}$$

Since I_1 and I_6 are both unknown when the circuit within the dashed line is considered as a node, this is as far as we can go. The numerical results obtained in the example do confirm that $I_1 + I_6 = 1$ mA.

Combination of Current Sources

We found in Sec. 4.3 (Example 4.3-6) that voltage sources in series could be combined into a single equivalent voltage source and that, in general, voltage sources were not connected in parallel. Current sources can be connected in parallel as illustrated in the following example.

EXAMPLE 4.7-3

In the circuits of Fig. 4.29, find the single equivalent current source which will replace the parallel combinations shown.

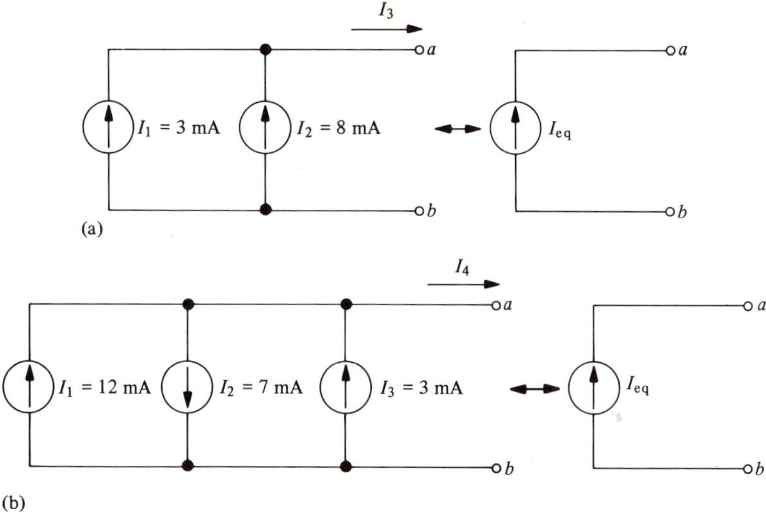

FIGURE 4.29
Circuits for Example 4.7-3.

Solution

Circuit *a*. Using KCL at node *a*

$$
\begin{aligned}
I_3 &= I_1 + I_2 \\
&= 3\text{ mA} + 8\text{ mA} = 11\text{ mA}
\end{aligned}
$$

Thus if we make $I_{eq} = 11$ mA, the single source on the right will be equivalent to the two parallel sources on the left.

Circuit *b*. Using KCL at node *a*.

$$
\begin{aligned}
I_4 + I_2 &= I_1 + I_3 \\
I_4 &= I_1 + I_3 - I_2 \\
&= 12\text{ mA} + 3\text{ mA} - 7\text{ mA} \\
&= 8\text{ mA}
\end{aligned}
$$

In this circuit, if $I_{eq} = 8$ mA, the single source will be equivalent to the three parallel sources. Note carefully that we cannot simply add the currents from the sources. They must be added algebraically using the sign convention that we have adopted for KCL.

• • •

Current sources cannot, in general, be connected in series as shown in Fig. 4.30 because KCL is violated unless the sources are identical. If we apply KCL at point *a* we have $I_2 - I_1 = 0$, which is only true if $I_1 = I_2$.

The examples in this section illustrate typical calculations the technologist is required to make during the analysis or design of electric or electronic systems.

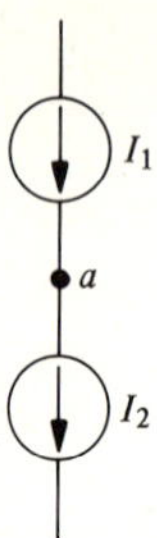

FIGURE 4.30
Current sources in series.

• • •

LEARNING EXERCISES FOR SEC. 4.7

1. Four wires are connected to a common junction. The currents in the first and third wires are toward the junction and measure 10 and 23 mA, respectively. The current in the second wire measures 82 mA away from the junction. What is the current in the fourth wire in a direction toward the junction?

2. You have available two current sources: 10 and 30 mA. They can be connected singly or in combination to a junction point *a*. List the different currents that can be supplied to point *a* and show a diagram for each.

Ans. ±10; ±40; 49; ±30; ±20

• • •

4.8 THE CURRENT DIVIDER

If the current into a pair of parallel resistors is known, the current into either of the resistors can be found without the intermediate step of finding the voltage by using the *current divider formula.* The circuit is drawn in Fig. 4.31a and the formula is derived by applying Ohm's law, noting that the voltage across all resistances is the same.

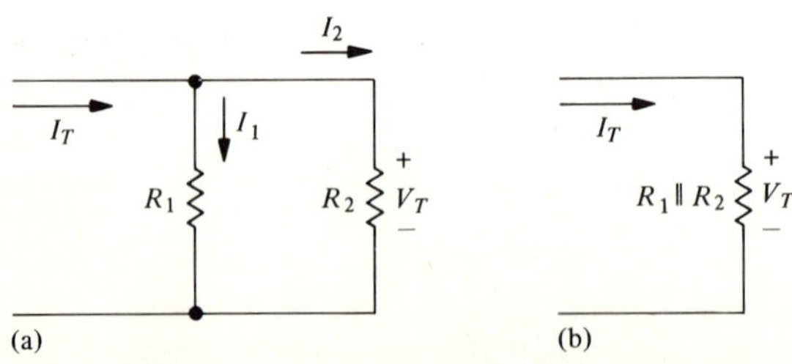

FIGURE 4.31
Current divider. (a) Circuit. (b) Equivalent circuit.

For the original circuit

$$V_T = R_1 I_1 \tag{4.8-1}$$

and

$$V_T = R_2 I_2 \tag{4.8-2}$$

For the equivalent circuit of Fig. 4.31b*

$$V_T = (R_1 \parallel R_2) I_T = \frac{R_1 R_2}{R_1 + R_2} I_T \tag{4.8-3}$$

Comparing Eqs. (4.8-1) and (4.8-3)

$$R_1 I_1 = \frac{R_1 R_2}{R_1 + R_2} I_T$$

Dividing through by R_1 in order to isolate I_1, we have the current divider formula

$$\boxed{I_1 = \frac{R_2}{R_1 + R_2} I_T} \tag{4.8-4}$$

and, comparing Eqs. (4.8-2) and (4.8-3),

$$\boxed{I_2 = \frac{R_1}{R_1 + R_2} I_T} \tag{4.8-5}$$

When we derived the voltage divider formula we used the relation $R_T = R_1 + R_2$. With this, the current divider formula can be written

$$I_1 = \frac{R_2}{R_T} I_T \tag{4.8-6}$$

*From this point on we will use the notation $R_1 \parallel R_2$ as an abbreviation for "R_1 in parallel with R_2" and we will understand that $R_1 \parallel R_2 = R_1R_2/(R_1 + R_2)$.

and

$$I_2 = \frac{R_1}{R_T} I_T \tag{4.8-7}$$

These relations should be compared with the voltage divider formula, Eq. (4.5-4), which states that $V_2 = (R_2/R_T)V_i$. Observe that in the current divider formula, the numerator contains the resistance for which we *do not* want the current, whereas the numerator in the voltage divider formula contains the resistance for which we *do* want the voltage. The denominator is the same in both formulas.

The *current transfer ratios* are found by dividing by I_T

$$A_{i1} = \frac{I_1}{I_T} = \frac{R_2}{R_T} \tag{4.8-8}$$

and

$$A_{i2} = \frac{I_2}{I_T} = \frac{R_1}{R_T} \tag{4.8-9}$$

These ratios, like the voltage transfer ratios, cannot be greater than 1, and neither I_1 nor I_2 can be larger than I_T. Also, as in the case of the voltage divider, R_1 and R_2 may each represent the equivalent resistance of a number of other resistors.

EXAMPLE 4.8-1 Current Divider Calculations

In the circuit of Fig. 4.31, $R_1 = 2.2\ \text{k}\Omega$, $R_2 = 6.8\ \text{k}\Omega$, and $I_T = 20$ mA. Find I_1, I_2, V_T, and the input resistance.

Solution

Here we apply the current divider formulas with $R_T = 2.2\ \text{k}\Omega + 6.8\ \text{k}\Omega = 9\ \text{k}\Omega$.

$$I_1 = \frac{R_2}{R_T} I_T = \frac{6.8\ \text{k}\Omega}{9\ \text{k}\Omega}(20\ \text{mA}) = 15.1\ \text{mA}$$

$$I_2 = \frac{R_1}{R_T} I_T = \frac{2.2\ \text{k}\Omega}{9\ \text{k}\Omega}(20\ \text{mA}) = 4.9\ \text{mA}$$

To find V_T we apply Ohm's law

$$V_T = R_1 I_1 = (2.2\ \text{k}\Omega)(15.1\ \text{mA}) = 33.2\ \text{V}$$

or

$$V_T = R_2 I_2 = (6.8\ \text{k}\Omega)(4.9\ \text{mA}) = 33.3\ \text{V}.$$

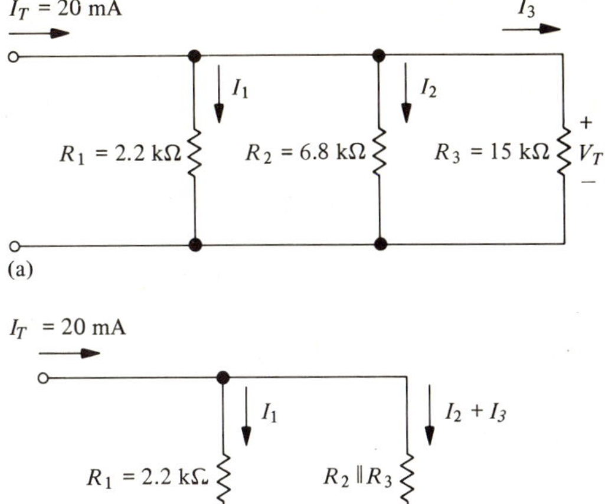

FIGURE 4.32
Example 4.8-2. (a) Original circuit. (b) Simplified circuit.

For this circuit, the input resistance is

$$R_{in} = R_1 \parallel R_2 = \frac{R_1 R_2}{R_1 + R_2} = \frac{(2.2\ \text{k}\Omega)\,(6.8\ \text{k}\Omega)}{2.2\ \text{k}\Omega + 6.8\ \text{k}\Omega} = 1.66\ \text{k}\Omega$$

In the voltage divider formula, the denominator is the same as the input resistance of the voltage divider circuit, $R_1 + R_2$. This is not true for the current divider formula.

• • •

EXAMPLE 4.8-2 Current Divider with Three Resistors

Circuits similar to that shown in Fig. 4.32 arise frequently in computer networks, where many satellite computers are connected to one main computer. Find I_1, I_2, I_3, V_T, and the input resistance.

Solution

In order to apply the current divider formula to this circuit we must recognize that the resistance for which we do not want the current consists of two resistors in parallel. Thus, in order to find I_1 we modify Eq. (4.8-4) as follows (see Fig. 4.32b).

$$I_1 = \frac{(R_2 \parallel R_3)}{R_1 + (R_2 \parallel R_3)}\, I_T$$

where

$$R_2 \parallel R_3 = \frac{(6.8\ \text{k}\Omega)(15\ \text{k}\Omega)}{6.8\ \text{k}\Omega + 15\ \text{k}\Omega} = 4.68\ \text{k}\Omega$$

Then

$$I_1 = \frac{4.68\ \text{k}\Omega}{2.2\ \text{k}\Omega + 4.68\ \text{k}\Omega}(20\ \text{mA}) = 13.6\ \text{mA}$$

There are a number of ways to find I_2 and I_3 once I_1 is known. Perhaps the easiest way is to use Ohm's law to find

$$V_T = R_1 I_1 = (2.2\ \text{k}\Omega)(13.6\ \text{mA}) = 29.9\ \text{V}$$

This voltage is then used with Ohm's law to find I_2 and I_3

$$I_2 = \frac{V_T}{R_2} = \frac{29.9\ \text{V}}{6.8\ \text{k}\Omega} = 4.4\ \text{mA}$$

$$I_3 = \frac{V_T}{R_3} = \frac{29.9\ \text{V}}{15\ \text{k}\Omega} = 1.99\ \text{mA}$$

The student should check to ensure that KCL is satisfied.

We check by again applying the current divider formula in the following way: If $I_1 = 13.6$ mA, then $20 - 13.6 = 6.4$ mA remains to divide between R_2 and R_3. Thus,

$$I_2 = \frac{R_3}{R_2 + R_3}(6.4\ \text{mA}) = \frac{15\ \text{k}\Omega}{6.8\ \text{k}\Omega + 15\ \text{k}\Omega}(6.4\ \text{mA}) = 4.4\ \text{mA}$$

and

$$I_3 = \frac{R_2}{R_2 + R_3}(6.4\ \text{mA}) = \frac{6.8\ \text{k}\Omega}{6.8\ \text{k}\Omega + 15\ \text{k}\Omega}(6.4\ \text{mA}) = 2\ \text{mA}$$

The input resistance is (using conductance)

$$G_{in} = G_1 + G_2 + G_3 = \frac{1}{2.2\ \text{k}\Omega} + \frac{1}{6.8\ \text{k}\Omega} + \frac{1}{15\ \text{k}\Omega} = 0.668\ \text{mS}$$

Using the reciprocal function on the calculator

$$R_{in} = \frac{1}{G_{in}} = 1.5\ \text{k}\Omega$$

As a further check we calculate $V_T = R_{in} I_T = (1.5\ \text{k}\Omega)(20\ \text{mA}) = 30\ \text{V}$

• • •

EXAMPLE 4.8-3 Design of a Current Divider

A phonograph amplifier input stage requires a current divider that has a current transfer ratio $A_i = 0.6$ and an input resistance greater than 10 kΩ.

Solution

The circuit is shown in Fig. 4.31a and we consider I_1 to be the output current. The current transfer ratio specification requires that

$$\frac{R_2}{R_1 + R_2} = 0.6 \qquad (4.8\text{-}10)$$

while the input resistance specification requires that

$$\frac{R_1 R_2}{R_1 + R_2} = 10^4\ \Omega \qquad (4.8\text{-}11)$$

We have here two equations in two unknowns, R_1 and R_2. These can be solved in several ways. One way is to note that both sides of Eq. (4.8-11) can be divided by R_1 to give

$$\frac{R_2}{R_1 + R_2} = \frac{10^4}{R_1}$$

The left side of this equation will be recognized as being the same as the left side of Eq. (4.8-10) and is thus equal to 0.6. Then

$$\frac{10^4}{R_1} = 0.6$$

from which

$$R_1 = 16.7\ \text{k}\Omega$$

This result is substituted into either Eq. (4.8-10) or (4.8-11) to yield $R_2 = 25$ kΩ.

If the divider is to be built with 20% resistors we could choose $R_1 = 15$ kΩ and $R_2 = 22$ kΩ, or $R_1 = 22$ kΩ and $R_2 = 33$ kΩ. The resulting current ratio and input resistance, based on the nominal values, would be for the first choice

$$\frac{I_1}{I_i} = \frac{22\ \text{k}\Omega}{15\ \text{k}\Omega + 22\ \text{k}\Omega} = 0.6$$

and

$$R_{in} = \frac{(15\ \text{k}\Omega)(22\ \text{k}\Omega)}{15\ \text{k}\Omega + 22\ \text{k}\Omega} = 8.9\ \text{k}\Omega$$

For the second choice

$$\frac{I_1}{I_i} = \frac{33\ \text{k}\Omega}{22\ \text{k}\Omega + 33\ \text{k}\Omega} = 0.6$$

$$R_{\text{in}} = \frac{(22\ \text{k}\Omega)\,(33\ \text{k}\Omega)}{22\ \text{k}\Omega + 33\ \text{k}\Omega} = 13.2\ \text{k}\Omega$$

Since the input resistance must be greater than 10 kΩ we specify $R_1 = 22$ kΩ and $R_2 = 33$ kΩ.

• • •

LEARNING EXERCISES FOR SEC. 4.8

1. Resistors of value 47 and 15 kΩ are connected as a current divider to a 5-mA source. Draw the circuit and use the current divider formula to find the current through each resistor. What is the voltage across the circuit?

2. Find the current transfer ratio for each resistor in the circuit of Exercise 4.8-1.

Ans. 57; 0.24; 1.2; 3.8; 0.76

• • •

4.9 THE EFFECTS OF ACCIDENTAL OPEN OR SHORT CIRCUITS

Technologists and engineers are sometimes required to troubleshoot and repair electric systems that have malfunctioned, or to "debug" systems which they have designed. There are many different types of malfunctions or "bugs" which can occur. The short circuit or open circuit are faults which occur quite frequently and we will consider in this section how each of these affects series and parallel circuits.

Series Circuit

Consider first the series circuit shown in Fig. 4.33. Assume that the resistance wire from which R_1 is constructed has been physically broken between *a* and *b*, causing R_1 to be an open circuit. If R_1 represents an electric heater, then this could happen because the resistance wire actually melted and broke apart due to overheating in one small area. One way to determine the effect of this break is to note that the resistance of the break is infinite for all practical purposes. Thus the current in the circuit must reduce to zero after the wire is broken. If we apply KVL around the loop we have

$$V_{ab} + R_1 I + R_2 I + R_3 I - V_i = 0$$

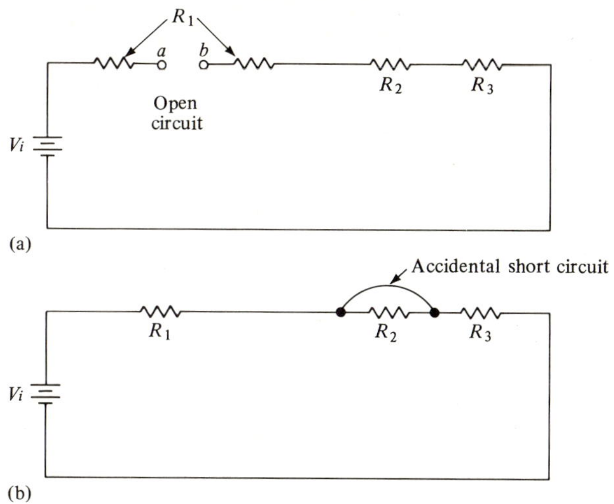

FIGURE 4.33
Investigation of an open circuit and a short circuit in a series string. (a) Open circuit. (b) Short circuit.

and since $I = 0$,

$$V_{ab} = V_i$$

In words, this says that the voltage across the open circuit is the same as the input voltage V_i. If V_i is, for example, 110 or 220 V, and we were to accidentally touch points a and b, we would receive a severe electric shock.

In order to check for an open circuit such as this we can measure the voltage across each resistance. All of these voltages will be zero except for the voltage across R_1, the resistance that has open-circuited, and this voltage will be the full source voltage.

Another fault which might occur would be the accidental short-circuiting of one of the resistors, for example, R_2 as shown in Fig. 4.33b. The effect of this malfunction is not quite as severe as the open circuit, but still requires investigation. Before the accidental short was applied the current was $V_i/(R_1 + R_2 + R_3)$. After the short is in place the current is increased to $V_i/(R_1 + R_3)$. If the possibility of such a short circuit exists, we must make sure that the increased current will not overload the source or cause the power dissipation in the remaining resistors to exceed their rated value.

EXAMPLE 4.9-1 Short Circuit Fault in a Series Circuit

The circuit of Fig. 4.33 has been designed to provide three equal voltages from a 220-V supply. In the circuit, $V_i = 220$ V, $R_1 = R_2 = R_3 = 12$ kΩ, 1/2-W resistors. Find the effect of an accidental short across one of the resistors.

Solution
Before the short, the current is

$$I = \frac{V_i}{R_1 + R_2 + R_3} = \frac{220\ \text{V}}{36\ \text{k}\Omega} = 6.1\ \text{mA}$$

Since the resistors are the same, the voltage across each is

$$V = \frac{V_i}{3} = \frac{220\ \text{V}}{3} = 73.3\ \text{V}$$

The power dissipation in each resistor is found from the relation derived in Sec. 2.7

$$P = VI = (73.3\ \text{V})(6.1\ \text{mA}) = 447\ \text{mW}$$

This is within the 1/2-W rating of the resistors.
If one of the resistors is shorted, the current becomes

$$I = \frac{V_i}{R_1 + R_2} = \frac{220\ \text{V}}{24\ \text{k}\Omega} = 9.17\ \text{mA}$$

and the voltage is 220 V/2 = 110 V.
The power dissipation in each of the two remaining resistors is then

$$P = VI = (110\ \text{V})(9.17\ \text{mA}) = 1010\ \text{mW} \approx 1\ \text{W}.$$

This exceeds the power rating of the remaining resistors, and they would undoubtedly overheat and be destroyed. If 1-W resistors were used and if the source could safely supply 9.17 mA, then no damage would be done.

• • •

Parallel Circuit

For the parallel circuit shown in Fig. 4.34a there are several possibilities for open-circuit faults. If the open circuit occurs between points *a* and *b*, in the main line, then current cannot reach any of the resistors and the circuit is dead. If the break occurs between points *c* and *d*, then resistor R_1 can receive current but the rest of the circuit is dead, and so on. On the other hand, if any of the resistors, such as R_1, suffer a break, the only effect would be that the current I_1 would become zero; the current in the other branches would be unaffected. This illustrates an advantage of having elements wired in parallel. If one element becomes open-circuited the other elements are unaffected.

A short-circuit fault in a parallel arrangement is much more serious. Consider what happens in the circuit of Fig. 4.34a if a short is accidentally

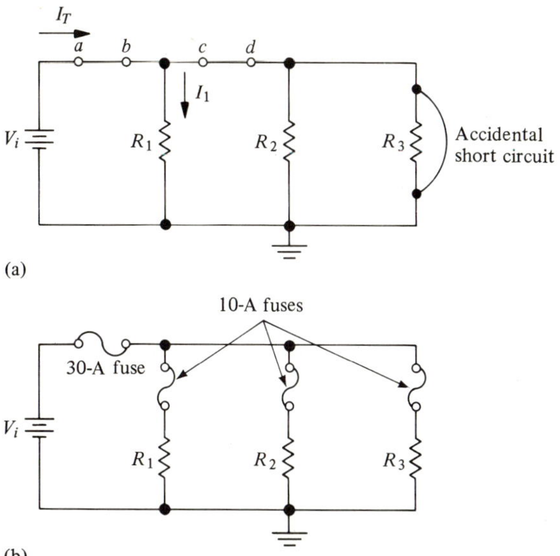

FIGURE 4.34
Investigation of an open circuit and a short circuit in a parallel arrangement. (a) Circuit. (b) Circuit with fuse protection.

placed across R_3. The conducting wires and the short circuit have essentially zero resistance. Therefore the total resistance of the circuit as seen from the battery is 0 Ω (recall from the formula for parallel resistors that 0 Ω in parallel with any resistance is 0 Ω). The current drawn from the battery will theoretically be $I = V_i/R_T = V_i/0 = \infty$. Because of the small resistance which actually is present, the current cannot be infinite. But the battery will try to supply a very high current, until it overheats or until the wires burn out. One way to protect against this kind of failure is to include fuses or circuit breakers in the system so that currents above a safe level will cause the individual branch or even the main line to be opened. It is interesting to note that the high current will flow only through the conducting wires up to the short and through the short itself. In Fig. 4.34a, with the short circuit in place, resistors R_1, R_2, and R_3 will draw essentially zero current (why?) and will not be damaged.

EXAMPLE 4.9-2 Effect of Short-Circuit in Parallel Network

The circuit of Fig. 4.34a represents an electric furnace in which three identical heaters are connected across the source. In the circuit, $R_1 = R_2 = R_3 = 33\ \Omega$ and $V_i = 220$ V. An accidental short occurs from point d to ground. The resistance seen from the source with the short in place is approximately 0.5 Ω. Find the current drawn from the source before and after the short.

Solution
Before the short

$$R_T = \frac{33\ \Omega}{3} = 11\ \Omega$$

$$I_T = \frac{V_i}{R_T} = \frac{220\ \text{V}}{11\ \Omega} = 20\ \text{A}$$

After the short

$$R_T \approx 0.5 \parallel 11 \approx 0.5\ \Omega$$

$$I_T = \frac{V_i}{R_T} \approx \frac{220\ \text{V}}{0.5\ \Omega} = 440\ \text{A}$$

This high current would undoubtedly quickly lead to overheating and failure of either the source or the wiring. In practice, a fuse or circuit breaker would be connected between points *a* and *b*. These devices are designed to break the circuit when the current exceeds a specified value. For this example, a 30-A fuse or circuit breaker would provide protection against a short-circuit fault. If additional protection is desired, a 10-A fuse may be provided for each heater, as shown in Fig. 4.34b.

• • •

LEARNING EXERCISES FOR SEC. 4.9

1. In Fig. 4.33, $V_i = 4400$ V, and R_1, R_2, and R_3 represent heating elements in a steel furnace. Each element has 2-Ω resistance and can safely dissipate 3 kW. Assuming that one element is short-circuited, find the current through the remaining elements and the power dissipated in each of them.

2. Figure 4.34 represents one circuit in an automobile powered by a 12-V battery. R_1 represents a stereo that has an effective resistance of 2 Ω, R_2 represents a cigarette lighter that requires 5 A, and R_3 represents headlights which draw 12 A. Find the battery current when all three loads are on. Assuming a conductor resistance of 0.08 Ω, how much current would the battery supply if the cigarette lighter short-circuited?

Ans. 23; 1100; 150; 2420

• • •

4.10 EFFECTS OF LOAD ON VOLTAGE AND CURRENT DIVIDERS

In this section we present examples of voltage and current dividers to which a load is connected. These circuits are combinations of the series and parallel circuits that we considered previously. This complicates the analysis because the resistance of the load must be taken into account.

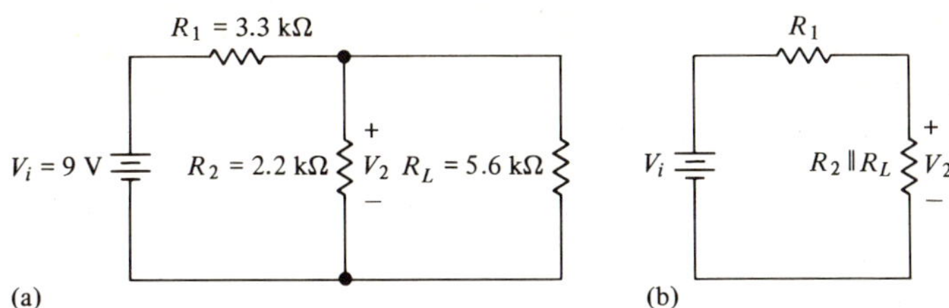

FIGURE 4.35
Loaded voltage divider. (a) Original circuit. (b) Equivalent circuit.

EXAMPLE 4.10-1 The Loaded Voltage Divider

The circuit of Fig. 4.35 represents a situation which arises frequently in electronic circuits. For example, R_1 and R_2 form a voltage divider across the 9-V supply. Resistor R_L, the load resistance, might represent the resistance of a meter connected to measure the voltage across R_2. As we will see, the fact that R_L is connected causes the voltage across R_2 to be different from what it would be without R_L connected. Thus we say that R_L "loads" the voltage divider. We will calculate the *loading effect* by finding the voltage across R_L.

Solution

We might be tempted to solve this problem by writing

$$V_2 = \frac{R_2}{R_T} V_i = \frac{2.2\ \text{k}\Omega}{2.2\ \text{k}\Omega + 3.3\ \text{k}\Omega}(9\ \text{V}) = 3.6\ \text{V}$$

This is the voltage output of the *unloaded* voltage divider. Unfortunately, this solution is wrong. The mistake arises because we have not considered the effect of R_L. Since R_L is in parallel with R_2 we must replace R_2 in the original formula by $R_2 \| R_L$. An equivalent two-resistor circuit is shown in Fig. 4.35b. Thus, $R_T = R_1 + (R_2 \| R_L)$ and the correct solution is

$$V_2 = \frac{(R_2 \| R_L)}{R_T} V_i = \frac{(2.2 \| 5.6)\text{k}\Omega}{3.3\ \text{k}\Omega + (2.2 \| 5.6)\text{k}\Omega}(9\ \text{V}) = 2.91\ \text{V}$$

Note that the effect of the load resistance is to reduce the output voltage compared to that of the unloaded divider because the effective R_2 is lowered by the presence of R_L.

This example points up the fact that any resistance in parallel with R_2 (or R_1) must be taken into account.

• • •

EXAMPLE 4.10-2 The Series-Dropping Resistor

Resistor R_1 in the circuit of Fig. 4.36a is often called a series-dropping resistor. It is used when a voltage source must supply one or more voltages of smaller value than

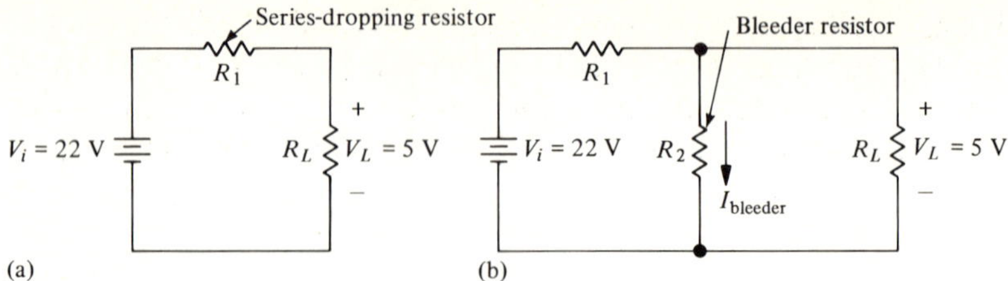

FIGURE 4.36
Series-dropping resistor. (a) Circuit for Example 4.10-2. (b) Circuit with bleeder resistor for Example 4.10-3.

the source voltage. For example, if we wish to operate a 5-V light bulb from a 22-V supply, we can use a dropping resistor to obtain the required 5 V. Find the value of this resistor if it is to drop a 22-V supply down to 5 V for a bulb that draws a current of 2 mA.

Solution
The data indicate that at 2-mA load current the load has a resistance

$$R_L = \frac{5\text{ V}}{2\text{ mA}} = 2.5\text{ k}\Omega$$

The circuit is shown in Fig. 4.36a. The only unknown is the value of R_1. This can be found in a number of different ways. One way is to use the voltage divider formula

$$\frac{V_L}{V_i} = \frac{R_L}{R_L + R_1}$$

Substituting the known values

$$\frac{5\text{ V}}{22\text{ V}} = \frac{2.5\text{ k}\Omega}{2.5\text{ k}\Omega + R_1}$$

from which we find $R_1 = 8.5$ kΩ. (The student should check this solution by applying KVL and Ohm's law directly. See Prob. 4.64.)

• • •

The use of the series-dropping resistor as in Example 4.10-2 has one disadvantage. Any change in load current will cause a change in load voltage because of the fact that the current flows through the dropping resistor. For example, assume that the load resistance in Example 4.10-2 changes so that the load current drops to 1 mA. Then the voltage across the 8.5-kΩ dropping resistor will be (8.5 kΩ) (1 mA) = 8.5 V and the load voltage will rise to 22 V − 8.5 V = 13.5 V. Thus, if the load resistance may vary and if the load voltage is to remain relatively constant, the series-dropping resistor does not provide a satisfactory solution.

A better solution is to connect a resistor in parallel with the load as shown in Fig. 4.36b. This resistor, called a "bleeder" resistor, is designed to draw a larger current than the load does, so that changes in load current will have a much smaller effect. This is illustrated in the next example.

EXAMPLE 4.10-3 Design with a Bleeder Resistor

In the circuit of Fig. 4.36b, the nominal load voltage is 5 V when the load current is 2 mA. However, the load resistance may rise to as much as twice its nominal value. Find the bleeder resistance R_2 and the series dropping resistor R_1 for the following two cases (use standard 20% tolerance resistors).

a. Bleeder current = 5 mA
b. Bleeder current = 10 mA

For each case find the total variation in load voltage as the load resistance varies over its indicated range.

Solution

a. Bleeder current = 5 mA. Since the voltage across the bleeder resistor is 5 V when the load current is 2 mA, and the bleeder current is to be 5 mA, the bleeder resistance must be $R_2 = 5\ \text{V}/5\ \text{mA} = 1\ \text{k}\Omega$. In order to find R_1 we note that it must drop $22\ \text{V} - 5\ \text{V} = 17\ \text{V}$ when the current through it is the bleeder current plus the load current ($5\ \text{mA} + 2\ \text{mA} = 7\ \text{mA}$). Thus $R_1 = 17\ \text{V}/7\ \text{mA} \approx 2.5\ \text{k}\Omega$ (we specify a 2.2-kΩ resistor).

In the circuit of Fig. 4.36b we now have the values $R_1 = 2.2\ \text{k}\Omega$, $R_2 = 1\ \text{k}\Omega$, and R_L may range from 2.5 to 5 kΩ. When $R_L = 2.5\ \text{k}\Omega$ ($I_L = 2\ \text{mA}$)

$$V_L = \frac{(R_2 \parallel R_L)}{R_1 + (R_2 \parallel R_L)} V_i = \frac{(1 \parallel 2.5)\ \text{k}\Omega}{2.2\ \text{k}\Omega + (1 \parallel 2.5)\ \text{k}\Omega}(22\ \text{V}) = 5.39\ \text{V}$$

When $R_L = 5\ \text{k}\Omega$ ($I_L = 1\ \text{mA}$)

$$V_L = \frac{(1 \parallel 5)\ \text{k}\Omega}{2.2\ \text{k}\Omega + (1 \parallel 5)\ \text{k}\Omega}(22\ \text{V}) = 6.04\ \text{V}$$

b. Bleeder current = 10 mA

For this case the bleeder resistance must be $R_2 = 5\ \text{V}/10\ \text{mA} = 0.5\ \text{k}\Omega$ (we specify a 0.47-kΩ resistor). The current through R_1 will be $10\ \text{mA} + 2\ \text{mA} = 12\ \text{mA}$. Thus $R_1 = 17\ \text{V}/12\ \text{mA} \approx 1.4\ \text{k}\Omega$ (we specify a 1.5-kΩ resistor).

We now have $R_1 = 1.5\ \text{k}\Omega$, $R_2 = 0.47\ \text{k}\Omega$, and R_L may range from 2.5 to 5 kΩ. When $R_L = 2.5\ \text{k}\Omega$

$$V_L = \frac{(0.47 \parallel 2.5)\ \text{k}\Omega}{1.5\ \text{k}\Omega + (0.47 \parallel 2.5)\ \text{k}\Omega}(22\ \text{V}) = 4.59\ \text{V}$$

When $R_L = 5\ \text{k}\Omega$

$$V_L = \frac{(0.47 \parallel 5)\ \text{k}\Omega}{1.5\ \text{k}\Omega + (0.47 \parallel 5)\ \text{k}\Omega}(22\ \text{V}) = 4.9\ \text{V}$$

Let us compare the results for the different bleeder currents. When the bleeder current is 5 mA the load voltage varies from 6.04 to 5.39 V, for a total variation of 0.65 V, while a bleeder current of 10 mA produces a load voltage variation of 4.9 to 4.59 V, a total variation of 0.31 V. The variation when there is no bleeder is more than 8 V. Thus we see that the load voltage variation decreases as we increase the bleeder current. This scheme has the disadvantage that the 22-V supply must furnish extra current and that power is dissipated in the bleeder resistor. However, this is sometimes a small price to pay for a relatively constant load voltage.

The student will have observed that the calculated nominal output voltages are not exactly 5 V. This is due to the fact that we used standard 20% tolerance resistance values for R_1 rather than the exact calculated values.

• • •

EXAMPLE 4.10-4 Double Current Divider

In the circuit of Fig. 4.37a find I_3.

Solution

The circuit requires two applications of the current divider rule. We first reduce the circuit by replacing the $R_3 - R_4$ parallel combination by its equivalent resistance, as shown in Fig. 4.37b. Then we use the current divider formula to find I_2. Since I_2 divides between R_3 and R_4 the current divider formula is then used again to find I_1.

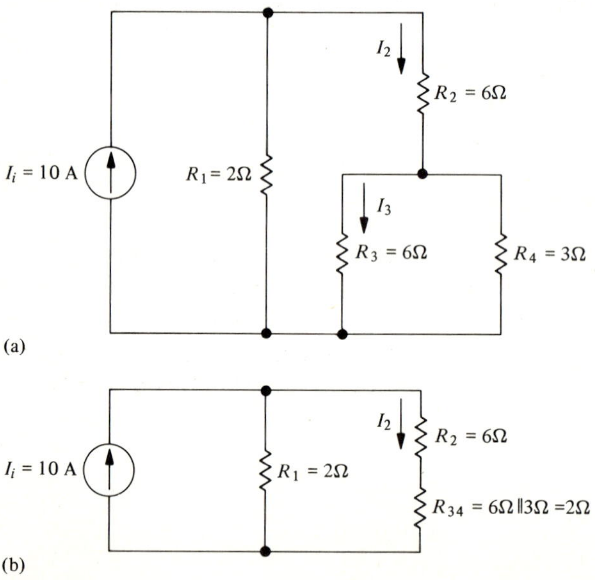

FIGURE 4.37
Example 4.10-4. (a) Original circuit. (b) Equivalent circuit.

The steps are as follows:
From Fig. 4.37b

$$I_2 = \frac{R_1 I_i}{(R_2 + R_{34}) + R_1} = \frac{2\ \Omega}{8\ \Omega + 2\ \Omega}(10\ \text{A}) = 2\ \text{A}$$

From Fig. 4.37a

$$I_3 = \frac{R_4}{R_3 + R_4} I_2$$

$$= \frac{3\ \Omega}{6\ \Omega + 3\ \Omega}(2\ \text{A})$$

$$= 0.67\ \text{A}$$

The student should complete this example by finding all currents and voltages and applying all possible checks using Ohm's law, KVL, and KCL.

• • •

LEARNING EXERCISES FOR SEC. 4.10

1. What series-dropping resistance is required to operate a 5-V light bulb from a 22-V battery if the bulb requires 150 mA?

2. In the circuit of Fig. 4.36b the supply is 110 V, the load R_L requires 5 V at 1 A, and the bleeder current is to be 2 A. Find R_1 and the bleeder resistance.

Ans. 35; 2.5; 113

• • •

4.11 POWER CALCULATIONS

In this section, we make use of the theory developed in Chap. 2 to calculate resistor power dissipation and source power requirements in the circuits discussed in this chapter. As a brief review, we note that the power delivered by or absorbed by a two-terminal element is simply the product of the voltage across the element and the current through it. Symbolically,

$$\boxed{P = VI} \qquad (4.11\text{-}1)$$

where P is in watts when V is in volts and I in amperes. The convention is that P is positive when the current enters the element at the terminal at which the actual voltage is positive, indicating that power is absorbed by (dissipated in) the element; P is negative otherwise, indicating that power is delivered by the element. When the element is a battery or current source, Eq. (4.11-1) can be applied directly. When the element is a resistor both voltage and current do not have to be known in order to calculate the power. We can make use of Ohm's law to eliminate either voltage or current from Eq. (4.11-1) as follows: Since $I = V/R$

$$P = VI = V\left(\frac{V}{R}\right)$$

$$\boxed{P = \frac{V^2}{R}} \qquad (4.11\text{-}2)$$

Also, since $V = RI$

$$P = VI = (RI)I$$

$$\boxed{P = RI^2} \qquad (4.11\text{-}3)$$

Thus, when the element is a resistor, we can use $P = V^2/R$ or $P = RI^2$ as appropriate.

Typical power calculations are given in the examples that follow.

EXAMPLE 4.11-1 Power in Resistances

The circuit of Fig. 4.38 represents the output circuit of the display amplifier in an electronic pinball machine. In the circuit, calculate the power dissipated in each resistor and check to see that it matches the power delivered by the battery.

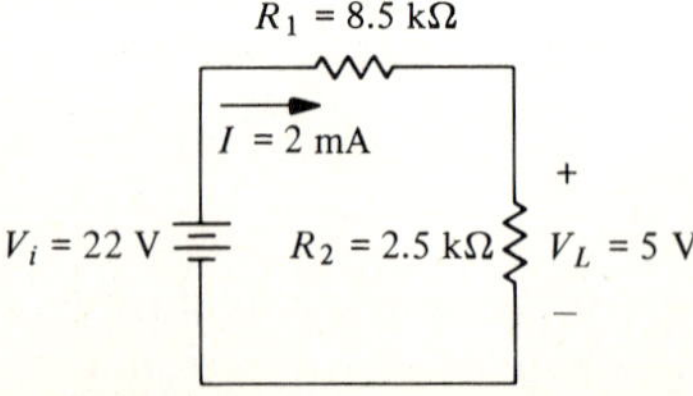

FIGURE 4.38
Circuit for Example 4.11-1.

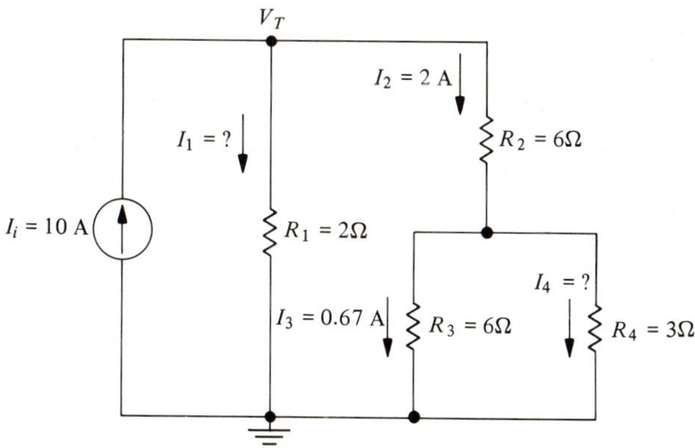

FIGURE 4.39
Circuit for Example 4.11-2.

Solution

The circuit with all voltages and currents is shown in Fig. 4.38. The resistor power dissipations are

$$P_1 = R_1 I^2 = (8.5 \times 10^3\ \Omega)(2 \times 10^{-3}\ \text{A})^2 = 34 \times 10^{-3}\ \text{W} = 34\ \text{mW}$$

$$P_2 = R_L I^2 = (2.5 \times 10^3\ \Omega)(2 \times 10^{-3}\ \text{A})^2 = 10 \times 10^{-3}\ \text{W} = 10\ \text{mW}$$

Thus the total power dissipated is

$$P_T = P_1 + P_2 = 34\ \text{mW} + 10\ \text{mW} = 44\ \text{mW}$$

The power delivered by the battery is

$$P_B = -V_i I = -(22\ \text{V})(2 \times 10^{-3}\ \text{A}) = -44 \times 10^{-3}\ \text{W} = -44\ \text{mW}$$

The power absorbed in the resistors is exactly equal in magnitude to the power delivered by the battery.

• • •

EXAMPLE 4.11-2 *Resistor Power Calculations*

In the circuit of Fig. 4.37a (see Example 4.10-4) find the power dissipated in each resistor and check to see that it matches the power delivered by the current source.

Solution

The circuit with values as found in Example 4.10-4 is shown in Fig. 4.39. In order to calculate the various powers we need I_1, I_4, and V_T. These are found as follows: Using KCL

$$I_1 = I_i - I_2 = 10\ \text{A} - 2\ \text{A} = 8\ \text{A}$$

Using Ohm's law

$$V_T = R_1 I_1 = (2\ \Omega)(8\ \text{A}) = 16\ \text{V}$$

Again using KCL

$$I_4 = I_2 - I_3 = 2\ \text{A} - 0.67\ \text{A} = 1.33\ \text{A}$$

The resistor power dissipations are

$$P_1 = R_1 I_1^2 = (2\ \Omega)(8\ \text{A})^2 = 128\ \text{W}$$

$$P_2 = R_2 I_2^2 = (6\ \Omega)(2\ \text{A})^2 = 24\ \text{W}$$

$$P_3 = R_3 I_3^2 = (6\ \Omega)(0.67\ \text{A})^2 = 2.7\ \text{W}$$

$$P_4 = R_4 I_4^2 = (3\ \Omega)(1.33\ \text{A})^2 = 5.3\ \text{W}$$

The total power absorbed by the resistors is

$$P_1 + P_2 + P_3 + P_4 = 160\ \text{W}$$

The power delivered by the current source is

$$P_S = -V_T I_i = -(16\ \text{V})(10\ \text{A}) = -160\ \text{W}$$

• • •

EXAMPLE 4.11-3

The circuit of Fig. 4.40 is part of the circuitry required to operate an olympic timer. Find the power dissipated in each resistance and choose a suitable power rating, keeping in mind that it is good engineering practice to include a safety factor of 1.5 or more when choosing power ratings. Available standard power ratings are 1/8, 1/4, 1/2, 1, 2, and 5 W.

Solution

We begin by finding the current in R_1 using Ohm's law

$$R_T = R_1 + (R_2 \parallel R_3)$$

$$= 0.05\ \text{k}\Omega + (2.2\ \text{k}\Omega \parallel 4.7\ \text{k}\Omega) = 1.5\ \text{k}\Omega$$

$$I_1 = \frac{V_1}{R_T} = \frac{80\ \text{V}}{1.5\ \text{k}\Omega} = 53\ \text{mA}$$

The power dissipated in R_1 is then

$$P_1 = R_1 I_1^2 = (47)(0.053)^2 = 0.13\ \text{W}$$

Applying a safety factor of 1.5 we have $1.5 \times 0.13 = 0.2$ W as the minimum power rating for R_1. The nearest standard value is 1/4 W, so that R_1 is specified as a 47-Ω, 1/4-W resistor.

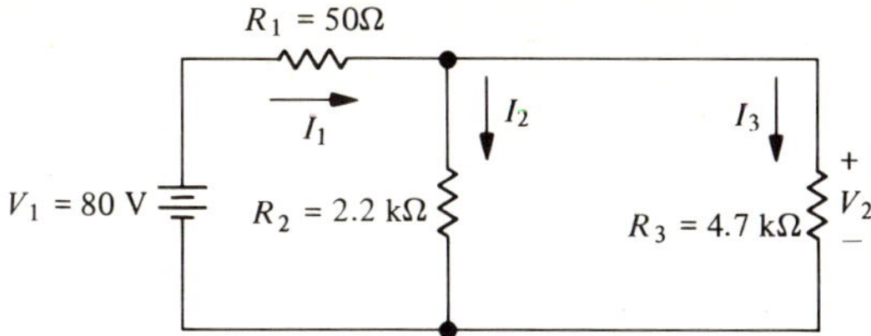

FIGURE 4.40
Circuit for Example 4.11-3.

In order to find the power dissipated in R_2 and R_3 we first find V_2 as follows

$$V_2 = V_1 - R_1 I_1 = 80 \text{ V} - (0.047 \text{ k}\Omega)(53 \text{ mA}) = 78 \text{ V}$$

Then

$$P_2 = \frac{V_2^2}{R_2} = \frac{78^2}{2200} = 2.8 \text{ W}$$

and

$$P_3 = \frac{V_2^2}{R_3} = \frac{78^2}{4700} = 1.3 \text{ W}$$

Applying the safety factor of 1.5 we find that R_2 must have a power rating of at least $1.5 \times 2.8 = 4.2$ W so a 5-W resistor is specified and R_3 must have $1.5 \times 1.3 = 2$ W, so a 2-W resistor is specified.

• • •

LEARNING EXERCISES FOR SEC. 4.11

1. A car stereo operates on 12 V and draws 6 A. How much power does it require?

2. A heater coil has a resistance of 12 Ω and requires 180 A to reach maximum heat. How much power must be supplied?

3. The display for a television hockey game has a resistance of 400 kΩ and is powered by a 900-V supply. How much power is required?

Ans. 2; 72; 390

• • •

4.12 ENERGY AND COST

In Chap. 2 we defined power as the rate at which energy is used or supplied. Thus

$$P = \frac{W}{t} \tag{4.12-1}$$

As we discussed earlier in this section, if P is positive then energy is being absorbed or used by the element, and the amount of energy is

$$\boxed{W = Pt} \tag{4.12-2}$$

where

W = energy in watt-seconds or joules
P = power in watts
t = time in seconds

The watt-second is used as the unit of energy for small devices such as electronic flash units. However, when we purchase energy from the power company, the watt-second is too small a unit for convenience, so when dealing with residential or commercial power, kilowatt-hours (kWh) are used. Equation (4.12-2) expressed in units of kilowatt-hours becomes

$$W = \frac{Pt}{1000} \tag{4.12-3}$$

where

W = energy in kilowatt-hours (kWh)
P = power in watts
t = time in hours (h)

With this formula we can calculate the total energy delivered to or by any electric device over a period of time.

If the power company charges D dollars per kilowatt hour ($/kWh), then the total cost of this energy is

$$\text{Cost} = WD = \frac{PtD}{1000}\ \$ \tag{4.12-4}$$

We illustrate with some examples based on the cost of energy during 1980.

EXAMPLE 4.12-1 Cost of Operating a Light Bulb

Find the cost of operating a 100-W light bulb for 1 year at an energy cost of 8¢/kWh.

Solution

Using cost = $PtD/1000$ we have

$$\text{Cost} = \frac{(100 \text{ W})(365 \text{ days})(24 \text{ h/day})(0.08 \text{ \$/kWh})}{(1000 \text{ W/kW})}$$

$$= \$70.08$$

• • •

EXAMPLE 4.12-2 *Cost of Cooking in an Electric Oven*

Find the cost of cooking a roast in a 220-V, 20-A electric oven. The total time required is 3 h; however, the oven is thermostatically controlled and is actually on for 70% of the time. Energy costs 8¢/kWh.

Solution

Again we use cost = $PtD/1000$ with

$$P = VI = (220 \text{ V})(20 \text{ A}) = 4400 \text{ W}$$

$$t = (3 \text{ h})(0.7) = 2.1 \text{ h}$$

Then

$$\text{Cost} = \frac{(4400 \text{ W})(2.1 \text{ h})(0.08 \text{ \$/kWh})}{1000 \text{ W/kW}} = \$0.74 = 74¢$$

• • •

A list of household appliances along with their power requirements is given in Table 4.1.

• • •

LEARNING EXERCISES FOR SEC. 4.12

1. A large air conditioner requires 10 kW of power. How much does it cost for 12 h of operation at 8¢/kWh?

2. A home computer is rated at 300 W. How much does it cost to run a long program which takes 3 h if the power cost is 8¢/kWh?

3. A 1200-W coffee maker takes 20 min to brew 8 cups of coffee. At 8¢/kWh how much does it cost to brew 1 cup?

Ans. 0.07; 9.60; 0.004

• • •

TABLE 4.1
Energy requirements of electric household appliances*

Appliance	Typical Power Requirements (W)	Annual Use (h)	Cost at 8¢/kWh ($)
Coffee maker	1 200	117	11.20
Dishwasher	1 200	302	29.00
Mixer	127	16	0.20
Range with oven	12 200	57	55.60
Toaster	1 200	34	3.30
Refrigerator	320	4 700	120.30
Clothes dryer	4 800	200	76.80
Washing machine	500	200	8.00
Central air conditioner (3 ton)	4 500	750	270.00
Room air conditioner	860	465	32.00
Electric blanket	180	830	12.00
Window fan	200	850	13.60
Hair dryer	600	42	2.00
Shaver	15	33	0.04
Radio	70	1 200	6.70
Color television	150	2 200	26.40
Clock	2	8 760	1.40
Vacuum cleaner	730	73	4.30

*Annual hours of use are based on a survey by the Edison Electric Institute and typical wattages are an average of various popular brands.

4.13 EFFICIENCY

The function of all electric systems is to transfer or transform energy, and the efficiency of a system is a quantitative measure of how good it is at this task. A typical system is shown schematically in Fig. 4.41. It is assumed that there are no energy sources within the system, that is, all energy is supplied via the input terminals. Since no system is perfect, there will always be some energy lost in the system. For example, in a power transmission system using wire conductors, there will be a power loss due to the current flow in the resistance of the conducting wires. Thus only a portion of the

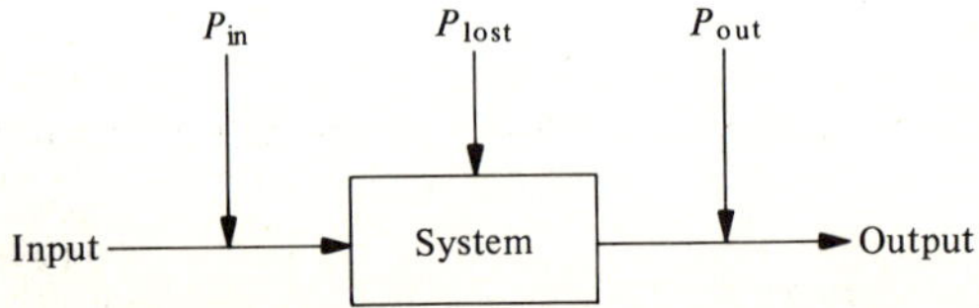

FIGURE 4.41
Electric system.

energy fed into the input terminals will reach the output. In terms of power, efficiency in percent is defined as

$$\eta = \frac{P_{out}}{P_{in}} \times 100\% \tag{4.13-1}$$

where η (eta) is the symbol usually used for efficiency.

The three powers are related by the law of conservation of energy.

$$P_{in} = P_{out} + P_{lost} \tag{4.13-2}$$

Using this, the basic equation for efficiency can also be written in the two forms

$$\eta = \frac{P_{out}}{P_{out} + P_{lost}} \times 100\% = \frac{P_{in} - P_{lost}}{P_{in}} \times 100\% \tag{4.13-3}$$

In all three formulas,

P_{out} = power ultimately transferred or transformed,
P_{in} = total input power to the system
P_{lost} = total power lost between input and output

All three quantities must, of course, be in the same units.

Since the power out cannot exceed the power in, Eq. (4.13-1) shows that the maximum possible efficiency is 100% and this occurs under ideal conditions when there are no losses. As the losses increase, the input power required to achieve a given output power increases and the efficiency decreases.

EXAMPLE 4.13-1 Efficiency of a Transmission System

In the power transmission system shown in Fig. 4.42a, the generator output is 1 MW, and this is the input to the transmission system $P_{T,in}$. The I^2R loss $P_{T,lost}$ in the transmission lines is 50 kW. Find the efficiency of the system.

Solution

From the problem statement, $P_{T,in} = 1$ MW and $P_{T,lost} = 50$ kW $= 0.05$ MW. Using the second form of Eq. (4.13-3)

$$\eta_T = \frac{P_{T,in} - P_{T,lost}}{P_{T,in}} \times 100 = \frac{1\text{ MW} - 0.05\text{ MW}}{1\text{ MW}} \times 100 = 95\%$$

• • •

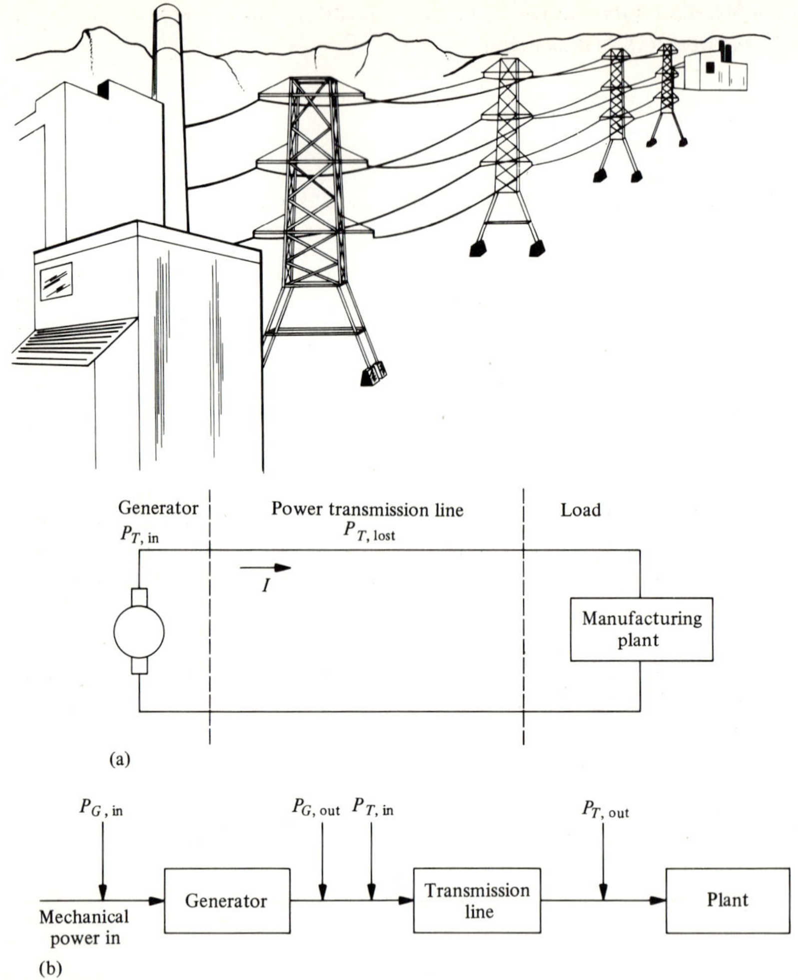

FIGURE 4.42
Power system for Examples 4.13-1 and 4.13-2. (a) System. (b) Cascaded systems.

EXAMPLE 4.13-2 Efficiency of Cascaded Systems

In Example 4.13-1 the mechanical power required to drive the generator is 1.23 MW. The generator system and transmission line system are said to be in *cascade* because the output of the generator is the input to the transmission line (see Fig. 4.42b). Find the generator efficiency and the overall efficiency.

Solution

The data indicate that $P_{G,\text{in}} = 1.23$ MW, and from Example 4.13-1 we have $P_{G,\text{out}} = P_{T,\text{in}} = 1$ MW. Then the generator efficiency is

$$\eta_G = \frac{P_{G,\text{out}}}{P_{G,\text{in}}} \times 100 = \frac{1\ \text{MW}}{1.23\ \text{MW}} \times 100 = 81.3\%$$

For the overall system the input power $P_{G,in}$ is 1.23 MW and the output power $P_{T,out}$ is 0.95 MW. Thus the overall efficiency is

$$\eta_{OV} = \frac{P_{T,out}}{P_{G,in}} = \frac{0.95\ \text{MW}}{1.23\ \text{MW}} \times 100 = 77.2\%$$

The overall efficiency can also be found by using the chain rule for systems in cascade

$$\eta_{OV} = \left(\frac{\eta_G}{100}\right)\left(\frac{\eta_T}{100}\right) = (0.813)(0.95) \times 100 = 77.2\% \tag{4.13-4}$$

The validity of this formula is proved in Prob. 4.90.

• • •

EXAMPLE 4.13-3 Motor Efficiency

An electric motor which drives a small rolling mill in a steel manufacturing plant operates on 220 V and provides a maximum mechanical output of 5 horsepower (hp) at an efficiency of 80%. Find the current into the motor under these conditions.

Solution

The horsepower (hp) is a United States Customary System unit of power used with electric motors. The SI conversion factor is

$$1\ \text{hp} = 746\ \text{W}$$

From the data, $P_{out} = 5\ \text{hp} \times 746\ \text{W/hp} = 3730\ \text{W} = 3.73\ \text{kW}$. Then

$$\eta = \frac{P_{out}}{P_{in}} \times 100\%$$

$$80\% = \frac{3730\ \text{W}}{P_{in}} \times 100\%$$

$$P_{in} = \frac{3730\ \text{W}}{80\%} \times 100\% = 4660\ \text{W}$$

Finally,

$$P_{in} = VI$$

$$4660\ \text{W} = (220\ \text{V})(I)$$

$$I = \frac{4660\ \text{W}}{220\ \text{V}} = 21.2\ \text{A}$$

• • •

LEARNING EXERCISE FOR SEC. 4.13

1. The electrical input to a steel mill motor is 4 kW and the mechanical power output is 5 hp. What is the efficiency? The 4 kW of electric power is obtained from a

generator that has an efficiency of 92%. How much power is required to drive the generator? What is the overall efficiency of the system?

Ans. 85.7; 93; 4.35

• • •

SUMMARY

In this chapter we have connected together resistors and sources to form circuits. The basic laws, Ohm's law and the two Kirchhoff laws, were applied to these circuits in order to determine currents and voltages. We then learned to design voltage divider and current divider networks when given appropriate specifications. In Chap. 5 we will extend these ideas to the analysis and design of more intricate circuits and systems.

Section

4.1 1. Ohm's law can be stated in three forms: $V = RI$; $I = V/R$; $R = V/I$.

2. The polarity of the voltage across a resistor is positive at the terminal at which the current enters.

4.2, 4.3 3. Kirchhoff's voltage law states that the algebraic sum of the voltages around any closed loop in a network is zero

$$\sum_{\text{Loop}} V_n = 0$$

4. In a series circuit: (a) the current is the same throughout the circuit; (b) the individual voltage drops across resistors are proportional to the resistance values; (c) the applied voltage equals the sum of the resistance RI voltage drops; (d) the total resistance $R_T = R_1 + R_2 + \ldots + R_n$.

4.5 5. The voltage divider formula states that

$$\frac{V_n}{V_i} = \frac{R_n}{R_T}$$

where the desired output voltage V_n appears across resistor R_n.

4.6, 4.7 6. Kirchhoff's current law states that the algebraic sum of the currents at any node is zero, $\Sigma_{\text{node}}\, i_n = 0$.

7. In a parallel circuit: (a) the voltage is the same across all branches; (b) the resistor currents are inversely proportional to the resistance values; (c) the total input current equals the sum of the resistor currents; (d) the input resistance can be determined using $R_T = R_1R_2/(R_1 + R_2)$, or if more than two resistors by using the conductance formula $G_T = G_1 + G_2 + \ldots + G_n$.

4.8 8. The current divider formula for resistors R_1 and R_2 in parallel states that

$$\frac{I_2}{I_i} = \frac{R_1}{R_T}$$

where the desired output current I_2 flows through R_2. If more than two resistors are in parallel all but the resistor R_2 through which the output current flows should be combined to form one equivalent resistance R_1. The formula above then applies.

4.10 9. When dropping a source voltage to a lower value using a series-dropping resistor, a bleeder resistor may be required to prevent large output voltage variations.

4.11 10. The power dissipated in a resistor can be found using any of the three forms $P = VI = RI^2 = V^2/R$.

4.12 11. The energy in kilowatt-hours delivered to or by an electric device over a period of t hours can be found from the formula $W = Pt/1000$.

4.13 12. (a) The efficiency of an electric system is defined as the ratio of the power out to the power in,

$$\eta = \frac{P_{out}}{P_{in}} \times 100\%$$

(b) For systems in cascade having individual efficiencies η_1, η_2, . . . , η_n the chain rule applies and the overall efficiency is $\eta_T = \eta_1\eta_2 \cdots \eta_n$.

QUESTIONS FOR REVIEW

Sec. 4.1

1. State Ohm's law in its three different forms.
2. What are the units appropriate for Ohm's law?
3. How is the polarity of the voltage drop across a resistor related to the direction of the current through the resistor?
4. Plot the *vi* characteristic for a 3.3-kΩ resistor.

Sec. 4.2

5. How can you tell that two elements are in series?
6. What rule governs the sum of the voltage drops around a closed loop?
7. What is the rule for finding the equivalent resistance of a number of series resistors?
8. State the general form of Kirchhoff's Voltage Law and describe how it is used.
9. The positive terminal of an 80-V battery is connected to the positive terminal of a 36-V battery. What voltage would be measured between the two negative terminals?

Sec. 4.3

10. KVL is applied to a loop in order to find an unknown voltage. The result turns out to be a negative number. What is the significance of the minus sign?

Sec. 4.4

11. What is the difference between reference polarity and actual polarity?
12. How can we determine actual polarity?
13. How is double-subscript notation used?
14. How is a voltmeter used to determine polarity?
15. What is the effect of changing the ground point in a circuit on the voltages across the elements in the circuit?

Sec. 4.5

16. What functions does a voltage divider circuit serve?
17. In a resistive voltage divider can the output voltage be greater than the input voltage?
18. What is the significance of the voltage transfer ratio?

Sec. 4.6

19. What determines whether two elements are in parallel?
20. State Kirchhoff's Current Law as applied to a parallel circuit.
21. In a parallel circuit, which resistance draws the larger current?
22. What is meant by equivalent resistance?
23. What is the equivalent resistance of two parallel resistors?

Sec. 4.7

24. State KCL in its general form.
25. A 4-A ideal current source is connected in series with a 2-A ideal current source. What is the resulting current?
26. The current sources of Question 25 are connected in parallel with both arrows pointing in the same direction. What is the resulting current?

Sec. 4.8

27. In what situation can we make use of a current divider circuit?
28. What is the significance of the current transfer ratio?

Sec. 4.9

29. Describe the effect of having one resistor in a series string of resistors open-circuited. Describe the effect of having one short-circuited.
30. Describe the effect on a circuit consisting of a number of parallel resistors if one resistor is open-circuited. What about short-circuited?
31. What is the function of fuses and circuit breakers?

Sec. 4.10

32. A voltage divider is designed to yield an output voltage of 2 V. The circuit is built and a load resistor comparable in value to the output resistor of the voltage divider is connected. The output voltage is much lower than the 2 V that the circuit was designed for. What went wrong?

33. What is the function of a dropping resistor?
34. What is the function of a bleeder resistor?

Sec. 4.11

35. State the basic relation for finding the power delivered to or from a two-terminal element.
36. State the alternate forms of the relation for power dissipated in a resistor.
37. In a closed circuit, what must be the sum of all the powers?

Sec. 4.12

38. Define power in terms of energy and time.
39. What are some typical units of electric energy?

Sec. 4.13

40. Define efficiency of a system in terms of power output, power input, and power loss.
41. Given the efficiencies of individual systems connected in cascade, how is the overall efficiency found?

PROBLEMS

Sec. 4.1

1. A current of 12 A flows through a resistance of 60 Ω. Find the voltage across the resistance and draw a circuit diagram showing current direction and voltage polarity.
2. Find the voltage across a 33-kΩ resistance that carries a current of 1.7 mA.
3. What is the current through a 75-Ω resistor if the voltage across it is 12 V?
4. A 15-Ω heater coil is connected to a 110-V power supply. How much current results?
5. A generator has an internal resistance of 0.4 Ω. What is the voltage across this resistance when the generator is supplying 150 A?
6. A 10-MΩ resistor has a voltage drop of 80 μV across it. How much current results?
7. A resistor in a transistor amplifier draws 25 mA from a 9-V battery. What is its resistance?
8. An electric furnace requires 83 A at 220 V. What is its resistance?
9. A fuse is rated at 25 mA and has a voltage drop of 500 μV before it overheats and interrupts the circuit. What is its resistance at this point?
10. Find the current in a 30-Ω relay coil when it is connected to 85 V.
11. A 6.8-kΩ resistor carries a current of 12.2 mA. What is the voltage drop across the resistor?
12. A heater coil has a resistance of 24 Ω. What voltage is required for the current in the coil to be 9.2 A?

Sec. 4.2

13. Resistors of 3.3 and 6.8 kΩ are connected in series. Find the equivalent resistance that can replace these two.

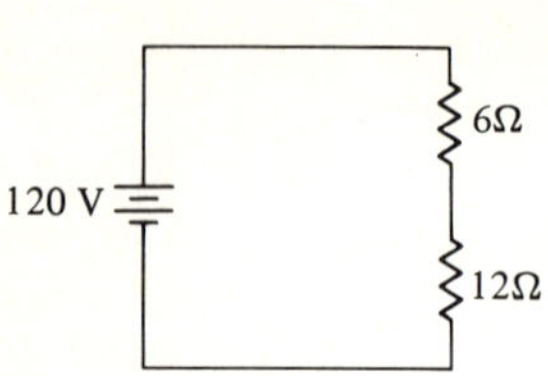

FIGURE 4.43

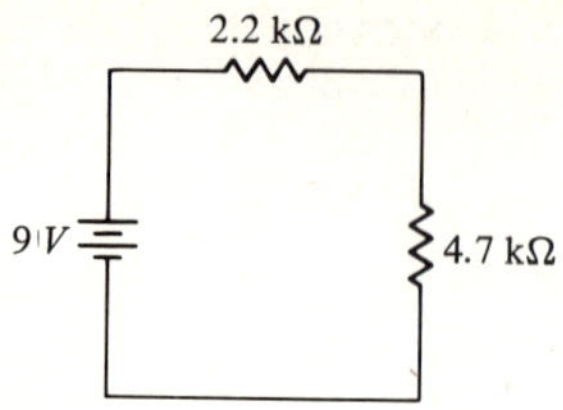

FIGURE 4.44

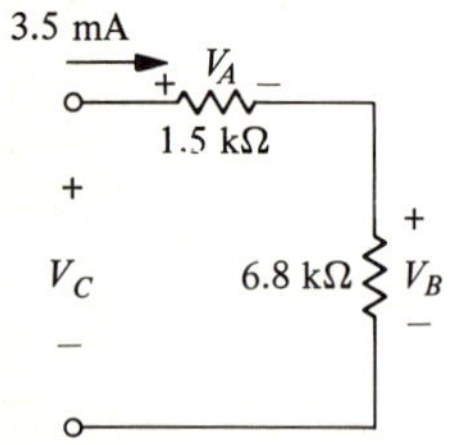

FIGURE 4.45

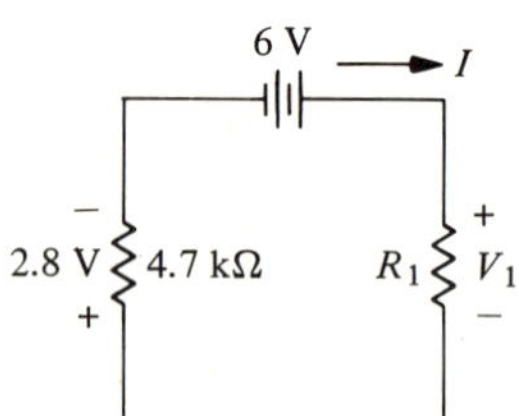

FIGURE 4.46

14. Find the current and resistor voltage drops in the circuit of Fig. 4.43. Indicate current direction and voltage polarity on all elements.
15. Repeat Prob. 14 for the circuit of Fig. 4.44.
16. Find V_A, V_B, and V_C in the circuit of Fig. 4.45.
17. In the circuit of Fig. 4.46 find I, V_1, and R_1.
18. In the circuit of Fig. 4.47 find V_1, V_2, and R_3.

Sec. 4.3

In Problems 19 through 22, the boxes contain batteries and/or resistors.

19. Find V_3 in the circuit of Fig. 4.48.
20. Find V_4 in the circuit of Fig. 4.49.
21. Find V_2 and V_4 in the circuit of Fig. 4.50.
22. In the circuits of Fig. 4.51 find the single source that will replace the series combinations of sources.
23. Find V_{ab} and V_4 in the circuit of Fig. 4.52.

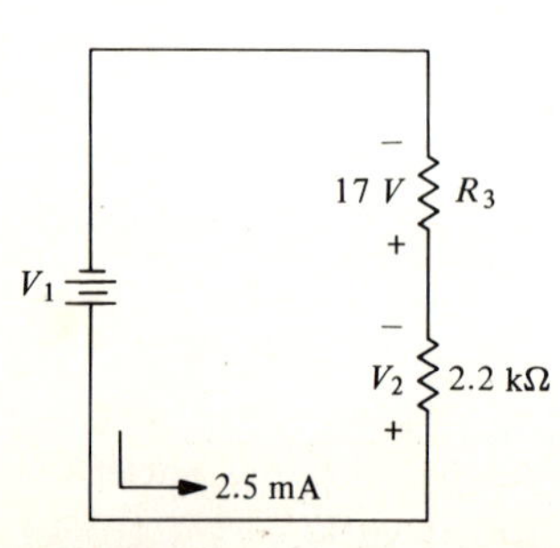

FIGURE 4.47

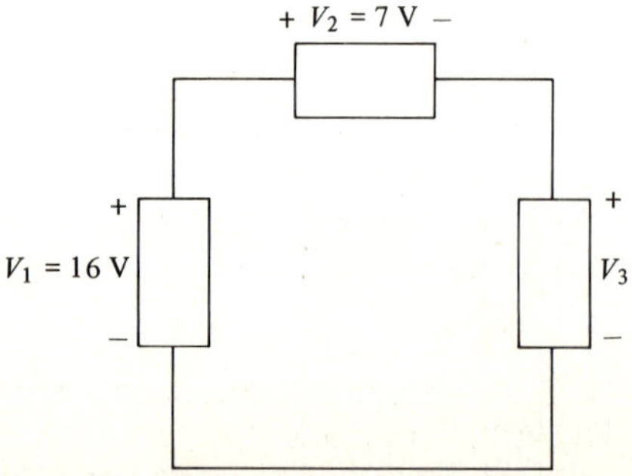

FIGURE 4.48

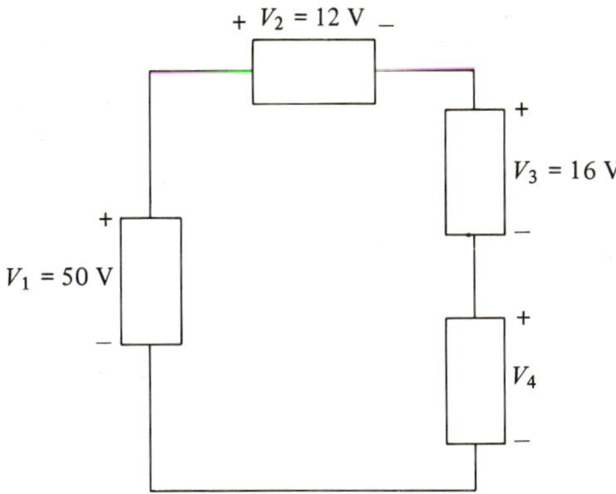

FIGURE 4.49

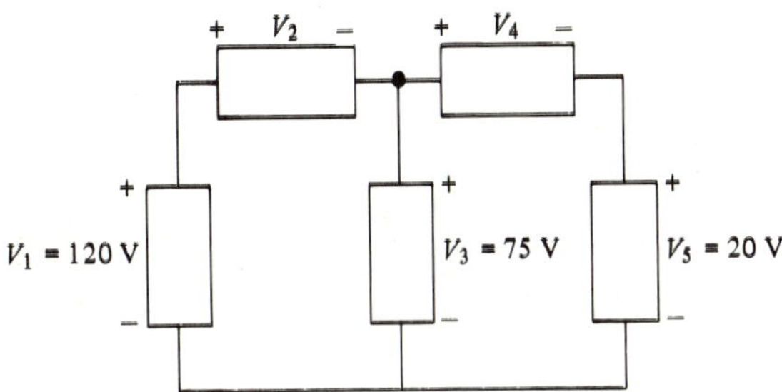

FIGURE 4.50

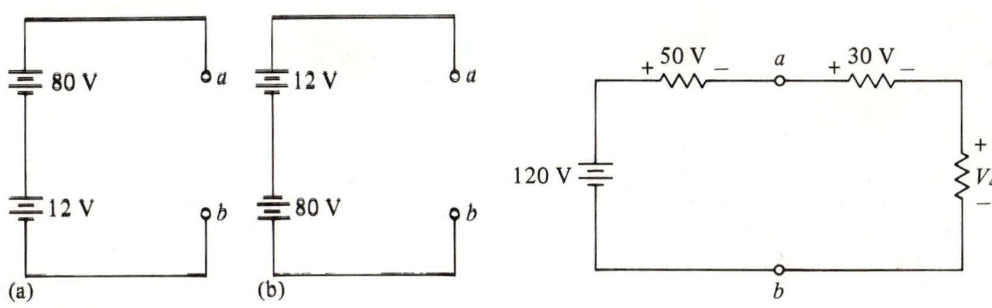

FIGURE 4.51

FIGURE 4.52

24. a. Find the current in the circuit of Fig. 4.53.
    b. Find the equivalent resistance seen by the battery.
    c. Find V_1 through V_4.
25. Find a circuit consisting of a single voltage source and a single resistor that will replace the circuits of Fig. 4.54.

Sec. 4.4

26. In the telephone distribution circuit of Fig. 4.55 it is found that $V_a = 120$ V, $V_b = 80$ V, $V_c = 37$ V, and $V_d = 12$ V. Find V_{ab}, V_{bc}, and V_{dc}.
27. a. In the circuit of Fig. 4.56 find all voltages if the ground is at point c.
    b. Repeat part a if the ground is moved to point b.

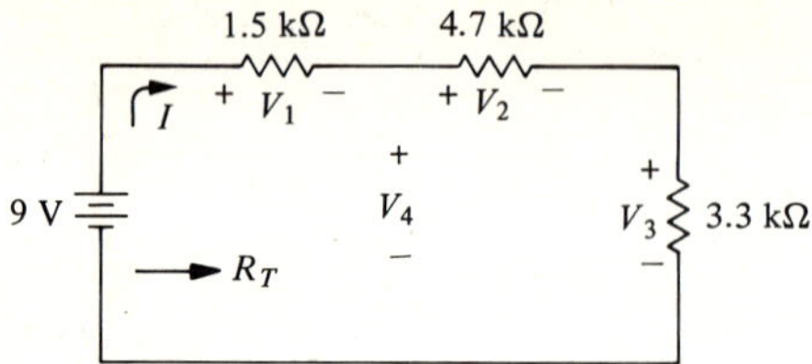

FIGURE 4.53

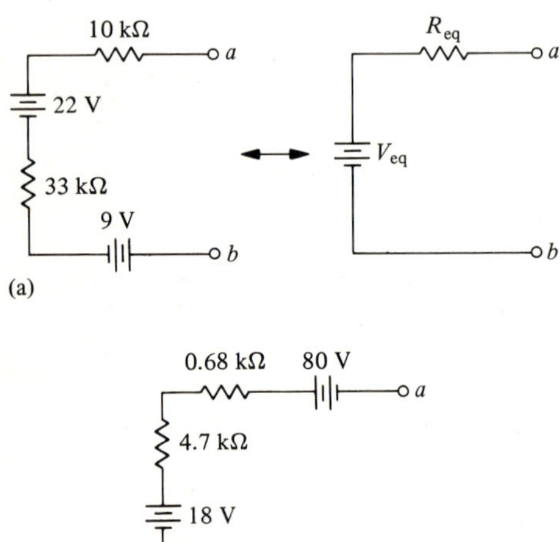

FIGURE 4.54

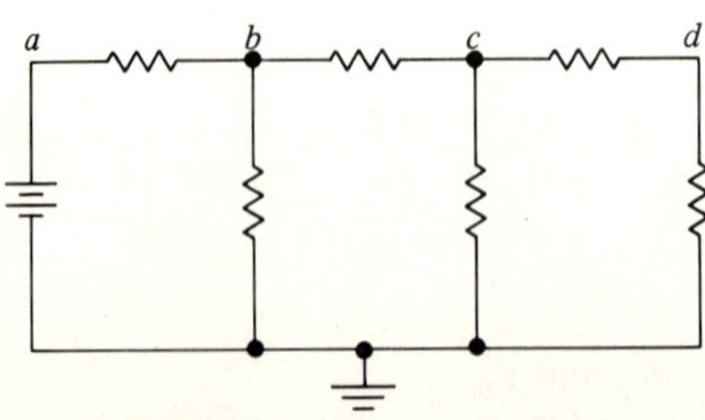

FIGURE 4.55

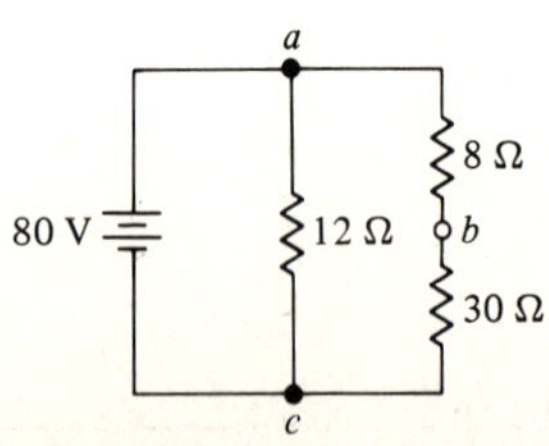

FIGURE 4.56

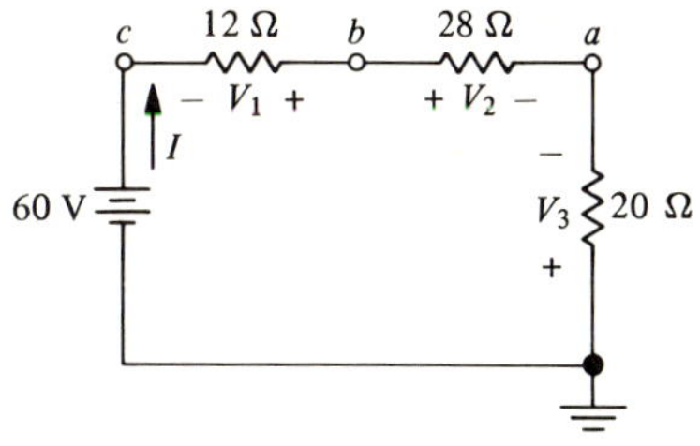

FIGURE 4.57

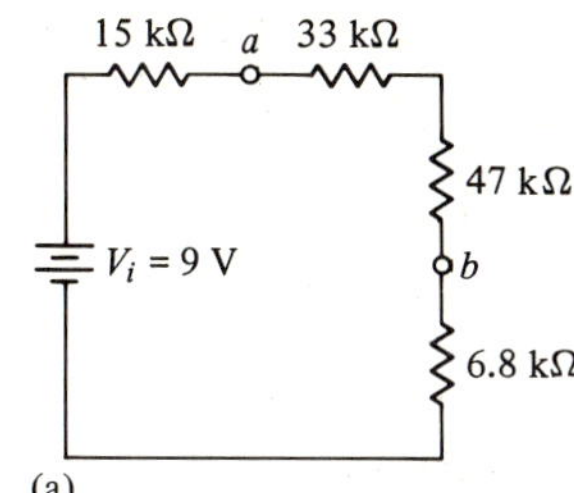

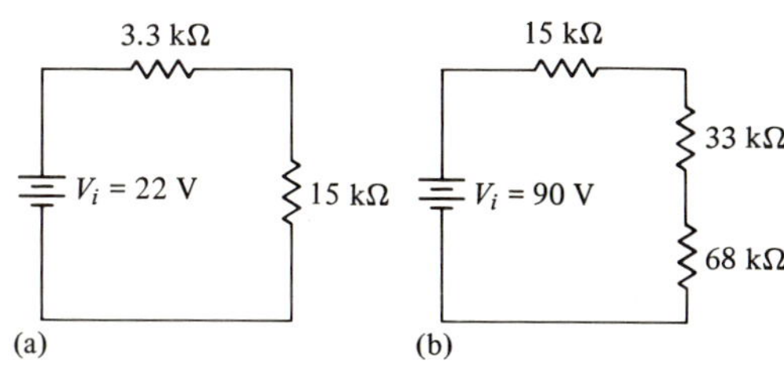

FIGURE 4.58

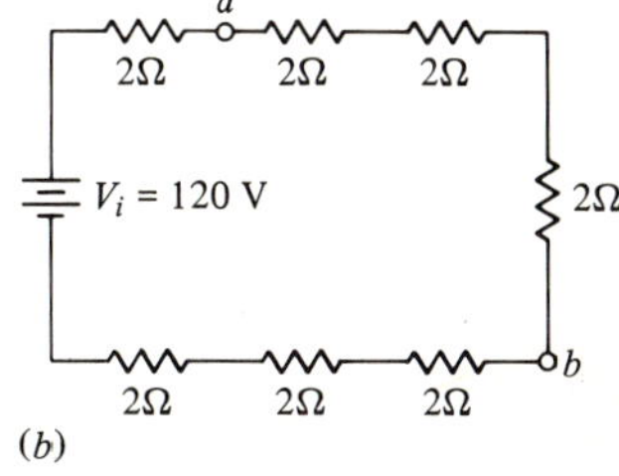

FIGURE 4.59

28. In the digital computer terminal circuit of Fig. 4.57 find I, V_1, V_2, V_3, V_{ab}, V_{bc}, V_a, and V_b. Use KVL to check as many ways as possible.

Sec. 4.5

29. Use the voltage divider formula to find all resistor voltages in the circuits of Fig. 4.58. Be sure to check using KVL whenever possible.
30. Use the voltage divider formula to find V_{ab} in the circuits of Fig. 4.59.
31. For the circuits in Fig. 4.59 find the voltage transfer ratios $A_v = V_{ab}/V_i$.
32. Prove that a voltage divider consisting of two equal resistors yields an output voltage exactly one-half of the input voltage.
33. Design a voltage divider for $A_v = 0.375$. The input resistance is to be greater than 50 kΩ.
34. Design a voltage divider that will provide an output of 5 V to operate a miniature light bulb when connected to a 22.5-V battery. The current drawn from the battery is to be less than 1 mA.
35. The circuit in Fig. 4.60 shows how a potentiometer (see Sec. 3.6) can be used as a variable voltage divider. In the circuit, x represents the fraction of the resistance R that appears between terminal b and the slider, terminal a. Thus x lies between 0 and 1. Find the voltage transfer ratio $A_v = V_{ab}/V_i$ in terms of x and R.

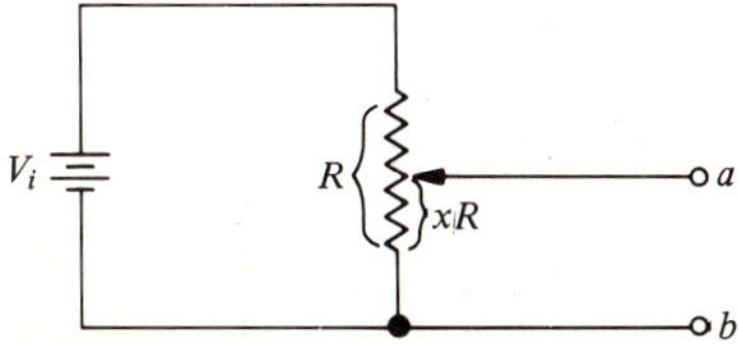

FIGURE 4.60

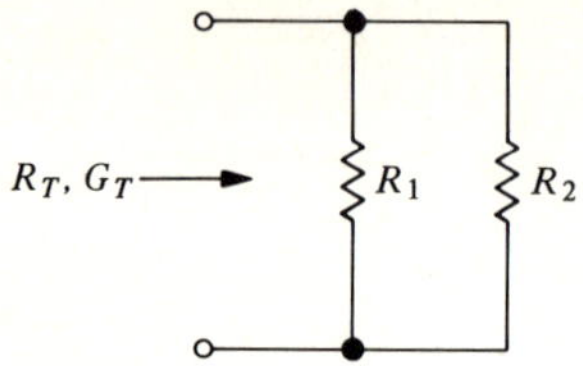

FIGURE 4.61

FIGURE 4.62

Sec. 4.6

36. Find the total resistance and conductance of the parallel circuit of Fig. 4.61 for the following values:
    a. $R_1 = 0$, $R_2 = 1.5\ \text{k}\Omega$
    b. $R_1 = 680\ \Omega$, $R_2 = 2.2\ \text{k}\Omega$
    c. $R_1 = 10\ \text{M}\Omega$, $R_2 = 1\ \text{k}\Omega$
    d. $R_1 = 6.8\ \text{k}\Omega$, $R_2 = 6.8\ \text{k}\Omega$
    e. $R_1 = 6.8\ \text{k}\Omega$, $R_2 = 22\ \Omega$
37. Find the total conductance and resistance of the parallel circuit of Fig. 4.62 for the following values:
    a. $R_1 = 1\ \Omega$, $R_2 = 100\ \Omega$, $R_3 = 10\ \text{k}\Omega$
    b. $R_1 = 6.8\ \text{k}\Omega$, $R_2 = 3.3\ \text{k}\Omega$, $R_3 = 15\ \text{k}\Omega$
    c. $R_1 = 10\ \text{M}\Omega$, $R_2 = 10\ \text{M}\Omega$, $R_3 = 10\ \text{k}\Omega$
    d. $R_1 = 150\ \text{k}\Omega$, $R_2 = 220\ \text{k}\Omega$, $R_3 = 100\ \text{k}\Omega$
38. Twenty-four resistors, each 150 kΩ, are connected in parallel. What is the equivalent resistance?

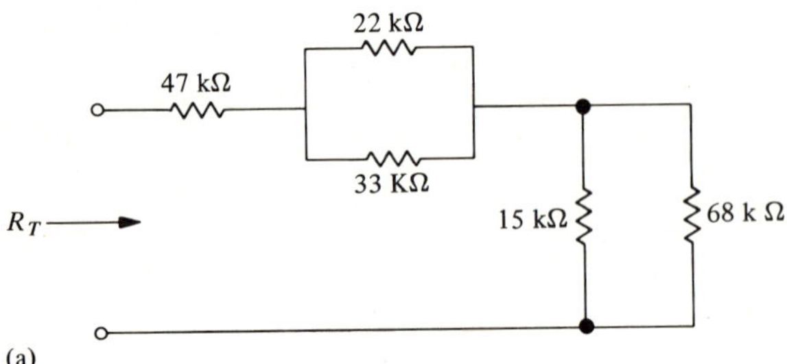

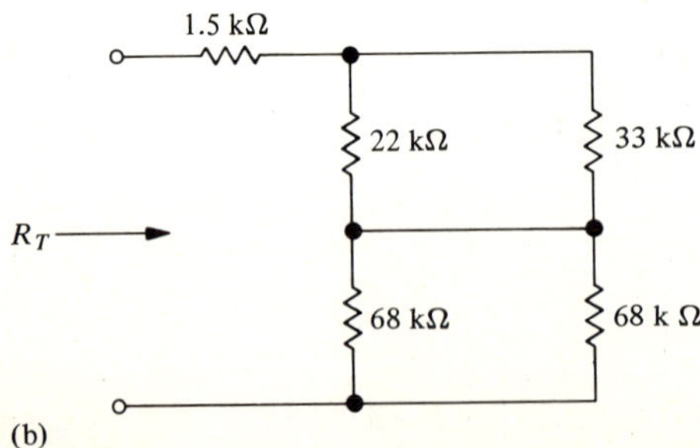

FIGURE 4.63

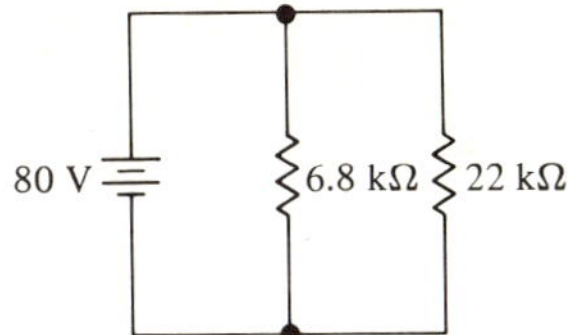

FIGURE 4.64

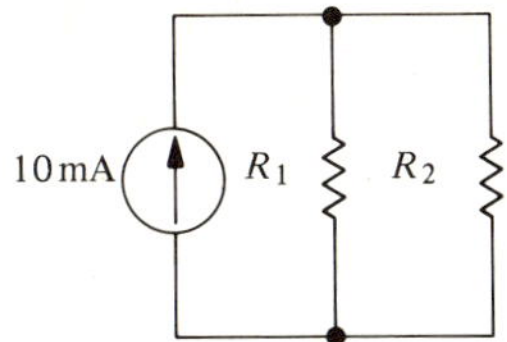

FIGURE 4.65

39. In the circuit of Fig. 4.61, $R_T = 16$ kΩ and $R_2 = 33$ kΩ. Find R_1.
40. In the circuit of Fig. 4.61, $G_T = 2$ mS, $R_1 = 1$ kΩ. Find R_2.
41. In the circuit of Fig. 4.62, $R_T = 100$ kΩ, $R_1 = 150$ kΩ, $R_2 = 330$ kΩ. Find R_3.
42. Calculate R_T in the circuits of Fig. 4.63.
43. In the circuit of Fig. 4.64 find all unknown currents and voltages.
44. In the circuit of Fig. 4.65 find all unknown currents and voltages for $R_1 = 1$ kΩ and $R_2 = 100$ kΩ, 10 kΩ, 1 kΩ, 100 Ω, and 10 Ω. Comment on the current division in relation to the resistance values.

Sec. 4.7

45. Find I_4 in the circuits of Fig. 4.66.
46. In the circuit of Fig. 4.67 find V_i, I_T, I_1, and R_2.

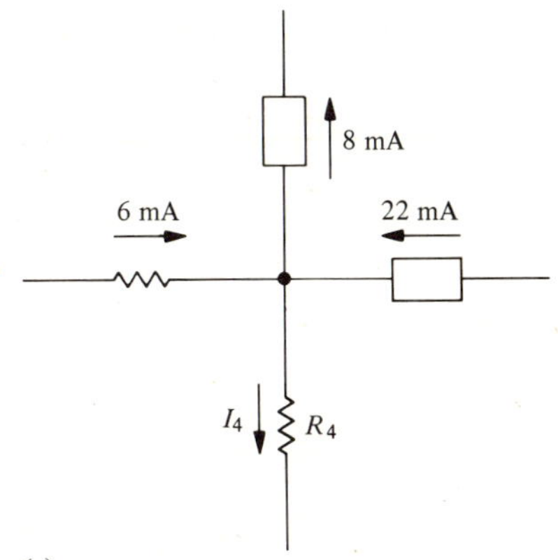

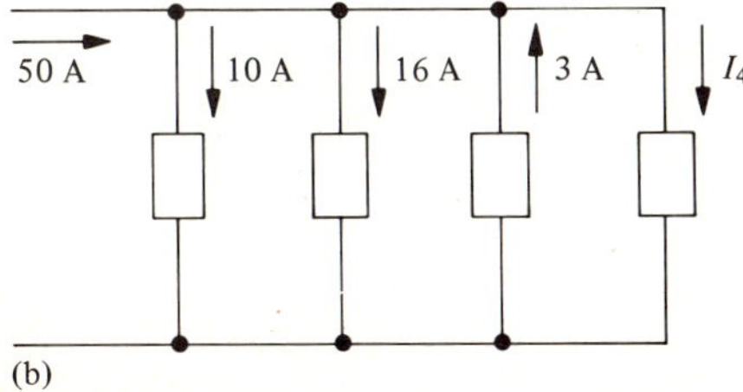

FIGURE 4.66

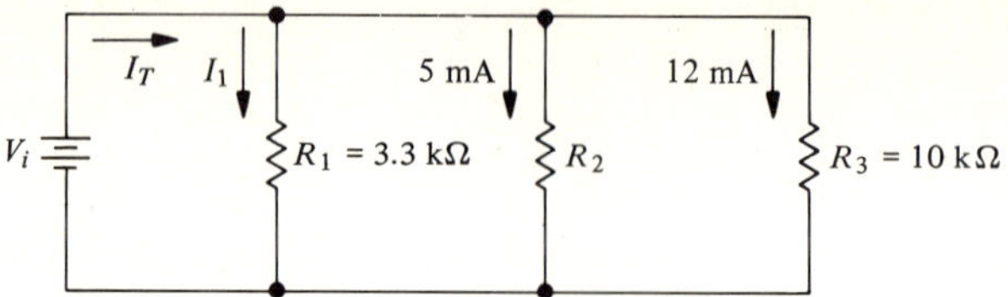

FIGURE 4.67

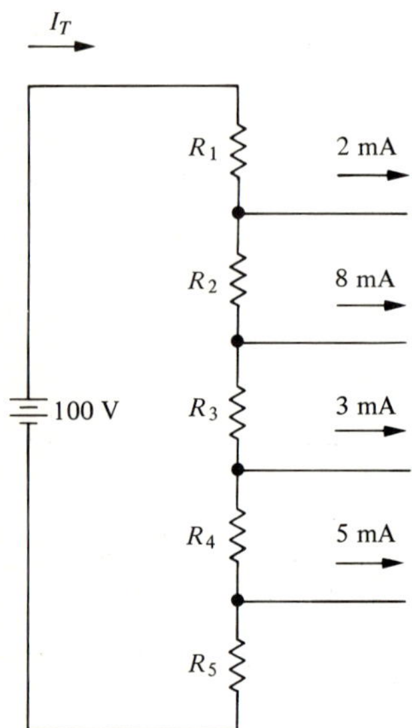

FIGURE 4.68

47. In the multiple voltage divider circuit of Fig. 4.68 find I_T and the voltage drop across each resistor.
48. Find the single current source that will replace each of the circuits shown in Fig. 4.69.
49. Find a circuit consisting of a single current source and a single resistor that will replace the circuits of Fig. 4.70.

Sec. 4.8

50. Use the current divider formula to find all currents in the circuits of Fig. 4.71. Be sure to check using KCL wherever possible.
51. Find the current transfer ratios A_{i1} and A_{i2} for the circuits of Fig. 4.71. Explain the fact that $A_{i1} + A_{i2} = 1$.
52. Use the current divider formula to find I_3 in the circuits of Fig. 4.72. Observe carefully the difference between the two circuits.
53. Prove that in a current divider consisting of two equal resistors, the current through each resistor is exactly one-half of the input current.
54. Find all currents in the circuits of Fig. 4.73.

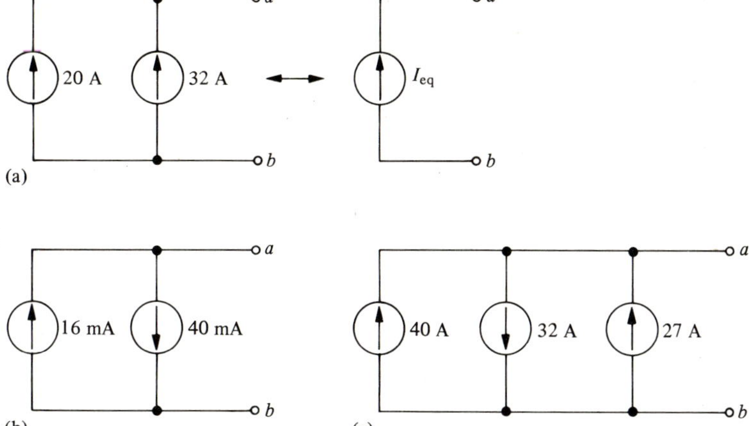

FIGURE 4.69

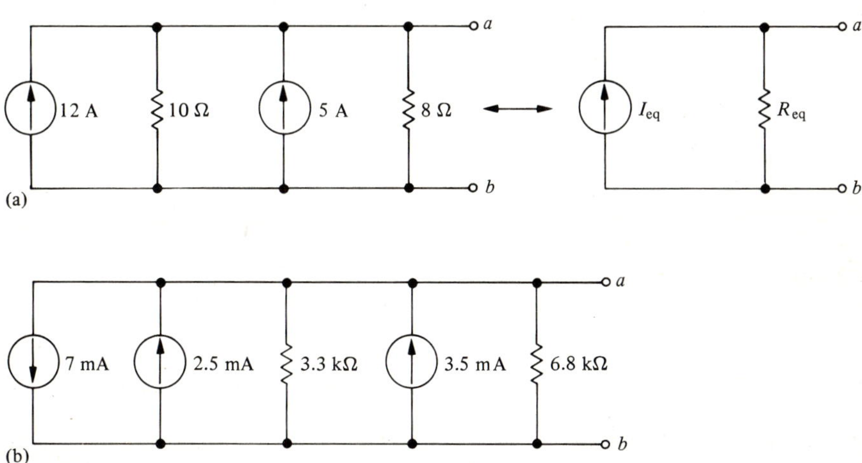

FIGURE 4.70

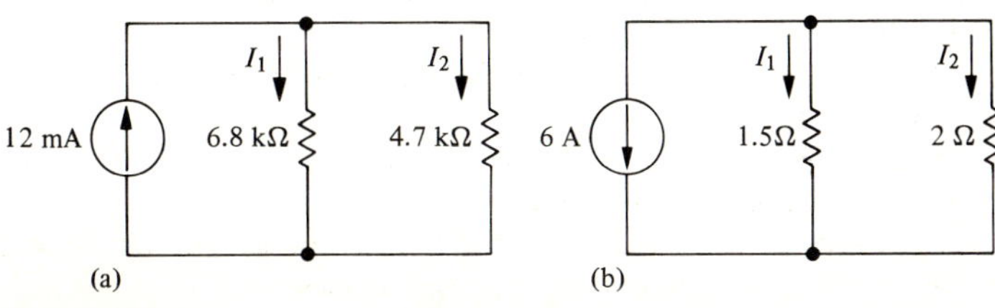

FIGURE 4.71

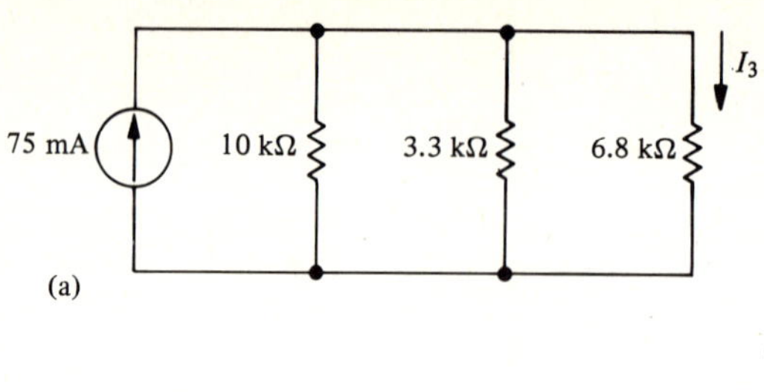

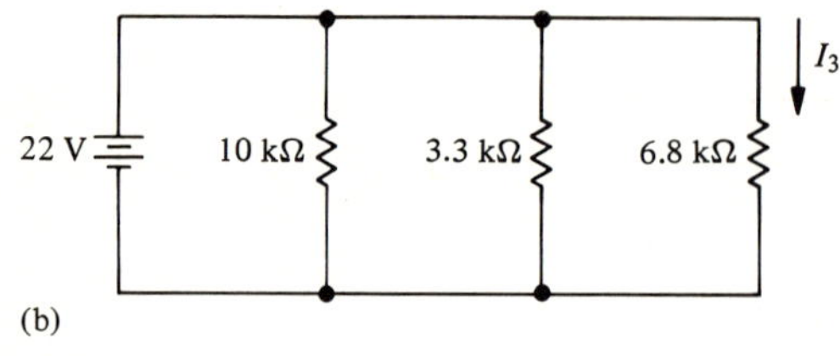

FIGURE 4.72

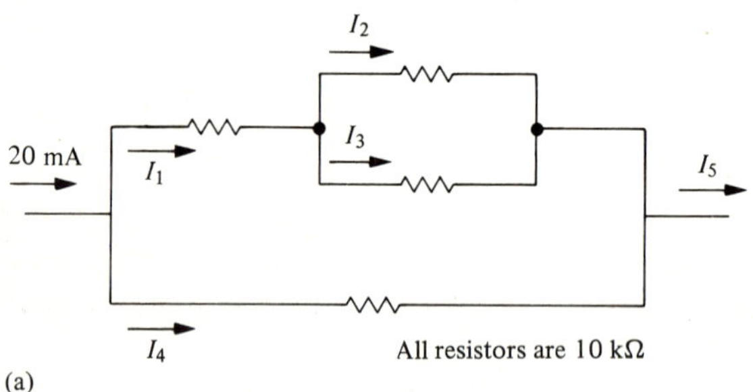

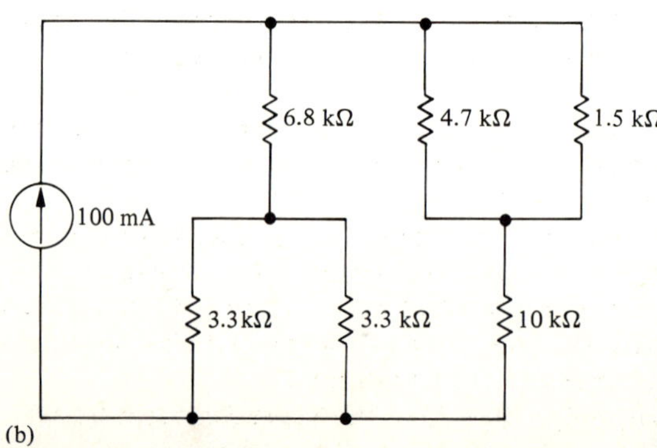

FIGURE 4.73

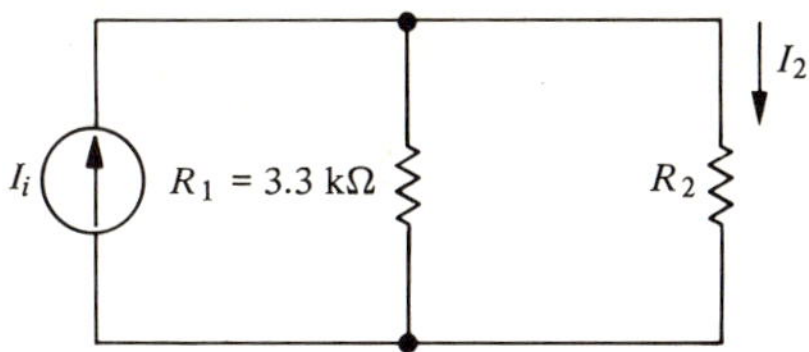

FIGURE 4.74

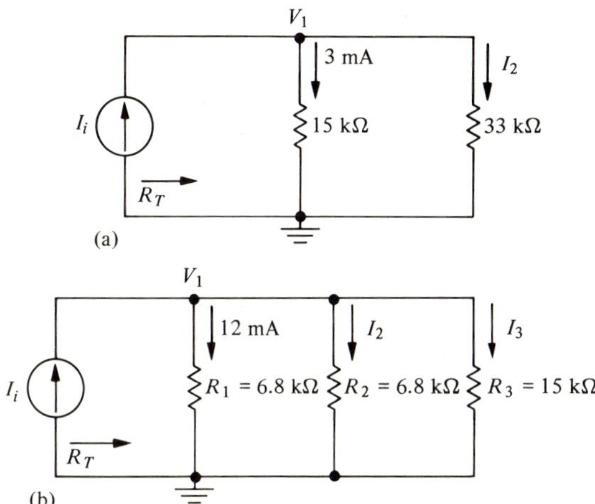

FIGURE 4.75

55. In the circuit of Fig. 4.74, find R_2 so that $A_{i2} = 0.375$.
56. Design a current divider for $A_{i1} = 0.4$. The input resistance is to be less than 1 kΩ.
57. Find the indicated unknowns in the circuits of Fig. 4.75.

Sec. 4.9

58. A circuit consists of twenty 6-Ω resistors in series with a fuse. The supply voltage is 120 V. How many of the resistors can be accidentally short-circuited before a 2-A fuse will open?
59. In the circuit of Fig. 4.33, $V_i = 110$ V, $R_1 = 33\ \Omega$, $R_2 = 68\ \Omega$, and $R_3 = 100\ \Omega$. What size fuse should be connected in series so that it will open up if any two resistors are acidentally short-circuited?
60. In the circuit of Prob. 59, find the power rating required for each resistor if they must not overheat when any one resistor is accidentally short-circuited.
61. In the residential lighting circuit of Fig. 4.76 what is the smallest fuse (current rating) which can be used between points *a* and *b* to protect the circuit in case of an accidental short circuit across one of the resistors?
62. In Fig. 4.34, $V_i = 110$ V and $R_1 = R_2 = R_3 = 20\ \Omega$. The line resistance is 0.34 Ω.
    a. Find the total current that must be supplied by the source.
    b. Find the total current if an accidental short circuit occurs across the load.

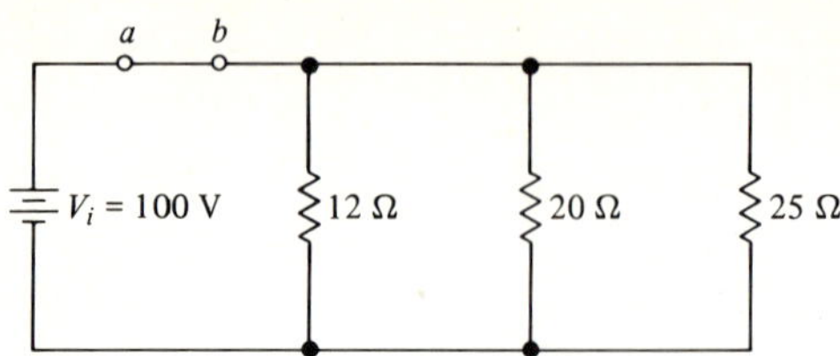

FIGURE 4.76

Sec. 4.10

63. In Fig. 4.35, $R_1 = 10$ kΩ, $R_2 = 3.3$ kΩ, and $R_L = 1.5$ kΩ.
    a. Find the voltage transfer ratio A_v.
    b. Find V_2 if $V_i = 35$ V.
64. In Example 4.10-2 find R_1 using KVL and Ohm's law.
65. A series-dropping resistor is to drop a 110-V supply down to 9 V for a transistor radio that draws 35 mA. Find the value of the resistance required.
66. A miniature soldering iron requires 2.2 A at 12 V. What size series-dropping resistor is required if the iron is to be used with a 220-V supply?
67. A transistor radio requiring 9 V at currents from a nominal value of 20 mA to a maximum value of 35 mA is to be operated from a 120-V source. Design a voltage divider with a bleeder resistance to supply the required 9 V at the nominal load current. The bleeder current is to be 50 mA. Use standard 20% resistors and check your design by finding the load voltage at both nominal and maximum load currents.
68. A fault has occurred in the circuit of Fig. 4.77. In order to locate the fault, voltage measurements are made. Specify the short- or open-circuit faults which would lead to the following measurements:
    a. $V_b = 0$, $V_{ab} = 110$ V
    b. $V_b = 110$ V, $V_{ab} = 0$
    c. $V_b = 0$, $V_{ab} = 0$
    d. $V_b = 45.4$ V
    e. $V_b = 26.6$ V
69. Find all currents and voltages in the circuits of Fig. 4.78.
70. Assume that a short or open circuit fault is possible in any of the four resistors in the circuit of Fig. 4.78a. If only one fault can occur at a time, find the effect of each possibility on V_3.

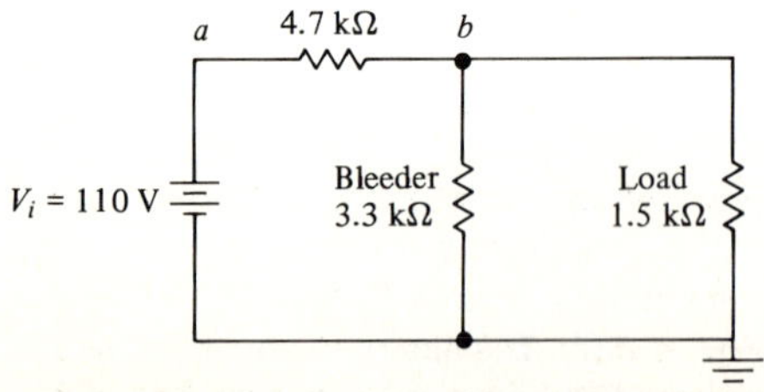

FIGURE 4.77

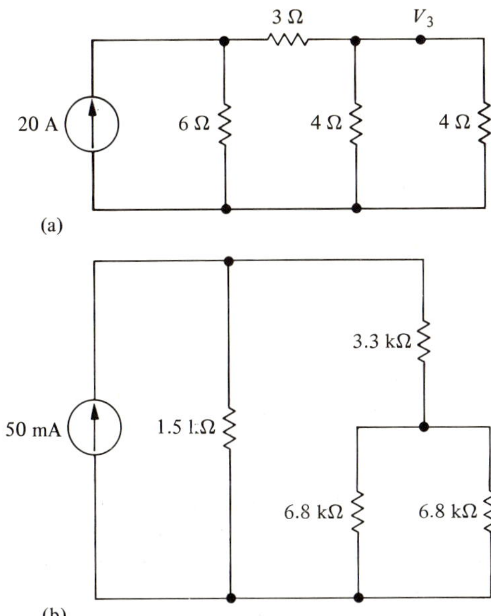

FIGURE 4.78

Sec. 4.11

71. A 4.7-kΩ resistor carries a current of 11.2 mA. How much power is dissipated?
72. A heater has a resistance of 7.3 Ω and carries a current of 15 A. Find the power dissipated.
73. A 1-MΩ resistor carries a current of 15 μA. How many microwatts are dissipated?
74. A 33-kΩ resistor is connected across 18 V. How much power is dissipated?
75. A light bulb has a resistance of 200 Ω when hot. How much power is dissipated when it is connected to 110 V?
76. A 150-W bulb operates on 120 V. How much current does it draw? What is its resistance?
77. Find the power dissipated in each element in the circuit of Fig. 4.43. Check to see that power delivered by the battery equals the power absorbed in the resistors.
78. Repeat Prob. 77 for the circuit of Fig. 4.56.
79. How much current can each of the following resistors safely carry?
    a. 10 kΩ, 1/2 W
    b. 4.7 kΩ, 2 W
    c. 33 Ω, 5 W
    d. 20 kΩ, 100 W
80. What power rating should each of the resistors in the circuit of Fig. 4.79 carry?

Sec. 4.12

In Problems 81 through 84 assume an energy cost of 8¢/kWh.

81. How much does it cost to operate a 3-ton central air conditioner for 8 h?
82. How much does it cost to operate a typical electric clock for a full year?

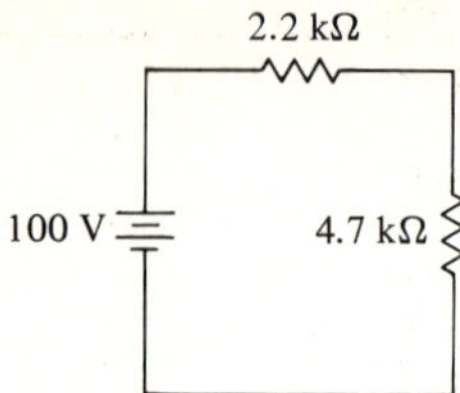

FIGURE 4.79

83. It takes 1.5 h to clean a house with a typical vacuum cleaner. What is the energy cost?
84. A color television set is used for 3 h per day, 310 days per year. What is the energy cost?

Sec. 4.13

85. The electrical input to a motor is 1100 W and the mechanical output is 1.2 hp. What is the efficiency of the motor?
86. The motor of Prob. 85 is connected to a 92% efficient gear train. How much power is available at the output shaft of the gear train?
87. A motor that drives an electric lawnmower is rated at 8 A, 120 V electrical input, 1 hp output. What is its efficiency?
88. An industrial drill press is 78% efficient. It can deliver 2.5 hp when the input voltage is 220 V. What is the input current?
89. In a transmission system the motor that drives the generator has a mechanical power output of 350 kW. When the motor is operating at full load the generator output is 300 kW at 440 V. The transmission line between the generator and load has an RI^2 loss of 25 kW at full load. Draw a block diagram and find the efficiency of each component and the overall efficiency.
90. Prove Eq. (4.13-4). [Hint: make use of Eq. (4.13-1).]

5
Resistive Circuit Analysis

OBJECTIVES

Upon completion of this chapter, the student should be able to:

Section

5.1
1. State the linearity principle.
2. State the superposition principle.
3. Apply the superposition principle to the solution of a multisource network.

5.2
4. State the steps required in the loop-current method.
5. Apply the loop-current method to two- and three-loop networks.

5.3
6. State the steps required in the node-voltage method.
7. Apply the node-voltage method to two and three node-pair networks.

5.4
8. Plot graphs of the terminal characteristics of ideal sources, practical sources, and resistors.
9. Determine the operating point of a circuit graphically using the terminal characteristic.
10. State the formula for percent regulation and use it to determine the regulation of a practical source.

5.5
11. Show a circuit that can be used to measure a terminal characteristic.

5.6
12. Find a Thevenin equivalent circuit that has a specified terminal characteristic.
13. Find the Thevenin equivalent of a complex network.
14. Find a Norton equivalent circuit which has a specified terminal characteristic.
15. Find the Norton equivalent of a complex network.

5.7
16. State the maximum power transfer theorem and apply it to an appropriate circuit.

17. Demonstrate that maximum power transfer is not always desirable.

5.8 18. Describe the standard ECAP branch.

19. Use ECAP to solve an appropriate network problem.

INTRODUCTION

All of the circuits analyzed in Chap. 4 consisted of resistors and sources connected in such a way that the circuits could be reduced to a single source and a single equivalent resistance. Sources are combined using KVL and KCL as shown in Secs. 4.3 and 4.7 and resistances are combined using the series and parallel resistance formulas. For example, consider the loaded voltage divider shown in Fig. 4.35. This is actually a series-parallel combination in which the resistances can be reduced to a single equivalent resistance by using the laws for combining series and parallel resistances. The tools required for complete analysis of such circuits are Ohm's law and KCL and KVL. When the circuits to be analyzed are more complex, and cannot be reduced to simple form, the basic tools are the same, but we must resort to more advanced network analysis techniques.

In this chapter we consider several of the most often-used of these techniques. In addition we will study graphical analysis and a computer circuit analysis program (ECAP) that is available on many computers and can completely solve extremely complex circuits in very little time.

5.1 LINEARITY AND SUPERPOSITION

The circuits we have been considering all have one thing in common. They are *linear* circuits consisting of resistors and independent sources. The resistance element is linear because its *vi* relation, Ohm's law ($V = RI$) plots as a straight line.

Consider that we are analyzing a linear network consisting of a number of resistors and two current sources, I_1 and I_2, as shown in Fig. 5.1a. In the circuit, the response of interest is the voltage V_1. Analysis of the circuit (which we do in detail in Example 5.1-1) will yield an equation of the form

$$V_1 = K_1 I_1 + K_2 I_2 \tag{5.1-1}$$

This is a linear equation similar to $V = RI$ for a single resistance. The difference is that in Eq. (5.1-1) the voltage depends on *two* currents. The coefficients K_1 and K_2 are constants that depend only on the resistances in the network.

One consequence of linearity is that the response V_1 is proportional to the source currents I_1 and I_2. Thus, if I_1 and I_2 are both doubled, the response is doubled. This is true not only of V_1, but of all other currents and

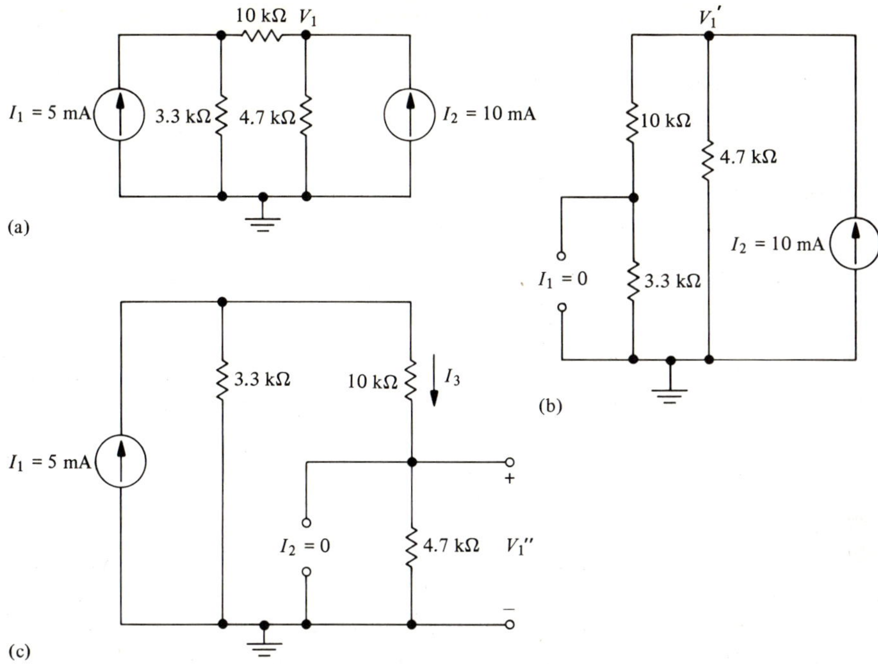

FIGURE 5.1
Circuit for Example 5.1-1. (a) Circuit. (b) Circuit for $I_1 = 0$. (c) Circuit for $I_2 = 0$.

voltages in the network. This principle can be extended to a linear network of any complexity. In the general case, the *linearity principle* can be stated as follows:

> *If all independent sources are multiplied by a constant, then all of the responses (currents and voltages) in the network will be multiplied by the same constant.*

Probably the most important consequence of linearity is the superposition theorem. In order to obtain it we return to the response Eq. (5.1-1) and consider each source separately. If we pretend to replace the independent source I_1 by an open circuit so that I_1 is zero and call the resulting response V_1', then

$$V_1' = V_1 |_{I_1=0} = K_2 I_2 \tag{5.1-2}$$

We next pretend to open-circuit source I_2 so that $I_2 = 0$ and call the resulting response V_1'' so that

$$V_1'' = V_1 |_{I_2=0} = K_1 I_1 \tag{5.1-3}$$

Now we add Eqs. (5.1-2) and (5.1-3) to get

$$V_1' + V_1'' = K_1 I_1 + K_2 I_2$$
$$= V_1 \qquad (5.1\text{-}4)$$

Thus the response V_1 can be found as the *sum* of two components, each due to one source *acting alone* while the other is set to zero. This is the principle of superposition applied to the network that led to Eq. (5.1-1). It is extremely useful because it reduces a complicated problem to a series of much simpler problems. A general statement of the *superposition principle* is as follows:

In a linear network containing more than one source (current or voltage), any response can be found by adding the response due to each source acting alone, all others being set to zero (short circuits for voltage sources, open circuits for current sources).

Thus, if there are five sources, we analyze five circuits, each containing only one source. Often the circuits containing only one source can be solved by inspection while the original circuit containing all of the sources would require the use of more complicated techniques and the solution of simultaneous equations. The utility of this theorem is demonstrated in the following examples.

EXAMPLE 5.1-1 Using Superposition

In the circuit of Fig. 5-1a find the response V_1 using superposition.

Solution

With this technique we analyze two circuits, each with only one source, and then simply add the results. The first circuit, with $I_1 = 0$, is shown in Fig. 5-1b. This is particularly easy to solve, since all we need to do is find the total resistance and then apply Ohm's law. Thus

$$V_1' = R_{eq} I_2 = [(10 + 3.3)\ \text{k}\Omega \parallel 4.7\ \text{k}\Omega] I_2$$
$$= \frac{(13.3)(4.7)}{(13.3 + 4.7)}\ \text{k}\Omega \times (I_2) = (3.47\ \text{k}\Omega)(I_2)$$

Since all resistance values are in kilohms, if I_2 is in mA then V_1' will be in volts and the constant $K_2 = 3.47\ \text{k}\Omega$ [see Eq. (5.1-1)].

The second circuit to be analyzed, with $I_2 = 0$, is shown in Fig. 5.1c. Here we use the current divider to find I_3 and then Ohm's law to find V_1''

$$I_3 = \frac{3.3\ \text{k}\Omega}{(10 + 4.7 + 3.3)\ \text{k}\Omega}(I_1) = 0.183 I_1$$

$$V_1'' = I_3(4.7\ \text{k}\Omega) = (0.183 I_1)(4.7\ \text{k}\Omega) = (0.86\ \text{k}\Omega)(I_1)$$

Since all resistance values are in kilohms, if I_1 is in milliamperes then V_1'' will be in volts and the constant $K_1 = 0.86\ \text{k}\Omega$.

Finally, the desired voltage is the sum of the voltages found in the two separate circuits.

$$\begin{aligned} V_1 &= V_1' + V_1'' \\ &= (3.47\ \text{k}\Omega)I_2 + (0.86\ \text{k}\Omega)I_1 \end{aligned}$$

Substituting the given values, $I_2 = 10$ mA and $I_1 = 50$ mA

$$\begin{aligned} V_1 &= 34.7\ \text{V} + 4.3\ \text{V} \\ &= 39\ \text{V} \end{aligned}$$

• • •

EXAMPLE 5.1-2 Using Superposition When Different Kinds of Sources Are Present

The circuit of Fig. 5.2a has both a voltage source and a current source. Find V_2 using superposition.

Solution

Let V_2' be the response to V_1 with $I_1 = 0$. The corresponding circuit is shown in Fig. 5.2b. Using the voltage divider formula

$$V_2' = \left(\frac{3.9\ \text{k}\Omega}{3.9\ \text{k}\Omega + 2.2\ \text{k}\Omega}\right)(9\ \text{V}) = 5.75\ \text{V}$$

V_2'' is the response to I_1 with $V_1 = 0$, and the corresponding circuit rearranged for clarity is shown in Fig. 5.2. Using Ohm's law

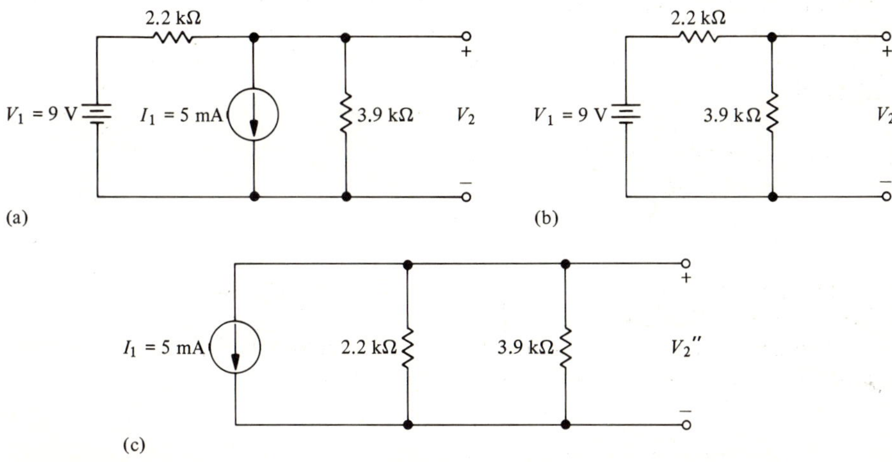

FIGURE 5.2
Circuit for Example 5.1-2. (a) Circuit. (b) Circuit for $I_1 = 0$. (c) Circuit for $V_1 = 0$.

$$V_2'' = -(2.2\text{ k}\Omega \parallel 3.9\text{ k}\Omega)(5\text{ mA}) = -(1.41\text{ k}\Omega)(5\text{ mA}) = -7.03\text{ V}$$

The negative sign is a consequence of the fact that, due to the direction of the 5-mA current source, the voltage polarity across the (2.2 kΩ ∥ 3.9 kΩ) equivalent resistance is + at the lower terminal.

Finally,

$$\begin{aligned} V_2 &= V_2' + V_2'' \\ &= 5.75\text{ V} - 7.03\text{ V} \\ &= -1.28\text{ V} \end{aligned}$$

• • •

The next example illustrates how the superposition theorem and two sets of measurements can be used to find the constants for an arbitrary network. Once the constants are known, the output in response to any inputs can be found.

EXAMPLE 5.1-3 Characterizing a Network Experimentally

In Fig. 5.3 we show a resistance network enclosed in a "black box" that has three pairs of external terminals. In use, a voltage source V_1 is connected to one pair of terminals, a current source I_1 is connected to another pair, and the output V_2 is taken from the third pair. Our problem is to characterize the network so that we can find the output V_2 for any values of the inputs V_1 and I_1. This can be done by measuring V_2 for two sets of values of V_1 and I_1 as follows. Assume that a voltmeter is connected to read V_2 and the following data are observed:

a. When $V_1 = 20$ V and $I_1 = 3$ A; $V_2 = 6$ V.
b. When $V_1 = 0$ V and $I_1 = 2$ A; $V_2 = 1$ V.

Find V_2 when $V_1 = 8$ V and $I_1 = 5$ A.

Solution

Since the network is linear, we can write an equation similar to Eq. (5.1-1)

$$V_2 = AV_1 + BI_1$$

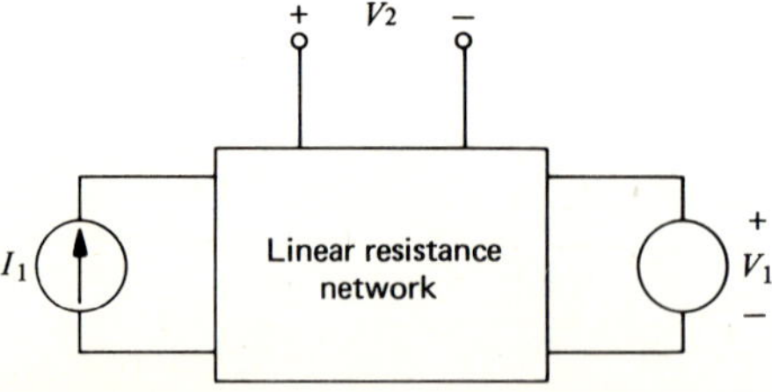

FIGURE 5.3
Circuit for Example 5.1-3.

The constants A and B are found from the experimental data. In order for the equation to be correct dimensionally, A must be dimensionless and B must have the dimensions of ohms. Substituting the measured values in volts and amperes

$$6 = 20A + 3B \tag{5.1-5}$$

$$1 = 0A + 2B \tag{5.1-6}$$

From Eq. (5.1-6) $B = (1/2)\Omega$. Substituting this value into Eq. (5.1-5)

$$6 = 20A + 3\left(\frac{1}{2}\right)$$

$$20A = 6 - \frac{3}{2} = \frac{9}{2}$$

$$A = \frac{9}{(2)(20)} = 0.225$$

Then

$$B = 0.5\Omega \qquad \text{and} \qquad A = 0.225$$

Thus, the response equation is

$$V_2 = 0.225V_1 + 0.5I_1$$

For the desired conditions, $V_1 = 8$ V and $I_1 = 5$ A, we obtain

$$V_2 = 0.225(8\text{ V}) + 0.5\ \Omega(5\text{ A}) = 4.3\text{ V}$$

• • •

LEARNING EXERCISE FOR SEC. 5.1

1. In the circuit of Fig. 5.4, use superposition to find V_1, I_1, and I_2.

Ans. −0.25; 0.75; 7.5

• • •

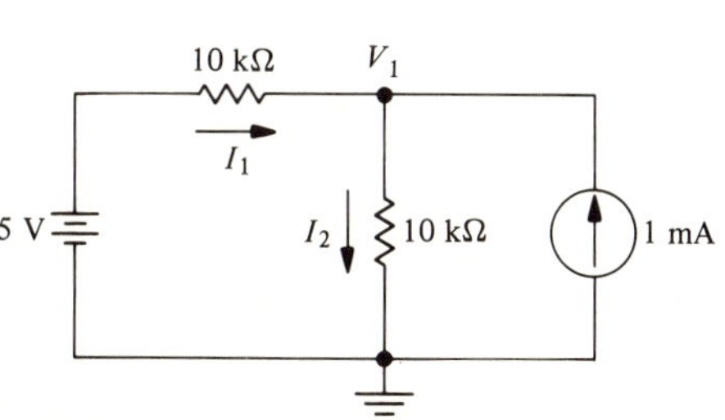

FIGURE 5.4
Circuit for Learning exercise 5.1-1.

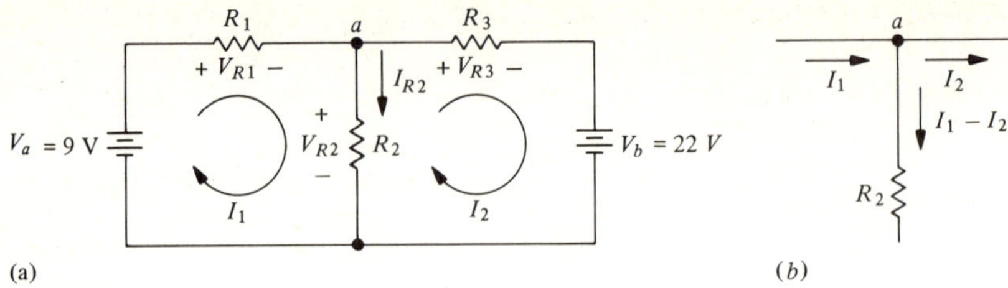

FIGURE 5.5
Two-loop network. (a) Circuit. (b) Magnified view of node *a*.

5.2 THE LOOP-CURRENT METHOD

In this section, we study a method based on KVL that is used in the analysis of complex networks. It is often called "mesh analysis." This method is highly organized in the sense that there is a specific sequence of steps that must be carried through. As we will see, the method leads to a set of simultaneous equations that must be solved in order to arrive at the solution.

To illustrate the method, consider the circuit of Fig. 5.5a. The idea is to write a KVL equation for each "window" of the network in terms of fictitious "loop" currents and the element *vi* relations. The number of equations required for a complete solution is equal to the number of loop currents. The sequence of steps to be followed is applied to the circuit of Fig. 5.5a as follows:

Step 1: If possible, draw the network in such a way that no conductors cross each other, then assign "loop" currents to each of the "windows" in the diagram. These currents should be given clockwise reference directions for uniformity.* In the circuit of Fig. 5.5a, there are two windows and the clockwise loop currents are designated I_1 and I_2.

Step 2: Assign temporary voltages across each resistor and mark them on the circuit diagram. In loop 1 of the circuit, these are V_{R1} and V_{R2}, while in loop 2 they are V_{R2} and V_{R3}. Note that V_{R2} appears in *both* loops.

Step 3: Write a KVL equation for each loop in terms of these temporary voltages. Start at a convenient point and trace around the loop in the direction of the loop current. For loop 1 we begin at the lower left-hand corner and obtain the following equation:

$$-V_a + V_{R1} + V_{R2} = 0 \tag{5.2-1}$$

Note carefully that when we trace through a voltage rise, the term takes a negative sign while the signs of voltage drops are positive.†

*These instructions are not mandatory, but are the safest way for beginning students to learn the method without unnecessary complications.

†An easy way to remember this is that the sign of the term is the same as the *first sign encountered* when tracing through the element.

For loop 2, starting at the ground end of R_2, we obtain

$$-V_{R2} + V_{R3} + V_b = 0 \tag{5.2-2}$$

Now we express the resistor voltages in terms of the loop currents as follows: Using Ohm's law

$$V_{R1} = R_1 I_1 \tag{5.2-3}$$

and

$$V_{R3} = R_3 I_2 \tag{5.2-4}$$

For V_{R2} we must be careful. The net current flowing in R_2 in the tracing direction is the *difference* between the two loop currents $I_1 - I_2$ (see Fig. 5.5b)

$$V_{R2} = R_2(I_1 - I_2) \tag{5.2-5}$$

Now we substitute these Ohm's law equations into the two KVL equations, (5.2-1) and (5.2-2). This yields

Loop 1:

$$-V_a + R_1 I_1 + R_2(I_1 - I_2) = 0 \tag{5.2-6}$$

Loop 2:

$$-R_2(I_1 - I_2) + R_3 I_2 + V_b = 0 \tag{5.2-7}$$

Rearranging Eqs. (5.2-6) and (5.2-7) we have the desired KVL equations

$$(R_1 + R_2)I_1 - R_2 I_2 = V_a \tag{5.2-8}$$

$$-R_2 I_1 + (R_2 + R_3)I_2 = -V_b \tag{5.2-9}$$

Note the symmetrical form of these equations. The coefficient of I_1 in Eq. (5.2-8) represents the total resistance connected in series around loop 1, that is, $R_1 + R_2$. The coefficient of I_2 represents the resistance common to loops 1 and 2, that is, R_2. The term $-R_2 I_2$ represents the effect on loop 1 of the current in loop 2 and is called the *coupling* or *mutual voltage*. When loop currents are all taken clockwise, the mutual terms always have a negative sign. Similar interpretations apply to the coefficients in Eq. (5.2-9). The right-hand sides of the equations are the net *voltage rise* of the sources in the tracing direction in each loop. While tracing around loop 1 in the clockwise direction, we go through a source voltage rise $V_a = +9$ V. On the other hand, while tracing loop 2 clockwise we go through a source voltage *drop* $V_b = 22$ V so that the voltage *rise* which appears on the right side of the equation is -22 V.

Step 4: Solve the simultaneous KVL equations for the unknown loop currents. Mutual branch currents and all voltage drops are then found as required. The method is illustrated by the following examples.

EXAMPLE 5.2-1 A Two-Loop Network

In the circuit of Fig. 5.2-1 $V_a = 9$ V, $V_b = 22$ V, $R_1 = 2.2$ kΩ, $R_2 = 15$ kΩ, and $R_3 = 4.7$ kΩ. Find all currents and voltages and check wherever possible.

Solution

Following the steps set down above we have the following:

Steps 1 and 2: Loop currents and the temporary voltages are shown on the circuit diagram.

Step 3: We substitute the given values directly into the final KVL equations using units of volts, kilohms, and milliamperes:*

Loop 1, Eq. (5.2-8)

$$(2.2 + 15)I_1 - 15I_2 = 9$$

Loop 2, Eq. (5.2-9)

$$-15I_1 + (15 + 4.7)I_2 = -22$$

Step 4: Solving with determinants:†

$$I_1 = \frac{\begin{vmatrix} 9 & -15 \\ -22 & 19.7 \end{vmatrix}}{\begin{vmatrix} 17.2 & -15 \\ -15 & 19.7 \end{vmatrix}} = \frac{177 - 330}{339 - 225} = \frac{-153}{114} = -1.34 \text{ mA}$$

$$I_2 = \frac{\begin{vmatrix} 17.2 & 9 \\ -15 & -22 \end{vmatrix}}{114} = \frac{-379 + 135}{114} = \frac{-243}{114} = -2.14 \text{ mA}$$

The negative signs indicate that the actual currents flow in directions opposite to the assumed clockwise loop currents. Using the results,

*From this point on we are going to occasionally indicate the units used in an equation separately, rather than to explicitly give the unit associated with each number.

†For students who have not studied determinants, a brief exposition is given in Appendix C.

$$I_{R2} = I_1 - I_2 = 0.8 \text{ mA}$$

$$V_{R2} = R_2 I_{R2} = (15)(0.8) - 12 \text{ V}$$

$$V_{R1} = R_1 I_1 = (2.2)(-1.34) = -2.95 \text{ V}$$

$$V_{R3} = R_3 I_2 = (4.7)(-2.14) = -10.1 \text{ V}$$

Checking with KVL we must have

$$V_a = V_{R1} + V_{R2} = 12 - 2.95 = 9.05 \text{ V} \approx 9 \text{ V}$$

and

$$V_b = -V_{R3} + V_{R2} = -(-10.1) + 12 = 22.1 \approx 22 \text{ V}$$

These check with sufficient accuracy for our purposes.

• • •

EXAMPLE 5.2-2 The Loop Current Method When a Current Source Is Present

The circuit of Fig. 5.6 contains a current source in addition to a voltage source. Use the loop current method to find all currents and voltages.

Solution

Again we follow the prescribed sequence of steps, indicating differences from the previous example because of the presence of the current source as we proceed:

Step 1: The loop currents are assigned as shown on the diagram. Note that loop current I_2 *must* be the same as the source current I_a by virtue of the fact that the current source is *ideal* and I_2 is the only loop current flowing through the branch containing the source.
Step 2: The temporary voltage polarities are indicated on the diagram.
Step 3: The KVL equations are written for loops that do not contain current sources; for this example, only loop 1. Thus the KVL equation for loop 1 is

$$-V_a + V_{R1} + V_{R2} = 0$$

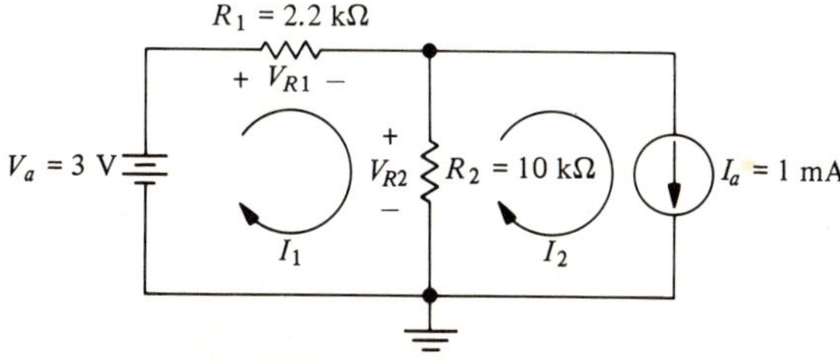

FIGURE 5.6
Circuit for Example 5.2-2.

where $V_{R1} = R_1I_1$ and $V_{R2} = R_2(I_1 - I_2)$. Substituting the given values with all resistances in kilohms and voltages in volts so that currents will be in milliamperes:

$$-3 + 2.2I_1 + 10I_1 - 10I_2 = 0$$

Collecting terms:

$$12.2I_1 - 10I_2 = 3$$

The loop containing the current source is often called a "dummy" loop and from it we write

$$I_2 = I_a = 1 \text{ mA}$$

This is substituted into the KVL equation for loop 1 and we have

$$12.2I_1 - (10)(1) = 3$$

$$12.2I_1 = 13$$

$$I_1 = 1.07 \text{ mA}$$

Thus even though at first glance the network appears to have two loops so that two simultaneous KVL equations will result, we end up with only one KVL equation because the current in the dummy loop is known. In effect, this circuit has only one *independent* loop. The rest of the currents and voltages are easily found. Thus

$$V_{R1} = R_1I_1 = (2.2)(1.07) = 2.35 \text{ V}$$

$$V_{R2} = R_2(I_1 - I_2) = 10(1.07 - 1) = 0.7 \text{ V}$$

Checking

$$V_a = V_{R1} + V_{R2} = 2.35 + 0.7 = 3.05 \text{ V}$$

• • •

As a final example of the loop-current method we consider a three-loop network.

EXAMPLE 5.2-3 Solving a Three-Loop Network

Use the loop-current method to find all currents in the circuit of Fig. 5.7.

Solution

We follow the steps set down previously.

Steps 1 and 2: The three clockwise window currents are shown on the diagram along with the temporary voltage polarities.

Step 3: The KVL equations in terms of the temporary voltages are (units throughout are volts, kilohms, and milliamperes):

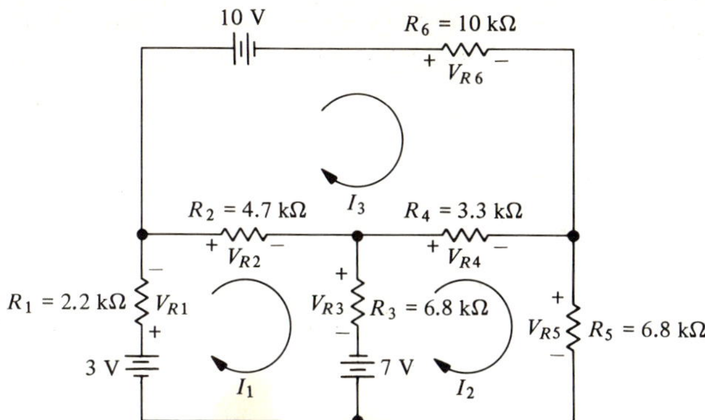

FIGURE 5.7
Three-loop network for Example 5.2-3.

Loop 1

$$-3 + V_{R1} + V_{R2} + V_{R3} - 7 = 0$$

Loop 2

$$+7 - V_{R3} + V_{R4} + V_{R5} = 0$$

Loop 3

$$+10 + V_{R6} - V_{R4} - V_{R2} = 0$$

The Ohm's law relations are:

$$V_{R1} = 2.2I_1$$
$$V_{R2} = 4.7(I_1 - I_3)$$
$$V_{R3} = 6.8(I_1 - I_2)$$
$$V_{R4} = 3.3(I_2 - I_3)$$
$$V_{R5} = 6.8I_2$$
$$V_{R6} = 10I_3$$

Substituting the Ohm's law relations into the KVL equations and collecting terms, we get

Loop 1

$$13.7I_1 - 6.8I_2 - 4.7I_3 = 10 \qquad (5.2\text{-}10)$$

Loop 2

$$-6.8I_1 + 16.9I_2 - 3.3I_3 = -7 \qquad (5.2\text{-}11)$$

Loop 3

$$-4.7I_1 - 3.3I_2 + 18I_3 = -10 \tag{5.2-12}$$

Step 4: These equations are solved using determinants. The actual evaluation of the determinants is a tedious task and the results are

$$I_1 = \frac{\begin{vmatrix} 10 & -6.8 & -4.7 \\ -7 & 16.9 & -3.3 \\ -10 & -3.3 & 18 \end{vmatrix}}{\begin{vmatrix} 13.7 & -6.8 & -4.7 \\ -6.8 & 16.9 & -3.3 \\ -4.7 & -3.3 & 18 \end{vmatrix}} = \frac{949}{2600} = 0.365 \text{ mA}$$

$$I_2 = \frac{\begin{vmatrix} 13.7 & 10 & -4.7 \\ -6.8 & -7 & -3.3 \\ -4.7 & -10 & 18 \end{vmatrix}}{2600} = \frac{-964}{2600} = -0.371 \text{ mA}$$

$$I_3 = \frac{\begin{vmatrix} 13.7 & -6.8 & 10 \\ -6.8 & 16.9 & -7 \\ -4.7 & -3.3 & -10 \end{vmatrix}}{2600} = \frac{-1374}{2600} = -0.528 \text{ mA}$$

It is left as a problem (Prob. 5.12) for the student to check these currents by finding the individual resistor voltage drops and checking to see that KVL is satisfied around each loop.

• • •

The Inspection Method for Writing Loop Equations

The examples done in this section all made use of temporary voltages and Ohm's law equations to arrive at the final sets of KVL equations such as Eqs. (5.2-8) and (5.2-9) or (5.2-10) through (5.2-12). This was done in order to emphasize that the method consists of nothing more than the basic relations correctly applied to the complicated network. The student will have observed that the final KVL equations can be written directly from observation of the circuit diagram without the intermediate step of using temporary voltages and Ohm's law equations. This is accomplished by making use of the coefficient interpretations given in connection with Eqs. (5.2-8) and

(5.2-9). Equation (5.2-10) is shown below with the circuit interpretation of each term.

Loop 1	$13.7I_1$	$-$	$6.8\ I_2$	$-$	$4.7I_3$	$=$	10	(5.2-10)
	Total resistance around loop 1 = 2.2 kΩ + 4.7 kΩ + 6.8 kΩ = 13.7 kΩ		Resistance common to loops 1 and 2, 6.8 kΩ with negative sign		Resistance common to loops 1 and 3, 4.7 kΩ with negative sign		Total voltage rise in tracing direction around loop 1 = 3 V + 7 V = 10 V	

For practice, Eqs. (5.2-11) and (5.2-12) should be derived directly from the circuit making use of the coefficient interpretations. We leave this as an exercise.

When the network contains more than three loops, the procedure is the same, except that there will be as many KVL equations as there are loops and the solution will involve evaluation of high-order determinants. The next example illustrates the inspection method of writing loop equations for a four-loop network.

EXAMPLE 5.2-4 Inspection Method Applied to Four-Loop Network

Write but do not solve the equations for the circuit shown in Fig. 5.8.

Solution

We use the coefficient interpretations outlined above. All resistance and voltage values are in kilohms and volts so currents will be in milliamperes.

Loop 1

$$(2.2 + 3.3 + 4.7)I_1 - (3.3)\ I_2 - (4.7)I_3 - (0)I_4 = 10$$

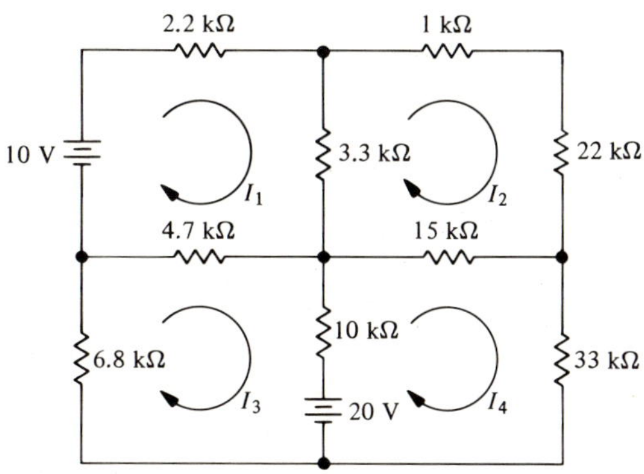

FIGURE 5.8
Circuit for Example 5.2-4.

Loop 2

$$-(3.3)I_1 + (1 + 22 + 15 + 3.3)I_2 - (0)I_3 - (15)I_4 = 0$$

Loop 3

$$-(4.7)I_1 - (0)I_2 + (6.8 + 4.7 + 10)I_3 - (10)I_4 = -20$$

Loop 4

$$(0)I_1 - (15)I_2 - (10)I_3 + (10 + 15 + 33)I_4 = 20$$

Collecting terms in the form of a table we get

	I_1	I_2	I_3	I_4	V
Loop 1	10.2	−3.3	−4.7	0	10
Loop 2	−3.3	41.3	0	−15	0
Loop 3	−4.7	0	21.5	−10	−20
Loop 4	0	−15	−10	58	20

The student should be sure that the origin of all of the numbers in the table is understood.

• • •

LEARNING EXERCISE

1. For the circuit of Fig. 5.9 find
   a. The total resistance around loop 1
   b. The total resistance around loop 2
   c. The resistance common to loops 1 and 2
   d. The total voltage rise in the tracing direction around loop 1
   e. The total voltage rise in the tracing direction around loop 2

Ans. 0; 21.7; 6.2; 18; 31.7

• • •

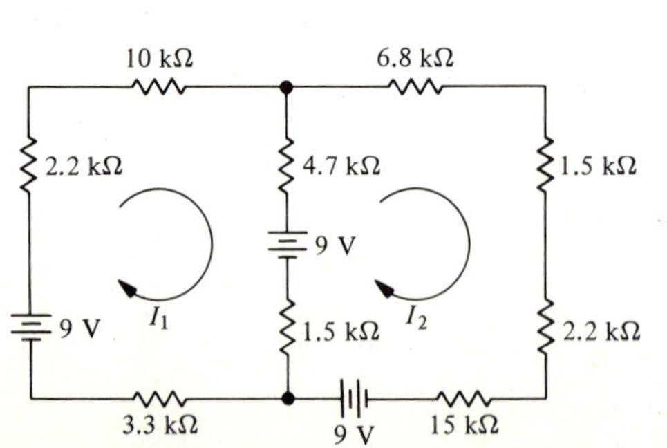

FIGURE 5.9
Circuit for Learning exercise 5.2-1.

5.3 THE NODE-VOLTAGE METHOD

When we are faced with the analysis of a complicated network, we must take advantage of any means available to ease the labor involved. The most laborious part of the process is the solution of the simultaneous equations. Thus if a method for writing circuit equations can be found that leads to fewer simultaneous equations, the labor involved will be decreased considerably. The node-voltage method to be described in this section is based on KCL and often leads to fewer equations than the KVL-based loop-current method. At the end of the section we discuss the problem of choosing between the methods.

Two Node-Pair Networks

Consider the network of Fig. 5.10, in which there are two independent nodes labeled *a* and *b*, and two independent current sources. The reference node is, in order to simplify the analysis, taken as the node that has the largest number of branches connected to it. Our problem is to find all voltages and currents in the network. For this circuit we use the node-voltage method in order to illustrate it. This method involves application of KCL at each independent node. The currents in the KCL equation are expressed in terms of the node voltages and element *vi* relations. The KCL equations are then solved for the node voltages which are, in turn, used to find all currents.

As in the loop-current method, there is a definite sequence of steps to be followed which we illustrate using the circuit of Fig. 5.10.

Step 1: Choose a reference node and identify and label the voltage at each independent node. For our example, there are two independent nodes, *a* and *b*, having node voltages V_a and V_b; both with respect to ground, which is assumed at zero voltage.

Step 2: In order to write KCL equations we assume branch currents through all of the resistances as shown in the figure. Currents through any resistors connected to ground are always assumed to flow in that direction, that is, from the node to ground. All other currents can be assumed to flow in any direction; for example, I_2 could have been chosen as flowing from node *b* to node *a*.

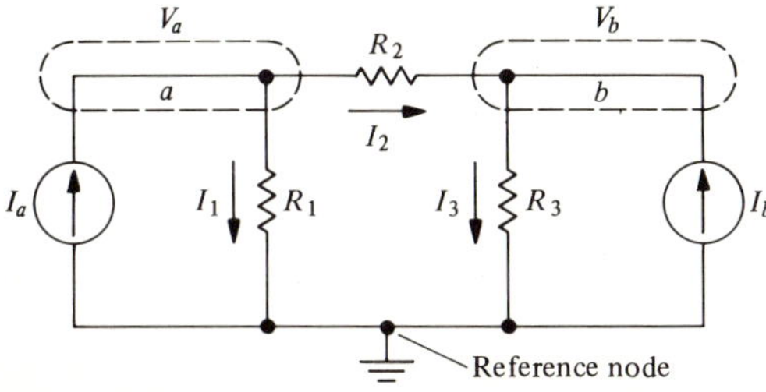

FIGURE 5.10
Two-node-pair network.

Step 3: Express the branch currents in terms of the node voltages. Here we have to be careful to apply Ohm's law correctly. Thus

$$I_1 = \frac{V_a}{R_1} \tag{5.3-1}$$

$$I_2 = \frac{V_a - V_b}{R_2} \tag{5.3-2}$$

$$I_3 = \frac{V_b}{R_3} \tag{5.3-3}$$

Step 4: Write a KCL equation for each independent node.

Node *a*:

$$I_1 + I_2 = I_a \tag{5.3-4}$$

Node *b*:

$$I_3 - I_2 = I_b \tag{5.3-5}$$

Step 5: Substitute the Ohm's law equations into the KCL equations to obtain a set of simultaneous equations in the unknown node voltages. This yields

$$\frac{V_a}{R_1} + \frac{V_a - V_b}{R_2} = I_a$$

$$\frac{V_b}{R_3} - \frac{V_a - V_b}{R_2} = I_b$$

Combining terms:

$$V_a\left(\frac{1}{R_1} + \frac{1}{R_2}\right) - V_b\left(\frac{1}{R_2}\right) = I_a \tag{5.3-6}$$

$$-V_a\left(\frac{1}{R_2}\right) + V_b\left(\frac{1}{R_2} + \frac{1}{R_3}\right) = I_b \tag{5.3-7}$$

This is the set of two simultaneous equations that describe the circuit. They must be solved for V_a and V_b using determinants or some other method.

The coefficients of the voltages in these equations are related to the physical structure of the network. The coefficient of V_a in Eq. (5.3-6), the equation for node *a*, represents the total conductance connected to node *a*, and the coefficient of $-V_b$ represents the conductance connected *between* nodes a and b. In Eq. (5.3-7), the equation for node b, the coefficient of V_b represents the total conductance connected to node b while the coefficient of $-V_a$ represents the conductance between b and a (which is obviously the same as that between a and b). These interpretations of the coefficients can be used as a check when equations are written for any network.

Step 6: Solve for the branch currents using Eqs. (5.3-1) through (5.3-3). This completes the solution. In a practical case it would usually not be necessary to solve for all of the unknown voltages and currents.

Step 7: Check all answers.

A numerical example follows.

EXAMPLE 5.3-1 Two Node-Pair Network

In the circuit of Fig. 5.10, $R_1 = \frac{1}{2}\,\Omega$, $R_2 = \frac{1}{3}\,\Omega$, $R_3 = \frac{1}{6}\,\Omega$, $I_a = 2$ A, and $I_b = 4$ A. Find all other currents and voltages.

Solution

We follow the sequence of steps previously set down.

Steps 1 and 2: As shown on the diagram.
Step 3: The Ohm's law equations are (units are volts, ohms, and amperes throughout)

$$I_1 = 2V_a$$

$$I_2 = 3(V_a - V_b)$$

$$I_3 = 6V_b$$

Step 4: Applying KCL at nodes a and b,

Node a:

$$I_1 + I_2 = 2$$

Node b:

$$I_3 - I_2 = 4$$

Step 5: Substituting from step 3 into step 4

$$2V_a + 3(V_a - V_b) = 2$$

$$6V_b - 3(V_a - V_b) = 4$$

Combining terms,

$$5V_a - 3V_b = 2$$

$$-3V_a + 9V_b = 4$$

Solving with determinants:

$$V_a = \frac{\begin{vmatrix} 2 & -3 \\ 4 & 9 \end{vmatrix}}{\begin{vmatrix} 5 & -3 \\ -3 & 9 \end{vmatrix}} = \frac{18 + 12}{45 - 9} = \frac{30}{36} = 0.833 \text{ V}$$

$$V_b = \frac{\begin{vmatrix} 5 & 2 \\ -3 & 4 \end{vmatrix}}{36} = \frac{20 + 6}{36} = 0.722 \text{ V}$$

Step 6

$$I_1 = 2(0.833) = 1.67 \text{ A}$$

$$I_2 = 3(0.833 - 0.722) = 0.333 \text{ A}$$

$$I_3 = 6(0.722) = 4.33 \text{ A}$$

Step 7: Checking KCL (step 4) we have

$$I_1 + I_2 = 1.67 + 0.333 = 2.00 \text{ A}$$

$$I_3 - I_2 = 4.33 - 0.333 = 4.00 \text{ A}$$

These results are seen to agree with the KCL equations in step 4. This type of checking should become a habit, since it often turns up numerical errors made along the way.

• • •

The next example illustrates the node-voltage method when a voltage source is present.

EXAMPLE 5.3-2 Node-Voltage Method with Voltage Source

The circuit of Fig. 5.11 contains a voltage source and a current source. Find all currents and voltages using the node-voltage method.

Solution

We again follow the sequence of steps given.

Step 1: The reference node is chosen as indicated on the diagram as that node with the largest number of branches connected to it and is marked with the ground symbol. The two independent nodes are marked *a* and *b*, respectively, with node voltages V_a and V_b. The node between the battery and R_1 is *not* independent because its voltage is *known*.

Step 2: Branch currents are assumed as shown in the diagram.

Step 3: The branch currents are (units in volts, kilohms, and milliamperes).

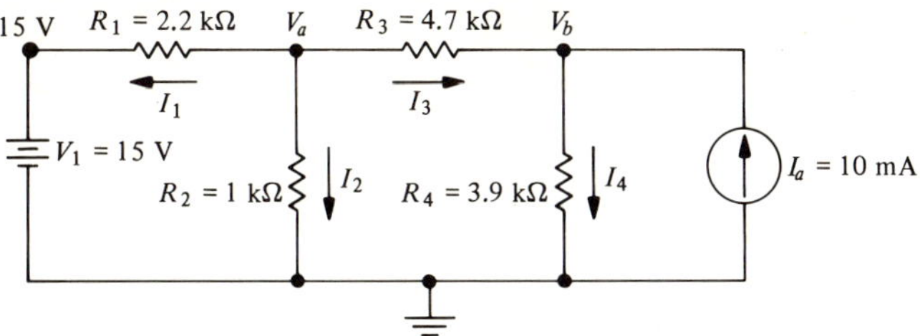

FIGURE 5.11
Two-node-pair network with voltage source.

$$I_1 = \frac{V_a - 15}{2.2}$$

$$I_2 = \frac{V_a}{1}$$

$$I_3 = \frac{V_a - V_b}{4.7}$$

$$I_4 = \frac{V_b}{3.9}$$

Step 4: The KCL equations are

Node *a*:

$$I_1 + I_2 + I_3 = 0$$

Node *b*:

$$I_4 - I_3 = 10$$

Step 5: The node-voltage equations are obtained by substituting the branch current equations into the KCL equations. This yields

$$\frac{V_a - 15}{2.2} + V_a + \frac{V_a - V_b}{4.7} = 0$$

$$\frac{V_b}{3.9} - \frac{V_a - V_b}{4.7} = 10$$

Combining terms and rearranging:

$$V_a\left(\frac{1}{2.2} + 1 + \frac{1}{4.7}\right) - \frac{V_b}{4.7} = \frac{15}{2.2}$$

$$-\frac{V_a}{4.7} + V_b\left(\frac{1}{3.9} + \frac{1}{4.7}\right) = 10$$

Simplifying the coefficients:

$$1.67V_a - 0.213V_b = 6.82$$

$$-0.213V_a + 0.469V_b = 10$$

Solving by the determinant method:

$$V_a = \frac{\begin{vmatrix} 6.82 & -0.213 \\ 10 & 0.469 \end{vmatrix}}{\begin{vmatrix} 1.67 & -0.213 \\ -0.213 & 0.469 \end{vmatrix}} = \frac{3.20 + 2.13}{0.782 - 0.045} = \frac{5.33}{0.737} = 7.23 \text{ V}$$

$$V_b = \frac{\begin{vmatrix} 1.68 & 6.82 \\ -0.213 & 10 \end{vmatrix}}{0.737} = \frac{16.7 + 1.45}{0.737} = 24.6 \text{ V}$$

Step 6: The individual branch currents are, from step 3:

$$I_1 = \frac{7.23 - 15}{2.2} = -3.53 \text{ mA}$$

$$I_2 = \frac{V_a}{1} = 7.23 \text{ mA}$$

$$I_3 = \frac{7.23 - 24.6}{4.7} = -3.70 \text{ mA}$$

$$I_4 = \frac{24.6}{3.9} = 6.31 \text{ mA}$$

Step 7: Checking the KCL equations from step 4:

$$I_1 + I_2 + I_3 = -3.53 + 7.23 - 3.70 = 0 \text{ mA}$$

$$I_4 - I_3 = 6.31 - (-3.70) = 10 \text{ mA}$$

• • •

The procedure is exactly the same regardless of the number of independent nodes. The number of simultaneous equations is equal to the number of independent node pairs. The next example illustrates the method for a network with three independent node pairs.

EXAMPLE 5.3-3 Three Node-Pair Network

Find V_1, V_2, and V_3 in the circuit of Fig. 5.12.

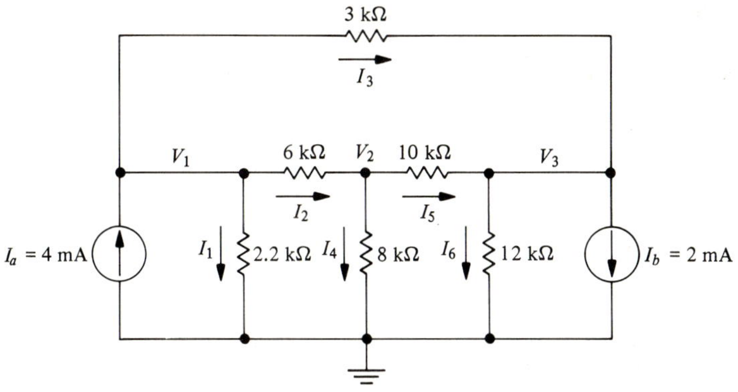

FIGURE 5.12
Circuit for Example 5.3-3.

Solution

Step 1: Choose a reference node and identify and label the voltage at each independent node. This is done on the circuit diagram. The voltages are V_1, V_2, and V_3.

Step 2: The *assumed* branch currents are shown in the figure.

Step 3: Expressing branch currents in terms of node voltages (units are volts, kilohms, milliamperes)

$$I_1 = \frac{V_1}{2.2}$$

$$I_2 = \frac{V_1 - V_2}{6}$$

$$I_3 = \frac{V_1 - V_3}{3}$$

$$I_4 = \frac{V_2}{8}$$

$$I_5 = \frac{V_2 - V_3}{10}$$

$$I_6 = \frac{V_3}{12}$$

Step 4: KCL equations at each node:

Node 1:

$$I_1 + I_2 + I_3 = I_a$$

Node 2:

$$-I_2 + I_4 + I_5 = 0$$

Node 3:

$$-I_5 + I_6 - I_3 = -I_b$$

The student should carefully check the sign of each term in these equations in order to be sure of completely understanding the sign convention.
Step 5: Substituting step 3 into step 4.

$$\frac{V_1}{2.2} + \frac{V_1 - V_2}{6} + \frac{V_1 - V_3}{3} = 4$$

$$\frac{-(V_1 - V_2)}{6} + \frac{V_2}{8} + \frac{V_2 - V_3}{10} = 0$$

$$\frac{-(V_2 - V_3)}{10} + \frac{V_3}{12} - \frac{(V_1 - V_3)}{3} = -2$$

Combining terms

$$V_1\left(\frac{1}{2.2} + \frac{1}{6} + \frac{1}{3}\right) - V_2\left(\frac{1}{6}\right) - V_3\left(\frac{1}{3}\right) = 4 \tag{5.3-8}$$

$$-V_1\left(\frac{1}{6}\right) + V_2\left(\frac{1}{6} + \frac{1}{8} + \frac{1}{10}\right) - V_3\left(\frac{1}{10}\right) = 0 \tag{5.3-9}$$

$$-V_1\left(\frac{1}{3}\right) - V_2\left(\frac{1}{10}\right) + V_3\left(\frac{1}{10} + \frac{1}{12} + \frac{1}{3}\right) = -2 \tag{5.3-10}$$

The student at this point should check the physical interpretation of each of the coefficients as outlined previously. For example, the coefficient of V_3 in Eq. (5.3-10) should be the total conductance connected to node 3. A glance at the circuit indicates that this is correct.
Step 6: Solve for V_1, V_2, and V_3. Simplifying the coefficients, we get the following set of equations:

$$0.955V_1 - 0.167V_2 - 0.333V_3 = 4$$

$$-0.167V_1 + 0.392V_2 - 0.1V_3 = 0$$

$$-0.333V_1 - 0.1V_2 + 0.517V_3 = -2$$

They are best solved by the determinant method if a computer is not available.

$$V_1 = \frac{\begin{vmatrix} 4 & -0.167 & -0.333 \\ 0 & +0.392 & -0.1 \\ -2 & -0.1 & +0.517 \end{vmatrix}}{\begin{vmatrix} 0.955 & -0.167 & -0.333 \\ -0.167 & +0.392 & -0.1 \\ -0.333 & -0.1 & +0.517 \end{vmatrix}} = \frac{0.476}{0.115} = 4.14 \text{ V}$$

V_2 and V_3 are found in a similar fashion.

$$V_2 = \frac{0.177}{0.115} = 1.54 \text{ V}$$

$$V_3 = \frac{-0.104}{0.115} = -0.904 \text{ V}$$

Step 7: Checking is left as an exercise. (See Prob. 5.18).

• • •

If the physical interpretation of the coefficients is well understood, it is not difficult to write the KCL equations (5.3-8, 9, and 10) directly from inspection of the circuit diagram. The student should try this shortcut procedure with the problems given at the end of the chapter.

As in the loop-current method, the tedious part of the process is the determinant evaluation. If the circuit leads to more than three simultaneous equations, a computer solution should be sought. However, if this is not possible, the procedure above will always lead to a solution.

Choice Between Node and Loop Methods

It is not possible to state a rule that can be used to determine the best method to be used for a specific circuit. The usual procedure is to use that method leading to the smallest number of equations that must be solved simultaneously. As an example, consider the network of Fig. 5.13. If node analysis is considered, we see that the circuit has two independent nodes, labeled 1 and 2 on the diagram. The node between the battery and R_1 and R_5 is *not* independent. Thus two node equations will suffice. On the other hand, there are four windows, one of which is a dummy, so that three loop equations are required. Thus, on the basis of the least number of simultaneous equations, the node-voltage method would be used for this circuit.

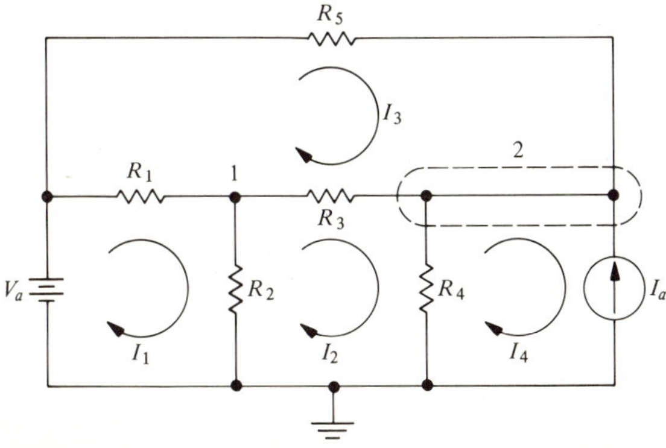

FIGURE 5.13
Circuit for loop or node analysis.

5.4 GRAPHICAL ANALYSIS USING TERMINAL CHARACTERISTICS

In this section we consider the solution of simple circuits using graphical analysis. We show how the terminal characteristic of the practical source can be used to find the output of the source for any value of load resistance. With these ideas available, we will go on to a discussion of equivalent circuits.

Let us first consider the graphical solution of the circuit shown in Fig. 5.14a. The graph of the terminal characteristic of the ideal source (usually called the "load line"), is shown in Fig. 5.14b. The equation for this characteristic is

$$v = V_T \tag{5.4-1}$$

The graph of the terminal characteristic of the resistor is shown in Fig. 5.14c. The equation for this characteristic is Ohm's law

$$v = R_L i \tag{5.4-2}$$

In order to solve this circuit graphically, we simply plot both graphs on the same set of axes as shown in Fig. 5.14d. We can do this because the

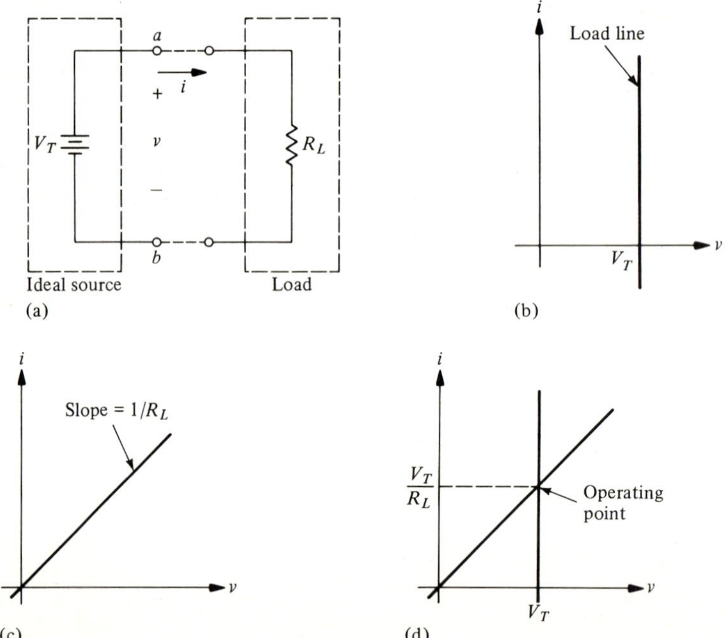

FIGURE 5.14
Graphical solution. (a) Circuit. (b) *vi* characteristic of source. (c) *vi* characteristic of resistor. (d) Intersection of characteristics.

variables v and i are the same for both elements when they are connected together. The point where the terminal characteristics intersect is the simultaneous solution of the equations for the terminal characteristics [Eqs. (5.4-1) and (5.4-2)], and represents the operating point of the circuit. If the value of the resistance is changed the slope of its characteristic changes, and the intersection with the source characteristic determines the new operating point. Note that the voltage across the load is constant and remains the same regardless of the value of the load resistance.

Now let us consider the graphical solution of the same circuit with a practical source. A practical source always has some internal resistance and can be represented for most practical purposes by an ideal source of voltage V_T in series with a resistance R_T as shown in Fig. 5.15a. Referring to the circuit, it is important to note that *only* terminals a and b are available. Terminal c, which is shown inside the practical source, is not available externally, and in fact may not exist at all, depending on the energy conversion method that generates the voltage.

We wish to plot the terminal characteristics as in Fig. 5.14. Thus we need the terminal characteristic of the practical source. We find it by writing KVL around the loop as follows:

$$-V_T + R_T i + v = 0 \tag{5.4-3}$$

Isolating the dependent variable v on the left-hand side

$$v = V_T - R_T i \tag{5.4-4}$$

This is the equation for the *vi* characteristic of the practical source. Observe that there are two unknowns in this equation, v and i. Therefore another equation is required before we can find a solution. The second equation comes about when a load is connected so that

$$v = R_L i \tag{5.4-5}$$

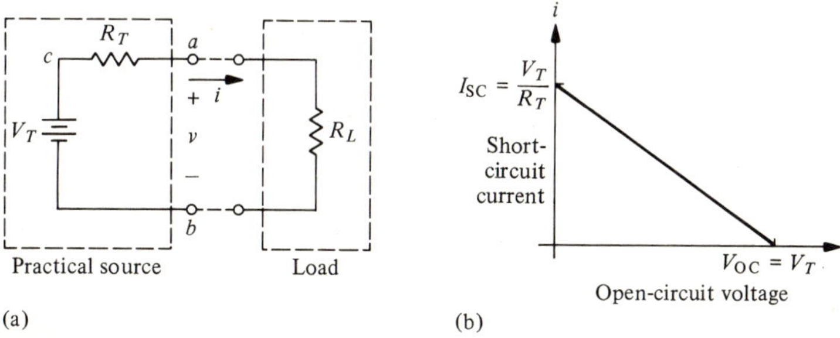

FIGURE 5.15
The practical source. (a) Circuit. (b) Terminal characteristic.

These last two equations involve the same two unknowns and thus they can be solved graphically in the same manner as for the ideal source. The graph of the practical source characteristic is a straight line as shown in Fig. 5.15b. In order to plot the graph we require only two points. Those most easy to obtain are the intersections on the axes, which have a useful physical interpretation. Consider the point on the voltage axis. At this point, the current is zero and there is no voltage drop across R_T so that

$$v|_{i=0} = V_{oc} = V_T \tag{5.4-6}$$

Thus the point represents the *open-circuit voltage.*

To find the point on the current axis we set $v = 0$. This can be done by placing a short circuit from a to b that forces the condition $v = 0$. When this is done, the *short-circuit current* is

$$i|_{v=0} = I_{sc} = \frac{V_T}{R_T} \tag{5.4-7}$$

This is the intercept on the current axis. From the above we have the important conclusion that the terminal characteristic of the practical source is a straight line that runs from the open-circuit voltage to the short-circuit current. The slope of the line can be found from Eq. (5.4-4) by isolating i on the left-hand side so that we have the slope-intercept form:

$$i = -\frac{v}{R_T} + \frac{V_T}{R_T} \tag{5.4-8}$$

Thus the slope of the line is $-1/R_T$. This terminal characteristic is called a *load line* when it arises in conjunction with the analysis of electronic devices. The load line will be studied in other courses.

The next step is to plot the terminal characteristic of the load on the same axes. This is shown in Fig. 5.16. The point where the two lines

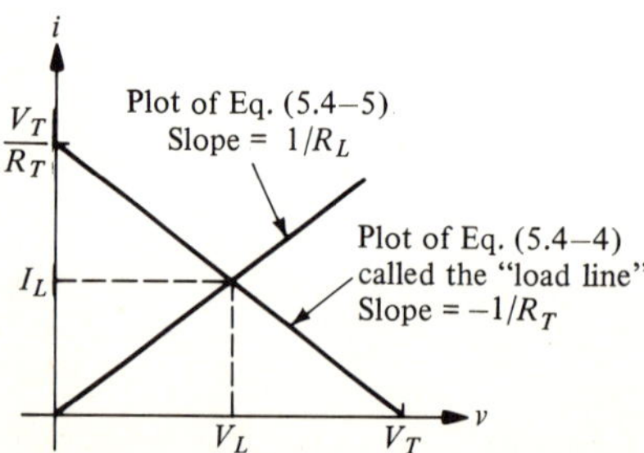

FIGURE 5.16
Graphical solution for practical source with load.

intersect is the operating point for the circuit. We can read directly from the graph the value of load voltage V_L and current I_L.

Now suppose that the value of the load resistance changes. The new value of load voltage and current is easily found by plotting the *vi* characteristic corresponding to the new value of resistance and observing the new point of intersection. It is important to observe that regardless of the value of the load resistance, the operating point *must* lie on the terminal characteristic of the source (the load line). This is illustrated in the following numerical example.

EXAMPLE 5.4-1 A Graphical Solution

An automobile battery has an open-circuit voltage $V_T = 12.8$ V, and an internal resistance $R_T = 0.15\ \Omega$. Using the graphical technique, find

a. The load voltage and current for load resistances of 0.2 and 0.5 Ω.
b. The load voltage when the load current is 60 A. What value of load resistance would lead to this condition?
c. Check all graphical answers using Ohm's law and KVL.

Solution

The graphs are shown in Fig. 5.17. The *vi* characteristic of the source runs from $V_{oc} = V_T = 12.8$ V to $I_{sc} = 12.8\ \text{V}/0.15\ \Omega = 85.3$A.

a. The *vi* characteristics corresponding to 0.2- and 0.5-Ω loads are plotted as shown. The load voltage and current values are read from the intersection points as

$$0.2\ \Omega: \quad V_L \approx 7.3\ \text{V}; \quad I_L \approx 37\ \text{A}$$

$$0.5\ \Omega: \quad V_L \approx 9.9\ \text{V}; \quad I_L \approx 19\ \text{A}$$

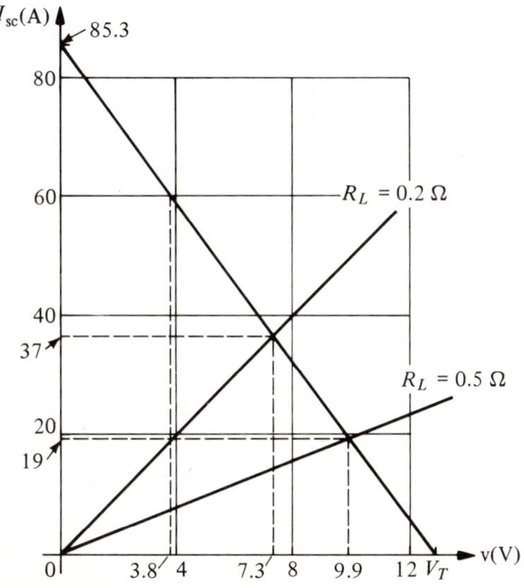

FIGURE 5.17
Graphs for Example 5.4-1.

b. At the point where $i = 60$ A, we read $v = V_L \approx 3.8$ V. The load resistance for this case is

$$R_L = \frac{V_L}{I_L} = \frac{3.8\ \text{V}}{60\ \text{A}} = 0.063\ \Omega$$

c. Checking:

$$0.2\ \Omega: \qquad I_L = \frac{V_T}{R_T + R_L} = \frac{12.8\ \text{V}}{0.15\ \Omega + 0.2\ \Omega} = 36.6\ \text{A}$$

$$V_L = R_L I_L = (0.2\ \Omega)(36.6\ \text{A}) = 7.32\ \text{V}$$

$$0.5\ \Omega: \qquad I_L = \frac{12.8\ \text{V}}{0.15\ \Omega + 0.5\ \Omega} = 19.7\ \text{A}$$

$$V_L = R_L I_L = (0.5\ \Omega)(19.7\ \text{A}) = 9.85\ \text{V}$$

$$60\ \text{A}: \qquad V_{R_T} = R_T I_L = (0.15\ \Omega)(60\ \text{A}) = 9\ \text{V}$$

$$V_L = V_T - V_{R_T} = 12.8\ \text{V} - 9\ \text{V} = 3.8\ \text{V}$$

$$R_L = \frac{V_L}{I_L} = \frac{3.8\ \text{V}}{60\ \text{A}} = 0.063\ \Omega$$

The answers read directly from the graph are seen to be quite close to the calculated answers.

• • •

Drop in Terminal Voltage as Load Current Increases

Earlier in this section we found that the terminal voltage of an ideal source did not change as the load current changed. The results of the last example can be used to point up the important fact that the voltage available at the terminals of a practical source *always* decreases as the load current increases. For the practical source the equation relating the terminal voltage to the load current is [see Eq. (5.4-4)]

$$v_L = V_T - R_T i_L \tag{5.4-9}$$

This equation was plotted in Fig. 5.15b with current as the dependent variable. In Fig. 5.18 we plot it again, this time with load voltage as the dependent variable. When the graph is plotted this way we see quite clearly how the load voltage decreases as the load current increases.

In practice, a practical source cannot supply currents beyond a certain maximum, or *full-load* value, without suffering physical damage. This is shown as I_{FL} on the graph along with the corresponding full-load voltage V_{FL}. The drop in voltage as the load varies from no-load to full-load is measured by a quantity called *regulation* which is defined as

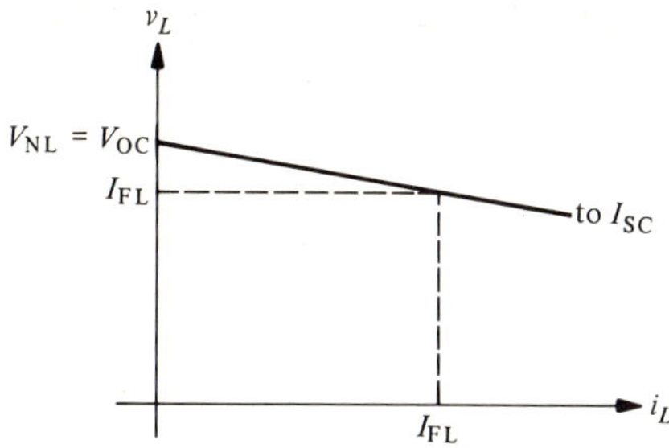

FIGURE 5.18
Illustrating the drop in terminal voltage of a practical source.

$$\% \text{ Regulation} = \frac{\text{no-load voltage} - \text{full-load voltage}}{\text{full-load voltage}} \times 100 \qquad (5.4\text{-}10)$$

where, in terms of the graph of Fig. 5.18, the conditions are:

At no-load:

$$v_L = V_{NL} = V_{oc} \text{(the open-circuit voltage)} \qquad \text{and} \qquad i_L = 0$$

At full-load:

$$v_L = V_{FL} \qquad \text{and} \qquad i_L = I_{FL}$$

An ideal source with $R_T = 0$ provides a constant voltage which is independent of the current so that $V_{FL} = V_{NL}$ and, using Eq. (5.4-10)

$$\% \text{ Reg} = \frac{V_{NL} - V_{FL}}{V_{FL}} \times 100 = 0\%$$

In a practical source the regulation is not zero and is a measure of the variation of the terminal voltage with load, smaller values of regulation indicating less terminal voltage variation.

It is interesting to determine the conditions that lead to a regulation of 100%. In order to do this, we write Eq. (5.4-10) in the form:

$$\% \text{ Reg} = \frac{V_{NL} - V_{FL}}{V_{FL}} \times 100 \qquad (5.4\text{-}11)$$

We can express V_{FL} in terms of the open-circuit voltage by applying the voltage divider relation to the circuit of Fig. 5.15a

$$v = V_{FL} = \frac{R_L}{R_L + R_T} V_{oc}$$

and since $V_{NL} = V_{oc}$, Eq. (5.4-11) becomes

$$\% \text{ Reg} = \frac{V_{oc} - \dfrac{R_L}{R_L + R_T} V_{oc}}{\dfrac{R_L}{R_L + R_T} V_{oc}} \times 100$$

After a little algebra this simplifies to

$$\% \text{ Reg} = \frac{R_T}{R_L} \times 100 \qquad (5.4\text{-}12)$$

Now when we set % Reg = 100 and solve we find that for 100% regulation, $R_L = R_T$. Thus when the load resistance is equal to the source resistance, the regulation is 100%. (Note also that the full-load voltage is exactly one-half of the open-circuit voltage at this point.) When the source resistance R_T is zero, the regulation is 0% and the source is ideal.

EXAMPLE 5.4-2 Regulation of a Battery

The automobile battery of Example 5.4-1 has a rated full-load current of 30 A. Find the regulation.

Solution

From the graph of Fig. 5.17 we read down from $i = I_{FL} = 30$ A to find $v = V_{FL} = 8.4$ V. Also, when $i = 0$, $v = V_{NL} = 12.8$ V. Then

$$\% \text{ Reg} = \frac{V_{NL} - V_{FL}}{V_{FL}} \times 100 = \frac{12.8 - 8.4}{8.4} \times 100 = \frac{4.4}{8.4} \times 100$$

$$= 52\%$$

• • •

LEARNING EXERCISE FOR SEC. 5.4

1. An aircraft generator has an output of 46 V at no load. When it is supplying full-load current of 26 A, the output voltage drops to 43.5 V. Assuming that the circuit is linear, find
   a. The internal resistance of the generator
   b. The voltage regulation in percent
   c. The output voltage when $I = 13$ A.

Ans. 0.096; 5.7; 44.8

• • •

5.5 TERMINAL CHARACTERISTICS AND EQUIVALENT CIRCUITS

Suppose we have a dc electric system that is completely enclosed except for two terminals at which we can make any electrical measurements we wish. We would like to determine as much as possible about the system inside the box from these measurements. In general, it is not possible to determine exactly what the unknown system is. However, if the system is *linear*, we can find a circuit that is exactly *equivalent*, as far as behavior at the terminals is concerned, from two voltage-current measurements. In what follows, we assume that the system is an electric circuit that consists of resistors and sources. However, all conclusions hold for any linear electric system or device.

Measurement of Terminal Characteristics

Since the circuit is linear, its terminal characteristic will be a straight line. Therefore, two different volt-ampere measurements will provide two points that are sufficient to determine the straight-line characteristic. These measurements can be made by connecting an adjustable voltage source in series with an ammeter across the terminals of the network as shown in Fig. 5.19a. The voltage source is adjusted to two different values and the current is measured for each of these voltages. (It is assumed that the voltage across the ammeter is negligible in this procedure.) Figure 5.19b shows two typical points along with the terminal characteristic drawn through them.

Linearity and Equivalence

We have described a way of finding the terminal characteristic of a linear network. It is important to note the relevance of the linearity of the circuit; as a consequence of this property it is possible to determine the complete terminal characteristic from two simple measurements. If a network is not known to be linear, we cannot conclude anything about its behavior except at those points where we have made measurements.

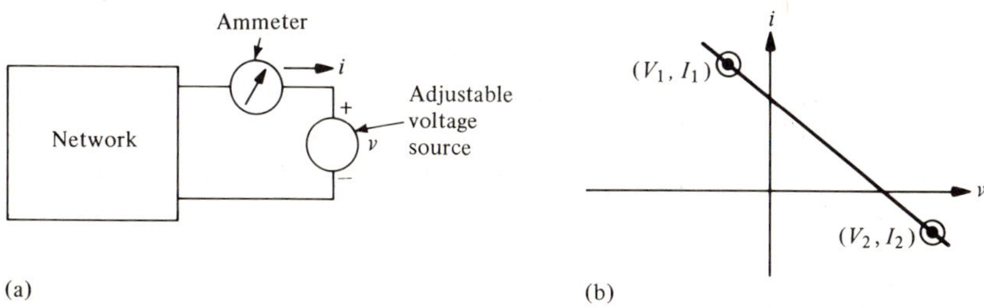

FIGURE 5.19
Measurement of network terminal characteristic. (a) Volt-ampere measurement. (b) Possible characteristic.

The original question we asked is how much can be determined about the network in the box. The following indicates that we can, in the case of a linear network, go only so far as determining an equivalent circuit. Let us assume that the terminal characteristic is that of a 4-Ω resistance. We cannot conclude that the network in the box is a 4-Ω resistance; it might be two 2-Ω resistances in series, a 3-Ω and a 1-Ω resistance in series, two 8-Ω resistances in parallel, or any one of an infinite number of other combinations that result in an *equivalent* resistance of 4 Ω.

Thus we conclude that

1. For a linear circuit only, we can find an equivalent circuit for the one in the box from two measured points.
2. The equivalent circuit will not necessarily be identical to the one in the box.
3. The circuits are equivalent only to the extent that they have the same volt-ampere characteristic at the terminal pair at which the measurements are made.

In the next section we will show how to find the equivalent circuit.

5.6 THEVENIN AND NORTON EQUIVALENT CIRCUITS

In this section we show how to find an equivalent circuit for any arbitrary linear network containing resistors and sources. In the last section we showed how the terminal characteristic of such a network could be found experimentally and pointed out that it will always be a straight line. In Sec. 5.4 we showed that the terminal characteristic of a practical source consisting of an ideal source and one resistor was a straight line. The circuit is repeated along with its terminal characteristic in Fig. 5.20. Since the arbitrary network and the practical source have the same straight-line terminal characteristic measured at terminals *ab*, we conclude that they will be exactly equivalent as far as terminals *ab* are concerned if we can find values for V_T and R_T that will make the two straight lines coincide. We will do this shortly.

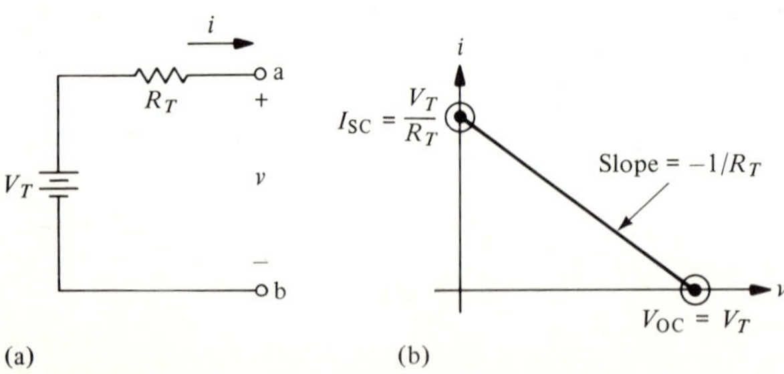

FIGURE 5.20
Thevenin equivalent circuit.

The two-element circuit we have called a practical source is known as the *Thevenin equivalent circuit,* after the man who first described its properties. Our next step is to find appropriate values for the Thevenin voltage V_T and the Thevenin resistance R_T. Before we do this, let us be certain that we understand the meaning of *equivalence.* As we stated before, two linear circuits are equivalent at a pair of terminals if they have exactly the same terminal characteristic measured at those terminals. Another way to look at this is the following: If the original circuit and its Thevenin equivalent are encased in boxes with only terminals a and b available, as shown in Fig. 5.21, then there is no way that we can tell the boxes apart by measurements.

We will often have occasion to replace a complicated network containing any number of resistors and sources by its Thevenin equivalent in order to simplify a problem in network analysis.

Finding the Thevenin Circuit Parameters

In the Thevenin equivalent circuit shown in Fig. 5.20a there are two parameters, V_T and R_T. Thus if we can find V_T and R_T from the original circuit, we will have completely determined the Thevenin equivalent. Another possibility exists if we examine the terminal characteristic in Fig. 5.20b. As discussed previously, this characteristic is most easily determined from two measurements: the open-circuit voltage,

$$V_{oc} = V_T \tag{5.6-1}$$

and the short-circuit current

$$I_{sc} = \frac{V_T}{R_T} = \frac{V_{oc}}{R_T} \tag{5.6-2}$$

Thus the short-circuit current is a third parameter.

Because of this last relation, we only have to find two of the three parameters, I_{sc}, V_{oc}, and R_T. The third parameter can then be found from the equation. Determining V_{oc} and I_{sc} from the original circuit is conceptually easy. In order to find V_{oc} we simply calculate the voltage across terminals ab or measure it using a high-resistance voltmeter so that the terminals are essentially open-circuited. To find I_{sc} we place a short circuit

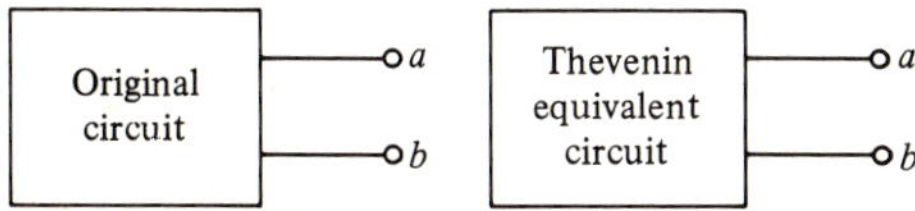

FIGURE 5.21
Illustrating equivalence.

from a to b and calculate (or measure) the resulting current through the short. If we are measuring the short-circuit current, we must make certain that the presence of the short circuit will not cause an overload in any of the circuit elements.

The Thevenin resistance R_T can also be found from the original circuit as follows:

If all of the ideal independent sources in the given circuit are replaced by zero values, the resulting network is a purely resistive circuit. The equivalent resistance, that is, the resistance that would be measured across terminals ab, is equal to the Thevenin resistance, R_T. This is so because the Thevenin equivalent circuit reduces to the Thevenin resistance when the Thevenin voltage source, V_T, is set to zero (see Fig. 5.20a). Remember, when using this method, that a voltage source is set to zero by replacing it by a short circuit while a current source is set to zero by replacing it by an open circuit.

Since the determination of the Thevenin equivalent from a circuit diagram is such a useful technique, we will illustrate it with several examples.

EXAMPLE 5.6-1

Find the Thevenin equivalent of the circuit of Fig. 5.22a across terminals ab.

Solution

The Thevenin voltage is the open-circuit voltage drop from terminal a to b. It is computed using the voltage divider formula:

$$V_T = V_{ab} = \frac{2\ \text{k}\Omega}{2\ \text{k}\Omega + 8\ \text{k}\Omega} \times 24\ \text{V} = 4.8\ \text{V} \tag{5.6-3}$$

The Thevenin resistance is found as the equivalent resistance across terminals ab with the voltage source replaced by a short circuit as shown in Fig. 5.22b. This resistance is computed using the formula for parallel resistances.

$$R_T = R_{ab} = \frac{(8\ \text{k}\Omega)(2\ \text{k}\Omega)}{8\ \text{k}\Omega + 2\ \text{k}\Omega} = 1.6\ \text{k}\Omega \tag{5.6-4}$$

The resulting Thevenin equivalent is shown in Fig. 5.22c.

As a check, we can compute the short-circuit current from terminal a to terminal b for the original circuit and the Thevenin equivalent.

In the circuit of Fig. 5.22d, which is the original circuit with the output shorted, the current I will all flow through the short circuit so that no current flows through the 2-kΩ resistance. (Why?) Thus $I = I_{ab}$ and Ohm's law yields

$$I_{sc} = I_{ab} = \frac{24\ \text{V}}{8\ \text{k}\Omega} = 3\ \text{mA} \tag{5.6-5}$$

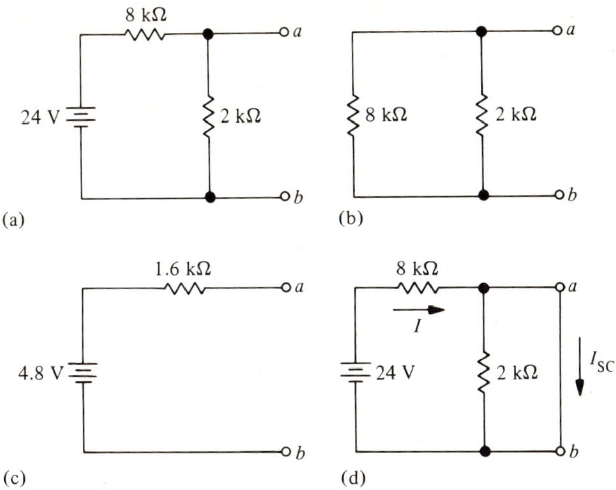

FIGURE 5.22
Circuits for Example 5.6-1. (a) Original circuit. (b) Circuit with voltage source set to zero. (c) Thevenin equivalent. (d) Circuit for calculating short-circuit current.

Applying Ohm's law to the Thevenin equivalent of Fig. 5.22c with a short circuit between points a and b we obtain

$$I_{sc} = I_{ab} = \frac{4.8 \text{ V}}{1.6 \text{ k}\Omega} = 3 \text{ mA} \tag{5.6-6}$$

in agreement with the short-circuit current for the original circuit.

• • •

EXAMPLE 5.6-2

Find the Thevenin equivalent at terminals ab of the circuit shown in Fig. 5.23a.

Solution

We will use superposition to compute the open-circuit voltage and short-circuit current due to each source acting alone, that is, with the other source replaced by a zero value. The total open-circuit voltage and short-circuit current are then obtained by taking the algebraic sum of the components due to each source.

Figure 5.23b illustrates the equivalent circuit with the current source replaced by an open circuit (recall that a zero current source is equivalent to an open circuit). The open-circuit voltage, V_{oc1} is easily computed using the voltage-divider relationship:

$$V_{oc1} = V_{ab1} = \frac{6 \text{ k}\Omega}{6 \text{ k}\Omega + 4 \text{ k}\Omega} \times 24 \text{ V} = 14.4 \text{ V} \tag{5.6-7}$$

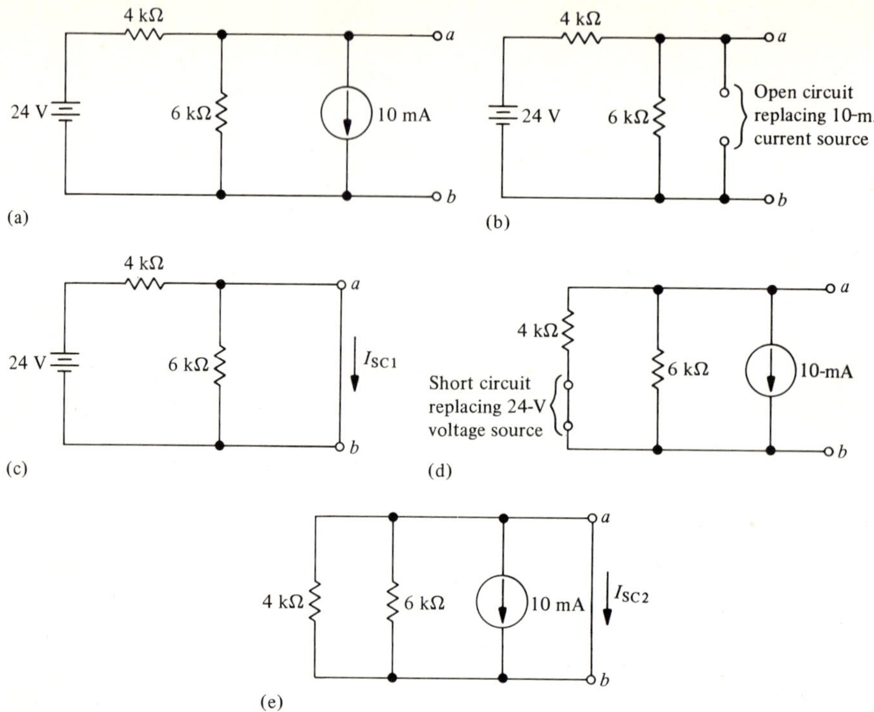

FIGURE 5.23
Example 5.6-2. (a) Circuit. (b) Zero current source equivalent circuit. (c) Circuit for finding I_{sc1}. (d) Zero voltage source equivalent circuit for finding V_{oc2}. (e) Circuit for fnding I_{sc2}.

The short-circuit current is computed using Ohm's law applied to the circuit but with a short circuit connected from a to b as shown in Fig. 5.23c. Note that the 6-kΩ resistance has no effect. (Why?) Using Ohm's law we find

$$I_{sc1} = I_{ab1} = \frac{24\ \text{V}}{4\ \text{k}\Omega} = 6\ \text{mA} \tag{5.6-8}$$

The next step is to compute the open-circuit voltage and short-circuit current due to the current source. The equivalent circuit with the voltage source replaced by a short circuit (recall that a zero voltage source is equivalent to a short circuit) is shown in Fig. 5.23d.

The parallel combination of the 4- and 6-kΩ resistances can be replaced by a single equivalent resistance having a value of

$$\frac{(4\ \text{k}\Omega)(6\ \text{k}\Omega)}{4\ \text{k}\Omega + 6\ \text{k}\Omega} = 2.4\ \text{k}\Omega \tag{5.6-9}$$

The open-circuit voltage is computed using Ohm's law:

$$V_{oc2} = V_{ab2} = -(2.4\ \text{k}\Omega)(10\ \text{mA}) = -24\ \text{V} \tag{5.6-10}$$

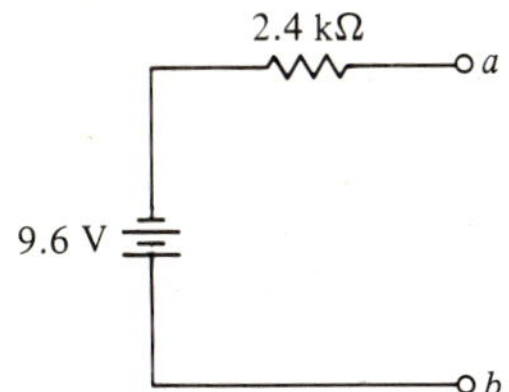

FIGURE 5.24
Thevenin equivalent circuit for Example 5.6-2.

Note that the current is flowing from the bottom to the top through the resistance. Thus the polarity of the *actual* output voltage is opposite to the *reference* polarity (drop from a to b); hence the minus sign.

To find the short-circuit current we place a short circuit from a to b as shown in Fig. 5.23e. All of the current from the 10-mA source *must* flow through the short circuit. (Why?) Thus

$$I_{sc2} = I_{ab2} = -10 \text{ mA}$$

Note the minus sign, which is a result of the difference between the *reference* direction for I_{ab2} (from a to b) and the *actual* direction of current due to the 10-mA source through the short circuit.

Finally, we compute the total open-circuit voltage and short-circuit current as

$$V_{oc} = V_{ab} = V_{ab1} + V_{ab2} = 14.4 - 24 = -9.6 \text{ V} \tag{5.6-11}$$

$$I_{sc} = I_{ab1} + I_{ab2} = 6 - 10 = -4 \text{ mA} \tag{5.6-12}$$

Therefore

$$V_T = V_{oc} = -9.6 \text{ V} \tag{5.6-13}$$

$$R_T = \frac{V_{oc}}{I_{sc}} = \frac{-9.6 \text{ V}}{-4 \text{ mA}} = 2.4 \text{ k}\Omega \tag{5.6-14}$$

The resulting circuit is shown in Fig. 5.24.

As a check we calculate R_T directly from the circuit. Figure 5.25 shows the circuit of Fig. 5.23a with both sources replaced by zero values. From this circuit it is

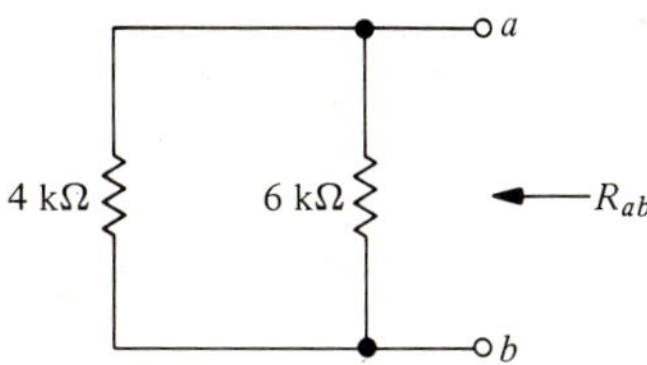

FIGURE 5.25
Equivalent circuit for R_T in Example 5.6-2.

clear that the resistance R_{ab} is the equivalent resistance of the parallel combination of the 4- and 6-kΩ resistances. Thus

$$R_{ab} = R_T = \frac{(4\ \text{k}\Omega)(6\ \text{k}\Omega)}{4\ \text{k}\Omega + 6\ \text{k}\Omega} = 2.4\ \text{k}\Omega \qquad (5.6\text{-}15)$$

• • •

EXAMPLE 5.6-3

Find the Thevenin equivalent across terminals *ab* for the circuit of Fig. 5.26a.

Solution

The first step in this solution is to determine which two of the three quantities (open-circuit voltage, short-circuit current, or equivalent resistance) should be computed. The short-circuit current is a good choice since the relatively complicated portion of the circuit connected across the output terminals is shorted out and therefore has no effect.

$$I_{sc} = I_{ab} = \frac{20\ \text{V}}{8\ \text{k}\Omega} = 2.5\ \text{mA} \qquad (5.6\text{-}16)$$

We will compute the Thevenin resistance as the second quantity. The resistive circuit with the source set to zero is shown in Fig. 5.26b. The parallel combination of the 6- and 4-kΩ resistances can be replaced by a single 2.4-kΩ equivalent resistance. The series combination of this resistance and the 4-kΩ resistance is a 6.4-kΩ resistance which is in parallel with the 8-kΩ resistance. The resulting resistance is

$$R_T = R_{ab} = \frac{(8\ \text{k}\Omega)(6.4\ \text{k}\Omega)}{8\ \text{k}\Omega + 6.4\ \text{k}\Omega} = 3.56\ \text{k}\Omega \qquad (5.6\text{-}17)$$

The Thevenin voltage is computed as

$$V_T = R_T I_{sc} = (3.56\ \text{k}\Omega)(2.5\ \text{mA}) = 8.9\ \text{V} \qquad (5.6\text{-}18)$$

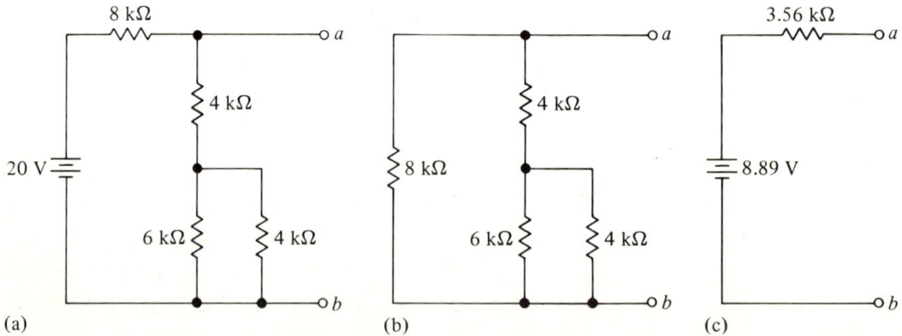

FIGURE 5.26
Circuit for Example 5.6-3. (a) Circuit. (b) Zero source equivalent. (c) Thevenin equivalent.

which together with the Thevenin resistance

$$R_T = 3.56 \text{ k}\Omega \tag{5.6-19}$$

define the Thevenin equivalent shown in Fig. 5.26c.

• • •

The Norton Equivalent Circuit

There is another form the equivalent circuit may take, called the Norton equivalent, after the engineer who first described it. The circuit is shown in Fig. 5.27, and it consists of a current source in parallel with a resistance.

The vi relation for this circuit is found by writing KCL at the upper node. This yields

$$i + \frac{v}{R_N} - I_N = 0$$

Solving for v we obtain

$$v = R_N I_N - R_N i \tag{5.6-20}$$

This is identical in form to Eq. (5.4-4), the vi characteristic of the Thevenin circuit. For the two circuits to be identical, the slopes and i intercepts must be the same, that is,

$$V_T = R_N I_N \tag{5.6-21}$$

and

$$R_T = R_N \tag{5.6-22}$$

When these two equations are satisfied the Thevenin and Norton circuits are equivalent at their external terminals and thus both are equivalent to the original linear circuit.

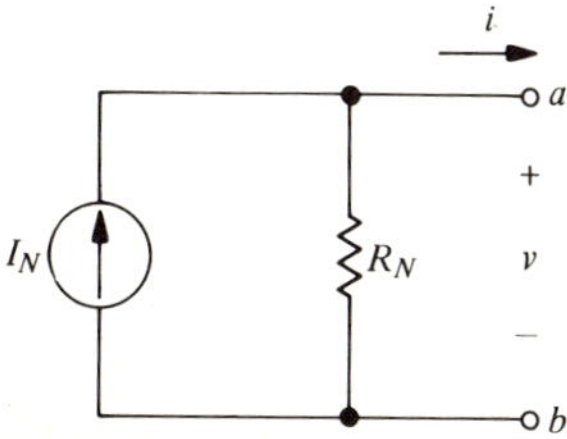

FIGURE 5.27
Norton equivalent circuit.

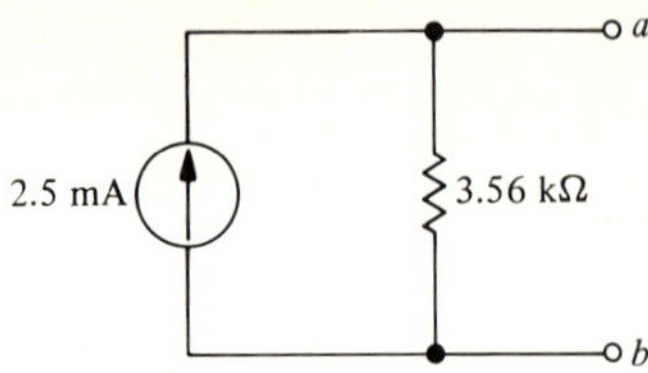

FIGURE 5.28
Norton equivalent circuit for Example 5.6-4.

The physical significance of the parameters of the Norton circuit is similar to that of the Thevenin circuit. For the Thevenin circuit, the voltage source represents the open-circuit voltage at the terminals of interest. For the Norton circuit all of the current I_N will flow through a short from a to b so that the current source represents the short-circuit current at these terminals. For both circuits the resistance represents the resistance of the original circuit with all sources dead (voltage sources replaced by short circuits and current sources by open circuits).

The Norton equivalent of a given circuit can be found using the same techniques as we used for the Thevenin circuit. An example follows.

EXAMPLE 5.6-4

Find the Norton equivalent of the circuit of Fig. 5.26a (the Thevenin equivalent was found in Example 5.6-3).

Solution

To find the Norton current, we connect a short circuit from a to b and calculate the resulting short-circuit current. Since no current will flow through the resistance network connected between a and b because of the short, the calculation simply involves Ohm's law. Thus

$$I_N = \frac{20\text{ V}}{8\text{ k}\Omega} = 2.5\text{ mA}$$

The Norton resistance R_N is the same as the Thevenin resistance found in Example 5.6-3, that is, $R_N = 3.56$ kΩ. The Norton circuit is shown in Fig. 5.28.

As a check, we find the open-circuit voltage of the Norton circuit to be (2.5 mA) (3.56 kΩ) = 8.9 V. This agrees with the value found in Example 5.6-3.

• • •

LEARNING EXERCISE FOR SEC. 5.6

1. In the circuit of Fig. 5.29 find V_T, R_T, and I_N.

Ans. 14.5; 0.85; 12.3

• • •

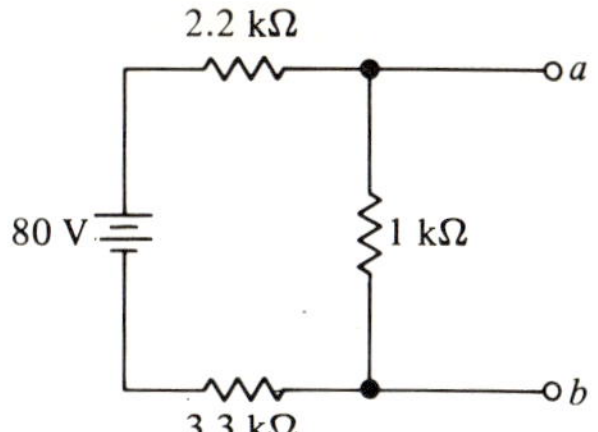

FIGURE 5.29
Circuit for Learning exercise 5.6-1.

5.7 THE MAXIMUM POWER TRANSFER THEOREM

As we found in Sec. 5.4, the terminal voltage of a practical source decreases as the load current increases. Often we are more interested in the power delivered to the load than the terminal voltage of the source, providing that the voltage stays within acceptable limits. Thus we are led to ask the following question: Given a practical source with fixed internal resistance, what is the maximum power that can be delivered to a load and what must be the characteristics of that load? The question can be answered by considering the circuit of Fig. 5.30 in which the source and its internal resistance are fixed while the load resistance can vary. Qualitatively, we can see that as R_L increases, the total resistance increases so that the current will decrease. At the same time the voltage across R_L increases so we cannot conclude anything about the power without actually making some calculations. To do this we begin by finding the current.

$$I_L = \frac{V_T}{R_L + R_T} \tag{5.7-1}$$

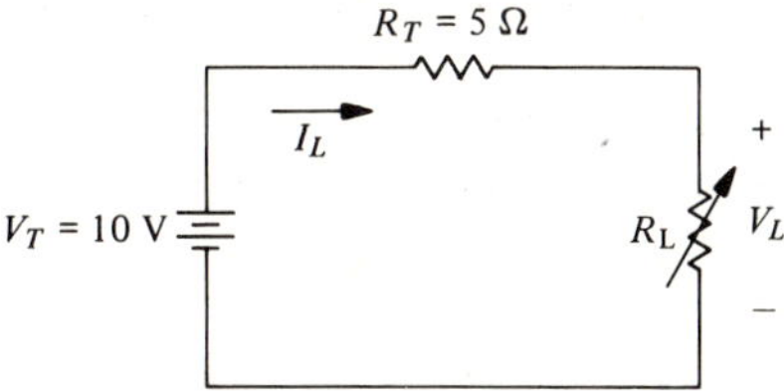

FIGURE 5.30
Circuit for maximum power calculation. V_T and R_T are fixed values that do not change.

Then the power is

$$P_L = R_L I_L^2$$
$$= \frac{V_T^2 R_L}{(R_L + R_T)^2}$$
$$= \frac{100R_L}{(R_L + 5)^2} \tag{5.7-2}$$

This is tabulated below and plotted in Fig. 5.31 for a range of values of R_L.

R_L, Ω	0	1	3	5	7	9	11	13	15
P_L, W	0	2.78	4.69	5	4.86	4.59	4.30	4.01	3.75

From the table and the graph we see that in this circuit maximum power is transferred to the load when the load resistance is equal to the internal resistance of the source. It can be proved using calculus that this is true in general. Thus the maximum power transfer theorem can be stated as follows:

> *Maximum power is transferred from a Thevenin circuit to a load when the load resistance is exactly equal to the Thevenin resistance.*

We often describe this by stating that the load is *matched* to the source. From the graph we see that some mismatch can be tolerated without serious loss of power output. For example, when $R_L = 2R_T = 10\ \Omega$ the power output is reduced to 4.44 W. When $R_L = (1/2)R_T = 2.5\ \Omega$ the power output again equals 4.44 W so that for load resistances from $(1/2)R_T$ to $2R_T$, the power

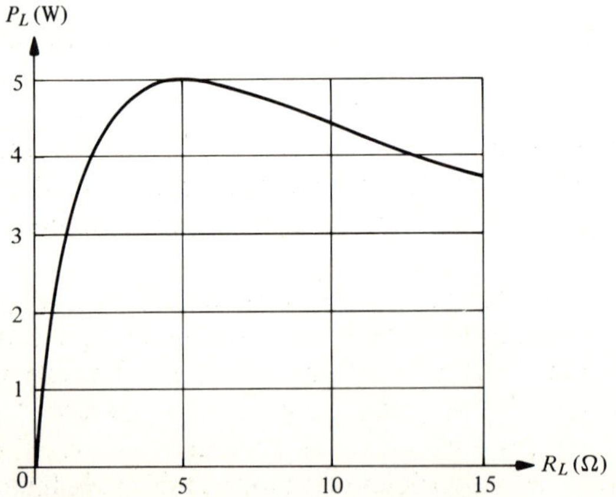

FIGURE 5.31
Graph of power versus load resistance.

output is within 11% of the maximum power available. Thus, in situations where this loss of available power can be tolerated, a mismatch by a factor of two will not be serious.

In order to complete the picture, we will find the load voltage, load current, and efficiency at maximum power when $R_L = R_T$. The load voltage is

$$V_L = \frac{R_L}{R_L + R_T} V_T = \frac{R_T}{2R_T} V_T = \frac{1}{2} V_T = 5 \text{ V} \tag{5.7-3}$$

Thus when maximum power is being delivered by a Thevenin circuit, the load voltage is exactly one-half of the open-circuit voltage.

The load current is

$$I_L = \frac{V_T}{R_L + R_T} = \frac{V_T}{2R_T} \tag{5.7-4}$$

The short-circuit (Norton) current is V_T/R_T. Thus when the circuit is delivering maximum power, the load current is one-half of the short circuit current.

In order to find the efficiency, we need the power supplied by the source. At maximum power, this is

$$P_T = \frac{V_T^2}{R_L + R_T} = \frac{V_T^2}{2R_T} = \frac{10^2}{(2)(5)} = 10 \text{ W} \tag{5.7-5}$$

We found above that the power delivered to the load was $P_L = 5$ W. Thus the efficiency is

$$\eta = \frac{P_L}{P_T} \times 100 = \frac{5}{10} \times 100 = 50\% \tag{5.7-6}$$

In some communications systems, this low efficiency can be tolerated because the cost of the lost power is not important. For example, the load on an antenna is often matched in order to obtain the maximum amount of signal power.

On the other hand, power companies must have equipment large enough to produce the power used plus the power lost and must also pay the cost of producing this power. Thus it is to their advantage to keep the system efficiency as high as possible, and an efficiency of 50% would be completely unacceptable. Power systems must be designed so that the internal resistance of the source is very much lower than the relatively fixed load resistance in order to keep the power loss low and the load voltage relatively constant.

In the previous discussion we considered a fixed supply (V_T and R_T constant) and found that the load resistance should equal R_T in order for the load power to be a maximum. Often we are faced with criteria other than the achievement of maximum power. For example, in many electronic circuits the Thevenin circuit is fixed but we wish to achieve a maximum voltage across the load. Theoretically, this will be the case if $R_L = \infty$, that is, an open circuit. However, for various reasons, the load resistor must usually be finite.

To put this on a quantitative basis, consider the voltage divider relation applied to the circuit of Fig. 5.30. We have

$$A_v = \frac{V_L}{V_T} = \frac{R_L}{R_L + R_T} \tag{5.7-7}$$

We tabulate below values of A_v for a variety of values of R_L.

R_L	0	R_T	$2R_T$	$4R_T$	$6R_T$	$10R_T$	∞
A_v	0	0.5	0.67	0.8	0.86	0.91	1.0

From the table we see that as long as R_L is greater than $4R_T$, the output voltage will be 80% or more of the maximum possible value. Thus as a rule of thumb we can state that for good voltage gain, we should have

$$R_L > 4R_T \tag{5.7-8}$$

Typical figures will be found in the following examples.

EXAMPLE 5.7-1 Maximum Power from a Norton Circuit

In the circuit of Fig. 3.32a, find the value of R_L that will result in maximum power transfer and find the maximum load power.

Solution

We convert the Norton circuit to its Thevenin equivalent shown in Fig. 5.32b. Maximum power is transferred to the load when $R_L = R_T = 12\ \Omega$.

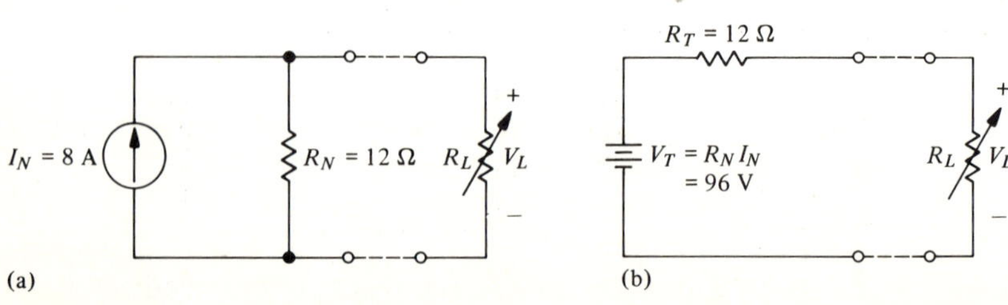

FIGURE 5.32
Example 5.7-1. (a) Norton circuit. (b) Thevenin circuit.

To find the load power, consider the Norton circuit. The load current is $I_L = (1/2)I_N - 4$ A. Then

$$P_L = R_L I_L^2 = (12\ \Omega)(4\ \text{A})^2 = 192\ \text{W}$$

This example shows that in order to apply the maximum power theorem to a Norton circuit we simply set the load resistance equal to the Norton resistance.

• • •

EXAMPLE 5.7-2 Maximum Power from an Arbitrary Circuit

In the circuit of Fig. 5.33a find R_L for maximum power and find the maximum power.

Solution

The easiest way to solve this problem is to convert the circuit to a Thevenin or Norton equivalent. For this circuit the Thevenin equivalent is easier to find. The Thevenin resistance is found by replacing the 72-V battery by a short circuit and calculating the resistance looking into terminals a–b

$$R_T = 1\ \Omega + (4\ \Omega \parallel 8\ \Omega) = 3.67\ \Omega$$

Thus the load resistance should be $R_L = 3.67\ \Omega$ for maximum power transfer.

The Thevenin voltage is the same as the voltage across the 4-Ω resistor with the load disconnected (why?). Using the voltage divider this is

$$V_T = \frac{4\ \Omega}{4\ \Omega + 8\ \Omega} \times 72\ \text{V} = 24\ \text{V}$$

The Thevenin equivalent circuit is shown in Fig. 5.32b. When $R_L = 3.67\ \Omega$ exactly 12 V appears across the load (why?) and

$$P_L = \frac{V_L^2}{R_L} = \frac{(12\ \text{V})^2}{3.67\ \Omega} = 39.2\ \text{W}$$

The student should observe that 39.2 W is also dissipated in the Thevenin resistance and that the Thevenin source supplies 2 × 39.2 W = 78.4 W. However,

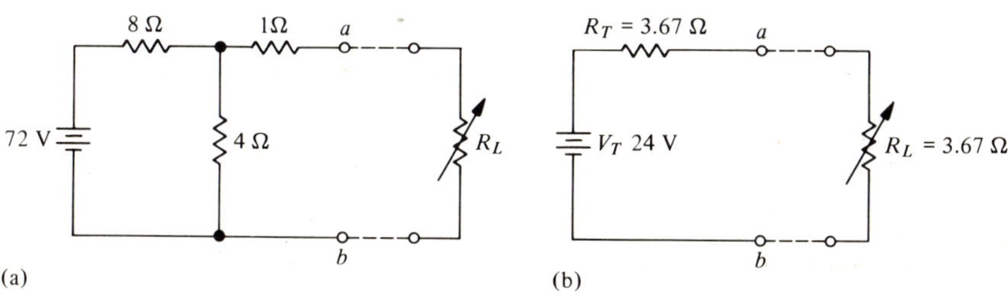

FIGURE 5.33
Example 5.7-2. (a) Original circuit. (b) Thevenin equivalent.

these figures do not necessarily correspond in any way to the power dissipated in the elements of the original circuit. The power in these elements is calculated in Prob. 5.38.

• • •

LEARNING EXERCISE FOR SEC. 5.7

1. In the circuit of Fig. 5.29 (p. 217) find
   a. The load resistance which will achieve maximum power transfer when connected between *a* and *b*
   b. The maximum power in the load under these conditions
   c. The minimum load resistance required across *a*–*b* for good voltage gain

Ans. 44.5; 0.85; 3.4

• • •

5.8 COMPUTER ANALYSIS OF NETWORKS—THE ECAP PROGRAM*

With computer services increasingly available, it is no longer necessary for the engineer or technologist to perform analysis of complex networks on paper. Computer programs are available to perform all kinds of complicated calculations. In this section, we describe one of the most popular of these programs, IBM's ECAP (Electronic Circuit Analysis Program). In practice, ECAP is used to solve complicated circuit problems; we will apply it to simpler problems in order to teach its use.

Briefly, when using ECAP, the technologist takes the schematic diagram of the circuit to be analyzed, and converts it to a standard ECAP equivalent circuit form. The information from the ECAP circuit is transferred to a coding sheet, using the ECAP language, then to punched cards, or directly to the computer via a typewriter terminal if the program is being used in a time-sharing system. When the computer receives the information from either the card deck or the typewriter, the ECAP program automatically formulates the network node equations, solves them, and prints or otherwise displays the results. The technologist does not have to go through the labor of setting up and solving equations; time can be spent more profitably interpreting and utilizing the results.

From the practical point of view, ECAP is even more useful because it can be used to determine changes in currents and voltages when circuit elements or signals vary. With this capability, the program can be used to find element values that will lead to a specified voltage or current and thus becomes an extremely powerful design tool. In this section we present a brief introduction to the language of ECAP and examples of its use in the

*This section may be omitted without loss of continuity if appropriate computer facilities are not available.

analysis of the types of networks considered in this chapter. It is important that the user learn the ECAP language thoroughly because communication with the computer must be exactly according to the ECAP rules or the computer will not recognize the commands.

The Simplified ECAP Standard Branch

The circuit *branch* is the basic unit used by ECAP. Each branch begins and ends on a node. The programmer specifies all of the circuit branches, and the program proceeds from this information to the solution.

A simplified form of the ECAP standard branch for dc analysis is shown in Fig. 5.34. This branch *must* contain a resistance and may, in addition, contain independent current and/or voltage sources as explained in detail below. The various features of the diagram are:

Branches: Each branch must contain a resistance R (or conductance G), which must be nonzero. If independent sources are present, they are designated by E and/or I as shown. Reference directions for these sources are discussed in Example 5.8-1. The circuit branches may be numbered in any order, but consecutive numbers beginning with 1 must be used. The branch number b is placed in a box, with an arrow on the box indicating the arbitrary reference direction from the initial node *d* to the final node e.

Nodes and node voltages: The network nodes (*d* and e in the figure) are numbered in any order using consecutive numbers beginning with 0. ECAP always assumes that node 0 is the reference (or ground) node. To distinguish node numbers from branch numbers, node numbers should be encircled. Node voltages in the figure are specified as NV*d* and NVe with respect to ground.

Element currents: The element current in branch b (called CAb by ECAP) is the current flowing through R from the initial node *d* to the final node e. Note that our original choice of direction determines which is the initial and which the final node.

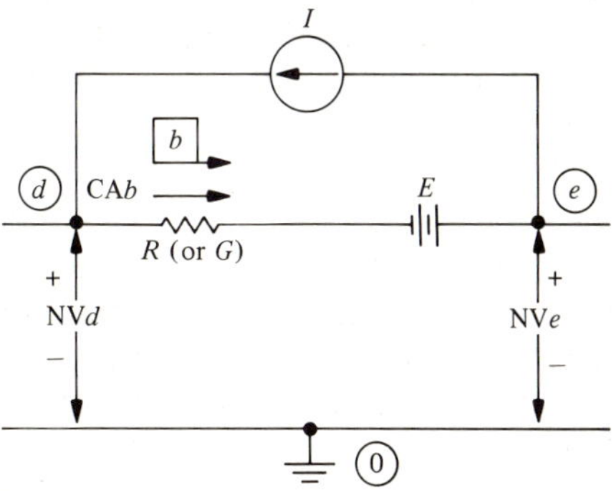

FIGURE 5.34
The simplified ECAP branch.

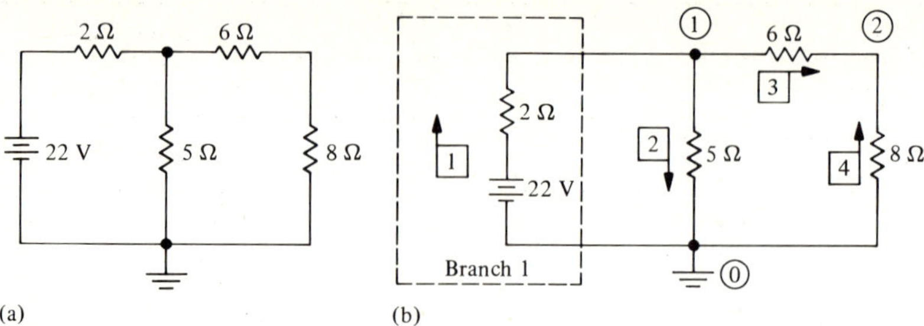

FIGURE 5.35
(a) Circuit for Example 5.8-1. (b) Circuit prepared for ECAP.

To prepare an ECAP circuit, branches and nodes are numbered and current directions are assumed. This information is transferred to a coding sheet using the ECAP language and then either to punched cards or a typewriter terminal. The procedure is illustrated in the following example.

EXAMPLE 5.8-1 Using ECAP with Punched Cards

Find all currents and voltages in the circuit of Fig. 5.35a using ECAP.

Solution

The ECAP circuit is shown in Fig. 5.35b. The circuit contains three nodes and four branches, one of which contains a voltage source. Note that all nodes and branches have been carefully numbered. The coding sheet is shown in Fig. 5.36 and an explanation of the numbered lines on the sheet follows.*

Line 1: Data required by the computer center to identify the user, the program (ECAP in this case), and so on. More than one line may be required.
Line 2: A command statement telling the ECAP program that a dc analysis is desired.
Line 3: The letter C specifies a comment. Any text following the C is printed as part of the output but has no other effect. Comments may be used anywhere in the program. A blank comment card or line is inserted after line 5 to provide a space in the printout.
Line 4: This line describes branch 1 to the program: Its initial node is 0, and its final node is 1. It contains a 2-Ω resistance and a 22-V voltage source. If the voltage source were reversed, we would write $E = -22$.

To determine the proper sign for a branch voltage source, pretend to short-circuit the branch. If the resulting current flow through the resistance is in the same direction as the preassigned reference arrow, the sign is positive; otherwise it is negative. For branch current sources (see Fig. 5.34) we pretend to disconnect the branch from the rest of the circuit. If the current from the source flows through the

*The column numbers on the coding sheet must be observed carefully if punched cards are used. Terminal input will be illustrated in the next example.

IBM FORTRAN Coding Form

PROGRAM | PUNCHING INSTRUCTIONS
PROGRAMMER | DATE

```
COMM. | STATEMENT NUMBER | CONT. | FORTRAN STATEMENT
1 →   AB$$   JØB758ECAP
2 →          DC ANALYSIS
3 →   C      EXAMPLE 5.8-1
4 →     B1   N(0,1), R=2, E=22
        B2   N(1,0), R=5
        B3   N(1,2), R=6
5 →     B4   N(0,2), R=8
      C
6 →          PRINT, NV, CA
7 →          EXECUTE
```

*A standard card form, IBM electro 888157, is available for punching statements from this form.

FIGURE 5.36
Coding sheet for Example 5.8-1. Each line represents one punched card.

resistance in the same direction as the preassigned reference arrow, it is positive; otherwise, it is negative.

Line 5: Specifies branch 4; initial node 0, final node 2, and resistance of 8 Ω.
Line 6: Tells ECAP that all node voltages (NV) and element currents (EC) or (CA) are to be printed as output.
Line 7: Command statement indicating that the program is to be executed.

Figure 5.37 shows the computer printout. The node voltages and element currents are listed in rows according to the numbering scheme used on the ECAP circuit. All numerical answers are in standard computer "floating point" form. The number following the E (sometimes D is used) indicates the power of 10 by which the number preceding the E is to be multiplied. Thus I_4 = EC 4 = −1.019 A, indicating that +1.019 A flows from node 2 to node 0.

• • •

```
ECAP        10:32          FRIDAY     I02



NV 1        NV 2        EC 1        EC 2        EC 3        EC 4
1.426E 01   8.148E 00   3.870E 00   2.852E 00   1.019E 00   -1.019E 00
```

FIGURE 5.37
Computer printout for Example 5.8-1.

EXAMPLE 5.8-2 Using ECAP with Terminal Input

Find all currents, voltages, and power dissipated in each branch in the circuit of Fig. 5.38a.

Solution

The ECAP circuit is shown in Fig. 5.38b and the typewriter terminal input in Fig. 5.39a. Note that at this installation line numbers must be used when the program is run, using a typewriter terminal for input.

In the program, the current in branch 1 is given with a negative sign in accordance with the procedure outlined in Example 5.8-1, and the PRINT statement calls for branch power (BP) in addition to voltages and currents. The final statement, RUN **ECAP, typed without a line number, calls the ECAP program. This may be somewhat different at other computer installations.

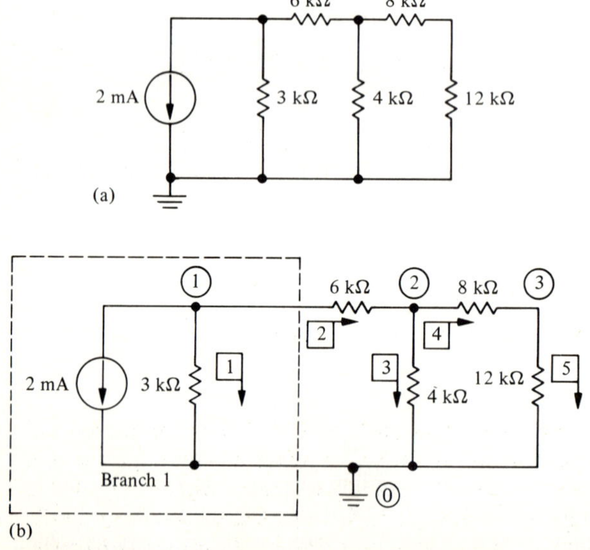

FIGURE 5.38
(a) Circuit for Example 5.8-2. (b) Circuit prepared for ECAP.

```
100   DC ANALYSIS
200C
300   B1 N(1,0),R=3E03,I=-2E-03
400   B2 N(1,2),R=6E03
500   B3 N(2,0),R=4E03
600   B4 N(2,3),R=8E03
700   B5 N(3,0),R=12E03
800C
900   PRINT,NV,CA,BP
1000  EXECUTE
1100  END

RUN **ECAP
```

(a)

```
ECAP        10:34                FRIDAY      102



 NODE VOLTAGES
  1- 3  -4.541E 00  -1.622E 00  -9.730E-01
 ELEMENT CURRENTS
  1- 5  -1.514E-03  -4.865E-04  -4.054E-04  -8.108E-05  -8.108E-05
 ELEMENT POWER LOSSES
  1- 5   6.872E-03   1.420E-03   6.574E-04   5.259E-05   7.889E-05


END OF RUN

PROCESSING     1 UNITS
```

(b)

FIGURE 5.39
(a) Typewriter terminal input for Example 5.8-2. (b) Computer printout.

The printout of results is shown in Fig. 5.39b. Verification of the results is left for the problems section.

• • •

EXAMPLE 5.8-3 A More Complicated Circuit

Use the ECAP Program to Find I_3 in the Circuit of Fig. 5.40a.

Solution

The ECAP circuit is shown in Fig. 5.40b. Note that the 10-V battery in the original circuit does not have a resistance in series with it and thus does not constitute a proper ECAP branch. In order to accommodate the battery we add a 0.001-Ω resistance in series with it. This is small enough so that it will not affect the solution but will satisfy the ECAP requirement that each branch have a nonzero resistance. The rest of the ECAP circuit follows the rules set down previously.

The terminal input is shown in Fig. 5.41a. The student should be convinced that the signs of the voltage and current sources are correct. The statement in line 1200

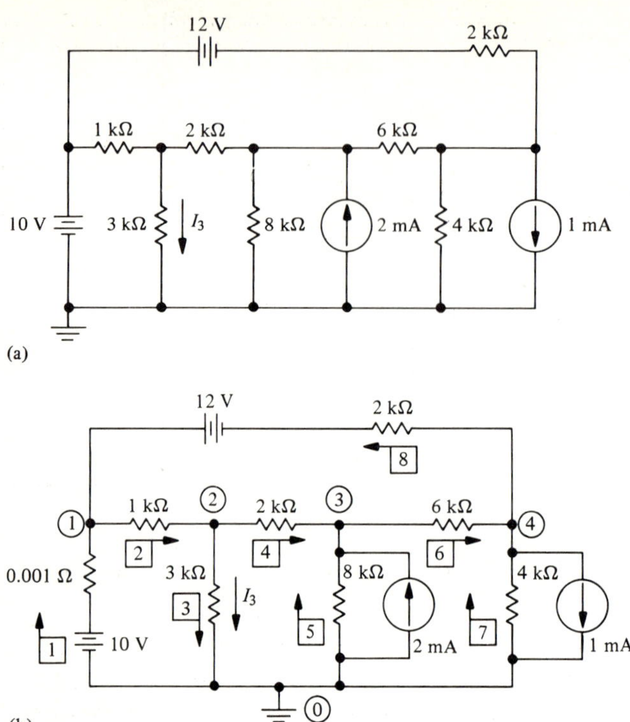

FIGURE 5.40
Example 5.8-3. (a) Original circuit. (b) ECAP circuit.

differs from previous examples in that it asks ECAP to print out only the value of element current 3, which is the desired answer. Otherwise, the program follows the previous examples.

Note that the original network has three independent nodes so that three node equations must be solved simultaneously for a solution. Additional branches would increase this number. With ECAP, each additional branch simply adds one line to the program.

The printout of the solution is shown in Fig. 5.41b. The output is indicated as EC3 (element current 3) and is seen to be 2.453 mA.

• • •

The MODIFY Routine

The MODIFY command allows us to repeat an immediately preceding DC ANALYSIS with modified parameter values. As we will see in the next example, it is necessary to add only the new parameter values. Values that remain the same need not be repeated.

The MODIFY routine can be used in a number of ways in both analysis and design. In analysis, it can be used to determine the effects of compo-

```
BKEX      10:35          03/26/80


100   DC ANALYSIS
200   C   EXAMPLE 5.8-3
300   B1 N(0,1), R=.001, E=10
400   B2 N(1,2), R=1E3
500   B3 N(2,0), R=3E3
600   B4 N(2,3), R=2E3
700   B5 N(0,3), R=8E3, I=-2E-3
800   B6 N(3,4), R=6E3
900   B7 N(0,4), R=4E3, I=1E-3
1000 B8 N(4,1), R=2E3, E=12
1100 C
1200 PRINT, CA3
1300 EXECUTE


RUN **ECAP
```

(a)

```
ECAP        10:40          03/26/80


EC  3

2.453E-03



END OF RUN
```

(b)

FIGURE 5.41
Example 5.8-3. (a) Terminal input. (b) Output.

nent tolerances on a particular current or voltage in the network. If a circuit has been designed and all component values are available, the MODIFY routine can be used to find the effect of changes in one or more circuit parameters on any or all of the currents and voltages in the network.

In design, one way that the MODIFY routine might be used occurs when all but a few components can readily be determined on paper. The undetermined components can then be found by using the MODIFY routine to scan appropriate ranges of these components until the desired response is obtained.

In using this routine a parameter is varied over a range by specifying the smallest value of the range, the largest value, and the total number of values required in the range. For example, the entry

100 B6 R = 1E3(9)10E3

tells the computer that the resistor in branch 6 starts at 1000 Ω and ends at 10 000 Ω. There are nine intervals in between, so that the resistance change per step is

$$\frac{10\,000 - 1000}{9} = 1000\ \Omega$$

Thus, the resistor starts at 1000 Ω and progresses in steps of 1000 Ω up to and including 10 000 Ω. When this is used in a MODIFY routine, a DC solution is produced for each value of the range.

The MODIFY routine is illustrated in the following example.

```
100  DC ANALYSIS
200  C  EXAMPLE 5.8-3
300  B1 N(0,1), R=.001, E=10
400  B2 N(1,2), R=1E3
500  B3 N(2,0), R=3E3
600  B4 N(2,3), R=2E3
700  B5 N(0,3), R=8E3, I=-2E-3
800  B6 N(3,4), R=6E3
900  B7 N(0,4), R=4E3, I=1E-3
1000 B8 N(4,1), R=2E3, E=12
1100 C
1200 PRINT, CA3
1300 EXECUTE
1310 MODIFY
1320 B3   R=1E3(4)9E3
1330 EXECUTE
```

(a)

```
RUN **ECAP


ECAP          10:57                03/26/80




 EC  3
2.453E-03


R     = 1.000E 03
 5.100E-03

R     = 3.000E 03
 2.453E-03

R     = 5.000E 03
 1.615E-03

R     = 7.000E 03
 1.204E-03

R     = 9.000E 03
 9.593E 03
```

(b)

FIGURE 5.42
Example 5.8-4. (a) MODIFY routine. (b) Output.

EXAMPLE 5.8-4 The MODIFY Routine

Use the MODIFY routine to find the variation in I_3 of the circuit of Fig. 5.40a if the 3-kΩ resistor in branch 3 varies from 1 to 9 kΩ in 2-kΩ steps.

Solution

Since the nominal solution of 3 kΩ was obtained in Example 5.8-3 we use the program in Fig. 5.41a for our nominal solution. The only additions required to take account of the parameter variation are the statements (1310, 1320, and 1330) shown in Fig. 5.42a. These statements are inserted after the EXECUTE statement of the original program as shown. (The procedure may be somewhat different on some computers.) When they are added to the original program a value of I_3 is calculated for each of the values of the modified resistor. The output is shown in Fig. 5.42b. Note that the nominal solution is given first and then repeated during the course of the parameter iteration because the 3-kΩ parameter value is included in the iteration range.

• • •

SUMMARY

In this chapter we have considered a number of the techniques used by technologists and engineers to analyze complex electric circuits. The important points are summarized below:

Section

5.1 1. The linearity principle states that in a linear circuit, if all independent sources in a network are multiplied by a constant, then all currents and voltages in the network are multiplied by the same constant.

2. The superposition principle states that a circuit response can be found by adding the response due to each source acting alone.

3. For linear networks, the response coefficients can be found experimentally by making as many measurements as there are coefficients.

5.2 4. In the loop-current method, a KVL equation is written for each window using Ohm's law for each resistor. The resulting set of simultaneous equations is solved for the currents.

5.3 5. In the node-voltage method, a KCL equation is written at each independent node using Ohm's law for each resistor. The resulting set of equations is solved for the voltages.

5.4 6. The terminal characteristics of ideal sources, practical sources, and resistances are straight lines.

7. Circuits can be solved graphically by plotting terminal characteristics of sections of the circuits on appropriate axes. The intersections of the characteristics give the operating point of the circuit.

8. The regulation of a source is a measure of the decrease in terminal voltage with increased load current. Smaller values of regulation mean less decrease in output from no load to full load.

5.5 9. For linear circuits, equivalent circuits can be found from two measured points.

5.6 10. The Thevenin equivalent circuit consists of a voltage source V_T in series with a resistance R_T.

11. The voltage V_T is the open-circuit voltage at the terminals of interest before the load is connected.

12. The resistance R_T is measured looking back into the terminals with all sources set to zero.

13. The Norton equivalent circuit consists of a current source I_N in parallel with a resistance R_N.

14. The current I_N is the current through a short circuit across the terminals of interest.

15. The resistance R_N is the same as the Thevenin resistance R_T.

16. The Norton and Thevenin sources are related by the equations $V_T = R_N I_N$ and $R_T = R_N$.

5.7 17. Maximum power is transferred from a Thevenin circuit to a load when the load resistance is equal to the Thevenin resistance.

18. Often maximum power transfer is not desirable. For example, when maximum voltage is required at the load, the load should be greater than $4R_T$.

5.8 19. The ECAP standard branch must contain a resistance, and may contain current and/or voltage sources.

20. In the ECAP circuit, all nodes are numbered and arbitrary current directions are specified. These are used to describe the circuit to the program.
21. The MODIFY routine is used to repeat solutions with modified parameter values.

QUESTIONS FOR REVIEW

Sec. 5.1

1. State the linearity principle.
2. State the superposition principle and describe how it is used to solve a circuit with more than one source.
3. Describe how the superposition principle can be used to characterize a circuit in a black box.

Sec. 5.2

4. Name the steps to be used in solving a circuit using mesh analysis.
5. Describe the inspection method of writing loop equations and give a physical explanation of the coefficients.

Sec. 5.3

6. Describe the steps to be used in solving a network using the node-voltage method.
7. What is the physical interpretation of the coefficients of the voltages in the node equations?
8. What is the usual criterion for selecting the node analysis method or the mesh analysis method?

Sec. 5.4

9. Sketch the graph of the terminal characteristic of an ideal 10-V voltage source.
10. Sketch the terminal characteristic of an ideal 2-A current source.
11. In Questions 9 and 10 a 2-Ω resistor is connected to each of the sources in turn. Show how the resulting circuits may be solved graphically.
12. What is the equation for the volt-amp characteristic of a practical source that has internal resistance R_i?
13. Describe the graphical method of solution for the case when a load resistance is connected to a practical source.
14. Why does the terminal voltage of a practical source change as the load current increases?
15. What is the meaning of *voltage regulation?*
16. What is the *no-load voltage?*
17. What is the *full-load voltage?*
18. What is the voltage regulation of an ideal source?

Sec. 5.5

19. What are the implications of the term *linear system?*
20. How many measurements are required to completely determine the terminal characteristic of a linear circuit?
21. What is meant by the term *equivalent circuit?*

22. Describe the Thevenin equivalent circuit.
23. What is the physical significance of the Thevenin voltage? The Thevenin resistance?
24. Describe several ways that can be used to find the Thevenin voltage.
25. Describe several ways that can be used to find the Thevenin resistance.
26. Describe the Norton equivalent circuit.
27. Define the Norton circuit parameters, R_N, and I_N.
28. Describe several ways to find the Norton resistance and the Norton current.

Sec. 5.7

29. When is maximum power transferred from a circuit to a load?
30. What load resistance do we use if we wish to achieve maximum voltage across the load when the source is a Thevenin circuit?
31. When maximum power transfer is achieved, what is the efficiency?
32. Indicate situations where maximum power transfer is not desirable.

Sec. 5.8

33. Describe the simplified ECAP standard branch.
34. In writing an ECAP program how do we determine the proper sign for a branch voltage source?
35. Describe the MODIFY routine.

PROBLEMS

Sec. 5.1

1. In the circuit of Fig. 5.43, find I_1 using superposition.
2. In the circuit of Fig. 5.43, find V_2 using superposition.
3. In the circuit of Fig. 5.44, find I_2 using superposition.

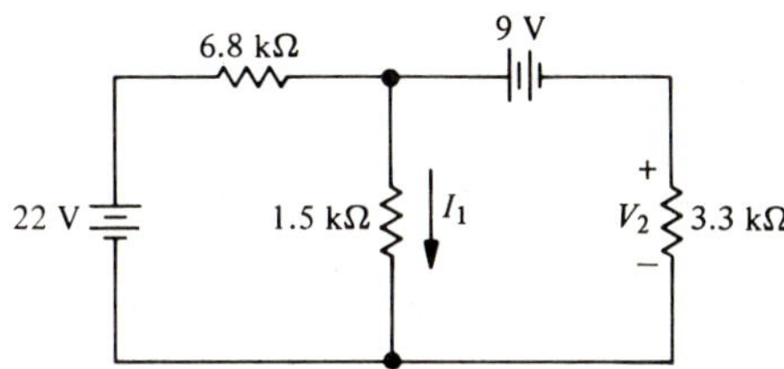

FIGURE 5.43

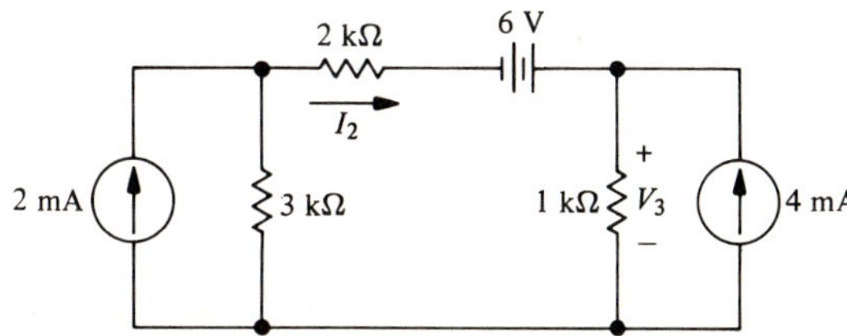

FIGURE 5.44

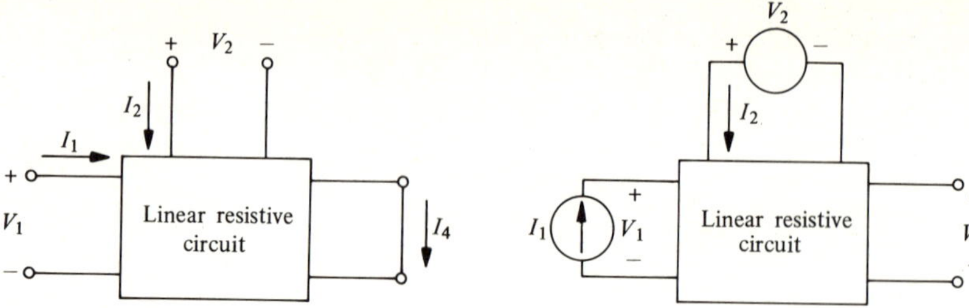

FIGURE 5.45

FIGURE 5.46

4. Use superposition to find V_3 in the circuit of Fig. 5.44.
5. In the circuit of Fig. 5.45 the following data are measured:
   When $I_1 = 3$ mA and $V_2 = 6$ V, $I_4 = 2$ mA.
   When $I_1 = 2$ mA and $V_2 = 8$ V, $I_4 = 7$ mA.
   Find the value of I_4 when $I_1 = 4$ mA and $V_2 = 2$ V.
6. In the circuit of Fig. 5.46 the following data are measured:
   When $I_1 = 6$ mA and $V_2 = 80$ V, $V_4 = 16$ V.
   When $I_1 = 0$ and $V_2 = 62$ V, $V_4 = 80$ V.
   Find V_4 when $I_1 = 12$ mA and $V_2 = 36$ V.

Sec. 5.2

7. Use the loop-current method to find all currents and voltages in the circuit of Fig. 5.47.

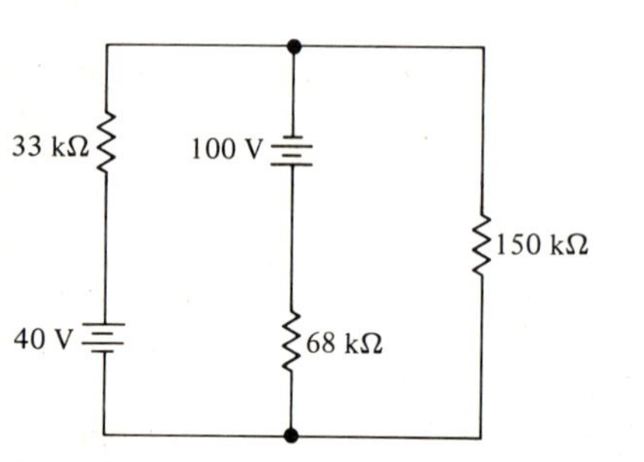

FIGURE 5.47

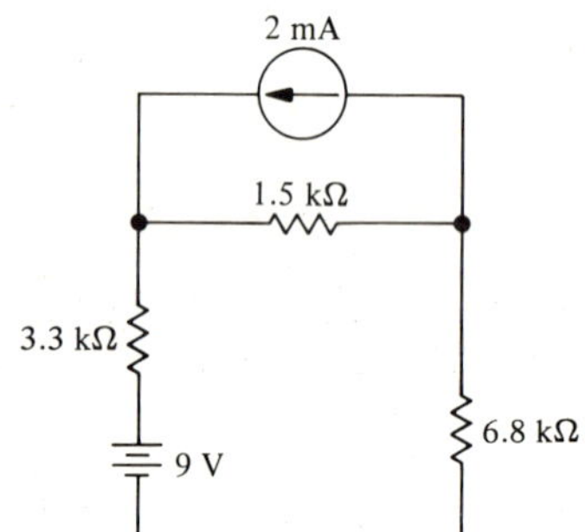

FIGURE 5.48

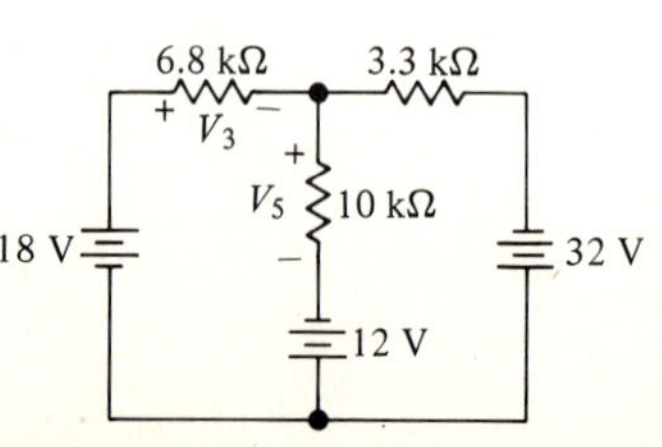

FIGURE 5.49

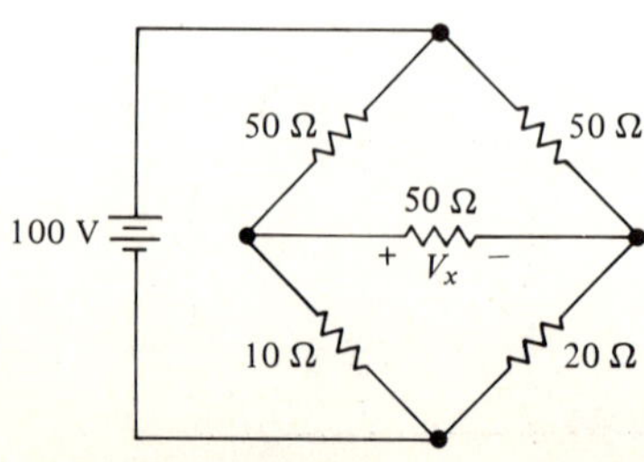

FIGURE 5.50

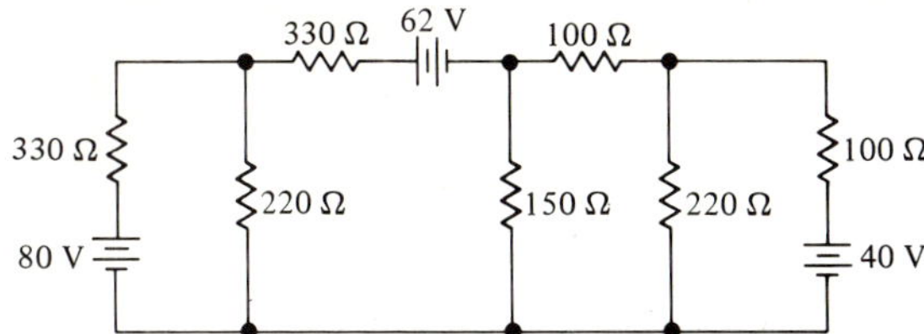

FIGURE 5.51

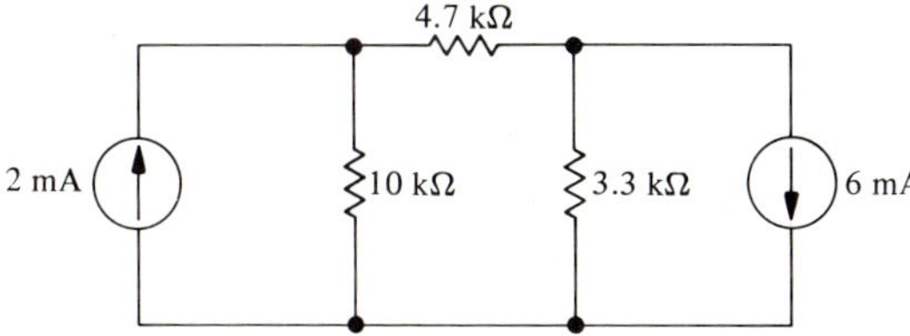

FIGURE 5.52

8. Find all currents and voltages in the circuit of Fig. 5.48.
9. Find V_3 and V_5 in the circuit of Fig. 5.49.
10. Find V_x in the bridge circuit of Fig. 5.50.
11. In the circuit of Fig. 5.51 write, but do not solve, a set of loop equations using the inspection method.
12. Check the currents found in Example 5.2-3 by finding the individual resistor voltage drops and using KVL around each loop.

Sec. 5.3

13. Use the node-voltage method to find all currents and voltages in the circuit of Fig. 5.52.
14. Repeat Prob. 13 for the circuit of Fig. 5.53.
15. Repeat Prob. 13 for the circuit of Fig. 5.50.
16. Find I_1 and I_3 in the circuit of Fig. 5.54.
17. In the circuit of Fig. 5.55 write, but do not solve, a set of node equations using the inspection method.
18. Check the results of Example 5.3-3 as many ways as you can.
19. In the circuits of Figs. 5.55, 5.54, 5.51, 5.50, 5.49, and 5.48 find the number of loop equations required for a solution and compare with the number of node equations required.

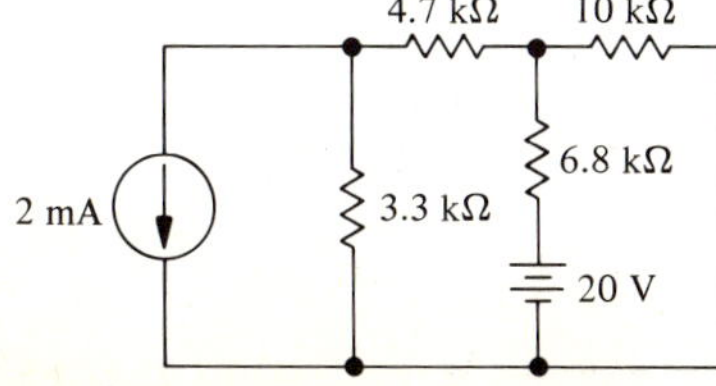

FIGURE 5.53

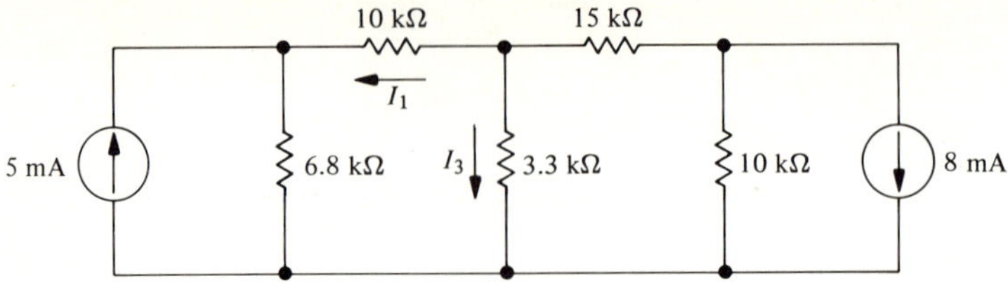

FIGURE 5.54

Sec. 5.4

20. The terminal characteristic of a particular battery is a straight line that intercepts the voltage axis at 6.4 V and the current axis at 180 A. Find V_T and R_T.
21. The battery of Prob. 20 can safely carry a full-load current of 45 A. Find the voltage regulation.
22. Three volt-ampere measurements are made at the terminals of a network. The measured values are 5 V, 1 mA; 11 V, 3 mA; 20 V, 5 mA. Is it possible to find an equivalent linear circuit for the given network? If yes, find it; otherwise explain why you can't.
23. Measurements on a laboratory power supply yield the following values:
    a. Open-circuit voltage = 16 V
    b. $V = 15.3$ V when $I = 8$ A

    Find R_T, the internal resistance of the supply.
24. Measurements on an automobile battery yield the following data:
    a. $V = 12.7$ V when $I = 1$ A
    b. $V = 12.3$ V when $I = 30$ A

    Find V_T, R_T, and the voltage regulation if the full-load current is 25 A.
25. A dc generator has an open-circuit voltage $V_T = 132$ V and an internal resistance $R_T = 0.32\ \Omega$. Find graphically
    a. The load voltage and current for load resistances of 0.5, 1, and 2 Ω
    b. The load voltage for currents of 380 and 50 A
    c. The load current when the load voltage is 125 V
    d. The rated full-load current is 100 A. Find the voltage regulation.

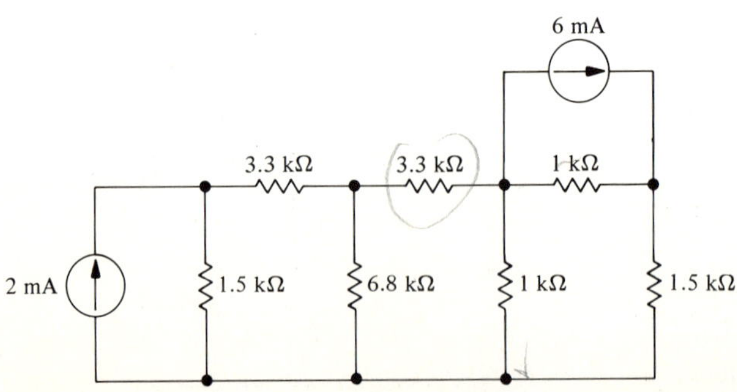

FIGURE 5.55

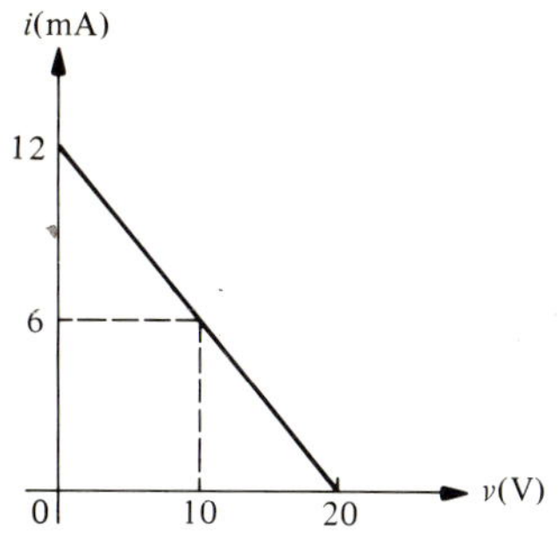

FIGURE 5.56

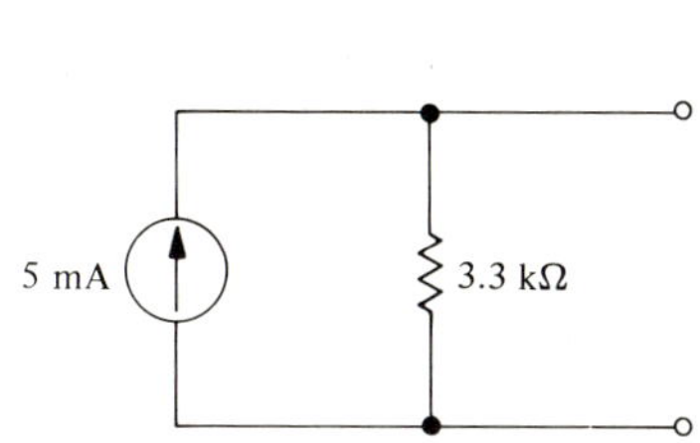

FIGURE 5.57

Sec. 5.6

26. Find the Thevenin equivalent for the network whose terminal characteristic is shown in Fig. 5.56.
27. Find the Thevenin equivalent for the network shown in Fig. 5.57.
28. Find the Thevenin equivalent for the network shown in Fig. 5.58.
29. Find the Thevenin equivalent for the circuit shown in Fig. 5.59.
30. Find the Thevenin and Norton equivalents for the circuit shown in Fig. 5.60.
31. Find the Thevenin and Norton equivalents for the circuit shown in Fig. 5.61.
32. Find the power delivered to the 5-Ω resistance in the circuit shown in Fig. 5.62. Hint: Replace everything but the 5-Ω resistor by a Thevenin circuit.
33. Find the voltage V_0 for the circuit shown in Fig. 5.63. Make use of Thevenin's theorem.
34. Find the Thevenin and Norton equivalents for the circuit of Fig. 5.64.
35. Find the Thevenin equivalent for the network shown in Fig. 5.65. Hint: Replace

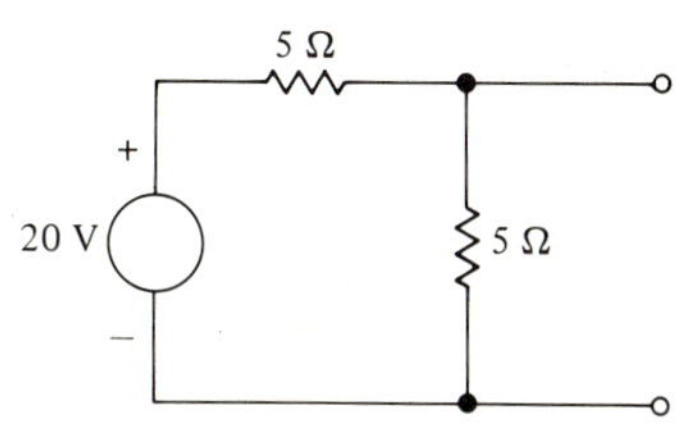

FIGURE 5.58

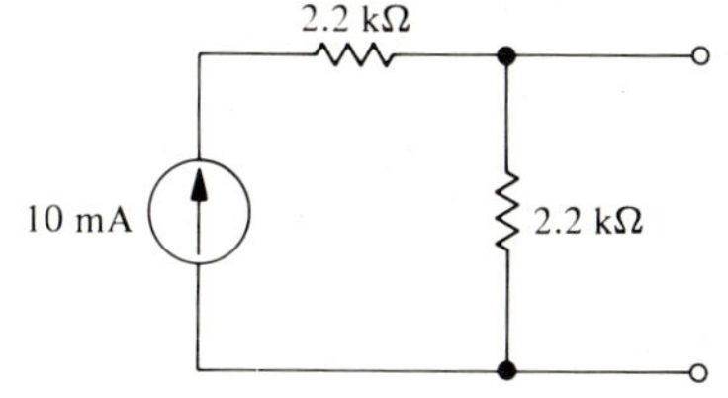

FIGURE 5.59

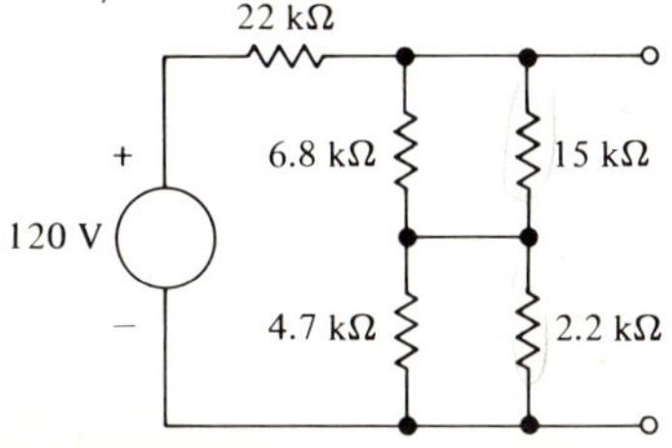

FIGURE 5.60

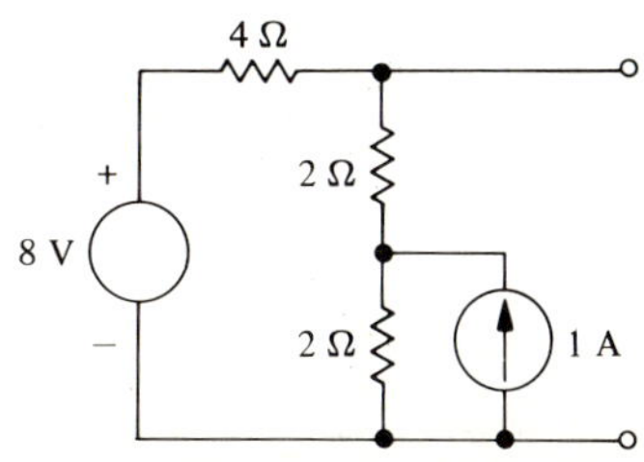

FIGURE 5.61

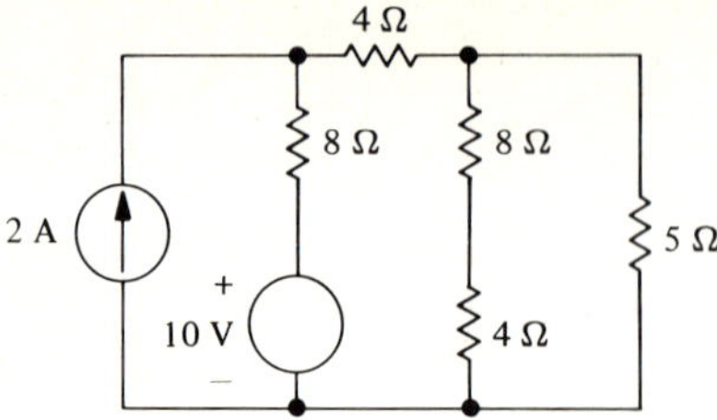

FIGURE 5.62

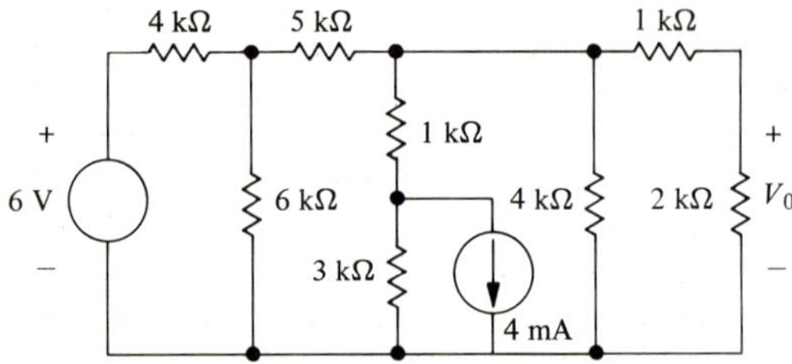

FIGURE 5.63

the series combination of sources by a single equivalent source. This may require some careful thought.

36. A transistor interstage network is shown in Fig. 5.66. The voltage V_B is to be $+2$ V and $R_1 \parallel R_2$ is to be 10 kΩ. Find R_1 and R_2 if (a) $I_B = 0$ and (b) $I_B = 225$ μA. In the diagram the -9V and $+16$V designations are voltages with respect to ground.

Sec. 5.7

37. In the circuit of Fig. 5.30, $V_T = 120$ V and $R_T = 20$ Ω. Find and plot V_L, I_L, P_L, and efficiency η vs R_L for values of R_L from 0.2 to 2000 Ω. Use a logarithmic scale for the *x*-axis (R_L) and a linear scale for the *y*-axis (V_L, I_L, P_L, and η).
38. In the circuit of Fig. 5.33a calculate the power dissipated in each of the resistors and the power delivered by the 72-V battery when the load $R_L = 3.67$ Ω for maximum power transfer. Compare with the power dissipated in R_T and the power delivered by the Thevenin source as found in Example 5.7-2.
39. An audio amplifier has an internal resistance of 8 Ω. The two-wire line to the

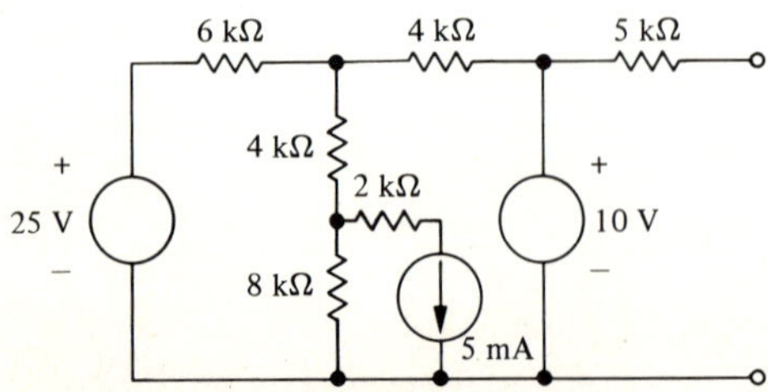

FIGURE 5.64

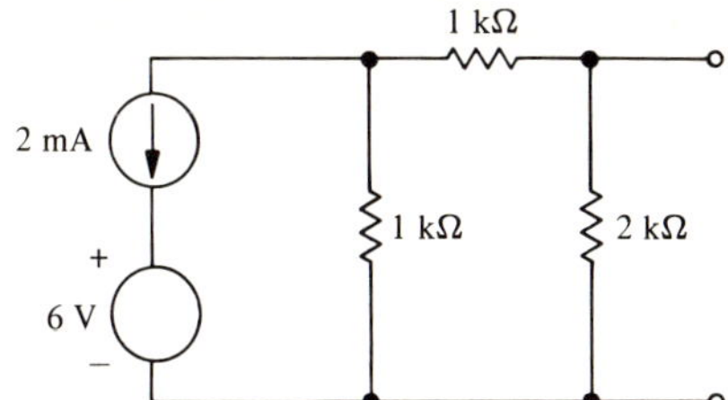

FIGURE 5.65

loudspeaker has a resistance of 1.5 Ω per wire. What should the loudspeaker resistance be for maximum power transfer?

40. A microphone has an internal resistance of 2 kΩ and its open-circuit voltage is 200 mV. It is connected to an amplifier which has an input resistance of 6.8 kΩ.
    a. How much voltage appears at the amplifier input?
    b. What is the maximum power output possible from the microphone?
41. An emergency generator has an open-circuit voltage of 125 V and an internal resistance of 0.35 Ω. What load resistance would dissipate 2.4 kW when connected to this generator? What is the maximum generator efficiency with this load?
42. An automobile battery has an open-circuit voltage of 13.2 V and an internal resistance of 0.03 Ω.
    a. Find the load voltage when a 1.2-Ω load is connected.
    b. Find the power dissipated in a 0.75-Ω load.
    c. A 0.22-Ω load is connected. Find the efficiency.
    d. What is the maximum power that this battery can deliver?
    e. What is the minimum load resistance for which the load voltage will be greater than 12.9 V?

Sec. 5.8

43. Write an ECAP program to find all currents and voltages in the circuit of Fig. 5.43. Run the program if you have a computer available.

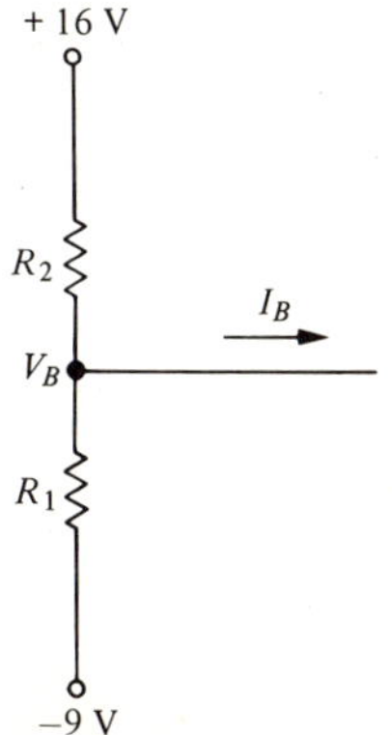

FIGURE 5.66

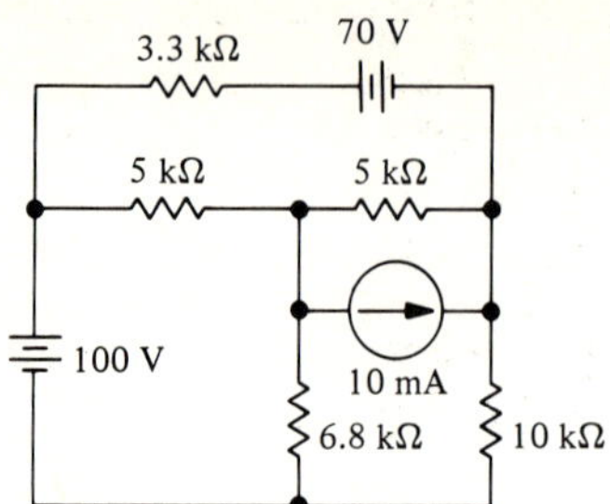

FIGURE 5.67

44. Repeat Prob. 43 for the circuit of Fig. 5.44.
45. Repeat Prob. 43 for the circuit of Fig. 5.50.
46. Repeat Prob. 43 for the circuit of Fig. 5.67.
47. In the circuit of Fig. 5.53, use ECAP to find the voltage across the 10-kΩ resistor. The 4.7-kΩ resistor may vary from 1 to 10 kΩ. Use the MODIFY routine to find the effect of this variation on the voltage across the 10-kΩ resistor.
48. In the circuit of Fig. 5.66, $R_1 = 6.8$ kΩ and $R_2 = 10$ kΩ. Use the MODIFY routine to find the effect on V_B of a range of values of I_B from 0 to 500 μA. Hint: In order to have a valid ECAP representation, I_B will have to be represented as a current source. The parallel resistance required for a valid ECAP branch can then have a very high value so that it will not affect the solution.

6
DC Measurements and Instruments

OBJECTIVES

Upon completion of this chapter, the student should be able to

Section

6.1 1. Describe the operation of the D'Arsonval movement.
6.2 2. Calculate the error introduced when a meter of known characteristics is used to measure current.
3. Calculate meter shunt resistance for a given movement.
6.3 4. Calculate the multiplier resistance required to convert a given movement to a voltmeter.
5. Define the current sensitivity of a D'Arsonval movement.
6. Calculate the error introduced when a meter of known characteristics is used to measure voltage.
6.4 7. Describe the operation of series and shunt ohmmeters.
8. Design a series ohmmeter for a given meter movement.
9. Design a shunt ohmmeter for a given meter movement.
6.5 10. Describe the operation of the Wheatstone bridge.
11. Calculate the unknown resistance from the parameters of a balanced bridge.
12. Calculate the range capability of a Wheatstone bridge.
6.6 13. Describe the operation of the dynamometer wattmeter.

INTRODUCTION

When working with actual electric circuits, we usually use indicating instruments to tell us what the voltages and currents are throughout the circuit. For dc measurements, such instruments come in two types. The first is classified as *analog*, where the output indication is continuously variable.

Often the position of a pointer gives the desired output indication. Sometimes a pen or stylus gives a permanent record of the variable being measured.

The second type of instrument is *digital* and is characterized by an output display that gives the measured value in decimal numeric values. Digital instruments are becoming more and more popular due to the availability and economy of integrated circuits. In general, for routine dc measurements, digital meters are much more accurate and reliable than their analog counterparts. However, the analog meters to be discussed in this chapter are often much less expensive than equivalent digital meters and so we can expect them to be around for some time to come.

In this chapter we consider the basic analog meter and its digital counterpart. Wherever possible we will develop the theory necessary to assess the effect of the connection of the instrument on the circuit variable being measured. This effect is usually called *loading* and its calculation will provide us with a practical application of the circuit analysis methods we have learned up to this point. Sufficient theory will be provided so that the student will be able to perform laboratory experiments involving these instruments.

6.1 THE D'ARSONVAL METER MOVEMENT

This movement is basically a dc current-sensitive electromagnetic device, and is used in both ammeters and voltmeters. A sketch of its principal parts is shown in Fig. 6.1. Current enters through the upper spring, goes through the moving coil, and exits through the lower spring. This current creates a magnetic field which interacts with the field produced by the permanent magnet. The result is that a torque is produced that tends to rotate the

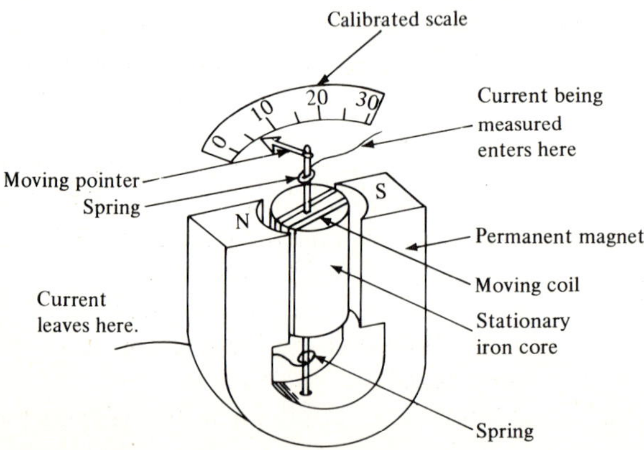

FIGURE 6.1
The D'Arsonval meter movement.

moving coil. This torque is opposed by a counter-torque provided by the springs. The instrument is designed so that when $I = 0$ the pointer will point to zero on the scale. The springs are adjusted so that the net torque produced when the current is at its maximum desired value will move the pointer so that it points to the desired full-scale value on the scale. A torque that moves the pointer upscale results only when current flows in the indicated direction. It can be shown that the pointer deflection is directly proportional to the current, so that the scale is linear.

D'Arsonval movements are manufactured to deflect to full scale for currents from 20 μA to 1 mA. Many of them have a voltage drop of 50 mV when full-scale current is flowing. The actual resistance of the moving coil is usually an odd amount. However, in order to achieve a 50-mV voltage drop with the desired full-scale current, the total movement resistance must lie between 50 and 2500 Ω. Thus a *calibrating resistor* constructed from precision resistance wire is added in series with the moving coil. If, for example, the full-scale current is 1 mA then the coil and calibrating resistances must add up to 50 mV/1 mA = 50 Ω. If the coil resistance is 19 Ω then the calibrating resistance must be 50 − 19 = 31 Ω.

The current required for full-scale deflection is called the *current sensitivity*. Along with the resistance it is used to specify the characteristics of the movement. It is important that the technologist understand how to determine the effect of connecting such a meter into a circuit. This is considered in Sec. 6.2.

6.2 AMMETERS

To use the D'Arsonval movement for dc current measurement, the meter is connected directly into the wire through which the desired current is flowing, that is, in *series*. All movements have a + sign on the terminal at which the current must enter. *Connecting the meter backward or in parallel instead of in series may result in physical damage* because the pointer will be forced against the mechanical stop usually provided to prevent excess motion of the coil.

It is convenient to model a real ammeter by a circuit consisting of an *ideal* ammeter having zero resistance in series with a resistance R_m which is equal to the resistance of the real meter. If we know the characteristics of the meter, we can predict its loading effect in a specific circuit.

Figure 6.2 shows a circuit in which an ammeter has been connected to measure the current in load resistor R_L. Note that the meter is connected in *series* with R_L, with the + sign where the current enters.

When full-scale current flows there will be a voltage drop across the meter

$$V_m = R_m I_m \tag{6.2-1}$$

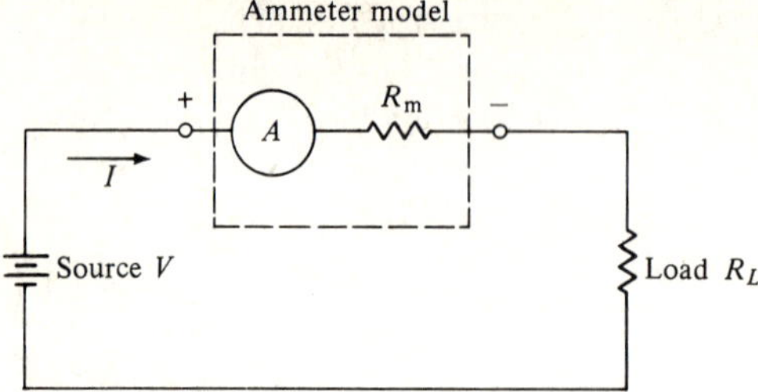

FIGURE 6.2
Ammeter used to measure current.

where

V_m = voltage drop at full scale, in volts
I_m = full-scale current (current sensitivity), in milliamperes
R_m = meter resistance, in kilohms

For most ammeters in most circuits, this voltage is small, and causes little error. However, there are situations where significant error is introduced, as in the following example.

EXAMPLE 6.2-1 Error Due to Ammeter

A 1-mA 200-Ω movement is to be used to measure the current in the circuit of Fig. 6.2 where we have a 1-V source feeding a 1-kΩ load.

a. Find the error introduced due to the insertion of the meter.
b. Find the error introduced if the source is increased to 10 V and the load to 10 kΩ.

Solution

The percent error is defined by the equation

$$\text{Percent error} = \frac{I_{\text{meas}} - I_{\text{true}}}{I_{\text{true}}} \times 100$$

where

I_{meas} = meter reading (measured value), in milliamperes
I_{true} = actual value (with meter not connected), in milliamperes

The numerator, $I_{\text{meas}} - I_{\text{true}}$, represents the difference between the measured value and the true value, and is called the *absolute error* or *absolute deviation.* If the presence of the meter did not alter the current being measured, this quantity would be zero.

Using Ohm's law, we find with the meter

$$I_{meas} = \frac{V}{R_m + R_L} \tag{6.2-2}$$

and without the meter

$$I_{true} = \frac{V}{R_L} \tag{6.2-3}$$

Substituting these equations into Eq. (6.2-1) we obtain an expression for the error in terms of the circuit parameters.

$$\text{Percent error} = \frac{I_{meas} - I_{true}}{I_{true}} \times 100$$

$$= \frac{V/(R_m + R_L) - V/R_L}{V/R_L} \times 100 \tag{6.2-4}$$

$$\text{Percent error} = \frac{-R_m}{R_m + R_L} \times 100$$

Before we substitute values into this equation we note that the percent error will always be negative, since $R_m/(R_m + R_L)$ will always be between 0 and +1. A little thought will show that this is quite reasonable, since a negative error indicates a *decrease* in the current in the series circuit. The decrease occurs because the presence of the meter has increased the total series resistance. Finally, substituting into Eq. (6.2-4) with all resistance values in kilohms:

a. $R_L = 1\ \text{k}\Omega$ $\quad \text{Percent error} = \frac{-0.2}{0.2 + 1} \times 100 = -16.7\%$

b. $R_L = 10\ \text{k}\Omega$ $\quad \text{Percent error} = \frac{-0.2}{0.2 + 10} \times 100 = -1.96\%$

For the higher resistance circuit the error is almost negligible. Thus, the criterion for small errors due to ammeter insertion is that the circuit resistance should be much larger than the meter resistance.

• • •

Meter Shunts

The maximum current a D'Arsonval movement can carry is equal to its current sensitivity I_m. In order to measure higher currents, we can connect a *shunt* (*parallel* resistance) across the meter so that only a part of the current will flow through the meter while the additional current flows through the shunt. This additional resistance is shown in Fig. 6.3a. The value of R_{sh} is calculated using full-scale values as follows:

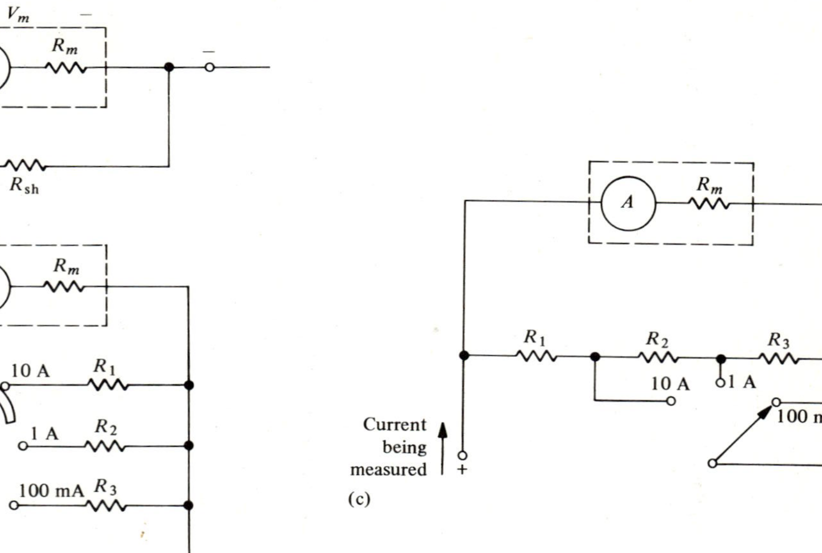

FIGURE 6.3
Ammeter (a) Shunt arrangement to increase range. (b) Multirange ammeter. (c) Meter with Ayrton shunt.

The current through the shunt I_s will be, using KCL

$$I_s = I_{fs} - I_m \tag{6.2-5}$$

where

I_{fs} = desired full-scale current
I_m = current sensitivity of meter movement

Since R_{sh} and R_m are in parallel, the voltage across each must be the same, that is,

$$R_m I_m = R_{sh} I_s \tag{6.2-6}$$

Usually, I_{fs}, I_m, and R_m are known and we have to find R_{sh}. We would like to solve these equations so that R_{sh} is given in terms of the known parameters. To do this we solve Eq. (6.2-6) for R_{sh}:

$$R_{sh} = \frac{R_m I_m}{I_s}$$

Next substitute for I_s from Eq. (6.2-5) to get

$$\boxed{R_{sh} = \frac{R_m I_m}{I_{fs} - I_m}} \tag{6.2-7}$$

which is the desired relation. Typical values are given in the following example.

EXAMPLE 6.2-2 Finding Required Shunt Resistance

It is desired to use a 1-mA 50-Ω D'Arsonval movement to measure 1 A full scale. Find the required shunt resistance in the circuit of Fig. 6.3a.

Solution

One way to find the required shunt resistance is to substitute the given values into Eq. (6.2-7). However, it is more instructive to arrive at a solution by applying basic principles so that we do not have to depend on memorized formulas. We will use this method in this example. Our overall plan is to take advantage of the fact that the voltage is the same across the meter and across the shunt because they are in parallel. The steps in the solution are as follows:

Step 1: The voltage across the meter at full scale is

$$V_m = R_m I_m = (50\ \Omega)(0.001\ \text{A}) = 0.05\ \text{V}$$

Step 2: If, at full scale, 1 A flows into the + terminal and 1 mA flows through the meter, then 999 mA must flow through the shunt.
Step 3: Since the voltage across the shunt is the same 0.05 V that appears across the meter, the shunt resistance must be

$$R_{sh} = \frac{V_m}{I_s} = \frac{0.05\ \text{V}}{0.999\ \text{A}} = 0.050\ \Omega$$

The student should observe that this solution did not require us to memorize the general formula given in Eq. (6.2-7). We used nothing more complicated than Ohm's law and KCL.

• • •

In practice, it is desirable to have a multistage ammeter, so that one can select from several different ranges. Various shunt arrangements are possible; one with separate shunts selected by a switch is shown in Fig. 6.3b. The switch must be of the *make-before-break** type, in order to prevent the possibility of excessive current through the meter movement during range switching. Each range might have its own scale, or one scale might serve with separate multiplying factors for each range. Each of the shunt resistances is determined exactly as R_2 for the 1-A range in Example 6.2-2. (Values for R_1 and R_3 are calculated in Prob. 6.3.) Another type of multirange ammeter is shown in Fig. 6.3c. In this configuration a series connection called an Ayrton shunt is used. As the meter is switched to lower current ranges, more of the shunt resistance is inserted by the switch. Note that in this circuit the meter is never without a shunt.

When using multirange ammeters, they should be set initially on their highest current range in order to avoid damage due to excessive current. A lower range can then be selected to give a suitable deflection. Calculation of the values in an Ayrton shunt are given in the following example.

EXAMPLE 6.2-3 Designing an Ayrton Shunt

The multirange ammeter of Fig. 6.3c is to be constructed using a 1-mA 50-Ω movement. Find the required values of R_1, R_2, and R_3.

Solution

The meter and the shunt are in parallel so the voltage across each is the same. The procedure here is to equate these voltage drops taking into account the fact that the resistance of the meter branch changes as we move the switch from one position to another.

Consider first that the switch is in the 100-mA position. Then

*The name arises from the fact that, for example, the 1-A terminal is connected to the moving contact *before* the 10-A terminal is disconnected. This avoids the possibility of the meter being without a shunt.

$$R_m I_m = R_{sh} I_s \tag{6.2-8}$$

where

$$R_{sh} = R_1 + R_2 + R_3 \tag{6.2-9}$$

In this equation we have $R_m = 50\ \Omega$, $I_m = 1$ mA, and $I_s = 100$ mA $- 1$ mA $= 99$ mA.

Substituting these data we have

$$(1\text{ mA})(50\ \Omega) = (99\text{ mA})(R_1 + R_2 + R_3)$$

$$R_1 + R_2 + R_3 = \frac{50\text{ mV}}{99\text{ mA}} = 0.505\ \Omega \tag{6.2-10}$$

When the switch is on the 1-A scale, the *effective* shunt resistance is $R_1 + R_2$ and the *effective* meter resistance is $R_m + R_3$. Equating voltage drops across the parallel branches, we get

$$I_m(R_m + R_3) = I_s(R_1 + R_2) \tag{6.2-11}$$

where $I_m = 1$ mA and $I_s = 1$ A $- 1$ mA $= 999$ mA.

This equation contains three unknowns, R_1, R_2, and R_3. In order to solve it we make use of Eq. (6.2-10) to write

$$R_1 + R_2 = 0.505 - R_3 \tag{6.2-12}$$

With this substitution Eq. (6.2-11) becomes

$$I_m(R_m + R_3) = I_s(0.505 - R_3) \tag{6.2-13}$$

in which the only unkown is R_3. Substituting the known values

$$(1\text{ mA})(50\ \Omega + R_3) = (999\text{ mA})(0.505\ \Omega - R_3)$$

When this is solved for R_3 we find

$$R_3 = 0.454\ \Omega \tag{6.2-14}$$

When the switch is on the 10-A scale, the shunt resistance is R_1 and the effective meter resistance is $R_m + R_2 + R_3$. Again equating voltage drops, we get

$$I_m(R_m + R_2 + R_3) = I_s R_1 \tag{6.2-15}$$

where $I_m = 1$ mA, $I_s = 10\,000$ mA $- 1$ mA $= 9999$ mA, $R_m = 50\ \Omega$, and $R_3 = 0.454\ \Omega$. Again we make use of Eq. (6.2-10) to write

$$R_2 + R_3 = 0.505 - R_1 \tag{6.2-16}$$

Using this in Eq. (6.2-15) we get

$$I_m(R_m + 0.505 - R_1) = I_s R_1$$

in which the only unknown is R_1. Substituting the known values

$$(1\text{ mA})(50.505\ \Omega - R_1) = (9999\text{ mA})R_1$$

Solving for R_1 we find

$$R_1 = 0.005\ 05\ \Omega$$

Finally, using Eq. (6.2-10) we have

$$\begin{aligned} R_2 &= 0.505 - R_1 - R_3 \\ &= 0.505 - 0.005\ 05 - 0.454 \\ &= 0.046\ \Omega \end{aligned}$$

This completes the design.

• • •

LEARNING EXERCISE FOR SEC. 6.2

1. A 50-μA, 250-Ω D'Arsonval movement is to be used to construct a meter that will measure 10, 25, and 100 A, full scale. Find the required shunt resistance values.

Ans. 500; 1250; 125

• • •

6.3 VOLTMETERS

The D'Arsonval movement is usually not suitable for the measurement of voltage without modification. The required modification consists of adding a series resistance R_s (called a *multiplier* resistance) as shown in Fig. 6.4a. This resistance is chosen so that full-scale current flows through the movement when the desired full-scale voltage V_{fs} is applied across the terminals. We can find the value of this resistance by noting that the voltage drop across it must be equal to the desired full-scale voltage minus the voltage drop across the meter. Thus, when full-scale voltage is applied,

$$V_{\text{multiplier}} = V_{fs} - V_m = V_{fs} - R_m I_m \qquad (6.3\text{-}1)$$

At full scale

$$V_{\text{multiplier}} = R_s I_m \qquad (6.3\text{-}2)$$

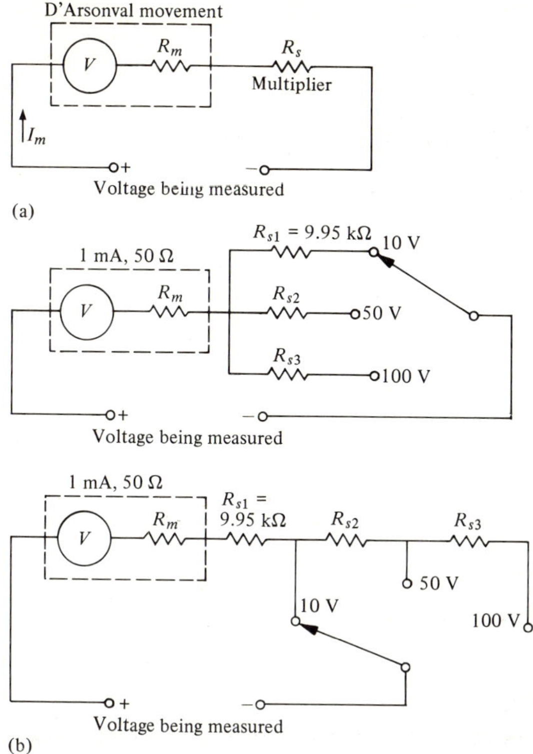

FIGURE 6.4
Voltmeters. (a) Multiplier. (b) Multirange switching.

Substituting this into Eq. (6.3-1) and solving for R_s we find

$$\boxed{R_s = \frac{V_{fs}}{I_m} - R_m} \qquad (6.3\text{-}3)$$

All of the quantities on the right-hand side would normally be given. Typical numerical values are included in the following example.

EXAMPLE 6.3-1 Finding a Multiplier Resistance

The 1-mA 50-Ω movement of Example 6.2-2 is to be used to measure 10 V full scale. Find the required multiplier resistance.

Solution

Equation (6.3-3) is used, being sure to keep all given quantities in proper units of volts, ohms, and amperes. Thus

$$R_s = \frac{10\ \text{V}}{0.001\ \text{A}} - 50\ \Omega$$

$$= 10\ 000\ \Omega - 50\ \Omega$$

$$= 9.95\ \text{k}\Omega$$

The student should check this result by using Ohm's law to find the current in the circuit and the voltages across R_s and R_m when $V_{fs} = 10$ V.

• • •

At this point we note that in contrast with the ammeter, the voltmeter is connected *across*, or *in parallel with*, the element or circuit for which the voltage is to be measured. The + terminal of the meter must be connected to the positive side of the voltage being measured in order to obtain an upscale reading.

In order to construct a multirange voltmeter, various switching arrangements can be used. Two possibilities are shown in Fig. 6.4b using the 1-mA 50-Ω movement of Example 6.4. Calculation of the multiplier resistors is done in Prob. 6.8.

As with multirange ammeters, precautions should be observed when connecting a multirange voltmeter across a circuit. The range switch should *always* be turned to the highest voltage range *before* connecting the meter. The range can then be lowered to obtain an adequate deflection.

Voltmeter Sensitivity—The Ohms-Per-Volt Rating

Most commercial voltmeters using the D'Arsonval movement will include an *ohms-per-volt* rating, which is usually printed on the face of the meter below the scales. This rating is extremely important, as it allows us to determine immediately the current sensitivity of the movement used in the voltmeter, and the total resistance of the meter for any of its ranges. This, in turn, allows us to readily calculate the *loading* effect of the meter on the original circuit.

The sensitivity of the meter, in ohms per volt, is simply the reciprocal of the current sensitivity I_m. (Recall that the units of I are volts per ohm.) Thus

$$\boxed{\text{Ohms per volt} = \frac{1}{I_m}} \qquad (6.3\text{-}3)$$

We can interpret this in the following way: Consider a 50-μA movement. Its rating is

$$\frac{1}{I_m} = \frac{1}{50 \times 10^{-6}\ \text{A}} = 20\ 000\ \text{ohms per volt}\ (\Omega/\text{V})$$

This means that regardless of the full-scale voltage that the meter is being designed for, we need a total resistance of 20 000 Ω for each volt of the full-scale value. The total meter resistance is then the *ohms-per-volt rating times the full-scale voltage* for the range being used. For example, if the full-scale voltage is to be 10 V, then the circuit resistance must be (20 000 Ω/V) × (10 V) = 200 000 Ω. This resistance is constant regardless of the magnitude of the voltage being measured on the 10-V range.

In order for a voltmeter to have little effect on the circuit being measured, it should draw as little current as possible. Thus, higher ohms-per-volt ratings are usually more desirable. Use of the sensitivity is illustrated in the following example.

EXAMPLE 6.3-2 Using the Sensitivity Rating

a. Find the sensitivity of the 1-mA 50-Ω movement of Example 6.3-1.
b. Find the total meter resistance for each of the ranges shown in Fig. 6.4b.

Solution

a. $$\text{Sensitivity} = \frac{1}{I_m} = \frac{1}{10^{-3}} = 1000\,\frac{\Omega}{\text{V}}$$

b. $$R_T = \frac{\Omega}{\text{V}} \times V_{\text{fs}}$$

10-V range:

$$R_T = \left(1000\,\frac{\Omega}{\text{V}}\right)(10\text{ V}) = 10\text{ k}\Omega$$

50-V range:

$$R_T = \left(1000\,\frac{\Omega}{\text{V}}\right)(50\text{ V}) = 50\text{ k}\Omega$$

100-V range:

$$R_T = \left(1000\,\frac{\Omega}{\text{V}}\right)(100\text{ V}) = 100\text{ k}\Omega$$

• • •

Meter Loading

The measurement error resulting because of the effect of the voltmeter on the circuit depends on the ratio of the circuit resistance to the voltmeter resistance. Consider Fig. 6.5. Clearly if R_2 and R_T are of the same order of magnitude, the error introduced will be high, because the original circuit without the voltmeter is considerably different from the circuit when the

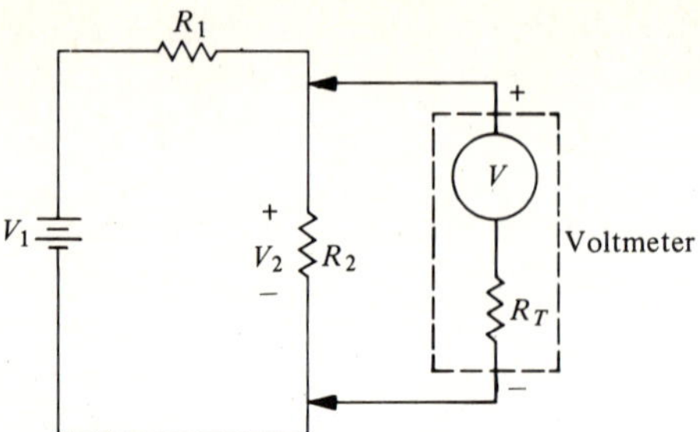

FIGURE 6.5
Circuit for determining voltmeter loading.

voltmeter is connected. On the other hand, if R_T is much greater than R_2, the disturbance will be small. The next example illustrates these points.

EXAMPLE 6.3-3 Calculation of Voltmeter Loading

In the circuit of Fig. 6.5, $R_1 = R_2 = 330\ \text{k}\Omega$, $V_1 = 100$ V, and the voltmeter uses a 50-μA movement and has a full-scale range of 50 V.

a. Find the sensitivity and total resistance of the meter.
b. Find the voltage across R_2 without the meter.
c. Find the voltage indicated by the meter when it is connected.
d. Find the percent error due to voltmeter loading.
e. Repeat (b), (c), and (d) if $R_1 = R_2 = 33\ \text{k}\Omega$.

Solution

a. $$\text{Sensitivity} = \frac{1}{I_m} = \frac{1}{50 \times 10^{-6}} = \frac{20\ \text{k}\Omega}{\text{V}}$$

$$R_T = \left(20\,\frac{\text{k}\Omega}{\text{V}}\right)(50\ \text{V}) = 1\ \text{M}\Omega$$

b. Using the voltage-divider formula,

$$V_2 = V_1\frac{R_2}{R_1 + R_2} = (100\ \text{V})\left(\frac{330\ \text{k}\Omega}{330\ \text{k}\Omega + 330\ \text{k}\Omega}\right) = 50\ \text{V}$$

c. The resistance of the parallel connection of R_2 and the meter is (keeping all resistances in units of megohms)

$$R_{eq} = \frac{R_T R_2}{R_T + R_2} = \frac{(1)(0.33)}{(1 + 0.33)} = 0.248\ \text{M}\Omega$$

Since the meter measures the voltage across R_{eq}

$$V_{2m} = V_1\frac{R_{eq}}{R_1 + R_{eq}} = 100\left(\frac{0.248}{0.330 + 0.248}\right) = 43\ \text{V}$$

d. The error is 43 V − 50 V = −7 V. Thus

$$\text{Percent error} = \frac{-7}{50} \times 100 = -14\%$$

e. When $R_1 = R_2 = 0.033\ \text{M}\Omega$,

$$V_2 = 100\left(\frac{0.033}{0.033 + 0.033}\right) = 50\ \text{V}$$

$$R_{eq} = \frac{(1)(0.033)}{1 + 0.033} = 0.032\ \text{M}\Omega$$

$$V_{2m} = 100\left(\frac{0.032}{0.033 + 0.032}\right) = 49\ \text{V}$$

$$\text{Percent error} = \frac{49 - 50}{50} \times 100 = -2\%$$

Clearly, the loading effect decreases as the ratio of meter resistance to circuit resistance increases. The student can become further convinced of this by repeating the calculations above for a 1-mA movement (Prob. 6.12).

• • •

Instrument Errors

In the previous example, the error was due totally to loading of the circuit by the meter. Readings may also depart from the true value because of inaccuracy of the movement itself. This is usually specified by the manufacturer and typical values are ±1, 2, 3, and 5% *of full scale.* This means that for a 50-V scale and ±2% accuracy the possible error is (±0.02)(50) = ±1 V at any point on the scale.

The effect of this error can be seen more clearly from the following example. Consider that when a particular voltage is applied to the meter, the reading is 50 V, that is, a full-scale reading. Then, since the possible error is ±1 V, the actual voltage applied to the meter lies between 49 and 51 V. For this full-scale value, the accuracy is ±2%. Now consider that a different voltage is applied to the meter and the reading is now 5 V, that is, a reading of about 10% of full scale. The possible error is still ±1 V so that the voltage applied to the meter lies between 4 and 6 V. If the actual voltage is 4 V the error is

$$\text{Percent error} = \frac{I_{meas} - I_{true}}{I_{true}} \times 100$$

$$= \frac{5 - 4}{4} \times 100$$

$$= 25\%$$

From this result we see that readings become less and less accurate as we move further away from full sale. We should always choose meter ranges so that readings are as close to full scale as possible. Readings below 10% of full scale are avoided whenever possible because of the large errors involved.

Other causes of error in measurements are called *systematic* errors which may be attributed to miscalibration, nonlinearity, undetected shift of the zero point, or bias on the part of the experimenter. This type of error usually tends to be in one direction and to have a relatively constant magnitude. The previously discussed errors were *random* in nature, as indicated by the ± specification.

These *instrument errors* should be carefully distinguished from *instrument loading*, which was illustrated in Example 6.3-2. Not too much can be done about instrument errors unless one has a very large budget with which to buy better instruments. However, instrument loading can be corrected for or minimized by careful design of the measuring method.

Using the Voltmeter for Troubleshooting

The voltmeter can be used to advantage for troubleshooting. Consider the circuit shown in Fig. 6.6. In the circuit an accidental open circuit has occurred and we are required to track it down so that it can be repaired. We can troubleshoot the circuit with the power on using a voltmeter to find the break. We begin by measuring the voltage from point *a* to ground. The reading is 220 V, the same as the applied voltage. This indicates that (1) there is an open circuit at some point that causes the current to be zero, so that there is no voltage drop across R_1 and (2) that the path from point *a* back to the source is *not* broken. We next move the + terminal of the voltmeter to point *b* where we again read 220 V. This indicates that no current is flowing between the source and point *b* and that the break has not been encountered yet. A measurement at point *c* yields the same result. However, when we move the voltmeter to point *d*, the reading is 0 V, indicating that the path back to the source has been broken and that the break is between points *c* and *d*.

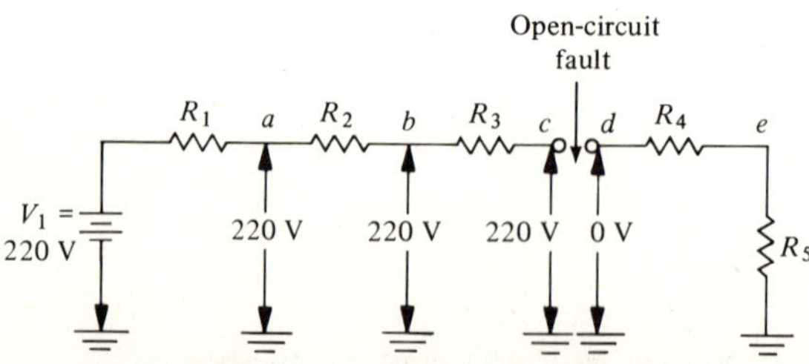

FIGURE 6.6
Using a voltmeter to find an open-circuit fault.

This example illustrated a very simple troubleshooting problem. Very often, such problems are not so easily solved, and considerable ingenuity is required to interpret the various readings. In all cases, the voltmeter should have a resistance much higher than the circuit resistances.

• • •

LEARNING EXERCISES FOR SEC. 6.3

1. What is the full-scale current of a meter movement rated at 50 000 Ω/V?

2. What multiplier resistances are required if a 50 000 Ω/V, 2.5-kΩ movement is to be used to measure 100- and 250-V full scale?

Ans. 12.5; 5; 20

• • •

6.4 RESISTANCE MEASUREMENT

The Voltmeter–Ammeter Method

In the absence of instruments that measure resistance directly, Ohm's law can be used with individual measurements of voltage across and current through the unknown resistance. These measurements are easily obtained using a voltmeter and ammeter set on appropriate ranges. There are two ways that meters can be connected, as shown in Fig. 6.7. In the configuration shown in Fig. 6.7a, the ammeter reading includes the current drawn by the voltmeter. If the voltmeter current is comparable to the current through the unknown resistor, then the ammeter reading will not be an indication of the true current. In the laboratory, this can be checked by temporarily disconnecting the voltmeter. If the ammeter reading changes by more than a small amount, then the voltmeter loading is not negligible. In this case, the circuit of Fig. 6.7b may be used. In this circuit the voltmeter reading

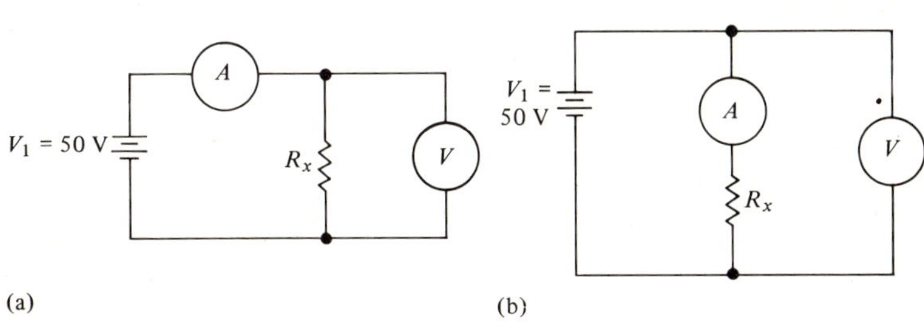

FIGURE 6.7
Voltmeter-ammeter measurement of resistance. (a) Ammeter reading includes voltmeter current. (b) Voltmeter reading includes ammeter voltage drop.

includes the voltage drop across the ammeter. If the ammeter resistance is much lower than R_x, then this voltage drop will be insignificant. This can be checked in the laboratory by temporarily placing a short circuit across the ammeter. If the voltmeter reading changes significantly, then the ammeter loading is not negligible.

If both methods will not work, then a high resistance voltmeter or lower resistance ammeter will have to be used. Or, if the meter resistances are known, the value of R_x can be calculated (see Prob. 6.19).

EXAMPLE 6.4-1 Voltmeter–Ammeter Loading in a Resistance Measurement

The circuits of Fig. 6.7 are used with $V_1 = 50$ V to measure a 33-kΩ resistance. A 5-mA full-scale, 50-Ω milliammeter is available, and two voltmeters are available. Both have 50-V scales and one is rated at 1000 Ω/V, the other at 20 000 Ω/V. For both circuits, using the ammeter and each voltmeter in turn, find the apparent resistance as calculated from the voltmeter and ammeter readings.

Solution

The circuits are shown in Fig. 6.8 with the meter resistances explicitly displayed. For the 1000 Ω/V voltmeter on the 50-V scale, the total resistance is 50 kΩ and for the 20 000 Ω/V voltmeter it is 1 MΩ.

If the ammeter were ideal (zero resistance) and the voltmeter were ideal (infinite resistance), the current would be

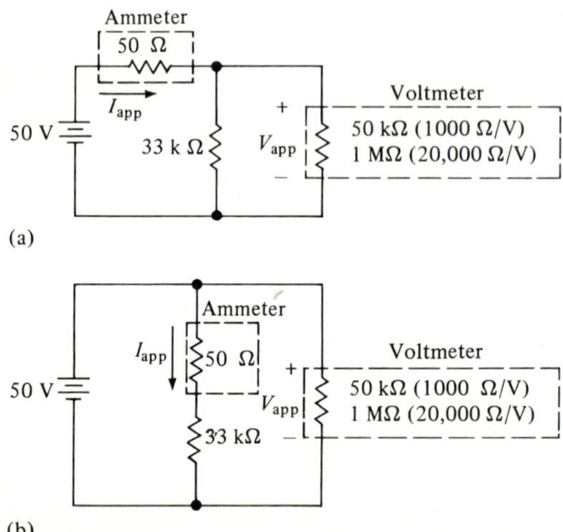

FIGURE 6.8
Example 6.4-1. In both circuits I_{app} = ammeter reading, V_{app} = voltmeter reading. (a) Ammeter reading includes voltmeter current. (b) Voltmeter reading includes drop across ammeter.

$$I_{true} = \frac{50\text{ V}}{33\text{ k}\Omega} = 1.52\text{ mA}$$

and the voltage would be $V_{true} = 50$ V.

With the meters present these values will change. For each different connection we will have an *apparent* current I_{app} indicated by the ammeter, and an *apparent* voltage V_{app} indicated by the voltmeter. In practice the value of R_x would not be known, and we would find the *apparent* resistance R_{app} as the ratio of the meter indications. The calculations follow.

Circuit (a): Using the 1000 Ω/V voltmeter, the meter readings would be as follows:

$$I_{app} = \frac{50\text{ V}}{0.05\text{ k}\Omega + (33\text{ k}\Omega \parallel 50\text{ k}\Omega)} = \frac{50\text{ V}}{0.05\text{ k}\Omega + 19.9\text{ k}\Omega}$$

$$= 2.51\text{ mA}$$

$$V_{app} = I_{app}(33\text{ k}\Omega \parallel 50\text{ k}\Omega) = (2.51\text{ mA})(19.9\text{ k}\Omega) = 49.9\text{ V}$$

Then

$$R_{app} = \frac{V_{app}}{I_{app}} = \frac{49.9\text{ V}}{2.51\text{ mA}} = 19.9\text{ k}\Omega \approx 20\text{ k}\Omega$$

$$\%\text{error} = \frac{R_{app} - R_{true}}{R_{true}} \times 100 = \frac{20 - 33}{33} \times 100 = -39.4\%$$

Using the 20 000 Ω/V voltmeter, the meter readings would be

$$I_{app} = \frac{50\text{ V}}{0.05\text{ k}\Omega + (33\text{ k}\Omega \parallel 1000\text{ k}\Omega)} = \frac{50\text{ V}}{0.05\text{ k}\Omega + 32\text{ k}\Omega}$$

$$= 1.56\text{ mA}$$

$$V_{app} = I_{app}(33\text{ k}\Omega \parallel 1000\text{ k}\Omega) = (1.56\text{ mA})(32\text{ k}\Omega) = 49.9\text{ V}$$

Then

$$R_{app} = \frac{V_{app}}{I_{app}} = \frac{49.9\text{ V}}{1.56\text{ mA}} = 32\text{ k}\Omega$$

$$\%\text{error} = \frac{R_{app} - R_{true}}{R_{true}} = \frac{32 - 33}{33} \times 100 = -3\%$$

Circuit (b): Using the 1000 Ω/V voltmeter, the meter readings are:

$$V_{app} = 50\text{ V}$$

$$I_{app} = \frac{50\text{ V}}{33\text{ k}\Omega} = 1.51\text{ mA}$$

$$R_{app} = \frac{50\text{ V}}{1.51\text{ mA}} = 33\text{ k}\Omega$$

The error here is negligible. Using the 20 000 Ω/V voltmeter leads to exactly the same result. (Why?) From this example we see that, for the typical instruments used, circuit (b) gives a much more accurate result. However, using circuit (a) with the 20 000 Ω/V voltmeter gives results which would be adequate for many applications.

The true value of the unknown resistance can be found from the apparent value of the unknown if the meter resistance is known. This is considered in Probs. 6.19 and 6.20.

• • •

The Series Ohmmeter

For routine measurement of resistance the *series* ohmmeter circuit of Fig. 6.9a is used. The accuracy is not impressive since it lies in the 10 to 20% range for most meters. However, for many applications this is adequate.

If the terminals are left open so that $R_x = \infty$, no current will flow, and the meter will not register. Thus the left end of the scale corresponds to $R_x = \infty$ as shown in Fig. 6.9b. When the terminals are shorted (connected directly together), $R_x = 0$ and the remaining resistance $R_s + R_z + R_m$ is chosen so that full-scale current flows. Thus the right end of the scale (full-scale) corresponds to $R_x = 0$. The required resistance can be found by noting that the meter current is

$$I = \frac{V}{R_s + R_z + R_m + R_x} \tag{6.4-1}$$

where R_z is a small variable resistance used as a *zero adjust* control to calibrate the ohmmeter and compensate for battery aging by bringing the pointer to zero when the input test leads are shorted. Thus, the full-scale current occurs when $R_x = 0$:

$$I_m = \frac{V}{R_s + R_z + R_m} \tag{6.4-2}$$

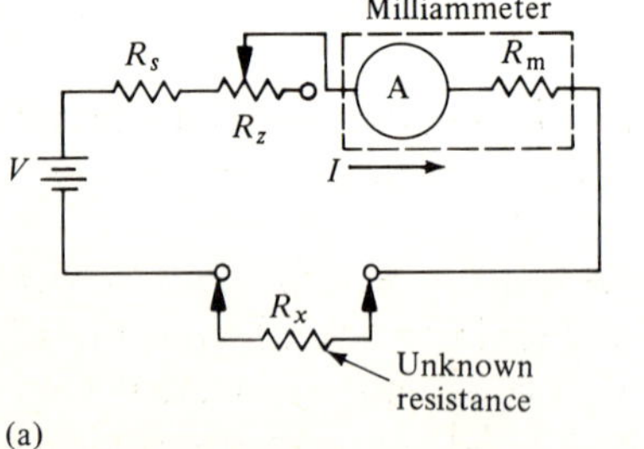

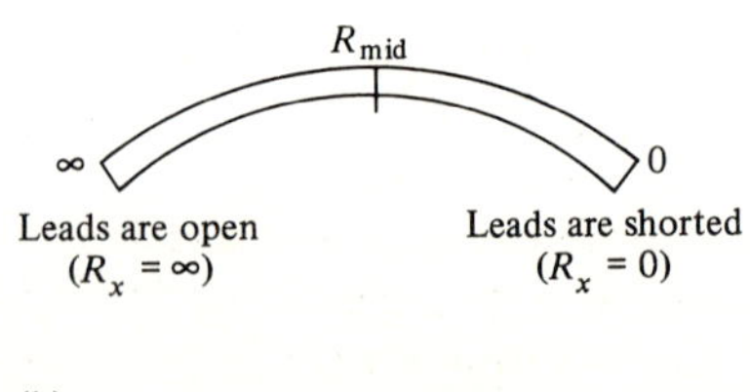

FIGURE 6.9
Series ohmmeter. (a) Circuit. (b) Scale.

Now let us factor Eq. (6.4-1) to get the form

$$I = \frac{V}{(R_s + R_z + R_m)\left(1 + \dfrac{R_x}{R_s + R_z + R_m}\right)}$$

Making use of Eq. (6.4-2) this can be written

$$I = \frac{I_m}{1 + \dfrac{R_x}{R_s + R_z + R_m}} \tag{6.4-3}$$

The current sensitivity of the movement I_m and the midscale resistance R_{mid} is often specified. When $R_x = R_{mid}$ the current will be one-half of its full-scale value, $I = I_m/2$. From Eq. (6.4-3) we see that this occurs when $R_x = R_{mid} = R_s + R_z + R_m$. Then we can write

$$I = \frac{I_m}{1 + R_x/R_{mid}}$$

Solving for R_x, we find

$$R_x = R_{mid}\left(\frac{I_m}{I} - 1\right) \tag{6.4-4}$$

This relation can be used to find the unknown resistance corresponding to any point on the meter scale.

In design, the battery voltage required to power the ohmmeter can be found from Eq. (6.4-2) by writing it in the form

$$V = (R_s + R_z + R_m)I_m = R_{mid} I_m$$

Since R_{mid} and I_m will be given, V can be found. R_s is chosen a small amount less than R_{mid} to allow for R_z and R_m. The design process will be illustrated in the next example.

One disadvantage of the series ohmmeter can be seen by considering the scale shown in Fig. 6.9b. The left half of the scale covers all values from R_{mid} to ∞, and the right half all values from 0 to R_{mid}. Thus the scale is very nonlinear and accuracy is poor, especially at the high resistance end of the scale where the values crowd together. Usually, ohmmeters are provided with multiple ranges so that different midscale values are available. In the following example, typical values are calculated.

EXAMPLE 6.4-2 Designing a Series Ohmmeter

A 50-μA 200-Ω meter movement is used in the series ohmmeter circuit of Fig. 6.9a.

a. Find the required battery voltage and suitable values for R_s and R_z so that the midscale resistance is 10 kΩ.

b. Find the midscale resistance if a 1.5-V battery is to be used and also find suitable values for R_s and R_z.

Solution

a. In order to have $R_{mid} = 10\ \text{k}\Omega$ we must have

$$R_s + R_z + R_m = 10\ \text{k}\Omega$$

Since $R_m = 200\ \Omega = 0.2\ \text{k}\Omega$

$$R_s + R_z = 9.8\ \text{k}\Omega$$

For the zero adjust, we assume that an adjustment of $\pm 25\%$ of full scale will be adequate. Thus if we use a 5-kΩ variable resistance for R_z and adjust R_s so that exactly full-scale current flows when the variable resistance R_z is at one-half of its maximum value, we will achieve the desired result.

Then,

$$R_s + \frac{5\ \text{k}\Omega}{2} = 9.8\ \text{k}\Omega$$

$$R_s = 7.3\ \text{k}\Omega$$

The final design values are then

R_s = 7.3-kΩ fixed resistance

R_z = 5-kΩ variable resistance

The required battery voltage is

$$V = R_{mid} I_m = (10\ \text{k}\Omega)(50\ \mu\text{A}) = 500\ \text{mV} = 0.5\ \text{V}$$

b. With the battery voltage specified, we find the midscale resistance

$$R_{mid} = \frac{V}{I_m} = \frac{1.5\ \text{V}}{50\ \mu\text{A}} = 0.03\ \text{M}\Omega = 30\ \text{k}\Omega$$

Then we must have

$$R_s + R_z + R_m = 30\ \text{k}\Omega$$

and since $R_m = 0.2\ \text{k}\Omega$

$$R_s + R_z = 29.8\ \text{k}\Omega$$

For the zero-adjust resistance, we choose the standard variable resistance value $R_z = 10\ \text{k}\Omega$. Then

$$R_s + \frac{10\ \text{k}\Omega}{2} = 29.8\ \text{k}\Omega$$

$$R_s = 24.8\ \text{k}\Omega$$

This completes the design.

• • •

The Shunt Ohmmeter

For relatively low values of resistance, a somewhat different arrangement, called a *shunt* ohmmeter, is used. The circuit is shown in Fig. 6.10a, where we see that it differs from the series ohmmeter in that the meter movement is directly in shunt (parallel) with the unknown resistance R_x. The switch is included so that power will not be drawn from the battery when the ohmmeter is not in use.

For this ohmmeter, the condition $R_x = 0$ places a direct short across the meter movement so that $R_x = 0$ corresponds to zero meter indication, as shown in Fig. 6.10b. When $R_x = \infty$ maximum current flows in the meter movement and R_s and R_z are adjusted so that this equals the current sensitivity of the meter movement.

Equation (6.4-2) holds for the shunt ohmmeter when $R_x = \infty$ with R_z now called an *infinity adjust* resistor. As contrasted to the series instrument, a midscale reading occurs when $R_x = R_m \parallel (R_s + R_z)$ so that the shunt ohmmeter has a relatively low midscale reading.

At this point we must warn the student that the ohmmeter (of any type) *must never be connected to a circuit in which sources are active*. The active circuit may try to force a current through the movement much larger than the movement's full-scale value, and thus cause permanent damage.

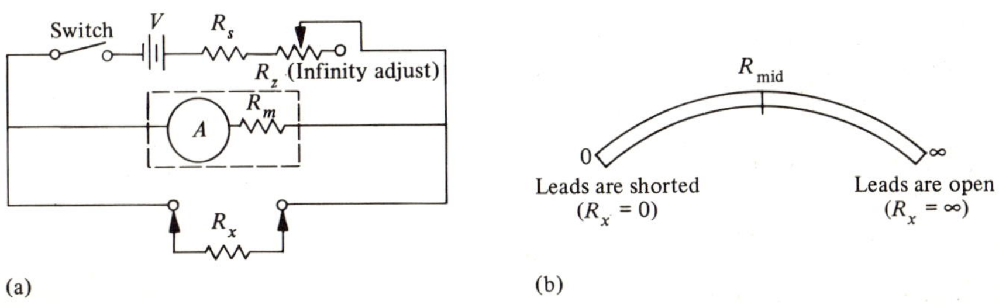

FIGURE 6.10
Shunt ohmmeter. (a) Circuit. (b) Scale.

EXAMPLE 6.4-3 Design of a Shunt Ohmmeter

The following components are to be used in a shunt ohmmeter:

1. 1-mA, 50-Ω movement
2. 200-Ω variable resistor for infinity adjustment
3. 2-V battery

a. Find a suitable value for R_s.
b. Find the unknown resistance R_x required for 3/4, 1/2, and 1/4 full-scale deflection.
c. Draw the scale

Solution

a. For the shunt ohmmeter of Fig. 6.10 the full-scale current of 1 mA flows through the meter when $R_x = \infty$. Using Ohm's law, we have, with $R_x = \infty$,

$$I_m = \frac{2\text{ V}}{R_s + R_z + 50\ \Omega} = 0.001\text{ A} \tag{6.4-5}$$

from which

$$R_s + R_z + 50\ \Omega = 2000\ \Omega$$

and

$$R_s + R_z = 1950\ \Omega \tag{6.4-6}$$

In order to have some adjustment available we assume R_z set at its midpoint so that $R_z = 200\ \Omega/2 = 100\ \Omega$. Then

$$R_s = 1850\ \Omega$$

b. Since we need to find several values it is convenient to derive an equation for R_x in terms of the desired meter current and the known resistance values and battery voltage. One way to do this is to first convert the battery branch into its Norton equivalent, as shown in Fig. 6.11a. The Norton circuit is then combined with the meter resistance and the unknown to give the reduced circuit of Fig. 6.11b. We apply the current divider formula to this circuit to get

$$\frac{I_{\text{meter}}}{1.03\text{ mA}} = \frac{R'_x}{0.05\text{ k}\Omega + R'_x} \tag{6.4-7}$$

where the units are milliamperes, kilohms, and volts, and

$$R'_x = 1.95\text{ k}\Omega \parallel R_x = \frac{1.95R_x}{1.95 + R_x}\text{ k}\Omega \tag{6.4-8}$$

When these equations are solved for R_x, we obtain, after some algebra (which the student should check),

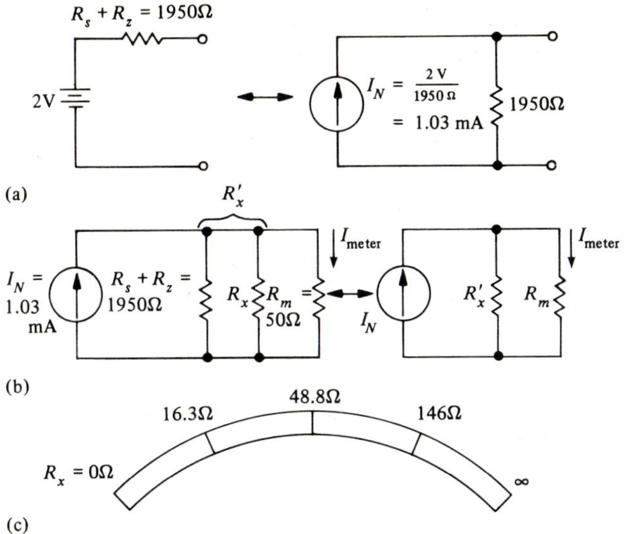

FIGURE 6.11
Example 6.4-3. (a) Conversion of battery branch. (b) Reduced circuit. (c) Meter scale calibrations.

$$R_x = \frac{0.048\,8 I_{\text{meter}}}{1 - I_{\text{meter}}}\ \text{k}\Omega \tag{6.4-9}$$

This is the desired equation relating the unknown resistance to the meter current. Let us check for the end-of-scale values, $I_{\text{meter}} = 0$ and 1 mA. When $I_{\text{meter}} = 0$,

$$R_x = 0$$

When $I_{\text{meter}} = 1$ mA,

$$R_x = \frac{0.048\,8}{1 - 1} \rightarrow \infty$$

For the intermediate values, at one-quarter of full scale, $I_{\text{meter}} = 0.25$ mA and

$$R_x = \frac{0.048\,8 \times 0.25}{1 - 0.25} = 0.016\,3\ \text{k}\Omega = 16.3\ \Omega$$

At one-half of full scale, $I_{\text{meter}} = 0.5$ mA and

$$R_x = \frac{0.048\,8 \times 0.5}{1 - 0.5} = 0.048\,8\ \text{k}\Omega = 48.8\ \Omega$$

At three-quarters of full scale, $I_{\text{meter}} = 0.75$ mA and

$$R_x = \frac{0.048\,8 \times 0.75}{1 - 0.75} = 0.146\ \text{k}\Omega = 146\ \Omega$$

The scale is drawn in Fig. 6.4-5c. In practice the divisions would be labeled at more convenient points, such as 10 Ω, 20 Ω, etc. Note the low mid-scale value of 48.8 Ω.

• • •

Using the Ohmmeter to Check Continuity

When troubleshooting malfunctioning electric circuits, we often suspect that a break exists in the circuit wiring. If the circuit can be checked with the power off, an ohmmeter can be used to check wiring continuity. When using an ohmmeter for this purpose, the lowest range should be used since the wire conductors that are not broken have very low resistance. To perform the continuity test we measure the resistance between different pairs of points on the conductors in the current flow path. If the path is continuous between the points the resistance will be essentially zero. If the break occurs between the points, then the resistance will be infinite. Sometimes a wire will break inside its insulating covering so that the break is not visible to the eye. When the ohmmeter is connected to the ends of the wire, it will register an open circuit rather than the very low resistance which we expect.

Sometimes bad connections occur at terminals. Here again, connecting the ohmmeter properly will indicate an open circuit and thus pinpoint the break.

The ohmmeter is also useful for checking continuity of resistors, but the results of such checks must be interpreted carefully. For example, when checking a parallel circuit the ohmmeter may indicate continuity when in fact a break exists in one of the parallel branches. This is shown in Fig.

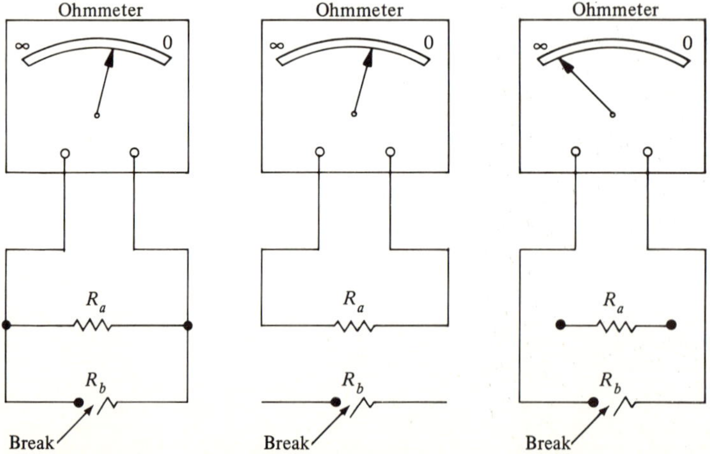

FIGURE 6.12
Continuity checking with the ohmmeter. (a) Meter reads finite resistance. (b) Meter again reads finite resistance with R_b disconnected. (c) Meter reads infinite resistance.

6.12a, where the ohmmeter indicates a noninfinite resistance. If it is suspected that this is the case, then each branch must be checked individually with the other branches disconnected. In Fig. 6.11b we show how branch *a* is checked with branch *b* disconnected. Here the finite resistance reading on the ohmmeter indicates that branch *a* is not open. In Fig. 6.12c we check branch *b* with branch *a* disconnected and find that the ohmmeter reads infinity ohms, indicating that branch *b* has an open circuit fault.

The ohmmeter can be used to locate individual wires in a multiconductor cable in which the wires are not individually color coded. This is shown in Fig. 6.13a. To begin, one terminal of the ohmmeter is connected to one wire and the other terminal is connected, in turn, to each of the wires at the other end. When the ohmmeter is connected to opposite ends of the same wire, the reading will be a very low value of resistance, as shown in the figure, depending on the wire size and the length of the cable. When the ohmmeter connections are to the ends of two different wires, then the reading will be infinity ohms.

When the cable is too long for the ohmmeter to reach the ends, another procedure is used to locate individual wires. We connect wires 1 and 2 together at the far end of the cable. Then the ohmmeter is used as shown in Fig. 6.13b to locate this pair of wires and we label them *a* and *b* at the ohmmeter end. The same procedure is used to locate pairs 3 and 4, which are labeled *c* and *d* at the ohmmeter end. The next step is to connect wires 1 and 3 at the far end and locate this pair with the ohmmeter. Assume that the ohmmeter indicates low resistance between wires *a* and *d*. Then *a* must be on the same wire as 1 and *d* on the same wire as 3. Also, by elimination, *b* and 2 are on the same wire, as are *c* and 4. The near end can now be relabeled with numbers to correspond with the far end.

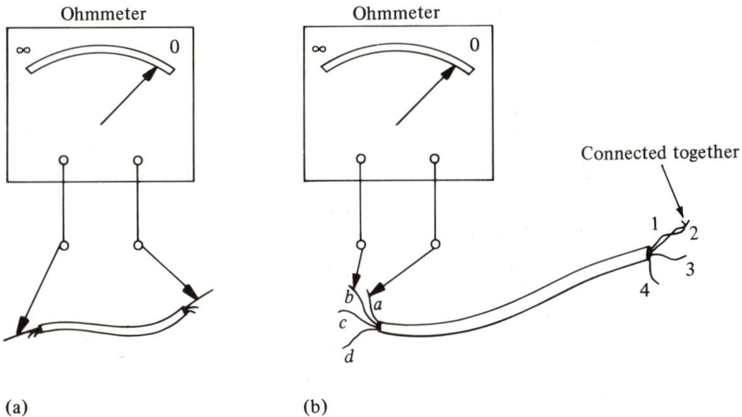

FIGURE 6.13
Checking multiconductor cable. (a) When Ohmmeter can connect to each end. (b) When cable is too long for ohmmeter to connect to each end.

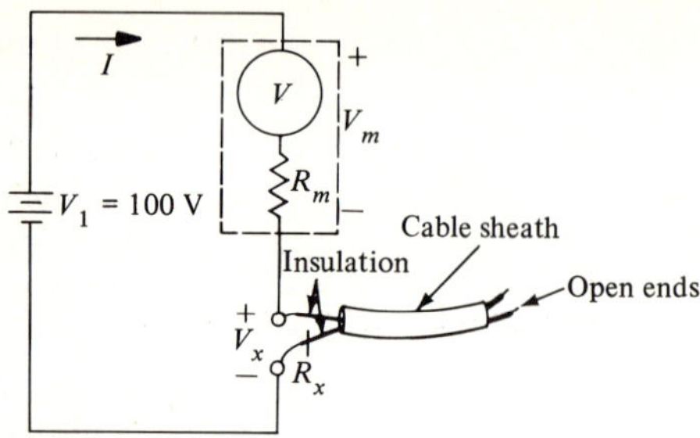

FIGURE 6.14
Using a voltmeter to measure high resistance.

Using the Voltmeter to Check High Resistance Values

In the standard ohmmeter, the high resistance values are crowded toward the left end of the scale and it is not possible to measure high resistances with any degree of accuracy. However, if voltage can be applied to the resistance to be measured, we can calculate its value from a simple voltage measurement by making use of our knowledge of circuit theory.

There are numerous instances where the measurement of resistance in the megohm range is required. One example is to find the leakage resistance of the insulation in a cable which has several conductors. This is illustrated in the following example.

EXAMPLE 6.4-4 Measuring Insulation Resistance Using a Voltmeter

In the circuit of Fig. 6.14, the voltmeter is a 20 000 Ω/V instrument that reads 15 V on the 100-V scale. The voltage source is 100 V. Show how the cable insulation resistance can be found from this voltage reading and knowledge of the resistance of the voltmeter.

Solution

In this configuration, the voltmeter is connected in *series* with the unkown high resistance to form a voltage divider. Using circuit theory we can then calculate the value of the unkown resistance. The high resistance is shown as a two-conductor cable (with the far ends open) so that we are measuring the resistance from one conductor to the other through the insulation.

Since the voltmeter and the unknown resistance are in series, they both carry the same current. Thus we can write

$$I = \frac{V_x}{R_x} = \frac{V_m}{R_m} \tag{6.4-10}$$

and, using KVL

$$V_x = V_1 - V_m \tag{6.4-11}$$

Combining these two equations we find

$$R_x = \left(\frac{V_1 - V_m}{V_m}\right) R_m \tag{6.4-12}$$

The data in the problem statement indicate that $V_1 = 100$ V, $V_m = 15$ V, and $R_m = (20\ 000\ \Omega/\text{V}) \times (100\ \text{V}) = 2\ \text{M}\Omega$. Substituting these values in Eq. (6.4-12) we find

$$R_x = \frac{(100 - 15)}{15} \times 2 = 11.3\ \text{M}\Omega.$$

• • •

LEARNING EXERCISE FOR SEC. 6.4

1. A series ohmmeter uses a 20-μA, 2.5-kΩ meter movement. If $R_s + R_z + R_m = 10$ kΩ find the values of unknown resistance for 1/4, 1/2, and 3/4 of full scale.

Ans. 10; 3.33; 30

• • •

6.5 BRIDGE MEASUREMENT OF RESISTANCE

When ohmmeter measurements do not provide the required accuracy we must turn to some sort of *bridge* measuring device, the *Wheatstone bridge* being one of the most popular types. The basic circuit is shown in Fig. 6.15, where R_1, R_2, and R_3 are standard resistors known very accurately, often to 0.01%, R_x is the unknown resistance, and G is a sensitive zero-center galvanometer (microammeter). Resistor R_s allows for a rough balance and provides for protection of the galvanometer.

In operation, one of the standard resistances is varied until the galvanometer current is zero when $R_s = 0$. This is called the *null*, or *balanced*, condition for the bridge. Since the meter current is zero at balance, the voltage across it is also zero. Then, using Ohm's law and KVL we obtain two

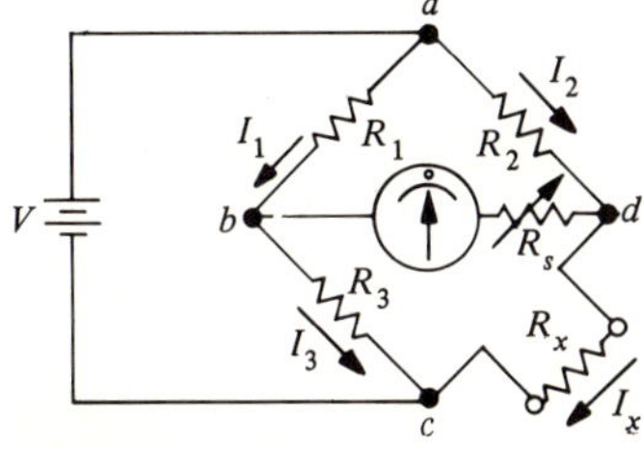

FIGURE 6.15
Wheatstone bridge circuit.

equations, as follows:

$$I_1 R_1 = I_2 R_2$$
$$I_3 R_3 = I_x R_x \qquad (6.5\text{-}1)$$

Noting that the meter current is zero, application of KCL at nodes *b* and *d* yields the current equations

$$I_1 = I_3$$
$$I_2 = I_x \qquad (6.5\text{-}2)$$

Now we divide the two equations in Eq. (6.5-1) to get

$$\frac{I_1 R_1}{I_3 R_3} = \frac{I_2 R_2}{I_x R_x}$$

Noting Eq. (6.5-2) the currents cancel and the balance condition becomes

$$\frac{R_1}{R_3} = \frac{R_2}{R_x} \qquad (6.5\text{-}3)$$

Finally, solving for R_x

$$\boxed{R_x = \frac{R_3 R_2}{R_1}} \qquad (6.5\text{-}4)$$

The value of the unknown can be calculated from Eq. (6.5-4) and is independent of the battery voltage V.

In some commercial versions of the Wheatstone bridge (see Fig. 6.16) resistors R_1 and R_3 are variable in steps of 10 and are connected together mechanically so that the ratio R_3/R_1 is either 1 or a positive power of 10. Resistor R_2 is continuously variable and has a calibrated scale from which we read the value of the unknown resistance. In use, the R_3/R_1 ratio is set so that the bridge can be balanced by adjustment of R_2. Thus the R_3/R_1 ratio and the maximum value of R_2 determine the range of unknowns that can be measured. Typical numbers will be found in the following example.

EXAMPLE 6.5-1 Wheatstone Bridge Range

Find the ranges available on a Wheatstone bridge for which the R_3/R_1 ratio can be set to 10, 1, and 1/10. Resistor R_2 is a 5-kΩ wire-wound rheostat.

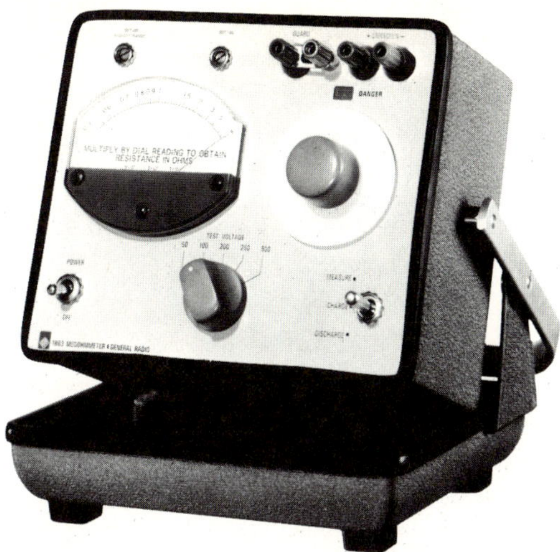

FIGURE 6.16
Commercial resistance bridge (courtesy of GenRad, Inc.).

Solution

The range is

$$\frac{R_3}{R_1} R_{2\min} \le R_x \le \frac{R_3}{R_1} R_{2\max}$$

Since $R_{2\max} = 5\ \text{k}\Omega$ and $R_{2\min} = 0$, this becomes

$$0 \le R_x \le \frac{5R_3}{R_1}\ \text{k}\Omega$$

Thus for $R_3/R_1 = 10$ the range is from 0 to 50 kΩ, for $R_3/R_1 = 1$ the range is from 0 to 5 kΩ, and for $R_3/R_1 = 1/10$ the range is from 0 to 500 Ω.

• • •

An interesting variation of the Wheatstone bridge is the *slide-wire* bridge shown in Fig. 6.17. In this circuit resistors R_1 and R_3 are replaced by a 1-m length of resistance wire chosen to have a very uniform resistance distribution along its length. The galvanometer connection between R_1 and R_3 is now a sliding contact that can move along the resistance wire. In use, the bridge is balanced by moving the sliding contact along the slide wire until the galvanometer reading is zero. At balance, Eq. (6.5-4) derived for the Wheatstone bridge also applies to the slide-wire bridge. It is repeated here for convenience.

$$R_x = \left(\frac{R_3}{R_1}\right) R_2 \tag{6.5-4}$$

For the slide wire, the resistivity ρ and cross-sectional area A are the same throughout, so $R_1 = (\rho/A)L_1$ and $R_3 = (\rho/A)L_3$. The balance equation now becomes

$$R_x = \frac{(\rho/A)L_1}{(\rho/A)L_3} R_2 = \frac{L_1}{L_3} R_2 \tag{6.5-5}$$

The values of L_1 and L_3 at balance are read from the meter scale to which the slide wire is attached.

EXAMPLE 6.5-2 The Slide-Wire Bridge

In the slide-wire bridge of Fig. 6.17, the bridge is balanced when $L_1 = 24$ cm with $R_2 = 80\ \Omega$. Find R_x.

Solution

If $L_1 = 24$ cm then $L_3 = 100 - 24 = 76$ cm and

$$R_x = \frac{24}{76} \times 80 = 25.3\ \Omega$$

• • •

LEARNING EXERCISE FOR SEC. 6.5

1. A Wheatstone bridge has $R_1 = 5\ \text{k}\Omega$, $R_3 = 10$ and $25\ \text{k}\Omega$, and R_2 can be varied from 2.5 to 20 kΩ. Find the range for each value of R_3.

Ans. 12.5; 5; 100; 40

• • •

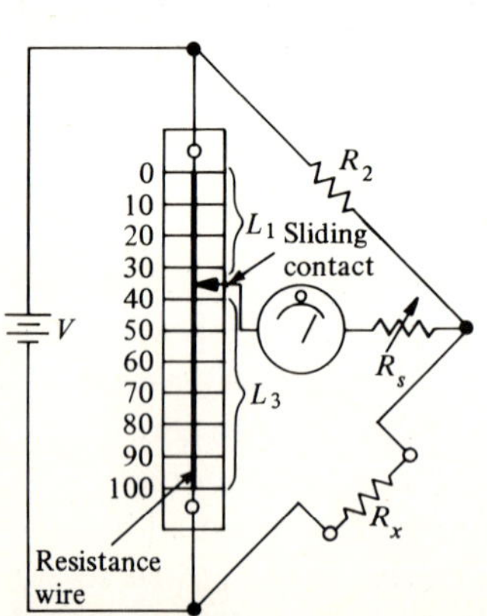

FIGURE 6.17
Slide-wire bridge.

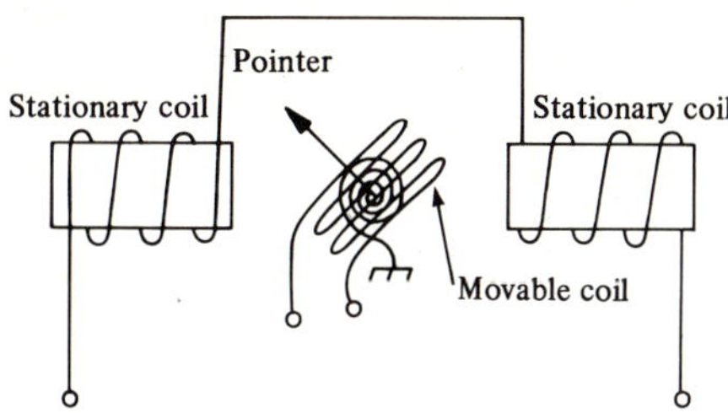

FIGURE 6.18
The electrodynamometer mechanism.

6.6 A WATTMETER: THE ELECTRODYNAMOMETER MOVEMENT

This type of movement is more expensive than the D'Arsonval type but can be used as a wattmeter for both dc and ac. The basic arrangement is shown in Fig. 6.18. The magnetic field in which the moving coil rotates is provided by the current flowing in the stationary coils. If the movable and stationary coils are in series, the same current flows through both, and since the torque is proportional to the product of the currents the movement of the pointer will also be proportional to this product, which represents the square of the current. The resulting scale, called a *square-law* scale, is nonlinear. Also, if the current reverses, it does so in both coils simultaneously so the direction of torque does not change. Thus the movement can be used for both dc and ac measurement.

When this movement is used as a wattmeter, the line current is passed through the stationary coil (called the current coil) and the line voltage is applied to the movable coil (the potential coil) through a multiplier resistance (see Fig. 6.19a). Thus the current in the potential coil is proportional to the line voltage and as noted in the previous paragraph the movement of the pointer is proportional to the product of the two currents. The meter reading is then proportional to the power. A cutaway drawing of a commercial electrodynamometer movement is shown in Fig. 6.20.

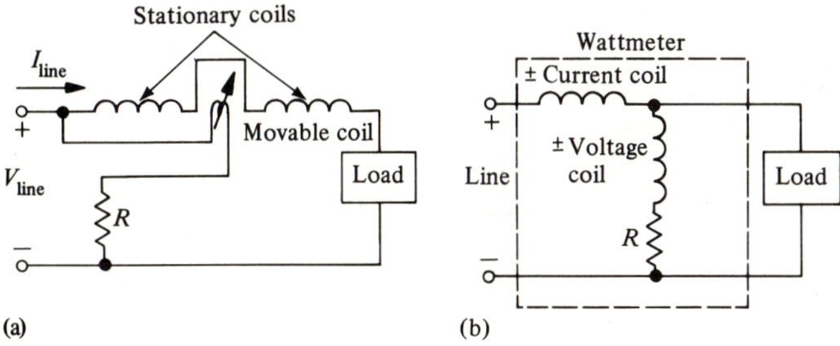

FIGURE 6.19
Wattmeter connections. (a) Pictorial schematic. (b) Conventional schematic.

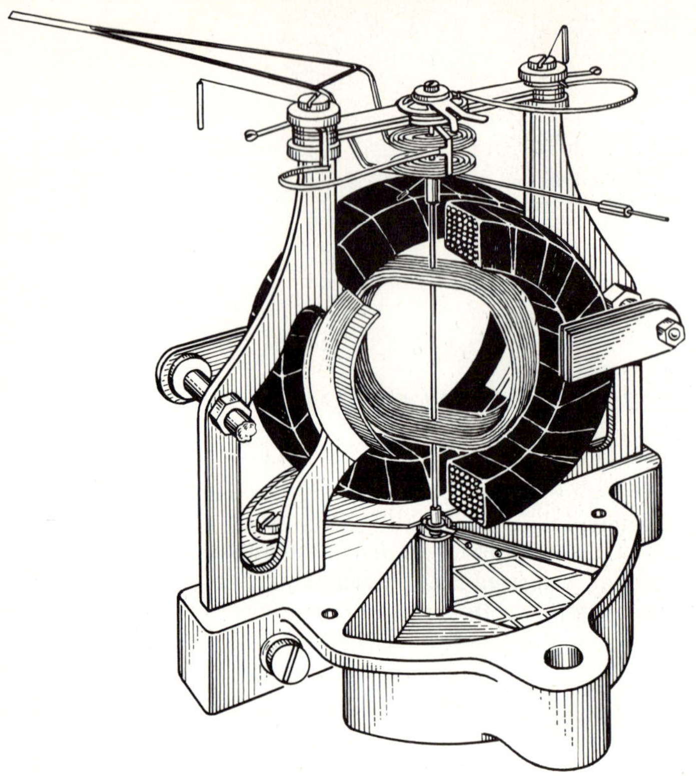

FIGURE 6.20
Electrodynamometer movement (courtesy of Weston Instruments, Inc.).

6.7 MULTIMETERS AND DIGITAL METERS

We have seen that the basic D'Arsonval meter movement can be used to measure current, voltage, and resistance by properly choosing the associated circuitry. With the use of gangs of switches, all three types of measurements, with several ranges for each type, can be performed with one instrument. Such instruments are called volt-ohm-milliammeters, usually abbreviated VOM. They usually have provisions for both ac and dc measurements and are battery operated for portability. A photograph of one popular VOM is shown in Fig. 6.21 and a schematic diagram is shown in Fig. 6.22. In the schematic, the Range switch is shown in the 500/1000 V position and the Function switch is at +dc. Tracing the schematic leads to the voltmeter circuit shown in Fig. 6.23. The circuits corresponding to other switch positions will be considered in the problems. Note that this VOM provides eight dc voltage ranges, six dc current ranges, and three resistance ranges in addition to various ac ranges. The rated accuracy for dc measurements is 2% of full scale and the meter sensitivity is 20,000 Ω/V.

FIGURE 6.21
Commercial VOM (courtesy of Simpson Electric Company).

One drawback of this VOM is its relatively low ohms-per-volt rating. Other, more expensive, VOMs utilize solid state devices called field-effect transistors (FETs) to increase the effective meter resistance to about 10 MΩ on *all* ranges. Thus the FET VOM will have negligible loading effect on almost any circuit to which it is connected.

The most sophisticated multimeters available at the time of this writing are digital multimeters, an example of which is shown in Fig. 6.24. From the picture, we see that the result of the measurement is displayed directly on a numerical display device and is very easy to read compared with the scale of an analog meter. In many digital multimeters (DMMs) range selection and decimal point placing are accomplished automatically, so that the user does not have to be concerned with these adjustments. The complex digital circuits which comprise the DMM provide extremely high accuracy, typically 0.05% of full scale in addition to input resistance of 1000 MΩ or more. As these instruments become cheaper, they are gradually displacing analog devices in most applications.

A number of different methods are used to convert a dc voltage to a direct-reading numerical display. The block diagram for one popular circuit is shown in Fig. 6.25. Operation is as follows: the voltage-controlled oscillator produces an output pulse train in which the number of pulses per second is proportional to the dc voltage being measured. Let us assume that

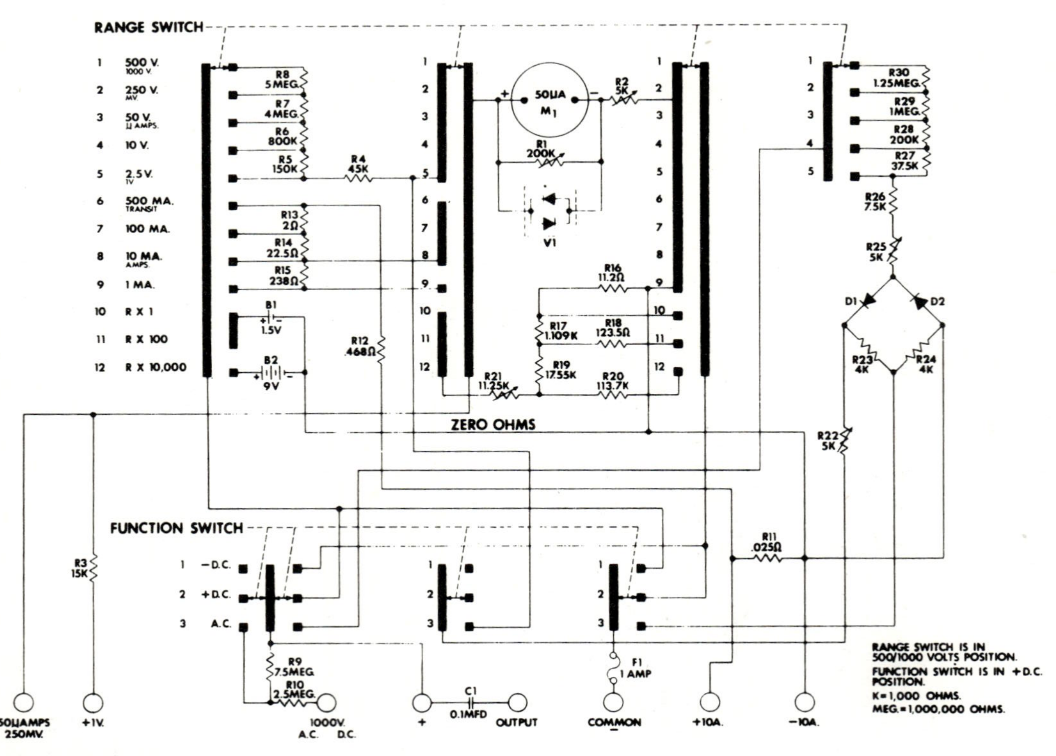

FIGURE 6.22
Schematic of commercial VOM (courtesy of Simpson Electric Company).

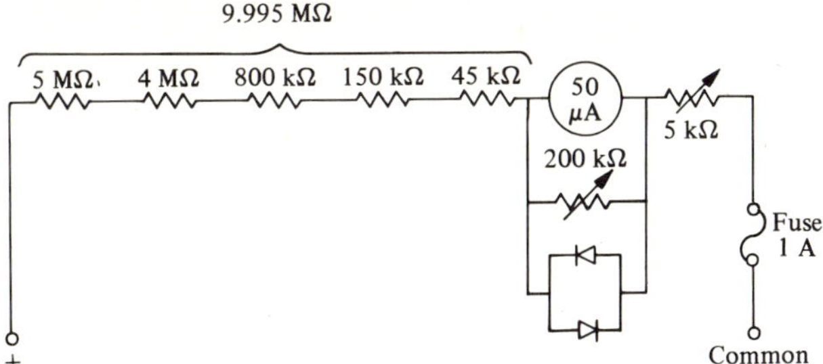

FIGURE 6.23
Voltmeter circuit of commercial VOM on 500-V range. Note: At 20,000 Ω/V, the total resistance on this range should be 20,000 Ω/V × 500 V = 10 MΩ.

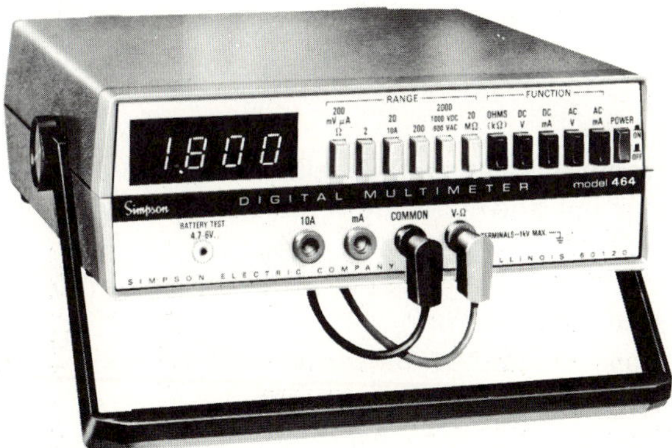

FIGURE 6.24
Digital multimeter (courtesy of Simpson Electric Company).

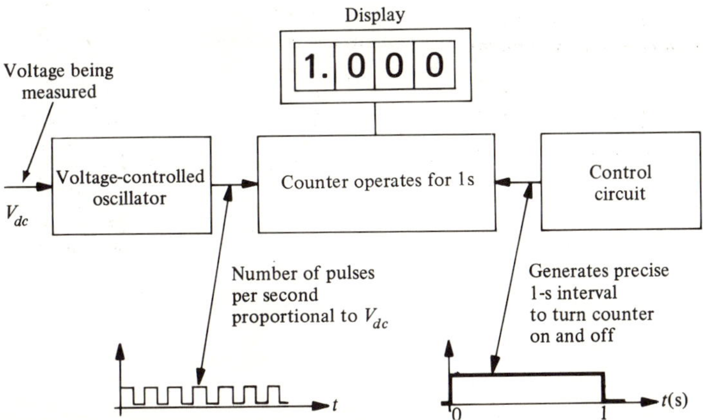

FIGURE 6.25
Basic digital voltmeter block diagram.

a dc output of 1 V produces a frequency of 1000 pulses per second. If the counter operates for exactly 1 s and then stops, 1000 pulses will have been counted. The circuit is designed so that the decimal point is placed properly and the display reads 1.000 V. If the input voltage is 1.5 V, the frequency will be 1500 pulses per second, 1500 pulses will be counted, and the display will read 1.500 V. The control block contains circuits that generate very accurate time intervals (1 s in our example), which are used as on-off signals for the counter.

SUMMARY

In this chapter we have considered the methods and instruments used to measure the current and voltage variables and resistance parameters that we studied in previous chapters. The important points concerning each of these are summarized below.

Section

6.1 1. The D'Arsonval movement is a current-sensitive device, used with shunts to measure current and with multipliers to measure voltage.

2. The sensitivity of a D'Arsonval movement is the current required for full-scale deflection.

6.2 3. An ammeter is always connected in *series*, and its resistance should be low compared to the circuit resistance.

4. A given movement can be used to measure currents greater than its current sensitivity by connecting a shunt in parallel. The extra current flows through the shunt.

6.3 5. A voltmeter is always connected in *parallel*, and its resistance should be high compared to the circuit resistance.

6. A given movement can be used to measure voltages greater than its full-scale voltage by connecting a multiplier resistor in series. The extra voltage appears across the multiplier.

7. The resistance of a given voltmeter is found by multiplying its sensitivity (ohms per volt) rating by the full-scale voltage.

6.4 8. Ohmmeters must *never* be connected to an energized circuit.

6.5 9. The Wheatstone bridge is used for precision measurement of resistance.

QUESTIONS FOR REVIEW

Sec. 6.1

1. Describe the operation of the D'Arsonval meter movement.
2. What is the *current sensitivity* of the D'Arsonval movement?

Sec. 6.2

3. How is the D'Arsonval movement used for dc current measurements?
4. How do we use a D'Arsonval movement to measure a current greater than its current sensitivity?
5. How can we construct a multirange ammeter?
6. What is the advantage of the Ayrton shunt?

Sec. 6.3

7. How is the D'Arsonval movement used to construct a voltmeter?
8. What is the ohms-per-volt rating of a meter movement?
9. What is meant by the term *meter loading?*
10. How are inaccuracies inherent in a meter movement specified?
11. Describe how a voltmeter can be used to find an accidental open circuit.

Sec. 6.4

12. Describe the series ohmmeter.
13. Why is the scale of a series ohmmeter very nonlinear?
14. Describe the shunt ohmmeter.
15. When is a shunt ohmmeter preferred over a series ohmmeter?
16. Describe the use of an ohmmeter to check continuity in a circuit.
17. Describe how a voltmeter can be used to measure very high resistances.

Sec. 6.5

18. Describe the *Wheatstone bridge.*
19. What is meant by the *range* of a Wheatstone bridge?
20. Describe the slide-wire bridge.

Sec. 6.6

21. Briefly describe the electrodynamometer wattmeter.

Sec. 6.7

22. Describe a typical VOM.

PROBLEMS

Sec. 6.1

1. A 1-mA 200-Ω movement reads 0.92 mA when exactly 0.9 mA is flowing. What is the percent error?
2. A 50-μA 1000-Ω D'Arsonval movement is to be used to measure current in the circuit of Fig. 6.2. In the circuit, $V = 4$ V and $R_L = 100$ kΩ. Find the error introduced due to the insertion of the meter.
3. Find values for R_1 and R_3 in Fig. 6.3b. (Refer to Example 6.2-2, in which the value of R_2 was found.)
4. A 50-μA 1000-Ω D'Arsonval movement is to be used with the Ayrton shunt shown in Fig. 6.3c. Find the required values of R_1, R_2, and R_3.
5. A 50-μA 1000-Ω movement is to be used in a milliammeter having ranges of 10, 25, 50, and 100 mA. Draw a circuit and show all component values. What voltage is across the meter when full-scale current flows for each range?
6. A 1-mA 100-Ω movement is to be used in a multirange ammeter to measure

full-scale currents of 1, 5, 10, and 50 A. Draw a circuit and show all component values.

7. A dc milliammeter has a full-scale reading of 5 mA. If the resistance on this scale is 2 Ω, what shunt resistance is required to extend the range to 5 A?

Sec. 6.3

8. Find suitable values for R_{s2} and R_{s3} in both circuits of Fig. 6.4b.
9. A voltmeter with a full-scale value of 30 V is to be made up using a 5-mA 20-Ω movement. Find the required multiplier resistance and the ohms-per-volt rating.
10. Repeat Prob. 9 for a 50-μA 1000-Ω movement.
11. A certain movement has a voltage of 100 mV across it at full scale. If the meter resistance is 10 Ω what value of multiplier resistance will convert it to a 100-V meter?
12. Repeat Example 6.3-3 for a 1-mA movement and compare the results.
13. A 50-V voltmeter has an accuracy of ±2% of full scale. When the meter reads 29 V, what is the range of the true value?
14. A voltmeter has ranges of 5, 10, and 50 V and its accuracy is rated at ±3% of full scale. A true voltage of 4 V is applied to the terminals. Find the possible error in the readings on the 5-, 10-, and 50-V scales. Which is the most accurate and why?
15. A multirange voltmeter has a sensitivity of 20 kΩ/V. Find the full-scale current and the resistance of each range if the ranges are 2.5, 10, 50, and 250 V.
16. A circuit consists of two 50-kΩ resistors in series connected to a 100-V source. A 100-V voltmeter with a resistance of 2 MΩ is used to measure the voltage across one of the resistors. Find the percent error in the measured value.
17. In the circuit of Fig. 6.26:

    a. With the meter disconnected, find the actual voltage across the 300-kΩ resistor.

    b. The meter is a 20 000 Ω/V unit with a range of 150 V full scale. What will it read?

    c. Another meter with a 1-mA movement and 150 V full-scale range is used. What will it read?
18. Could you use a voltmeter to track down the open-circuit fault in the circuit of Fig. 6.27?

Sec. 6.4

19. a. In the voltmeter-ammeter method for measuring resistance (see Example 6.4-1), the true value of the unknown resistance can be found from the apparent value of the resistance if the voltmeter resistance R_{vm} is known. Show that for the circuit of Fig. 6.7a, the true resistance is

$$R_x = \frac{R_{app}}{1 - R_{app}/R_{vm}}$$

(Hint: Note that $R_{app} = R_x \parallel R_{vm}$.)

b. Show that the true resistance for the circuit of Fig. 6.7b is

$$R_x = R_{app} - R_{am}$$

where R_{am} is the ammeter resistance.

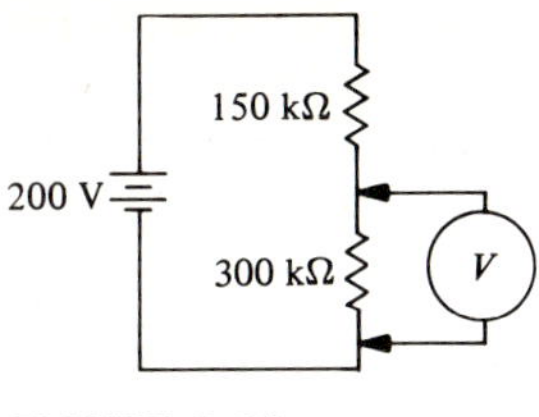

FIGURE 6.26

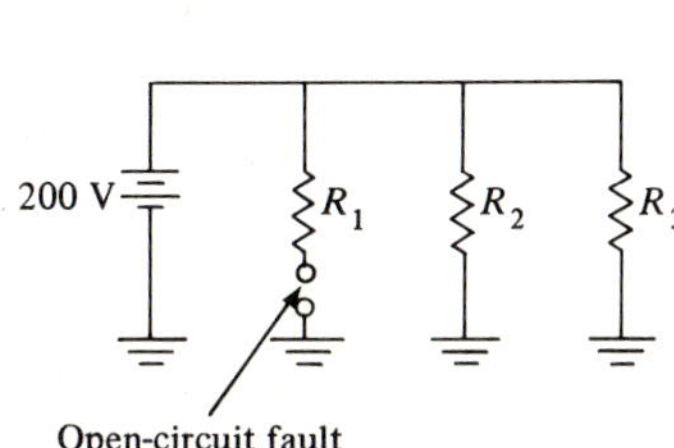

FIGURE 6.27

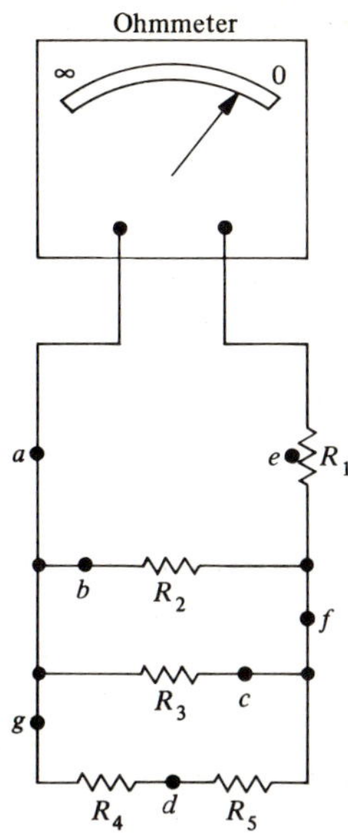

FIGURE 6.28

20. In the circuit of Fig. 6.7a V_1 is unknown, the ammeter is rated at 10 mA, 200 Ω, and the voltmeter is rated at 50 V, 1000 Ω/V. The ammeter reads 8 mA and the voltmeter reads 46 V. Find R_{app} and R_x.
21. A series ohmmeter is to use the following:
    1. A 100-μA 1000-Ω movement
    2. A zero-adjust variable resistor of 2 kΩ
    3. A 3-V battery
    4. A series resistor, value to be determined

    a. Find $R_s + R_z$ and the resistance R_x required for full, three-quarter, one-half, and one-quarter scale deflection.

    b. Draw the scale to be used with the meter.
22. For the shunt ohmmeter, prove that the midscale reading occurs when $R_x = R_m \parallel (R_s + R_z)$.
23. The components of Prob. 21 are to be used to construct a shunt ohmmeter.

    a. Find a suitable value for R_s.

    b. Find the unknown resistance R_x required for three-quarters, one-half, and one-quarter of full-scale deflection.

    c. Draw the scale.
24. Design a series ohmmeter using a 50-μA 1000-Ω movement and a 1.5-V battery. The midscale resistance must be 2 kΩ. (Hint: You may require a voltage divider to obtain an effective battery voltage less than 1.5 V.)
25. In a long two-wire cable is it possible to identify the individual wires using only an ohmmeter? Explain.
26. Repeat Prob. 25 for a three-wire cable and a five-wire cable.
27. In the circuit of Fig. 6.28, breaks can occur at different times at points *a* through *g*. For each possible break find the ohmmeter reading. In the circuit $R_1 = 100\ \Omega$, $R_2 = R_3 = R_4 = R_5 = 1\ \text{k}\Omega$.

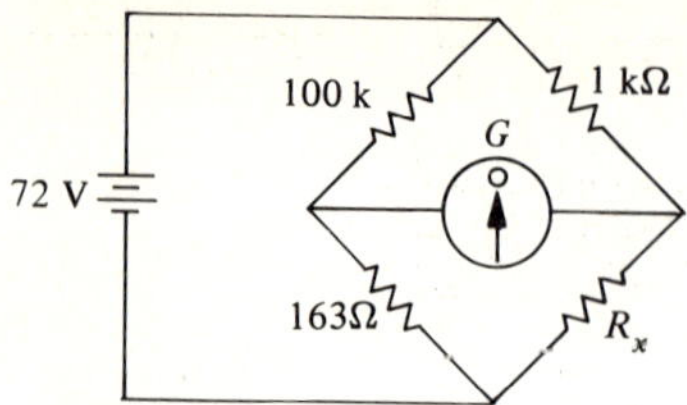

FIGURE 6.29

28. In the circuit of Fig. 6.14, the voltmeter has a sensitivity of 5000 Ω/V and reads 32 V on the 150-V scale. Find the insulation resistance of the cable.
29. In Fig. 6.14, the voltmeter has a sensitivity of 20 000 Ω/V and the cable resistance is 6 MΩ. What would a suitable scale be for the voltmeter in order that it read a voltage greater than 25% of full scale?

Sec. 6.5

30. The Wheatstone bridge shown in Fig. 6.29 is balanced. Find R_x.
31. In the Wheatstone bridge of Fig. 6.15 $R_1 = 10$ kΩ, $R_3 = 100$ Ω, and $R_x = 1632$ Ω. What value must R_2 have to obtain balance?
32. The Wheatstone bridge of Fig. 6.15 is balanced. The current through R_1 is 200 mA, the voltage across R_3 is 2.2 V, and the battery current is 280 mA. Find R_x.
33. A Wheatstone bridge is to have ranges 0 to 1, 0 to 10, and 0 to 100 kΩ. A variable 7.5-kΩ precision resistor is to be used for R_2. Find suitable R_3/R_1 ratios for the different ranges.
34. In the slide-wire bridge of Fig. 6.17, find R_x if $L_1 = 37$ cm and $R_2 = 53$ Ω.
35. In the slide-wire bridge of Fig. 6.17, $R_2 = 120$ Ω and $R_x = 153$ Ω. Find L_1 and L_3.

Sec. 6.7

36. In the multimeter circuit of Fig. 6.22, trace the 10-V dc voltmeter range and draw the resulting circuit.
37. Repeat Prob. 36 for the 1-mA dc range.
38. Repeat Prob. 36 for the $R \times 100$ range.

7
Magnetism and Magnetic Circuits

OBJECTIVES

Upon completion of this chapter the student should be able to

Section

7.1
1. Describe the magnetic effects that may be observed using iron filings and a permanent magnet.
2. State that a magnetic field can be thought of as a set of invisible lines of force, which interact with magnets and certain metals.

7.2
3. State that magnetic flux is synonymous with total number of lines of force.
4. State that flux density measures magnetic field strength in a small area.
5. Determine flux density in a region when appropriate data are provided.

7.3
6. Describe the magnetic effects that may be observed when iron filings are placed in the vicinity of a current-carrying conductor.
7. Using the right-hand rule, determine the direction of flux lines around a current-carrying conductor.
8. Using the solenoid right-hand rule, find the direction of the flux through the center of a solenoid.
9. State that magnetic flux can be established in a magnetic circuit only when a magnetomotive force is applied.
10. Find the electric analog circuit that corresponds to a given magnetic circuit.
11. Find the reluctance of a magnetic circuit given its dimensions and permeability.
12. State that the magnetic field intensity is the magnetomotive force per unit length of the magnetic circuit and is independent of the material.

	13.	State the formula that relates flux density to magnetic field intensity.
	14.	Use the BH curve for a magnetic material to find flux density or magnetic field intensity given appropriate data.
7.4	15.	Analyze a single loop magnetic circuit in which the magnetic path lies in one material.
7.5	16.	Analyze series magnetic circuits in which the magnetic path goes through two or more different materials.
	17.	Analyze a series magnetic circuit with an air gap.
7.6	18.	Analyze a series-parallel magnetic circuit with or without an air gap.
7.7	19.	Describe the operation of a typical electromagnetic relay.
7.8	20.	State Faraday's law for the voltage induced in a conductor when it is in motion and cuts across flux lines.
	21.	Make use of Faraday's law to find induced voltages.
7.9	22.	Compute the force on a current-carrying conductor in a magnetic field.

INTRODUCTION

One of the first things that we perceive is the relative motion of the objects around us. This motion is initiated, maintained, and terminated by the action of forces. One such force with which we are all familiar is the gravitational force that keeps us from tumbling about in space in the weightless state experienced by astronauts. The gravitational force is classed as a *field force*, which is a force that acts without any physical contact.

Magnetism is another field phenomenon with which all of us are familiar. We may have experienced the invisible forces that exist between permanent magnets, or felt the magnetic force in a motor when we tried to stop it from rotating. Magnetic force fields can arise in two ways. One way is from permanent magnets and the other is due to current flow in conductors. The analysis of magnetic circuits in many ways parallels that of electric circuits and we will exploit the electric analog as we progress.

We come in contact with many practical applications of magnetism every day. These include motors and generators of all kinds, audio recording on magnetic tape, video recording on magnetic tape or disc, floppy disc memories used in home computers, magnetic bubble memories used in large computers, and many others.

In this chapter we study the relationships between electricity, magnetism, and magnetic forces. These ideas will then be used to explain the operation of typical electromagnetic devices such as the relay and D'Arsonval movement. In a later chapter we describe the operation of the ac generator and the transformer.

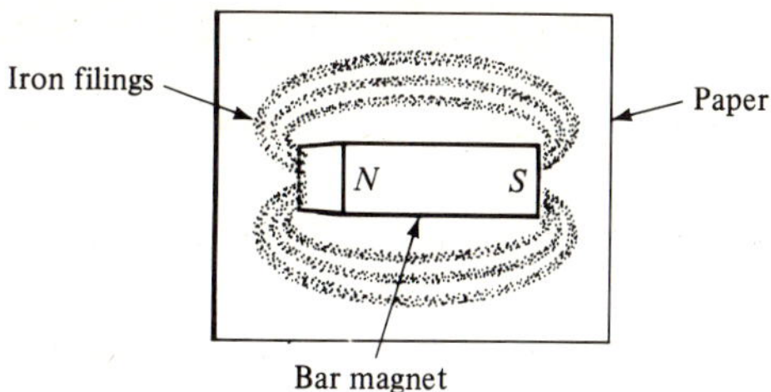

FIGURE 7.1
Pattern of field lines produced by iron filings sprinkled over a bar magnet.

7.1 THE MAGNETIC FIELD

We stated in the Introduction that magnetic force is a *field* phenomenon. This means that there exists a magnetic field in which a magnetic material will be acted upon by a magnetic force. The phenomenon of magnetism results in these forces, and as we will see later, induced voltages. Both the forces and induced voltages are relatively easy to measure and provide us with means to identify the presence of the field.

The magnetic field can be thought of as a set of invisible lines of force that interact with magnets and certain metals. These lines of force can be measured by the force they exert on a small magnet used as a test probe; a compass needle is frequently used for this purpose. Iron filings sprinkled on a piece of paper under which there is a magnet align themselves along these lines of force, providing a picture of the field as shown in Fig. 7.1.

The areas at the ends of the magnet where the filings are most concentrated are called the *poles* of the magnet. The north pole is the one that will point to the earth's north pole when the magnet is suspended and the direction of the lines of force is considered to be from north to south. The lines of force outlined by the iron filings are shown schematically in Fig. 7.2. They can also be detected by placing a small compass in the field as shown in the figure. The compass needle will align itself with the magnetic lines of force existing in the area it is placed in.

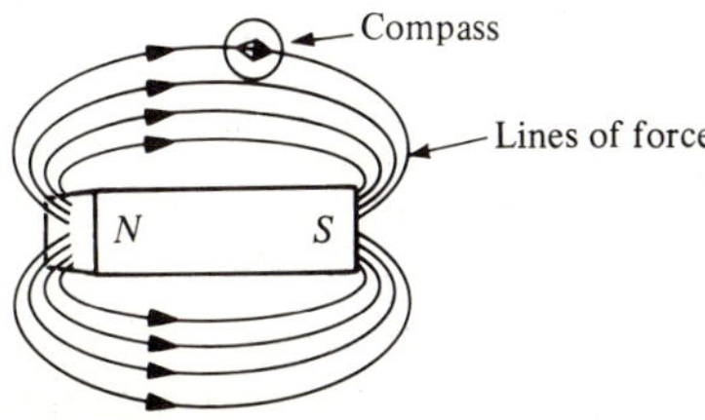

FIGURE 7.2
The magnetic field of the bar magnet.

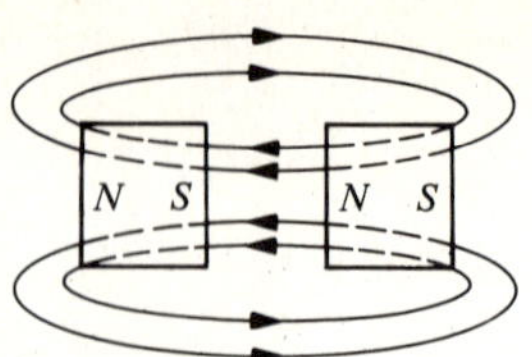

FIGURE 7.3
The magnetic field when a bar magnet is cut in two.

An interesting phenomenon occurs if we cut the bar magnet in two as shown in Fig. 7.3. Using iron filings or a compass needle, we find that lines of magnetic force have formed in the gap between the two halves and that new north and south poles have formed on each side of the gap. If we now cut the magnet again we find the same behavior in the new gaps. This occurs regardless of the number of times that the magnet is cut. Thus we conclude that the magnetic lines of force are continuous, and are connected between the south pole and north pole *inside* the magnet in order to form completely closed loops.

Another interesting phenomenon occurs if we place a piece of ferromagnetic material such as iron in the field of a permanent magnet. We find that the iron becomes *magnetized,* that is, it exhibits the properties of a magnet. This is illustrated in Fig. 7.4. At the end at which the direction of the magnetic force lines is toward the iron, a south pole appears, and a north pole appears at the opposite end. The piece of iron has become a temporary magnet by *induction.* When it is removed from the field of the permanent magnet it will lose most of its magnetism.

The magnetic effect we have all observed, that like poles repel each other, while unlike poles attract each other, can be explained by the fact that magnetic lines of force repel each other. This can be seen in Fig. 7.5, which shows the field resulting when two north poles are brought close together. The fact that the lines of force have a repelling effect on each

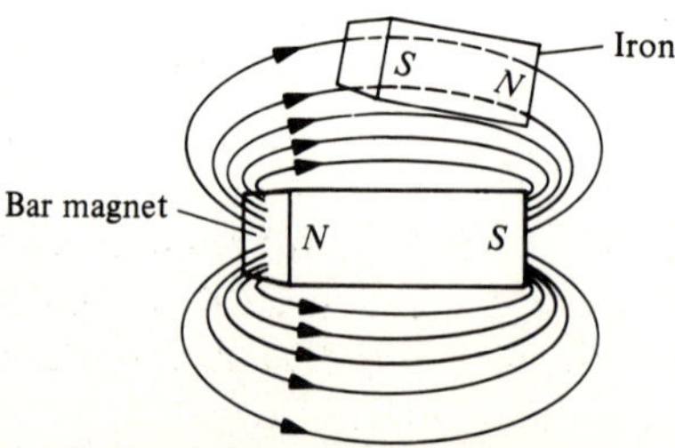

FIGURE 7.4
Ferromagnetic material in a magnetic field becomes magnetized.

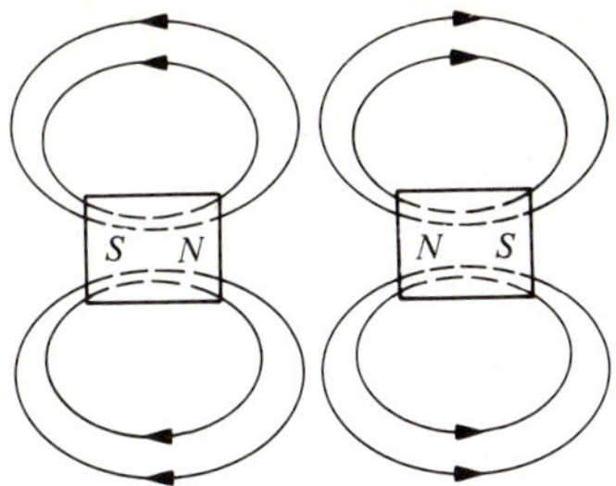

FIGURE 7.5
The magnetic field when like poles are placed close together.

other also indicates that they cannot intersect, but always form individual closed loops.

The characteristics described above all lead to the form of the magnetic field around a permanent magnet as shown in Fig. 7.1. These magnetic fields are found to exist in the vicinity of current-carrying conductors as well as near permanent magnets. The theory that follows is based on numerous experiments and provides a unified explanation for all of the observed results.

• • •

LEARNING EXERCISES FOR SEC. 7.1

Fill in the blanks.

1. Magnetic force is a ______ force.
2. Magnetic lines of force form ______ loops.
3. In the field outside a permanent magnet, the direction of the flux lines is from the ______ pole to the ______ pole.
4. Unlike magnetic poles ______ each other.

Ans. Closed; attract; north, south; field

• • •

7.2 MAGNETIC FLUX

In the previous section we used the term *magnetic field* in connection with the distribution of lines of force around a permanent magnet. Actually magnetic field refers to a number of different but related quantities. The first of these which we will consider is the *magnetic flux*.

Magnetic flux is a measure of the *total* number of lines of force associated with the field. Therefore, *magnetic flux* and *total number of lines of force* are synonymous. The symbol used for flux is the Greek letter ϕ

(phi), and the unit is the *weber* (Wb). As we found for the lines of force, lines of magnetic flux always form closed loops. If we look at the flux lines associated with a magnet they seem to start on one pole and terminate on the other; the lines continue inside the magnet, however, and form a closed loop. These lines do not have an actual physical existence; we cannot isolate them and say there is a line passing through this particular point in space. However, they may be compared to lines tracing the flow of electrons around an electric circuit, especially if they are confined to a relatively small area. Using this analogy, we can refer to the closed loop paths of the flux lines as a *magnetic circuit*. In the case of the magnetic circuit there is no actual *flow*, or motion of particles, as there is in the case of the electrons that form the current in the electric circuit. However, we will find that the similarities between the two enable us to make use of the electric circuit theory developed in previous chapters.

Magnetic Flux Density

We have seen that magnetic fields are distributed throughout regions of space. The field distribution, however, does not have to be uniform throughout the region. The *flux density* is a measure of the field strength over small regions rather than over the entire region. For example, in Fig. 7.6a we show a *uniform* field in which the flux lines are evenly distributed throughout the region. If we place an area A perpendicular to the field at any point in the region and then move it to another point, the area will always intersect the same number of flux lines. Thus, for this uniform field we can write, using the letter B to symbolize flux density, that

$$\boxed{B = \frac{\phi}{A}} \quad \frac{\text{Wb}}{\text{m}^2} \text{ (webers per square meter)} \tag{7.2-1}$$

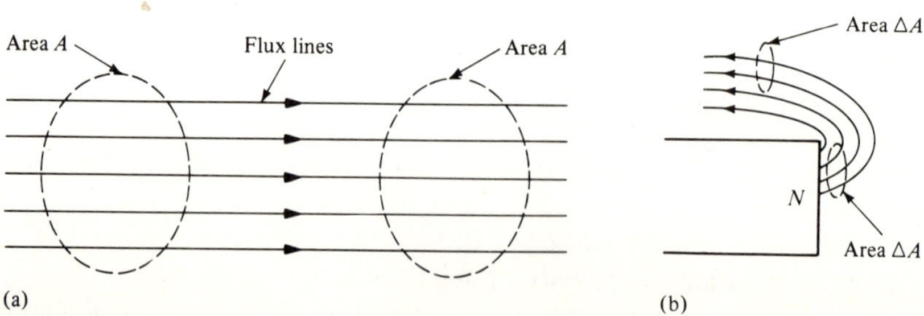

FIGURE 7.6
Flux density. (a) For a uniform field. (b) For a nonuniform field.

where ϕ is the flux in webers intercepted by area A square meters. In the SI the unit for flux density is the tesla (T) so that we have

$$1\ \text{T (tesla)} = 1\ \frac{\text{Wb}}{\text{m}^2}$$

In Fig. 7.6b we show a small region near the north pole of a permanent magnet. From the figure we see that a small area ΔA located near the pole of the magnet will intersect more flux lines than the same area located farther away from the pole. Thus the flux density (number of flux lines per unit area) is much greater as we move closer to the pole of the magnet. The reason is that the field is not *uniform* over the region, so the flux density will vary from point to point. Here we are faced with a situation similar to that which arose when we defined current and voltage in Chap. 2. By analogy, we are then led to a more general definition of flux density,

$$B = \frac{\Delta\phi}{\Delta A} \tag{7.2-2}$$

In most of the applications that we will consider, the flux will be distributed uniformly, so that Eq. (7.2-1) will suffice.

EXAMPLE 7.2-1 *Finding Flux Density*

The total flux inside a bar magnet is found to be 3.3 mWb. The area is 30 cm^2. Find the flux density, assuming that the flux is uniformly distributed throughout the cross section.

Solution

We first convert the area to square meters:

$$A = 30\ \text{cm}^2 \left(\frac{1\ \text{m}^2}{10^4\ \text{cm}^2}\right) = 3 \times 10^{-3}\ \text{m}^2$$

Then

$$B = \frac{\phi}{A} = \frac{3.3 \times 10^{-3}\ \text{Wb}}{3 \times 10^{-3}\ \text{m}^2} = 1.1\ \text{T}$$

• • •

EXAMPLE 7.2-2 *Finding the Total Flux*

The flux density inside the magnet of Fig. 7.2 is uniform with a value of 5000 T. The radius of the magnet's circular cross section is 0.5 cm. Find the total flux in the air around the magnet.

Solution

The *total flux* in the air is the same as the flux in the magnet since the flux lines are closed loops. (Note, in contrast, that the *flux density* is not the same inside and outside. Why?) Therefore, since the field is uniform,

$$\phi = BA = 5000\,\frac{\text{Wb}}{\text{m}^2} \times \pi(0.5\text{ cm})^2 \times \left(\frac{1\text{ m}}{100\text{ cm}}\right)^2$$

$$= 0.393\text{ Wb}$$

• • •

The Magnetic Field Due to a Current

We stated before that magnetic fields are found in the vicinity of current-carrying conductors as well as permanent magnets. This gives us the capability of exerting control over the magnetic field since it is easy to turn a current on and off or to change its strength, something we cannot easily do with permanent magnets. In this section we consider the magnetic field around a current-carrying conductor.

If we set up the experiment shown in Fig. 7.7a, we find that the iron filings sprinkled on the paper remain in the same random positions they assumed originally. However, when we close the switch, so that a current is established, we find that the filings arrange themselves in concentric circles around the conductor, indicating the presence of concentric magnetic lines of force. As long as the current is constant, the magnetic field is stationary; however, if we increase the current, the field expands, indicating that it is becoming stronger.

The fact that the flux lines indicated by the distribution of the iron filings have direction can be demonstrated by placing a small compass needle in the field. When the current is turned on, the needle will point in a certain direction. If the current is then reversed, it is found that the compass needle reverses its position, indicating that the field does indeed have

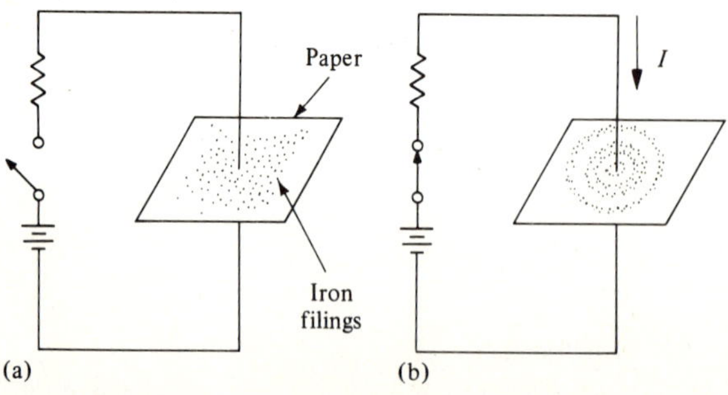

FIGURE 7.7
Magnetic field around a current-carrying conductor. (a) No current. (b) Current applied.

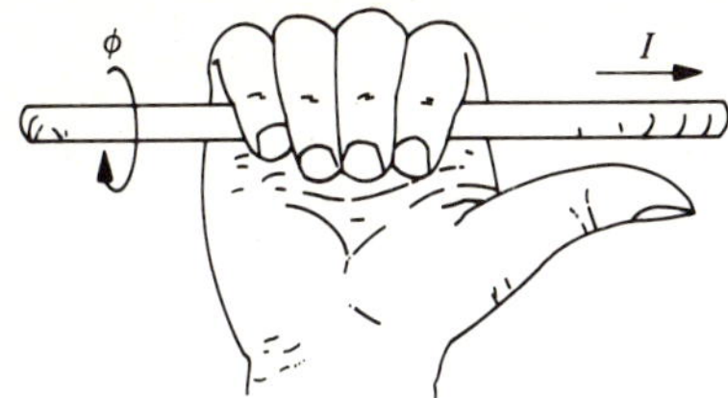

FIGURE 7.8
Right-hand rule for finding direction of flux lines around a current-carrying conductor.

direction. The actual direction can be determined by using the right-hand rule illustrated in Fig. 7.8. The rule can be stated in words as follows:

> *If the conductor is held in the right hand so that the thumb points in the direction of the current, then the flux lines around the conductor are in the same direction as the fingers circling the conductor.*

THE SOLENOID

When using magnetic fields we often want to concentrate the flux in a small area. This can be done in the following way: Consider that the current-carrying conductor is formed into a loop as shown in Fig. 7.9. Then, using the right-hand rule, we find that all of the flux lines pass through the center of the loop in the same direction. Thus, because the total number of lines has not changed, the flux density inside the loop will be greater than that outside. The field can be concentrated still further by using a large number of loops. This is done by winding the conductor in a cylindrical form as shown in Fig. 7.10. This configuration is called a *solenoid.* As shown on the diagram, the current in each turn of the coil is flowing in the same direction around the form so that the solenoid acts like a single loop of a multiple conductor cable, each conductor of which carries a current equal to the battery current. The magnetic field that results is similar to that of the bar magnet of Fig. 7.2, with the flux density greatest in the interior of the cylinder. Thus, the solenoid can be called an *electromagnet.*

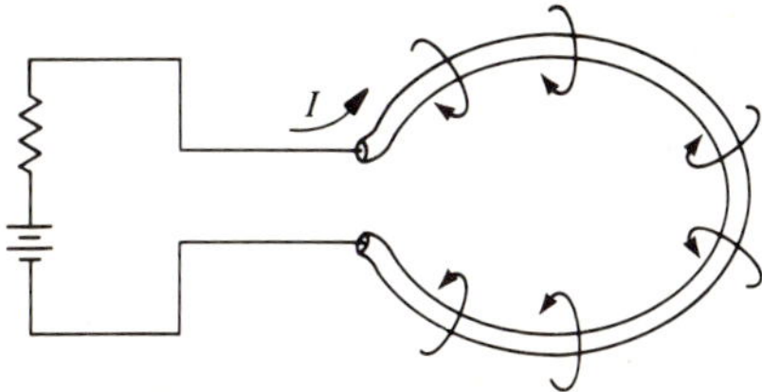

FIGURE 7.9
Illustrating the flux distribution inside a wire loop.

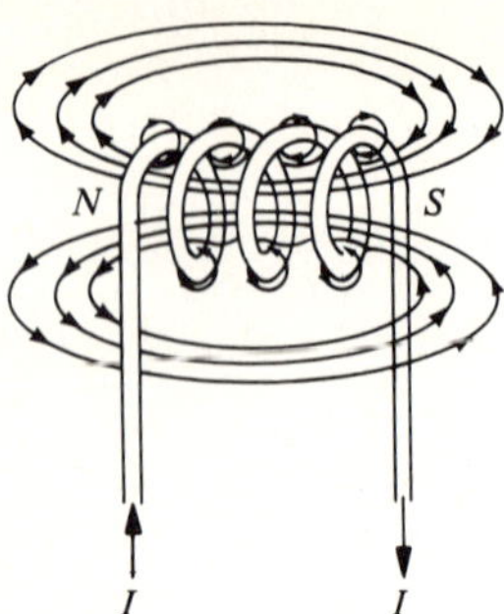

FIGURE 7.10
Magnetic field of a solenoid.

The direction of the flux lines can be found by another version of the right-hand rule, shown in Fig. 7.11.

If the fingers of the right hand are placed around the solenoid in the direction of the current, then the thumb will point in the direction of the magnetic flux lines though the center of the solenoid.

By analogy to the bar magnet in Fig. 7.2, we see that in this case the thumb will point to the north pole of the electromagnet.

Another interesting experiment is shown in Fig. 7.12a. Here we have arranged a single loop circuit so that two conductors carry the same current in opposite directions through our iron filing magnetic field detector. Using the right-hand rule we find that the flux directions are as shown and that between the two conductors, the flux lines all have the same direction. Since the flux lines exert a repelling force on one another, the two conductors will experience a force of repulsion. If we now rearrange the circuit so that the currents are in the same direction through the iron filings, we find that the flux lines between the conductors have opposite directions as shown in Fig. 7.12b. This results in a force of attraction between the two conductors. The magnitude of the force depends on the magnitude of the current in the conductors and on their length and spacing.

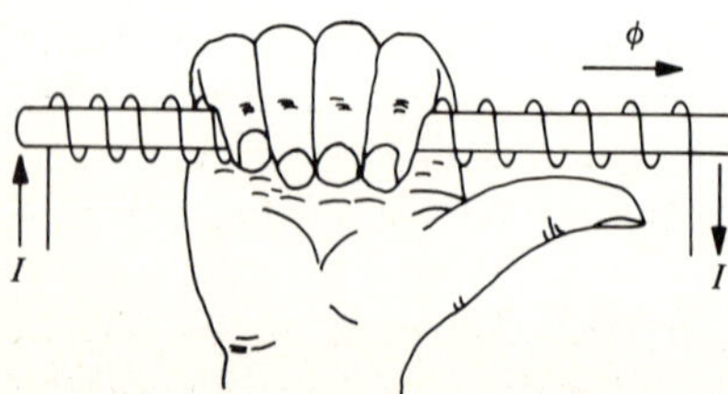

FIGURE 7.11
Right-hand rule for solenoids.

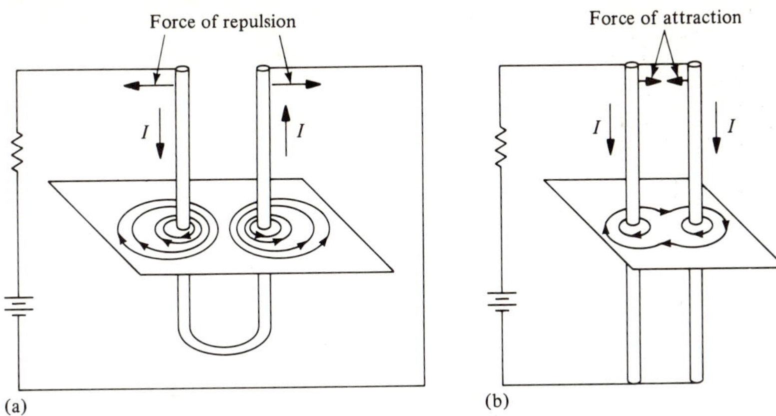

FIGURE 7.12
Illustrating the forces between current-carrying conductors. (a) Repulsion. (b) Attraction.

The fact that electric currents and magnetic fields interact to produce mechanical forces leads to such devices as the D'Arsonval meter movement, which we studied in Chap. 6, and others, such as the electric motor.

• • •

LEARNING EXERCISES FOR SEC. 7.2

1. The magnetic flux density inside the deflection yoke of a TV picture tube is 1.2 T and is uniform over an area of 5.2 cm^2. Find the total flux through this area in milliwebers.

2. In a nuclear particle accelerator, the total flux is 2.3 Wb concentrated in an area of 1.6 cm^2. What is the flux density in kiloteslas?

Ans. 0.624; 14.4

• • •

7.3 MAGNETOMOTIVE FORCE

The magnetic flux described in Sec. 7.2 is directly analogous to current in the electric circuit. In the typical electric circuit, we found that current was produced by application of a suitable voltage or *electromotive force*. Similarly, in a magnetic circuit a flux cannot be established until a *magnetomotive force* is applied.

When the magnetic field is due to an electric current, as in the case of the solenoid, the flux produced in the core depends on the current and the number of turns in the coil. The magnetomotive force (MMF) for which we use the letter symbol F_m is

$$F_m = NI \tag{7.3-1}$$

where

I = coil current, amperes
N = number of turns of wire in the coil
F_m = magnetomotive force in ampere-turns (At)

Reluctance

We now have two of the parameters required to set up a magnetic circuit–electric circuit analog. These are the flux ϕ, which is analogous to electric current, and the magnetomotive force F_m, which is analogous to battery voltage. The third parameter, which represents the proportionality constant between the applied MMF and the resulting magnetic flux, is the *reluctance* of the magnetic circuit. We use the letter symbol R_m for reluctance and can now set down Ohm's law for magnetic circuits (analogous to $V = RI$ for electric circuits)

$$F_m = R_m \phi \tag{7.3-2}$$

where

F_m = magnetomotive force in ampere-turns
ϕ = magnetic flux in webers
R_m = reluctance in ampere-turns per weber

A diagram illustrating the analogy between electric circuits and magnetic circuits is shown in Fig. 7.13. The solution of the electric circuit is usually extremely simple because the resistance element is *linear*, that is, its resistance value does not in any way depend on the applied voltage or the resulting current. Unfortunately, this is seldom true in the case of reluctance. The reluctance of practical magnetic circuits that contain ferromagnetic materials changes with variations in MMF and flux so that the magnetic circuit is *nonlinear*. Solution of such circuits often involves trial-and-error procedures and requires data on the magnetic properties of the materials involved. Before we consider this type of circuit some additional terms must be defined.

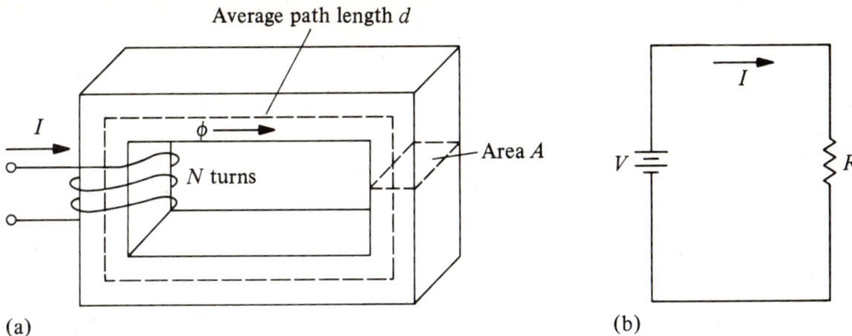

FIGURE 7.13
Single-loop magnetic circuit and electric analog. (a) Magnetic circuit. (b) Electric analog.

Permeability

In Chap. 3 we found that the resistance of a conductor to the flow of electricity is directly proportional to its length and inversely proportional to its cross-sectional area. The same holds true for magnetic reluctance, that is,

$$\boxed{R_m = \frac{d}{\mu A}} \tag{7.3-3}$$

where

R_m = reluctance in ampere-turns per weber
d = length of the magnetic path in meters
A = cross-sectional area in square meters
μ = permeability in webers per ampere-meter

The quantity μ is called the *permeability* of the material.* It is a measure of the ease with which a magnetic flux can be established in the material and is very similar to conductivity in electric circuits. Often, magnetic fields exist in air or vacuum. The permeability of free space (vacuum) is used often and is symbolized by μ_0. Its numeric value is

$$\mu_0 = 4\pi \times 10^{-7} \text{ Wb/Am} \tag{7.3-4}$$

*This is the Greek mu, the same symbol as the prefix for *micro*. Micro seldom arises in connection with magnetic circuits.

The permeability of *nonmagnetic* materials such as air, glass, wood, brass, and aluminum is essentially the same as that for vacuum. Magnetic materials, which include iron, steel, cobalt, and their alloys, have permeabilities many times that of free space and they are referred to as ferromagnetic materials.

Somctimes the relative permeability μ_r is used. This is defined as

$$\mu_r = \frac{\mu}{\mu_0}$$

μ_r is dimensionless and for free space $\mu_r = 1$.

For most materials of interest μ is not constant and values must be obtained from experimental data provided by the manufacturer of the material. This will be discussed in the next section.

Magnetic Field Intensity

We observed in Fig. 7.7 that magnetic fields exist around conductors carrying electric currents, as well as around permanent magnets. We would like to relate the strength of the magnetic field surrounding a current-carrying conductor to the current that gives rise to it. The field depends not only on the current, however, but on the material in which the field exists as well. Furthermore, the relationship between the current and magnetic flux density is nonlinear for many materials.

The *magnetic field intensity* is introduced to separate the effects of the current from the effects of the material on the magnetic field caused by the current. It is sometimes called magnetizing force or magnetic field strength and is defined as the magnetomotive force per unit length of the magnetic circuit. The letter symbol is H and, by definition,

$$\boxed{H = \frac{F_m}{d}} \tag{7.3-6}$$

where

F_m = MMF in ampere-turns
d = length of magnetic circuit in meters
H = magnetic field intensity in ampere-turns per meter

Using $F_m = NI$ [see Eq. (7.3-1)], we have

$$H = \frac{NI}{d} \tag{7.3-7}$$

Some numbers are given in the following example.

EXAMPLE 7.3-1 Finding the Magnetizing Force

In the magnetic circuit of Fig. 7.14, the current is 3 A, there are 20 turns, and the mean path length is 0.4 m. Find the magnetizing force.

Solution

$$H = \frac{NI}{d} = \frac{20 \times 3}{0.4} = 150 \text{ At/m}$$

• • •

It is important to note that the magnetizing force H is independent of the type of material in the magnetic path. It is determined only by the current, number of turns, and length of the path.

In order to complete our analysis we need a relation between magnetic field intensity and flux density. This can be derived as follows:

From Eqs. (7.3-2) and (7.3-3) we have

$$R_m = \frac{F_m}{\phi} = \frac{d}{\mu A} \tag{7.3-8}$$

Manipulating the last two forms so as to isolate μ, we find

$$\mu = \frac{\phi}{A} \times \frac{d}{F_m} \tag{7.3-9}$$

But $\phi/A = B$ [see Eq. (7.2-1)] and $F_m/d = H$ [see Eq. (7.3-6)]. Substituting these two relations we have

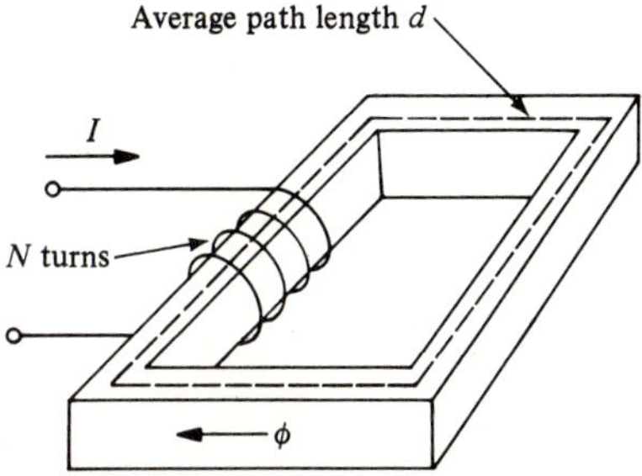

FIGURE 7.14
Magnetic circuit for Example 7.3-1.

$$\mu = \frac{B}{H} \tag{7.3-10}$$

which is rearranged as

$$\boxed{B = \mu H} \tag{7.3-11}$$

According to this equation, materials with higher permeabilities will have higher flux densities induced for a given applied magnetic field intensity.

As stated previously, most magnetic materials are nonlinear. The value of μ for most magnetic materials is not constant; it decreases with increasing values of H. For this reason the relationship between B and H for a specific material is often presented by an experimentally determined BH curve. Several curves are shown in Fig. 7.15. Notice that as H gets larger and larger, B tends to approach a constant value, called the "saturation level." In other words, there is a point of diminishing returns beyond which adding more ampere-turns will not noticeably increase the flux density. The value of μ for a given material and value of H is found by dividing the corresponding value of B by the value of H; graphically, it is the slope of the

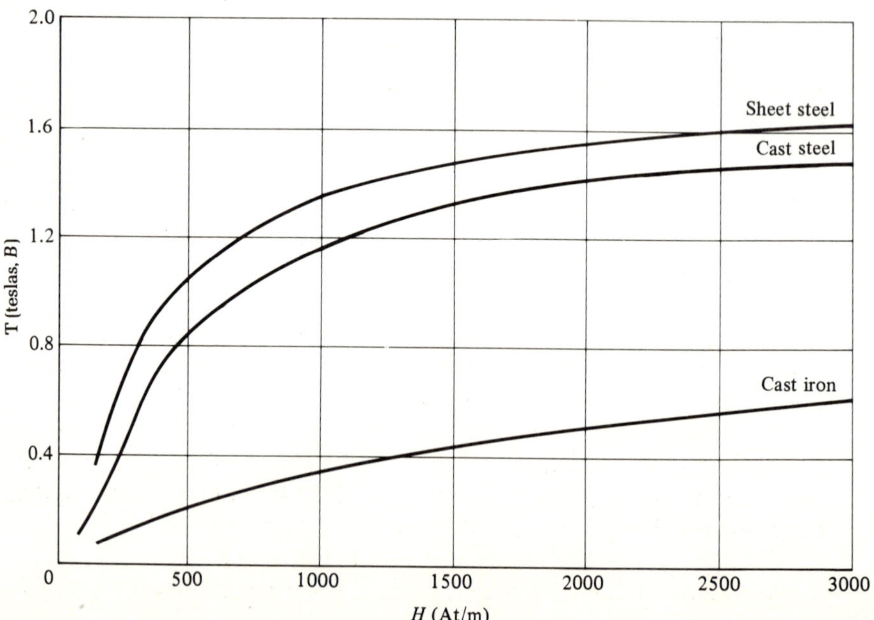

FIGURE 7.15
BH curves for some common ferromagnetic materials.

line from the origin to the point on the curve corresponding to the given values of B and H.

EXAMPLE 7.3-2 Using the BH Curve

The magnetic field intensity in cast steel is 300 At/m. Find the flux density. Find the new flux densities if H is first doubled and then quadrupled.

Solution

From Fig. 7.15 we find the following values:

H (At/m)	B (Wb/m^2)
300	0.55
600	0.93
1200	1.22

Note the saturation effect. It is evidenced by the fact that a 300 At/m change in H (from 300 to 600 At/m) results in a change of 0.38 Wb/m^2 in B, while a 600 At/m change in H (from 600 to 1200 At/m), which is double the first change, results in a smaller change in B of only 0.29 Wb/m^2.

• • •

SUMMARY

At this point we will pause and summarize what has been presented thus far. For convenience we list the steps in the development in sequence.

1. A magnetic field may be produced by a permanent magnet or by an electric current.
2. The field may be detected by the force it exerts on a test magnet or by using iron filings.
3. This force is explained by assuming that there are invisible lines of force, called magnetic flux ϕ. These lines always form closed loops.
4. The number of lines per unit area at a point in space is called the flux density B.
5. A new quantity called the magnetic field intensity H is introduced. It isolates the effect of the current on the field from the effect of the particular material in which the field exists.
6. The effect of the material is accounted for by making use of the relation between B and H for that material. The relation is usually nonlinear and is often available graphically. It provides the required link relating the magnetic flux ϕ to the current I which causes it.

In the next section we make use of these relations to solve typical magnetic circuits.

• • •

LEARNING EXERCISES FOR SEC. 7.3

1. In the electric circuit analog, ______ is analogous to current, ______ to voltage, and ______ to resistance.
2. The magnetic circuit of Fig. 7.14 is to be used in an automatic door opener. The mean path length is 8 cm and the coil has 50 turns. What current is required to establish a magnetizing force of 220 At/m?
3. How many ampere-turns per meter are required in cast iron and in cast steel to establish a flux density of 0.5 T?

Ans. 300; 0.35; 2000; flux, MMF, reluctance.

• • •

7.4 MAGNETIC CIRCUITS; ONE MATERIAL

Many of the everyday devices we think of as electrical depend on proper magnetic design as much as they do on proper electrical design. For example, in loudspeakers, motors, generators, and relays the magnetic circuit is often more important than the electric circuit. The electric circuit serves only to provide the MMF required to establish a certain flux density, which in turn provides the required mechanical forces. Some devices use permanent magnets for this purpose.

The solution of magnetic circuits is more complex than the solution of electric circuits, primarily because of the nonlinear *BH* curve. This complicates the evaluation of the answer; the circuit theory remains the same, and here we can make good use of the Ohm's law analogy.

Problems are basically of two types. In one case, the required flux is given and the MMF required to achieve this flux must be found. This type of problem occurs in the design of motors, generators, and transformers. In the other type, the MMF is given and the flux is to be found. This type of problem is more difficult when more than one kind of material is contained in the magnetic path and often requires a trial-and-error solution.

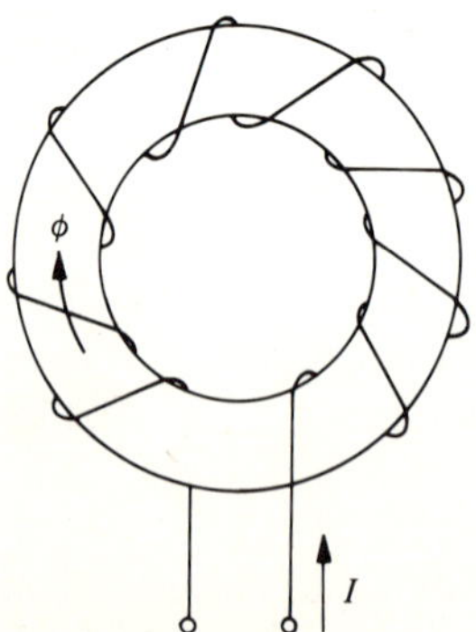

FIGURE 7.16
Toroidal coil.

The next example illustrates the simplest magnetic circuit. It consists of a circular core called a toroid around which is wound a coil of wire as shown in Fig. 7.16.

EXAMPLE 7.4-1 Calculating the Flux in a Toroidal Coil

The core in Fig. 7.16 has a mean path length of 50 cm and a cross-sectional area of 2 cm^2. The coil has 200 turns and carries a current of 2 A. Find the flux if the core material is (a) brass, and (b) cast steel.

Solution

For both materials, the MMF is

$$F_m = NI = 200 \times 2 = 400 \text{ At}$$

The reluctance is

$$R_m = \frac{d}{\mu A} = \frac{0.5 \text{ m}}{\mu \times 2 \times 10^{-4} \text{ m}^2} = \frac{2500}{\mu} \text{ At/Wb}$$

a. When the material is brass, the permeability is essentially the same as that of free space and

$$R_m = \frac{2500}{\mu_0} = \frac{2500}{4\pi \times 10^{-7}} = 2 \times 10^9 \text{ At/Wb}$$

and, using $\phi = F_m/R_m$

$$\phi_{\text{brass}} = \frac{F_m}{R_m} = \frac{4 \times 10^2 \text{ At}}{2 \times 10^9 \text{ At/Wb}} = 0.2 \times 10^{-6} \text{ Wb}$$

b. When the material is cast steel, we use the magnetization curves of Fig. 7.15. For $F_m = 400$ At and $d = 0.5$ m we have

$$H = \frac{400 \text{ At}}{0.5 \text{ m}} = 800 \text{ At/m}$$

From the BH curve for cast steel we read, for this value of H,

$$B = 1.1 \text{ T}$$

Then

$$\mu = \frac{B}{H} = \frac{1.1 \text{ Wb/m}^2}{800 \text{ At/m}} = 1.38 \times 10^{-3} \text{ Wb/Am}$$

$$R_m = \frac{2500}{\mu} = \frac{2500}{1.38 \times 10^{-3}} = 1.81 \times 10^6 \text{ At/Wb}$$

and

$$\phi_{\text{steel}} = \frac{F_m}{R_m} = \frac{4 \times 10^2}{1.81 \times 10^6} = 0.22 \text{ mWb}$$

The last three steps were included in order to illustrate the similarity to Ohm's law. However, once B is known, we can simply multiply by the core area to obtain

$$\phi = BA = (1.1 \text{ Wb/m}^2)(2 \times 10^{-4} \text{ m}^2) = 0.22 \text{ mWb}$$

Note that much more flux is established in the steel than in the brass.

• • •

LEARNING EXERCISE FOR SEC. 7.4

1. A toroidal memory core has a mean path length of 1.2 cm and is wound with 10 turns that carry a current of 50 mA. The cross-sectional area is 0.6 cm^2. Find the magnetizing force, flux density, reductance, and total flux if μ = 0.07 Wb/Am for the material.

Ans. 2.9; 41.7; 0.17×10^{-3}; 2.94×10^3

• • •

7.5 SERIES MAGNETIC CIRCUITS

In order to be able to analyze circuits in which the flux passes through different materials as in Fig. 7.17, we need some additional rules. If we assume that all of the flux is confined to the ferromagnetic material and none "leaks" into the surrounding air, then a line of force will make a complete loop passing through both materials. Thus the two sections are

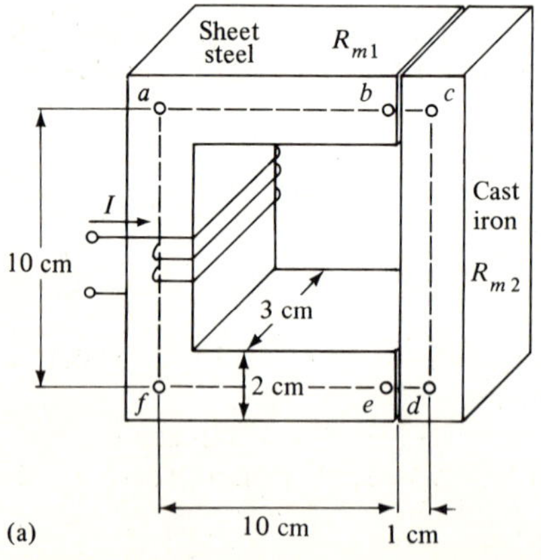

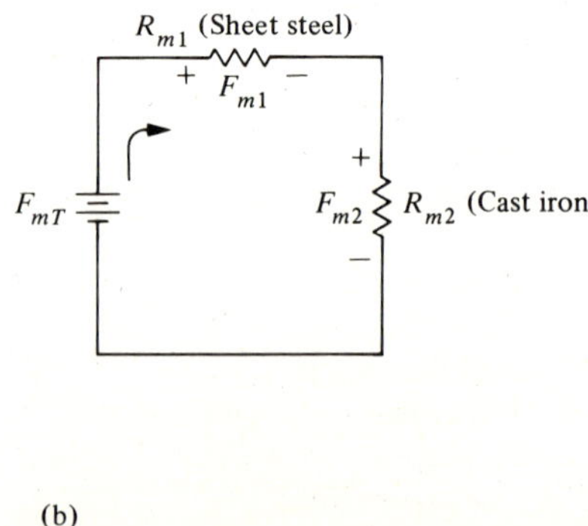

FIGURE 7.17
Series magnetic circuits with two different materials. (a) Magnetic circuit. (b) Electric analog.

effectively in *series* and just as the total resistance in a series electric circuit is the sum of the individual resistances, the total reluctance of the series magnetic circuit is the sum of the individual reluctances. Then

$$R_{mT} = R_{m1} + R_{m2} \tag{7.5-1}$$

The equivalent electric circuit is shown in Fig. 7.17b. The additional law required to solve series magnetic circuits is Ampere's Circuital Law, which is the analog of KVL.

The total magnetomotive force in a series magnetic circuit is the sum of the MMFs required for each section.

Applying this to the magnetic circuit of Fig. 7.17a we have

$$F_{mT} = F_{m1} + F_{m2} \tag{7.5-2}$$

where

F_{mT} = applied MMF = NI
F_{m1} = MMF "drop" in U-shaped sheet steel section = $R_{m1}\phi$
F_{m2} = MMF drop in cast iron section = $R_{m2}\phi$

EXAMPLE 7.5-1 MMF Required to Produce a Specified Flux

The relay frame shown in Fig. 7.17 is a typical series magnetic circuit. In the circuit the flux is to be 0.3 mWb. Find the current required to establish this flux if the coil has 100 turns.

Solution

The steps in the solution are as follows:

1. Find the flux density for each section using $B = \phi/A$.
2. From the magnetization curves (Fig. 7.15), using the values of B found in step 1, find the magnetic field intensity H for each material.
3. Using these values, determine the MMF = Hd required for each section.
4. Apply Ampere's Circuital Law (KVL) to find the total MMF required to establish the desired flux.
5. Using $F_m = NI$ find the required current.

From the diagram, we see that the flux path is divided into three straight-line segments in each material. In the sheet steel, the three segments are *b–a*, *a–f*, and *f*–*e*, each of which is 10 cm long, leading to a total path length in the sheet steel of d_{SS} = 30 cm = 0.3 m. In the cast iron, the segments are *b*–*c*, *c*–*d*, and *d*–*e*, which add up to a total path length of d_{CI} = 12 cm = 0.12 m. The area is constant throughout at 6 $\text{cm}^2 = 6 \times 10^{-4}\ \text{m}^2$. Next we follow the steps set down previously.

1. $B = \dfrac{\phi}{A} = \dfrac{0.3 \times 10^{-3}\ \text{Wb}}{6 \times 10^{-4}\ \text{m}^2} = 0.5\ \text{T}$

2. From Fig. 7.15 we find that a flux density of 0.5 T requires the following magnetizing forces:

   Sheet steel $\quad H_{SS} = 200\ \text{At/m}$
   Cast iron $\quad H_{CI} = 1850\ \text{At/m}$

3. The MMF required for each section is:

   Sheet steel $\quad H_{SS}d_{SS} = (200\ \text{At/m}) \times (0.3\ \text{m}) = 60\ \text{At}$
   Cast iron $\quad H_{CI}d_{CI} = (1850\ \text{At/m}) \times (0.12\ \text{m}) = 222\ \text{At}$

4. Using Ampere's Circuital Law,

   $$NI = H_{SS}d_{SS} + H_{CI}d_{CI} = 60 + 222 = 282\ \text{At}$$

5. Since N = 100 turns,

   $$100I = 282\ \text{At}$$

   $$I = \frac{282}{100} = 2.82\ \text{A}$$

Note that the bulk of the MMF is required for the path through the cast iron.

• • •

The Effects of an Air Gap in the Magnetic Circuit

Very often, the flux path in a magnetic circuit is not completely within magnetic material. Often an air gap is required for the operation of the device, as in the electromagnetic relay. Consider the magnetic circuit of Fig. 7.17. If we introduce an air gap at some point in the circuit, all of the lines of force in the ferromagnetic material must pass through the air gap. Thus the gap is in series with the remainder of the circuit and the solution is the same as in Example 7.5-1 with the addition of the reluctance due to the air gap. However, because magnetic lines of force tend to repel one another, their passage through the air will result in a spreading out of the lines as shown in Fig. 7.18. This effect is called *fringing*, and if the length of the air gap is small compared with its cross section, we can approximately account for the fringing by calculating an effective cross-sectional area in which the length

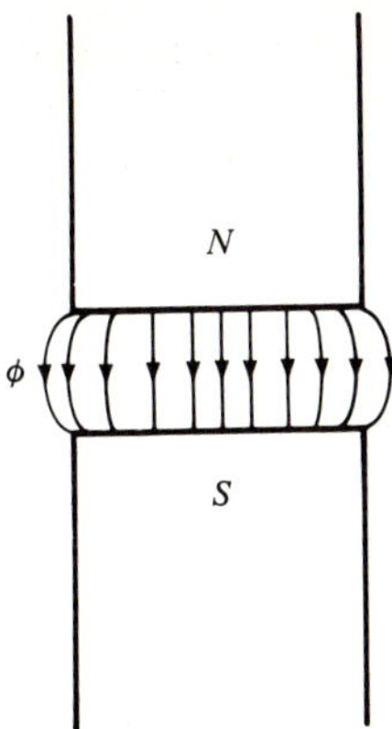

FIGURE 7.18
Fringing effect at an air gap.

of the air gap is added to each cross-sectional dimension. This is illustrated in the following example.

EXAMPLE 7.5-2 *Effect of Air Gap*

In the magnetic circuit of the relay shown in Fig. 7.17 assume that the relay is open so that a 2-mm air gap is introduced at point b. Find the current required to establish a flux of 0.3 mWb. As in Example 7.5-1 the coil has 100 turns.

Solution

The steps in the solution are identical to Example 7.5-1 with the addition of the air gap. For the air gap, we add the length 2 mm to each cross-sectional dimension. Then

$$A_G = (2\text{ cm} + 0.2\text{ cm})(3\text{ cm} + 0.2\text{ cm}) = 7.04\text{ cm}^2 = 7.04 \times 10^{-4}\text{ m}^2$$

The flux density in the air gap is

$$B_G = \frac{\phi}{A_G} = \frac{3 \times 10^{-4}\text{ Wb}}{7.04 \times 10^{-4}\text{ m}^2} = 0.426\text{ T}$$

The magnetic field intensity in the air gap is

$$H_G = \frac{B_G}{\mu_0} = \frac{0.426}{4\pi \times 10^{-7}} = 0.34 \times 10^6\text{ At/m}$$

The magnetomotive force required for the air gap is

$$F_{mG} = H_G d_G = (3.4 \times 10^5\text{ At/m})(0.002\text{ m}) = 680\text{ At}$$

The total MMF is (see step 4 of Example 7.5-1)

$$
\begin{aligned}
NI &= H_{SS}d_{SS} + H_{CI}d_{CI} + H_G d_G \\
&= 60 + 222 + 680 \\
&= 962 \text{ At}
\end{aligned}
$$

Finally

$$I = \frac{962 \text{ At}}{100 \text{ t}} = 9.62 \text{ A}$$

This should be compared with the current of 2.82 A found in Example 7.5-1 for the same flux without the air gap. Note that most of the MMF required to establish the flux is due to the nonmagnetic air gap.

• • •

In the next example we consider the more interesting problem of finding the flux due to a given coil current.

EXAMPLE 7.5-3 Finding the Flux When NI is Given

In the magnetic circuit of Fig. 7.19 the coil current and number of turns are given. Find the resulting flux.

Solution

This problem requires a trial-and-error solution. The steps in the solution are as follows:

1. Assume that a large fraction of the applied MMF appears across the air gap.
2. Use this value of MMF to find H_G, B_G, and ϕ_G.
3. The flux will be the same in the cast steel, so we use it to find the MMF required for the cast steel.

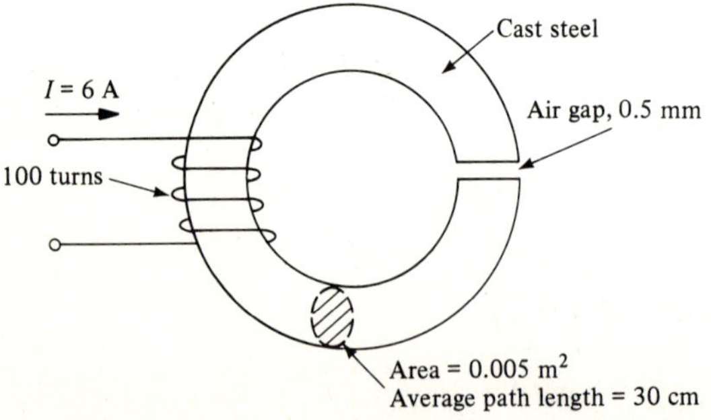

FIGURE 7.19
Toroid with an air gap for Example 7.5-3.

4. The MMFs for both materials are added, and the results compared with the given applied MMF.
5. If this result is close enough to the applied MMF we have our answer. If not, we change the fraction in step 1 and repeat the procedure.
6. Step 5 is repeated until the desired accuracy is obtained.

Applying these steps to the magnetic circuit of Fig. 7.19:

1. We assume that 80% of the applied MMF appears across the gap. Then

$$(NI)_G = 0.8 \times 100 \text{ turns} \times 6 \text{ A} = 480 \text{ At}$$

2. Using $H_G d_G = (NI)_G$ we have

$$H_G = \frac{(NI)_G}{d_G} = \frac{480 \text{ At}}{0.5 \times 10^{-3} \text{ m}} = 0.96 \times 10^6 \text{ At/m}$$

Then

$$B_G = \mu_0 H_G = (4\pi \times 10^{-7} \text{ Wb/Am}) \times (9.6 \times 10^5 \text{ At/m}) = 1.2 \text{ T}$$

and the flux is (we neglect fringing)

$$\phi_G = B_G A_G = (1.2 \text{ T}) \times (0.005 \text{ m}^2) = 6 \times 10^{-3} \text{ Wb}$$

3. We assume $\phi_{CS} = \phi_G = 6 \times 10^{-3}$ Wb. Then since the area is the same, $B_{CS} = 1.2$ T. From Fig. 7.15 we find the corresponding value of H_{CS} to be 1100 At/m. The MMF required for the cast steel is then

$$(NI)_{CS} = H_{CS} d_{CS} = (1100 \text{ At/m}) \times (0.3 \text{ m}) = 330 \text{ At}$$

4. The total MMF is

$$\begin{aligned}(NI)_T &= (NI)_{CS} + (NI)_G \\ &= 330 + 480 \\ &= 810 \text{ At}\end{aligned}$$

This is about 35% greater than the actual applied MMF of 600 At, so we will need a second trial. For the second trial we assume 70% of the applied MMF appears across the air gap. Thus

1. $(NI)_G = 0.7 \times 600 = 420$ At

2. $H_G = \dfrac{(NI)_G}{d_G} = \dfrac{420 \text{ At}}{0.5 \times 10^{-3} \text{ m}} = 0.84 \times 10^6 \text{ At/m}$

$$B_G = \mu_0 H_G = (4\pi \times 10^{-7} \text{ Wb/Am}) (8.4 \times 10^5 \text{ At/m}) = 1.06 \text{ T}$$

$$\phi_G = B_G A_G = (1.06 \text{ T}) (0.005 \text{ m}^2) = 5.3 \text{ mWb}$$

3. $B_{CS} = B_G = 1.06$ T

From Fig. 7.15, $H_{CS} = 760$ At/m. Then

$$(NI)_{CS} = H_{CS}d_{CS} = (760 \text{ At/m})(0.3 \text{ m}) = 228 \text{ At}$$

4. $$(NI)_T = (NI)_{CS} + (NI)_G$$
$$= 228 + 420$$
$$= 648 \text{ At}$$

This is within 10% of the actual value of 600 At. If a closer estimate were required we would repeat the process using 67% as the fraction in step 1. The final result is that the flux in the magnetic circuit is

$$\phi \approx 5.3 \text{ mWb}$$

• • •

This type of problem can become quite tedious if several materials with different lengths and cross-sectional areas are involved. When this is the case, calculations should be organized in tabular form in order to facilitate checking.

7.6 SERIES-PARALLEL MAGNETIC CIRCUITS

In many magnetic devices, the flux paths are similar to current paths in a series-parallel electric circuit. For example, in the magnetic circuit of Fig. 7.20 the flux ϕ_T due to the current in the coil splits at junction a into two component fluxes, ϕ_1 and ϕ_2. The electric analog of this magnetic circuit is shown in Fig. 7.20b. In order to analyze this circuit, we need to know how the component fluxes are related to the total flux. This behavior is the direct magnetic analog of KCL, that is, at junction a

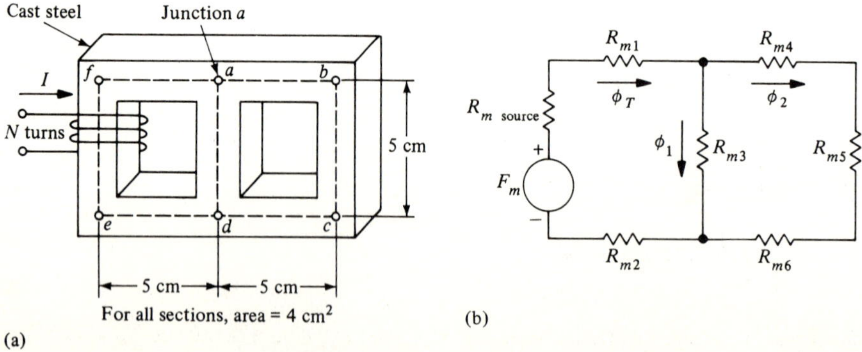

FIGURE 7.20
Series-parallel magnetic circuit. (a) Pictorial diagram. (b) Electric analog.

$$\phi_T = \phi_1 + \phi_2 \qquad (7.6\text{-}1)$$

With this rule available, we consider a numerical example.

EXAMPLE 7.6-1 MMF in a Series-Parallel Magnetic Circuit

Figure 7.20 shows the magnetic circuit of a multiwinding telephone transformer. Find the current required to establish a flux in the outer leg $\phi_2 = 0.1$ mWb.

Solution

The procedure for this example is to start with the outer leg in which ϕ_2 flows and work back to the coil. The steps are as follows.

1. Knowing ϕ_2 find the flux density in the ϕ_2 path *abcd*

$$B_{abcd} = \frac{\phi_2}{A} = \frac{0.1 \times 10^{-3}\ \text{Wb}}{4 \times 10^{-4}\ \text{m}^2} = 0.25\ \text{T}$$

2. Using the magnetization curves of Fig. 7.15, find the magnetizing force H for this path: From the curve, $H_{abcd} = 160$ At/m
3. The path length is $d_{abcd} = 5\ \text{cm} + 5\ \text{cm} + 5\ \text{cm} = 15\ \text{cm} = 0.15\ \text{m}$.
   Thus the magnetomotive force is

$$H_{abcd}d_{abcd} = (160\ \text{At/m}) \times (0.15\ \text{m}) = 24\ \text{At}$$

4. Use Ampere's Circuital Law around loop *abcda* to find $H_{ad}d_{ad}$.
   The appropriate form of the law is

$$\sum_{abcda} F_m = 0$$

For the sum, we break the paths into two segments, *abcd* and *ad*, then, taking account of "polarity" as we would with KVL, we have

$$H_{abcd}d_{abcd} - H_{ad}d_{ad} = 0$$
$$24 - H_{ad}d_{ad} = 0$$
$$H_{ad}d_{ad} = 24\ \text{At}$$

5. Find the magnetizing force for path *ad*

$$H_{ad} = \frac{24\ \text{At}}{0.05\ \text{m}} = 480\ \text{At/m}$$

6. From the magnetization curves find B_{ad}. We read from the curves $B_{ad} = 0.83$ T.
7. Find ϕ_1.

$$\phi_1 = B_{ad}A = 0.83 \times 4 \times 10^{-4} = 0.332\ \text{mWb}$$

8. Find ϕ_T

$$\phi_T = \phi_1 + \phi_2$$
$$= 0.332 + 0.1$$
$$= 0.432 \text{ mWb}$$

9. Find the flux density in sections *de*, *ef*, and *fa*.

$$B_{defa} = \frac{\phi_T}{A} = \frac{0.432 \times 10^{-3}}{4 \times 10^{-4}} = 1.08 \text{ T}$$

10. From the *BH* curves, find H_{defa}. We read from the curve

$$H_{defa} = 800 \text{ At/m}$$

11. Find the MMF for the path *defa*

$$H_{defa}d_{defa} = (800 \text{ At/m})(0.15 \text{ m}) = 120 \text{ At}$$

12. Apply Ampere's Circuital Law around loop *fadef*. This yields the total ampere turns required to establish the desired flux.

$$NI = H_{defa}d_{defa} + H_{ad}d_{ad}$$
$$= 120 + 24$$
$$= 144 \text{ At}$$

13. Find the current. Since $N = 100$

$$I = \frac{144 \text{ At}}{100 \text{ t}} = 1.44 \text{ A}$$

This problem is much more complicated than the equivalent electrical problem due to the nonlinear relation between *B* and *H*. However, with high-speed computers available, the technologist should know how to set the problem up, and can leave the tedious details of solution to the computer.

• • •

7.7 DESIGN OF THE MAGNETIC CIRCUIT OF A RELAY

The relay is an electromechanical switch; basically a switch that is operated by an electromagnet whose input is an electric signal. A diagram illustrating the construction of a typical relay is shown in Fig. 7.21. When current starts in the coil a magnetic flux is established in the core, causing it to become magnetized. The magnetic force causes the movable part of the core, called the *armature*, to be attracted to the fixed part, thereby closing the contacts.

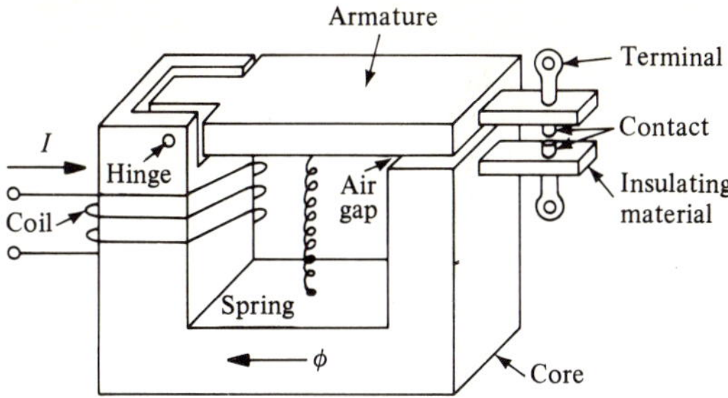

FIGURE 7.21
Construction of a typical relay.

When the current is shut off, the magnetic flux becomes zero and there is no magnetic force. The spring then pushes the armature of the core open, opening the contacts.

The problem considered here is the determination of the number of turns required in the coil to cause the relay to close for a given current. Suppose the specifications are as follows:

1. The mean path length of the core and armature is 8 cm
2. The cross section of the core and armature is 1 × 1.5 cm
3. The length of the air gap is 2 mm
4. The material is cast steel throughout
5. The flux required to overcome the spring force and close the air gap is 0.12 mWb.

Find the number of turns required if the pull-in current, the current required to close the relay, is to be 200 mA. Then determine the release current, the current at which the relay opens.

The steps required in the determination of the number of turns are

1. Compute the cross-sectional area of the core

$$A_C = 1 \text{ cm} \times 1.5 \text{ cm} \times \left(\frac{1 \text{ m}}{100 \text{ cm}}\right)^2 = 0.15 \times 10^{-3} \text{ m}^2$$

2. Find the flux density in the core:

$$B_C = \frac{\phi}{A_C} = \frac{0.12 \times 10^{-3} \text{ Wb}}{1.5 \times 10^{-4} \text{ m}^2} = 0.8 \text{ Wb/m}^2$$

3. Find the magnetic field intensity in the core from the BH curve in Fig. 7.15

$$H_C = 450 \text{ At/m}$$

4. Find the effective area of the air gap. This is greater than the area of the core, due to fringing (discussed in Sec. 7.5). If the length of the air gap is relatively small compared with the dimensions of its cross section, then a reasonable approximation to its effective cross-sectional area is found by adding its length to each cross-sectional dimension of the core. Thus

$$A_G = (1 + 0.2)\text{ cm} \times (1.5 + 0.2)\text{ cm} \times \left(\frac{1\text{ m}}{100\text{ cm}}\right)^2 = 0.204 \times 10^{-3}\text{ m}^2$$

5. Compute the flux density in the air gap

$$B_G = \frac{\phi}{A_G} = \frac{0.12 \times 10^{-3}\text{ Wb}}{2.04 \times 10^{-4}\text{ m}^2} = 0.588\text{ Wb/m}^2$$

6. Calculate the magnetic field intensity in the air gap

$$H_G = \frac{B_G}{\mu_0} = \frac{0.588\text{ Wb/m}^2}{4\pi \times 10^{-7}\text{ Wb/Am}} = 0.468 \times 10^6\text{ At/m}$$

7. Find the total MMF required using Ampere's circuital law:

$$\text{MMF} = H_C d_C + H_G d_G = (0.08)(450) + (2 \times 10^{-3})(0.468 \times 10^6)$$
$$= 36 + 936 = 972\text{ At}$$

8. The number of turns required is found by equating the MMF to the product of the pull-in current and the number of turns, N.

$$N = \frac{\text{MMF}}{I_P} = \frac{972\text{ At}}{0.200\text{ A}} = 4860\text{ turns}$$

With this number of turns and a current of 200 mA, the force produced by the electromagnet will be sufficient to close the air gap. When this happens the contacts will come together, closing the switch. Most of the MMF appears across the air gap when it is open. However, when the gap closes, less MMF (fewer NI) is required to maintain the same flux. Thus, once the air gap is closed, we no longer need 200 mA in the coil. The minimum current required to maintain the desired flux of 0.12 mWb is called the *release* current. It is found by following the same sequence of steps except that there is no air gap since the relay is closed and in the last step the MMF is divided by $N = 4860$ turns to compute I_R. The steps are:

9. The magnetic field intensity in the core is the same as calculated for the pull-in current in step 3: $H_C = 450$ At/m

10. The MMF is

$$\text{MMF} = H_C d_C = (0.08)(450) = 36 \text{ At}$$

When the gap is closed this is the total MMF.

11. The release current is

$$I_R = \frac{\text{MMF}}{N} = \frac{36 \text{ At}}{4860 \text{ t}} = 7.4 \text{ mA}$$

Thus the current required to hold the relay closed once it has been closed is much less than that required to close it when it is open. This is due to the absence of the air gap when the relay is closed.

7.8 ELECTROMAGNETIC INDUCTION

In 1831 Faraday discovered that a voltage was *induced* within a conductor when it moved through a magnetic field, as shown pictorially in Fig. 7.22. In the diagram a conducting rod is shown being moved through the magnetic field of a permanent magnet. The fact that a voltage is induced between the ends of the conducting rod is indicated by the zero-center galvanometer. As the conductor is moved upward through the magnetic field, the galvanometer needle will deflect away from zero in one direction. When the motion is downward, the galvanometer needle will deflect in the other direction. It is important to note that the only time that a voltage will be induced so that a

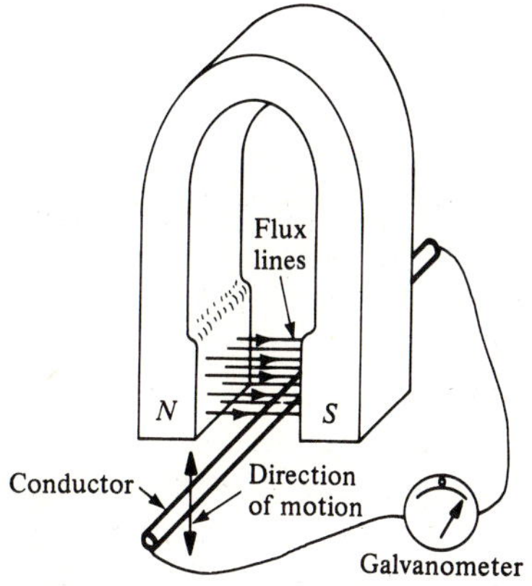

FIGURE 7.22
Electromagnetic induction when a conductor moves through a magnetic field.

reading appears on the galvanometer is when the conductor is *moving through* the magnetic field and thereby cutting across magnetic flux lines. The same effect is obtained if the conductor is stationary and the magnetic field moves. This discovery is of prime importance in the history of electricity, for as a direct result, Faraday was able to build the first electric *generator* to convert mechanical power input to electric power output.

The type of voltage induction described above involves mechanical motion of either the conductor or the apparatus producing the magnetic field. This is called *electromagnetic induction.* It is possible to induce voltages without requiring mechanical motion by using the scheme shown in Fig. 7.23. Here we achieve the same effect as motion between the flux lines and the conductors on the secondary winding without requiring mechanical motion of either the winding or the core. This is achieved by varying the current in the primary winding so that the flux in the core changes with time. As we increase the voltage of the variable power supply, the current flowing in the primary winding increases. This in turn increases the number of flux lines in the core. Since the secondary winding is on the same core, the number of flux lines through this winding is increasing and we find that the galvanometer needle moves from its zero position during the time that the dc power supply voltage is being changed. In this configuration most of the flux lines are confined to the iron core (there is very little leakage) and we say that the flux *links* the secondary winding. If we now decrease the voltage of the power supply, we find that the galvanometer needle deflects in the opposite direction, showing that the induced voltage now has the reverse polarity. The only time that the galvanometer deflects is when the current in the primary is changing due to a change in the dc power supply. This phenomenon, the generation of a voltage in the secondary winding due to a *changing* current in the primary winding, is called *mutual induction.*

For either type of electromagnetic induction the magnitude of the voltage induced is proportional to N, the number of turns in the coil ($N = 1$ for a straight conductor) and the rate at which the flux lines are *changing*

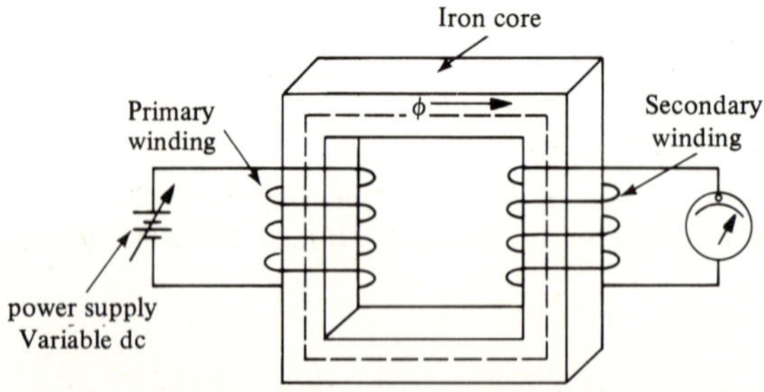

FIGURE 7.23
Mutual induction in which the magnetic field changes with time.

with time (the number of flux lines per second crossed by the conductor). The formula for induced voltage according to Faraday's law is

$$v = -N\frac{\Delta\phi}{\Delta t} \tag{7.8-1}$$

where

v = voltage induced at the terminals of the coil or conductor, in volts. (We use a lower-case v because the voltage may not be constant. The negative sign is a consequence of Lenz's law, which will be explained shortly.)
N = number of turns in the coil ($N = 1$ for a straight conductor)
$\Delta\phi/\Delta t$ = number of webers per second cut at right angles by the coil or conductor when mechanical motion is involved, or number of webers per second linked by the coil in the case of mutual induction.

It is interesting to consider this equation from the dimensional point of view. This is

$$\text{Volts} = \frac{\text{webers}}{\text{seconds}}$$

Isolating webers on the left side, we have

$$\text{Webers} = \text{volts} \times \text{seconds}$$

Thus the weber can be expressed in terms of other units as volts × seconds (V·s) and can be defined in terms of Faraday's law as follows:

One volt is induced in a conductor that cuts 1 weber of magnetic flux in 1 second.

We illustrate Faraday's law with the following examples.

EXAMPLE 7.8-1. Finding the Number of Turns

The flux linking a receiver coil mounted on a satellite changes from 0 to 0.28 mWb in 140 ms. During this time, −3.2 V is measured at the coil terminals. How many turns are there in the coil?

Solution

We rearrange Faraday's law in order to isolate the number of turns N. This yields

$$N = \frac{-v}{\Delta\phi/\Delta t}$$

From the data,

$$\frac{\Delta\phi}{\Delta t} = \frac{0.28 \text{ mWb}}{140 \text{ ms}} = 2 \text{ mWb/s}$$

and

$$N = \frac{-(-3.2) \text{ V}}{2 \times 10^{-3} \text{ Wb/s}} = 1.6 \times 10^3 = 1600 \text{ turns}$$

• • •

EXAMPLE 7.8-2 Voltage Induced by a Time-Varying Flux

The flux linking a 1000-turn coil in a radar receiver varies as shown in Fig. 7.24a. Find the waveform of the induced voltage.

Solution

The student will recognize that this problem is similar to Example 2.3-4, where we found the current due to a time-varying charge. As in that example we divide the flux curve into regions, in each of which the flux variation is linear. The voltage calculation in each region is then done separaty as follows:

Region 1

$$\frac{\Delta\phi}{\Delta t} = \frac{(10 - 0) \text{ mWb}}{(5 - 0) \text{ s}} = 2 \text{ mWb/s}$$

$$v = -N\frac{\Delta\phi}{\Delta t} = -10^3 \times 2 \times 10^{-3} = -2 \text{ v} \qquad 0 < t < 5 \text{ s}$$

Region 2

$$\Delta\phi = 0 \qquad \text{therefore } v = 0 \qquad 5 \text{ s} < t < 10 \text{ s}$$

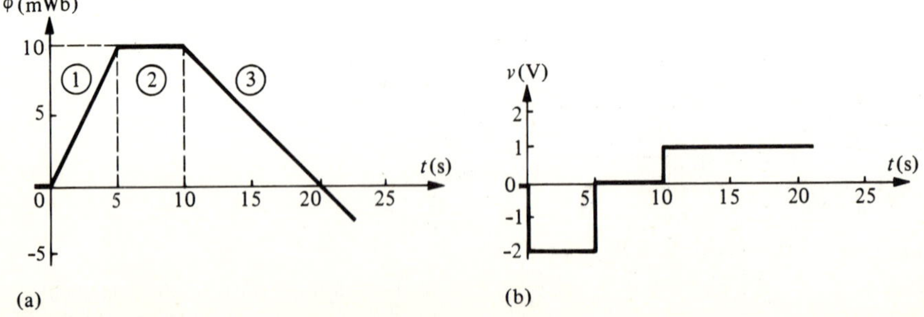

FIGURE 7.24
Flux and voltage for Example 7.8-2. (a) Flux. (b) Voltage.

Region 3

$$\frac{\Delta\phi}{\Delta t} = \frac{(0 - 10)\ \text{mWb}}{(20 - 10)\ \text{s}} = -1\ \text{mWb/s}$$

$$v = -10^3(-1 \times 10^{-3}) = +1\ \text{v} \qquad 10\ \text{s} < t < \infty$$

A graph of v vs. time is shown in Fig. 7.24b.

• • •

The negative sign in Faraday's law is a consequence of *Lenz's law*, which states that

> *The polarity of an induced voltage produces a current such that the magnetic field resulting from this current* opposes *the change in flux which gave rise to the original induced voltage.*

This law follows from the law of conservation of energy. If the flux from the current resulting from the induced voltage were in the same direction as the original flux they would add together, leading eventually to an infinite voltage.

An illustration of Lenz's law is shown in Fig. 7.25. In the diagram a coil of wire is wound over a cardboard tube. A galvanometer is connected to the coil so that there is a complete circuit for the current. The current will be indicated by the galvanometer. When the permanent bar magnet shown in the diagram is stationary, no flux will be changing in the coil, no voltage will be induced, and the galvanometer will read zero. When the bar magnet is moved toward the coil, flux from the magnet will be increasing in the coil in the direction from left to right. According to Lenz's law, the direction of the flux caused by the induced current must be from right to left as shown. In order to produce this flux according to the right-hand rule, the current in the coil must be in the direction shown. Note that this current will also effectively produce a north pole at the left end of the coil. This north pole opposes the motion of the north pole of the bar magnet which is moving

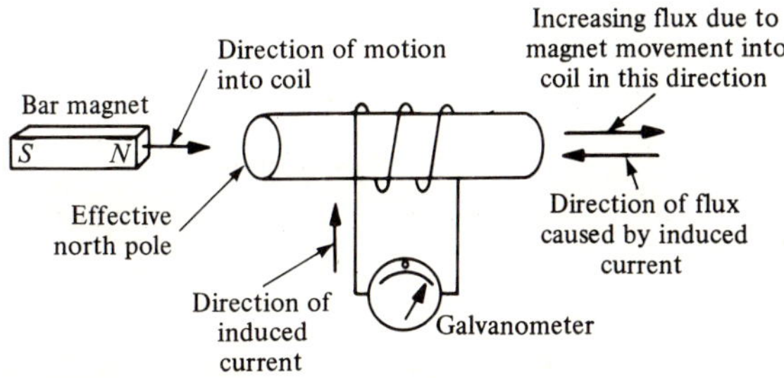

FIGURE 7.25
Lenz's law.

toward it. Thus we must do work against this force of repulsion in order to push the bar magnet into the coil. The work expended in moving the bar magnet is the source of energy for the current induced in the coil.

When the induced voltage is due to motion of the conductor or the field, Faraday's law takes on a different form. For the case when the magnetic field is uniform of flux density B and the conductor or field is moving at constant velocity S meters per second (m/s), the voltage induced in a conductor of length d is

$$\boxed{v = BdS} \tag{7.8-2}$$

This formula is correct only when B, d, and S are mutually perpendicular. The polarity of the induced voltage can be obtained by the *right-hand* or *generator* rule:

> *When the first three fingers of the right hand are extended at right angles to each other with the thumb pointing in the direction of motion and the forefinger pointing in the direction of the magnetic flux, the second finger points in the direction of current flow.*

This is illustrated pictorially in Fig. 7.26.

EXAMPLE 7.8-3 Voltage Induced in an Airplane Wing

An airliner has a wing span of 65 m and flies at 250 m/s. The component of the flux density of the earth's magnetic field, which is cut at right angles by the wing is 50 μT. How much voltage is induced between the tips of the wings?

Solution

The induced voltage is

$$\begin{aligned} v &= BdS \\ &= (50 \times 10^{-6})(65)(250) \\ &= 812 \text{ mV} \end{aligned}$$

• • •

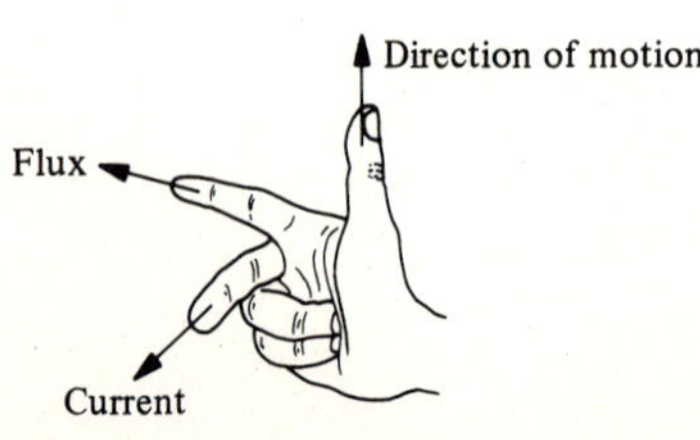

FIGURE 7.26
The right-hand generator rule.

LEARNING EXERCISES FOR SEC. 7.8

1. The voltage induced in a 30-turn receiving antenna coil for a model airplane is 7.3 μV. At what rate is the flux linking the coil changing?

2. The flux linking a 500-turn AM receiver antenna coil is decreasing at the rate of 0.02 mWb/s. What voltage is applied to the receiver?

3. A model airplane has a wing span of 0.75 m. If the component of the flux density of the earth's magnetic field cut at right angles by the wing is 40 μT, how fast must the plane fly in order for the voltage induced between the wing tips to be 1 μV?

Ans. 0.033; −0.243; 10

• • •

7.9 FORCE ON A CURRENT-CARRYING CONDUCTOR

Devices such as the D'Arsonval meter movement and the electric motor are possible because of the fact that current-carrying conductors situated in a magnetic field are acted on by electromagnetic forces. When a conductor of length d carries a current I, and is situated in a uniform magnetic field of flux density B, as shown in Fig. 7.27, there will be a force acting on the conductor whose magnitude is

$$\boxed{F = BId} \tag{7.9-1}$$

where

F = force, in newtons (N)

B = flux density, in webers per square meter

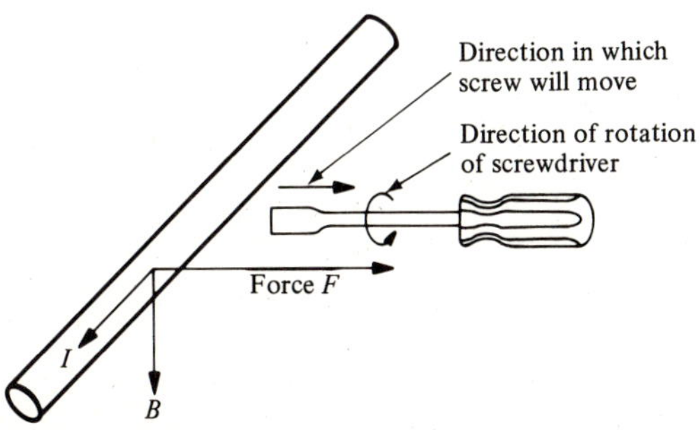

FIGURE 7.27
Direction of force on a current-carrying conductor in a magnetic field.

I = current, in amperes

d = length, in meters

In order for this formula to be valid, the conductor must be in a direction perpendicular to the flux lines. Then the direction of the force is perpendicular to the plane formed by B and I. The direction of the force can be found by applying the *right-hand screw rule*. To apply this rule, refer to Fig. 7.27 and imagine that we have a large screwdriver blade turning a right-hand screw. Place the blade of the screwdriver parallel to I and then rotate it 90° until it is parallel to B. The force F will then be in the direction that the screw moves. This will always be perpendicular to the plane formed by B and I.

In the general case, B and I are not perpendicular. The formula for the force in this case is

$$F = BId \sin \theta \tag{7.9-2}$$

where θ is the angle less than 180° between a line in the direction of B and a line in the direction of I. The direction of F is again perpendicular to the plane containing B and I and is given by the right-hand screw rule. In this case the screwdriver blade is rotated through the angle θ instead of 90° and the force is in the direction that the screw moves.

EXAMPLE 7.9-1 Flux Density from Force Measurements

A test wire 2 cm long carrying 10 A is placed in a uniform magnetic field in a direction perpendicular to the flux lines. The force on the test wire is 0.05 N. Find the flux density.

Solution

$$B = \frac{F}{Id} = \frac{0.05\ \text{N}}{(10\ \text{A})(0.02\ \text{m})} = 0.25\ \text{T}$$

• • •

The D'Arsonval Meter Movement

The diagram of a simple D'Arsonval meter movement is shown in Fig. 7.28a. This device was described in Sec. 6.1 in connection with dc measurements. It consists of an armature pivoted between two bearings with a coil of wire wrapped parallel to the axis of the armature and a permanent magnet that provides a magnetic field. The armature is restrained from turning by a spring.

When current is carried by the armature coil, a force F acts on each side of the coil as shown in Fig. 7.28b. Since B and I are perpendicular, this force can be found from Eq. (7.9-1), $F = BId$, where d is the length of the armature. The directions of the forces F are given by the right-hand rule

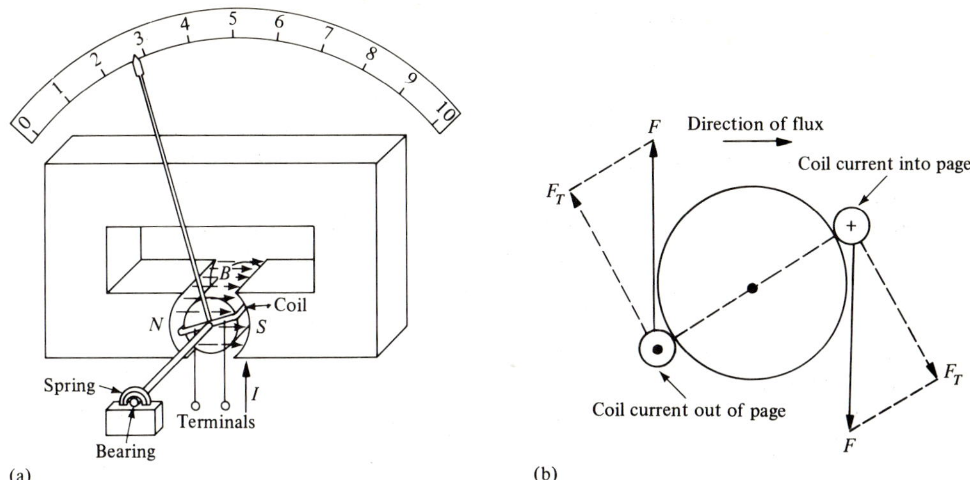

FIGURE 7.28
D'Arsonval meter movement. (a) Pictorial diagram. (b) Diagram of forces.

and are as shown in Fig. 7.28b. Each of these forces has two components: one in the radial direction, which is not shown, and one in the tangential direction, shown as F_T in the diagram. The two tangential components provide a torque (rotational force), which rotates the armature clockwise; this rotation is resisted by the restraining force of the spring. The greater the current, the greater is the force tending to rotate the armature clockwise; the more the spring is stretched the greater is its force tending to rotate the armature *counterclockwise*. The armature rotates clockwise until these forces are balanced, at which point it stops moving. The dial is calibrated to indicate the amount of current in the armature winding.

• • •

LEARNING EXERCISES FOR SEC. 7.9

1. A wire being tested for a dc power transmission system carries 1250 A. It is 2.2 m in length and is situated perpendicular to the flux lines in a field which has a flux density of 0.13 T. What is the force on the wire?

2. The wire in Exercise 7.9-1 is reoriented at an angle of 60° to the flux lines. What is the force now?

Ans. 310; 358

• • •

SUMMARY

Many of the devices and systems with which technologists are concerned depend on magnetic effects for their operation. In this chapter we have studied the most important of these magnetic phenomena and have learned how to calculate their effect on the electric system of which they are a part. The important points to be learned from each section are listed below.

Section

7.1
1. Magnetism results in measurable magnetic forces and induced voltages.
2. Magnetic fields of force exist in the vicinity of permanent magnets and current-carrying conductors.

7.2
3. Magnetic flux is a measure of the total number of lines of force associated with a field. It is measured in webers (Wb). Lines of magnetic flux always form closed loops.

7.3
4. Magnetic flux density is the flux per unit area $B = \Delta\phi/\Delta A$ Wb/m^2. The unit for B is the tesla (T) which is 1 Wb/m^2.
5. The direction of the magnetic field in the vicinity of a current-carrying conductor can be found using the right-hand rule.
6. A magnetic field can be concentrated in a small area by using a solenoid coil. The direction of the concentrated flux in the center of the solenoid is determined by the right-hand rule.
7. Magnetomotive force is the source of magnetic flux. For a coil of N turns, carrying a current I A, it is MMF $= NI$ At (ampere-turns).
8. The MMF "drop" in a material is found from Ohm's law for magnetic circuits to be $F_m = R_m\phi$, where R_m is the reluctance (At/Wb) of the magnetic circuit.
9. Reluctance is analogous to resistance in an electric circuit and is found from the formula $R_m = d/\mu A$, where μ is the magnetic permeability of the material in webers per ampere-meter, and d and A are length and area in meters and square meters.
10. The permeability of free space is $\mu_0 = 4\pi \times 10^{-7}$ Wb/Am. The permeability of most nonmagnetic materials is approximately the same as that of free space. For magnetic materials, the permeability is found from curves supplied by the manufacturer.
11. The magnetic field intensity H is determined by the applied MMF and the length of the magnetic path d according to the relation $H = NI/d$ At/m.
12. The flux density in a material is related to the magnetizing force by the relation $B = \mu H$.
13. For ferromagnetic materials, the relation between B and H is nonlinear and is given on graphs provided by manufacturers.

7.4
14. In a magnetic circuit constructed from one material the flux can be found by first calculating the reluctance and then using Ohm's law for magnetic circuits.

7.5
15. When more than one type of material is involved in a magnetic circuit we make use of the fact that reluctances add in series just

as resistances in series in an electric circuit. In addition, Ampere's circuital law $F_{mT} = \Sigma F_m$ is used in place of KVL.

16. When an air gap is present, the fringing effect may be accounted for by adding the length of the air gap to each cross-sectional dimension.
17. When the MMF is given in a magnetic circuit containing ferromagnetic materials, a trial-and-error procedure is required to find the flux.

7.6 18. To solve series-parallel magnetic circuits we use the magnetic analog of KCL, that is, at a junction of magnetic materials, $\phi_T = \Sigma \phi$.

7.7 19. In an electromagnetic relay, a larger current is required to close the relay than to hold it closed because of the presence of the air gap when it is open.

7.8 20. According to Faraday's law, the voltage induced in a conductor or coil of N turns in which the magnetic flux is changing with time is $v = -N\,\Delta\phi/\Delta t$. The negative sign is a consequence of Lenz's law.

21. When the induced voltage is due to motion of the conductor or the field, Faraday's law becomes $v = BdS$, where B is the flux density in webers per square meter, d is the length of the conductor in meters, and S is the relative velocity in meters per second.

7.9 22. The force on a current-carrying conductor in a magnetic field is $F = BId$ newtons, where B is the flux density in webers per square meter, I is the current in amperes, and d is the length of the conductor in meters.

QUESTIONS FOR REVIEW

Sec. 7.1

1. What is meant by the statement "magnetic force is a *field* phenomenon"?
2. Sketch a diagram of the magnetic field around a bar magnet.
3. What are *magnetic lines of force* and how can they be detected?
4. How is a north pole different from a south pole?
5. Describe how a piece of iron can be magnetized.
6. How do we know that magnetic lines of force form closed loops?

Sec. 7.2

7. Define *magnetic flux* and *flux density.*
8. What is a *magnetic circuit*?
9. Describe the magnetic field around a current-carrying conductor.
10. How do we find the direction of the flux lines around a current-carrying conductor?
11. How can we use a compass needle to find the direction of current in a conductor?
12. Why does one wind a conductor into the form of a solenoid?

13. How do we find the direction of flux lines through the center of a solenoid?
14. If two current-carrying conductors are placed close together, how can we determine the direction of the force on them?

Sec. 7.3

15. What is *magnetomotive force*?
16. What is *reluctance*?
17. State *Ohm's law* for magnetic circuits.
18. Draw a simple magnetic circuit and its electric analog. What is the difference between these two if the magnetic circuit contains ferromagnetic materials?
19. Define *permeability* and state the relation between reluctance and permeability.
20. What is the permeability of free space?
21. Define *relative permeability.*
22. What is *magnetic field intensity*?
23. How is flux density related to magnetizing force?
24. Describe the *BH* curve for typical ferromagnetic materials.
25. What is the significance of *saturation*?

Sec. 7.4

26. Describe how you would calculate the flux in a continuous magnetic circuit that consists of a single material.

Sec. 7.5

27. Repeat Question 26 for a circuit consisting of two different materials.
28. Describe how you would find the MMF required to produce a specified flux in a series magnetic circuit.
29. What is the effect of an air gap in a magnetic circuit?

Sec. 7.6

30. Describe how you would find the required MMF in a series-parallel magnetic circuit.

Sec. 7.7

31. Describe the construction and operation of an *electromagnetic relay.*
32. What is the meaning of *pull-in current*? *Release current*?

Sec. 7.8

33. Describe *electromagnetic induction.*
34. How is it possible to have voltage induced in a conductor without any mechanical motion?
35. Describe *mutual induction.*
36. What is *Faraday's law*?
37. What is *Lenz's law*?
38. What is Faraday's law as applied to the voltage induced in a conductor due to motion of the conductor or the field?

Sec. 7.9

39. How is the force on a current-carrying conductor related to the magnitude and direction of the flux density and the length of the conductor?
40. Describe the construction and operation of a D'Arsonval meter movement.

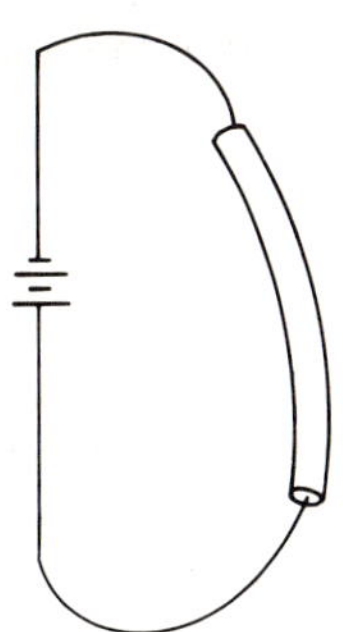

FIGURE 7.29

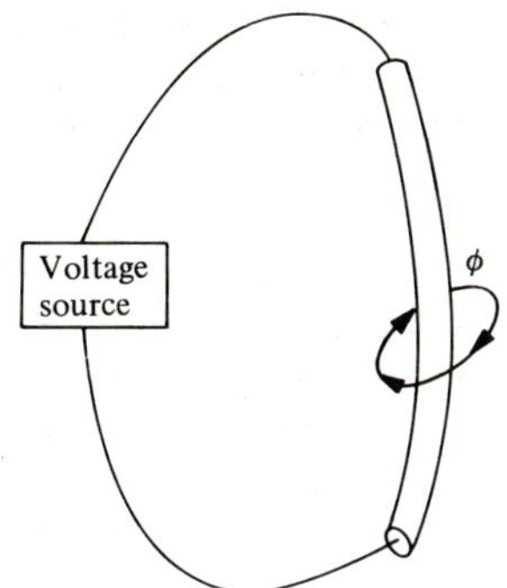

FIGURE 7.30

PROBLEMS

Sec. 7.2

1. The flux perpendicular to a surface whose cross section is 0.06 m^2 is known to be 32 mWb. Find the flux density B.
2. The magnetic field of a dc motor is measured and found to be 240 μWb over a 0.0027-m^2 area. What is the flux density?
3. In the field of a large horseshoe magnet the flux density is known to be 0.042 Wb/m^2. What will be the flux through a surface whose area is 6 cm^2?
4. The field in a permanent magnet motor has a flux density of 3.12 T over a rectangular area 0.13 by 0.16 m. Find the flux.
5. The flux density in the magnetic field of a solenoid is measured at 25 T and the flux through a particular area is 340 μWb. What is the area?
6. The field in a magnetic core has a uniform flux density of 15 Wb/m^2. Its cross section is a 2×5 cm rectangle. Find the total flux ϕ.
7. A doughnut-shaped cast-steel magnetic core has a circular cross section whose radius is 1.5 cm. The diameter of the entire core from outer edge to outer edge is 10 cm. The total flux through it is 0.8 mWb. Find the flux density.
8. Sketch the magnetic force lines around the conductor in Fig. 7.29 and show their direction.
9. In Fig. 7.30 the direction of the flux lines around a current-carrying conductor is shown. Find the direction of the current in the conductor.
10. In Fig. 7.31, show the direction of current in the coil.
11. Sketch the flux lines in the vicinity of the coil in Fig. 7.32, show their direction, and indicate the location of the induced north and south poles.

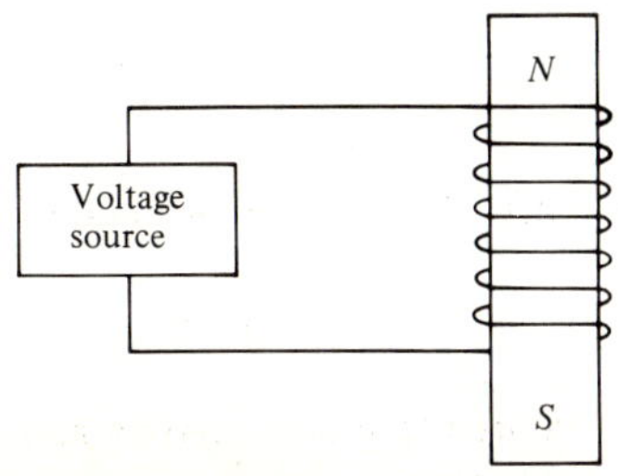

FIGURE 7.31

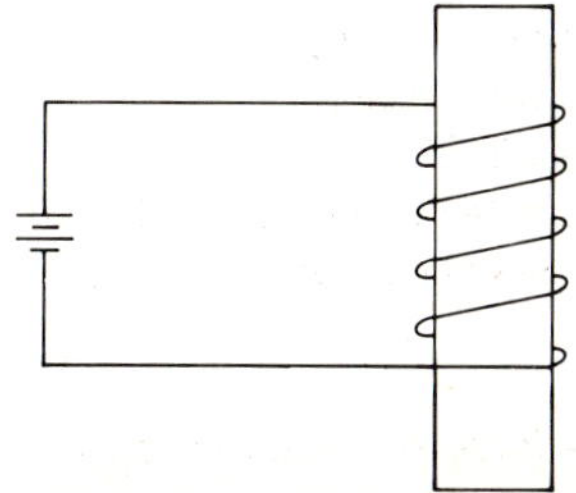

FIGURE 7.32

Sec. 7.3

Note: Whenever necessary, the *BH* curves of Fig. 7.15 should be used for the following problems.

12. Find the magnetomotive force produced by 2.5 A through a 200-turn coil.
13. An MMF of 1250 At is to be produced by a 300-turn coil. What current is required?
14. What is the reluctance of a magnetic circuit in which a flux of 2.5 mWb is measured when 3.5 A is carried by a 200-turn coil supplying the MMF?
15. A 450-turn relay coil is to produce a flux of 1.5 mWb in a magnetic circuit whose reluctance is 2.3×10^6 At/Wb. What current must be established in the coil?
16. Find the reluctance of a magnetic circuit if a flux of 0.042 μWb is established by an impressed MMF of 480 At.
17. An MMF of 680 At is impressed on a magnetic circuit that has a reluctance of 0.62×10^6 At/Wb. What is the flux?
18. The magnetic circuit of a transformer has a uniform cross-sectional area of 5.2 cm^2 and an average flux path length of 0.3 m. Its reluctance is 2.3×10^6 At/Wb. Find the permeability of the material.
19. A solenoid has a core that is in the form of a brass cylinder 0.12 m in length with a diameter of 0.023 m. Find its reluctance.
20. In a large electromagnet, the total flux is 68 mWb, the reluctance is 500 kAt/Wb. The current in the coil is 0.5 A. How many turns of wire are there on the coil?
21. A piece of iron 1.5 cm $\times$ 1.5 cm $\times$ 10 cm has a reluctance of 1300 kAt/Wb in the 10-cm direction. Find the permeability of the iron.
22. Find the reluctance of the air gap of a relay that has a square cross section of 1.25 cm on a side and a spacing of 1 mm.
23. A material used in computer memories is characterized by a relative permeability $\mu_r = 2500$. Find its permeability.
24. A flux density of 1.0 Wb/m^2 is established in a cast steel core.
    a. Find the magnetic field intensity, *H*.
    b. Determine the permeability, μ.
    c. If the cross section of the core is 1 $\times$ 4 cm, find the flux, ϕ.
25. Find the reluctance of a 10-cm length of the core described in Prob. 24.
26. A cast-steel magnetic core has a circular cross section with a diameter of 1 cm. Its mean length is 20 cm. It is wrapped with a 100-turn coil carrying 250 mA. Find
    a. The applied MMF
    b. The magnetic field intensity
    c. The flux density
    d. The total flux
27. A 500-turn coil is wrapped around the core described in Prob. 7. Find the flux if the current is 1 A.
28. In the magnetic circuit of a motor a field intensity of 2500 At/m produces a flux density of 1.2 T. What is the permeability of the material at this flux density?
29. Find the flux density created by a magnetizing force of 2200 At/m in (a) cast iron; (b) cast steel; (c) sheet steel.
30. A flux density of 1.3 T is required in an automobile generator fabricated from cast steel. What magnetic field intensity will produce this flux density?
31. The magnetizing force in a relay is 220 At/m. The coil has 130 turns and a current of 1.7 A. Find the flux path length.

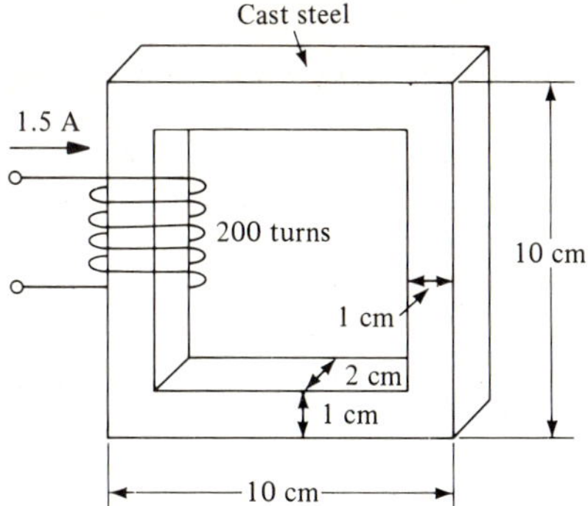

FIGURE 7.33

32. A particular flux path is 45 cm long, has an area of 1.2 cm^2, and a relative permeability of 100. Calculate the reluctance of this flux path.
33. A steel core has a relative permeability of 2500. It has a cross-sectional area of 7.2 in^2 and is 28 in long. The applied MMF is 75 At. Find
    a. The reluctance of the magnetic circuit
    b. The flux in the core
    c. The magnetizing force
    d. The flux density
34. A cast-iron toroidal core has a cross-sectional area of 2.5×10^{-3} m^2 and an average length of 1.2 m. A coil wound on the core has 250 turns. If the coil current is 5 A, find
    a. The MMF
    b. The flux density
    c. The flux in the core
    d. The reluctance

Sec. 7.4

35. In the magnetic circuit of Fig. 7.33 the material is cast steel. Find the flux, repeat if the core material is (a) aluminum, (b) cast iron. (Note: Be sure to use

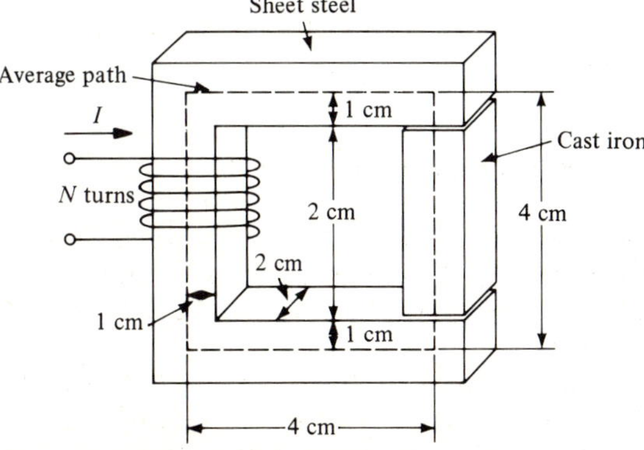

FIGURE 7.34

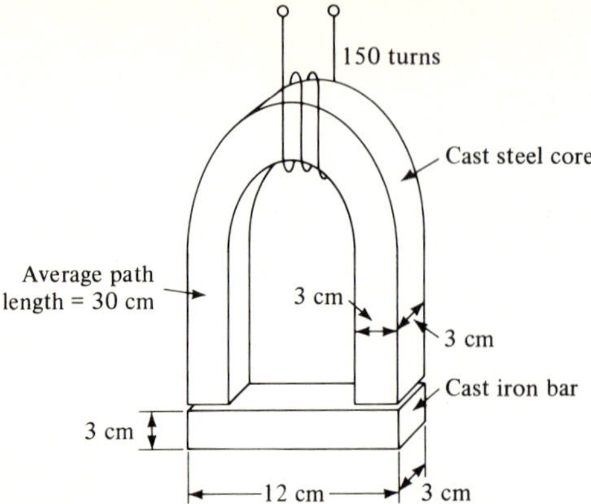

FIGURE 7.35

the average path length which is one-half the sum of the outer perimeter and the inner perimeter.)

36. In the magnetic circuit of Fig. 7.33 find the current required to establish a flux of 0.25 mWb if the material is cast steel.
37. The toroidal core in Fig. 7.16 has an outside diameter of 5 cm and a cross-sectional diameter of 1.25 cm. If I = 50 mA, μ_r = 800, and ϕ = 12.5 μWb, find
    a. The MMF
    b. The number of turns in the coil
    c. The reluctance of the coil
38. A current of 500 mA is established in a 1000-turn coil wound on a toroidal cast-steel core which has a cross-sectional area of 15 cm^2 and an average path length of 0.3 m. Find the total flux in the core.

Sec. 7.5

39. In the magnetic circuit of Fig. 7.34 find the NI required to establish a flux of 0.24 mWb. Find the permeability of each material.
40. The horseshoe electromagnet of Fig. 7.35 has the dimensions shown on the

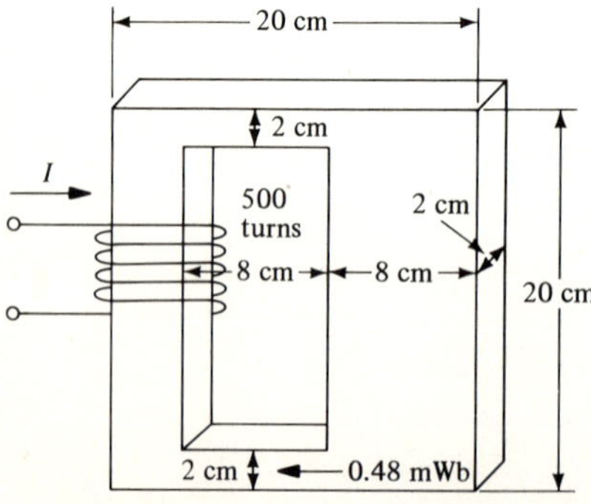

FIGURE 7.36

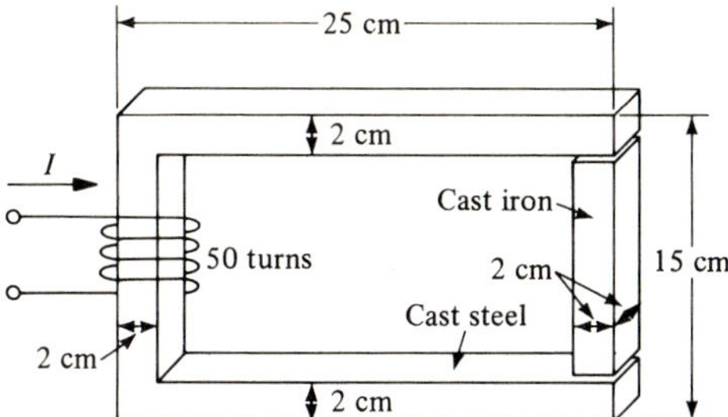

FIGURE 7.37

diagram. What current in the coil will develop a flux of 0.54 mWb in the cast-iron bar being held by the magnet?

41. A sheet of brass 1 mm thick is placed between the magnet and the cast iron in Fig. 7.35. Find the current required to develop 1.2 mWb in the cast iron. (Hint: Treat the brass as two air gaps.) Use the result of Prob. 40 as your initial guess.
42. In the magnetic circuit of Fig. 7.35 find the flux in the cast iron if the current in the coil is 375 mA. (A trial-and-error procedure will be required.)
43. In Problem 7.42 the current is 1.5 A. Find the total flux in the cast iron.
44. The cast-steel magnetic core shown in Fig. 7.36 has a uniform thickness of 2 cm. Find the current *I*.
45. Find the current required in a 500-turn coil to produce a flux of 0.1 mWb in a cast-steel toroidal core if a 0.1-cm air gap is cut in it. It has a circular cross section with a radius of 1.5 cm and the outer diameter of the toroid is 10 cm.
46. Find the current required in the magnetic circuit of Fig. 7.37 to produce a flux density of 0.5 Wb/m^2.

Sec. 7.6

47. Repeat Example 7.6-1 if the flux in the outer leg is to be 0.05 mWb.
48. In the magnetic circuit of Fig. 7.20 an air gap 0.1 mm thick is cut in the center leg. Find the current required to establish a flux of 50 μWb in the outer leg.
49. In the sheet-steel magnetic circuit of Fig. 7.38 the flux in the center leg is to be 1.5 mWb. How many turns of wire must be wound on this leg if the current is to be 100 mA?

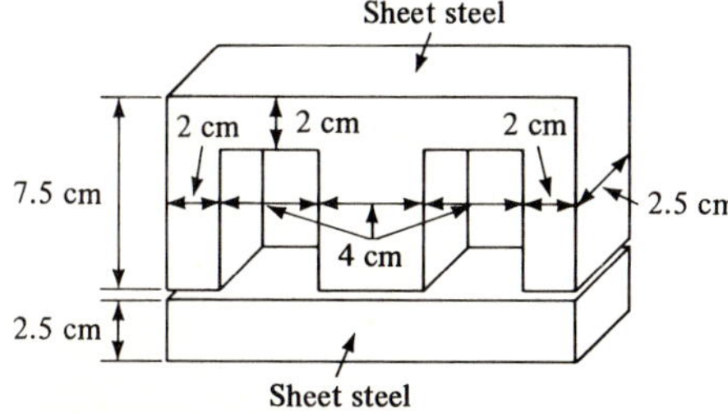

FIGURE 7.38

Sec. 7.8

50. A time-varying flux changes from 0 to 4 mWb in 0.5 s. This flux links a coil that has 2500 turns. How much voltage would be measured at the coil terminals during the time that the flux is changing?
51. A voltage of -20 mV is measured at the terminals of a coil. The coil is linked by a flux which changes from 3 to 6.2 mWb in 0.25 s. How many turns are there in the coil?
52. A voltage of $+2$ V is measured at the terminals of a 150-turn coil situated in a changing magnetic field. Find the rate at which the flux is changing.
53. Flux linking a 250-turn coil varies as shown in Fig. 7.39. Find the waveform of the induced voltage.
54. The total flux at the end of a long bar magnet is 300 μWb. The magnet is pulled through a 500-turn coil in 10 ms. What voltage appears across the terminals of the coil during this time?
55. A conductor 50 cm long moves through a uniform magnetic field at a velocity of 30 m/s. The flux density in the field is 0.15 T. Find the voltage induced in the conductor.
56. A conductor 2.5 m long passes through a magnetic field of 0.3 T at a velocity of 10 m/s. Find the voltage induced in the conductor.
57. Repeat Example 7.8-3 for a small plane with a wing span of 60 ft that flies at 120 mi/h.
58. In Fig. 7.22 the flux density in the air between the pole faces is 1.5 T. The pole faces are square, 5 cm on a side. A voltage of 5 mV is to be produced between the ends of a conductor moving through the air gap. How fast must the conductor be moving?

Sec. 7.9

59. A wire 2.6 cm long is oriented perpendicular to the flux lines in a uniform field. The current in the wire is 14 A and the force is measured as 0.068 N. Find the flux density.
60. A test wire 2 cm long carrying 10 A is placed in a uniform field at an angle of 55° to the flux lines. The force on the wire is measured to be 0.075 N. Find the flux density.
61. A conductor 50 cm long is placed perpendicular to a field in which the flux density is 1.2 T. The current in the conductor is 15 A. Find the force on the conductor.

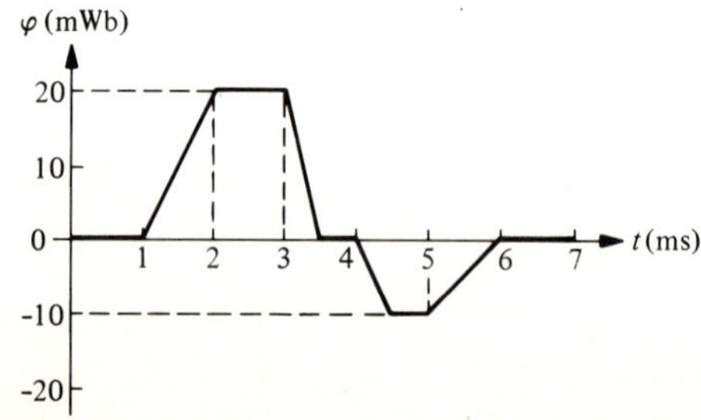

FIGURE 7.39

62. A conductor 30 cm long carries a current of 18 A and is placed at an angle of 60° to a magnetic field of strength 0.2 Wb/m^2.
    a. Find the magnitude of the force on the conductor.
    b. If the conductor lies in the *xy* plane and the flux lines lie in the positive *x* direction, determine the direction of the force.
63. A wire 250 cm long is placed on the *x* axis in a magnetic field that has a flux density of 2.8 T with the flux direction along the negative *y* axis. The current in the wire is 5 A in the positive *x* direction. Find the magnitude of the force on the wire and its direction.

8
Inductance and Capacitance

OBJECTIVES

Upon completion of this chapter, the student should be able to

Section

Section		Objective
8.1	1.	Find the inductance of a coil given values for the appropriate parameters.
	2.	State the volt-ampere relation for the inductance element.
	3.	State that the voltage across an inductor is zero unless the current is changing with time.
	4.	Use the inductance *vi* relation to find the voltage response to various linear current waveforms.
8.2	5.	Find the equivalent inductance for series and parallel combinations of inductors.
8.3	6.	Describe mutual magnetic coupling.
	7.	Explain the dot convention for coupled coils.
	8.	Find the equivalent inductance for series inductors when magnetic coupling is present.
	9.	Write KVL equations for circuits with magnetic coupling.
8.4	10.	State the relation between charge and voltage in a capacitor.
	11.	State the volt-ampere relation for the capacitance element.
	12.	State that current flows through a capacitor only when the voltage is changing with time.
	13.	Use the capacitance *vi* relation to find the current response to various linear voltage waveforms.
8.5	14.	Find the capacitance of a parallel-plate capacitor given values for the appropriate parameters.
	15.	Discuss dielectric strength.
	16.	Describe nonideal capacitors and parasitic capacitance.

8.6 17. Find the capacitor voltage required to achieve a specified constant current.

18. Describe the unit step function and find the response of a capacitor to a current step.

8.7 19. Find the equivalent capacitance of series and parallel combinations of capacitors.

20. Find all charges and voltages in a series-parallel capacitor network.

21. Design a capacitive voltage divider.

8.8 22. State the relations for energy stored in capacitors and inductors and use them to find the stored energy.

INTRODUCTION

Up to this point all of the circuits we have considered have consisted solely of resistors and dc sources. In all cases the circuit currents and voltages were dc and we were only required to calculate their relative levels. We are now ready to add to our catalog of linear elements, two for which this behavior is no longer true. These are the inductor and the capacitor.

To give you an idea of what is to come, we note that a capacitor is an element in which no current flows unless the voltage across it is changing with time, and an inductor is an element which has no voltage across its terminals unless the current through it is changing with time. The capacitor opposes changes in the voltage across its terminals while the inductor opposes changes in the current through it. These properties are used to advantage in many signal-shaping circuits and in all radio and TV receivers. Because of the fact that changes with time are involved, we are going to have to make considerable use of the concept of rate of change, which we studied in Chap. 1 and again in Chap. 7.

In this chapter we consider the fundamental behavior of these elements and their effect on basic waveforms.

8.1 INDUCTANCE

In this section we describe the inductance element and derive its *vi* relation. The form usually taken by the inductance element is that of a long cylindrical coil of many turns of wire as shown pictorially in Fig. 8.1a. Often the coil is wound over a core of magnetic material, and the actual physical form practical inductors take consists of many varieties, some of which are shown in Fig. 8.1b. The circuit symbol usually used is shown in Fig. 8.1c. Variations on this symbol sometimes indicate the presence of a magnetic core or a slug of magnetic material, which can be moved in or out to vary the properties of the element. These variations are shown in Fig. 8.1d.

To derive the *vi* relation for the inductance we begin with Faraday's law (see Sec. 7.8). Faraday's law states that a coil of N turns of wire placed

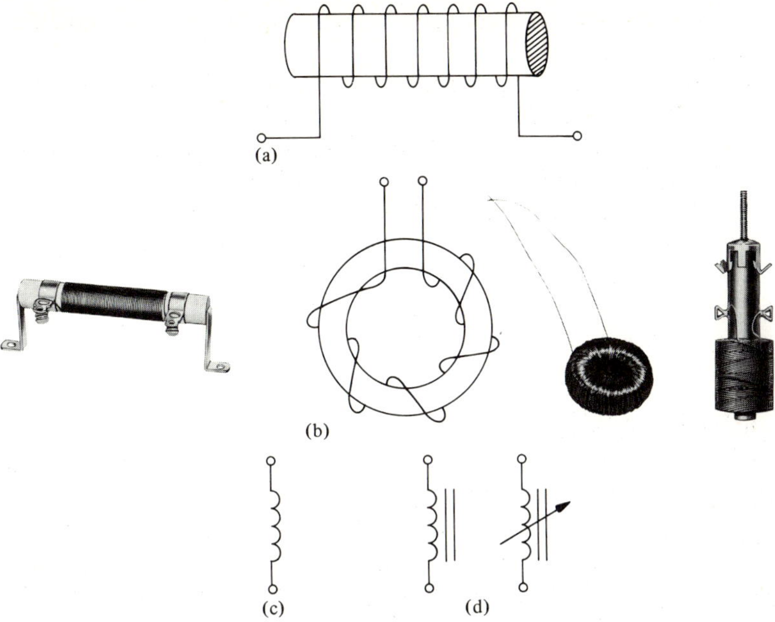

FIGURE 8.1
Inductance. (a) Cylindrical coil. (b) Types of inductors: ceramic core inductor, toroidal inductor, variable inductor with movable core. (c) Circuit symbol. (d) Circuit symbol when magnetic core or movable plug is present.

in a magnetic flux field that is changing with time will have a voltage induced across its terminals of a magnitude

$$v(t) = N\frac{\Delta\phi}{\Delta t} \tag{8.1-1}$$

where

$v(t)$ = magnitude of the voltage at the coil terminals, in volts (V)
N = number of turns
$\Delta\phi/\Delta t$ = rate of change of flux linked by the coil, in webers per second (Wb/s).

At this point a word of explanation is in order concerning notation. The voltage in Eq. (8.1-1) is written as $v(t)$. This is called *functional* notation and it does *not* mean *v times t*. It should be read as "*v as a function of time* equals $N\Delta\phi/\Delta t$." It will always be clear from the context that a lower-case v followed by t in parentheses is meant to be functional notation. We write it this way in order to emphasize the fact that the voltage will, in general, vary with time. Often the t in parentheses is omitted and only the lower-case letter v is used. From this point on we will understand that lower-case v is to be interpreted as $v(t)$. The same holds for current i, charge q, and any other quantity which may vary with time. Sometimes the words "instantaneous voltage v" are used. This also means that the voltage v may vary with time. As noted in previous chapters, when a *capital* V appears, it refers to a constant, or dc, voltage.

Now consider the circuit shown in Fig. 8.2. When the switch is closed, the current begins to increase, and since flux ϕ is proportional to current i, the flux will also begin to increase. According to Faraday's law a voltage of *self-induction* will be induced across the coil because the flux is changing. This voltage will be in such a direction as to *oppose* the change in current, according to Lenz's law (Sec. 7.8). Because of this opposition to changes in current, it is not possible for the current in such a circuit to rise instantly.

We can relate the voltage across the coil to the current by noting that, from Eq. (7.3-2), the flux is

$$\phi = \frac{Ni}{R_m} \tag{8.1-2}$$

where

ϕ = instantaneous flux, in webers (Wb)
N = number of turns
i = instantaneous current, in amperes (A)
R_m = reluctance, in ampere-turns per weber (At/Wb)

N is constant for a given coil and if we make the reasonable assumption that R_m is constant we can write

$$\Delta\phi = \frac{N}{R_m}\,\Delta i \tag{8.1-3}$$

Dividing both sides by Δt we get

$$\frac{\Delta\phi}{\Delta t} = \frac{N}{R_m}\frac{\Delta i}{\Delta t} \tag{8.1-4}$$

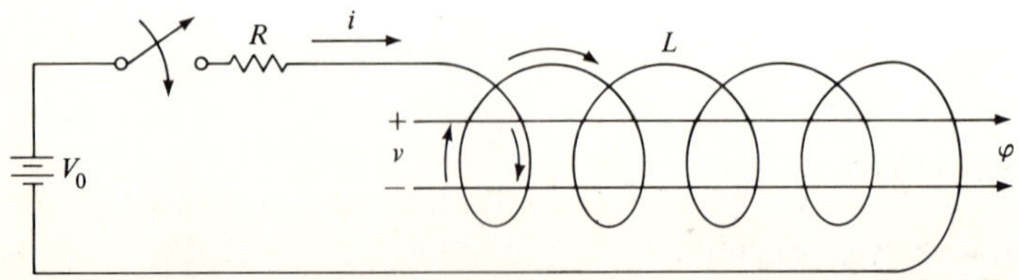

FIGURE 8.2
Inductive circuit.

Next we substitute $\Delta\phi/\Delta t$ from Eq. (8.1-4) into Eq. (8.1-1). This yields

$$v(t) = \frac{N^2}{R_m}\frac{\Delta i}{\Delta t} \tag{8.1-5}$$

The constant N^2/R_m is called the inductance of the coil, symbolized by L, and the unit of inductance is the henry (H), in honor of Joseph Henry, an American physicist. Using this definition, Eq. (8.1-5) becomes

$$\boxed{v(t) = L\frac{\Delta i}{\Delta t}} \tag{8.1-6}$$

This equation is the *vi* relation for the inductance. Note that if the current is constant, then $\Delta i/\Delta t = 0$ and no voltage appears across the terminals. Thus the only time that voltage will appear across the terminals of an inductance is when the current is changing with time. This represents a major difference from the resistive circuits we have studied up to now. This difference can best be appreciated by considering several examples.

EXAMPLE 8.1-1 Inductance of a Toroidal Coil

An inductance coil for a special space experiment is constructed in the form of a toroid on a cast-steel ring. There are 1800 turns of wire, the cross-sectional area is 3.5 cm^2, and the average path length is 30 cm. At the rated value of current, the permeability μ is 800 μWb/Am. Find the inductance of the coil.

Solution

From Eqs. (8.1-5) and (8.1-6) we have

$$L = \frac{N^2}{R_m} \tag{8.1-7}$$

and using $R_m = d/\mu A$ [see Eq. (7.3-3)]

$$\boxed{L = \frac{N^2\mu A}{d}} \tag{8.1-8}$$

Substituting the given values,

$$L = \frac{(1800)^2(800 \times 10^{-6})(3.5 \times 10^{-6})}{(0.3)}$$

$$= 3.02 \text{ H}$$

When using Eq. (8.1-8) to calculate inductance, we must keep in mind that it is based on the assumption that all of the flux links all of the turns. Thus it is most accurate when applied to iron-core coils and toroids. For air-core coils where there is appreciable flux leakage, other more accurate formulas derived experimentally are available in handbooks.

• • •

EXAMPLE 8.1-2 Voltage Due to a Linearly Increasing Current

In a TV picture-tube circuit, the current through a 250-mH inductor changes linearly from 5 A at $t_1 = 1$ s to 65 A at $t_2 = 3$ s. Find the voltage across the coil during this interval and plot the current and voltage.

Solution

A graph of the current is shown in Fig. 8.3. Since it is increasing linearly (that is, with constant slope) the voltage will be constant. Using the inductance *vi* relation [Eq. (8.1-6)] we have

$$v = L\frac{\Delta i}{\Delta t} = L\frac{i(t_2) - i(t_1)}{t_2 - t_1} = 0.25\,\frac{65 - 5}{3 - 1} = 0.25 \times 30 = 7.5 \text{ V}$$

This is plotted in Fig. 8.3. Note that the graphs cannot extend beyond the 1 to 3 s interval because we are not told what the current variation is outside of this interval.

• • •

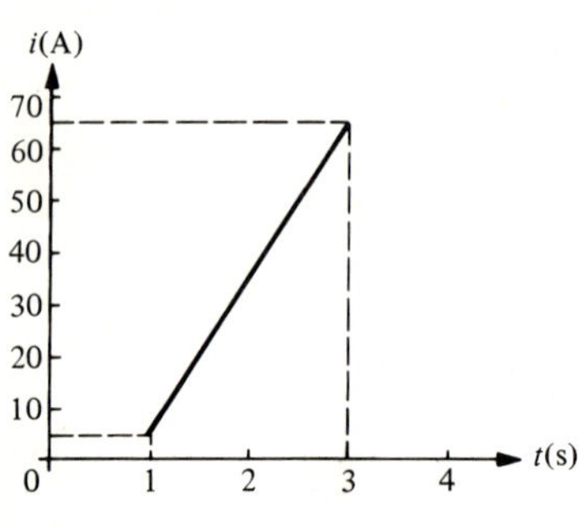

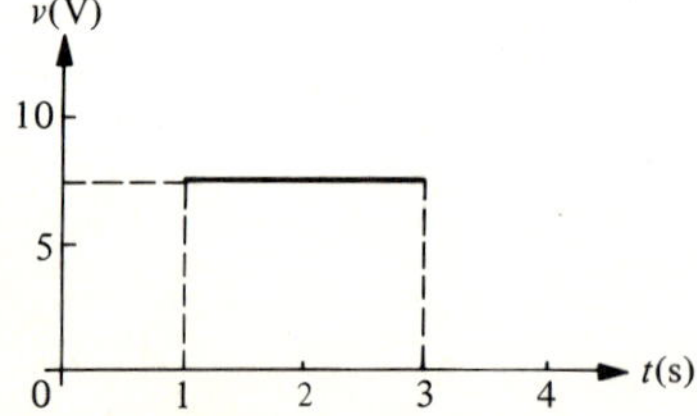

FIGURE 8.3
Current and voltage for Example 8.1-2.

EXAMPLE 8.1-3 Current Required for Constant Voltage

A 20-mH inductor is to have a constant voltage across it of 3 V. (a) At what rate must the current through it be changing in order for this voltage to be maintained across its terminals? (b) Sketch several current waveforms that will yield the desired voltage.

Solution

a. Rearranging the inductor *vi* relation so that the rate of change of current is isolated on the left, we have

$$\frac{\Delta i}{\Delta t} = \frac{v}{L} = \frac{3\text{ V}}{0.02\text{ H}} = 150\text{ A/s}$$

b. Several graphs of current vs. time are shown in Fig. 8.4. Note that each of these currents will produce 3 V across the 20-mH coil even though all three appear different. The point is that all three have the same *slope* or *rate of change* of 150 A/s.

• • •

EXAMPLE 8.1-4 Voltage Waveform in an RL *Circuit*

The current waveform of Fig. 8.5b is applied to the *RL* circuit of Fig. 8.5a. Find the voltages across *L* and *R*, and the total voltage.

Solution

Because of the nature of the current waveform, it is convenient to treat each segment separately. For $t < 0$, the current is zero, so $v(t) = 0$. For the first line segment we can represent the current by the straight-line equation

$$i(t) = t \qquad 0 < t < 2\text{ s}$$

Then using Eq. (8.1-6),

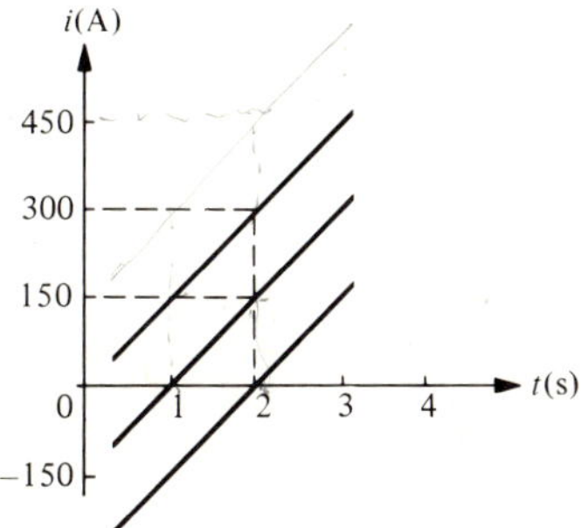

FIGURE 8.4
Current waveforms which produce 3 V across a 20-mH coil.

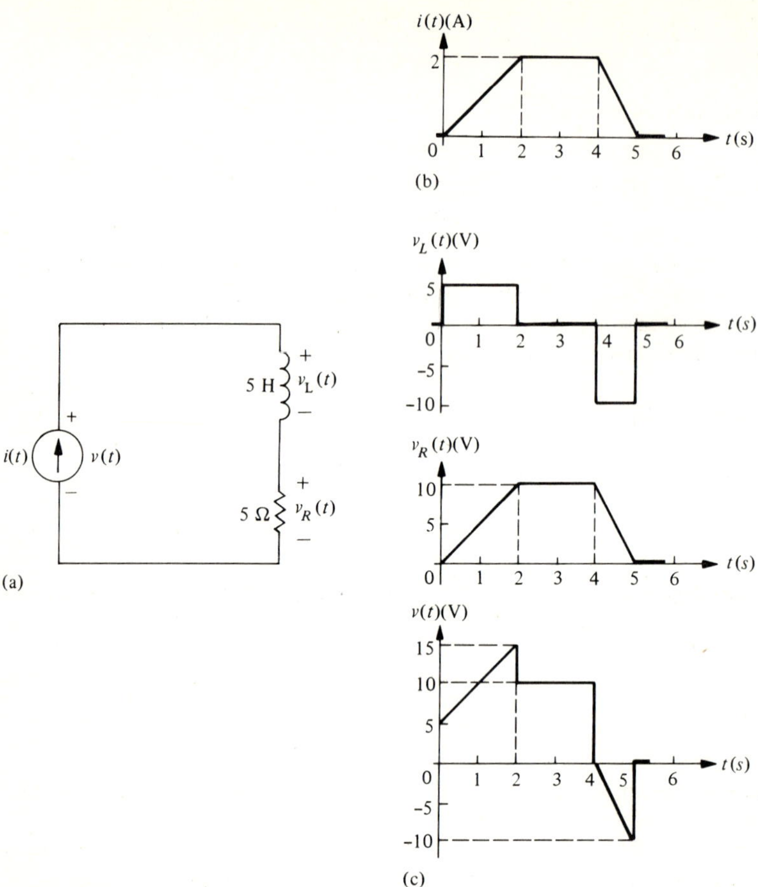

FIGURE 8.5
Example 8.1-4. (a) Circuit. (b) Input current waveform. (c) Voltage waveforms.

$$v_L(t) = L\frac{\Delta i}{\Delta t} = (5\text{ H})(1\text{ A/s}) = 5\text{ V} \qquad 0 < t < 2\text{ s}$$

Thus, at $t = 0$, $v(t)$ jumps to 5 V where it remains until $t = 2$ s.

For the interval from $t = 2$ s to $t = 4$ s, $i(t)$ is a constant so that $v_L(t) = 0$ for $2 < t < 4$ s. Thus the voltage drops from 5 to 0 V at $t = 2$ s, where it remains until $t = 4$ s.

For the final interval, from $t = 4$ s to $t = 5$ s, the straight-line equation is (be sure to verify this)

$$i(t) = 10 - 2t \qquad 4 < t < 5\text{ s}$$

For this line, the slope is $m = -2$ A/s. Thus [see Eq. (1.6-12)]

$$m = \frac{\Delta i}{\Delta t} = -2 \text{ A/s}$$

and the voltage is

$$v_L(t) = L\frac{\Delta i}{\Delta t} = (5 \text{ H})(-2 \text{ A/s}) = -10 \text{ V} \qquad \text{for } 4 < t < 5 \text{ s}$$

Thus at $t = 4$ s the voltage drops from 0 to -10 V where it remains until $t = 5$ s when it returns to zero. The inductor voltage waveform is plotted in Fig. 8.5c.

The resistor voltage waveform is also plotted in Fig. 8.5c. It is obtained simply by noting that Ohm's law applies, that is,

$$v_R(t) = Ri(t) = 5i(t)$$

The total voltage is obtained by using KVL,

$$v(t) = v_L(t) + v_R(t)$$

This is the final graph in Fig. 8.5c. The student should take careful note of the difference between the resistor voltage waveform and the inductor voltage waveform.

• • •

LEARNING EXERCISES FOR SEC. 8.1

1. Pressure is symbolized in general by the letter *P*. How would pressure as a function of time be symbolized?
2. An air-core toroidal coil for a low-power transmitter has 30 turns and an area of 0.8 cm^2. The average path length is 5 cm. Find the inductance.
3. A 38-mH coil is used in a control circuit. During one 3-ms cycle of operation the current changes linearly from 5 to 18 mA. Find the voltage across the coil during this interval.
4. The current in a TV deflection circuit is changing at 320 A/s. What inductance must this current pass through for a voltage of 350 V to appear across the terminals?

Ans. 0.165; p(t); 1.81; 1.1

• • •

8.2 SERIES AND PARALLEL INDUCTORS

In this section our objective is to find a single equivalent inductance L_{eq} that will replace a series or parallel combination of two or more inductances.

If two inductors L_1 and L_2 are connected in series as shown in Fig. 8.6 in such a way that none of the flux from L_1 links any of the turns of L_2, and vice versa, then according to KVL, we must have, at any instant of time,

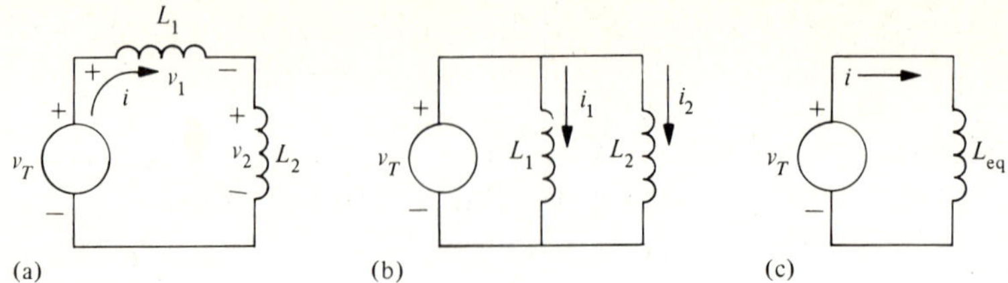

FIGURE 8.6
Inductors. (a) In series. (b) In parallel. (c) The equivalent circuit.

$$v_T = v_1 + v_2 \tag{8.2-1}$$

Since the current is the same throughout the circuit,

$$v_1 = L_1 \frac{\Delta i}{\Delta t} \quad \text{and} \quad v_2 = L_2 \frac{\Delta i}{\Delta t} \tag{8.2-2}$$

Substituting into Eq. (8.2-1)

$$v_T = L_1 \frac{\Delta i}{\Delta t} + L_2 \frac{\Delta i}{\Delta t} = (L_1 + L_2) \frac{\Delta i}{\Delta t} \tag{8.2-3}$$

For the equivalent circuit of Fig. 8.6c we have

$$v_T = L_{eq} \frac{\Delta i}{\Delta t} \tag{8.2-4}$$

Comparing Eqs. (8.2-3) and (8.2-4) the equivalent inductance is simply

$$\boxed{L_{eq} = L_1 + L_2} \tag{8.2-5}$$

This has exactly the same form as the equation for the equivalent resistance of two series resistors.

If the same two inductors are connected in parallel as shown in Fig. 8.6b, again with no common flux linkages, the equivalent inductance can be shown to be (Prob. 8.13)

$$\boxed{L_{eq} = \frac{L_1 L_2}{L_1 + L_2}} \tag{8.2-6}$$

It is evident from Eqs. (8.2-5) and (8.2-6) that the equivalent inductance of inductors in series or parallel is found in exactly the same way as the equivalent resistance of series or parallel combinations of resistors.

When the inductors are not magnetically isolated as above, the effect of the flux from each coil acting on the other must be taken into account. This is considered in the next section.

Coil Resistance

Practical inductors are made of numerous turns of wire, which has finite resistance. This resistance cannot actually be separated from the ideal inductance of the coil, and analysis taking account of this effect becomes very difficult. To simplify the analysis, we replace the actual coil by a *model*, in which the inductance and resistance are in series as shown in Fig. 8.7. This model will provide results that are sufficiently accurate for most purposes. When the resistance is negligible, the coil behaves like a pure inductance. We have made this assumption in the development of this section. In succeeding sections, the coil resistance is assumed to be negligible unless otherwise stated.

• • •

LEARNING EXERCISE FOR SEC. 8.2

1. A laboratory bench drawer contains two magnetically shielded inductors marked 30 and 40 mH. What values of inductance can be realized if they are used singly or in combination?

Ans. 17; 30; 70; 40

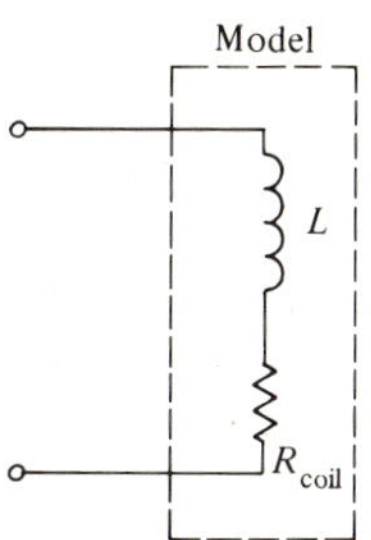

FIGURE 8.7 Practical model for inductance coil.

8.3 MUTUAL INDUCTION

When two coils are placed so that some or all of the flux produced by the current in one coil passes through (links) the second coil, a voltage will be induced across the second coil due to changes in the flux produced by the first coil. This effect is called *mutual induction,* and is accounted for by an element called *mutual inductance.* In this section we consider several aspects of mutual inductance coupling.

In the circuit of Fig. 8.8 we have two coils placed close together. The coils need not have an iron core. The primary coil has N_1 turns and self-inductance L_1; the secondary coil has N_2 turns and self-inductance L_2. When voltage is applied to the primary, current will flow and primary flux ϕ_1 will be produced as shown. According to Faraday's law [see Eqs. (8.1-1) and (8.1-6)],

$$v_1 = N_1 \frac{\Delta\phi_1}{\Delta t} = L_1 \frac{\Delta i_1}{\Delta t} \tag{8.3-1}$$

A portion of this flux will link the secondary and if it is changing with time a secondary voltage will be induced according to

$$v_2 = N_2 \frac{\Delta\phi_m}{\Delta t} \tag{8.3-2}$$

where ϕ_m is the portion of ϕ_1 that links the secondary.

To measure the degree of coupling, a *coefficient of coupling* k is defined as follows:

$$k = \frac{\phi_m}{\phi_1} \tag{8.3-3}$$

If the coupling is perfect so that all of the primary flux links the secondary, then $\phi_m = \phi_1$ and $k = 1$. Values of k close to 1 are realized in iron-core

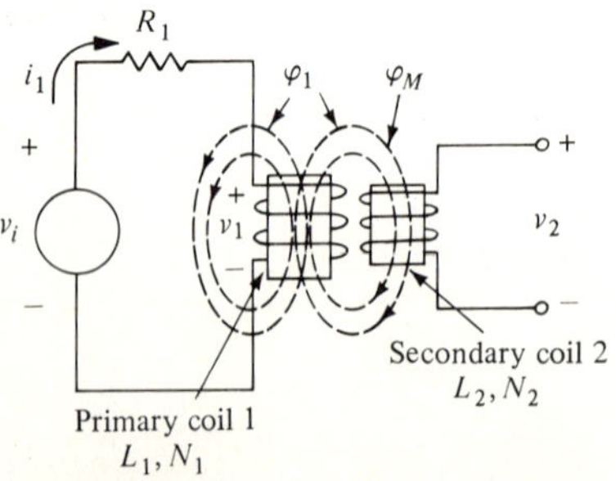

FIGURE 8.8
Mutually coupled coils.

transformers, and in configurations where the secondary is wound directly over the primary. Values of $k \ll 1$ (0.01 to 0.2) are typical of air-core coils.

Using the definition of Eq. (8.3-3), the secondary voltage of Eq. (8.3-2) can be written

$$v_2 = kN_2 \frac{\Delta \phi_1}{\Delta t} \tag{8.3-4}$$

Noting Eq. (8.3-1)

$$\frac{\Delta \phi_1}{\Delta t} = \frac{v_1}{N_1} \tag{8.3-5}$$

When this is substituted into Eq. (8.3-4) the result is

$$v_2 = k\left(\frac{N_2}{N_1}\right) v_1 \tag{8.3-6}$$

In this equation, k, N_2, and N_1 are constants. Since N_2 is easily made greater than N_1 and with an iron core $k \approx 1$, it is possible to have v_2 greater than v_1. This is the basis for the *electrical transformer* which we study in more detail in a later chapter. For now, it is important to keep in mind that this effect occurs only when the current i_1 is changing with time, otherwise v_2 is zero.

Mutual Inductance

The *mutual inductance* M, measured in henries, between coils 1 and 2 is given by either of two formulas:

$$M_{12} = N_2 \frac{\Delta \phi_m}{\Delta i_1} = kN_2 \frac{\Delta \phi_1}{\Delta i_1}$$

or (8.3-7)

$$M_{21} = N_1 \frac{\Delta \phi_m}{\Delta i_2} = kN_1 \frac{\Delta \phi_2}{\Delta i_2}$$

In all cases $M_{12} = M_{21} = M$. Thus Eq. (8.3-7) becomes

$$M = kN_2 \frac{\Delta \phi_1}{\Delta i_1}$$

$$M = kN_1 \frac{\Delta \phi_2}{\Delta i_2} \tag{8.3-8}$$

Now we multiply these two equations together to get

$$M^2 = k^2 N_1 N_2 \frac{\Delta\phi_1}{\Delta i_1} \frac{\Delta\phi_2}{\Delta i_2}$$

Rearranging,

$$M^2 = k^2 \left(N_1 \frac{\Delta\phi_1}{\Delta i_1}\right)\left(N_2 \frac{\Delta\phi_2}{\Delta i_2}\right) \tag{8.3-9}$$

Using this equation, the mutual inductance can be related to the primary and secondary inductances as follows. From Eq. (8.3-1) the self-inductance of each coil can be written as

$$L_1 = N_1 \frac{\Delta\phi_1}{\Delta i_1} \quad \text{and} \quad L_2 = N_2 \frac{\Delta\phi_2}{\Delta i_2} \tag{8.3-10}$$

Using these relations Eq. (8.3-9) becomes

$$M^2 = k^2 L_1 L_2$$

Taking the square root of both sides,

$$\boxed{M = k\sqrt{L_1 L_2}} \tag{8.3-11}$$

This is an important relation. Typical calculations are shown in the following examples.

EXAMPLE 8.3-1 *Finding Primary and Secondary Voltages in Coupled Coils*

In the coupled coils of Fig. 8.8 the following values apply: $k = 0.6$, $L_1 = 0.2$ H, $N_1 = 100$ turns, $L_2 = 0.4$ H, $N_2 = 200$ turns. Find

a. The mutual inductance M
b. The induced voltages across the primary and the secondary if the primary flux changes at the rate of 700 mWb/s

Solution

a. From Eq. (8.3-11),

$$M = k\sqrt{L_1 L_2} = 0.6\sqrt{0.2 \times 0.4} = 0.6\sqrt{0.08} = 0.17 \text{ H}$$

b. From Eq. (8.3-1),

$$v_1 = N_1 \frac{\Delta\phi_1}{\Delta t} = (100)(0.7) = 70 \text{ V}$$

From Eq. (8.3-4),

$$v_2 = kN_2 \frac{\Delta\phi_1}{\Delta t} = (0.6)(200)(0.7) = 84 \text{ V}$$

• • •

Series Connection of Coupled Coils

The coupled coils of Fig. 8.8 may be connected in series in some applications. When this is the case, the total inductance is not simply $L_1 + L_2$ but is complicated by the presence of the mutual inductance. Because of the series connection, the same current flows in both coils as shown in Fig. 8.9a and b and the voltage induced in coil 1 will consist of a term due to the self-inductance L_1 and a term due to the current i flowing in coil 2 which induces a voltage in coil 1 through the mutual inductance M. The voltage is

$$v_1 = L_1 \frac{\Delta i}{\Delta t} \pm M \frac{\Delta i}{\Delta t}$$

$$= (L_1 \pm M) \frac{\Delta i}{\Delta t} \qquad (8.3\text{-}12)$$

Now we must determine the sign to be used with the mutual term. In Fig. 8.9a we see that the current i flowing in coil 2 produces a flux ϕ_2 which is in the same direction as ϕ_1 and will thus simply add to ϕ_1. The mutual voltage induced in coil 1 will then have the same polarity as the self-induced voltage $L_1\Delta i/\Delta t$ and Eq. (8.3-12) becomes

$$v_1 = (L_1 + M) \frac{\Delta i}{\Delta t} \qquad (8.3\text{-}13)$$

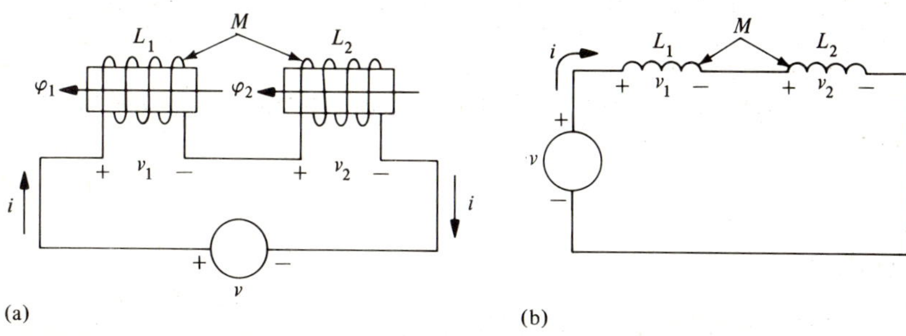

FIGURE 8.9
Series-coupled coils. (a) Pictorial diagram. (b) Circuit.

If the winding direction of one of the coils is changed or if the wires going to one coil are simply interchanged, the components of flux due to the two coils would be in opposite directions, and the sign of the mutual term would be negative. This case will be considered shortly.

For the voltage across coil 2 we have, using similar reasoning,

$$v_2 = (L_2 + M)\frac{\Delta i}{\Delta t} \tag{8.3-14}$$

The total voltage v is, using KVL,

$$v = v_1 + v_2 = (L_1 + M)\frac{\Delta i}{\Delta t} + (L_2 + M)\frac{\Delta i}{\Delta t}$$

$$= (L_1 + L_2 + 2M)\frac{\Delta i}{\Delta t}$$

This is the same as the vi relation of an equivalent inductance L_{eq}, where

$$L_{eq} = L_1 + L_2 + 2M \tag{8.3-15}$$

If one coil is reversed, this becomes

$$L_{eq} = L_1 + L_2 - 2M \tag{8.3-16}$$

It is interesting to note that these two equations can be used to measure M. We would proceed by first measuring L_{eq} both ways. M is then found by subtracting Eq. (8.3-16) from Eq. (8.3-15). This yields

$$L_{eq+} - L_{eq-} = L_1 + L_2 + 2M - (L_1 + L_2 - 2M)$$

$$= 4M$$

Thus

$$M = \tfrac{1}{4}(L_{eq+} - L_{eq-}) \tag{8.3-17}$$

where L_{eq+} is the larger of the measured equivalent inductances and L_{eq-} is the smaller.

EXAMPLE 8.3-2 Series Coils with Mutual Inductance

Two coils are wound so that mutual inductance is present between them. The self-inductances are 10 and 9 mH and the mutual inductance is 6 mH. Find the total inductance and coefficient of coupling if the coils are connected in the two possible arrangements of Fig. 8.9b.

Solution
For the case where the fluxes add,

$$L_{eq} = L_1 + L_2 + 2M = 10 + 9 + 12 = 31 \text{ mH}$$

The coefficient of coupling is

$$k = \frac{M}{\sqrt{L_1 L_2}} = \frac{6}{\sqrt{90}} = 0.63$$

For the other connection,

$$L_{eq} = L_1 + L_2 - 2M = 10 + 9 - 12 = 7 \text{ mH}$$

The coefficient of coupling is the same, $k = 0.63$.

• • •

In the derivation above we determined the sign of the mutual term by considering the actual coil-winding directions. It is seldom feasible to do this in practice because the normal circuit diagram does not show the winding direction pictorially, so a system of dot markings is used to indicate the winding direction and, as a result, the sign of the mutual terms.

To illustrate the dot convention, consider Fig. 8.10a where we have shown a pair of coupled coils in series. The significance of the dots is as follows:

> *When current flows in at the dot on one winding, a voltage is induced across the other winding, which is positive at the dotted end. When current flows out at the dot on one winding, the voltage across the other winding is negative at the dotted end.*

With the dots as shown in part a of the figure the current i flows *in* at the dot in coil 1 so the mutual voltage induced in coil 2 is more positive at the dotted

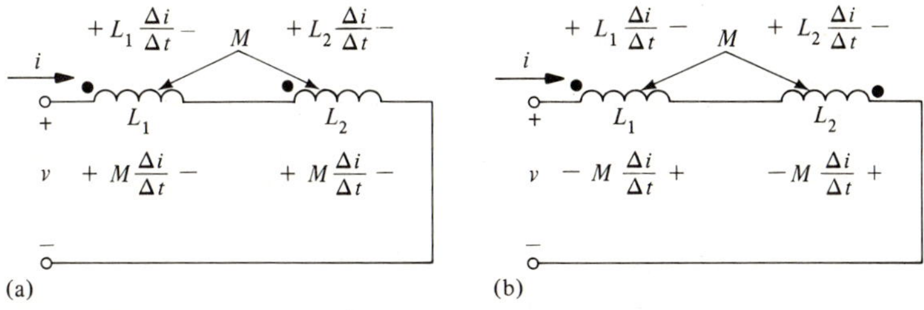

FIGURE 8.10
Dot convention for coupled coils. (a) Series connection, *M* terms positive. (b) *M* terms negative.

end. Also, the current is flowing in at the dot in coil 2 so the mutual voltage induced in coil 1 is more positive at the dotted end. Since the self-inductance voltage across each coil is positive at the end at which the current enters, we see that the mutual voltages add, and the M terms are then positive. The KVL equation for the circuit with an explanation of each term is as follows:

$$v = L_1 \frac{\Delta i}{\Delta t} + M \frac{\Delta i}{\Delta t} + L_2 \frac{\Delta i}{\Delta t} + M \frac{\Delta i}{\Delta t} \tag{8.3-18}$$

where

$L_1(\Delta i/\Delta t)$ = voltage across coil 1 due to current i in coil 1
$M(\Delta i/\Delta t)$ = voltage across coil 1 due to current i in coil 2
$L_2(\Delta i/\Delta t)$ = voltage across coil 2 due to current i in coil 2
$M(\Delta i/\Delta t)$ = voltage across coil 2 due to current i in coil 1

For the other dot arrangement, shown in Fig. 8.10b, the induced voltages in one coil due to the current flowing in the other are seen to have a polarity opposed to the self-inductance drops. Thus the KVL equation is identical to Eq. (8.3-18) except that the M terms are negative.

Establishing Dot Markings

When the winding directions of the coils are not known and cannot be determined, the dot markings are easily established using the test circuit of Fig. 8.11. The battery is usually of relatively low voltage, and the momentary switch can be simply two pieces of bare wire. When the switch is closed the primary current will flow in the direction shown so a dot is placed at the end of the primary coil into which the current flows. If the voltmeter kicks in the positive direction, then the dot is placed where the positive terminal of the voltmeter is connected to the secondary. If the voltmeter kicks

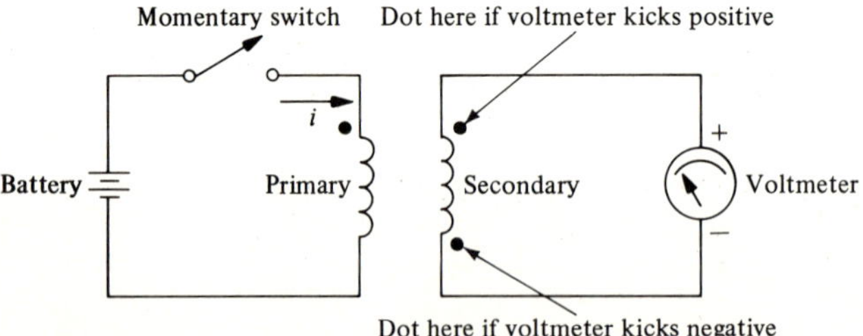

FIGURE 8.11
Circuit for establishing dot positions.

negative, the dot is placed at the other end of the secondary. Use of the dots in writing circuit equations when coupled coils are present is illustrated in the following example.

EXAMPLE 8.3-3 Loop Equations with Mutual Inductance

Write the KVL equations for the radio transmitter input circuit of Fig. 8.12.

Solution

This is a two-loop network, so we use the loop-current method of Sec. 5.2. Loop currents i_1 and i_2 are assumed to flow clockwise as shown. The KVL equation for loop 1 is then

$$v_1 = R_1 i_1 + L_1 \frac{\Delta i_1}{\Delta t} - M \frac{\Delta i_2}{\Delta t}$$

The sign of the mutual term is found by noting that i_2 flows into coil 2 at the end *opposite* the dot. Thus the mutual voltage induced in coil 1 due to i_2 in coil 2 is more positive at the end of coil 1 opposite the dot. This is a voltage *rise* in the clockwise direction in which we are tracing loop 1 so the sign of the voltage is negative in accordance with our convention that voltage drops take a positive sign and voltage rises take a negative sign.

For loop 2 the KVL equation is

$$0 = -M \frac{\Delta i_1}{\Delta t} + L_2 \frac{\Delta i_2}{\Delta t} + R_2 i_2$$

The student should be sure to understand the reasoning behind the negative sign of the mutual term in this equation.

• • •

EXAMPLE 8.3-4 Equivalent Inductance of Series Coils

Determine the equivalent inductance that can replace the series circuits of Fig. 8.13a and b.

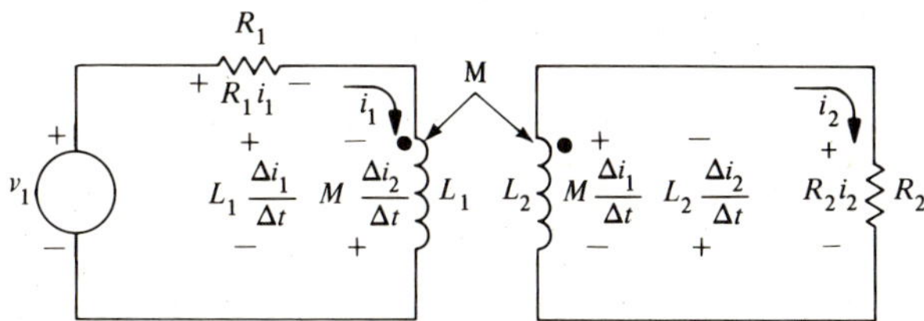

FIGURE 8.12
Circuit for Example 8.3-3. For clarity, the various voltage drops are shown explicitly.

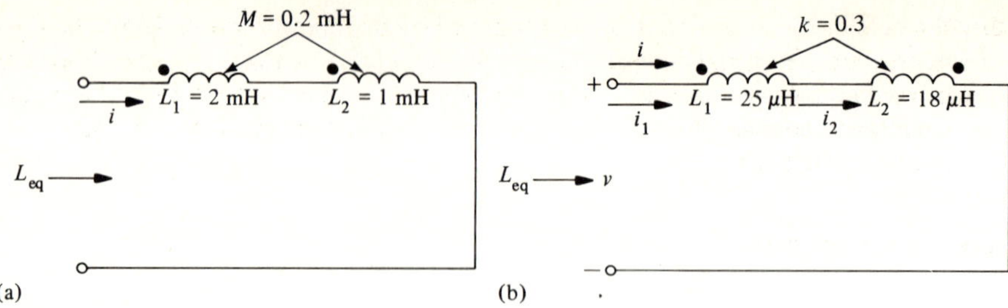

FIGURE 8.13
Circuits for Example 8.3-4.

Solution

Circuit a: We find L_{eq} by pretending to write KVL around the loop. Since the current is in the same direction into the dots on both coils, the mutual voltages will have the same polarities as the self-inductance drops. Thus the mutual terms will be positive and

$$L_{eq} = L_1 + L_2 + 2M = 2 + 1 + 2(0.2) = 3.4 \text{ mH}$$

Circuit b: For this circuit we are given the coupling coefficient. The mutual inductance is then

$$M = k\sqrt{L_1 L_2} = 0.3\sqrt{25 \times 18} = 6.4 \ \mu\text{H}$$

If i_1 is the current in L_1 and i_2 the current in L_2, then the KVL equation taken clockwise around the loop is

$$v = L_1 \frac{\Delta i_1}{\Delta t} - M\frac{\Delta i_2}{\Delta t} + L_2 \frac{\Delta i_2}{\Delta t} - M\frac{\Delta i_1}{\Delta t}$$

The negative sign of the $M\Delta i_2/\Delta t$ term arises because i_2 flows into coil 2 at the end *opposite* the dot. This makes the voltage induced in coil 1 due to i_2 positive at the end opposite the dot. Thus the voltage induced in coil 1 due to i_2 is a rise in the tracing direction and requires a negative sign. Similar reasoning leads to the negative sign in the $M\Delta i_1/\Delta t$ term. Clearly, in this circuit $i_1 = i_2 = i$ so that

$$v = (L_1 + L_2 - 2M)\frac{\Delta i}{\Delta t}$$

and

$$L_{eq} = L_1 + L_2 - 2M = 25 + 18 - 2(6.4) = 30.2 \ \mu\text{H}$$

• • •

LEARNING EXERCISES FOR SEC. 8.3

1. An audio transformer is closely coupled with $k = 0.98$. Find M if $L_1 = 300$ mH and $L_2 = 80$ mH.

2. Two separate coils are wound on the same nonmagnetic core. When they are connected in series the inductance is measured and found to be 16 μH. When one of the coils is reversed the same measurement yields 22 μH. Find the mutual inductance between the two coils and the coefficient of coupling if the inductance of one coil is 5 μH.

Ans. 1.5; 0.18; 0.15

• • •

8.4 CAPACITANCE

The inductance element considered in Sec. 8.1 differs from the previously studied resistance element in several ways. One difference is that while voltage and current are directly proportional in the resistance, voltage across an inductance is zero unless the current is changing with time. Another difference concerns energy; the resistance element dissipates energy in the form of heat while the inductance element cannot dissipate energy but can only store it in its magnetic field (this will be discussed in Sec. 8.8).

The capacitance element is closely related to the inductance. Current will flow through it only when the voltage across it is changing with time and it cannot dissipate energy but can only store it in its electric field. An actual capacitor usually consists of a pair of metal plates or foil separated by an insulating material, called a *dielectric*, as shown in Fig. 8.14a along with the circuit symbol for capacitance. To achieve reasonable values of capacitance, a large surface area is required so the assembly is often rolled into cylindrical form to conserve space. Various types of capacitors are shown in Fig. 8.14b. The *ceramic capacitor* consists of a ceramic disk of dielectric, on each side of which a thin layer of silver is applied. Wires are attached to the silver and an outer insulating coating is applied. The *paper capacitor* consists of aluminum foil and wax-impregnated paper dielectric rolled into cylindrical form. *Plastic-film dielectric capacitors* using Mylar, polyethylene, or polycarbonate films are also manufactured in cylindrical form. The *mica capacitor* consists of a stack of alternate layers of mica dielectric and conducting foil or silver fired onto the mica surface. *Variable capacitors* usually consist of sets of aluminum plates separated by air as the dielectric. Alternate plates are movable so that the capacitance can be varied by varying the area. *Electrolytic capacitors* utilize aluminum foil as one plate and an alkaline electrolyte as the other. Application of a dc voltage during manufacture causes a thin layer of aluminum oxide to be formed, which acts as a dielectric. This type of capacitor exhibits very high values of capacitance but has the disadvantage that the voltage across it can only be of one polarity.

The color-coded band marking system for tubular paper capacitors is shown in Fig. 8.14c. For the capacitance value, given by the first three

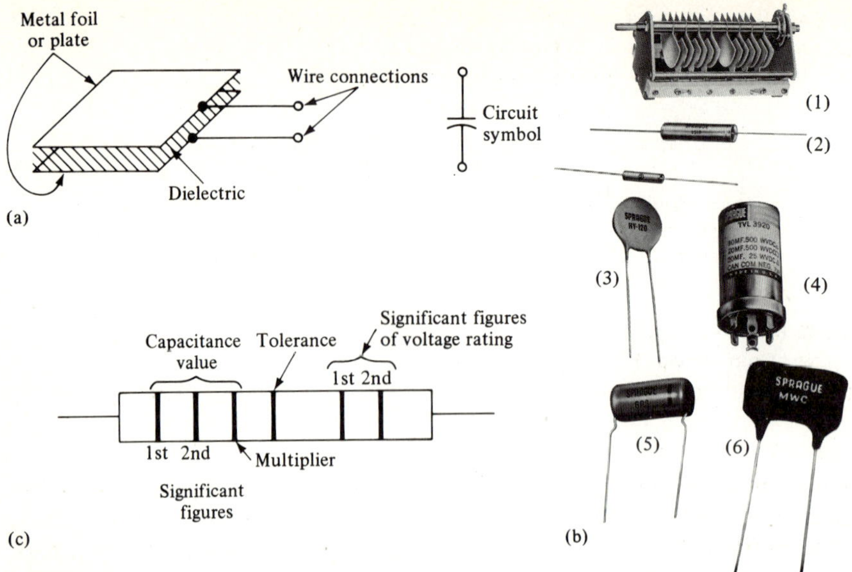

FIGURE 8-14
Capacitance. (a) Pictorial sketch of a capacitor and circuit symbol. The curved line represents the plate which is connected to the lowest voltage point. (b) (1) Variable air capacitor (courtesy of James Millen Manufacturing Co., Inc.); (2) Tantalum capacitors; (3) Ceramic disc capacitor; (4) Electrolytic capacitor; (5) Plastic film capacitor; (6) Mica capacitor (courtesy of Sprague Electric Co.). (c) Color-code bands for tubular capacitors.

bands, the code is the same as for resistors given in Fig. 3.15. For the fourth (tolerance) band, green denotes 5%, white 10%, black 20%, orange 30%, and yellow 40%. The voltage rating is identified by a single-digit number (fifth band) up to 900 V. The rating is obtained by multiplying the fifth band digit by 100. If the rating is above 900 V, a sixth band is added and the resulting two-digit number is multiplied by 100. Other types of capacitors may use dots rather than bands and the significance of the bands and dots varies somewhat for different types.

The Electric Field

Consider the circuit of Fig. 8.15. When the switch is open and the plates are not initially charged there will be no net positive or negative charge on either plate. When the battery voltage is applied to the capacitor by closing the switch, electrons cannot move through the air between the plates. However, electrons are drawn from the upper plate to the positive terminal of the battery so that the upper plate has less negative charge and thus becomes positively charged. At the same time, electrons are repelled by the negative terminal of the battery, and move to the lower plate at the same

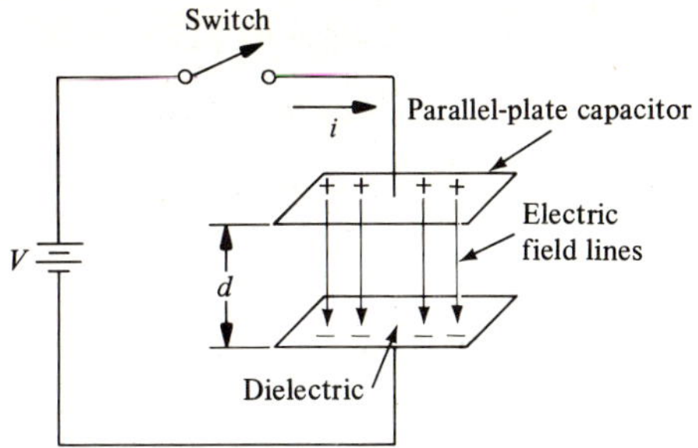

FIGURE 8.15
Electric field lines in a parallel-plate capacitor.

rate that they are being drawn to the positive battery terminal from the upper plate. This motion of electrons forms a transient current, which continues until the voltage across the plates is exactly equal to the battery voltage. After this initial transient period no current flows and the plate connected to the positive battery terminal carries a charge $+Q$ while the opposite plate carries a charge $-Q$. At this time a steady *electric field* appears between the plates. The electric field is very similar to the magnetic field that we studied in Chap. 7. However, lines of electric force differ from magnetic lines of force in that they are not continuous, but always begin on positive charges and end on negative charges. Thus the electric force lines between the plates of the capacitor begin on the positive charges on the upper plate and end on the negative charges on the lower plate.

In contrast to magnetic fields, where it is not possible to isolate magnetic poles, electric charges do exist in isolated form. The electric force between two small electric charges is directly proportional to the product of the charges and inversely proportional to the square of the distance between them

$$F = k\frac{Q_1Q_2}{d^2} \tag{8.4-1}$$

where

Q_1 and Q_2 = charges, in coulombs (C)
d = distance between charges, in meters (m)
k = a constant equal to 9.0×10^9 Nm^2/C^2
F = electric force in newtons (N)

The strength of the electric field, called the *field intensity*, is defined as the electric force per unit charge on a small test charge at a particular point in space. The symbol used for field intensity is E so that

$$E = \frac{F}{Q} \tag{8.4-2}$$

where

F = electric force, in newtons (N)
Q = test charge, in coulombs (C)
E = electric field intensity, in newtons per coulomb (N/C)

For the case of two charged plates that we are considering, the field intensity is uniform between the plates, and can be shown to be

$$E = \frac{V}{d} \tag{8.4-3}$$

where

V = voltage between plates, in volts (V)
d = distance between plates, in meters (m)
E = electric field intensity, in volts per meter (V/m)

For the configuration of Fig. 8.15, the field between the plates will be essentially uniform, that is, linear. This means that if the battery voltage is 100 V and if we had an instrument capable of measuring the voltage in the space between the plates, we would find that the voltage from the lower plate to any point between the plates is proportional to the distance to the point of measurement. For example, the voltage from the lower plate to the half-way point between the plates is 50 V.

Capacitance

Returning to the circuit of Fig. 8.15, the charge Q on the plates is found to be directly proportional to the voltage even if the voltage is varying with time and the current is not zero. Thus the defining relation for the *ideal* capacitance is

$$\boxed{q = Cv} \tag{8.4-4}$$

where*

*We are often going to use lower-case letters for variables from this point on, in order to indicate that they may vary with time.

q = charge on plates, in coulombs (C)
v = voltage across plates, in volts (V)
C = constant of proportionality, called capacitance, measured in farads (F)

The farad is an extremely large unit and most actual capacitors range from picofarads (10^{-12} F) to microfarads (10^{-6} F). Typical values of mica capacitors range from 25 to 500 pF, while paper capacitors range from about 0.0005 to 2 μF. Electrolytic capacitors range from about 1 to 2500 μF with maximum allowable voltages from 3 to 450 V.

Referring to Eq. (8.4-4) we note that charge is usually an inconvenient variable to work with in circuit analysis. By returning to the definition of current we can convert the equation to the more convenient current and voltage variables. To accomplish this, we write Eq. (8.4-4) in the form

$$\frac{\Delta q}{\Delta t} = C\frac{\Delta v}{\Delta t} \tag{8.4-5}$$

From Sec. 2.3 we have the definition of current: $i = \Delta q/\Delta t$. Thus

$$\boxed{i = C\frac{\Delta v}{\Delta t}} \tag{8.4-6}$$

This is the desired *vi* relation for the capacitor, and we see that the current is zero unless the voltage is varying with time. It is instructive to compare this with the *vi* relation for the inductor:

$$v = L\frac{\Delta i}{\Delta t} \tag{8.4-7}$$

If we interchange v and i, and replace L by C or C by L in either of these relations, we arrive at the other one. This is an example of *duality*. A useful consequence of duality is that all of the theory associated with Eq. (8.1-6) in Sec. 8.1 can be applied to Eq. (8.4-6) by merely interchanging v and i, and replacing L by C.

When we studied the inductance element in Sec. 8.1, we found that inductance opposed changes in current and it was not possible for current to change instantaneously in an inductance. A dual conclusion holds for capacitance. Since the voltage across a capacitance is proportional to the charge, any change in the voltage must be brought about by a redistribution of the charge on the plates. Electric charges (such as the electron), even though extremely small, do have finite mass, so it is impossible for them to move any distance in zero time. Thus we conclude that the voltage across a capacitor cannot change instantaneously. We will make use of these properties of inductance and capacitance in the next chapter.

At this point it will be useful to list for review some of the properties of the three elements: resistance, inductance, and capacitance.

1. The resistance element opposes the flow of current. It dissipates energy by converting it to heat.
2. The capacitance element opposes any change in the voltage across its terminals. It stores energy in its electric field (this is discussed in Sec. 8.8).
3. The inductance element opposes any change in the current through it. It stores energy in its magnetic field (Sec. 8.8).

• • •

LEARNING EXERCISE FOR SEC. 8.4

1. The negative terminal of a 22.5-V lantern battery is connected to the bottom plate of a parallel-plate capacitor. The positive terminal is connected to the top plate. The plate spacing is 0.5 cm and the total capacitance is 5 pF.
   a. Find the voltage 1 mm and 3 mm from the bottom plate.
   b. Find the total charge.

Ans. 13.5; 4.5; 112

• • •

8.5 THE PARALLEL-PLATE CAPACITOR

In this section we present the formula for the capacitance of the parallel-plate capacitor shown in Fig. 8.16 in terms of its physical dimensions. This will then be applied in several examples in order to illustrate the actual dimensions involved in obtaining practical capacitance values.

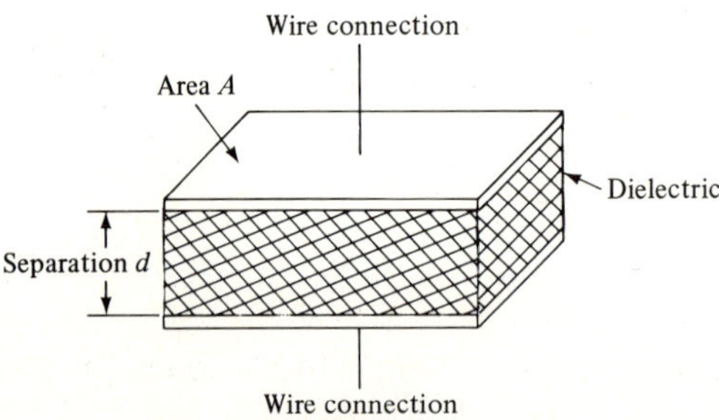

FIGURE 8.16
Parallel-plate capacitor.

We can arrive at the way in which the capacitance varies with the dimensions of the plates by some simple physical reasoning. If we isolate C in the formula $Q = CV$ we have

$$C = \frac{Q}{V} \tag{8.5-1}$$

Thus, for a fixed voltage between the plates, capacitance is directly proportional to charge. Since electric charges such as electrons do have finite dimensions, only a certain amount of charges will fit in a given area A. If we double the area, keeping the voltage the same, we can accommodate twice as many charges on the plate, and the capacitance will be doubled. We therefore conclude that the capacitance of the parallel-plate capacitor is directly proportional to its area. This conclusion is similar to that found in Chap. 3, where the resistance of a conductor is found to be inversely proportional to its cross-sectional area, so that its conductance is directly proportional to the area.

For the conductor, conductance is *inversely* proportional to length and similarly, the capacitance of parallel plates is found to be inversely proportional to the distance d between them. We then write the formula for the capacitance of a pair of parallel plates as

$$\boxed{C = \epsilon \frac{A}{d}} \tag{8.5-2}$$

In this formula the constant of proportionality ϵ (epsilon) is analogous to conductivity σ, which is the reciprocal of resistivity ρ. It depends on the dielectric material between the plates and is called the *permittivity* of the dielectric. If C is in farads (F), A in meters squared, and d in meters, then the units of ϵ must be farads/meter (F/m).

The permittivity of vacuum is symbolized by ϵ_0 and has the value 8.85×10^{-12} F/m. It is convenient to define a *relative permittivity* ϵ_r as the ratio of the permittivity of the dielectric in question to that of vacuum. Thus

$$\epsilon_r = \frac{\epsilon}{\epsilon_0} \tag{8.5-3}$$

ϵ_r is usually called the *dielectric constant* of the material. Some typical values are given in Table 8.1.

It is convenient to express Eq. (8.5-2) in terms of the dielectric constant. We do this by solving Eq. (8.5-3) to get $\epsilon = \epsilon_r\epsilon_0$ so that Eq. (8.5-2) becomes

TABLE 8.1
Dielectric constant of various materials

Dielectric	ϵ_r
Vacuum	1
Air	1.0006
Teflon	2
Mylar	3
Paper (paraffin-coated)	4
Mica	5
Tantalum oxide	25
Ceramic	15 to 10 000

$$C = \frac{\epsilon_r \epsilon_0 A}{d} \tag{8.5-4}$$

Substituting $\epsilon_0 = 8.85 \times 10^{-12}$ F/m

$$\boxed{C = \frac{8.85 \times 10^{-12}\, \epsilon_r A}{d} \text{ F}} \tag{8.5-5}$$

It must be remembered that this formula applies only to parallel-plate capacitors in which the plate dimensions are much larger than their separation. Formulas for other configurations can be found in electronics handbooks.

We illustrate the use of this formula with an example.

EXAMPLE 8.5-1 Capacitance of Parallel Plates

A parallel-plate capacitor used in a physics experiment has an area A of 100 cm^2 and a separation $d = 1$ mm. Find the capacitance if the dielectric is

a. Air
b. Paraffin-coated paper
c. Ceramic, $\epsilon_r = 250$

Solution

We use Eq. (8.5-5) with

$$A = 100 \text{ cm}^2 \times \left(\frac{1 \text{ m}}{100 \text{ cm}}\right)^2 = 10^{-2} \text{ m}^2$$

and

$$d = 1 \text{ mm} \times \left(\frac{1 \text{ m}}{1000 \text{ mm}}\right) = 10^{-3} \text{ m}$$

Then

$$C = \frac{8.85 \times 10^{-12} \times \epsilon_r \times 10^{-2}}{10^{-3}} = 88.5\, \epsilon_r \times 10^{-12}\ \text{F} \tag{8.5-6}$$

a. For air, $\epsilon_r = 1.0006$, and

$$C = 88.5 \times 1.0006 \times 10^{-12}\ \text{F} = 88.6\ \text{pF}$$

(recall that 10^{-12} F = 1 pF)

b. For paraffin-coated paper, $\epsilon_r = 4$, and

$$C = 88.5 \times 4 \times 10^{-12}\ \text{F} = 354\ \text{pF}$$

c. For ceramic, $\epsilon_r = 250$, and

$$C = 88.5 \times 250 \times 10^{-12}\ \text{F} = 22\,000\ \text{pF} = 0.022\ \mu\text{F}$$

(recall that 1 μF = 10^6 pF)

Thus by simply changing the dielectric material with the same plate dimensions and spacing we achieve a capacitance change of 250 to 1.

• • •

In the last example, the plate area was 100 cm² (approximately 15 in²). For practical reasons, it is not possible to allocate this much space in a system for one capacitor. One way to reduce the space requirement is to form the capacitor from two long rectangular flexible plates, for example, 2 × 50 cm, as shown in Fig. 8.17a. The dielectric in flexible form is inserted between the plates and the whole assembly is rolled into cylindrical form as shown in Fig. 8.17b.

Dielectric Strength

In modern circuit design we try to keep all components as small as possible. Thus, in the design of capacitors we try to obtain maximum capacitance in minimum volume. If we consider the basic formula for capacitance, C =

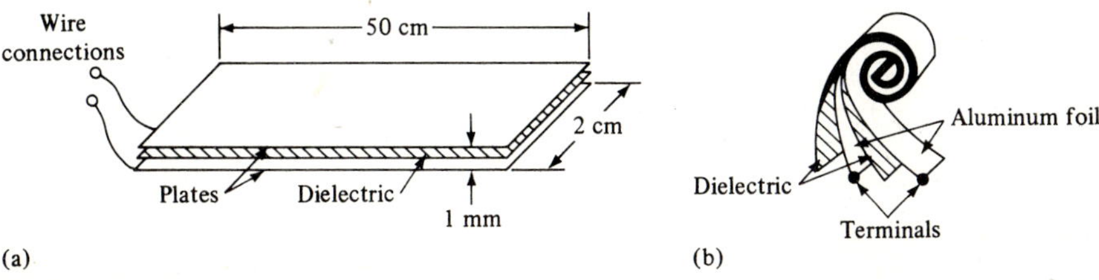

FIGURE 8.17
Parallel-plate capacitor. (a) Assembly. (b) Final rolled-up form.

$\epsilon A/d$, we see that one way to increase C for a given area and dielectric is to decrease d, the separation between the plates. This can only be carried so far until other problems are encountered. The limits on how small the separation can be made are set by the maximum voltage to be applied to the capacitor and the *dielectric strength* of the material between the plates. Every dielectric material is characterized by a maximum field intensity E [see Eq. (8.4-2)], which is the field intensity it can withstand before breaking down. This voltage per unit length is called the *dielectric strength* and is expressed in kilovolts per millimeter (kV/mm). Some typical values are given in Table 8.2.

When breakdown occurs, the dielectric acts like a pure conductor, essentially a short circuit. An interesting example of dielectric breakdown is lightning. If we think of the clouds as one plate of a capacitor and the earth the other plate, then the atmosphere between the two is the dielectric. When the voltage between the two is so large that the dielectric strength of the atmosphere is exceeded, a lightning bolt passes from the clouds to the earth.

The significance of the dielectric strength can best be understood in terms of the following example. Consider a capacitor built with a ceramic dielectric having a dielectric strength of 3 kV/mm. If we use a dielectric thickness of 0.3 mm the capacitor should be rated at a maximum voltage of (3 kV/mm)(0.3 mm) = 0.9 kV = 900 V. If we reduce the spacing to 0.1 mm the capacitor can be rated at only 300 V. Thus a trade-off may be necessary between capacitance and breakdown voltage. Typical values will be found in the following examples.

EXAMPLE 8.5-2 Characteristics of Parallel-Plate Capacitors

A capacitor for use in radio and TV circuits is to be manufactured in the form of a circular disk. The dielectric is to be 1 cm in diameter and 0.2 mm in thickness. The plates are to be formed from a coating of silver plated onto both sides of the dielectric, which is to be either mica ($\epsilon_r = 5$) or ceramic ($\epsilon_r = 5000$). For each of these materials find

TABLE 8.2
Dielectric strength of various materials

Dielectric	Average Dielectric Strength (kV/mm)
Air	3
Ceramic	3
Paper (paraffin-coated)	30
Teflon	60
Glass	120
Mica	200

a. The capacitance
b. The maximum voltage rating
c. The charge on the plates when maximum voltage is applied

Solution

a. Mica ($\epsilon_r = 5$):

$$C = \frac{8.85 \times 10^{-12}\, \epsilon_r A}{d}$$

$$= \frac{8.85 \times 10^{-12} \times 5 \times \pi \times (5 \times 10^{-3})^2}{2 \times 10^{-4}}$$

$$= 17.4 \times 10^{-12}\ \text{F} = 17.4\ \text{pF}$$

Ceramic ($\epsilon_r = 5000$): Since C is directly proportional to ϵ_r and the ϵ_r for ceramic is 1000 times the ϵ_r for mica, we see that $C = 17.4 \times 10^{-9}\ \text{F} = 0.0174\ \mu\text{F}$.

b. Mica (dielectric strength = 200 kV/mm):

$$\text{Voltage rating} = 200\ \text{kV/mm} \times 0.2\ \text{mm} = 40\ \text{kV}$$

Ceramic (dielectric strength = 3 kV/mm):

$$\text{Voltage rating} = 3\ \text{kV/mm} \times 0.2\ \text{mm} = 0.6\ \text{kV} = 600\ \text{V}$$

c. Mica:

$$\begin{aligned} Q &= CV \\ &= 17.4 \times 10^{-12}\ \text{F} \times 40 \times 10^3\ \text{V} \\ &= 700 \times 10^{-9}\ \text{C} \\ &= 0.7\ \mu\text{C} \end{aligned}$$

Ceramic:

$$\begin{aligned} Q &= 17.4 \times 10^{-9}\ \text{F} \times 600\ \text{V} \\ &= 10\,440 \times 10^{-9} \\ &= 10.4\ \mu\text{C} \end{aligned}$$

• • •

Nonideal Capacitors

Consider the circuit of Fig. 8.15. When the switch is closed the capacitor will charge to the battery voltage. If the capacitor is ideal and the switch is now opened, the charge will remain on the plates indefinitely. Such conditions are, of course, never achieved, and in a practical situation, the charge will "leak" from the plates until the capacitor is completely discharged. The time taken for this discharge depends on the *effective resistance* of the capacitor. This resistance is in parallel with the ideal capacitance, as shown in Fig. 8.18. The circuit is a *model* which approximately accounts for the various nonideal characteristics of actual capacitors.

One way that leakage can occur is directly through the dielectric. Since every dielectric has some small amount of free electrons, application of voltage will set them into motion causing a small current. This current flow is accounted for by an effective resistance, which is 1000 MΩ or more for the most commonly used dielectric materials. It is such a large resistance that in most cases it can be considered an open circuit. Other possibilities for leakage exist. For example, leakage current can flow through the material in which the capacitor is encapsulated. In electrolytic capacitors, leakage currents can be relatively high, and the effective resistance low enough so that it sometimes must be taken into account.

Parasitic Elements

Let us extend the idea of nonideal capacitors a little further. None of the actual physical elements that we use in a circuit is ideal. However, many times they approach close enough to the ideal so that the design calculations that we make based on ideal characteristics yield useful results. It is not possible to set down rules that can be used to predict when the nonideal character of an element must be taken into account in a design. However, at this time we will point out several situations that do arise frequently.

Consider first the pair of conducting wires shown in Fig. 8.19a. They are metal conductors separated by air and thus there is capacitance between them, as indicated in dashed lines on the diagram. This capacitance can be calculated using formulas found in handbooks. The important point is that this capacitance is something that we would rather not have around. When we study alternating current, we will find that at very high frequencies, this capacitance will actually make it appear that alternating

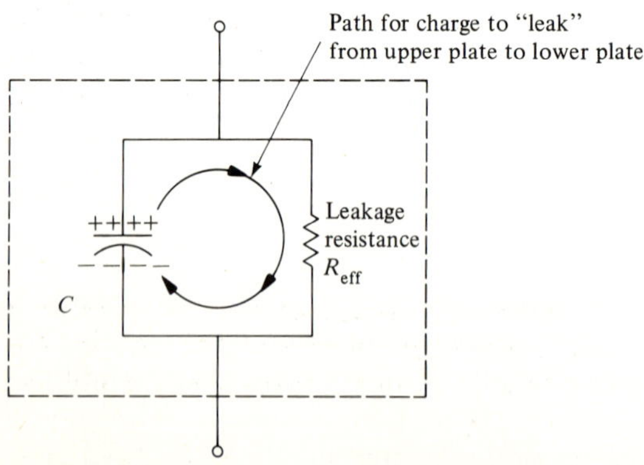

FIGURE 8.18
Nonideal capacitor.

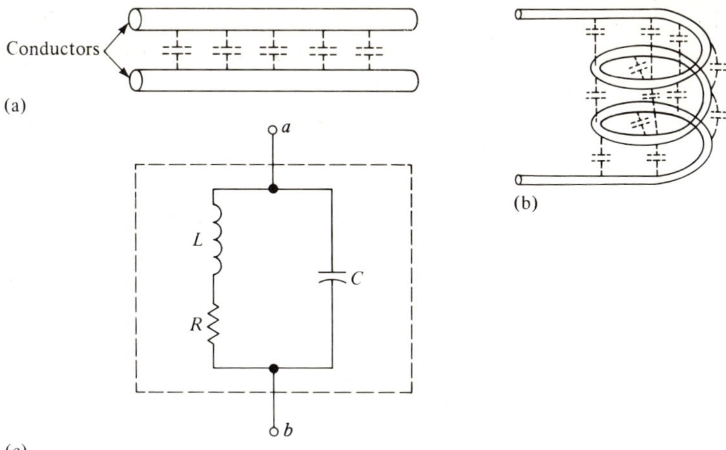

FIGURE 8.19
Parasitic capacitance. (a) Between conductors. (b) In a coil. (c) Coil with models for parasitic elements.

current is flowing through the air between the wires, an undesirable effect. This is then a *parasitic* phenomenon due to the parasitic capacitance between the wires.

Another example occurs in an inductance coil. Consider the several turns of wire of an inductance coil as shown in Fig. 8.19b. We pointed out in Sec. 8.2 that practical inductors always have parasitic resistance associated with the inductance. Now we find that parasitic capacitance is also present. This capacitance is a combination of the many tiny capacitances between turns and from one side to the other of a single turn. It is usually modeled by a single capacitor in parallel with the coil, as shown in Fig. 8.19c. It must be clearly understood that R, L, and C in this circuit cannot be separated physically and that the only terminals available are a and b.

Many other parasitic effects exist in all electric elements. Fortunately, most of them can be ignored in a preliminary design. If the design must be refined further, experience will dictate which parasitic elements must be taken into account.

• • •

LEARNING EXERCISE FOR SEC. 8.5

1. A 5-pF parallel-plate capacitor is required for an X-ray machine. The dielectric is to be mica which has $\epsilon_r = 5$ and a dielectric strength of 200 kV/mm. The maximum voltage which the capacitor must withstand is 50 kV. Find the minimum spacing and the required area.

Ans. 28; 0.25

• • •

8.6 RESPONSE OF CAPACITORS

In this section we present examples that illustrate certain aspects of the voltage-current behavior of the capacitance element. In later chapters we consider the response of circuits that contain capacitors along with other elements.

EXAMPLE 8.6-1 Voltage Required to Achieve Constant Current

a. A 10-μF capacitor is to carry a constant current of 1 mA. At what rate must the voltage across it change in order that it carry this current?

b. A constant current of 1 mA is applied to the positive terminal of a 10-μF capacitor that has an initial voltage of 10 V across it at $t = 0$. Find the capacitor voltage at $t = 100$ ms.

Solution

a. In the equation $i = C\Delta v/\Delta t$ we isolate $\Delta v/\Delta t$ to get

$$\frac{\Delta v}{\Delta t} = \frac{i}{C} = \frac{10^{-3}\text{ A}}{10 \times 10^{-6}\text{ C}} = 100\text{ V/s}$$

Possible waveforms are shown in Fig. 8.20a.

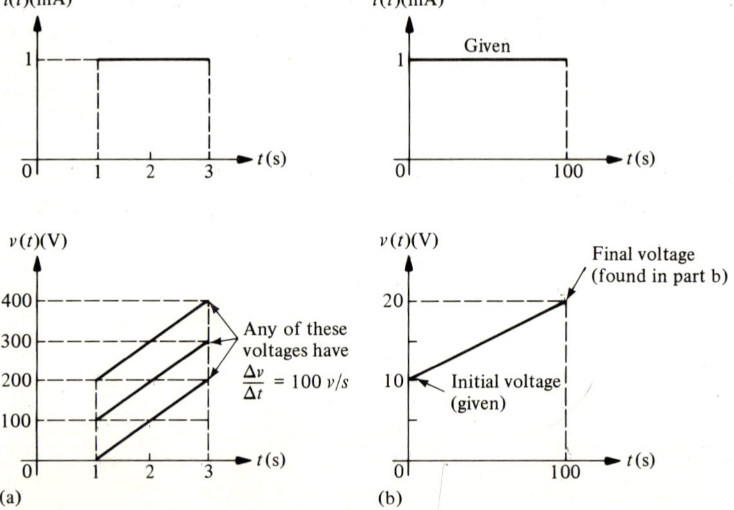

FIGURE 8.20
Example 8.6-1. (a) Possible voltage waveforms for part a. (b) Waveforms for part b.

b. We solve the capacitor vi equation for Δv:

$$\Delta v = \frac{i}{C} \Delta t$$

In this equation, $i = 1$ mA, $C = 10\ \mu F$, and $\Delta t = 100 - 0 = 100$ ms. Substituting, we get

$$\Delta v = \frac{10^{-3}\ \text{A} \times 0.1\ \text{s}}{10 \times 10^{-6}\ \text{F}} = 10\ \text{V}$$

This is *not* the desired answer. It represents the *change*, Δv, from the initial voltage V_0, which is given as 10 V. Then

$$v = V_0 + \Delta v = 10 + 10 = 20\ \text{V}$$

The waveforms are shown in Fig. 8.20b.

Note that in part (a) of this example the voltage can begin at any value and as long as it rises at the rate of 100 V/s, the current will be 1 mA. In part (b), the voltage started at a specified initial value, so that there is only one possible voltage.

• • •

In Sec. 8.4 we stated that the voltage across a capacitor could not change suddenly. However, the current through the capacitor can change suddenly. Let us investigate the effect of such a change. The graph shown in Fig. 8.21a illustrates a dimensionless *unit step function* that is used to represent sudden changes. From the figure we see that

$$\begin{aligned} u(t) &= 0 \qquad t < 0 \\ u(t) &= 1 \qquad t > 0 \end{aligned} \tag{8.6-1}$$

$u(t)$ is undefined at $t = 0$ where the jump takes place. This function may be used to represent a sudden change in current voltage, charge, or any other

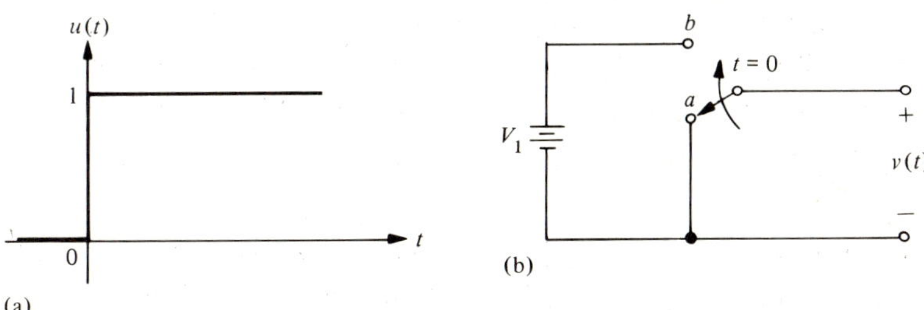

FIGURE 8.21
The unit step function *u*(*t*). (a) Graph of *u*(*t*). (b) Circuit which generates $V_1 u(t)$.

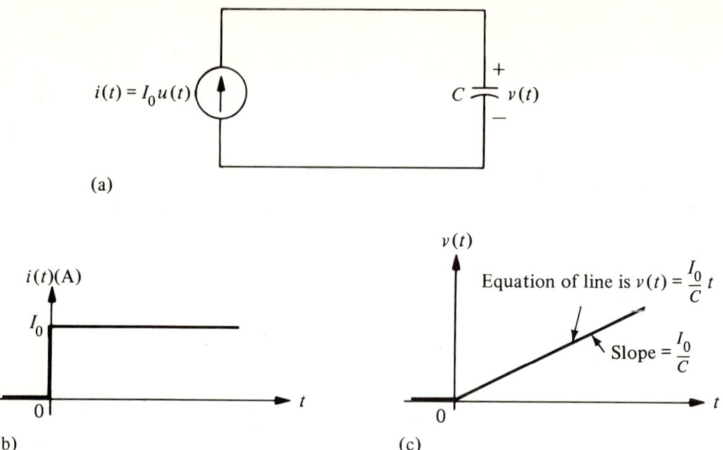

FIGURE 8.22
Step-function response. (a) Circuit. (b) Input current. (c) Output voltage.

variable in the following way. Consider that we want to write an expression to represent a sudden change in voltage in which the voltage goes from 0 to V_1 volts at $t = 0$. We then write

$$v(t) = V_1 u(t) \text{ V} \tag{8.6-2}$$

The circuit shown in Fig. 8.21b generates this voltage. When the switch, which is assumed to be ideal, is thrown from a to b at $t = 0$ the voltage across the output terminals jumps from 0 to V_1 volts. Equation (8.6-2) is the mathematical expression which represents this jump. Electronic circuits can be arranged to generate step functions of voltage or current. Let us assume that we have such a circuit in the form of a current source $i(t)$, which we connect to a capacitor as shown in Fig. 8.22a. The waveform of the current from the source is then shown in Fig. 8.22b and we wish to find the resulting voltage waveform. If we assume that the capacitor is initially uncharged, we can proceed as follows: Consider the time before $t = 0$ as the first interval and the time after $t = 0$ as the second interval. Then during the first interval, since $i = 0$ for $t < 0$

$$\Delta v = \frac{i}{C}\Delta t = 0 \qquad \text{for } t < 0 \tag{8.6-3}$$

During the second interval $i = I_0$ so that

$$\Delta v = \frac{I_0}{C}\Delta t$$

Dividing both sides by Δt, we get

$$\frac{\Delta v}{\Delta t} = \frac{I_0}{C} \qquad \text{for } t > 0 \tag{8.6-4}$$

Now $\Delta v/\Delta t$ represents the *slope* of the voltage-time curve, and since I_0/C is a constant we conclude that the graph of voltage vs. time is a straight line of slope I_0/C, that is,

$$v(t) = \frac{I_0}{C} t \qquad t > 0 \tag{8.6-5}$$

This is plotted in Fig. 8.22c and the waveform is called a *ramp* function. It is evident that the ramp is nonphysical because it theoretically increases without bound. However, ramp functions which last for a finite time are easily generated and are used to form "sawtooth" waveforms, which are often used in electronic circuits.

• • •

LEARNING EXERCISES FOR SEC. 8.6

1. What capacitance is required to achieve a constant current of 3 μA when the voltage across it is changing at the rate of 12 V/s?
2. At what rate must the voltage be changing if 6 mA is to be maintained through a 50-μF capacitor?
3. The voltage across a 330-pF capacitor is changing at the rate of 2 kV/s. Find the current.

Ans. 0.66; 0.25; 120

• • •

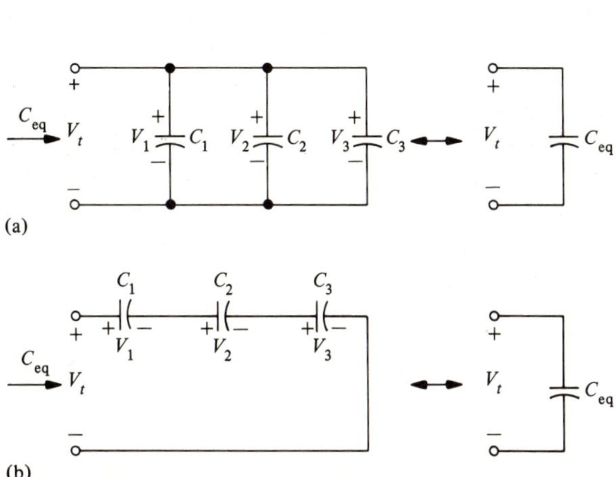

FIGURE 8.23
Equivalent capacitance. (a) Parallel. (b) Series.

8.7 SERIES AND PARALLEL CAPACITORS

In this section we find the equivalent capacitance for series and parallel capacitors. Consider first the parallel connection in Fig. 8.23a. This has the same effect as increasing the area of the plates of a single capacitor. The charge on the individual capacitors is

$$Q_1 = C_1V_1 \qquad Q_2 = C_2V_2 \qquad Q_3 = C_3V_3$$

However, because of the parallel connection

$$V_1 = V_2 = V_3 = V_t$$

The total charge is then

$$Q_t = C_{eq}V_t = Q_1 + Q_2 + Q_3 = C_1V_t + C_2V_t + C_3V_t$$

Dividing by V_t we get

$$\boxed{C_{eq} = C_1 + C_2 + C_3} \tag{8.7-1}$$

Thus any number of *parallel* capacitances add directly in the same way that *series* resistances do and we have the rule

> *The equivalent capacitance for a parallel combination of any number of capacitors is the sum of the individual capacitances.*

Next we consider the series connection of Fig. 8.23b. Using KVL we see that the sum of the voltages across the individual capacitors will be equal to the terminal voltage V_t

$$V_t = V_1 + V_2 + V_3 \tag{8.7-2}$$

where

$$V_1 = \frac{Q_1}{C_1} \qquad V_2 = \frac{Q_2}{C_2} \qquad V_3 = \frac{Q_3}{C_3} \tag{8.7-3}$$

The series connection is similar to increasing the plate separation d of a single capacitor. Since $C = \epsilon A/d$, the result is a *decrease* in the total capacitance. However, the combination will be able to withstand a higher terminal voltage than any of the individual capacitors.

To find the total capacitance, we note that in a series connection of capacitors, the charge on each must be the same so that $Q_t = Q_1 = Q_2 = Q_3$. Thus

$$V_t = \frac{Q_t}{C_{eq}} = \frac{Q_t}{C_1} + \frac{Q_t}{C_2} + \frac{Q_t}{C_3}$$

Dividing by Q_t,

$$\boxed{\frac{1}{C_{eq}} = \frac{1}{C_1} + \frac{1}{C_2} + \frac{1}{C_3}} \qquad (8.7\text{-}4)$$

It is evident that the formula for capacitors in series is similar to that for resistors in parallel. For the important case of two capacitors in series

$$\boxed{C_{eq} = \frac{C_1 C_2}{C_1 + C_2}} \qquad (8.7\text{-}5)$$

As with parallel resistors, the equivalent series capacitance is *always less than* any of the individual capacitances and the rule can be stated

For series capacitors, the reciprocal of the equivalent capacitance is equal to the sum of the reciprocals of the individual capacitances.

EXAMPLE 8.7-1 Series and Parallel Capacitors

Two capacitors, 0.02 and 0.05 μF, are first connected in parallel and then in series. Find the equivalent capacitance for each case.

Solution

Equivalent parallel capacitance

$$C_p = C_1 + C_2 = 0.02\ \mu F + 0.05\ \mu F = 0.07\ \mu F$$

Equivalent series capacitance

$$C_s = \frac{C_1 C_2}{C_1 + C_2} = \frac{(0.02\ \mu F)(0.05\ \mu F)}{0.02\ \mu F + 0.05\ \mu F} = 0.014\ \mu F$$

• • •

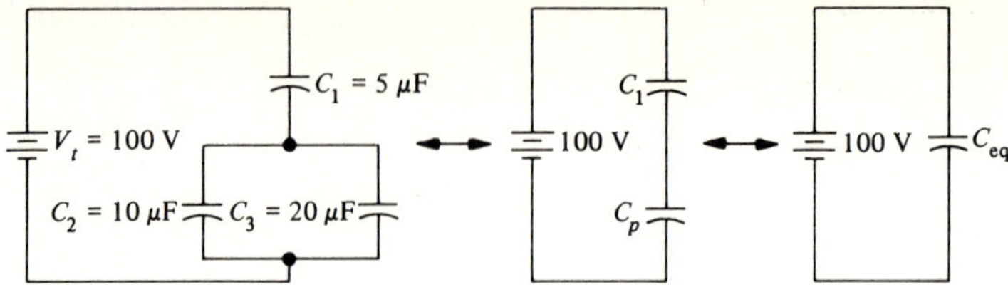

FIGURE 8.24
Circuit for Example 8.7-2.

EXAMPLE 8.7-2 Finding Equivalent Capacitance, Voltage, and Charge

For the circuit of Fig. 8.24 find

a. The equivalent capacitance
b. The charge on each capacitor
c. The voltage across each capacitor

Solution

a. We first combine C_2 and C_3 into C_p

$$C_p = C_2 \parallel C_3 = 10 + 20 = 30\ \mu\text{F}$$

Next we combine C_1 and C_p into C_{eq}

$$C_{eq} = \frac{C_1 C_p}{C_1 + C_p}$$

$$= \frac{(5)(30)}{5 + 30} = 4.3\ \mu\text{F}$$

b. To find the charge on each capacitor, we note that the total charge is

$$Q_t = C_{eq} V_t = (4.3 \times 10^{-6})(100)$$

$$= 430\ \mu\text{C}$$

Since C_1 is in series with $(C_2 + C_3)$, we have

$$Q_t = Q_1 = Q_2 + Q_3 = 430\ \mu\text{C}$$

To find Q_2 and Q_3 we need the capacitor voltages.

c. $$V_1 = \frac{Q_1}{C_1} = \frac{430 \times 10^{-6}}{5 \times 10^{-6}} = 86\ \text{V}$$

Using KVL

$$V_2 = V_3 = V_t - V_1 = 100 - 86 = 14\ \text{V}$$

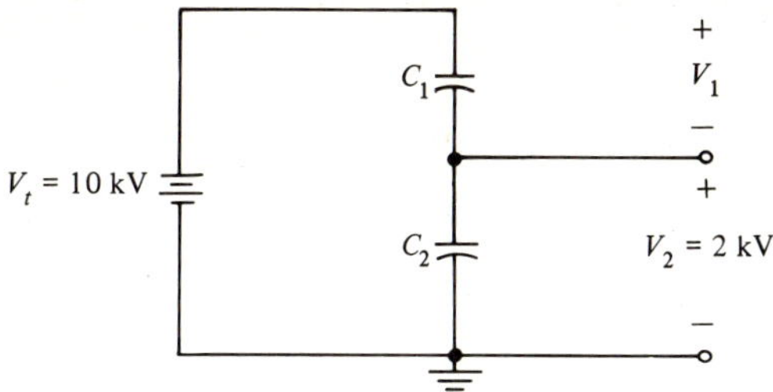

FIGURE 8.25
Capacitive voltage divider for Example 8.7-3.

Thus

$$Q_2 = C_2V_2 = (10 \times 10^{-6})(14) = 140\ \mu C$$
$$Q_3 = C_3V_3 = (20 \times 10^{-6})(14) = 280\ \mu C$$

Note that the *larger* voltage appears across the *smaller* capacitance in the series connection.

• • •

EXAMPLE 8.7-3 Design of a Capacitive Voltage Divider

In high-voltage dc circuits in which very little current is drawn from the high-voltage source, capacitors rather than resistors are used to form voltage dividers. Design a capacitive voltage divider to reduce a 10-kV supply to 2 kV for use in a dental X-ray machine. Specify the capacitance and voltage rating of each capacitor.

Solution

The circuit is shown in Fig. 8.25. Since the capacitors are in series, the charge on each is the same. Then $Q_1 = Q_2$ where $Q_1 = C_1V_1$ and $Q_2 = C_2V_2$ so that

$$C_1V_1 = C_2V_2$$

This is rearranged to yield

$$\boxed{\frac{V_1}{V_2} = \frac{C_2}{C_1}} \tag{8.7-6}$$

This formula tells us that the ratio of the two capacitor voltages equals the inverse ratio of the capacitances. It is instructive to find the voltage transfer ratio from input to output, $A_v = V_2/V_t$. Using KVL, $V_1 = V_t - V_2$. This is substituted into Eq. (8.7-6)

$$\frac{V_t - V_2}{V_2} = \frac{V_t}{V_2} - 1 = \frac{C_2}{C_1}$$

From which

$$\frac{V_t}{V_2} = \frac{C_2}{C_1} + 1 = \frac{C_2 + C_1}{C_1}$$

and

$$\boxed{A_v = \frac{V_2}{V_t} = \frac{C_1}{C_2 + C_1}} \qquad (8.7\text{-}7)$$

If we choose $C_2 = 1\ \mu F$ arbitrarily we can find C_1 using either Eq. (8.7-6) or Eq. (8.7-7). We use Eq. (8.7-6) with $V_2 = 2$ kV, $V_1 = 10 - 2 = 8$ kV, and $C_1 = 1\ \mu F$. Then

$$\frac{8}{2} = \frac{1}{C_1}$$

and

$$C_1 = 0.25\ \mu F$$

Finally, the required capacitances and their voltage ratings are:

$C_1 = 0.25\ \mu F$, 8 kV dc
$C_2 = 1\ \mu F$, 2 kV dc

• • •

LEARNING EXERCISE FOR SEC. 8.7

1. An experimenter's kit contains two capacitors, 0.01 and 0.02 μF. What values of capacitance can be obtained using them singly or in combination?

Ans. 0.0067; 0.01; 0.03; 0.02

• • •

8.8 ENERGY STORAGE

In earlier sections of this chapter it was mentioned that energy is stored in the magnetic field of an inductor in which current is flowing and also in the electric field of a capacitor that has a voltage across its terminals. If a capacitor is charged to a finite voltage by connecting it to a source of energy and then the source is suddenly removed, it is found that the voltage that existed across the capacitor at the instant when the source was removed is maintained for long periods of time, as long as there is negligible leakage between the plates of the capacitor. During this time, the energy that was drawn from the source is stored in the electric field of the capacitor. If a resistance is connected across the plates of a previously charged capacitor, the stored energy will be dissipated as heat in the resistance. The energy stored in the electric field of a capacitor that has V_1 volts across it is

$$W_C = \frac{1}{2} C V_1^2 \tag{8.8-1}$$

where W_C is in joules (J), C is in farads (F), and V_1 is in volts (V).

The fact that this energy can be stored for long periods makes the capacitor useful where large currents are required for short periods of time, as in electronic flash guns and electric spot welders. In these devices, storage capacitors are charged at low current through relatively high resistances. When the current is required for the flash tube or weld the charged capacitor is discharged through the low resistance of the flash tube or welding material. The current during the discharge is very high.

In the inductance, the energy stored in the magnetic field is

$$W_L = \frac{1}{2} L I_1^2 \tag{8.8-2}$$

where W_L is in joules (J) when L is in henries (H) and I_1 is in amperes (A). This formula is clearly the *dual* of Eq. (8.8-1) for the capacitor.

The long-time energy storage feature of the capacitance element is not available in the inductance. This is because the current I_1 must be maintained in order to maintain the magnetic field. In order to maintain the current *without* an energy source, a zero resistance path for the current is required. Since every material has resistance (except at temperatures at or near absolute zero) this is not practical. If the current in an inductive circuit is interrupted by opening a switch, the magnetic field collapses very rapidly due to the sudden interruption. This rapidly collapsing field can generate very high voltages across the opening switch contacts, causing arcing and burning of the contacts. This problem can often be eliminated either by connecting a capacitor across the switch contacts to oppose the voltage change or by connecting a discharge resistor in parallel with the inductive circuit to absorb the energy in the collapsing field.

EXAMPLE 8.8-1 Electronic Photoflash

An electronic flash gun has a 2000-μF storage capacitor, which is charged to 48 V for operation. The flash duration in use is 1/1000 s.

a. Find the charge on the capacitor and the energy stored.
b. Assuming that all of the charge passes through the flashtube in the 1/1000-s flash duration, find the current through and the power delivered to the tube.
c. After the flash the capacitor must be recharged. How long will this take if the charging power supply delivers a constant current of 100 mA?

Solution

a. The charge is given by

$$Q = CV = (2 \times 10^{-3})(48) = 0.096 \text{ C}$$

The energy stored is

$$W_C = \frac{1}{2}CV^2 = \left(\frac{1}{2}\right)(2 \times 10^{-3})(48)^2 = 2.3 \text{ J}$$

b. Using the definition of current,

$$I = \frac{\Delta Q}{\Delta t} = \frac{0.096 \text{ C}}{0.001 \text{ s}} = 96 \text{ A}$$

The power delivered is the time rate of change of energy (see Sec. 2.7):

$$P = \frac{\Delta W}{\Delta t} = \frac{2.3 \text{ J}}{0.001 \text{ s}} = 2300 \text{ W}$$

Note that during the flash, the current and power are very high.

c. The charging time is found from the definition of current

$$I_{\text{charge}} = \frac{\Delta Q}{\Delta t}$$

where ΔQ is the required charge of 0.096 C. Since the charging current is 100 mA, this gives

$$0.1 = \frac{0.096}{\Delta t}$$

from which

$$\Delta t = \frac{0.096}{0.1} = 0.96 \text{ s}$$

This should be compared to the 1 ms taken to discharge the flashtube.

• • •

LEARNING EXERCISE FOR SEC. 8.8

1. In communication circuits, parallel combinations of inductance and capacitance are used frequently. In one such circuit $L = 0.3$ mH and $C = 2.2\ \mu$F. The voltage across the circuit is 5.2 V and the current in the inductance is 8 mA. Find the energy stored in each element.

Ans. 0.0096; 30

• • •

SUMMARY

In this chapter we have considered the fundamental behavior of those elements whose responses depend on time. The next step will be to consider the response waveforms that result when inductance and capacitance are combined with resistance to form practical combinations as used in radio, TV, and computer circuits.

Section

8.1 1. The inductance of a solenoid coil is given by

$$L = \frac{N^2 \mu A}{d}\ \text{H}$$

2. The volt-ampere relation for the inductance is

$$v(t) = L\frac{\Delta i}{\Delta t}$$

3. The current through an inductor cannot change instantaneously. Therefore, the inductance opposes any change in the current flowing through it.
4. The nonideal nature of practical inductors can be accounted for by including a resistance in series with an ideal inductance in the circuit model for the element.

8.2 5. The total inductance of any number of series inductors without any mutual magnetic coupling is the sum of the individual inductances

$$L_{eq} = L_1 + L_2 + \cdots + L_n$$

6. Inductors in parallel with no mutual induction present are combined in exactly the same way as parallel resistors. For two inductors in parallel

$$L_{eq} = \frac{L_1 L_2}{L_1 + L_2}$$

8.3 7. The voltages across a pair of magnetically coupled coils are related by the equation

$$\frac{v_2}{v_1} = k\frac{N_2}{N_1}$$

8. The mutual inductance between coupled coils can be found from the equation

$$M = k\sqrt{L_1 L_2}$$

9. When two magnetically coupled coils are connected in series, the equivalent inductance is found from the equation

$$L_{eq} = L_1 + L_2 \pm 2M$$

10. Dot markings can be used to indicate the polarity of mutual voltages in magnetically coupled coils. The significance of the dots is that current flowing in at the dotted end of one winding induces a voltage in the other winding, which is positive at the dotted end.

8.4 11. The electric force between two charges is

$$F = k\frac{Q_1 Q_2}{d^2}$$

12. The electric field intensity between two points is

$$E = \frac{V}{d}$$

13. The defining relation for capacitance is

$$q = Cv$$

14. The *vi* relation for the capacitor is

$$i(t) = C\frac{\Delta v}{\Delta t}$$

15. The voltage across a capacitor cannot change instantaneously; therefore, the capacitor *opposes* changes in its terminal voltage.

8.5 16. For a parallel-plate capacitor the capacitance is

$$C = \frac{8.85 \times 10^{-12} \epsilon_r A}{d} \text{ F}$$

17. The dielectric strength of a material is a measure of the maximum voltage per unit length that can be applied across the material before breakdown will occur.
18. The nonideal nature of actual capacitors can often be accounted for by including a high resistance in parallel with an ideal capacitor in the circuit model for the element.

8.6 19. The unit step function is zero for $t < 0$ and is 1 for $t > 0$. It is used to represent switching operations.

20. Capacitors in parallel add in the same way as resistors in series, that is, for any number of parallel capacitors

$$C_{eq} = C_1 + C_2 + \cdots + C_n$$

21. *Series* capacitors combine exactly as if they were *parallel* resistors. For any number of series capacitors

$$\frac{1}{C_{eq}} = \frac{1}{C_1} + \frac{1}{C_2} + \cdots + \frac{1}{C_n}$$

For two capacitors in series

$$C_{eq} = \frac{C_1 C_2}{C_1 + C_2}$$

22. In a capacitive voltage divider,

$$\frac{V_2}{V_t} = \frac{C_1}{C_1 + C_2}$$

This differs from the resistive voltage divider formula in that the numerator contains the capacitor across which the output voltage *does not* appear.

23. The energy stored in the electric field of a capacitor is

$$W_c = \frac{1}{2} CV^2$$

24. The energy stored in the magnetic field of an inductance is

$$W_L = \frac{1}{2} LI^2$$

QUESTIONS FOR REVIEW

Sec. 8.1

1. Explain the meaning of functional notation and give examples.
2. How does the inductance element differ from the resistance element in terms of its voltage-current behavior?
3. State the *vi* relation for inductance.
4. State the formula for the inductance of a coil in which the flux links all of the turns.
5. Explain why the current through an inductance cannot change instantaneously.

Sec. 8.2

6. Given two inductors that do not interact magnetically, how do you find the equivalent inductance if they are first connected in series and then in parallel?

Sec. 8.3

7. Describe mutual induction.
8. What is the coefficient of coupling?
9. How is the mutual inductance between a pair of magnetically coupled coils related to the individual coil inductance?
10. Describe the dot marking system for coupled coils.
11. How can dot placement be determined if the coils are encapsulated?

Sec. 8.4

12. Describe the physical configuration of a typical capacitor.
13. Describe the color-code band marking system for tubular capacitors.
14. What is the meaning of the term *electric field?*
15. How are charge and voltage related in a capacitor?
16. State the *vi* relation for a capacitor.
17. Why does a capacitor oppose changes in voltage?

Sec. 8.5

18. Explain in physical terms how capacitance varies with plate spacing and separation.
19. State the formula for the capacitance of a parallel-plate capacitor.
20. Describe some typical dielectric materials.
21. What is the meaning of *dielectric strength?*
22. What capacitor specification depends on the dielectric strength of the material between the plates?
23. Describe how the nonideal nature of a capacitor can be taken into account by a circuit model.

Sec. 8.6

24. Describe the unit step function.
25. Show a circuit that can generate a unit step function using a 1-V battery and a SPDT switch.

Sec. 8.7

26. State the formula for the equivalent capacitance of two capacitors connected first in series and then in parallel.

27. What is the advantage of a capacitive voltage divider in high voltage dc circuits?

Sec. 8.8

28. Describe the capacitive element and the inductive element in terms of energy storage.
29. State the formula for energy stored in a capacitor or an inductor.
30. Describe several ways in which stored energy is used.

PROBLEMS

Sec. 8.1

1. A coil consists of 250 turns of wire wound on a closed steel core for which $\mu_r = 1000$. The length of the core is 0.2 m and it has a cross-sectional area of 0.04 m^2. Find the inductance of the coil.
2. An amateur radio operator winds 100 turns of wire around a pencil 0.5 cm in diameter and 10 cm long. Find the inductance.
3. An air core coil for a TV receiver is to have an inductance of 20 μH and is to be 3 cm long and 3 cm in diameter. How many turns are required?
4. A coil has 800 turns and an inductance of 160 mH. If turns are removed from the coil without changing its dimensions, how many turns must be removed to reduce the inductance to 105 mH?
5. A toroidal air-core coil for a radar set has 300 turns, an area of 1.5 cm^2, and an effective length of 0.1 m. Find its inductance.
6. Find the voltage induced in a 400-turn coil if the flux changes by 0.5 mWb in 0.05 s.
7. It is desired to induce a voltage of 6 V in a coil when the flux changes by 5 mWb in 0.02 s. How many turns should the coil have?
8. Current in a 10-mH coil changes uniformly from 10 to 100 mA in 5 ms. What is the voltage across the coil during this interval?
9. The current waveform of Fig. 8.26 is applied to a 10-mH inductance. Find and plot the resulting voltage.
10. Repeat Prob. 9 for the current of Fig. 8.27.

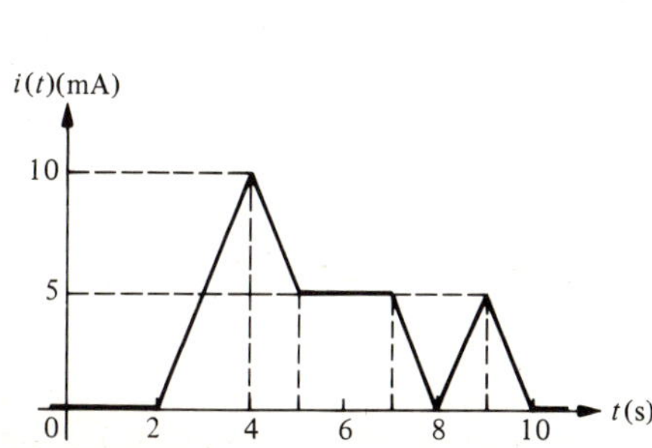

FIGURE 8.26

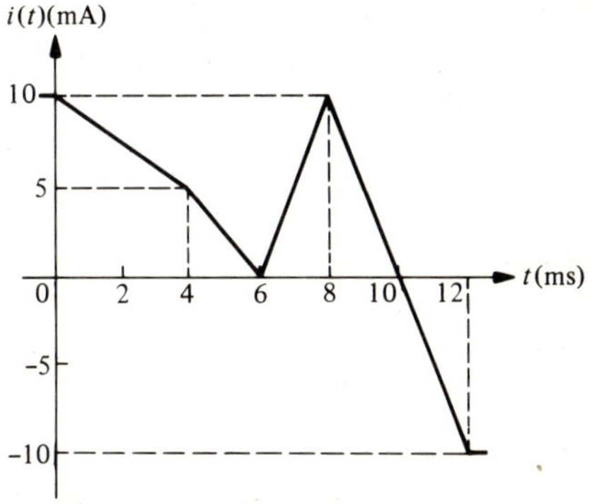

i(t)(mA)
10
5
-5
-10
0
2
4
6
8
10
12
t(ms)

FIGURE 8.27

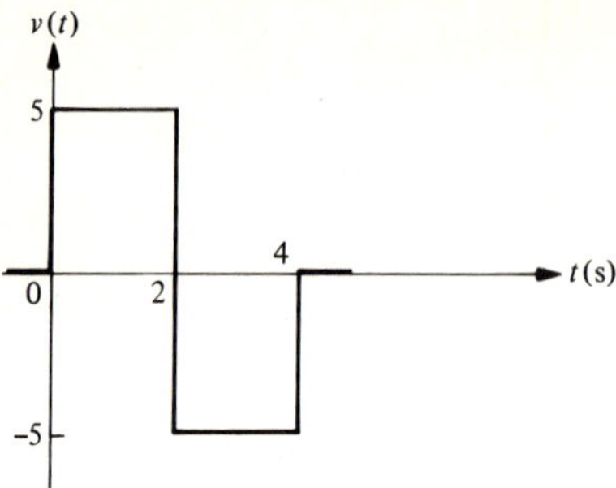

FIGURE 8.28

11. Sketch two currents that will yield the voltage waveform of Fig. 8.28 when applied to a 5-H inductance.
12. The current waveform of Fig. 8.26 is applied to a series combination of resistance and inductance with $R = 5\ \Omega$, $L = 4$ H. Find and sketch the voltage across L and R separately, and the total voltage.

Sec. 8.2

13. Show that the equivalent inductance of two inductors connected in parallel, with no common flux linkages, is

$$L_{eq} = \frac{L_1 L_2}{L_1 + L_2}$$

14. Find the equivalent inductance of the two circuits of Fig. 8.29.
15. Three inductors that are isolated magnetically are mounted on a circuit board. If the inductance values are 2, 3, and 5 mH, find all possible values that can be found using series, parallel, and series-parallel arrangements of one, two, or all three of the coils.
16. Find the inductance required in parallel with 90 μH to reduce the equivalent inductance to 40 μH.

Sec. 8.3

17. The coupled coils of Fig. 8.8 are wound over an iron core so that the coupling is very "tight" with $k = 0.99$. The primary has 100 turns and a self-inductance of 0.5 H. The secondary has 350 turns and an inductance of 3 H.

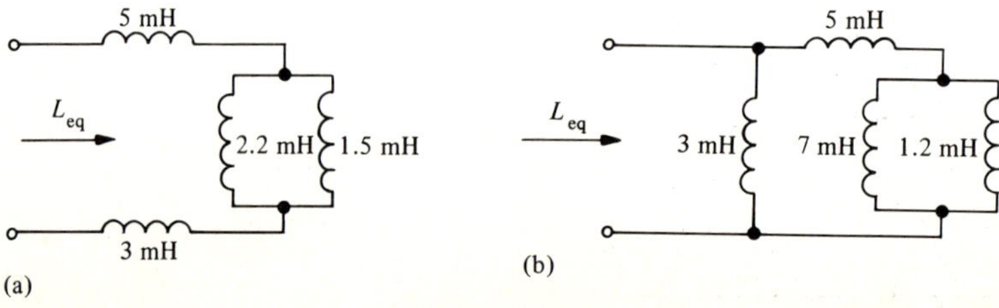

FIGURE 8.29

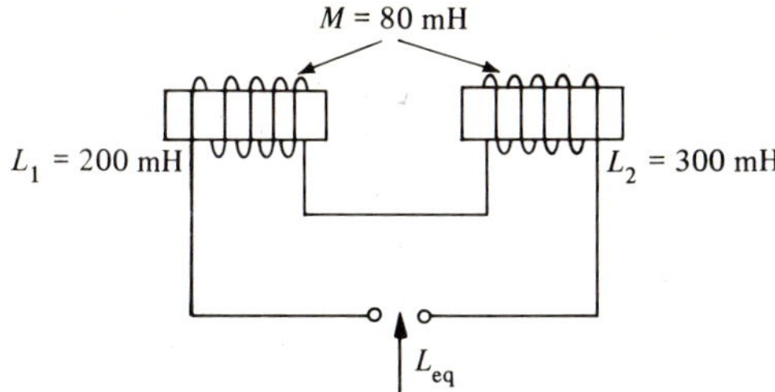

FIGURE 8.30

a. Find M.
b. Find the primary and secondary voltages if the primary flux is changing at the rate of 0.85 Wb/s.

18. In the coupled coils of Fig. 8.8, $k = 0.8$, the primary has 320 turns and 0.5 H inductance, while the secondary has 80 turns and 0.2 H inductance. Find:
    a. The mutual inductance M.
    b. The primary and secondary voltages if the primary current is changing at the rate of 20 A/s.
19. Repeat Prob. 18 if k is changed to 0.2. Compare.
20. Two coils have 300 and 550 turns each. They are placed close together so that a change of 1.2 A in the 550-turn coil produces a flux change of 0.12 mWb in the 550-turn coil and 0.1 mWb in the 300-turn coil. Find:
    a. The self-inductance of the 550-turn coil
    b. The coefficient of coupling
    c. The mutual inductance
21. For the circuit of Fig. 8.30 find the equivalent inductance and the coefficient of coupling. Show the dot placement for each coil.
22. For the series connection and dot placement of Fig. 8.31 find the equivalent inductance if the coefficient of coupling is (a) 1, (b) 0.5, (c) 0.
23. Repeat Prob. 22 if the dots are reversed on one of the coils.
24. Two coupled coils in series have a measured equivalent inductance of 1.6 H. One coil is reversed and the inductance is measured at 1.2 H. If one coil has a self-inductance of 0.5 H find the inductance of the other coil and the mutual inductance.
25. Find the equivalent inductance for the circuit of Fig. 8.32.
26. In the circuit of Fig. 8.12 the components have the following values: $R_1 = R_2 = 500\ \Omega$, $L_1 = 1.2$ H, $L_2 = 0.3$ H, $k = 0.8$. When i_1 is 5 mA and is changing at the rate of 3 A/s, the input voltage v_1 is 10 V. Find i_2 and the rate at which it is changing.

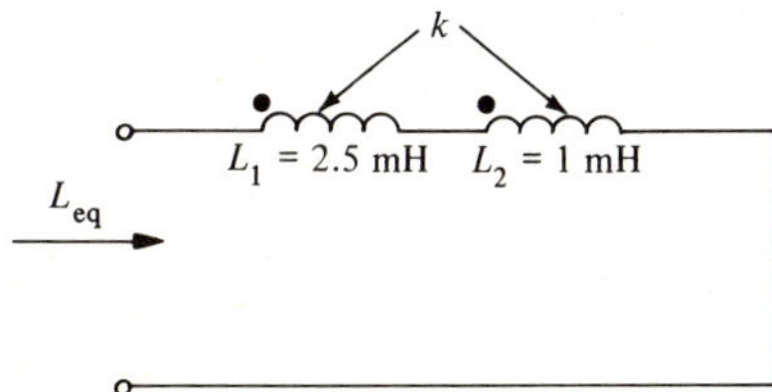

FIGURE 8.31

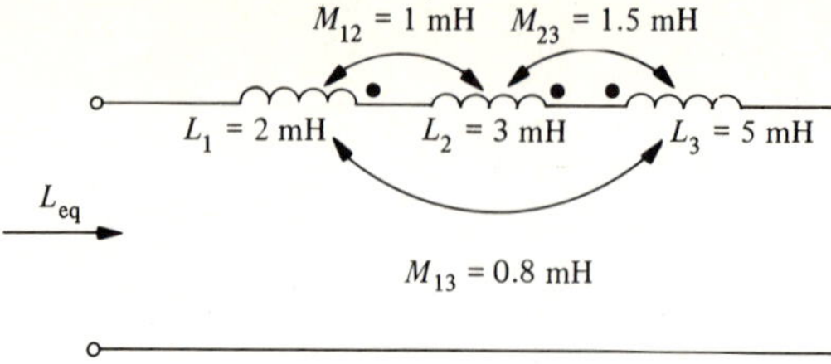

FIGURE 8.32

Sec. 8.4

27. Find the force acting on a single electron located 0.6 cm from a point charge of $+6\ \mu$C. Is this a force of attraction or repulsion?
28. Two point charges of 6 μC are separated in air by a distance of 1 m. Find the force acting on either charge.
29. Find the electric field intensity 1 m from a charge of 12 C.
30. A pair of parallel plates is 1.5 cm apart. If 400 V are applied to the plates, find the electric field intensity between the plates.
31. A 10-μF capacitor in a power supply carries a charge of 450 μC. What is the voltage across the capacitor?
32. A capacitor carries a charge of 20 mC and the voltage across it is 60 V. Find the capacitance.
33. A 0.001-μF motor-starting capacitor has a voltage of 220 V across it. How much charge is present?
34. A constant current of 2 mA is to be carried by a 0.05-μF capacitor. At what rate must the voltage across it be changing?
35. The voltage across a 5-μF capacitor decreases linearly from 10 to 2 V in 0.01 s. Find the current during this interval.
36. A constant current of 1 mA is to be maintained through a capacitor across which the voltage is changing at the rate of 500 V/s. Find the capacitance.
37. A 5000-μF capacitor used in low-voltage power supplies has a voltage rating of 15 V. What is the maximum charge that can be stored on it?
38. A 50-nF parallel plate capacitor has 220 μC of charge on its plates. Find the electric field intensity between the plates if they are 3 mm apart.

Sec. 8.5

39. A capacitor is constructed from two square plates each 2 $\times$ 2 cm separated by a distance of 0.1 cm. Find the capacitance if the dielectric is
    a. air b. mica c. ceramic ($\epsilon_r = 750$)
40. A parallel-plate air capacitor has $C = 2.6$ pF. If the plate spacing is doubled, what is the new capacitance?
41. Silver plates are deposited on both sides of a mica sheet that measures 3 cm $\times$ 1 cm $\times$ 0.03 mm. Find the capacitance between the silver plates.
42. A cloud with an area of 2 km^2 situated 1.5 km above the earth has collected a charge of $+300$ C. Assuming that the ground acts as a parallel plate, find the capacitance from the cloud to the earth and the voltage between them. What is the breakdown voltage for this configuration?
43. An air dielectric capacitor has C = 20 pF. When the capacitor is immersed in oil the capacitance increases to 54 pF. What is the dielectric constant of the oil?

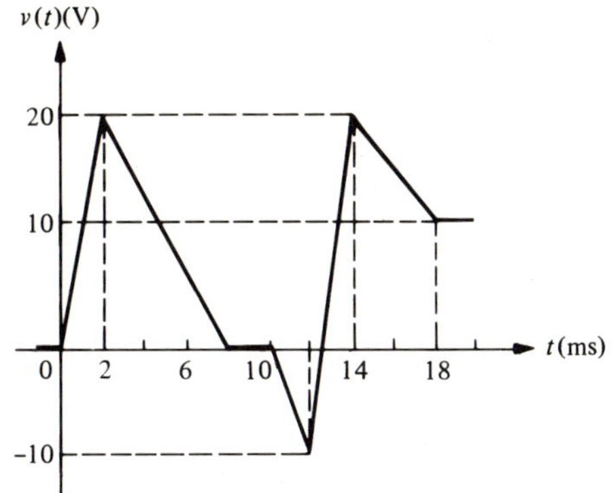

FIGURE 8.33

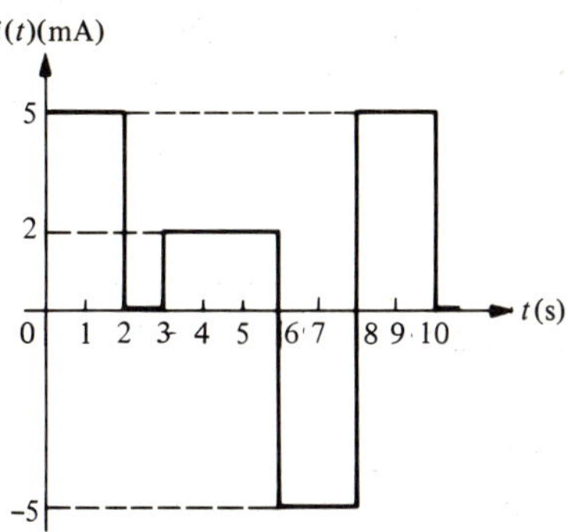

FIGURE 8.34

44. A tubular capacitor for general electronic use is formed by rolling two aluminum sheets 3 cm wide and 1 m long separated by paraffinned paper 0.1 mm thick. Find the capacitance and voltage rating.
45. Find the breakdown voltages for the capacitors of Prob. 39.
46. Find the breakdown voltage of the capacitor in Prob. 41.

Sec. 8.6

47. An initially uncharged 0.5-μF capacitor carries a constant current of 600 μA. Find the voltage across the capacitor at $t = 350$ ms.
48. A current $i(t) = 150u(t)$ μA is applied to an uncharged 0.05-μF capacitor. Find and sketch the voltage waveform in the interval -10 ms $< t <$ 100 ms.
49. Repeat Prob. 48 with an initial charge of -10 V on the capacitor.
50. The voltage waveform of Fig. 8.33 is applied to a 0.001-μF capacitor. Find and plot the current waveform.
51. The current waveform of Fig. 8.34 is flowing in an initially uncharged 2.5-μF capacitor. Find and sketch the resulting voltage waveform.
52. The voltage waveform shown in Fig. 8.35 is applied to a capacitor.
    a. Find and sketch the current waveform.
    b. Reduce the duration of the ramp to $\epsilon/2$ and repeat part (a).
    c. Reduce the duration of the ramp to $\epsilon/4$ and repeat part (a).

    As the duration approaches zero, the current approaches an *impulse* function.

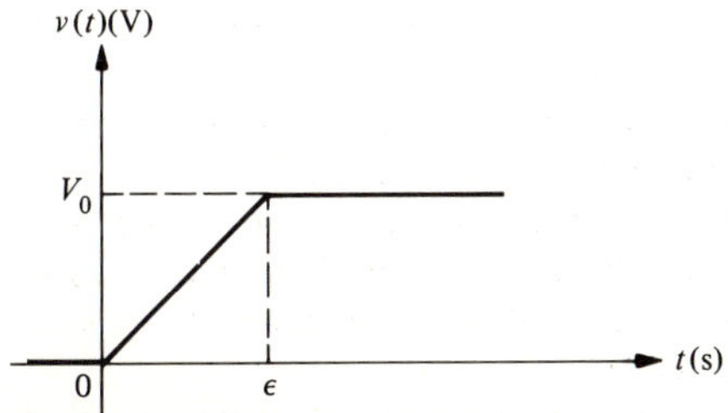

FIGURE 8.35

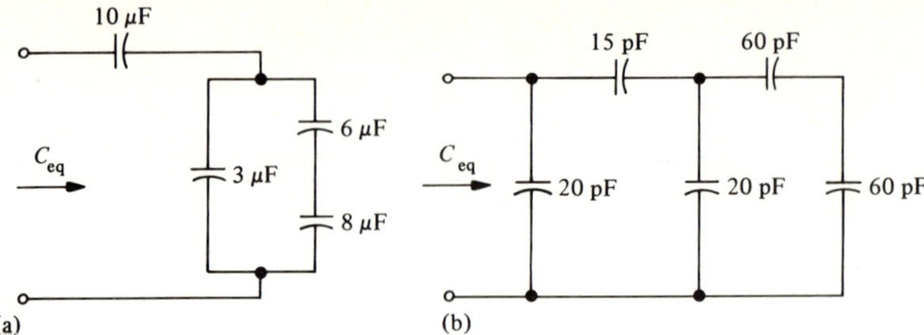

FIGURE 8.36

Sec. 8.7

53. Find the equivalent capacitance of the circuits of Fig. 8.36.
54. Find the voltage across and the charge on each capacitor in the circuits of Fig. 8.37.
55. Three capacitors of value 0.1, 0.22, and 0.5 μF are available. How many different series, parallel, and series-parallel combinations using one, two, or three of these capacitors are possible? Draw the circuit and find the capacitance for each possibility.
56. Two capacitors are connected in series. One is rated at 0.05 μF, 400 V and the other at 0.022 μF, 200 V. What is the maximum voltage that can be safely applied?
57. Find the output voltage in the high-voltage circuit of Fig. 8.38.
58. Design a capacitive voltage divider to reduce 15 kV to 3 kV for an X-ray machine. Specify voltage ratings for the capacitors.

Sec. 8.8

59. Find the energy stored in a 20-μF capacitor which has 120 V across its terminals.
60. The energy stored in a 0.1-μF capacitor is 4.3 J. Find the voltage across the capacitor and the charge on each plate.
61. A 10-μF and a 5-μF capacitor are connected in series across a 120-V source. Find the energy stored and the charge on each capacitor.
62. In order to spot-weld certain materials, 5 J of energy is required in a time of 10 ms. A 100-μF storage capacitor is available.
    a. To what voltage must the capacitor be charged in order to supply the required energy?
    b. What is the current during the weld?

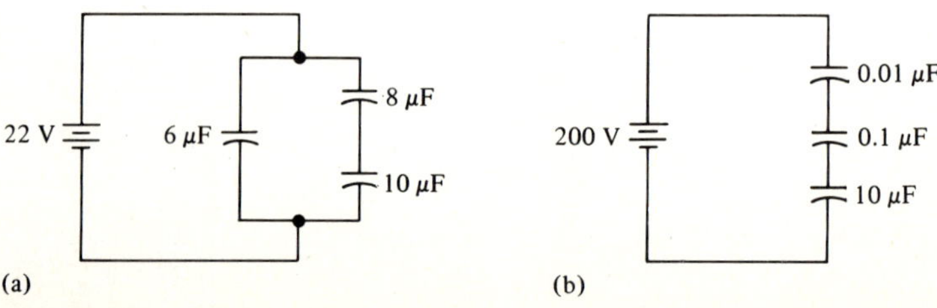

FIGURE 8.37

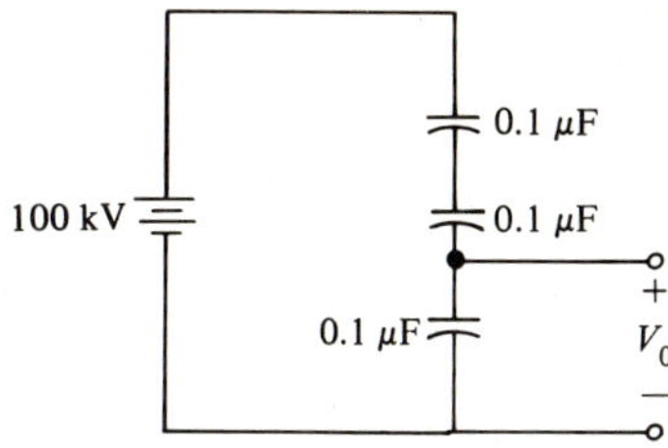

FIGURE 8.38

63. In a certain electromagnet, 60 A is required for the coil, which has an inductance of 8.3 H. Find the energy stored in the magnetic field when the magnet is on.
64. A relay coil carries a current of 125 mA. If it is required to store 12 J in its magnetic field, find the required inductance.
65. A certain dc motor has a field winding with an inductance of 85 H and the current is 1.2 A. When the field circuit is accidentally opened, an arc flashes across the open circuit. Find the power dissipated in the arc if it lasts 5 ms.
66. Two capacitors, 0.05 and 0.01 μF, are connected in parallel across a 120-V source. Calculate the energy stored in each.
67. The capacitors of Prob. 66 are connected in series across the 120-V source. Find the energy stored in each.
68. A 20-μF capacitor is charged to 500 V. If it is later discharged to 450 V, how much energy was removed?
69. A TV receiver power supply uses a 500-pF high-voltage filter capacitor. How much energy does it store when the voltage is 15 kV?
70. The flash tube in the electronic flash gun of Example 8.8-1 is replaced by one that has a flash duration of 1/1250 s at the same energy level, that is, 2.3 J. Repeat the example for these conditions.

9
Charge and Discharge

OBJECTIVES

Upon completion of this chapter, the student should be able to

Section

9.1 1. Use Kirchhoff's laws and the R, L, and C *vi* relations to write the equations for RL and RC circuits.
9.2 2. Solve a simple RL or RC circuit using the step-by-step method.
9.3 3. Describe the process of capacitor charge and discharge.
4. Write the basic equations for capacitor charge and discharge.
5. State that the RC time constant determines the relative speed of the charge or discharge.
9.4 6. Find values of the exponential function for arbitrary exponents using a scientific calculator.
7. State that the exponential decays to 37% of its initial value after one time constant and less than 1% after five time constants.
8. State that a rising exponential reaches 63% of its final value after one time constant and 99% after five time constants.
9.5 9. Describe buildup and decay of current in an inductance.
10. Write the equation for the buildup and decay of current in an inductance and plot appropriate graphs.
9.6 11. State the general solution for a circuit containing one inductor or one capacitor and define the three constants that completely specify the response.
12. Find initial conditions in RC and RL circuits.
9.7 13. Describe the physical significance of the forced response, the natural response, and the total response.
14. State that the natural response takes the response from the initial condition to the final value.

9.8 15. Use the general solution to find the response of RL and RC circuits to pulses.

16. Find the rise time of a typical RC circuit.

9.9 17. Use the general solution to find the response of circuits when initial conditions are not zero.

18. Find the time constant of a network from an oscilloscope trace of its transient response.

INTRODUCTION

In previous chapters we introduced the three linear passive elements: resistance, capacitance, and inductance, and their *vi* relations. In this chapter, we will begin to study the response of circuits containing two or more different elements.

Since the *vi* relations of inductors and capacitors contain rates of change such as $\Delta v/\Delta t$, our circuit equations will contain such terms. The circuit responses that are the solutions of these equations will vary with time, and we will consider the important properties of these solutions. Because of the known nature of these time variations, such circuits are often used to generate precise time intervals, which are used for various applications. These include timing circuits in digital computers, automotive control circuits, radar systems, and many others.

In the circuits of this chapter, the sources will be constant voltage or constant current sources applied to the circuit through switches as shown in Fig. 9.1. In this type of circuit, immediately after a switching action takes place, an unsettled state exists during which currents and voltages change with time. This is called the *transient state*. After sufficient time has elapsed, the transient state disappears, and all currents and voltages are constant (dc); these are called final, or *steady-state*, values. Some typical transient waveforms are shown in Fig. 9.1. The transient states are usually associated with the opening and closing of switches, but will also occur if the input voltage changes abruptly, as for example when a step function is

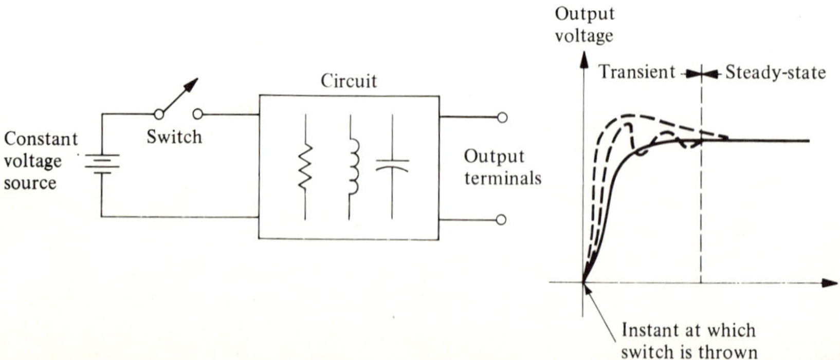

FIGURE 9.1
The transient and steady states.

applied. When considering the response to such signals, the procedure we discuss in this chapter is often called *time-domain* analysis.

Most transient states are characterized by currents and voltages that vary *exponentially*, and our aim in this chapter is to characterize these exponentials in terms of the circuit elements and inputs.

9.1 FORMULATING THE PROBLEM

In this section we consider the formulation of the equations for circuits in which transients are occurring. Consider first the circuit of Fig. 9.2a. The battery is switched across the series RC combination at $t = 0$ as indicated by the arrow on the switch. In the circuit of Fig. 9.2b a step-function source (see Sec. 8.6) replaces the battery and switch. Either circuit can be used for the analysis if the capacitor is uncharged before the switching. This means that the voltage across it is zero just before the switch is thrown. We wish to find the voltage across the capacitor after the switch is closed, that is, for $t > 0$ (see Fig. 9.2c). In order to do this we will need the value of the voltage across the capacitor immediately *after* the switch is thrown; we are given the fact that the capacitor is uncharged ($v_C = 0$) *before* the switch is thrown. In problems of this type, it is important to distinguish between the time just before the switch is thrown and the time just after the switch is thrown, because the circuit configuration changes as a result of the switching. We will designate $t = 0^-$ as the time just prior to the switching and $t = 0^+$ as the time just after the switching.

If the capacitor voltage were to change instantaneously, that is, "jump," the graph of voltage vs. time would contain a vertical line at the point where the change occurs, as shown in Fig. 9.3. Since $i = C\,\Delta v/\Delta t$, the current at any time is proportional to the *slope* of the voltage-vs.-time curve. From the graph, we see that the slope is infinite at the point where the change occurs,

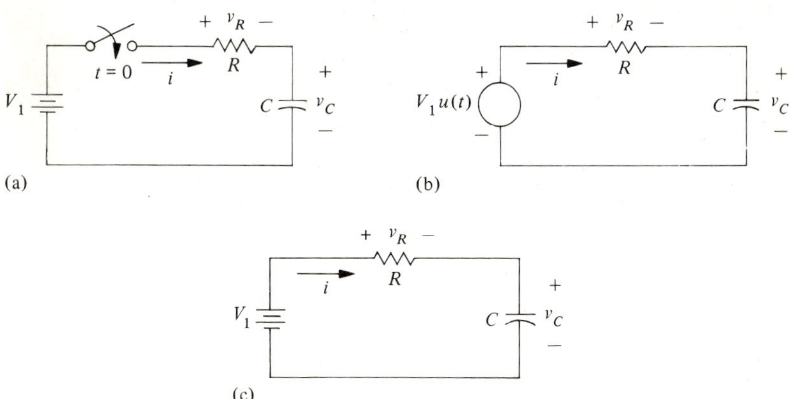

FIGURE 9.2
Transient circuit. (a) Switched source. (b) Step-function source. (c) Circuit after switch is thrown ($t > 0$).

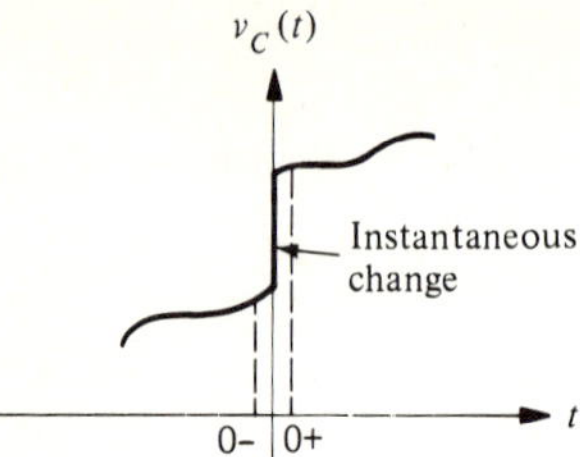

FIGURE 9.3
An Instantaneous change in voltage.

so that an infinite current would be required. This, in turn, would require that charges move a finite distance in zero time. As pointed out in Sec. 8.4 this is impossible, so we conclude that the voltage across a capacitor cannot change instantaneously, and the voltage across the capacitor in our circuit just *after* the switch is thrown remains at 0. Symbolically, $v_C(0^+) = v_C(0^-) = 0$ V. The value of v_C just *after* the switch is thrown is called the *initial* condition and the fact that the capacitor voltage cannot change instantaneously is known as a continuity condition.*

To analyze the network, we apply KVL *after* the switch is thrown. This yields

$$\begin{aligned} V_1 &= v_R + v_C \\ &= Ri + v_C \end{aligned} \tag{9.1-1}$$

where i, v_R, and v_C are unknown functions of t. When a voltage such as v_C depends on time, t, it should be written in the form $v_C(t)$ where the t in parentheses (called the *argument* of the function) explicitly shows that v_C *depends on t*. However, we often write it as v_C, assuming that the reader will understand that this is shorthand for $v_C(t)$. When we wish to indicate a value of v_C at a specific time, say t_1, we replace the t in the parentheses by t_1. Thus $v_C(2\text{ s})$ refers to the instantaneous value of v_C when $t = 2$ s. For the voltage at $t = 0^+$, we write $v_C(0^+)$. These points are shown in Fig. 9.4 for an arbitrary waveform. Throughout the remainder of this text, we will indicate values of variables at specific times as discussed above, for example, $i(2\text{ s})$, $v_R(0^+)$, $v_C(2\text{ s})$. However, we will not always explicitly indicate that a variable is a function of t. If the variable is symbolized by a lower-case letter (v, i, q) we will *understand* that it is, in general, a function of time.

Returning to the circuit, for the capacitor we have the vi relation [see Eq. (8.4-6)]

$$i = C\frac{\Delta v_C}{\Delta t} \tag{9.1-2}$$

*The continuity condition for capacitors applies only to voltage, *not* to current.

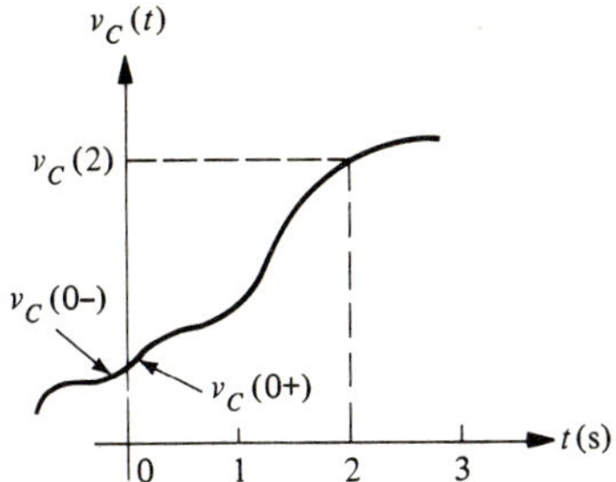

FIGURE 9.4
Functional notation.

When this is substituted into Eq. (9.1-1) we obtain

$$RC\frac{\Delta v_C}{\Delta t} + v_C = V_1 \tag{9.1-3}$$

This equation holds *only* for $t > 0$. It relates the capacitor voltage v_C to the circuit elements R and C and the input voltage V_1. The equation alone is not sufficient to determine the entire solution for $v_C(t)$; additional information is required. In this case the additional information is most conveniently expressed as the *initial condition* $v_C(0^+) = 0$.

The solution to this equation will be considered in the next section. We can, however, find some of its properties in advance.

Consider Eq. (9.1-1) immediately after the switch is thrown. We know that $v_C = 0$ at this time from the continuity condition. Thus, for $t = 0^+$

$$\begin{aligned} V_1 &= Ri(0^+) + v_C(0^+) \\ &= Ri(0^+) + 0 \end{aligned}$$

and therefore

$$i(0^+) = \frac{V_1}{R} \tag{9.1-4}$$

Thus the current immediately after the switch is thrown is found by using Ohm's law and by considering the capacitor as a short circuit. Note that the current changes instantaneously from zero just prior to the switching at $t = 0^-$ to V_1/R just after the switching at $t = 0^+$.

When the source voltage is constant, as in this circuit, the capacitor voltage will eventually become constant. Then $\Delta v_C/\Delta t = 0$ and we have from Eq. (9.1-3),

$$v_C(\infty) = V_1 \tag{9.1-5}$$

where $v_C(\infty)$ is used to indicate the value of v_C after a long time has elapsed.

From Eq. (9.1-2) we see that $i(\infty) = 0$ indicating that the capacitor "looks like" an *open circuit* in the steady state for dc sources. This fact can be used to simplify steady-state calculations in capacitive circuits with dc sources.

We have found the initial values $v_C(0^+) = 0$ and $i(0^+) = V_1/R$, and the final values $v_C(\infty) = V_1$ and $i(\infty) = 0$ from physical considerations; the values in between, which determine the transient state, will be found when we solve the equation. Application of these ideas to circuits containing capacitance and inductance will be found in the following examples.

EXAMPLE 9.1-1 *Initial and Final Values in a Capacitive Circuit*

In the circuit of Fig. 9.2a $V_1 = 120$ V, $R = 10$ kΩ, and $C = 10$ μF. Just before the switch is closed at $t = 0$, the voltage across the capacitor is 25 V.

Find:

a. $v_C(0^+)$
b. $i(0^+)$
c. $v_C(\infty)$
d. $i(\infty)$

Solution

a. The data tell us that $v_C(0^-) = 25$ V. Since the capacitor voltage cannot change instantaneously, we must have

$$v_C(0^+) = v_C(0^-) = 25 \text{ V}$$

b. Immediately after the switch is thrown, the voltage across the resistor is $V_1 - v_C(0^+)$. Therefore

$$i(0^+) = \frac{V_1 - v_C(0^+)}{R} = \frac{120 \text{ V} - 25 \text{ V}}{10 \text{ k}\Omega} = 9.5 \text{ mA}$$

c. After a long time, the capacitor will charge up to the source voltage, so that

$$v_C(\infty) = V_1 = 120 \text{ V}$$

d. When $v_C(\infty) = 120$ V, there is no further motion of charge and

$$i(\infty) = 0$$

• • •

EXAMPLE 9.1-2 *An Inductive Circuit*

Write the equation for the circuit of Fig. 9.5 and find the initial and final values of the current and voltages.

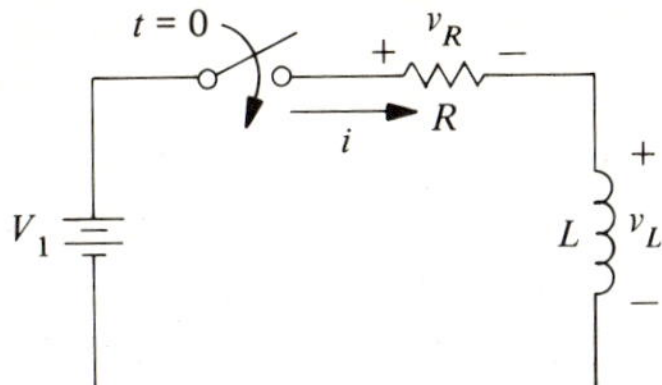

FIGURE 9.5
Circuit for Example 9.1-2.

Solution
Using KVL, and assuming that the switch has been thrown so that $t > 0$,

$$\begin{aligned} V_1 &= v_L + v_R \\ &= v_L + Ri \end{aligned} \tag{9.1-6}$$

Using the vi relation for the inductor, $v_L = L\Delta i/\Delta t$,

$$L\frac{\Delta i}{\Delta t} + Ri = V_1 \tag{9.1-7}$$

This is the desired circuit equation for $t > 0$. In order to find the initial and final values we need the continuity condition for inductance. For the current in an inductance to change instantaneously, an infinite voltage is required. The explanation of this fact is the *dual* of that for the capacitor. Since real voltages cannot be infinite, the current through an inductor cannot change instantaneously. Thus, because the open switch forces the current to be 0 before $t = 0$, the current must also be 0 immediately after the switch is closed and we have $i(0^+) = i(0^-) = 0$. Since $v_R = Ri$, we also have $v_R(0^+) = 0$ and consequently $v_L(0^+) = V_1$. However, we note that $v_L(0^-) = 0$.

For the final values we note that after a long time the current must become constant so that for large t,

$$v_L(t \to \infty) = L\frac{\Delta i}{\Delta t} \to 0$$

Thus $v_L(\infty) = 0$ and consequently $v_R(\infty) = V_1$. Finally,

$$i(\infty) = \frac{V_R(\infty)}{R} = \frac{V_1}{R}$$

The fact that $v_L(\infty) = 0$ indicates that in the steady state, for dc sources, the inductor looks like a *short circuit*. Note the duality to the capacitor, which looks like an *open circuit* under the same conditions.

• • •

LEARNING EXERCISES FOR SEC. 9.1

1. The equation of a voltage that is a linear function of t in seconds is $v(t) = 2t + 6$ V. Find $v(0^+)$, $v(2)$, and $v(26)$. Draw a graph of $v(t)$ and show these points on it.

Ans. 6; 10; 58

2. In the circuit of Fig. 9.2a, the capacitor is initially uncharged and $V_1 = 22$ V, $C = 12\ \mu$F, and $R = 47$ kΩ. Find $v_C(0^+)$, $v_C(\infty)$, $i(0^+)$, and $i(\infty)$.

Ans. 0; 22; 0; 0.47

• • •

9.2 STEP-BY-STEP SOLUTION

In this section we will demonstrate a method of solution of the equation for the RC circuit which is very similar to the way in which the solution would be computed by a suitably programmed digital computer. The equation to be solved is Eq. (9.1-3), which describes the buildup of voltage across the capacitor in the circuit of Fig. 9.2. It is repeated for convenience:

$$RC\frac{\Delta v_C(t)}{\Delta t} + v_C(t) = V_1 \tag{9.2-1}$$

This holds for $t > 0$ and is subject to the initial condition $v_C(0^+) = 0$.

Note that each term in this expression must be given in volts. In order for the first term to be in volts the units of RC must be *time* (seconds). The product RC in this type of circuit is given the symbol τ (tau) and is called the *time constant;* it is discussed in detail in Sec. 9.4. Using this definition, Eq. (9.2-1) can be written in the form

$$\frac{\Delta v_C(t)}{\Delta t} + \frac{v_C(t)}{\tau} = \frac{V_1}{\tau} \tag{9.2-2}$$

The graph of a typical $v_C(t)$ waveform is shown in Fig. 9.6. From the graph [also see Eq. (1.6-10)] we see that the increment $\Delta v_C(t)$ can be expressed in the form

$$\Delta v_C(t) = v_C(t + \Delta t) - v_C(t) \tag{9.2-3}$$

Using this, Eq. (9.2-2) becomes

$$\frac{v_C(t + \Delta t) - v_C(t)}{\Delta t} + \frac{v_C(t)}{\tau} = \frac{V_1}{\tau} \tag{9.2-4}$$

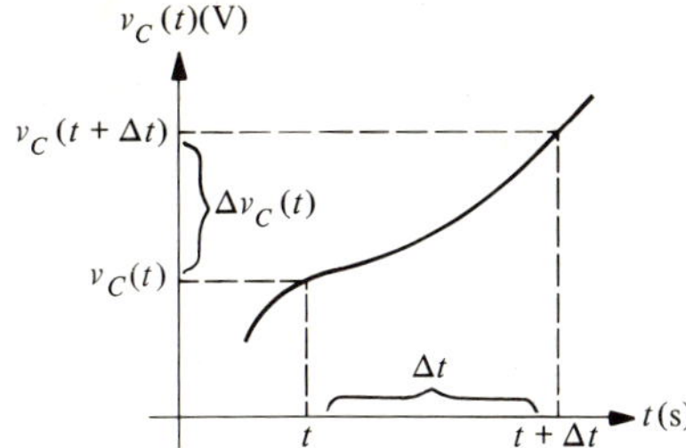

FIGURE 9.6
Typical waveform.

Rearranging, we get

$$v_C(t + \Delta t) = v_C(t)\left(1 - \frac{\Delta t}{\tau}\right) + V_1 \frac{\Delta t}{\tau} \tag{9.2-5}$$

This is the desired relation, which is called a *recursion formula*. It expresses v_C at a time $t + \Delta t$ in terms of v_C at time t (Δt seconds *earlier*) and in terms of the circuit parameters R, C, and V_1. If we can find v_C for *any time t*, we can use the equation to find its approximate value Δt seconds *later*. This value can then be used to find the value another Δt seconds later, and so on in steps of Δt seconds. We illustrate by using this procedure for the values $R = 1\ \text{M}\Omega$, $C = 2\ \mu\text{F}$, and $V_1 = 10$ V. Now

$$\tau = RC = 10^6 \times 2 \times 10^{-6} = 2\ \text{s}$$

so that Eq. (9.2-5) becomes

$$v_C(t + \Delta t) = v_C(t)\left(1 - \frac{\Delta t}{2}\right) + 10 \frac{\Delta t}{2} \tag{9.2-6}$$

The value chosen for the time step Δt should be one-tenth or less of the time constant; however, in order to simplify the illustration of the method we choose $\Delta t = 1$ s (this makes $\Delta t/\tau = 1/2$). Then Eq. (9.2-6) becomes

$$v_C(t + \Delta t) = \frac{1}{2} v_C(t) + 5 \tag{9.2-7}$$

This holds for $t > 0$ and the initial condition is $v_C(0^+) = 0$. It is the initial condition that gives us the starting value for our step-by-step solution. We proceed as follows: At $t = 0^+$

$$v_C(0^+) = 0\ \text{V} \qquad \text{(initial condition)}$$

Δt seconds later, Eq. (9.2-7) becomes

$$v_C(\Delta t) = \frac{1}{2} v_C(0^+) + 5 = 5 \text{ V}$$

$2\Delta t$ seconds later,

$$\begin{aligned} v_C(2\Delta t) &= \tfrac{1}{2} v_C(\Delta t) + 5 \\ &= \tfrac{1}{2}(5) + 5 \\ &= 7.5 \text{ V} \end{aligned}$$

$3\Delta t$ seconds later,

$$\begin{aligned} v_C(3\Delta t) &= \tfrac{1}{2} v_C(2\Delta t) + 5 \\ &= \tfrac{1}{2}(7.5) + 5 \\ &= 8.75 \text{ V} \end{aligned}$$

Continuing in this way, we get the results tabulated below:

t	$v_C(t)$
0	0
Δt	5
$2\Delta t$	7.5
$3\Delta t$	8.75
$4\Delta t$	9.375
$5\Delta t$	9.6875
$6\Delta t$	9.8438
$7\Delta t$	9.9219
$8\Delta t$	9.9609

These data are plotted in Fig. 9.7. The points are not joined together because our solution holds only at points Δt seconds apart. It must be emphasized that this solution is only an approximation which becomes better as $\Delta t/\tau$ is

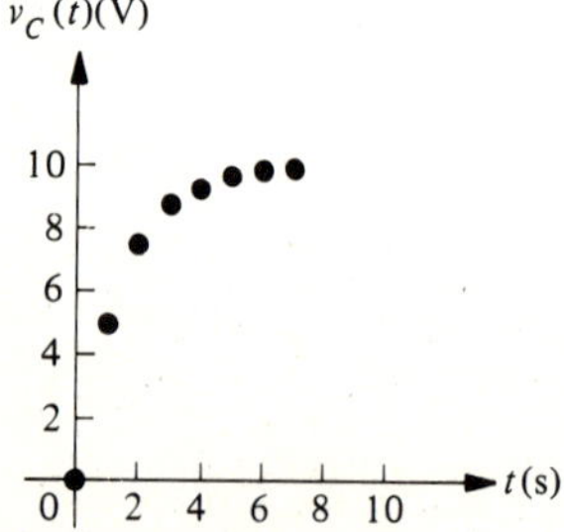

FIGURE 9.7
Stepwise solution of capacitive charging circuit.

taken smaller; however, even our choice of $\Delta t/\tau = 0.5$ leads to a solution remarkably similar to the actual solution, as we will see in the next section.

The approximate solution does show how the capacitor voltage increases as the capacitor charges. As this increase is taking place, the resistor voltage is decreasing. This may be seen from the KVL equation for $t > 0$,.

$$v_R(t) = V_1 - v_C(t)$$

Since the resistor voltage is decreasing, the current is also decreasing. This causes a decrease in the *rate* at which the capacitor voltage is increasing [recall that $i = C(\Delta v_C/\Delta t)$]. As this rate of change approaches zero, all currents and voltages approach constants, that is, steady-state values. In Sec. 9.3 we set down the actual solution in general form and show how it can be applied to more complicated circuits. Additional examples of the stepwise method will be found in the end-of-chapter problems.

9.3 CAPACITOR CHARGE AND DISCHARGE

In the last section we saw how the voltage builds up across the capacitor when a battery is switched across an RC circuit. This process is called *charging* and we say that the capacitor *charges up* to the final (steady-state) voltage. If a charged capacitor is disconnected from the charging circuit, it will ideally hold its charge indefinitely. If a resistor is connected across a charged capacitor, the difference in voltage between the plates will cause current to flow through the resistor and the energy stored in the electric field of the capacitor will be dissipated as heat in the resistor. This process is referred to as *discharging* the capacitor.

Consider the circuit of Fig. 9.8a. Assume that the switch has been in position 1 long enough so that the capacitor is charged to V_1 volts. At a time designated (for convenience) as $t = 0$, the single-pole double-throw switch disconnects the battery and connects the resistor and capacitor together as

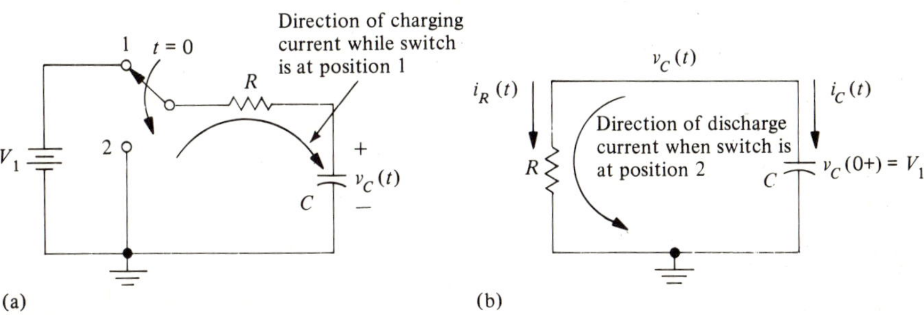

FIGURE 9.8
Discharging a capacitor. (a) Circuit. (b) Circuit for $t > 0$.

shown in part b of the figure. After $t = 0$, the only energy in the circuit is that stored in the electric field of the capacitor. The response under these conditions is called the *natural* response, since the response is determined by the network alone, that is, with no external input. As noted in Sec. 9.2, the capacitor voltage cannot change instantaneously, and since it has been charged to V_1 volts before $t = 0$, it will remain at V_1 volts immediately after the switching and the initial condition for the circuit is $v_C(0^+) = V_1$. After $t = 0^+$ the capacitor begins to *discharge*, that is, the voltage across it begins to decrease. While the capacitor voltage is decreasing, the energy stored in the electric field is being dissipated as heat in the resistor. Since the voltage across the resistor is the same as the voltage across the capacitor, the resistor voltage and consequently the current in the circuit will also decrease. This action continues until the capacitor is completely discharged, at which point the voltage across it is reduced to zero. At the same time the current is reduced to zero. The circuit is now said to be in the steady state.

To find the equation for the natural response so that we can plot it, we first write the circuit equation. Using KCL at the single independent node,

$$i_C(t) + i_R(t) = 0 \tag{9.3-1}$$

Using the *vi* relations for C and R,

$$C\frac{\Delta v_C}{\Delta t} + \frac{v_C}{R} = 0$$

Dividing through by C and setting $\tau = RC$,

$$\frac{\Delta v_C}{\Delta t} + \frac{v_C}{\tau} = 0 \tag{9.3-2}$$

This equation holds for $t > 0$ and is subject to the initial condition $v_C(0^+) = V_1$. Its exact solution can be shown to be, for $t > 0$,

$$v_C(t) = v_C(0^+)e^{-t/\tau} \tag{9.3-3}$$

$$= V_1 e^{-t/\tau} \tag{9.3-4}$$

This is an *exponential* function, which is discussed in detail in Sec. 9.4. It describes the way in which a capacitor discharges. A plot of the equation is shown in Fig. 9.9a. From the graph, we see that the final value of the capacitor voltage, $v_C(\infty)$, is zero. Thus it decays from the initial value V_1 to zero. While the capacitor voltage is decreasing from V_1 to zero, the resistor voltage is decreasing in exactly the same way because they are in parallel. Therefore the current will be decreasing from its initial value $i_R(0^+) = V_1/R$ to zero, as shown in Fig. 9.9b. It turns out that the natural response of *any* circuit containing only one capacitor or one inductor and any number of resistors has the same form, that is, an exponential decay.

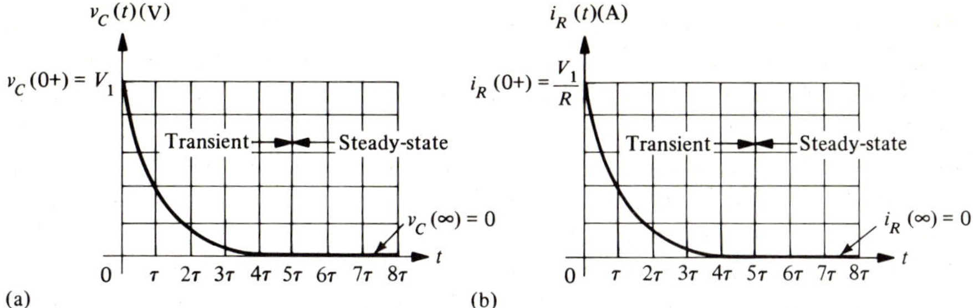

FIGURE 9.9
Natural response of RC circuit. (a) Capacitor and resistor voltage. (b) Resistor current.

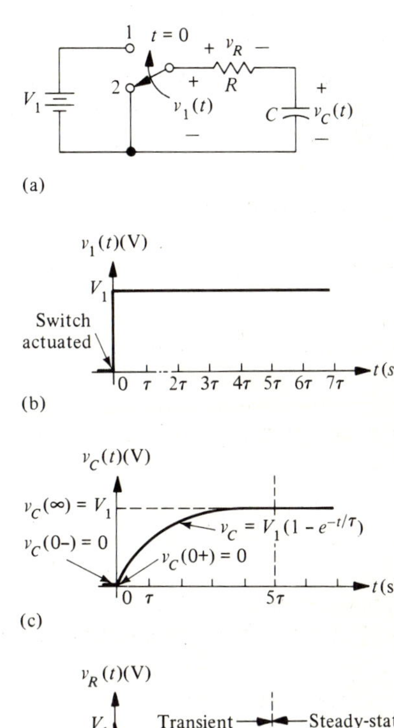

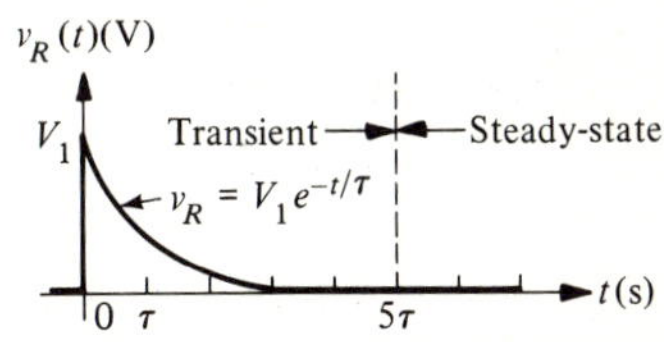

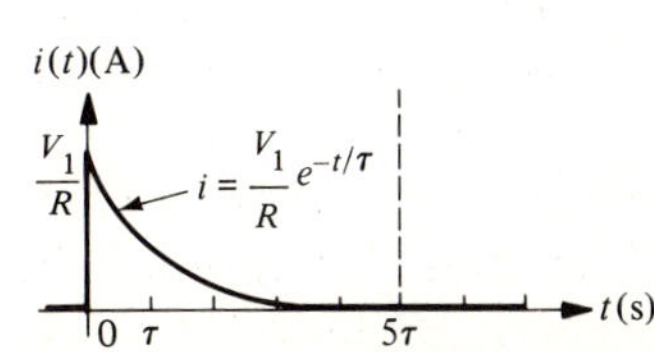

FIGURE 9.10
Charging the capacitor. (a) Circuit. (b) Input voltage. (c) Capacitor voltage. (d) Resistor voltage. (e) Current.

Charging the Capacitor

Consider the circuit of Fig. 9.10a and assume that the switch has been in position 2 long enough for the capacitor to completely discharge so that the voltage across it is zero. Now let us assume that the switch is returned to position 1, at a time which we designate as $t = 0$, so that the capacitor is charging. The voltage applied to the circuit is shown in Fig. 9.10b. The equation governing the charging process was found in Sec. 9.1-1 and is Eq. (9.1-3). An *approximate* solution to this equation was found in Sec. 9.2 and a graph shown in Fig. 9.7. The *exact* exponential solution is

$$\begin{aligned} v_C(t) &= v_C(\infty)(1 - e^{-t/\tau}) \\ &= V_1(1 - e^{-t/\tau}) \end{aligned} \tag{9.3-5}$$

The corresponding graph is shown in Fig. 9.10c, and we see that the charging process takes the capacitor voltage in the circuit of Fig. 9.10a from its initial value of zero volts to its final value $v_C(\infty) = V_1$. As the capacitor gradually charges, the voltage across it increases and energy is stored in its electric field. The increasing voltage across the capacitor opposes the battery voltage so that the voltage across the resistor decreases as the capacitor voltage increases. Since the resistor voltage is proportional to the current, the current in the circuit will gradually decrease as the capacitor voltage gradually increases. The reduction in resistor voltage continues until the capacitor voltage equals the battery voltage, at which point the resistor voltage and the circuit current are reduced to zero. At this point the circuit has reached the steady state. The resistor voltage can be found using KVL. From the circuit of Fig. 9.10a, we have, for $t > 0$,

$$\begin{aligned} v_R(t) &= V_1 - v_C(t) \\ &= V_1 e^{-t/\tau} \end{aligned} \tag{9.3-6}$$

The current is

$$i(t) = \frac{v_R(t)}{R} = \frac{V_1}{R} e^{-t/\tau} \tag{9.3-7}$$

These equations are plotted in Fig. 9.10d and e.

EXAMPLE 9.3-1 A Capacitive Circuit

Figure 9.8a represents the input circuit of a logic gate in a microcomputer. In the circuit, $V_1 = 5$ V, $R = 600\ \Omega$, and $C = 33$ pF.

a. The switch is in position 2 for a time long enough for the capacitor to be completely discharged. At a time designated as $t = 0$ the switch is moved to position 1. Find $v_C(t)$ for $t > 0$.

b. The switch is in position 1 for a time long enough for the capacitor to be fully charged. At a time designated as $t = 0$ the switch is moved to position 2. Find $v_C(t)$ for $t > 0$.

Solution

a. For this case the capacitor voltage is zero just before the switch is moved and, because it cannot change instantaneously, it will remain at 0 V through the switching process. The capacitor will then charge up to the battery voltage of 5 V. The charging process is described by Eq. (9.3-5). In the equation

$$V_1 = 5 \text{ V}$$
$$\tau = RC = (600\ \Omega)(33 \times 10^{-12}\text{ F}) \approx 20\,000 \text{ ps} = 20 \text{ ns}$$

Then, with t in nanoseconds,

$$v_C(t) = 5(1 - e^{-t/20}) \text{ V}$$

The graphs of Fig. 9.10 apply directly to this equation.

b. For this case the capacitor voltage is 5 V just before the switch is moved and will remain at 5 V through the switching. After $t = 0$ it will discharge to zero as in Eq. (9.3-4), in which $V_1 = 5$ V and $\tau = 20$ ns. Then, with t in nanoseconds

$$v_C(t) = 5e^{-t/20} \text{ V}$$

The graphs of Fig. 9.9 apply for this case.

• • •

9.4 THE EXPONENTIAL FUNCTION

The exponential function in Eq. (9.3-4) is extremely important in engineering. Most beginning technology students, however, will not have encountered it in previous math courses. In order to introduce it we begin by considering a slight variation of the scientific notation studied in Chap. 1, that is,

$$y = 10^t \tag{9.4-1}$$

Here y is the dependent variable, and the independent varible t, which is the exponent, represents time. Only positive values of time are considered in this section. This function is tabulated and plotted in Fig. 9.11a. On

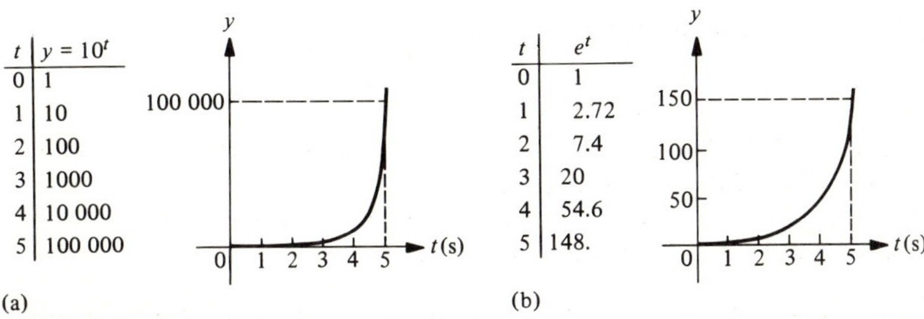

FIGURE 9.11
Exponential growth functions. (a) 10^t. (b) e^t.

the linear scale used, the function is almost coincident with the t axis up to about $t = 4$ s and rises very steeply thereafter.

The exponential function of most interest to us uses the base of the natural (Naperian) system of logarithms instead of 10. This irrational number is $e = 2.71828\ldots$, which we can round off to 2.72. Its importance stems from the fact that it turns up in the solutions of many problems involving natural events. A few of these are population growth, growth and decay of current and voltage in inductive and capacitive circuits, and motion of masses and springs restrained by viscous friction.

We next consider

$$y = e^t = 2.72^t \tag{9.4-2}$$

This is tabulated and plotted in Fig. 9.11b. The values in the table are easily obtained on a scientific calculator. We enter the value of t and then depress the key marked e^x.* The display then reads the value of e^t. One phenomenon described by this curve is the growth of population under the simplifying assumption that the growth at any time is proportional to the population existing at that time. The steep rise predicted by the curve is what causes such grave concern to statesmen and demographers.

The two functions just considered were growth functions. In electric circuit theory, we are more concerned with decay functions in which the exponent is negative. Consider

$$y = 10^{-t} \tag{9.4-3}$$

This is plotted in Fig. 9.12a. Note that the function decays very rapidly from its initial value of 1. In fact, it is down to 1% of its initial value when $t = 2$.

Next consider

$$y = e^{-t} \tag{9.4-4}$$

This decay function is plotted in Fig. 9.12b. The waveform describes the decay process that takes place when a capacitor is discharged through a resistor, as we saw in Sec. 9.3. We will now point out those features of the curve that are used to characterize it. To do this, we consider a more general form of the exponential function, expressed as a voltage for convenience

$$v(t) = V_1 e^{-t/\tau} \tag{9.4-5}$$

Here $v(t)$ is the voltage, V_1 is a constant called the *amplitude*, t is the independent variable, time, and τ (lower-case Greek letter tau) is the *time*

*Some calculators require that the INV key be depressed first and then the LNX key. If neither method works on your calculator, consult the instruction manual for the proper method to be used.

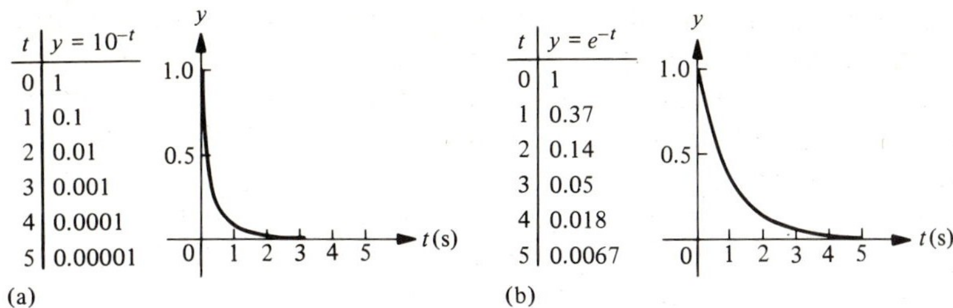

FIGURE 9.12
Exponential decay functions. (a) 10^{-t}. (b) e^{-t}.

constant, to be explained shortly. Note that the exponent $-t/\tau$ is dimensionless. This will always be true when the exponential represents the solution of a physical problem. Equation (9.4-5) is plotted in Fig. 9.13 in normalized form. This curve is universal and applies to any system that has an exponential response as given by Eq. (9.4-5). All that is needed to draw the actual curve are values of τ and V_1.

From the curve we can explain the significance of the time constant. Consider the value of $v(t)$ when $t = \tau$. This value is

$$v(\tau) = V_1 e^{-1} = \frac{V_1}{2.72} = 0.37V_1 \qquad (9.4\text{-}6)$$

Since the decay curve starts off at $v(t) = V_1$ when $t = 0$, Eq. (9.4-6) indicates that it will decay to 37% of its initial value at a time equal to the time constant. Thus the time constant can be used as a measure of the speed of the decay. For brevity, an exponential decay waveform is often specified by

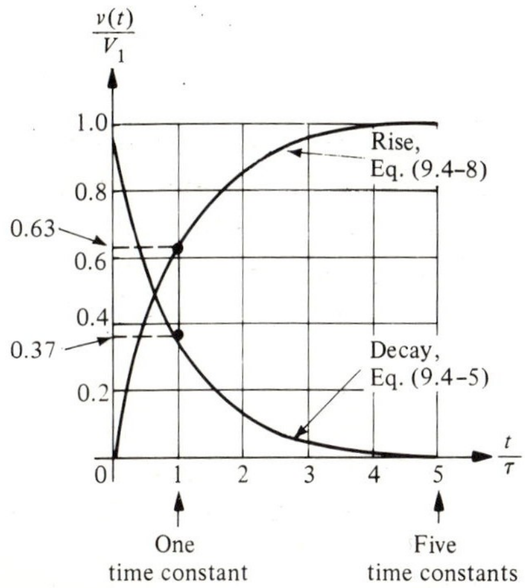

FIGURE 9.13
Universal exponential curves.

simply stating its amplitude and time constant. This will be illustrated in the next example.

Another point on the exponential decay that is worth remembering is the point at which $t = 5\tau$:

$$v(5\tau) = V_1 e^{-5} = 0.007V_1 \tag{9.4-7}$$

In words, this says that when $t = 5\tau$, the waveform has decayed to less than 1% of its initial value, and for all practical purposes, the decay is complete. Values of e^{-t} for intermediate values of t can be found using the calculator or the table of exponentials in Appendix D.

In addition to the exponential decay, which is characteristic of the discharge of a capacitor, we often encounter an exponential rise. This is different from the exponential growth considered previously, and is characteristic of the charging of a capacitor. The rise is governed by the equation

$$v(t) = V_1(1 - e^{-t/\tau}) \tag{9.4-8}$$

This is also plotted in Fig. 9.13. For this curve, we have, after one time constant,

$$v(t) = V_1(1 - e^{-1}) = 0.63V_1 \tag{9.4-9}$$

Thus the significance of the time constant for the exponential rise is that it represents the time at which the waveform has risen to 63% of its final value.

For the point where $t = 5\tau$, we find

$$v(t) = V_1(1 - e^{-5}) = 0.99V_1 \tag{9.4-10}$$

So after five time constants the exponential rise is up to 99% of its final value. In the following examples, we consider both types of exponentials with actual numerical values.

EXAMPLE 9.4-1 Graph of the Exponential Decay

An exponentially decaying voltage waveform is characterized by an amplitude of 20 V and a time constant of 3 ms. Write the equation for this voltage and sketch it to suitable scales.

Solution

Using Eq. (9.4-5) we note that $V_1 = 20$ V and $\tau = 3 \times 10^{-3}$ s. Thus

$$v(t) = 20e^{-t/3\times10^{-3}}\ \text{V} \tag{9.4-11}$$

In order to plot Eq. (9.4-11) we tabulate below some significant values.

t(ms)	$v(t)$(V)
0	20
3	7.4
6	2.7
15	0.13

The student should verify the computed values using a scientific calculator. The key strokes required at $t = 6$ ms on most calculators are as follows:

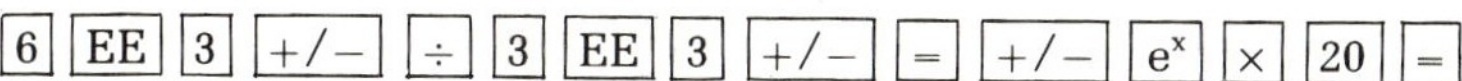

The waveform is shown in Fig. 9.14.

• • •

EXAMPLE 9.4-2 The Effect of Different Time Constants

An exponentially rising current has an amplitude of 10 mA. Write the equation for the waveform with a time constant of 10 μs and plot it for time constants of 10, 20, and 30 μs.

Solution

Referring to Eq. (9.4-8) with $I_1 = 10$ mA in place of V_1, $\tau = 10\ \mu$s, and replacing $v(t)$ with $i(t)$, we have

$$i(t) = 10(1 - e^{-t/10\times10^{-6}})\ \text{mA} \qquad (9.4\text{-}12)$$

For the other time constants the exponent is changed accordingly. Pertinent values of $i(t)$ are tabulated in Table 9.1 and plotted in Fig. 9.15. Note that it is the time

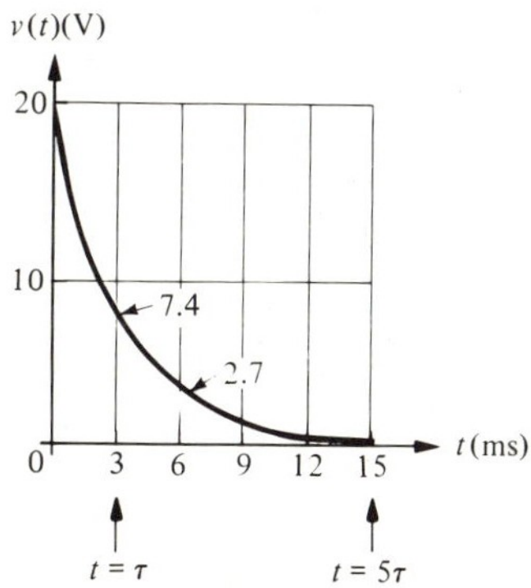

FIGURE 9.14
Waveform for Example 9.4-1.

TABLE 9.1

	$i(t)$(mA)		
t (μs)	$\tau = 10\ \mu$s	$\tau = 20\ \mu$s	$\tau = 30\ \mu$s
0	0	0	0
10	6.3		
20		6.3	
30			6.3
50	9.9		
100		9.9	
150			9.9

constant, not the individual value of R or C, that determines the speed with which the waveform rises toward its final value. Longer time constants lead to circuits requiring more time to acquire their final charge.

If desired, intermediate values can be filled into the table using the calculator or the tables of the function e^{-t} in Appendix D. The student should complete the table for practice.

• • •

LEARNING EXERCISES FOR SEC. 9.4

1. A 100-μF capacitor is charged through a 10-kΩ resistor to a voltage of 12 V. If the charging begins at $t = 0$ find $v_C(0^+)$, $v_C(0.5)$, $v_C(1)$, $v_C(2)$, and $v_C(5)$. All values of t are in seconds.

Ans. 10.4; 0; 4.72; 11.9; 7.59

2. A 0.22-μF capacitor is charged to 180 V and, at $t = 0$, is connected across a 10-kΩ resistor. Find the time constant and the voltage after 0, 1, and 5 time constants.

Ans. 2.2; 66; 1.2; 180

• • •

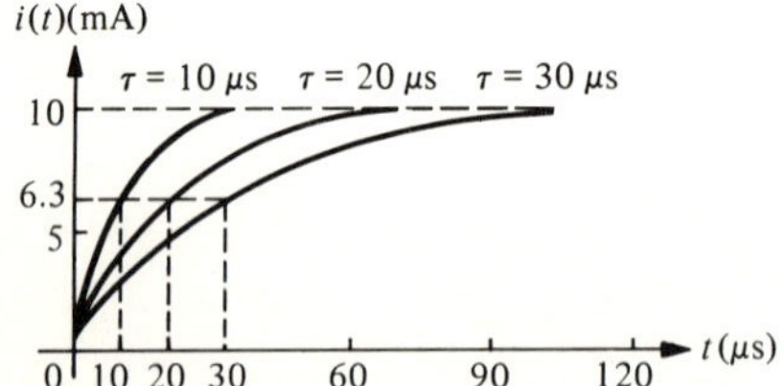

FIGURE 9.15
Waveforms for Example 9.4-2.

9.5 BUILDUP AND DECAY OF CURRENT IN AN INDUCTANCE

Inductive Current Buildup

A circuit containing resistance and inductance is shown in Fig. 9.16a. When the switch is moved from position 2 to position 1 at $t = 0$, the current builds up in the inductance in exactly the same way that the voltage builds up across a charging capacitor. The equation for the buildup of current in the inductor is

$$i_L(t) = i_L(\infty)(1 - e^{-t/\tau}) \tag{9.5-1}$$

A graph of this solution will be found in Fig. 9.17. The important points on the graph are as follows:

1. The initial value $i_L(0^+)$ is found by applying the continuity condition for inductance, which states that current through an inductance cannot change instantaneously. Thus if the current is zero before $t = 0$, it will remain zero through the switching and the circuit for a short instant after $t = 0$ takes the form shown in Fig. 9.16b, that is, the inductance "looks like" an open circuit for this instant.
2. After the initial instant the current builds up according to Eq. (9.5-1) and the circuit is shown in Fig. 9.16c.
3. The final value $i_L(\infty)$ is found by noting that after a long time has elapsed

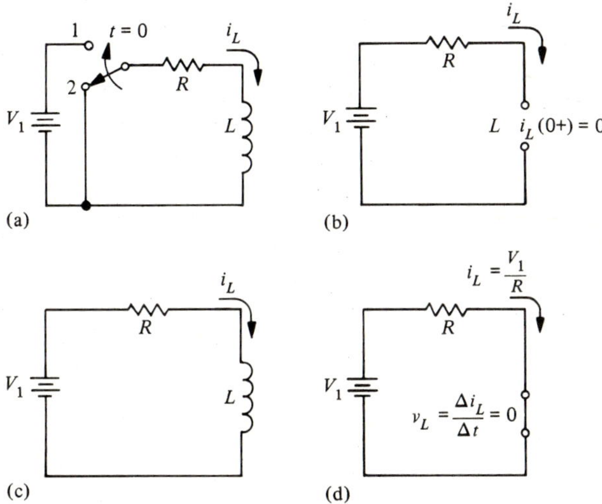

FIGURE 9.16
The *RL* circuit. (a) Original circuit. (b) Circuit just after $t = 0$. (c) circuit for $t > 0$. (d) Circuit for $t \to \infty$.

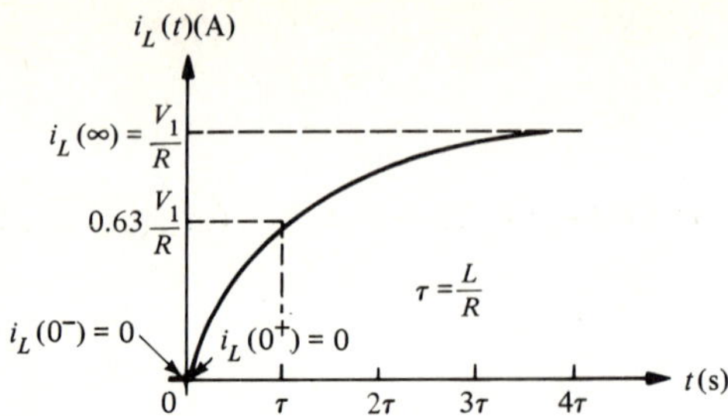

FIGURE 9.17
Buildup of current in an inductance.

the transient will have died down and the current will no longer be changing. Then $\Delta i_L/\Delta t = 0$ and

$$v_L = L\frac{\Delta i_L}{\Delta t} = 0 \qquad (9.5\text{-}2)$$

Since the voltage across the inductor is zero, the circuit appears as in Fig. 9.16d and all of the battery voltage appears across the resistor so that $i_L(\infty) = V_1/R$.

4. The time constant τ for the RL circuit is

$$\tau = \frac{L}{R} \qquad (9.5\text{-}3)$$

If L is in henrys and R is in ohms, then τ will be in seconds. Some typical numbers will be found in the following example.

EXAMPLE 9.5-1 Inductive Current Buildup

The deflection circuit of a TV picture tube has an inductance of 230 mH and a resistance of 30 Ω. A voltage of 90 V is switched across the combination. Find the inductor current.

Solution

The circuit is that of Fig. 9.16a and the current is given by Eq. (9.5-1). In the equation

$$i_L(\infty) = \frac{V_1}{R} = \frac{90\text{ V}}{30\ \Omega} = 3\text{ A}$$

$$\tau = \frac{L}{R} = \frac{230\text{ mH}}{30\ \Omega} = 7.7\text{ ms}$$

Then, with t in milliseconds,

$$I_L(t) = 3(1 - e^{-t/7.7})\ \text{A}$$

A graph of the current is shown in Fig. 9.17. On the graph, $i_L(\infty) = 3$ A and $\tau = 7.7$ ms.

• • •

Inductive Current Decay

The decay of current in an inductive circuit has the same form as the decay of voltage across a discharging capacitor. In the circuit of Fig. 9.16a, let us assume that the switch is in position 1 for a time long enough for the current in the coil to have reached its steady-state value, V_1/R. If we now move the switch to position 2, at a time which we designate as $t = 0$, the current will decay to zero according to the equation

$$i_L(t) = i_L(0^+)e^{-t/\tau} \tag{9.5-4}$$

In order to find $i_L(0^+)$ we must make use of the continuity condition for inductance. If the current is V_1/R at $t = 0^-$ just before the switching, it will not change during the switching process and $i_L(0^+) = i_L(0^-) = V_1/R$. The time constant is the same as before, $\tau = L/R$, so that

$$i_L(t) = \frac{V_1}{R}e^{-t/\tau} \tag{9.5-5}$$

The graph for this equation is shown in Fig. 9.18.

EXAMPLE 9.5-2 Inductive Current Decay

The deflection coil in Example 9.5-1 is short-circuited. Find the current.

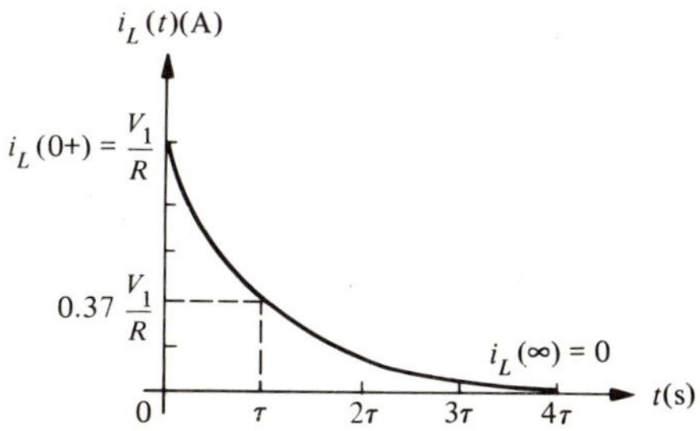

FIGURE 9.18
Decay of current in an inductance.

Solution

A short circuit across the coil is the same as moving the switch in Fig. 9.16a from position 1 to position 2, and the current will decay from an initial value of 3 A according to Eq. (9.5-5). Then, with t in milliseconds

$$i_L(t) = 3e^{-t/7.7} \text{ A}$$

• • •

LEARNING EXERCISE FOR SEC. 9.5

1. In the circuit of Fig. 9.16a, $V_1 = 80$ V, $R = 6.8$ kΩ, and $L = 4.2$ H. The switch is in position 2 for a long time before being moved to position 1 at $t = 0$. Find $i_L(0^+)$, $i_L(\infty)$, and the time constant τ.

Ans. 11.8; 0.62; 0

• • •

9.6 A GENERAL SOLUTION

In this section we present a general solution for a large class of transient responses. The responses are the currents and voltages in single energy-storage element circuits. For our purposes, such circuits consist of either one capacitor or one inductor, any number of resistors, and one source. The source may be a switched battery or constant current source, or a step function voltage or current source. All such circuits lead to a circuit equation similar to Eq. (9.1-3).

The general solution, which holds for $t > 0$ is*

$$r(t) = r(\infty) + [r(0^+) - r(\infty)]e^{-t/\tau} \qquad (9.6\text{-}1)$$

where

$r(t)$ = response at time t, either current or voltage
$r(\infty)$ = final value of response, either a constant or zero
$r(0^+)$ = initial value of response just after switching or step, either a constant or zero
$e^{-t/\tau}$ = exponential factor
τ = circuit time constant, to be explained shortly

The response is thus completely characterized by three numbers, $r(\infty)$, $r(0^+)$, and τ. These numbers are obtained from the circuit, using basic laws that are explained here, and demonstrated in the examples that follow.

*The student should check this by substituting $t = 0^+$ and $t = \infty$ to see that it gives the expected solutions.

1. $r(\infty)$ is the response a long time after the switching or step has taken place. This is found, as noted in Sec. 9.1, by considering capacitors as open circuits and inductors as short circuits.
2. $r(0^+)$ is the initial condition, which has been discussed in connection with each example thus far. It is found by finding the response just prior to the switching and then applying the appropriate continuity condition to find the initial condition just after switching.
3. τ is the circuit time constant. For RC circuits

$$\tau = R_T C \tag{9.6-2}$$

For RL circuits

$$\tau = \frac{L}{R_T} \tag{9.6-3}$$

L and C represent the inductance or capacitance. The Thevenin resistance R_T is found by setting the source to zero (short circuit for independent voltage sources, open circuit for independent current sources) and calculating the resistance looking into the circuit from the terminals of the inductor or capacitor. This is illustrated in the following examples.

EXAMPLE 9.6-1 Capacitor Step Function Response

The capacitor is initially uncharged in the circuit of Fig. 9.19a and the source is a 10-V step. Find

a. $v_C(t)$
b. $i_C(t)$

Solution

a. The general equation for $r(t) = v_C(t)$ becomes

$$v_C(t) = v_C(\infty) + [v_C(0^+) - v_C(\infty)]e^{-t/\tau} \tag{9.6-4}$$

The constants in this equation are as follows:

1. After a long time the capacitor will charge to 10 V so that $v_C(\infty) = 10$.
2. Since the capacitor is initially uncharged, $v_C(0^-) = 0$ before the step begins. The continuity condition requires that $v_C(0^+) = 0$ immediately after the step begins. Thus the capacitor "looks like" a short circuit in the time interval between $t = 0^-$ and $t = 0^+$.
3. For this circuit $\tau = RC = 2$ s by inspection.

Using these values in Eq. (9.6-4) we have

$$\begin{aligned} v_C(t) &= 10 + (0 - 10)e^{-t/2} \\ &= 10(1 - e^{-t/2})\text{ V} \end{aligned} \tag{9.6-5}$$

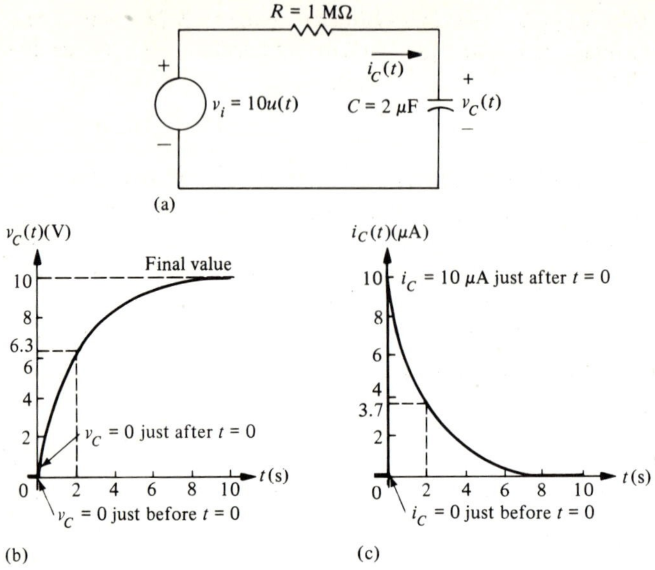

FIGURE 9.19
Example 9.6-1. (a) Circuit. (b) Capacitor voltage. (c) Current.

This equation is plotted in Fig. 9.19b. Its important characteristics were discussed in Sec. 9.4. Briefly, they are that the voltage rises to 63% of the final value in a time equal to one time constant (2 s) and to 99% of the final value in five time constants (10 s).

b. The general equation for $r(t) = i_C(t)$ becomes

$$i_C(t) = i_C(\infty) + [i_C(0^+) - i_C(\infty)]e^{-t/\tau} \tag{9.6-6}$$

The constants are as follows:

1. After a long time the capacitor appears as an open circuit so that $i_C(\infty) = 0$ A.
2. Immediately after the step begins we have $v_C(0^+) = 0$ V. Using Ohm's law,

$$i_C(0^+) = \frac{v_i(0^+) - v_C(0^+)}{R} = \frac{10 - 0}{10^6} = 10\ \mu\text{A}$$

3. The time constant is the same as in part a, $\tau = 2$ s.

Substituting these values in Eq. (9.6-6) we obtain

$$\begin{aligned} i_C(t) &= 0 + (10 - 0)e^{-t/2}\ \mu\text{A} \\ &= 10e^{-t/2}\ \mu\text{A} \end{aligned} \tag{9.6-7}$$

This is plotted in Fig. 9.19c. Note that the current is zero just before the step begins, and that the current *is* discontinuous (jumps) at $t = 0$. This does not violate the continuity condition, which applies only to voltage for a capacitor. The exponential decay is to 37% of its initial value after one time constant and is down to less than 1% after five time constants.

• • •

EXAMPLE 9.6-2 Finding Intermediate Values on a Charging Curve

Refer to the circuit of Fig. 9.2a. If $R = 39\ \text{k}\Omega$, $C = 0.05\ \mu\text{F}$, and $V_1 = 22$ V, find

a. The voltage across the capacitor at $t = 4$ ms,
b. The time required for the voltage across the capacitor to reach 17 V.

Solution

a. The capacitor voltage is [from Eq. (9.6-4)]

$$v_C(t) = V_1(1 - e^{-t/\tau})$$

where $\tau = RC = 39 \times 10^3 \times 0.05 \times 10^{-6} = 1.95$ ms. Then, at $t = 4$ ms,

$$\begin{aligned} v_C(t) &= 22(1 - e^{-4/1.95}) \\ &= 22(1 - e^{-2.05}) \end{aligned}$$

Using the $\boxed{e^x}$ key on the calculator we find $e^{-2.05} = 0.129$ so that

$$\begin{aligned} v_C(t) &= 22(1 - 0.129) \\ &= 19.2 \text{ V} \end{aligned}$$

b. $v_C(t) = 17 = 22(1 - e^{-t/\tau})$

Solving for t/τ,

$$1 - e^{-t/\tau} = \frac{17}{22}$$

$$e^{-t/\tau} = 1 - \frac{17}{22} = \frac{5}{22} = 0.227 \qquad (9.6\text{-}8)$$

To find the exponent, $-t/\tau$, we enter 0.227 on the calculator and depress the $\boxed{\text{LNX}}$ key. The display then reads -1.48, which means that

$$e^{-1.48} = 0.227 \qquad (9.6\text{-}9)$$

Comparing Eq. (9.6-9) with Eq. (9.6-8) we have

$$-\frac{t}{\tau} = -1.48$$

from which $t = 1.48\tau = (1.48)(1.95 \text{ ms}) = 2.89$ ms.

Thus it takes 2.89 ms for the capacitor to charge to 17 V.

• • •

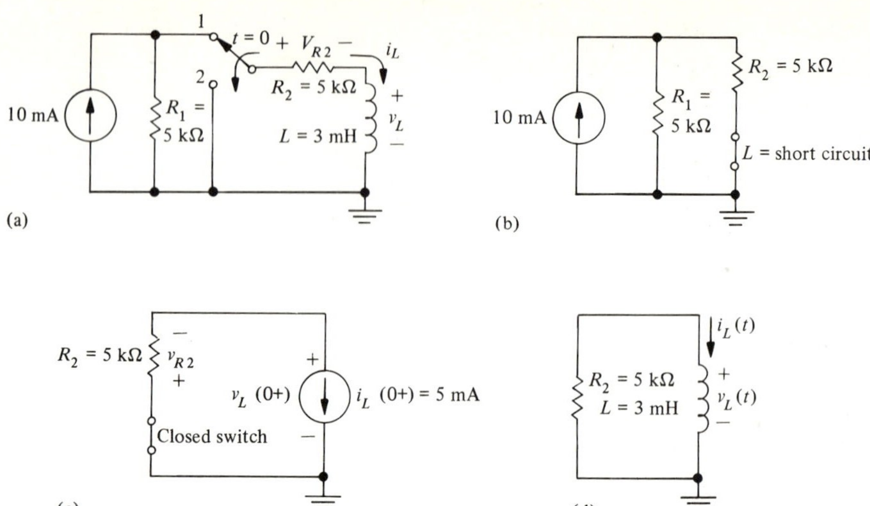

FIGURE 9.20
Example 9.6-3. (a) Circuit. (b) Circuit prior to switching ($t = 0-$). (c) Circuit just after switching ($t = 0+$). (d) Circuit for $t > 0$.

EXAMPLE 9.6-3 An Inductive Circuit

In the circuit of Fig. 9.20a, the switch is moved to position 2 at $t = 0$ after having been in position 1 for a long time. Find and sketch $i_L(t)$, $v_L(t)$, and $v_{R2}(t)$ for $t > 0^-$.

Solution

Consider first the instant before the switch is thrown, $t = 0^-$. At this time the circuit is as shown in Fig. 9.20b with the inductor appearing as a short circuit. Thus, just prior to the switching, the 10 mA from the current source will divide equally between the two 5-kΩ resistors so that $i_L(0^-) = 5$ mA and $v_L(0^-) = 0$ V.

At the instant after the switch is closed $(t = 0^+)$, the continuity condition on inductor current tells us that the current will remain at 5 mA $[i_L(0^+) = 5 \text{ mA}]$, and the circuit will then appear as shown in Fig. 9.20c. In the circuit we have replaced the inductor by a 5-mA current source because the circuit only holds for the $t = 0^+$ instant just after the switching has taken place. The voltage across the inductor, $v_L(0^+)$, is the same as the voltage across R_2. This is $(-5 \text{ mA})(5 \text{ k}\Omega) = -25$ V because the initial inductor current, $i_L(0^+) = 5$ mA, is flowing into R_2 at the ground terminal. Thus $v_L(0^+) = -25$ V while $v_L(0^-) = 0$. The continuity condition is not violated here because it does not apply to inductor voltage.

From $t = 0^+$ to ∞ the circuit appears as shown in Fig. 9.20d. Since there is no independent energy source the current will decay to zero $[i_L(\infty) = 0]$ and consequently the voltage v_L will also decay to zero $[v_L(\infty) = 0]$. The equation for the current is

$$i_L(t) = i_L(0^+)e^{-t/\tau}$$

where

$$i_L(0^+) = 5 \text{ mA}$$

$$\tau = \frac{L}{R_2} = \frac{3 \text{ mH}}{5 \text{ k}\Omega} = 0.6\ \mu\text{s}$$

The inductor voltage is

$$v_L(t) = v_L(0^+)e^{-t/\tau}$$

where

$$v_L(0^+) = -25 \text{ V}$$
$$\tau = 0.6\ \mu\text{s}$$

The resistor voltage is

$$\begin{aligned} v_{R2}(t) &= R_2 i_L(t) \\ &= 5 \text{ k}\Omega \times 5 \text{ mA} \times e^{-t/0.6} \\ &= 25e^{-t/0.6} \text{ V} \end{aligned}$$

The waveforms are plotted in Fig. 9.21. Observe that the resistor and inductor voltages add up to zero for $t > 0^+$ and to 25 V for $t = 0^-$.

• • •

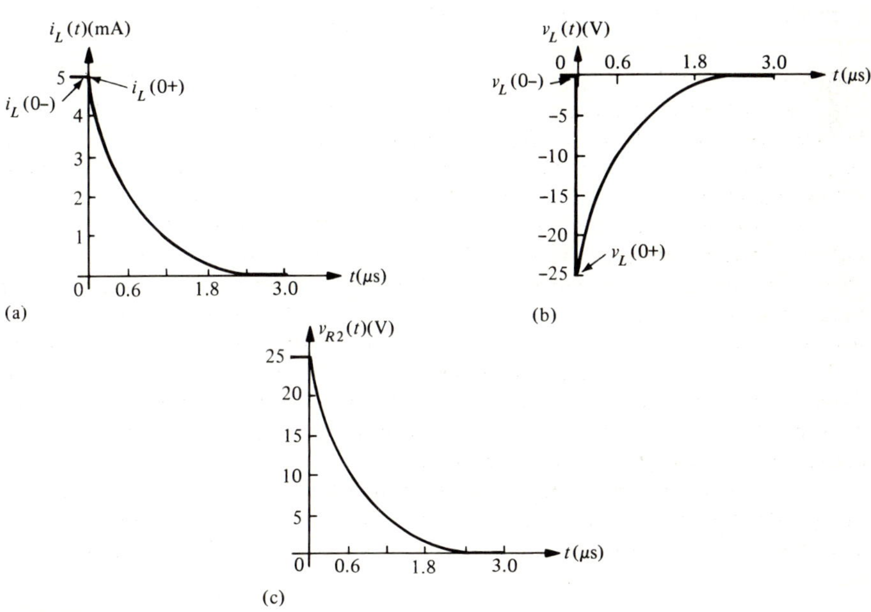

FIGURE 9.21
Waveforms for Example 9.6-3. (a) Inductor current. (b) Inductor voltage. (c) Resistor voltage.

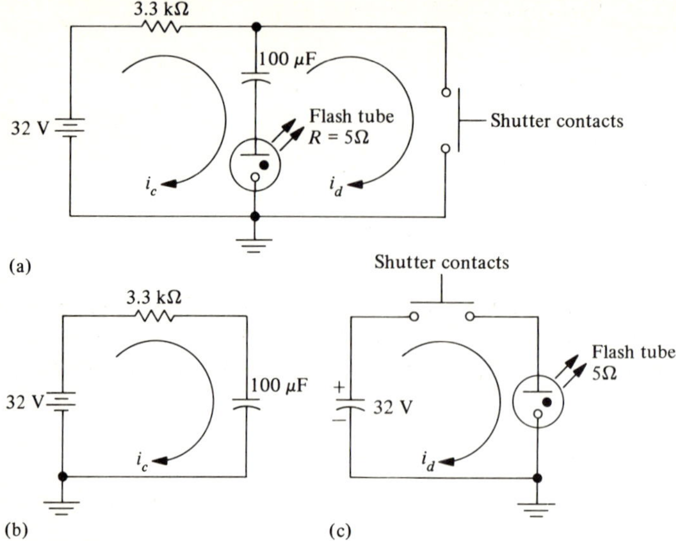

FIGURE 9.22
Electronic flash operation. (a) Circuit. (b) Circuit for charging cycle. (c) Circuit for discharge cycle.

EXAMPLE 9.6-4 Operation of an Electronic Flash Gun

In Fig. 9.22a we show a simplified version of the circuit of an electronic flash gun. For proper operation the flash tube requires a very high current for a very short period of time. It is not good practice to use batteries to supply such current surges, so we utilize the energy storage capability of a capacitor. Operation of the circuit is as follows: when the shutter contacts are open, the circuit appears as in Fig. 9.22b in which the 5-Ω flash tube resistance is neglected in comparison with the 3.3-kΩ resistance. During the time that the shutter contacts are open, the 100-μF capacitor will charge toward 32 V, with the charging current supplied by the battery. If the contacts are open long enough, it will charge fully, and in a practical flash gun a "ready light" will be activated. The current drawn from the battery is limited by the 3.3-kΩ resistor during the charging cycle.

When the photographer closes the shutter contacts, the circuit appears as in Fig. 9.22c. The capacitor discharges through the 5-Ω flash tube and the discharge takes place much more rapidly than the previous charge. All of the discharge current is supplied by the capacitor.

Find and sketch the waveforms of the current through the flash tube during the charge and discharge cycles.

Solution

During the charge cycle we symbolize the current as i_c. Then

$$i_c = i_c(\infty) + [i_c(0^+) - i_c(\infty)]e^{-t/\tau}$$

The constants are:

$$\tau = RC = (3.3 \times 10^3\ \Omega)(100 \times 10^{-6}\ \mathrm{F}) = 0.33\ \mathrm{s}$$

$$i_c(\infty) = 0\ \mathrm{A}$$

$$i_c(0^+) = \frac{32\ \mathrm{V}}{3.3\ \mathrm{k\Omega}} = 9.7\ \mathrm{mA}$$

Thus

$$i_c(t) = 9.7e^{-t/0.33}\ \mathrm{mA}$$

If the time constant is used as a measure of charging rate then it takes 0.33 s for the capacitor to charge between flashes and the peak charging current is 9.7 mA.

During the discharge cycle we symbolize the current shown in Fig. 9.22c as i_d. Then

$$i_d = i_d(\infty) + [i_d(0^+) - i_d(\infty)]e^{-t/\tau}$$

For this cycle the constants are

$$\tau = RC = (5\ \Omega)(100 \times 10^{-6}\ \mathrm{F}) = 500 \times 10^{-6}\ \mathrm{s} = 0.5\ \mathrm{ms}$$

$$i_d(\infty) = 0$$

$$i_d(0^+) = \frac{32\ \mathrm{V}}{5\ \Omega} = 6.4\ \mathrm{A}$$

Thus

$$i_d(t) = 6.4e^{-t/0.5\times 10^{-3}}\ \mathrm{A}$$

The discharge through the flash tube thus takes 0.5 ms as compared to the 330 ms required for the charge cycle. The peak discharge current supplied by the capacitor is 6.4 A, while the peak charging current supplied by the battery is limited to 9.7 mA by the 3.3-kΩ resistor. The waveforms are sketched in Fig. 9.23.

• • •

LEARNING EXERCISES FOR SEC. 9.6

1. In the circuit of Fig. 9.19a, $v_i = 20u(t)$ V, $R = 330$ kΩ, $C = 0.1\ \mu$F, and $v_C(0^-) = 6$ V. Find $v_C(0^+)$, $v_C(\infty)$, $i(0^-)$, $i(0^+)$, and $i(\infty)$.

Ans. 0.042; −0.018; 0; 20; 6

• • •

9.7 NATURAL AND FORCED RESPONSE

Let us return to the expression, given by Eq. (9.3-5), for the voltage rise across a capacitor:

$$v_C(t) = V_1(1 - e^{-t/\tau}) \qquad (9.3\text{-}5)$$

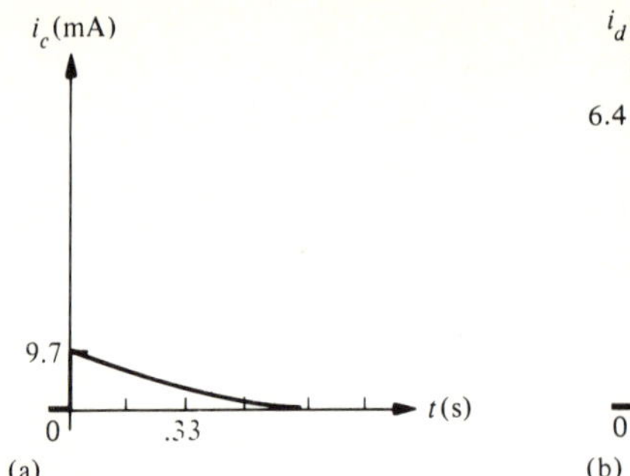

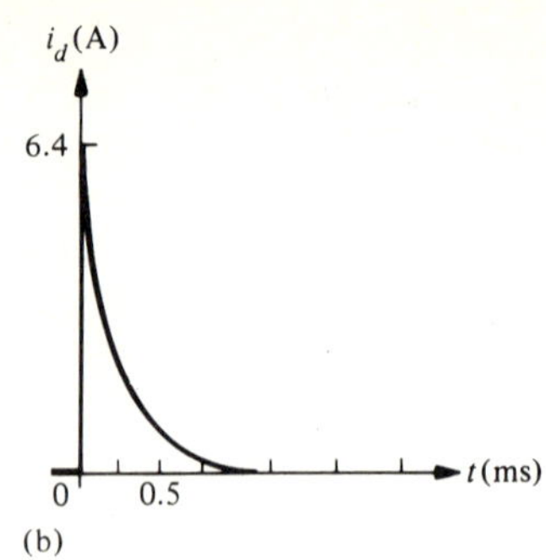

FIGURE 9.23
Waveforms for Example 9.6-4. (a) Charge. (b) Discharge. Note the difference in the scales of the two graphs. If they were plotted to the same scale, the discharge curve would appear as a sharp spike, and the charge curve would appear to coincide with the time axis.

In order to interpret this solution physically, we separate it into two parts. The first part consists of the constant term V_1. This is called the *forced* part of the solution, and we write

$$v_f(t) = V_1 \tag{9.7-1}$$

The second part of the solution, called the *natural response,* is

$$v_n(t) = -V_1 e^{-t/\tau}$$

These are plotted separately in Fig. 9.24 for $V_1 = 10$ V and $\tau = 2$ s.

The physical significance of the forced response is that it is the value which the output voltage assumes as $t \rightarrow \infty$. Thus, since the initial voltage is known to be zero and the final voltage, or forced response, is known to be 10 V, we see that the natural response *takes the solution from the initial condition to the final value.*

All circuit problems leading to linear equations like Eq. (9.3-2) are characterized by this same type of solution, that is, a forced response plus a natural response that provides the transition from the initial condition to the final value. In our example, the forced response is a constant, and has the same form as the input voltage. It is found that the forced response often has the same form as the input signal, whether it be a step function, sine wave, or some other wave form.

In single energy-storage element circuits, the natural response is a decaying exponential of the form $Ae^{-t/\tau}$. In more complicated circuits it has the form $Ae^{-t/\tau}\cos(\omega t + \theta)$ or even a sum of terms of this type, the number of terms depending on the complexity of the circuit.

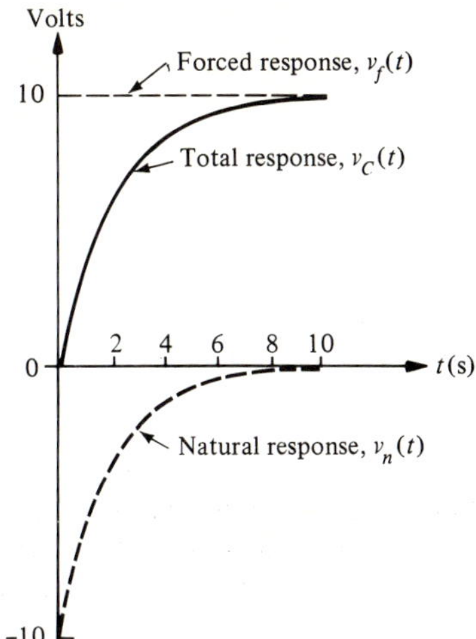

FIGURE 9.24
Components of response of *RC* circuit to step-function input.

9.8 PULSE RESPONSE AND RISE TIME

The pulse signals used in digital systems are often the input signals to circuits of the form shown in Fig. 9.25a. The capacitance in the circuits is usually *parasitic* in nature and is unintentionally introduced into the system because of the physical construction of the device. It is important that the pulse signals not be seriously distorted (that is, made different from the input) in passing through these circuits. Thus it becomes important to

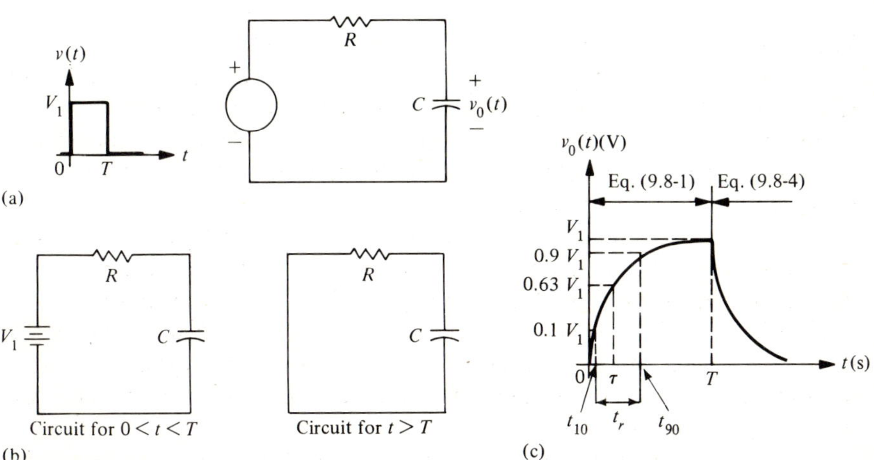

FIGURE 9.25
Pulse response. (a) Circuit. (b) Circuits during and after pulse. (c) Response waveform.

characterize the distortion of the pulse in some measurable way as it passes through the circuit. A quantity called *rise time* is used as a measure of this distortion and is a very important digital system parameter.

In order to define rise time, we consider the circuit of Fig. 9.25a with a pulse input as shown. The output voltage is found by first finding the output voltage due to a step function of V_1 volts beginning at $t = 0$. The response equation is then used to find the output voltage at $t = T$ when the pulse switches off. After $t = T$, the input voltage is zero, so the input source appears as a short circuit (Fig. 9.25b). Thus after $t = T$ the capacitor will discharge through the resistor. The voltage across the capacitor at $t = T$ is the initial condition for the discharge phase.

For the charging phase, we have, from Eq. (9.3-5),

$$v_0(t) = V_1(1 - e^{-t/\tau}) \tag{9.8-1}$$

This expression is valid *only* while the pulse is on, that is, for $0 < t < T$.

At time T the capacitor is charged to a voltage

$$v_0(T) = V_1(1 - e^{-T/\tau}) \tag{9.8-2}$$

If $T > 5\tau$ the exponential factor will be less than 0.01 and

$$v_0(t) \approx V_1 \tag{9.8-3}$$

Thus the voltage across the capacitor just *before* the pulse turns off is V_1 volts. After the pulse has turned off, the circuit shown in Fig. 9.25b for $t > T$ holds. The capacitor voltage immediately after the pulse has turned off will be V_1 volts since it cannot change instantaneously, and this is the initial condition for the circuit for $t > T$. Since there is no source in this circuit the capacitor will discharge through the resistor and its voltage will decay according to the relation

$$v_C(t) = v_C(T)e^{-(t-T)/\tau} \tag{9.8-4}$$

This is valid *only* for $t > T$. Note that the exponent is in terms of a "shifted" time $(t - T)$ since the time origin for this part of the calculation is at $t = T$. When $t = T$

$$v_C(t) = v_C(T)e^{-(T-T)/\tau} = v_C(T)e^0 = v_C(T)$$

as we would expect.

Equations (9.8-1) and (9.8-4) are plotted in Fig. 9.25c and it is evident that the output waveshape differs from the input waveshape due to the curvature of both the leading and trailing edges. In a digital system we are usually concerned with the *speed* with which $v_0(t)$ rises toward its final value. We know that the rising exponential reaches 63% of its final value after one time constant has elapsed, so that the time constant can be used as a measure of the speed of response of this circuit.

The criterion more often used for such circuits is the time it takes for $v_0(t)$ to rise from 10 to 90% of its final value. This is called the *rise time* t_r, and we can find it as follows.

If we call t_{10} the time at which $V_0 = 0.1V_1$, then from Eq. (9.8-1)

$$v_0(t_{10}) = 0.1V_1 = V_1(1 - e^{-t_{10}/\tau})$$

Solving this yields

$$e^{-t_{10}/\tau} = 0.9$$

Using the [LNX] key on a scientific calculator, ln 0.9 ≈ −0.11. Thus

$$\frac{-t_{10}}{\tau} = -0.11 \qquad \text{and} \qquad t_{10} = 0.11\tau$$

If t_{90} is the time at which $v_0 = 0.9V_1$ the steps above yield

$$e^{-t_{90}/\tau} = 0.1$$
$$t_{90} = 2.3\tau$$

From the definition of rise time (see Fig. 9.25c),

$$t_r = t_{90} - t_{10} = 2.3\tau - 0.11\tau \approx 2.2\tau \tag{9.8-5}$$

Finally, for the *RC* circuit,

$$t_r = 2.2RC \tag{9.8-6}$$

This relates the 10 to 90% rise time to the element values of the circuit.

EXAMPLE 9.8-1 Rise Time

The circuit of Fig. 9.26a is a model for the input to one of the amplifiers in a video monitor. The 20 pF represents parasitic capacitance across the input terminals. Find the rise time of the circuit and sketch the waveform applied to the amplifier if V_1 = 10 mV and T = 10 μs.

Solution

We begin by finding the Thevenin circuit as seen from the terminals of the capacitor. The Thevenin resistance is

$$R_T = 68\ \text{k}\Omega \parallel 100\ \text{k}\Omega = 40\ \text{k}\Omega$$

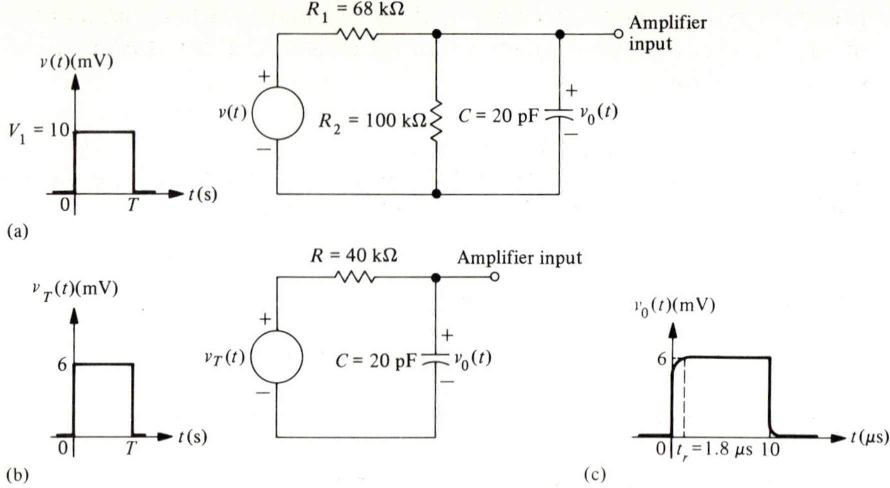

FIGURE 9.26
Example 9.8-1. (a) Circuit. (b) Thevenin equivalent. (c) Waveform at amplifier input.

and the Thevenin voltage is

$$V_T = \frac{100 \text{ k}\Omega}{100 \text{ k}\Omega + 68 \text{ k}\Omega} \times 10 \text{ mV} = 6 \text{ mV}$$

This is shown in Fig. 9.26b, and since it is identical to Fig. 9.25a we have for the rise time,

$$\begin{aligned} t_r = 2.2RC &= 2.2(40 \times 10^3 \ \Omega)(20 \times 10^{-12} \text{ F}) \\ &= 1.8 \ \mu\text{s} \end{aligned}$$

The pulse waveform at the amplifier input is shown in Fig. 9.26c. In an actual amplifier, there would be additional circuits in the signal path so that additional distortion would be introduced before the signal reached the output.

• • •

LEARNING EXERCISE FOR SEC. 9.8

1. In the circuit of Fig. 9.25a, R = 47 kΩ, C = 0.22 μF, and V_1 = 20 mV. Find the output voltage at t_{10} and t_{90} and the rise time.

Ans. 18; 2; 23

• • •

9.9 THE EFFECT OF NONZERO INITIAL CONDITIONS

When the capacitor in a transient circuit has acquired a charge before the switching has taken place or the step has been applied, care must be exercised in applying the general solution. The same is true for an inductor in which current flows before the switching instant. This is illustrated in the examples that follow.

EXAMPLE 9.9-1 A Capacitive Circuit

In the circuit of Fig. 9.25a, $R = 100\ \text{k}\Omega$, $C = 0.005\ \mu\text{F}$, $V_1 = 10$ V, $T = 0.75$ ms, and the capacitor is initially uncharged. Find and graph $v_0(t)$ and $i(t)$.

Solution

We proceed as in Sec. 9.8. For the circuit

$$\tau = RC = 100 \times 10^3 \times 0.005 \times 10^{-6} = 0.5 \times 10^{-3}\ \text{s} = 0.5\ \text{ms}$$

While the pulse is on, the capacitor charges according to Eq. (9.8-1)

$$v_0(t) = V_1(1 - e^{-t/\tau}) = 10(1 - e^{-t/0.5})\ \text{V}$$

where t is in milliseconds.

When $t = T = 0.75$ ms the input pulse goes to zero and the capacitor has charged to [see Eq. 9.8-2)],

$$v_0(0.75) = 10(1 - e^{-0.75/0.5}) = 10(1 - e^{-1.5}) = 7.77\ \text{V}$$

This is the final voltage for the charging phase, which becomes the initial condition for the discharge phase

$$v_0(T) = 7.77\ \text{V}$$

Observe that the capacitor has not become fully charged when the input pulse goes to zero.

The equation that describes the discharge is Eq. (9.8-4)

$$v_0(t) = 7.77e^{-(t-0.75)/0.5}\ \text{V}$$

The results are plotted in Fig. 9.27a. For the discharge phase, the voltage after one time constant is 37% of the initial value. Thus the voltage at $t = 0.75 + 0.5 = 1.25$ ms is $7.77 \times 0.37 = 2.87$ V.

The current can be found by several methods. We will use the general equation for exponential response. The student should check our results using the voltages found above to find the voltage across the resistor at $t = 0.25$, 0.5, 0.75, and 1.25 ms. Ohm's law is then used to find the current.

The general equation for the current during the time the pulse is on is

$$i(t) = i(0^+)e^{-t/\tau}$$

where

$$i(0^+) = \frac{V_1}{R} = \frac{10}{10^5} = 100\ \mu\text{A}$$

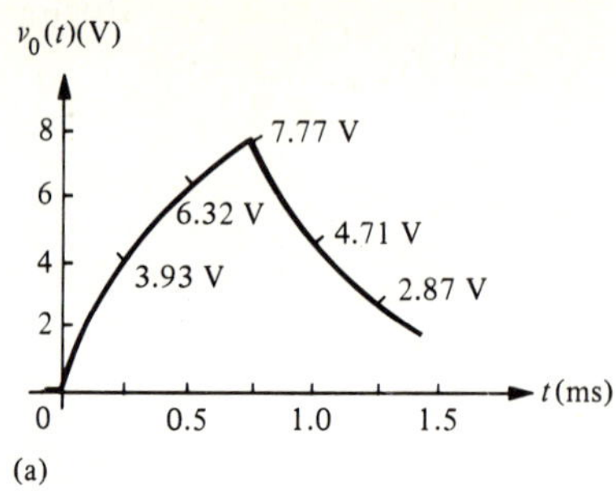

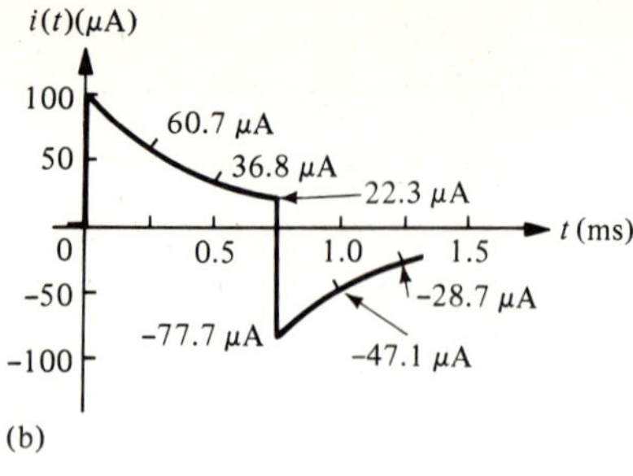

FIGURE 9.27
Example 9.9-1. (a) Output voltage waveform. (b) Current waveform.

Thus

$$i(t) = 100e^{-t/0.5}\ \mu\text{A}$$

where again t is in milliseconds.

An instant before the pulse goes off ($t = 0.75^-$) we have

$$i(0.75^-) = 100e^{-0.75/0.5} = 22.3\ \mu\text{A}$$

When the pulse goes off ($t = 0.75^+$) the input voltage is zero and the capacitor voltage is 7.77 V (see the circuit for $t > T$ in Fig. 9.25b). The voltage across the resistor is then -7.77 V, so that the current is

$$i(0.75^+) = \frac{0 - 7.77}{R} = -77.7\ \mu\text{A}$$

From $t = 0.75^+$ on, the current will decay exponentially to zero according to

$$i(t) = -77.7e^{-(t-0.75)/0.5}\ \mu\text{A}$$

The current is plotted in Fig. 9.27b.

• • •

EXAMPLE 9.9-2 *An Inductive Circuit*

In the circuit of Fig. 9.28a the switch is in position A for a long time. At $t = 0$ it is moved to position B. Find and plot $i_L(t)$ for $t > 0$.

Solution

The general response equation for $i_L(t)$ is

$$i_L(t) = i_L(\infty) + [i_L(0^+) - i_L(\infty)]e^{-t/\tau}$$

For the circuit when $t > 0$, $R_T = (R_1 + R_3) \parallel R_2$, and

$$\tau = \frac{L}{R_T} = \frac{10 \times 10^{-3}}{[(2.2 + 2.2) \parallel 4.7] \times 10^3} = \frac{10 \times 10^{-3}}{2.27 \times 10^3} = 4.4 \times 10^{-6}\ \text{s} = 4.4\ \mu\text{s}$$

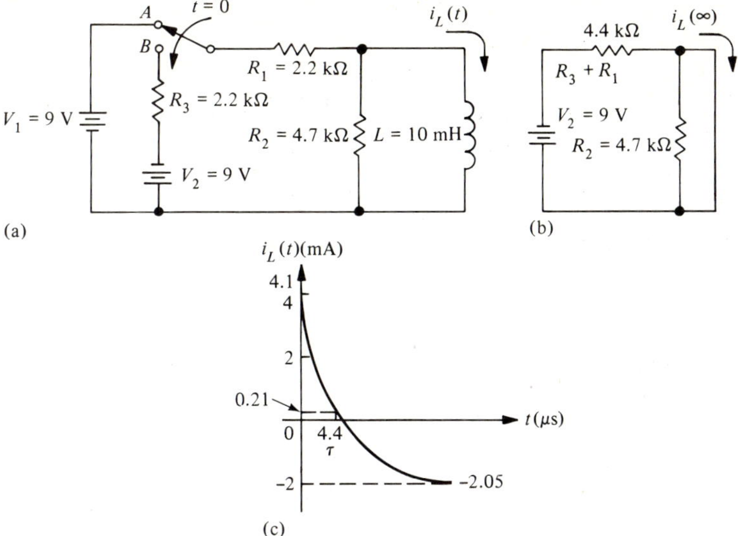

FIGURE 9.28
Example 9.9-2. (a) Circuit. (b) Circuit for $t \rightarrow \infty$. (c) Response.

Just before the switch is thrown the current through the coil will be constant and the coil will appear as a short circuit because the switch has been in position A for a long time and all transients will have disappeared. This current is

$$i_L(0^-) = \frac{V_1}{R_1} = \frac{9 \text{ V}}{2.2 \text{ k}\Omega} = 4.1 \text{ mA}$$

The continuity condition for inductance requires that this current not change instantaneously. Thus, the current immediately after the switching is $i_L(0^+) = 4.1$ mA. This is our initial condition.

After the switch has been in position B for a long time the transient will have disappeared and the circuit for the calculation of $i_L(\infty)$ will be as shown in Fig. 9.28b. From this circuit

$$i_L(\infty) = \frac{-9 \text{ V}}{4.4 \text{ k}\Omega} = -2.05 \text{ mA}$$

The response is then, for $t < 0$,

$$i_L(t) = 4.1 \text{ mA}$$

and for $t > 0$,

$$\begin{aligned} i_L(t) &= -2.05 + [4.1 + 2.05]e^{-t/4.4} \\ &= -2.05 + 6.15e^{-t/4.4} \text{ mA} \end{aligned}$$

with t in microseconds. These results are plotted in Fig. 9.28c.

• • •

LEARNING EXERCISE FOR SEC. 9.9

1. In the graph of Fig. 9.27b find the value of the current when t = 0.1, 0.6, 0.8, and 1.5 ms.

Ans. 81.9; −70.3; −17.3; 30.1

• • •

9.10 SYSTEM TIME CONSTANT IDENTIFICATION FROM EXPERIMENTAL DATA

It is sometimes necessary to determine an unknown system time constant from experimental observation of the system transient response. This can be done by several different methods. Perhaps the easiest way to get a ballpark estimate of an unknown time constant is to apply pulses or a square wave to the system under study and to observe the response waveform on an oscilloscope. The pulse width or square wave frequency is varied until a pattern similar to that shown in Fig. 9.29a is obtained, where it

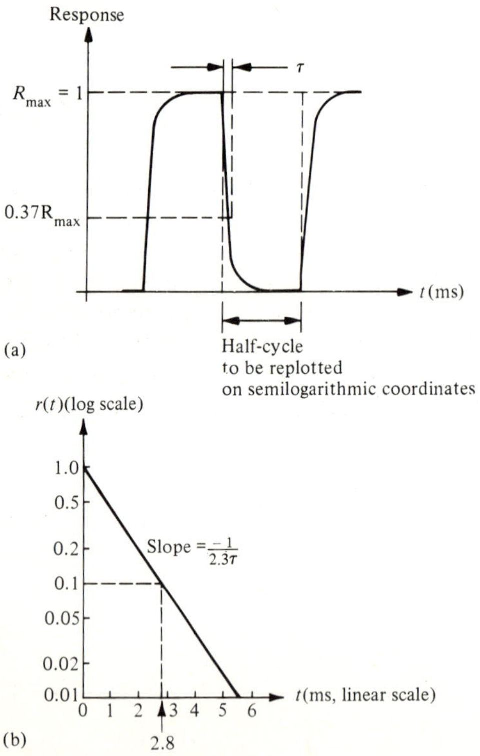

FIGURE 9.29
Experimental determination of time constant. (a) Square-wave response. (b) Semilog plot of one-half cycle.

is clear that the transient is completed within a half-cycle. The calibrated amplitude scale on the oscilloscope is then used to determine the point where the transient has risen (or decayed) 63% of the difference between its final and initial values. The time up to this point is one time constant, which can be read from the oscilloscope time axis calibration as indicated on the figure. The accuracy of this process is about 5% at best.

If more accuracy is desired, a plot on semilog paper may be used. This method has the advantage of averaging out errors made in individual readings.

Consider an exponential decay normalized to unity amplitude

$$r(t) = e^{-t/\tau} \tag{9.10-1}$$

This is plotted on linear scales in Fig. 9.9. Taking the natural log of both sides,

$$\ln r(t) = \frac{-t}{\tau} \ln e = \frac{-t}{\tau} \tag{9.10-2}$$

Since $\ln x = 2.3 \log x$ we have

$$\log r(t) = -\frac{t}{2.3\tau} \tag{9.10-3}$$

If $r(t)$ is plotted versus t on semilog paper as shown in Fig. 9.29b, the result is a straight line, which has a slope $-1/2.3\tau$.

If we have an experimental curve of $r(t)$ vs. t such as a tracing of an oscilloscope pattern and we replot the curve on semilog paper, we should be able to fit a straight line to the points if the system being tested can be adequately described by a single time constant. Once we have fitted a straight line to the experimental points the easiest way to find τ is to note times t_2 and t_1 at which $r(t_2)$ and $r(t_1)$ are a decade apart, that is, $r(t_2) = 10r(t_1)$. Then, using Eq. (9.10-3),

$$\log r(t_2) = -\frac{t_2}{2.3\tau} \tag{9.10-4}$$

and

$$\log r(t_1) = -\frac{t_1}{2.3\tau} \tag{9.10-5}$$

Next we subtract Eq. (9.10-5) from Eq. (9.10-4) to get

$$\log r(t_2) - \log r(t_1) = \frac{1}{2.3\tau}(t_1 - t_2)$$

Since $\log x - \log y = \log (x/y)$, this becomes

$$\log \frac{r(t_2)}{r(t_1)} = \frac{1}{2.3\tau}(t_1 - t_2)$$

But

$$\frac{r(t_2)}{r(t_1)} = 10$$

Thus

$$\log 10 = 1 = \frac{1}{2.3\tau}(t_1 - t_2)$$

and

$$\tau = \frac{t_1 - t_2}{2.3} \tag{9.10-6}$$

It is usually convenient to take $t_2 = 0$ so that finally

$$\tau = \frac{t_1}{2.3} \tag{9.10-7}$$

where t_1 is the time at which the response is one-tenth of its value at $t = 0$. For the graph of Fig. 9.29b, $t_1 = 2.8$ ms so that $\tau = 2.8/2.3 = 1.22$ ms.

It is important to note that the amplitude of the exponential does not affect the *slope* of the semilog plot. If the amplitude is other than 1, the straight line will be shifted up or down but the slope will not be affected.

This technique can also be applied to rising exponentials. In this case it is convenient to subtract the final value before plotting on semilog paper. Additional examples will be found in the problems section.

9.11 TRANSIENT RESPONSE OF COMPLEX NETWORKS

When a network contains more than one capacitor or one inductor, the transient solution becomes much more complicated and the mathematics involved is beyond the scope of this book. However, as an illustration we will describe the response of the circuit of Fig. 9.30a qualitatively. This circuit contains one of each type of element: R, L, and C. The capacitor is charged to a voltage V_0 at time $t = 0$. The response of interest is the current for $t > 0$. The *form* of this response depends on the ratio $(R/2)\sqrt{(C/L)}$. If the resistance is chosen so that this ratio is greater than one, the response is said to be *overdamped* and will appear as shown in Fig. 9.30b. If the ratio is less than one, the circuit is said to be *underdamped,* or oscillatory, and the

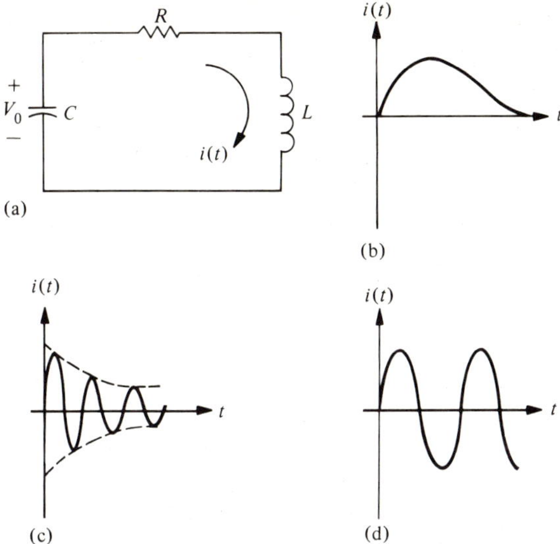

FIGURE 9.30
Transient network. (a) Circuit. (b) Overdamped response. (c) Underdamped response. (d) Undamped response.

response is a damped sine wave, as shown in Fig. 9.30c. As the resistance is further reduced toward zero, the response approaches a pure sine wave, as shown in Fig. 9.30d, and the circuit is said to be *undamped.* The oscillations occur because energy is being interchanged between the electric field of the capacitor and the magnetic field of the inductor. These oscillations are eventually reduced to zero (unless $R = 0$) because some energy will be dissipated in the resistance during each cycle. In transistor oscillators, the transistors supply the lost energy, thus producing sustained oscillations. This type of circuit is considered further in Chap. 13 in terms of frequency response.

SUMMARY

In this chapter we studied charging and discharging of capacitors and inductors. Circuits that make use of these responses are an integral part of many electric and electronic devices, ranging all the way from the flashing lights at a model railroad crossing to large-scale digital computers.

Section

9.1 1. Equations for circuits containing R and C or R and L elements involve rates of change of voltage or current. Initial conditions are required in order to complete the solution.

9.2 2. The step-by-step solution is completed by using a recursion formula derived from the original equation.

3. When a capacitor and resistor in series are switched across a battery, the voltage across the capacitor rises slowly, the speed depending on the *RC* product.

9.3 4. The equation that describes the charging of a capacitor is $v_C(t) = v_C(\infty)(1 - e^{-t/\tau})$.

5. The equation that describes the discharge of a capacitor is $v_C(t) = v_C(0^+)e^{-t/\tau}$.

9.4 6. Exponentials are found using the $\boxed{e^x}$ or the $\boxed{\text{LNX}}$ keys on a scientific calculator.

7. An exponential discharge decays to 37% of its initial value in one time constant, and less than 1% in five time constants.

8. An exponential charge increases to 63% of its final value in one time constant and more than 99% in five time constants.

9.5 9. The equations for the buildup and decay of current in an inductance are identical to the capacitor charge and discharge equations. The time constant for an inductive circuit is L/R.

9.6 10. The general solution of the equation for a circuit containing one capacitor or one inductor and any number of resistors is $r(t) = r(\infty) + [r(0^+) - r(\infty)]e^{-t/\tau}$. $r(0^+)$ is the initial value and $r(\infty)$ is the final value. The time constant for *RC* circuits is R_TC and for *RL* circuits it is L/R_T.

11. To find initial values, the circuit condition just before the switching is found, and then the appropriate continuity condition is applied.

12. To find final values remember that as $t \rightarrow \infty$ in this type of circuit, inductors "look like" short circuits, while capacitors "look like" open circuits.

9.7 13. The total solution goes from the initial value to the final value. The natural response provides the transition.

9.8 14. The pulse response of an *RC* circuit is found by applying the general solution to find the initial rise and the values just before the pulse goes off as initial conditions to find the fall of the output.

15. The rise time of an *RC* circuit is $t_r = 2.2R_TC$, where R_T is the resistance seen by the capacitor with all sources set to zero.

9.10 16. An unknown time constant is found from a straight-line plot of the exponential decay on semilog paper using the equation $\tau = t_1/2.3$ where t_1 is the time at which the response is one-tenth of its value at $t = 0$.

QUESTIONS FOR REVIEW

Sec. 9.1

1. Explain why the voltage across a capacitor cannot change instantaneously.
2. Explain why the current through an inductance cannot change instantaneously.
3. Describe the behavior of capacitors and inductors in circuits with dc sources as $t \rightarrow \infty$.

Sec. 9.2

4. What is the meaning of the term "initial condition"?

Sec. 9.3

5. Describe the charging and discharging processes in an *RC* circuit.
6. State the formula for a typical exponential discharge and define each of the constants.
7. State the formula for a typical exponential charge and define each of the constants.

Sec. 9.4

8. Use your calculator to prove that $e^{-2.46} = 0.085$.
9. What is the significance of the term "amplitude"?
10. What is the significance of the time constant?
11. Describe the charge and discharge curves in terms of the values after 1 and 5 time constants.

Sec. 9.5

12. How is the time constant of an inductive circuit related to the element values?
13. Describe what happens when a dc voltage is switched across an *RL* circuit.

Sec. 9.6

14. State the general solution that applies to both charge and discharge.
15. Define $r(\infty)$, $r(0^-)$, and $r(0^+)$.
16. How is the time constant determined in circuits where more than one resistance is present?
17. Describe how the value of $r(0^+)$ is found, given the value of $r(0^-)$.

Sec. 9.7

18. Define forced response, natural response, and total response.
19. How is the natural response related to the initial condition and the final value?

Sec. 9.8

20. Define rise time.
21. State the formula for rise time of an *RC* circuit.
22. What is the effect of the finite rise time on the pulse response of an *RC* circuit?

Sec. 9.9

23. Describe how nonzero initial conditions are taken into account when solving *RC* and *RL* networks.

Sec. 9.10

24. Describe how the semilog plot of an exponential can be used to find the time constant.

PROBLEMS

Sec. 9.1

1. Formulate the equation for $v_C(t)$ in the circuit of Fig. 9.31.
2. Formulate the equation for $i_L(t)$ in the circuit of Fig. 9.32.

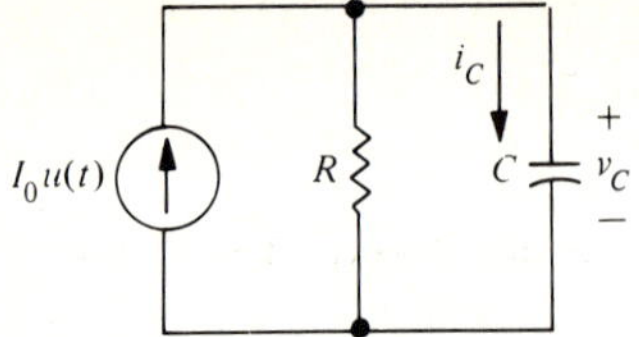

FIGURE 9.31

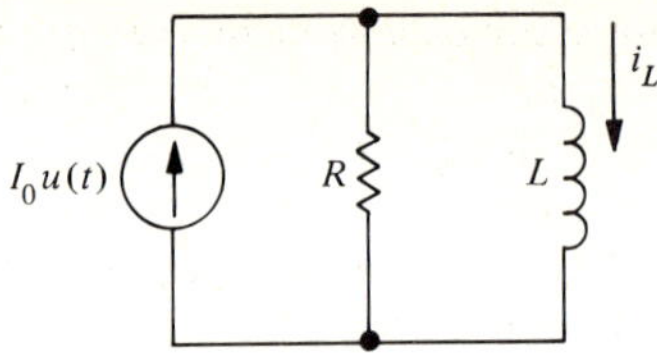

FIGURE 9.32

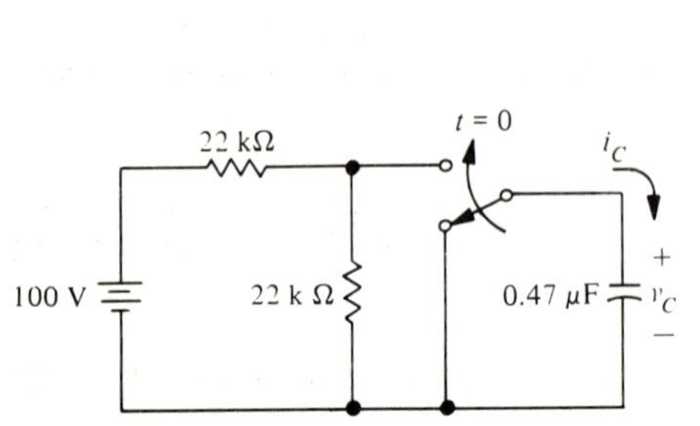

FIGURE 9.33

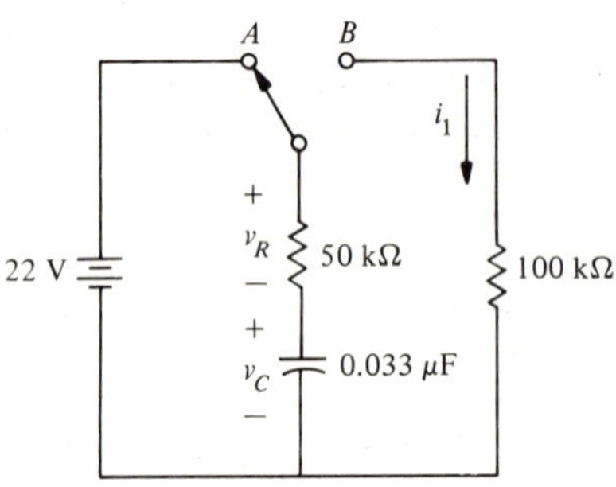

FIGURE 9.34

3. In Fig. 9.31, $I_0 = 2$ mA, $R = 4.7$ kΩ, $C = 0.1$ μF, and $v_C(0^-) = 0$. Find all initial and final conditions.
4. In Fig. 9.32, $I_0 = 2$ mA, $R = 2.2$ kΩ, $L = 10$ mH, and $i_L(0^-) = -3$ mA. Find all initial and final conditions.

Sec. 9.2

5. For the values in Prob. 3 solve the equation using the step-by-step method. Choose $\Delta t = \tau/5$ and plot your results up to $t = 5\tau$.

Sec. 9.3

6. Find and sketch $v_C(t)$ and $i_C(t)$ in the circuit of Fig. 9.33.
7. In Prob. 6 find the time for the capacitor voltage to reach 25 V.
8. In the circuit of Fig. 9.34 the switch is at A for a long time. At $t = 0$ it is moved to B. Find and plot $v_C(t)$ for $t > 0$.
9. In the circuit of Fig. 9.34, find and plot i_1 and v_R for $t > 0$.

Sec. 9.4

10. Use a calculator to find $e^{0.1}$, $e^{1.3}$, $e^{-0.12}$, $e^{-2.4}$, $e^{-7.6}$.
11. A certain voltage is described by the equation $v(t) = 30e^{-t/(20\times10^{-6})}$ V. What are the amplitude and time constant? Neatly sketch the voltage using graph paper.
12. An exponentially decaying current has an initial value (amplitude at $t = 0$) of 8 mA and a time constant of 10 μs. Using graph paper, neatly sketch the waveform.
13. An exponentially decaying voltage has a time constant of 5 ms. When $t = 5$ ms, $v(t) = 8$ V. Find the amplitude of the voltage at $t = 0$ and write the equation.
14. An exponentially decaying current has an initial value of 10 mA. At $t = 3$ ms, $i(t) = 1$ mA. Find the time constant and write the equation for $i(t)$.

15. An exponentially rising voltage is described by

$$v(t) = 10(1 - e^{-t/3\times 10^{-3}})\text{ V}.$$

What are its initial value, final value, and time constant? Sketch the waveform on graph paper.

16. A certain current is described by $i(t) = 3 - 7e^{-t/2}$ A. Is this a rise or a decay? Sketch the waveform on graph paper.
17. The equation of a voltage is $v(t) = 10 + 15e^{-t/3}$ V. Is this a rise or a decay? Sketch the waveform on graph paper.
18. An exponentially rising current has an initial value of zero, a final value of 10 mA, and a value of 2 mA after 2 ms. Find the time constant and write the equation for $i(t)$.

Sec. 9.5

19. In the circuit of Fig. 9.35, the switch is moved to position 1 at $t = 0$ after having been in position 2 for a long time. Find and plot i_L for $t > 0$.
20. Repeat Prob. 19 for v_L and v_R.
21. In the circuit of Fig. 9.36 the switch is opened at $t = 0$. Find and sketch i_L for $t > 0$.
22. Repeat Prob. 21 for v_L and i_R.
23. In the circuit of Fig. 9.35 the switch is moved to position 2 at $t = 0$ after having been in position 1 for a long time. Find and plot i_L for $t > 0$.
24. Repeat Prob. 23 for v_L and v_R.

Sec. 9.6

25. In the circuit of Fig. 9.37 the switch is closed at $t = 0$. The capacitor voltage at $t = 0^-$ is 2 V. Use the general relation to find and sketch v_C for $t > 0$.
26. Repeat Prob. 25 for i_1.
27. In the circuit of Fig. 9.38 the switch is closed at $t = 0$. Use the general relation to find and sketch i_L for $t > 0$.

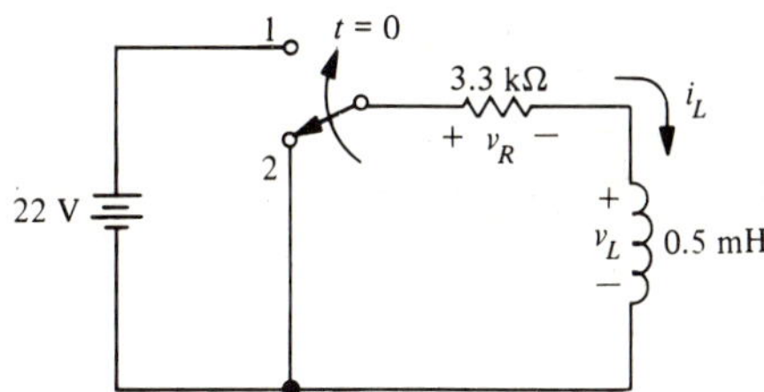

FIGURE 9.35

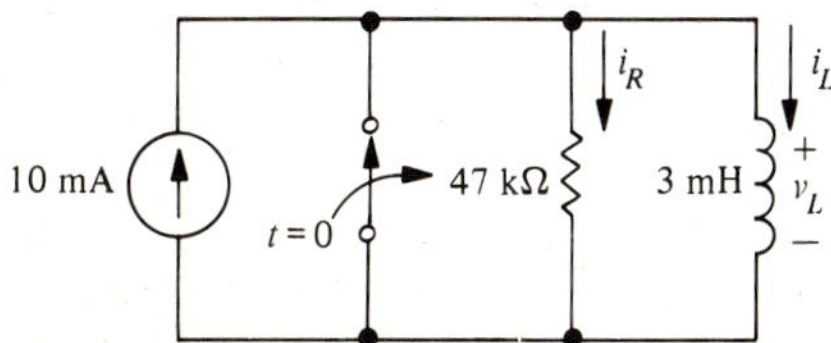

FIGURE 9.36

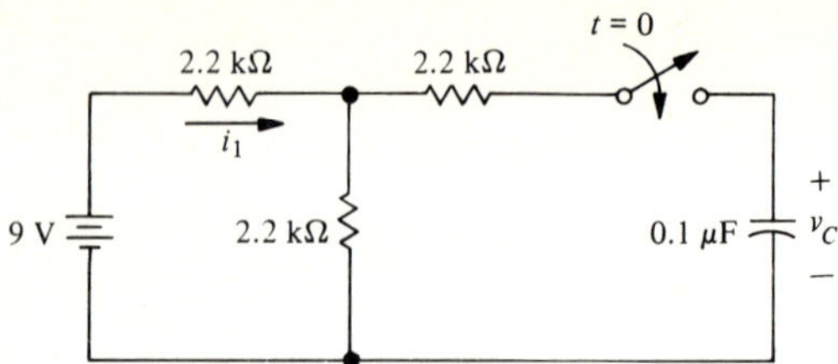

FIGURE 9.37

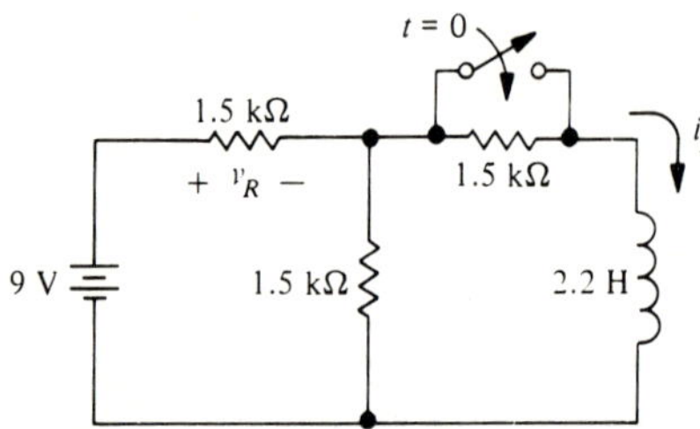

FIGURE 9.38

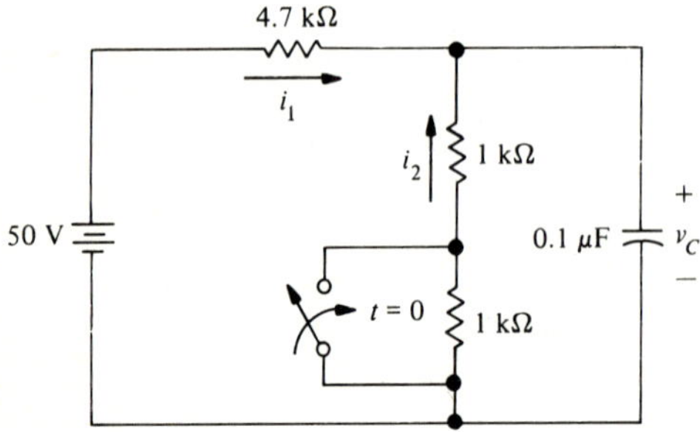

FIGURE 9.39

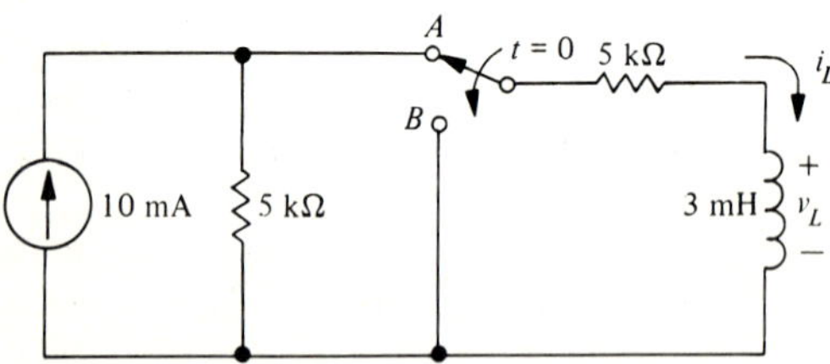

FIGURE 9.40

28. Repeat Prob. 27 for v_R.
29. In Example 9.6-4 the capacitor must charge to 28 V before the ready light is actuated. Find the time that the photographer must wait between flashes.
30. In the circuit of Fig. 9.39 the switch is closed at $t = 0$ after having been open for a long time. Find and plot v_C for $t > 0$.
31. Repeat Prob. 30 for i_1 and i_2.

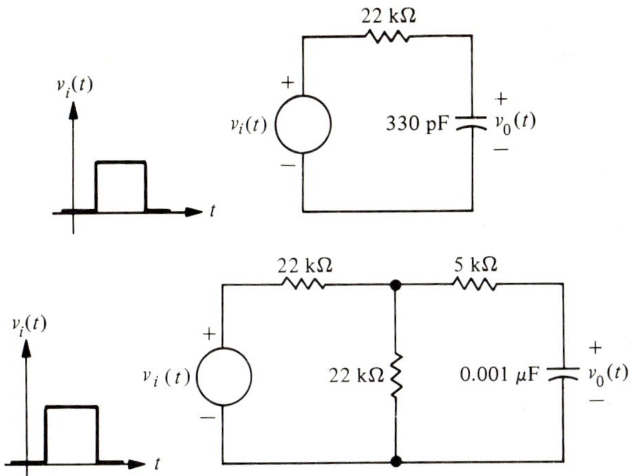

FIGURE 9.41

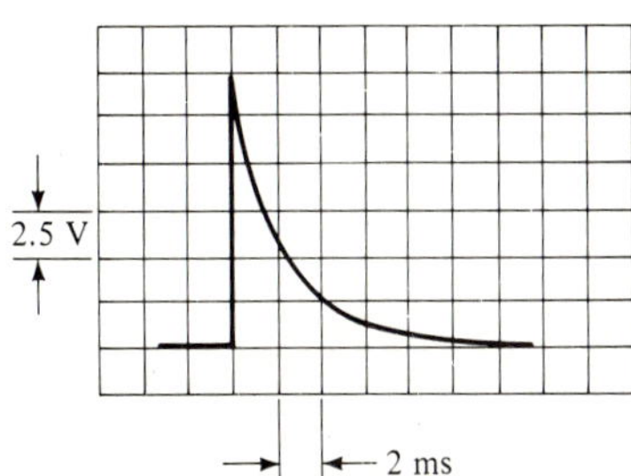

FIGURE 9.42

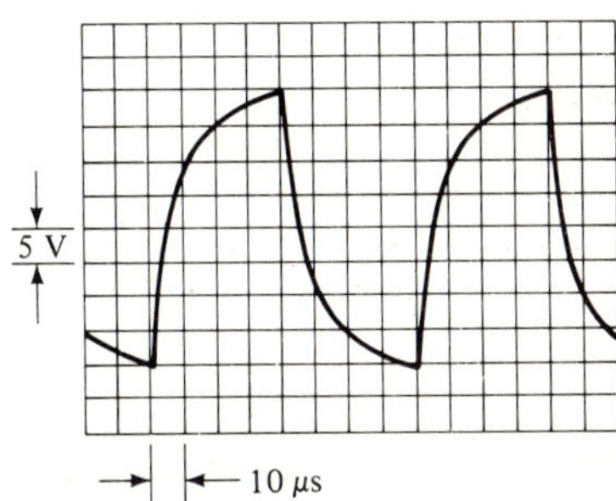

FIGURE 9.43

Sec. 9.8

32. In the circuit of Fig. 9.25a $T = 1$ ms, $V_1 = 5$ V, $R = 1.5$ kΩ, $C = 250$ pF. Find the rise time and sketch the output voltage waveform. Repeat if $R = 1.5$ MΩ, 15 MΩ.
33. In the circuit of Fig. 9.34 the switch is at A for a long time when it is moved to B at $t = 0$. At $t = 5$ ms it is moved back to A. Find and plot $v_C(t)$ for $0 < t < 30$ ms.
34. In the circuit of Fig. 9.31 the capacitor is charged to -10 V prior to $t = 0$. If $I_0 = 2$ mA, $R = 5$ kΩ, and $C = 0.01$ μF, find and plot $v_C(t)$ and $i_C(t)$.
35. In the circuit of Fig. 9.40 the switch goes from A to B at $t = 0$. At $t = 0.3$ μs it goes back to A and at $t = 0.6$ μs it goes back to B where it remains. Find and plot $v_L(t)$ and $i_L(t)$ for $0 < t < 10$ μs.
36. Find the rise time of the circuits of Fig. 9.41.

Sec. 9.9

37. Repeat Example 9.9-1 if the capacitor has an initial charge of 5 V.

Sec. 9.10

38. Replot the oscilloscope trace of Fig. 9.42 on semilog paper and determine the time constant.
39. Repeat Prob. 38 for the trace of Fig. 9.43.

10 Sinusoidal Signals and Phasors

OBJECTIVES

Upon completion of this chapter, the student should be able to

Section

Section		
10.1	1.	Write the equation for the general form of a sinusoid.
	2.	Define the amplitude, period, frequency, and phase angle.
	3.	Sketch the graph of a typical sinusoid.
	4.	Convert angles from degrees to radians and vice versa.
10.2	5.	Define the *j* operator and describe the complex plane.
	6.	State that the coordinate of a point along the real axis in the complex plane is the real part of the complex number and the coordinate of the point along the y axis is the imaginary part of the complex number.
	7.	Plot complex numbers given in either polar or rectangular form on the complex plane.
10.3	8.	Convert a given complex number from polar to rectangular form and vice versa.
	9.	Add, subtract, multiply, and divide complex numbers.
	10.	Define the complex conjugate of a given complex number.
10.4	11.	Find the phasor representation of a given sinusoidal signal using the effective value.
	12.	Show the correspondence between the phasor domain and time domain representations.
	13.	Convert from the phasor domain to the time domain and vice versa.
	14.	Add together two phasors using complex addition.
10.5	15.	Calculate phasor currents and voltages in resistive circuits.
10.6	16.	State that the formulas for average ac power in a resistance are

identical to the formulas for dc power when effective values are used.

17. Define effective value.

INTRODUCTION

In this chapter we are going to consider the *sinusoid*, the most basic ac waveform. As noted previously, the letters ac are an abbreviation for *alternating current* and this abbreviation is loosely used to describe currents and voltages that *alternate* between positive and negative values. Sinusoidal voltages are easily generated by rotating a coil of wire in a uniform magnetic field, as shown in Fig. 10-1a. As the coil rotates, the induced voltage changes according to the sine of the angle that the coil makes with the direction of the field. A given point on the coil first cuts the lines of force in one direction, and one-half revolution later cuts the same lines of force in the opposite direction. This causes the voltage to reverse polarity for one-half of each revolution as shown in the load voltage waveform in Fig. 10-1b. In order to transfer the voltage to the load, the ends of the rotating coil are connected to *slip rings*. The external circuit is connected to *brushes* (often made of carbon) held in contact with the slip rings by springs which are not shown in the diagram. Because of the ease of generation, the sinusoid is the waveform used in power systems and thus is the waveform of the ac voltage present at the electric outlets in our homes. It is also used extensively in radio and TV systems. For example, the signal transmitted by an AM radio station when no information is present is a pure sinusoid called a *carrier*. In order to transmit information, a property of the carrier is *modulated* by the information signal. When this is done the transmitted signal is no longer a pure sinusoid because of the modulation.

In Chap. 9 we considered the responses of RC and RL circuits to step-function inputs. The step is important because it is a standard signal for

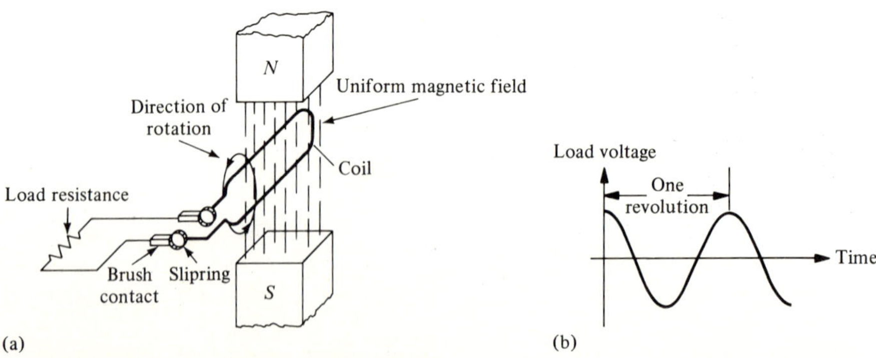

FIGURE 10.1
Generation of sinusoidal voltage. (a) Coil rotating in a magnetic field. (b) Waveform of voltage across load.

testing the *transient* response of a circuit, that is, the time response to an abrupt change in the input signal. The steady-state response to a sinusoidal signal is another very important characteristic of a network. It is important because it determines a network's frequency response, which we study in a later chapter, and because electric power is generated and transmitted in sinusoidal form. This means that all network computations in power distribution systems involve sinusoidal analysis.

The analysis of circuits in which the sources are sinusoidal is greatly simplified by the use of *phasors*. These are complex numbers that are manipulated using the rules of the complex number algebra to be introduced in this chapter.

10.1 CHARACTERISTICS OF SINE AND COSINE WAVES

The sinusoid that includes both the sine and cosine is the most basic of a class of signals that are called *periodic*. The property that all periodic waveforms share is that they repeat themselves every T seconds, where T is called the *period*. Some typical periodic waveforms are shown in Fig. 10.2. In mathematical terms, periodicity is expressed simply as

$$v(t) = v(t + T) \tag{10.1-1}$$

In words, this equation states that the value of the time-varying voltage v is exactly the same at time $t + T$ as it was at time t. The largest nonrepeating portion of a periodic waveform is called a *cycle*. According to Eq. (10.1-1), one cycle occurs every T seconds.

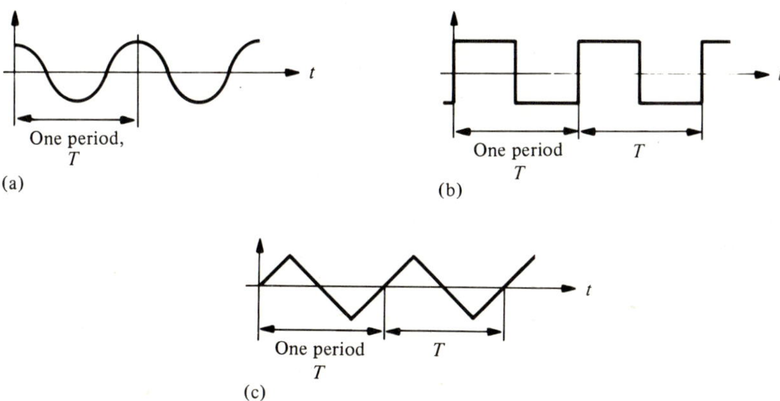

FIGURE 10.2
Periodic waveforms. (a) Cosine wave. (b) Square wave. (c) Triangular wave.

Cosine Waves

A graph of $\cos\alpha$ is shown in Fig. 10.3 with scales on the horizontal axis showing the angle α in both degrees and radians.* For those who may not be familiar with radian measure, a circle contains a total angle of 2π radians, which is therefore equal to 360°. Thus, given an angle in degrees, ϕ_{deg}, we can find the corresponding angle in radians, ϕ_{rad}, from the formula

$$\boxed{\phi_{rad} = \frac{\pi}{180} \times \phi_{deg}} \tag{10.1-2}$$

Conversely, given the angle in radians, we can convert to degrees using

$$\boxed{\phi_{deg} = \frac{180}{\pi} \times \phi_{rad}} \tag{10.1-3}$$

Also from the graph, one cycle, the largest nonrepeating portion of the cosine curve, occupies $360° = 2\pi$ rad on the α axis.

Let's find the angle in radians corresponding to some common angles in degrees.

180°:

$$\phi_{rad} = \frac{\pi}{180} \times 180 = \pi \text{ rad}$$

*A unit called the *gradient* (grad) is also being used for the measurement of angles and appears on many calculators. For this unit 100 grad = 90°.

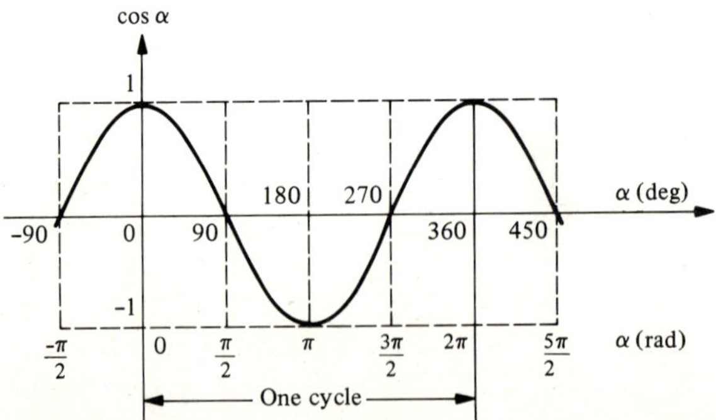

FIGURE 10.3
Graph of cos α.

90°:

$$\phi_{\text{rad}} = \frac{\pi}{180} \times 90 = \frac{\pi}{2} \text{ rad}$$

45°:

$$\phi_{\text{rad}} = \frac{\pi}{180} \times 45 = \frac{\pi}{4} \text{ rad}$$

30°:

$$\phi_{\text{rad}} = \frac{\pi}{180} \times 30 = \frac{\pi}{6} \text{ rad}$$

For some common angles given in radians:

1 rad:

$$\phi_{\text{deg}} = \frac{180}{\pi} \times 1 = 57.3°$$

$\pi/3$ rad:

$$\phi_{\text{deg}} = \frac{180}{\pi} \times \frac{\pi}{3} = 60°$$

$3\pi/2$ rad:

$$\phi_{\text{deg}} = \frac{180}{\pi} \times \frac{3\pi}{2} = 270°$$

Sinusoidal Voltage or Current

In the introduction to this chapter we described briefly how a sinusoidal voltage is generated by rotating a coil in a magnetic field. This generated voltage is shown in Fig. 10.4. In the figure we have chosen $t = 0$ at the point where the voltage is a maximum. This choice leads naturally to the description of the voltage by the cosine function rather than the sine function. The choice of cosine rather than sine is dictated by the fact that the cosine is the base for the phasor method which will be introduced later.

The voltage in Fig. 10.4 is represented by the equation

$$v(t) = |V_m| \cos \omega t \tag{10.1-4}$$

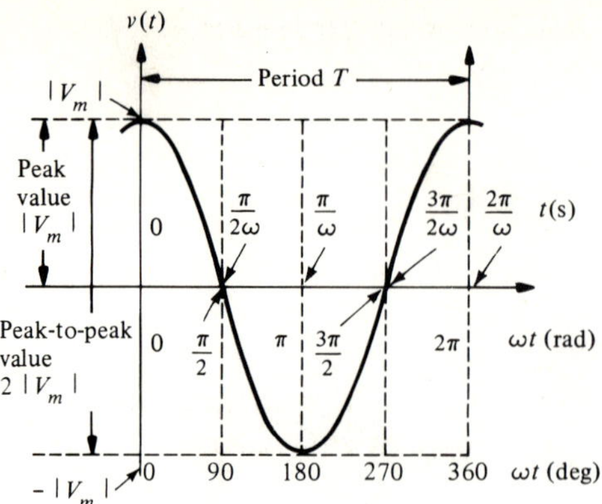

FIGURE 10.4
Basic cosine waveform.

$v(t)$ is the *instantaneous* value of the voltage v at time t. It is seen to oscillate between the maximum positive value $|V_m|$ and the maximum negative value $-|V_m|$. The two parameters that completely define $v(t)$ are:

1. The *maximum*, or *peak*, value $|V_m|$. Since the cosine function has a maximum value of 1, $|V_m|$ is the maximum value that $v(t)$ can attain. The symbolism $|V_m|$ means "the absolute value of V_m" or "the magnitude of V_m" and we use it at this point in order to be consistent with the complex phasors to be introduced later. On the graph of Fig. 10.4, the maximum positive value $|V_m|$ is attained at points on the radian ωt scale where $\omega t = 0, 2\pi, 4\pi, 6\pi \ldots$ rad and the maximum negative value, $-|V_m|$, at points where $\omega t = \pi, 3\pi, 5\pi, \ldots$ rad. Sometimes the peak-to-peak value of the waveform is required. For the cosine wave this is seen from Fig. 10.4 to be $2|V_m|$.
2. The *angular frequency* ω radians per second (rad/s). The cosine wave goes through a complete cycle when its angle goes through 360° or 2π rad. Since one cycle takes place in T seconds, the angle of the cosine changes at the rate of 2π radians every T seconds. Therefore the *angular frequency*, ω, is

$$\boxed{\omega = \frac{2\pi}{T} \text{ rad/s}} \tag{10.1-5}$$

More often we use the *cyclic frequency* f, which is the number of cycles per second. Since the period T is the number of *seconds per cycle* its reciprocal will be the number of *cycles per second*. The unit for cycles per second is the hertz (Hz), so that

$$f = \frac{1}{T}\,\text{Hz} \tag{10.1-6}$$

Using Eq. (10.1-6) in Eq. (10.1-5) the angular frequency becomes

$$\omega = 2\pi f\ \text{rad/s} \tag{10.1-7}$$

This relates the angular and cyclic frequencies. On the graph of Fig. 10.4, the horizontal axis is scaled with ωt in radians and degrees, and time in seconds. The radian and degree scales are universal and apply to any cosine wave. The time scale requires a knowledge of the frequency before actual times can be added.

General Form for a Sinusoidal Voltage

As shown in Fig. 10.5, the $t = 0$ point does not always occur precisely at the maximum of the voltage wave. To take account of this possibility the general form for a cosine voltage waveform is

$$v(t) = |V_m| \cos(\omega t + \phi) \tag{10.1-8}$$

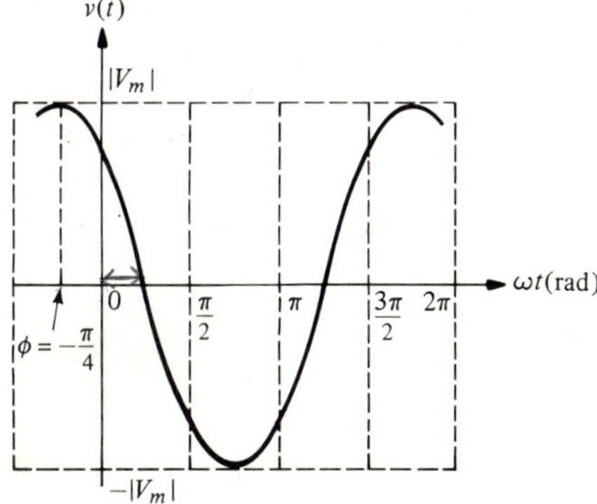

FIGURE 10.5
Cosine wave at an arbitrary phase angle ϕ.

The angle ϕ is called the *phase angle*. In Fig. 10.5 we see that the *phase angle* is the angle between the positive maximum closest to $t = 0$ and the point $t = 0$. When this maximum occurs to the left of $t = 0$, the angle takes the plus sign in the expression, otherwise it takes a minus sign. For the cosine wave in Fig. 10.5, where $\phi = -\pi/4$ rad $= -45°$, the expression would be $|V_m|\cos(\omega t + 45°)$.

The examples which follow illustrate some typical calculations involving sinusoids.

EXAMPLE 10.1-1 Sine Wave Calculations

For the voltage $v(t) = 150 \cos(377t - 72°)$ V, find

a. The amplitude
b. The angular frequency
c. The cyclic frequency
d. The period
e. The phase angle
f. Draw a graph of $v(t)$, showing all important parameters. The time scale is to be in seconds.

Solution

Comparing with the general form, Eq. (10.1-8), we have

a. $|V_m| = 150$ V

b. $\omega = 377$ rad/s

c. $f = \dfrac{\omega}{2\pi} = \dfrac{377}{6.28} = 60$ Hz

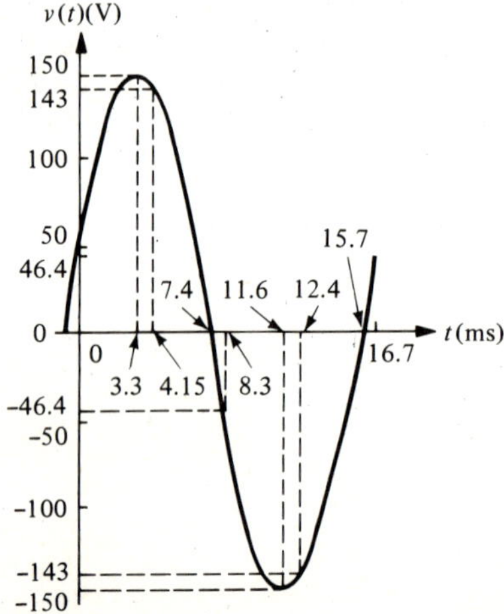

FIGURE 10.6
Graph for Example 10.1-1.

TABLE 10.1

ϕ_{deg}	$t = 46 \times 10^{-3}\,\phi_{deg}$ (ms)	$v(t) = 150 \cos(\phi_{deg} - 72°)$(V)
0	0	46.4
72	3.3	150
90	4.15	143
162	7.4	0
180	8.3	−46.4
252	11.6	−150
270	12.4	−143
342	15.7	0
360	16.7	46.4

d. $T = \frac{1}{f} = \frac{1}{60} = 0.0167 \text{ s} = 16.7 \text{ ms}$

e. $\phi = -72°$

f. The graph is shown in Fig. 10.6. In order to express the horizontal scale in units of time, we note that $\omega t = 377t$ is in units of (rad/s) × s = rad. To convert to degrees we use Eq. (10.1-3). This yields

$$\phi_{deg} = \frac{180}{\pi} \times 377t$$

Solving for t we get

$$t = \frac{\pi}{180 \times 377} \times \phi_{deg} = 46 \times 10^{-6}\,\phi_{deg} \text{ s}$$

In order to check this we note that the period is 16.7 ms = 0.0167 s. Thus if $\phi_{deg} = 360$ is substituted in the formula the result should be $t = 0.0167$ s, which it is. Values of t and $v(t)$ for various values of ϕ_{deg} are tabulated in Table 10.1 and shown on the graph. All calculations were done on a scientific calculator and students are urged to check several of the points on their own calculators.*

• • •

EXAMPLE 10.1-2 More Sine Wave Calculations

A voltage is described by the equation

$$v(t) = 60 \cos 377t \text{ V}$$

Find $v(t)$ when $t = 0.001$ s.

*For those without scientific calculators, tables of trigonometric functions are included in Appendix D.

Solution

$$\omega t = (377)(0.001) = 0.377 \text{ rad}$$

Most scientific calculators have a switch that allows the user to enter the angle in either degrees or radians. If we set the switch to radians, enter 0.377 and depress the cos x key, the display reads 0.93. Thus

$$\cos(0.377 \text{ rad}) = 0.93$$

Alternately, we can use Eq. (10.1-3) to convert radians to degrees, that is,

$$0.377 \times \frac{180}{\pi} = 21.6°$$

Finally,

$$v(0.001) = 60 \cos(0.377 \text{ rad}) = 60 \cos(21.6°) = 60(0.93) = 55.8 \text{ V}$$

A graph of $v(t)$ is shown in Fig. 10.7. The student will find that it is sometimes more convenient to use radians than degrees.

• • •

EXAMPLE 10.1-3 More Sine Wave Calculations

Find $v(t)$ at $t = 0.001$ s for the voltage of Example 10.1-2 if it has a phase angle of $+60°$.

Solution

With this phase angle, the equation of the voltage is

$$v(t) = 60 \cos(377t + 60°) \text{ V}$$

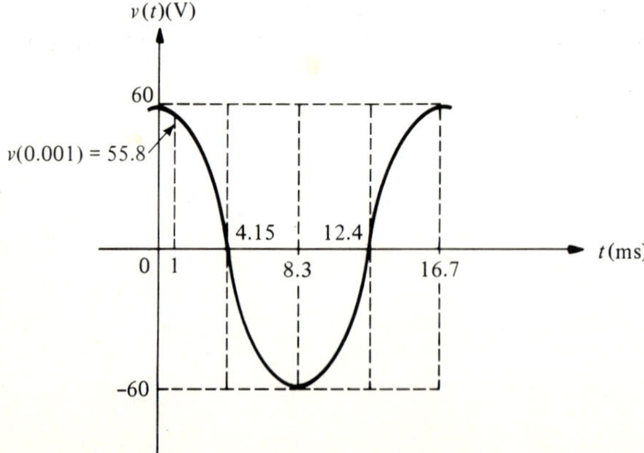

FIGURE 10.7
Graph for Example 10.1-2.

Note that the argument of the cosine has two terms, $377t$ having units of radians and 60° having units of degrees. It is very bad practice to mix units this way, but it is usually done in this case. Degrees are always written with the degree symbol (°). Before any calculations are made both terms must be converted to the same units.

Since we have, from Example 10.1-2, that $\omega t = 21.6°$, the argument of the cosine is $21.6° + 60° = 81.6°$. Then

$$v(0.001) = 60 \cos(81.6°) = 60(0.146) = 8.76 \text{ V}$$

A graph is shown in Fig. 10.8.

• • •

LEARNING EXERCISE FOR SEC. 10.1

1. The current through an antenna coil is

$$i(t) = 3 \cos(10^6 t + 22°)\ \mu\text{A}$$

Find

a. The amplitude
b. The angular frequency
c. The cyclic frequency
d. The period
e. The phase angle

Ans. 10^6; 3; 159 000; 22; 6.28

• • •

10.2 COMPLEX NUMBERS

Steady-state sinusoidal analysis requires extensive use of complex numbers. Since most technology students will not have studied this subject in their math courses we will devote this section and the next to a discussion of those topics required for the circuit theory to follow.

In order to define complex numbers we note first that the real numbers

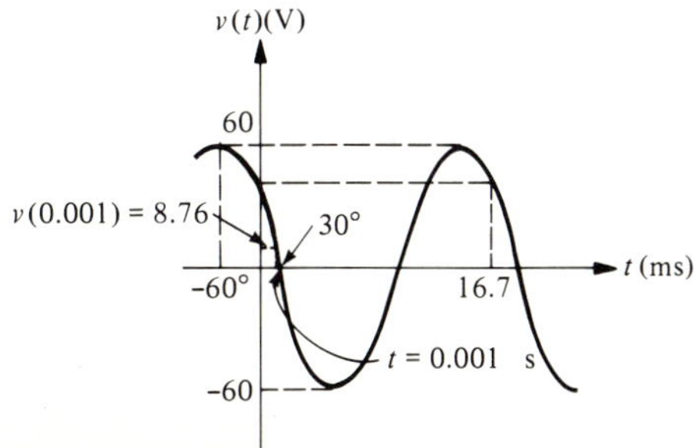

FIGURE 10.8
Graph for Example 10.1-3.

may be represented graphically by the points along a straight line as shown in Fig. 10.9a. A point on the line is chosen as the origin and is associated with the number zero. Every other point on the line is associated with a numerical value proportional to its distance from the origin; the value is positive if it is to the right of the origin and negative if it is to the left.

Consider the number $+3.2$. This can be represented by a line from the origin to the point $+3.2$ on the real axis shown on the figure. Next we ask the question, is it possible to perform an operation on the number $+3.2$ such that it will rotate 180° counterclockwise and become -3.2? The answer is yes; all we need to do is multiply by -1. This works for any number on the positive real axis as well as numbers on the negative real axis. Thus we may consider the number -1 as an operator that produces 180° counterclockwise rotation of any number it multiplies.

Now let us define an operator that produces 90° counterclockwise rotation of any number it multiplies. This operator will be symbolized by the letter j. In Fig. 10.9b the number $+3.2$ is shown along with the result obtained when it is multiplied by the j operator so that it becomes $j3.2$, which is at an angle of 90° to the original number.

Now let us operate on $j3.2$ with the operator j. The result is $j^2 \times 3.2$, which is 90° counterclockwise from $j3.2$ as shown in Fig. 10.9c. If we compare Fig. 10.9c with Fig. 10.9a, we see that the lines representing $j^2 \times 3.2$ and -3.2 are identical so that

$$j^2 \times 3.2 = -3.2$$

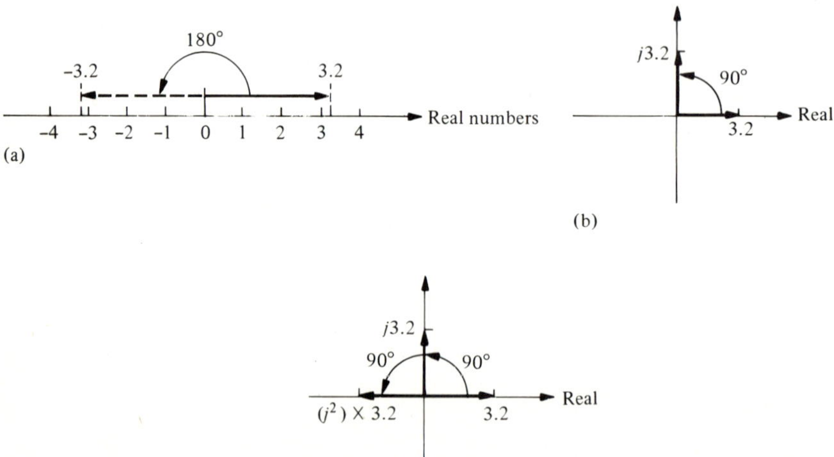

FIGURE 10.9
Graphical representation of numbers. (a) Real-number axis. (b) Operating on +3.2 with *j*. (c) Operating on *j*3.2 with *j*.

Dividing both sides by 3.2 we get

$$j^2 = -1 \tag{10.2-1}$$

from which*

$$j = \sqrt{-1} \tag{10.2-2}$$

where we have used only the positive square root.

When we consider the square root operation, that is, finding a number which when multiplied by itself produces a product equal to the given number, a problem arises; there is no *real number* which when multiplied by itself results in a negative product. However, we have just found that j when multiplied by itself yields -1. For this reason j is called an *imaginary* number.

The imaginary numbers are represented graphically by points along the line perpendicular to the real axis passing through its origin, as shown in Fig. 10.10. This line is called the imaginary axis. The real axis is customarily shown as a horizontal line; the imaginary axis is then a vertical line. Positive real values are to the right of the origin, negative real values to the left,

*The symbol i is used in math texts for $\sqrt{-1}$; however, in electrical technology i is always used for current, so j is used for $\sqrt{-1}$.

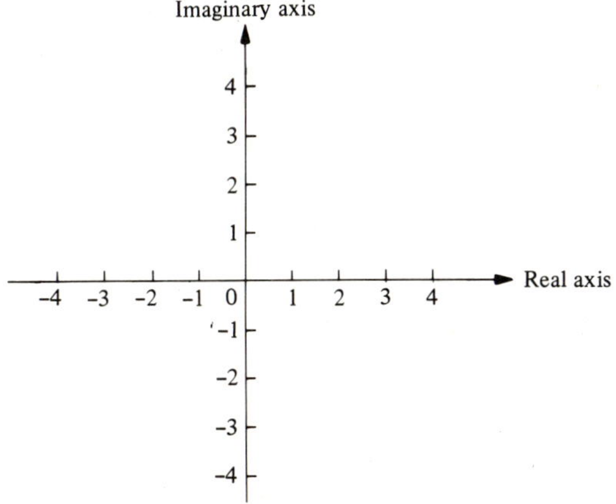

FIGURE 10.10
Real and imaginary axes of the complex plane.

positive imaginary values are above the origin, and negative imaginary values are below the origin.

Summarizing this discussion, the imaginary property of j causes the graphical representation of a number multiplied by j to be rotated 90° counterclockwise (CCW) from the original number. Thus since $+1$ is one unit to the right of the origin, $j \times 1 = j$ is one unit straight up, a rotation of 90° CCW; $j \times j = -1$ is one unit to the left, another 90° CCW rotation; $j \times -1 = -j$ is straight down, another 90° CCW rotation; and finally, $j \times -j = -j^2 = +1$ is again one unit to the right, the result of a fourth 90° CCW rotation. It is useful to think of the j operator as a number that rotates any number it multiplies by 90° CCW.

Complex numbers are a combination of real and imaginary numbers lying in the complex plane formed by the intersection of the real and imaginary axes shown in Fig. 10.10. They are specified using one of two forms; rectangular or polar. The rectangular form can be understood by observing the point corresponding to the arbitrary complex number z shown in Fig. 10.11. The coordinate of this point along the real axis, x, is called the *real part* of z, symbolized

$$x = \text{Re } z \tag{10.2-3}$$

while the coordinate along the imaginary axis, y, is called the *imaginary part* of z, symbolized

$$y = \text{Im } z \tag{10.2-4}$$

Note that the imaginary part of a complex number is a real number; it does not include the j by which it is multiplied in the complex number. The complex number is the sum of its real part and j times its imaginary part:

$$\boxed{z = x + jy} \tag{10.2-5}$$

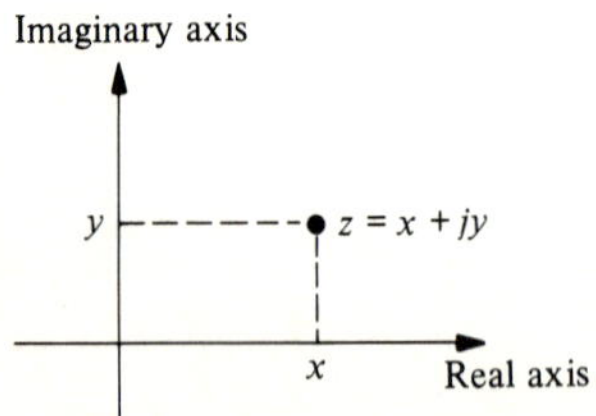

FIGURE 10.11
The rectangular coordinates *x* and *y* of a complex number z.

Graphically, z is obtained by drawing a rectangle, one of whose sides is of length x along the real axis while a perpendicular side is of length y along the imaginary axis. The rectangle is completed by drawing the other two sides as shown in Fig. 10.11 and z is at the corner of the rectangle diagonally across from the origin.

EXAMPLE 10.2-1 Locating Points on the Complex Plane.

a. Find the real and imaginary parts of the complex number a shown in Fig. 10.12a.

b. Plot the point $b = 2 - j3$ on the same set of axes. Find Re b and Im b.

Solution

a. From the graph we read the values

$$\text{Re}\, a = 5 \qquad \text{Im}\, a = 4$$

Thus $a = 5 + j4$

b. The complex number b is shown plotted in Fig. 10.12b. In order to plot the point we have used the values

$$\text{Re}\, b = 2 \qquad \text{Im}\, b = -3$$

• • •

The coordinates of a point representing a complex number are its real part and imaginary part. A complex number written as the sum of the real and j times the imaginary coordinates, as in Eq. (10.2-5), is said to be expressed in *rectangular*, or *cartesian*, form.

The position of a point in a plane can also be specified in polar form. The point representing the complex number z is shown in Fig. 10.13 in terms of its *polar* coordinates. The distance r from the origin to the point z is

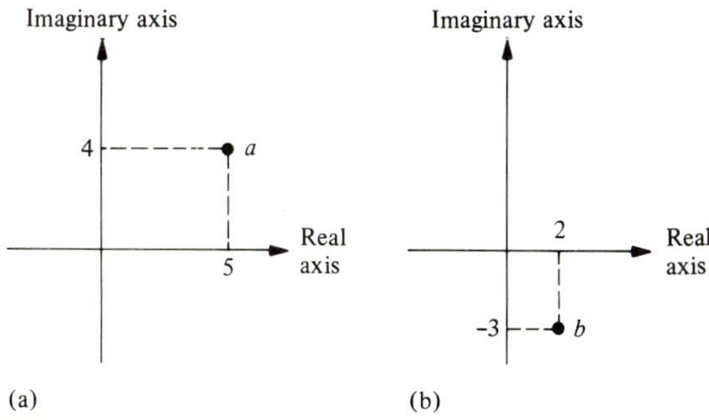

FIGURE 10.12
Example 10.2-1. (a) Complex number. (b) Complex number $b = 2 - j3$.

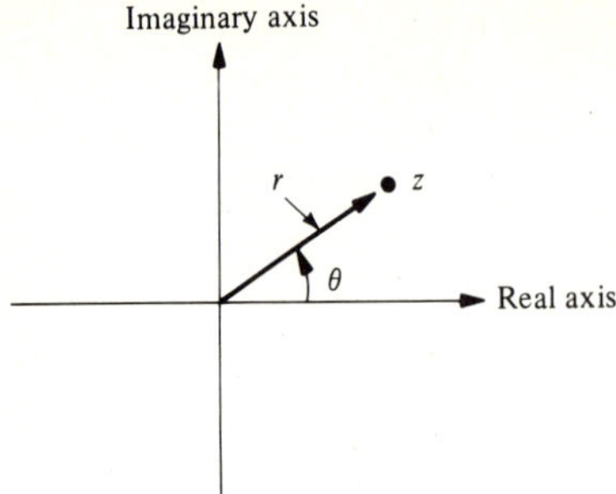

FIGURE 10.13
Complex number in terms of polar coordinates.

called the *magnitude*, or *absolute value*, of z and is symbolized by a pair of vertical bars surrounding z:

$$r = |z| \tag{10.2-6}$$

The absolute value is always a positive real number.

The angle, θ, measured counterclockwise from the positive real axis to the line connecting the point z and the origin is called the *argument* of z:

$$\theta = \arg z \tag{10.2-7}$$

θ is a real number and is generally chosen between 0 and 360°, that is,

$$0^\circ \leq \theta < 360^\circ \qquad \text{or} \qquad 0 \leq \theta < 2\pi \text{ rad} \tag{10.2-8}$$

The magnitude r and argument θ of a complex number z are the polar coordinates of the point representing z. The polar form of z is written*

$$\boxed{z = r\underline{/\theta}} \tag{10.2-9}$$

This is to be read "z equals r at angle θ."

EXAMPLE 10.2-2 The Polar Form

Plot the following complex numbers:

a. $z_1 = 2\underline{/35^\circ}$
b. $z_2 = 4\underline{/-35^\circ}$
c. $z_3 = 3\underline{/120^\circ}$

*This is shorthand for Euler's exponential form $z = re^{j\theta}$.

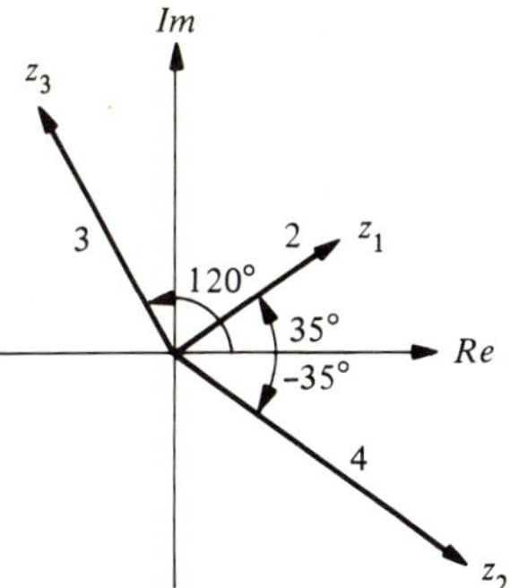

FIGURE 10.14
Example 10.2-2.

Solution

a. A line is drawn at an angle of 35° to the real axis in the complex plane. The point two units from the origin shown in Fig. 10.14 represents the complex number z_1. Complex numbers z_2 and z_3 are shown on the same diagram.

• • •

LEARNING EXERCISES FOR SEC. 10.2

1. Simplify as far as possible: j^5, j^7, j^8.

Ans. $+1; j; -j$

2. a. Given the complex number $z_4 = 6 - j11$, find Re z_4 and Im z_4.
   b. Given the complex number $z_5 = 20\underline{/82°}$, find r_5 and θ_5.

Ans. 20; 6; -11; 82

• • •

10.3 OPERATIONS WITH COMPLEX NUMBERS

Polar-to-Rectangular Conversion

It is frequently necessary to convert between the polar and rectangular forms for a given complex number. The polar coordinates r and θ of a complex number z are shown in Fig. 10.15a. The line connecting the point z with the origin is the hypotenuse of a right triangle whose legs are the real and imaginary parts of the complex number. Using the trigonometric formulas for the right triangle we have

$$\boxed{x = r \cos \theta} \qquad (10.3\text{-}1)$$

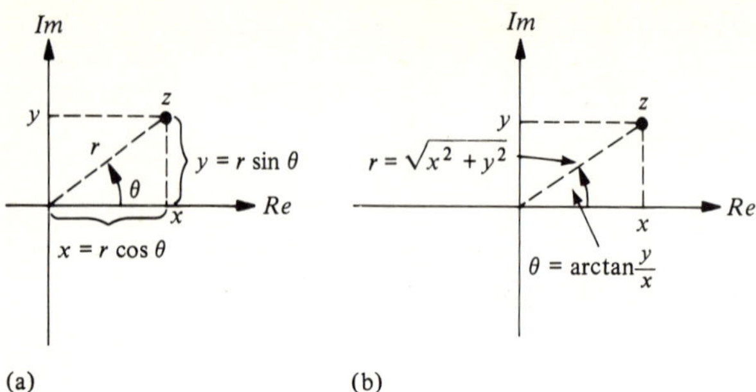

(a) (b)

FIGURE 10.15
Conversion between polar and rectangular forms of a complex number. (a) Conversion from polar to rectangular. (b) Conversion from rectangular to polar.

and

$$y = r \sin \theta \tag{10.3-2}$$

z can now be written in the form

$$z = x + jy = r(\cos \theta + j \sin \theta) \tag{10.3-3}$$

With these formulas, we can convert any complex number from polar to rectangular form.

Rectangular-to-Polar Conversion

The rectangular coordinates of a complex number are shown in Fig. 10.15b. This time we use the formulas for a right triangle to obtain the polar coordinates. The pythagorean formula is used to find the hypotenuse of the triangle, which is the magnitude of the complex number:

$$r = |z| = \sqrt{x^2 + y^2} \tag{10.3-4}$$

The angle (argument) of z is given by

$$\theta = \arg z = \arctan \frac{y}{x} \tag{10.3-5}$$

where arctan y/x means "the angle whose tangent is y/x." These two formulas enable us to convert from rectangular to polar form.

Calculator Conversion of Complex Numbers

Scientific calculators such as those shown in Fig. 1.1 can execute direct conversion of complex numbers in either direction, that is, polar-to-rectangular or rectangular-to-polar. Thus Eq. (10.3-1) through (10.3-5) may not be required. However, the geometrical concepts involved in these conversions are very important in ac circuit theory. For this reason, the conversions in the following examples are done using the equations.

EXAMPLE 10.3-1 Rectangular-to-Polar Conversions

Find the polar forms for the complex numbers

a. $6 + j8$
b. $-3.2 + j12$
c. $-3 - j6$

Solution

In this type of conversion, determination of the angle can sometimes be a problem. However, if a rough freehand sketch is drawn of the components of the complex number, this difficulty can be avoided. This is illustrated below.

a. The number $6 + j8$ is shown in Fig. 10.16a. The magnitude is found from Eq. (10.3-4)

$$r = \sqrt{x^2 + y^2} = \sqrt{6^2 + 8^2} = \sqrt{100} = 10$$

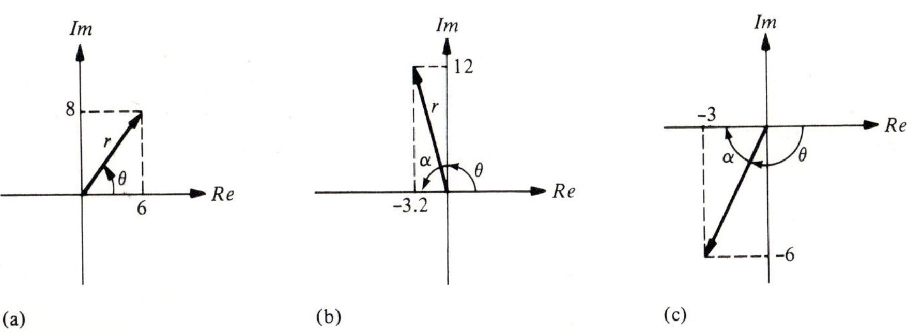

FIGURE 10.16
Example 10.3-1. (a) $6 + j8$. (b) $-3.2 + j12$. (c) $-3 - j6$.

The angle is found from Eq. (10.3-5)

$$\theta = \arctan \frac{8}{6} = 53°$$

This example is straightforward. However, the student should note that the answers can be estimated directly from the freehand sketch if it is made with a little care. This estimate provides a rough check on calculations and often helps to catch decimal point errors in magnitude or quadrant errors in angle.

b. $-3.2 + j12$ is sketched in Fig. 10.16b. From the sketch we estimate that the amplitude will be slightly more than 12 and the angle slightly over 90°. Using Eq. (10.3-4)

$$r = \sqrt{x^2 + y^2} = \sqrt{(-3.2)^2 + (12^2)} = 12.4$$

In order to find the angle, we note that on most calculators the arctan function provides angles from +90° to −90°. Thus we can use the calculator to find the angle α shown on the diagram. This is

$$\alpha = \arctan \frac{12}{3.2} = 75.1°$$

The desired angle θ is then the *supplement* of α, that is,

$$\theta = 180° - \alpha = 180° - 75.1° = 104.9°$$

The freehand sketch is invaluable for resolving questions in situations like this.

c. $-3 - j6$ is sketched in Fig. 10.16. We estimate the magnitude to be about 7 and the angle θ to be about −120°. Using Eq. (10.3-5) to find the angle α, we have

$$\alpha = \arctan \frac{6}{3} = 63.4°$$

Then

$$\theta = -(180° - \alpha) = -117°$$

The magnitude is found from Eq. (10.3-4)

$$r = \sqrt{x^2 + y^2} = \sqrt{(-3)^2 + (-6)^2} = 6.7$$

Note that the magnitude can also be found from Eq. (10.3-1) or Eq. (10.3-2) once the angle has been determined. Using Eq. (10.3-2)

$$r = \frac{y}{\sin \theta} = \frac{-6}{\sin(-117°)} = \frac{-6}{-0.891} = 6.7$$

The student should check these examples using the direct conversion capability of a scientific calculator.

• • •

EXAMPLE 10.3-2 Polar-to-Rectangular Conversions

Find the rectangular forms for the complex numbers

a. $12\underline{/42°}$
b. $36\underline{/2.25 \text{ rad}}$
c. $14.2\underline{/-80°}$

Solution

a. $12\underline{/42°}$ is sketched in Fig. 10.17a. Using Eqs. (10.3-1) and (10.3-2)

$$x = r\cos\theta = 12\cos 42° = 8.92$$
$$y = r\sin\theta = 12\sin 42° = 8.03$$

b. $36\underline{/2.25}$ Here the angle is given in radians. We can convert to degrees ($\theta = 2.25 \times 180°/\pi = 129°$) or set the calculator's DEG/RAD switch to RAD. The sketch is shown in Fig. 10.17b. We estimate $x \approx -20$ and $y \approx 30$. To check our estimate we have

$$x = 36\cos 2.25 = -22.6$$
$$y = 36\sin 2.25 = 28$$

c. $14.2\underline{/-80°}$ This is sketched in Fig. 10.17c. We estimate $x \approx 2$ and $y \approx -14$. The calculated values are

$$x = 14.2\cos(-80°) = 2.47$$
$$y = 14.2\sin(-80°) = -14$$

Again, these examples should be checked using the direct conversion capability of the scientific calculator.

• • •

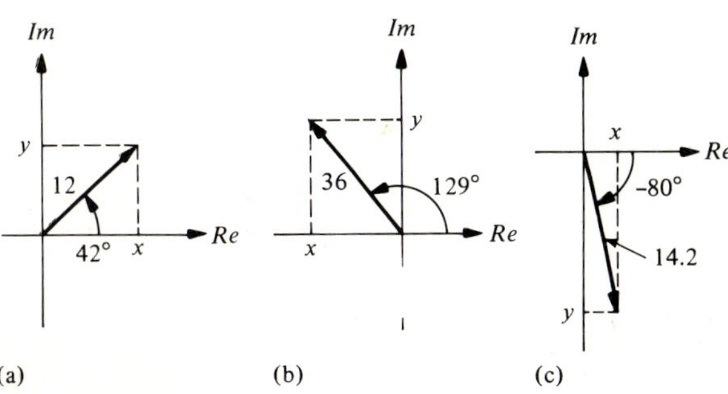

FIGURE 10.17
Example 10.3-2. (a) $12\underline{/42°}$. (b) $36\underline{/2.25}$. (c) $14.2\underline{/-80°}$.

We cannot overemphasize the importance of the free-hand sketches that should be drawn before doing such operations with complex numbers. They provide the student with a quick graphical check on signs, magnitudes, and angles. It will be found that most of the errors made in this type of calculation can be detected by careful checking of the calculated results against the sketch.

It is important to note that the calculations in Examples 10.3-1 and 10.3-2 could all have been carried out graphically using graph paper, a scale, and a protractor to measure angles. This will be considered in the problems.

• • •

LEARNING EXERCISES FOR SEC. 10.3

1. Convert the following to polar form:
   a. $6.2 + j8.7$
   b. $6.2 - j8.7$
   c. $-6.2 + j8.7$

Ans. $10.7\underline{/-55^\circ}$; $10.7\underline{/55^\circ}$; $10.7\underline{/125^\circ}$

2. Convert the following to rectangular form:
   a. $12\underline{/68^\circ}$
   b. $12\underline{/-68^\circ}$
   c. $12\underline{/112^\circ}$

Ans. $4.5 + j11$; $-4.5 + j11$; $4.5 - j11$

• • •

Addition of Complex Numbers

Two complex numbers, $z_1 = x_1 + jy_1$ and $z_2 = x_2 + jy_2$, are shown in Fig. 10.18a. The two are graphically added by moving one of them, z_2 in our example, so that it starts at the end of the other one rather than at the origin; its length and direction remain unchanged. This is shown in Fig. 10.18b, where the sum $z = z_1 + z_2$ is obtained by drawing a line from the origin to the end of the line representing z_2. From the figure we see that

$$\begin{aligned} z = z_1 + z_2 &= (x_1 + jy_1) + (x_2 + jy_2) \\ &= (x_1 + x_2) + j(y_1 + y_2) \end{aligned} \tag{10.3-6}$$

This shows that the real parts are added together and the imaginary parts are added together separately.

If z_1 or z_2 is in polar form it should be converted to rectangular form before performing the addition. If the result is to be in polar form, it should be converted after the addition.

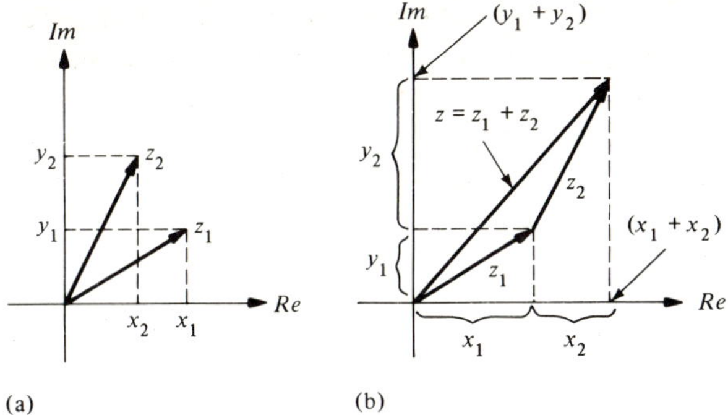

FIGURE 10.18
Addition of complex numbers. (a) Complex representation of numbers. (b) Graphical addition.

Subtraction of Complex Numbers

Subtraction is performed by separately subtracting the real and imaginary parts, just as they were separately added in addition. Thus

$$
\begin{aligned}
z = z_1 - z_2 &= (x_1 + jy_1) - (x_2 + jy_2) \\
&= (x_1 - x_2) + j(y_1 - y_2)
\end{aligned}
\tag{10.3-7}
$$

A graphical representation of this process is shown in Fig. 10.19, which indicates that the difference $z_1 - z_2$ is obtained by adding $-z_2$ to z_1. Note that $-z_2$ has the same magnitude as z_2 but the opposite direction. The addition and subtraction of complex numbers are illustrated in the following example.

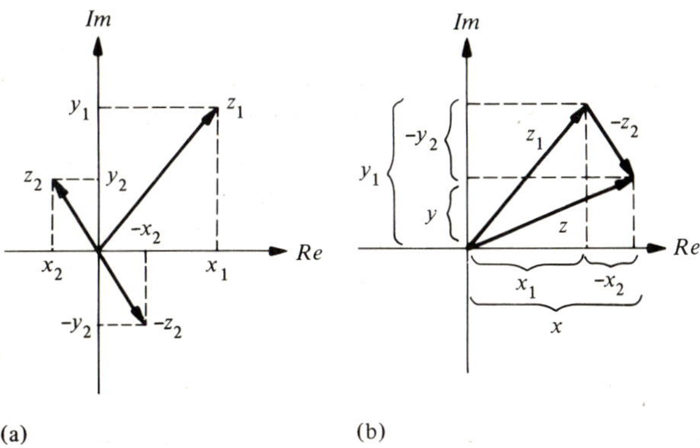

FIGURE 10.19
Subtraction of complex numbers. (a) The numbers in the complex plane. (b) The graphical subtraction process in which $-z_2$ is added to z_1.

EXAMPLE 10.3-3 Addition and Subtraction of Complex Numbers

Perform the indicated arithmetic operations on the following complex numbers, expressing the result in the same form as the operands in each case. Check the results graphically.

a. $(2) + (3 + j7) =$

b. $(5 + j6) - (7 + j2) =$

c. $2\angle 90° + 4\angle 30° =$

d. $5\angle 60° - 2\angle 300° =$

The reader should try to solve these before looking at the solution.

Solution

The first two are done using Eqs. (10.3-6) and (10.3-7):

$$\text{a.}\quad \begin{array}{r} 2 + j0 \\ +\,3 + j7 \\ \hline =5 + j7 \end{array} \qquad\qquad \text{b.}\quad \begin{array}{r} 5 + j6 \\ -\,7 - j2 \\ \hline =-2 + j4 \end{array}$$

The graphical solutions are shown in Fig. 10.20.

c. Here, we must first convert the polar forms to rectangular, perform the addition, and convert the result back to polar.

$$\begin{array}{rcl} 2\angle 90° = 2(\cos 90° + j\sin 90°) = & 0 & + j2 \\ +\;4\angle 30° = +4(\cos 30° + j\sin 30°) = & +3.5 & + j2 \\ \hline & 3.5 & + j4 = 5.3\angle 49° \end{array}$$

The graphical solution is shown in Fig. 10.21a.

d. The steps are the same as the previous problem except that the operation is subtraction.

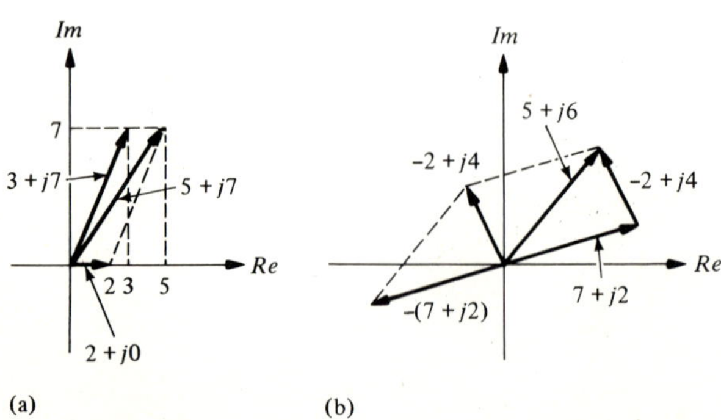

FIGURE 10.20
Example 10.3-3. (a) (2 + *j*0) + (3 + *j*7). (b) (5 + *j*6) − (7 + *j*2).

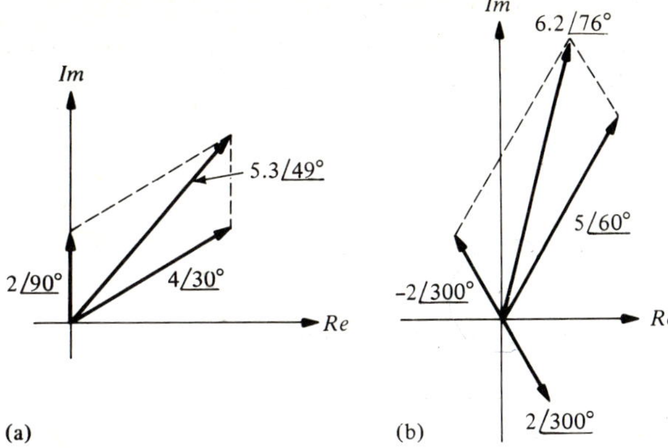

FIGURE 10.21
Example 10.3-3. (a) $2\angle 90° + 4\angle 30°$. (b) $6\angle 60° - 2\angle 300°$.

$$
\begin{array}{rlr}
5\angle 60° = 5(\cos 60° + j \sin 60°) = & 2.5 + j4.3 & \\
-\ 2\angle 300° = -2(\cos 300° + j \sin 300°) = & -1 \quad + j1.7 & \\
\hline
& 1.5 + j6 & = 6.2\angle 76°
\end{array}
$$

The graphical solution is shown in Fig. 10.21b.

• • •

LEARNING EXERCISE FOR SEC. 10.3

3. Given the following complex numbers

$z_1 = 3 + j4$
$z_2 = 5 - j2$
$Z_3 = -1 + j6$

Find

a. $z_1 + z_2$
b. $z_1 - z_2$
c. $z_1 + z_2 - z_3$

Ans. $9 - j4$; $-2 + j6$; $8 + j2$

8 + j2

• • •

Multiplication and Division of Complex Numbers

Complex numbers are most easily multiplied or divided if they are in polar form. The product of two complex numbers is

$$
\begin{aligned}
z = z_1 z_2 &= (r_1\underline{/\theta_1})(r_2\underline{/\theta_2}) \\
&= r_1 r_2 \underline{/\theta_1 + \theta_2}
\end{aligned}
\tag{10.3-8}
$$

Thus, the magnitude of the product of two complex numbers is the product of their magnitudes, while the angle of the product of two complex numbers is the sum of their angles.

The quotient of two complex numbers is

$$
\begin{aligned}
z = \frac{z_1}{z_2} &= \frac{r_1\underline{/\theta_1}}{r_2\underline{/\theta_2}} \\
&= \frac{r_1}{r_2}\underline{/\theta_1 - \theta_2}
\end{aligned}
\tag{10.3-9}
$$

Therefore, the magnitude of the quotient of two complex numbers is the quotient of their magnitudes while the argument of the quotient of two complex numbers is the difference between their arguments, that is, the numerator angle minus the denominator angle. Multiplication and division of complex numbers are illustrated in the following example.

EXAMPLE 10.3-4 Multiplication and Division of Complex Numbers

Perform the indicated operations, expressing the result in the same form as the operands.

a. $25\underline{/\pi} \times 3.5\underline{/-\pi/4} =$
b. $10\underline{/20°} \div 4\underline{/70°} =$
c. $(3 + j4) \times (1 + j) =$
d. $(5 - j2.5) \div (4 - j3) =$

Solution

The first two parts are done using Eqs. (10.3-8) and (10.3-9):

a. $25\underline{/\pi} \times 3.5\underline{/-\pi/4} = (25 \times 3.5)\underline{/\pi - \pi/4} = 87.5\underline{/3\pi/4}$

b. $10\underline{/20°} \div 4\underline{/70°} = \frac{10}{4}\underline{/20° - 70°} = 2.5\underline{/-50°}$

The third and fourth parts require conversions so the operations may be performed on the polar form:

c. $(3 + j4) \times (1 + j) = 5\underline{/53°} \times 1.41\underline{/45°} = 7.1\underline{/98°}$
$= -1 + j7$

d. $(5 - j2.5) \div (4 - j3) = 5.59\underline{/-26.6°} \div 5\underline{/-37°}$
$= \frac{5.59}{5}\underline{/-26.6° + 37°}$
$= 1.12\underline{/10.4°} = 1.10 + j0.202$

• • •

Although it is easiest to multiply and divide numbers in polar form, these operations may also be done with numbers expressed in rectangular form. If both operands and the result are to be in rectangular form it may be easier to do the somewhat more complex arithmetic and not have to be bothered with the three conversions that would be required to use the polar form.

The formula for multiplication follows directly from the rules of algebra:

$$(x_1 + jy_1)(x_2 + jy_2) = x_1x_2 + jx_1y_2 + jx_2y_1 + j^2y_1y_2$$

Since $j^2 = -1$, this results in

$$(x_1 + jy_1)(x_2 + jy_2) = (x_1x_2 - y_1y_2) + j(x_1y_2 + x_2y_1) \tag{10.3-10}$$

The use of this formula is illustrated by the following example.

EXAMPLE 10.3-5 Multiplication in Rectangular Form

Multiply $3 + j4$ by $1 + j$.

Solution

$$(3 + j4) \times (1 + j) = (3 \times 1 - 4 \times 1) + j(3 \times 1 + 4 \times 1)$$
$$= (3 - 4) + j(3 + 4) = -1 + j7$$

Compare this with Example 10.3-4c.

• • •

The division of complex numbers in rectangular form is presented in the following section because it requires the use of complex conjugate numbers.

Complex Conjugate Numbers

The complex conjugate of a complex number z is denoted z*. Graphically, it is the mirror image of z in the complex plane with the real axis being the axis of symmetry as shown in Fig. 10.22. From this figure we find that if $z = r\underline{/\theta} = x + jy$, then

$$z^* = r\underline{/-\theta} = x - jy \tag{10.3-11}$$

The product of a complex number with its complex conjugate is the square of the magnitude of the complex number, as we prove in the following:

$$zz^* = (r\underline{/\theta})(r\underline{/-\theta}) = r^2\underline{/\theta - \theta} = r^2 = |z|^2 \tag{10.3-12}$$

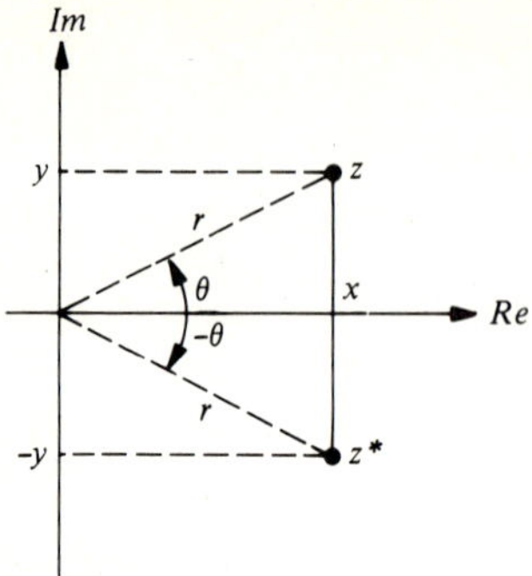

FIGURE 10.22
Complex conjugate of a complex number.

The same result may be obtained using the rectangular form:

$$zz^* = (x + jy)(x - jy) = [x^2 - (-y^2)] + j(-xy + xy)$$
$$= x^2 + y^2 = |z|^2$$

Using this result we may derive the formula for the quotient of two complex numbers expressed in rectangular form; the method is based on elimination of the complex denominator by multiplying it by its complex conjugate. In order to not change the value of the fraction, the numerator is multiplied by the same value as follows:

$$\frac{z_1}{z_2} = \frac{z_1}{z_2} \times \frac{z_2^*}{z_2^*} = \frac{z_1 z_2^*}{|z_2|^2} = \frac{(x_1 + jy_1)(x_2 - jy_2)}{x_2^2 + y_2^2}$$
$$= \frac{x_1 x_2 + y_1 y_2}{x_2^2 + y_2^2} + j\frac{x_2 y_1 - x_1 y_2}{x_2^2 + y_2^2}$$

Division of complex numbers in rectangular form is illustrated in the following example.

EXAMPLE 10.3-6 Division in Rectangular Form

Divide $5 - j2.5$ by $4 - j3$.

Solution

$$\frac{5 - j2.5}{4 - j3} = \left(\frac{5 - j2.5}{4 - j3}\right)\left(\frac{4 + j3}{4 + j3}\right)$$
$$= \frac{(5 \times 4 - j^2\, 2.5 \times 3) + j(5 \times 3 - 2.5 \times 4)}{4^2 + 3^2}$$

$$= \frac{(20 + 7.5) + j(15 - 10)}{16 + 9}$$

$$= \frac{27.5 + j5}{25} = 1.1 + j0.2$$

Compare this result with that of Example 10.3-4d.

• • •

LEARNING EXERCISE FOR SEC. 10.3

4. Given the following complex numbers:
   $z_1 = 6\angle 24°$
   $z_2 = 8\angle 102°$
   $z_3 = 3\angle 67°$
   Find

   a. $z_1 \times z_2$ b. $\frac{z_1}{z_2}$

   c. $z_1 \times z_3^*$ d. $\frac{z_3}{z_2^*}$

Ans. $18\angle -43°$; $0.75\angle -78°$; $0.375\angle 169°$; $48\angle 126°$

• • •

10.4 PHASORS

The use of phasors for ac circuit analysis provides us with a shorthand method for handling sinusoidal network calculations. In this method, a complex number is substituted for each sine wave. The algebra of complex numbers is then used to combine the sine waves as needed in the calculation. This is far easier than using the trigonometric formulas that are required if the sine waves are to be manipulated in the standard form of Eq. (10.1-8). After we have defined phasors, we will give an example that shows the phasor calculation side-by-side with the trigonometric calculation. The difference will then be quite apparent.

In order to define the phasor, consider Fig. 10.23a. In this figure, the complex plane has been rotated 90° from its usual orientation so that the real axis is vertical. In this complex plane we have shown a vector of length $|V_m|$ on the real axis. We let this vector rotate counterclockwise at constant angular velocity ω radians per second with the rotation beginning at $t = 0$. The graph next to the complex plane shows the projection of the tip of the vector on the positive real axis as the vector rotates. The resulting curve is symbolized as $v(t)$. Now, in order to find a mathematical expression for $v(t)$

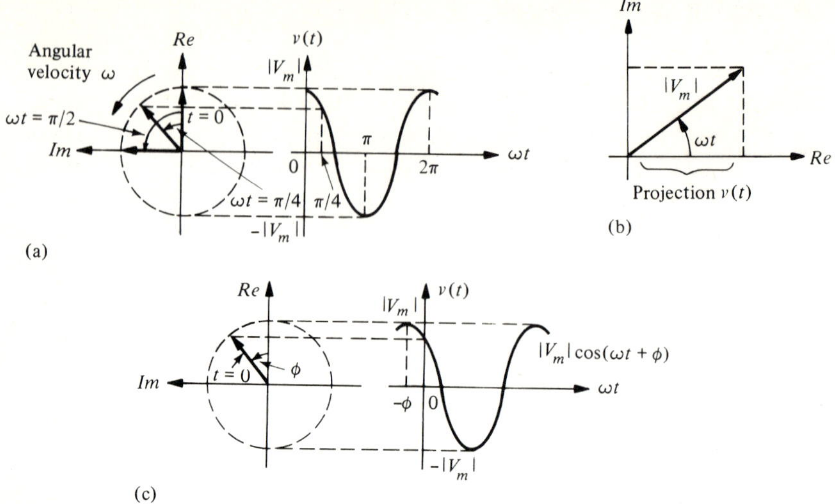

FIGURE 10.23
Projection of rotating vector. (a) Zero phase angle. (b) Position of phasor at time *t*. (c) Arbitrary phase angle.

consider Fig. 10.23b in which the vector is frozen at one instant of time, t. The angle that the vector makes with the real axis is ωt radians and from the right triangle we have

$$\cos \omega t = \frac{v(t)}{|V_m|}$$

from which

$$v(t) = |V_m| \cos \omega t \qquad (10.4\text{-}1)$$

This shows that the projection of the rotating complex plane vector does indeed yield the cosine function. Next consider the rotating vector in Fig. 10.23c which starts rotating at $t = 0$ but with an initial phase angle ϕ. For this case the angle at any time t becomes $\omega t + \phi$ so that the projection on the real axis is

$$v(t) = |V_m| \cos(\omega t + \phi) \qquad (10.4\text{-}2)$$

This equation is identical to Eq. (10.1-8), which gives the standard form for the cosine wave. In linear circuits, when the source is a cosine wave of one frequency, all of the currents and voltages in the circuit are cosine waves of the *same* frequency but different amplitudes and phase angles. The complex plane vector is called a *phasor* and it contains all of the information required to specify the cosine wave $v(t)$ except for the angular frequency ω.

The standard form for the sinusoid, Eq. (10.4-2), is referred to as the *time domain* form, while the phasor designation is called the *frequency*, Fourier, or *phasor domain* form. When dealing with phasors it is more convenient to use the *effective* value $|V|$, which is defined in Sec. 10.6. The relationship between the peak and the effective value is

$$|V| = \frac{|V_m|}{\sqrt{2}} \tag{10.4-3}$$

A typical phasor voltage would be written in the form*

$$V = |V| \angle \phi \tag{10.4-4}$$

where $|V|$ is the magnitude of the effective phasor V and ϕ is its phase angle.

The correspondence between the two representations is shown here.

Phasor representation	Equivalent time function
$V = \|V\| \angle \phi$	$v(t) = \sqrt{2}\|V\| \cos(\omega t + \phi)$
	$= \|V_m\| \cos(\omega t + \phi)$

EXAMPLE 10.4-1 Phasor-to-Time Domain Conversions

Convert the following phasors to the time domain and plot both phasors and time functions. The frequency is 60 Hz.

a. $V = 50\angle 0°$ V
b. $I = 12\angle 30°$ mA

Solution

a. The peak amplitude of the voltage is $50\sqrt{2} = 70.7$ V and the phase angle is 0°. The angular frequency is $\omega = 2\pi \times 60 = 377$ rad/s. Thus the phasor $V = 50\angle 0°$ V is

*A word about notation: In the past, boldfaced letters were often used to symbolize complex numbers. However, current practice as set forth in USAS Standard Y10.5-1968, "Letter Symbols for Quantities Used in Electrical Science and Electrical Engineering" dictates that ordinary italic type be used when the complex number represents a phasor. Thus the typeface cannot be used to distinguish between a complex quantity or a real quantity. However, it should always be clear from the context and if the quantity is complex, then the arithmetic operations detailed in this chapter must be used.

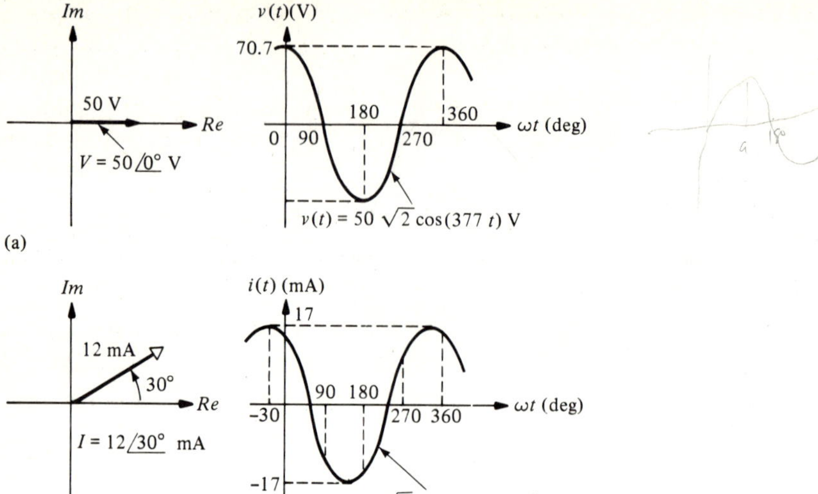

FIGURE 10.24
Phasor-time domain equivalences. (a) A typical voltage. (b) A typical current.

equivalent to the time function $v(t) = 70.7 \cos 377t$ V. The graphs for both of these are shown in Fig. 10.24a.

b. The peak amplitude of the current is $12\sqrt{2} = 17$ mA and the phase angle is 30°. Thus the phasor $I = 12\underline{/30°}$ mA is equivalent to $i(t) = 17 \cos(377t + 30°)$ mA. The graphs are shown in Fig. 10.24b.

• • •

EXAMPLE 10.4-2 *Time-to-Phasor Domain Conversions*

Convert the following time domain functions to the phasor domain.

a. $i(t) = 75 \cos(377t + 82°)$ mA

b. $v(t) = 22 \cos\left(\omega t + \dfrac{\pi}{4}\right)$ mV

Solution

a. The effective current is $75/\sqrt{2} = 53$ mA. Then the time domain form $i(t) = 75 \cos(377t + 82°)$ mA becomes the phasor $I = 53\underline{/82°}$ mA.
b. The effective voltage is $22/\sqrt{2} = 15.6$ mV. The time function $v(t) = 22 \cos[\omega t + (\pi/4)]$ mV is equivalent to the phasor $V = 15.6\underline{/\pi/4}$ mV.
The student should draw the graphs for this example.

• • •

LEARNING EXERCISES FOR SEC. 10.4

1. For each of the following currents, find the effective current and the cyclic frequency.
   a. $i(t) = 600 \cos(2000\pi t - 38°)$ mA
   b. $i(t) = 7.3 \cos(377t + 22°)$ A

Ans. 1000; 424; 5.16; 60

2. For each of the following phasors, find the peak value of the time function and the phase angle in degrees.
   a. $V = 311\underline{/-\pi/6}$ V
   b. $I = 100\underline{/3\pi/4}$ mA

Ans. 141; −30; 440; 135

• • •

Addition of Sine Waves

In ac circuit analysis we often have to add or subtract sinusoids of the same frequency but different phase angles. This sum (or difference) is always a sinusoid of the same frequency but different amplitude and phase, which can be found by finding the sum (or difference) of their phasor representations, a considerably easier process than using the trigonometric form, as shown in the following example.

EXAMPLE 10.4-3 Comparing Phasor and Time Domain Addition

Given the two voltages,

$$v_1(t) = 10 \cos(\omega t + 30°) \text{ V}$$

$$v_2(t) = 15 \cos(\omega t + 45°) \text{ V}$$

Find $v(t) = v_1(t) + v_2(t)$ using (a) trigonometric identities and (b) phasors.

Solution

a. Since $\cos(x + y) = \cos x \cos y - \sin x \sin y$,

$$\begin{aligned} v_1(t) &= 8.7 \cos \omega t - 5 \sin \omega t \\ v_2(t) &= 10.6 \cos \omega t - 10.6 \sin \omega t \\ \hline v(t) &= 19.3 \cos \omega t - 15.6 \sin \omega t \end{aligned}$$

To combine these terms we use the identity

$$a \cos \omega t + b \sin \omega t = \sqrt{a^2 + b^2} \cos\left(\omega t - \arctan \frac{b}{a}\right)$$

Thus

$$v(t) = \sqrt{(19.3)^2 + (15.6)^2}\cos\left(\omega t - \arctan\frac{-15.6}{19.3}\right)$$

$$= 24.8\cos(\omega t + 39°)\text{ V}$$

b. Using phasors, we write

$$\mathbf{V} = \mathbf{V}_1 + \mathbf{V}_2 = \frac{10}{\sqrt{2}}\angle 30° + \frac{15}{\sqrt{2}}\angle 45° = 7.07\angle 30° + 10.6\angle 45°$$

$$= 6.1 + j3.5 + 7.5 + j7.5$$

$$= 13.6 + j11$$

$$= 17.5\angle 39°\text{ V}$$

Recall that it is understood that 17.5 V is the effective value. Then, converting back to the time domain,

$$v(t) = 17.5\sqrt{2}\cos(\omega t + 39°)$$

$$= 24.8\cos(\omega t + 39°)\text{ V}$$

The graphical addition process is shown in Fig. 10.25. The procedure for adding the two sine waves in part a of the figure is to add instantaneous values of v_1 and v_2 at a number of points. The resultant curve for $v(t)$ is then drawn through these points. This point-by-point procedure is obviously quite tedious.

In the phasor method shown in part b of the figure, we draw the phasors $\mathbf{V}_1$ and $\mathbf{V}_2$, complete the parallelogram, and draw the diagonal which represents the sum $\mathbf{V}$. The resultant phasor is scaled to find the amplitude, and its angle is measured with a protractor to give the final numerical answer.

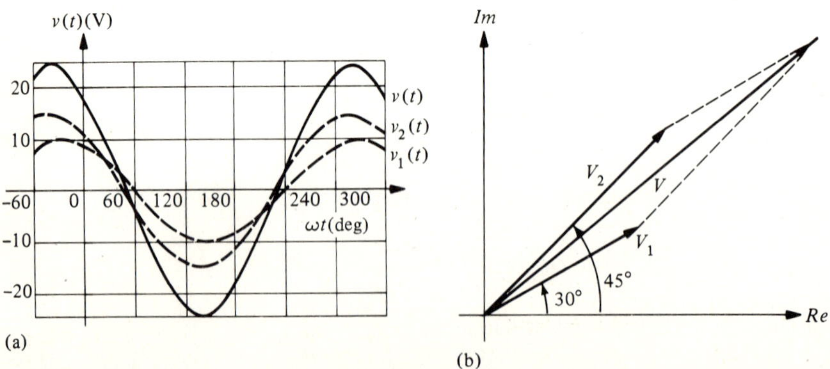

FIGURE 10.25
Example 10.4-3. (a) Point-by-point addition of sinusoids. (b) The corresponding phasor addition.

Clearly, the phasor method is easier to apply. It can be done graphically, with a carefully scaled diagram, or calculated quickly using the conversion capability of the scientific calculator.

• • •

10.5 THE BASIC LAWS IN PHASOR FORM

In this section we illustrate the phasor form of the basic laws as applied to ac networks.

Consider that the voltage across the resistance in the circuit of Fig. 10.26a is a sinusoid as shown by the solid curve in Fig. 10.26c. Then, at any particular instant of time, say t_1, the instantaneous voltage has a certain magnitude $v(t_1)$. If we restrict ourselves only to this particular instant, then at t_1 the circuit is identical to the dc circuit shown in Fig. 10.26b. The difference between the two circuits is that at another instant of time the magnitude of the ac voltage will change, while the voltage in the dc circuit remains constant. Thus, for any particular instant of time, all of the dc circuit laws that we have learned up to this point apply to the ac circuit. With this in mind, Ohm's law in terms of the instantaneous current and voltage takes on the general form

$$i(t) = \frac{v(t)}{R} \tag{10.5-1}$$

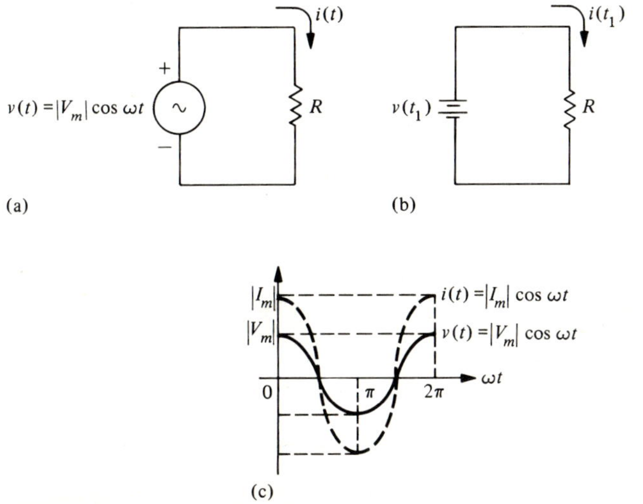

(a) $v(t) = |V_m| \cos \omega t$, $i(t)$, R

(b) $v(t_1)$, $i(t_1)$, R

(c) $|I_m|$, $|V_m|$, 0, π, 2π, ωt, $i(t) = |I_m| \cos \omega t$, $v(t) = |V_m| \cos \omega t$

FIGURE 10.26
Current and voltage in a resistance. (a) Circuit. (b) Circuit at one particular time, t_1. (c) Waveforms.

Since R is a constant, this equation shows that if the voltage is a cosine wave, then the current must also be a cosine wave with the same phase as shown in Fig. 10.26c. The instantaneous current will reach its peak value at the same instant as the instantaneous voltage and will otherwise follow exactly the same form as the voltage. Since $v(t)$ and $i(t)$ are both cosine waves with the same phase angle (zero degrees in this case), they are said to be "in phase." This "in phase" condition only holds when the element is a resistance.

If $v(t)$ is given by

$$v(t) = |\mathrm{V}_m| \cos \omega t \tag{10.5-2}$$

then

$$i(t) = \frac{|\mathrm{V}_m|}{R} \cos \omega t = |I_m| \cos \omega t \tag{10.5-3}$$

where

$$|I_m| = \frac{|\mathrm{V}_m|}{R} \tag{10.5-4}$$

This relates the *magnitude* of the voltage across a resistance to the *magnitude* of the current through it. We can also write Ohm's law in phasor form:

$$\boxed{\mathrm{V}_m = RI_m} \tag{10.5-5}$$

where it is understood that the resistance parameter is a real number with zero phase angle. For example, a 10-kΩ resistor would be designated as $R = 10\underline{/0°}$ kΩ. In addition, we must remember that V_m and I_m are phasor quantities, with both magnitude and phase angle.

Using effective phasors, Ohm's law looks exactly like its dc counterpart:

$$\boxed{\mathrm{V} = RI} \tag{10.5-6}$$

EXAMPLE 10.5-1 Phasor Voltage and Current in a Resistor

An effective current of $3.7\underline{/41°}$ mA flows through a 2.2-kΩ resistor. Express the voltage across the resistor in both phasor and time domain form.

Solution

Using Eq. (10.5-6) with current in milliamperes and resistance in kilohms the phasor voltage is

$$
\begin{aligned}
V &= RI \\
&= 2.2\underline{/0^\circ}\ \text{k}\Omega \times 3.7\underline{/41^\circ}\ \text{mA} \\
&= 8.1\underline{/41^\circ}\ \text{V}
\end{aligned}
$$

The circuit and phasor diagram are shown in Fig. 10.27. In the phasor diagram we use an open arrowhead for the current phasor and a closed arrowhead for the voltage phasor. This convention will be used for all phasor diagrams.

In time domain form the voltage is

$$
\begin{aligned}
v(t) &= \sqrt{2} \times 8.1 \cos(\omega t + 41^\circ) \\
&= 11 \cos(\omega t + 41^\circ)\ \text{V}
\end{aligned}
$$

• • •

The Kirchhoff laws can also be expressed in phasor form. For the voltage law, in terms of instantaneous values the general form is

$$v_1(t) + v_2(t) + v_3(t) + \cdots = 0 \tag{10.5-7}$$

where the sum on the left-hand side of the equation contains all of the voltages around the closed loop. In words, this says that, *at any instant of time t*, the voltages around the closed loop must add algebraically to zero. In terms of effective phasors this is

$$\boxed{V_1 + V_2 + V_3 + \cdots = 0} \tag{10.5-8}$$

Example 10.4-3 illustrated this type of phasor addition. From the example, we see that magnitudes of phasors *cannot* be added, that is,

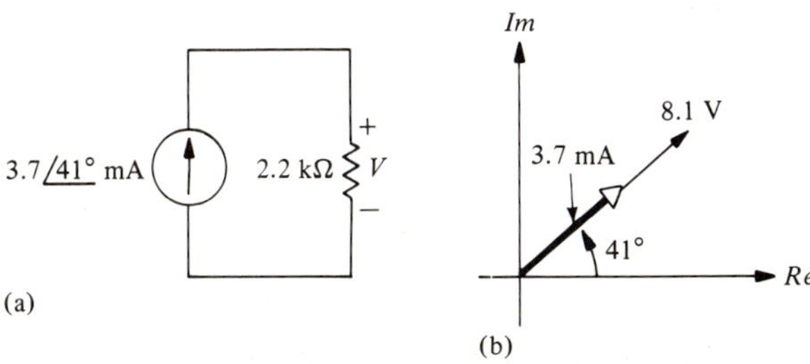

FIGURE 10.27
Example 10.5-1. (a) Circuit. (b) Phasor diagram.

$$|V_1| + |V_2| + |V_3| + \cdots \neq 0 \tag{10.5-9}$$

Each phasor will, in general, have a different angle and the angles *must be taken into account* when the phasors are added or subtracted.

By analogy, KCL can be written in phasor form as

$$\boxed{I_1 + I_2 + I_3 + \cdots = 0} \tag{10.5-10}$$

where all of the currents entering or leaving the node must be included with their proper algebraic sign.

We illustrate the phasor form of the basic laws in the following example.

EXAMPLE 10.5-2 A Resistive Circuit

Figure 10.28a shows the input circuit of a phonograph preamplifier. Find all unknown currents and voltages.

Solution

This problem is solved in much the same manner as a dc circuit problem. We begin by finding the resistance seen by the source.

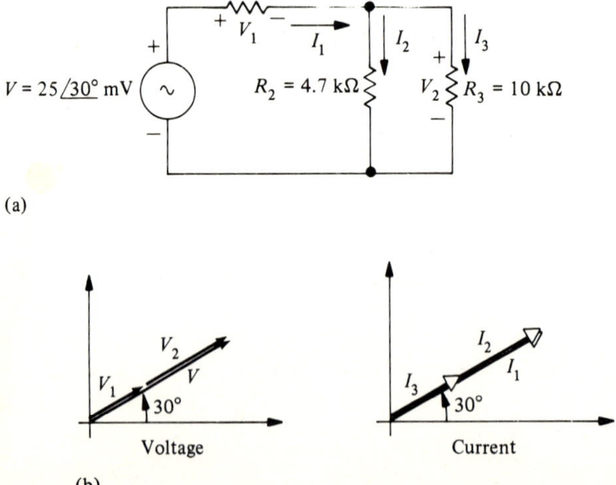

FIGURE 10.28
Example 10.5-2. (a) Circuit. (b) Phasor diagrams.

$$R_T = R_1 + (R_2 \parallel R_3)$$
$$= 2.2\ \text{k}\Omega + (4.7\ \text{k}\Omega \parallel 10\ \text{k}\Omega)$$
$$= 2.2\ \text{k}\Omega + 3.2\ \text{k}\Omega$$
$$= 5.4\ \text{k}\Omega$$

Then, using Ohm's law in phasor form we find

$$I_1 = \frac{V}{R_T} = \frac{25\underline{/30^\circ}\ \text{mV}}{5.4\underline{/0^\circ}\ \text{k}\Omega} = 4.6\underline{/30^\circ}\ \mu\text{A}$$

The voltage across R_1 is then

$$V_1 = R_1 I_1$$
$$= 2.2\underline{/0^\circ}\ \text{k}\Omega \times 4.6\underline{/30^\circ}\ \mu\text{A}$$
$$= 10\underline{/30^\circ}\ \text{mV}$$

The voltage across R_2 and R_3 is found using KVL:

$$V_2 = V - V_1$$
$$= 25\underline{/30^\circ}\ \text{mV} - 10\underline{/30^\circ}\ \text{mV}$$

When both phasors are at the same angle we do not have to find their real and imaginary components in order to perform the indicated subtraction. Since they lie in a straight line we can subtract their magnitudes to get

$$V_2 = 15\underline{/30^\circ}\ \text{mV}$$

The currents through R_2 and R_3 are now found using Ohm's law:

$$I_2 = \frac{V_2}{R_2} = \frac{15\underline{/30^\circ}\ \text{mV}}{4.7\underline{/0^\circ}\ \text{k}\Omega} = 3.2\underline{/30^\circ}\ \mu\text{A}$$

$$I_3 = \frac{V_2}{R_3} = \frac{15\underline{/30^\circ}}{10\underline{/0^\circ}\ \text{k}\Omega} = 1.5\underline{/30^\circ}\ \mu\text{A}$$

The phasor diagrams are shown in Fig. 10.28b. We can check using KCL, which requires that

$$I_1 = 4.6\underline{/30^\circ} = I_2 + I_3 = 3.2\underline{/30^\circ}\ \mu\text{A} + 1.5\underline{/30^\circ}\ \mu\text{A} = 4.7\underline{/30^\circ}\ \mu\text{A}.$$

The slight difference is due to rounding.

• • •

The calculations in this example were quite straightforward. In the next chapter we include capacitance and inductance and we find that the phasors no longer lie at the same angle. Calculations then become more complicated.

• • •

LEARNING EXERCISE FOR SEC. 10.5

1. Resistors of values 4.7 and 6.8 kΩ are connected in series across a test oscillator which generates a 1-kHz sinusoid with 1.2-V peak amplitude and a phase angle of 62°. Find the phasor current and the voltage across each resistor.

Ans. $0.5\angle 62°$; $74\angle 62°$; $0.35\angle 62°$

• • •

10.6 POWER AND ENERGY IN AC CIRCUITS

In Chap. 4 we studied the distribution of power in resistive circuits powered by dc sources. In this section we consider the same problem, but the circuits are powered by sinusoidal ac sources.

Consider the circuit of Fig. 10.29a. The instantaneous value of the current fluctuates in accordance with the cosine curve:

$$i(t) = |I_m| \cos \omega t \qquad (10.6\text{-}1)$$

At any instant of time the dc formula for power holds so that the instantaneous power will also fluctuate. The formula for instantaneous power is

$$\boxed{p(t) = Ri^2(t)} \qquad (10.6\text{-}2)$$

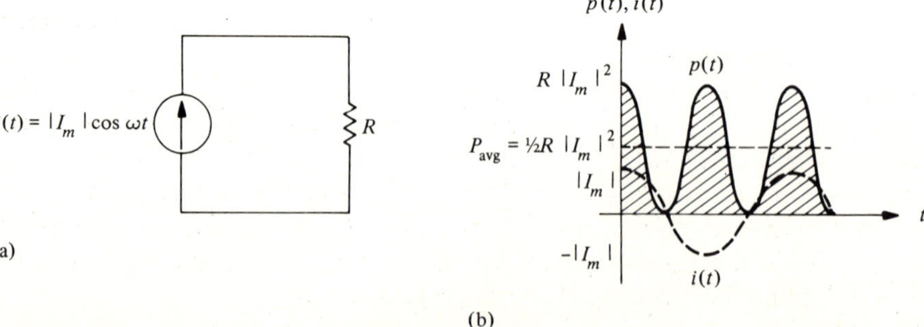

FIGURE 10.29
Power in ac circuits. (a) Circuit. (b) Waveforms.

where $i(t)$ is the instantaneous current through the resistance. Equation (10.6-2) indicates that the instantaneous power varies as a cosine-squared wave. In order to determine what this waveform looks like, we substitute Eq. (10.6-1) into Eq. (10.6-2) to get

$$p(t) = R\,|I_m|^2 \cos^2 \omega t \tag{10.6-3}$$

We could draw a graph of $p(t)$ directly from this equation. However, it is more instructive to simplify it using the trigonometric identity

$$\cos^2 x = \frac{1}{2}(1 + \cos 2x) \tag{10.6-4}$$

Using this, $p(t)$ becomes

$$\begin{aligned} p(t) &= \frac{R\,|I_m|^2}{2}(1 + \cos 2\omega t) \\ &= \frac{R\,|I_m|^2}{2} + \frac{R\,|I_m|^2}{2}\cos 2\omega t \end{aligned} \tag{10.6-5}$$

This is plotted in Fig. 10.29b along with the current waveform. Because of the squaring of the cosine wave the instantaneous power is always positive and, as we pointed out in Sec. 2.7, positive power is converted into heat in a resistor. The graph shows that the instantaneous power swings from zero to its maximum value at a frequency exactly twice that of the current. Thus, as far as the heating effect of the cosine wave of current is concerned, one pulse of energy is converted into heat on the positive half-cycle of current, and another pulse of energy is converted into heat on the negative half-cycle. If the current is at the standard power frequency of 60 Hz, then the instantaneous power pulsates at 120 Hz. Sometimes electric lights appear to flicker due to these pulsations of the instantaneous power.

The user of electric power is seldom concerned with its instantaneous value. It is the average power over a large number of cycles that is of most interest. For this reason, we define the *effective value* of a sinusoidal current as the value of direct current that produces the same amount of heat as the alternating current in a given resistance. In general, this must be measured over a large number of cycles of the ac. Since the amount of heat energy delivered by the resistance is proportional to the average power dissipated in the resistance, we can find the effective value of the current by comparing the dc power to the average ac power.

The dc power is

$$P_{dc} = RI^2 \tag{10.6-6}$$

The average ac power can be found by observation of the instantaneous power curve of Fig. 10.29b. Since the instantaneous power oscillates *symmetrically* between 0 and $R|I_m|^2$ it appears that its average value is exactly one-half of the maximum power. Thus

$$\text{avg } P_{ac} = \frac{R|I_m|^2}{2} \tag{10.6-7}$$

The effective value is that direct current which will produce the same power loss in the resistor as the sinusoidal current. Equating the powers

$$P_{dc} = \text{avg } P_{ac}$$

$$RI^2 = \frac{R|I_m|^2}{2}$$

Cancelling R and taking the square root of both sides, we find

$$\boxed{I = \frac{|I_m|}{\sqrt{2}} = 0.707|I_m|} \tag{10.6-8}$$

This is the effective value* and we see that it is 0.707 times the peak value of the alternating current. It is important to note that this was derived for sine waves and applies only to that waveform.

A development similar to that above will show that the effective value of a sinusoidal voltage has the same form as the current

$$\boxed{|V| = \frac{|V_m|}{\sqrt{2}} = 0.707|V_m|} \tag{10.6-9}$$

As noted earlier in this chapter it is more convenient to work with effective values in ac circuit analysis. Therefore, from this point on, all ac currents and voltages will be stated as effective values, and the symbols $|V|$ or $|I|$ will be used to denote the magnitude of the phasors V or I. Peak values, if required, will always be identified by the subscript m.

*This is often called the rms (root-mean-square) value because we found it by taking the square *root* of the *mean* (average) of the *square* of the current.

In terms of the effective current, the average power dissipated in the resistance in Fig. 10.29a is (we use the letter P for average power from this point on)

$$P = R|I|^2 \tag{10.6-10}$$

where

$$|I| = |I_m|/\sqrt{2}$$

In terms of voltage,

$$P = \frac{|V|^2}{R} \tag{10.6-11}$$

Finally, in terms of $|V|$ and $|I|$ the formula is

$$P = |V||I| \tag{10.6-12}$$

EXAMPLE 10.6-1 Effective Value

What is the time-domain form of a sinusoidal voltage that will produce the same heat in a 10-Ω resistor as a 4-V dc voltage?

Solution

The same heat will be produced by an ac voltage having an *effective* value of 4 V. Its peak amplitude will be

$$|V_m| = 4\sqrt{2} = 5.66 \text{ V}$$

and its time-domain form is

$$v(t) = 5.66 \cos(\omega t + \theta) \text{ V}$$

• • •

EXAMPLE 10.6-2 *Power in a Resistor*

A current $i(t) = 250 \cos \omega t$ A passes through a 2-Ω ballast resistor.

a. How much dc current will produce the same heat in the resistor?
b. How much power is dissipated?

Solution

a. The amount of direct current required is the same as the effective value

$$I_{dc} = |I| = \frac{|I_m|}{\sqrt{2}} = \frac{250}{\sqrt{2}} = 177 \text{ A}$$

b. We use the RI^2 form for power

$$P = R|I|^2 = (2)(177)^2 = 63 \text{ kW}$$

• • •

EXAMPLE 10.6-3 *Finding Current and Resistance*

A 120-V light bulb is rated at 150 W. If it can be considered to be purely resistive, find:

a. The effective value of the current
b. The resistance of the bulb

Solution

a. The effective value of the source is 120 V. Therefore, using $P = |V||I|$, we have

$$|I| = \frac{P}{|V|} = \frac{150}{120} = 1.25 \text{ A}$$

b. The resistance can be found from Ohm's law,

$$R = \frac{|V|}{|I|} = \frac{120}{1.25} = 96 \ \Omega$$

• • •

LEARNING EXERCISES FOR SEC. 10.6

1. The standard residential voltage in the United States is 120 V at 60 Hz. What is the peak value of this voltage?

2. A stage set is illuminated by eight 500-W spotlights. If they operate on 220-V circuits, how much current is required?

3. The sinusoidal output of an AM radio transmitter is measured on an oscilloscope and found to be 375 V from peak to peak. What is the effective value of the voltage?

Ans. 132; 170; 18.2

• • •

SUMMARY

In this chapter we have considered the fundamentals of ac circuit analysis, that is, the sinusoid, its phasor representation, the basic laws in phasor form, and the concept of effective value. In the next chapter we use these ideas to find the current, voltage, and power distribution in networks containing R, L, and C elements when the source is sinusoidal.

Section

10.1

1. A sinusoidal signal is a periodic signal characterized by an amplitude, period, and phase angle.
   a. Its amplitude is the peak value it attains.
   b. Its period T is the time interval for one complete cycle. Its frequency f, measured in hertz (Hz), is the reciprocal of the period measured in seconds (s). Its angular frequency is $\omega = 2\pi f$ and is measured in radians per second.
   c. Its phase angle is a measure of its position along the t axis.
2. The basic sinusoidal signal is

   $$v(t) = |V_m| \cos(\omega t + \phi)$$

   where $v(t)$ is the instantaneous value, $|V_m|$ is the peak value, ω is the angular frequency, and ϕ is the phase angle.
3. Multiply degrees by $\pi/180$ to get radians.
4. Multiply radians by $180/\pi$ to get degrees.

10.2

5. The imaginary operator is $j = \sqrt{-1}$.
6. The rectangular form of a complex number is

   $$z = x + jy$$

   where $x = \text{Re } z$ and $y = \text{Im } z$
7. The polar form of a complex number is

   $$z = r\underline{/\theta}$$

   where $r = |z|$ and $\theta = \arg z$

10.3

8. To convert complex numbers from one form to the other, use

   $$x = r\cos\theta \qquad r = \sqrt{x^2 + y^2}$$
   $$y = r\sin\theta \qquad \theta = \arctan(y/x)$$

9. To add (subtract) complex numbers, add (subtract) the real and imaginary parts separately to get the real and imaginary parts of the sum (difference).
10. To multiply (divide) complex numbers, multiply (divide) their

magnitudes to get the magnitude of the product (quotient) and add (subtract) the phase angles.

11. The conjugate of z is z^* where $z^* = x - jy = r\angle{-\theta}$.

10.4 12. A phasor is a vector in the complex plane which, when rotated at ω rad/s, projects a sinusoid on the real axis.

13. The effective phasor $V = |V|\angle\theta$ corresponds to the sinusoid $v(t) = \sqrt{2}|V|\cos(\omega t + \theta)$.

14. Addition (subtraction) of sinusoids can be performed by finding the sum (difference) of their complex phasor representations.

10.5 15. Ohm's law, KVL, and KCL have exactly the same form when applied to effective phasors as when applied to dc networks. The difference is that the phasors have associated angles, which complicate computations.

10.6 16. The effective value of a sinusoid is the dc value that produces the same amount of heat over a period of time as the ac. It is $1/\sqrt{2}$ times the peak value.

17. The average power dissipated in a resistor in an ac circuit is

$$P = |V||I| = R|I|^2 = \frac{|V|^2}{R}$$

where $|V|$ and $|I|$ are effective values.

QUESTIONS FOR REVIEW

Sec. 10.1

1. Write the equation for the instantaneous value of a typical sinusoidal voltage.
2. What is the significance of the maximum, or peak, value?
3. Compare the angular frequency and the cyclic frequency.
4. How is the phase angle represented?
5. How do we convert from radians to degrees and vice versa?
6. Describe the symbolism used to represent the absolute value of a phasor.

Sec. 10.2

7. Plot an arbitrary point in the complex plane and show its real part, imaginary part, magnitude, and angle.
8. Simplify j^n as far as possible where $n = 0, 1, 2, 3$, and 4.

Sec. 10.3

9. Explain how a complex number in cartesian form is converted to polar form.
10. Explain how a complex number in polar form is converted to cartesian form.
11. Explain the procedure for adding two complex numbers graphically by geometrical construction.
12. Repeat the previous question for the subtraction process.
13. Can the multiplication or division of two complex numbers be carried out by geometrical construction?
14. What is the geometrical significance of the conjugate of a complex number?

Sec. 10.4

15. Explain the correspondence between the projection of the rotating phasor on the real axis of the complex plane to the sinusoidal function.
16. What piece of information concerning the sinusoid is missing from the phasor representation?
17. In a linear circuit the source is a sinusoid of frequency 100 kHz. How can we find the frequencies of all of the currents and voltages?
18. Explain how the phasor $I = |I| \angle \theta$ is converted to time domain form.
19. Explain how the time domain form $v(t) = |V_m| \cos(\omega t + \theta)$ is converted to the phasor domain.
20. What is the advantage of using phasor quantities when sinusoids must be added?

Sec. 10.5

21. What is Ohm's law in terms of instantaneous values?
22. What can be deduced when the statement is made that a voltage and current are "in phase"?
23. Write KVL and KCL in phasor form.

Sec. 10.6

24. State the formula for instantaneous power.
25. The instantaneous power in a resistance is always a positive quantity. Why?
26. Describe the fluctuations in the instantaneous power in a resistance.
27. How do we define the *effective value* of a sinusoidal current?
28. How is the effective value related to the peak value of a sinusoidal waveform?
29. State the three forms of the formula for power in a resistance in terms of effective values.

PROBLEMS

Sec. 10.1

1. The sinusoidal voltage output of an audio oscillator is described by the equation

   $$v(t) = 50 \cos(6280t - 32°) \text{ mV}$$

   Find its amplitude, angular frequency, cyclic frequency, period, and phase angle. Sketch the wave on graph paper, being sure to scale the axes properly.
2. a. Convert the angles 32°, 68°, 88°, 130°, 222°, and 310° to radians.
   b. Convert the angles $\pi/8$ rad, 0.1 rad, 1.0 rad, 2.4 rad, and 6 rad to degrees.
3. Sketch the following functions on the same set of axes.
   a. $v(t) = 10 \cos 377t$
   b. $v(t) = 10 \cos(377t + \pi/6)$
   c. $v(t) = 10 \cos(377t + \pi/3)$
4. A cosine voltage waveform has a maximum value of 440 V and zero phase angle. Find $v(t)$ when
   a. $\omega t = 30°$ b. $\omega t = 0.2$ rad c. $\omega t = 220°$

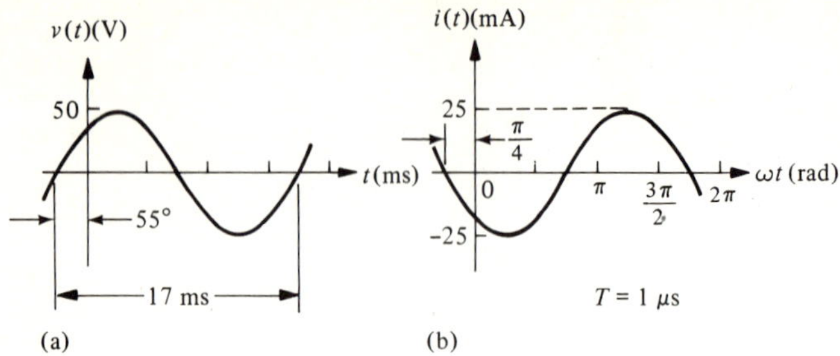

FIGURE 10.30

5. A current is given by $i(t) = 200 \sin(628t + 80°)$ mA. Find its instantaneous value when
   a. $t = 5$ ms b. $t = 50$ ms
6. A cosine voltage waveform has a phase angle of 30° and an instantaneous value of 130 V when $\omega t = \pi/4$. What is its peak value?
7. Write the equation for the sinusoids shown in Fig. 10.30 using the cosine function.
8. A 60-Hz generator has a peak voltage of 171 V and a phase angle of 0°. What is the voltage after 2 ms? 5 ms? 100 ms?
9. The instantaneous voltage from a 500-Hz source is -3 V when $t = 1.3$ ms. Find the instantaneous voltage when $t = 1.7$ ms.
10. Write the equation for the output of a 440-Hz aircraft alternator that has a peak voltage of 200 V.

Sec. 10.2

11. Find both the rectangular and the polar forms for the complex numbers shown in Fig. 10.31.
12. Make a labeled sketch showing the complex plane representation of each of the following:
    a. $p = 5 - j8$ b. $i = 10\underline{/90°}$ c. $z = j10$ d. $v = 25\underline{/120°}$
13. A complex number has Re $z = -3.2$ and Im $z = 12$. Write the number in rectangular form.
14. A complex number has a magnitude of 12.4 and an angle of $3\pi/2$ rad. Write the number in polar form.
15. A complex number has Re $z = 4$ and an angle of 30°. Using a ruler and protractor on graph paper, find Im z and $|z|$.
16. A complex number has Im $z = 6$ and $|z| = 8$. Using a ruler, compass, and protractor, find Re z and arg z.

Sec. 10.3

17. Convert the following to rectangular form (be sure to draw a freehand sketch to check each one).
    a. $10\underline{/\pi}$ b. $20\underline{/-30°}$ c. $100\underline{/225°}$ d. $0.062\underline{/102°}$
18. Repeat Prob. 17 for the following:
    a. $12\underline{/3\pi/2}$ b. $60\underline{/-72°}$ c. $150\underline{/290°}$ d. $0.08\underline{/110°}$

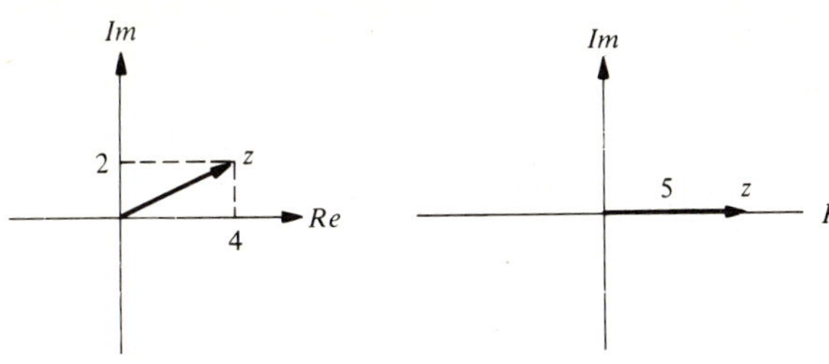

(a) (b)

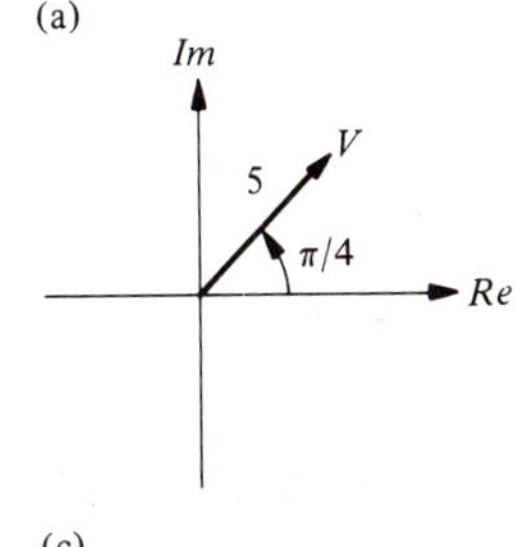

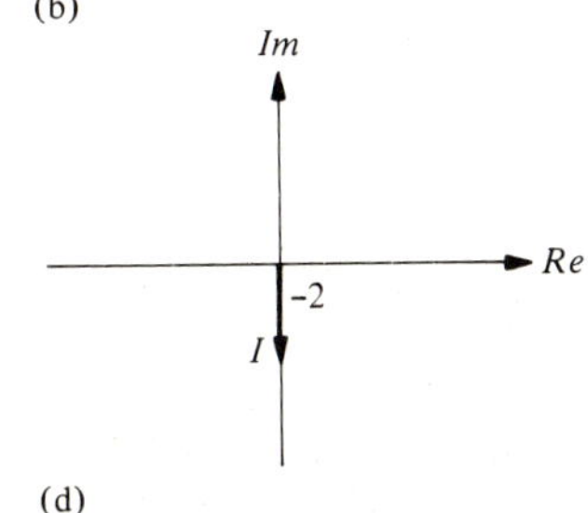

(c) (d)

FIGURE 10.31

19. Convert the following to polar form (be sure to check each one with a freehand sketch).
    a. $-3 + j4$ b. $-10 - j10$ c. $20 + j50$ d. $j16$
20. Repeat Prob. 19 for the following:
    a. $3 + j4$ b. $-14 - j8$ c. $0.8 - j0.4$ d. $0.2 + j0$
21. Using only a ruler, protractor, compass, and graph paper, convert the following rectangular (polar) complex numbers to polar (rectangular) forms.
    a. $10 + j4$ b. $-6 - j9.2$ c. $12\angle 50^\circ$ d. $16\angle \pi/4$
22. Repeat Prob. 21 for the following:
    a. $16 - j10$ b. $3.5 + j7$ c. $10\angle 70^\circ$ d. $12\angle -\pi/3$
23. Perform the indicated additions and subtractions; check your results graphically.
    a. $(2 + j3) + (5 - j4)$ b. $25\angle 90^\circ + 100\angle 143^\circ$
    c. $(10 - j4) - (5 + j5)$ d. $10\angle 57.3^\circ - 8\angle -85.9^\circ$
24. Repeat Prob. 23 for the following:
    a. $(6 - j2) + (3 + j3)$ b. $12\angle 240^\circ + 16\angle 45^\circ$
    c. $(6 + j2) - (12 + j12)$ d. $8\angle 60^\circ - 6\angle -90^\circ$
25. Perform the indicated multiplications and divisions:
    a. $(3 + j4)(2.5 - j1)$ b. $18\angle 20^\circ \times 4.5\angle 60^\circ$
    c. $5\angle 180^\circ \div 40\angle -45^\circ$ d. $(10 + j15) \div (6 + j2)$
26. Repeat Prob. 25 for the following:
    a. $(2.5 + j5)(3 - j6)$ b. $16\angle 45^\circ \times 2\angle 45^\circ$
    c. $5\angle 170^\circ \div 20\angle -10^\circ$ d. $(6 + j12) \div (6 + j2)$
27. Evaluate the following with $z_1 = 6\angle 30^\circ$, $z_2 = 9\angle 70^\circ$

$$z = \frac{z_1 z_2}{z_1 + z_2}$$

28. Repeat Prob. 27 for $z_1 = 3 + j4$ and $z_2 = 6 - j12$.

29. Find the complex conjugates of the following numbers:
    a. $-4 - j2$ b. $2\angle -30°$
30. Represent each of the following in terms of $|z|$, arg z, Re z, and Im z:
    a. zz^* b. $z + z^*$ c. $z - z^*$

Sec. 10.4

31. Find the phasor representation of the following:
    a. $i(t) = 5\cos(200\pi t + 30°)$ mA
    b. $v(t) = 171\cos(377t + 80°)$ mV
    c. $i(t) = 12.5\sin(10^4\pi t + 20°)$ μA
    Hint: You may need the identity $\sin x = \cos(x - 90°)$.
32. Repeat Prob. 31 for the following:
    a. $v(t) = 622\cos(377t + 10°)$ V
    b. $i(t) = 3.2\cos(10^5\pi t - 10°)$ μA
    c. $v(t) = 16\sin(100t + 60°)$ mV
    Hint: You may need the identity $\sin x = \cos(x - 90°)$.
33. Find the time domain equations for the following:
    a. $V = 10\angle 30°$ V; $f = 1$ MHz
    b. $I = 2.6\angle 110°$ mA; $f = 50$ Hz
    c. $V = 100\angle 30°$ mV; $\omega = 600$ rad/s
34. Repeat Prob. 33 for the following:
    a. $I = 16\angle 45°$ μA; $f = 2$ MHz
    b. $V = 120\angle 30°$ V; $f = 60$ Hz
    c. $I = 2\angle -45°$ A; $\omega = 100$ rad/s
35. Given $i_1(t) = 22\cos(\omega t + 15°)$ mA; $i_2(t) = 16\cos(\omega t + 80°)$ mA, find $i(t) = i_1(t) + i_2(t)$ using phasors.
36. Repeat Prob. 35 for $i_1(t) = 23.8\cos(377t + 25°)$ A; $i_2(t) = 16\cos(377t - 12°)$ A.

Sec. 10.5

37. The voltage across a 470-kΩ resistor is measured and found to be $64\angle -21°$ V. Find the current in both phasor and time domain form.
38. The current through a 6.8-kΩ resistor is $12\angle 32°$ mA. Find the voltage across the resistor in both phasor and time domain form.
39. In the circuit of Fig. 10.32 $I = 16\angle 30°$ mA. Find I_1, I_2, and V.
40. In the circuit of Fig. 10.32, $I_2 = 2\angle 18°$ mA. Find V, I_1, and I.
41. In the circuit of Fig. 10.33, $V = 60\angle 10°$ V. Find I, V_1, and V_2.
42. In the circuit of Fig. 10.33, $V_1 = 20\angle 30°$ V. Find I, V_2, and V.
43. In the circuit of Fig. 10.34, $I = 6\angle 20°$ mA. Find all currents and voltages.

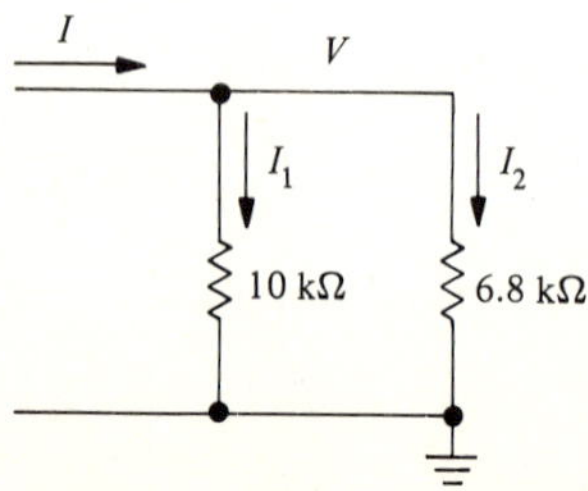

FIGURE 10.32

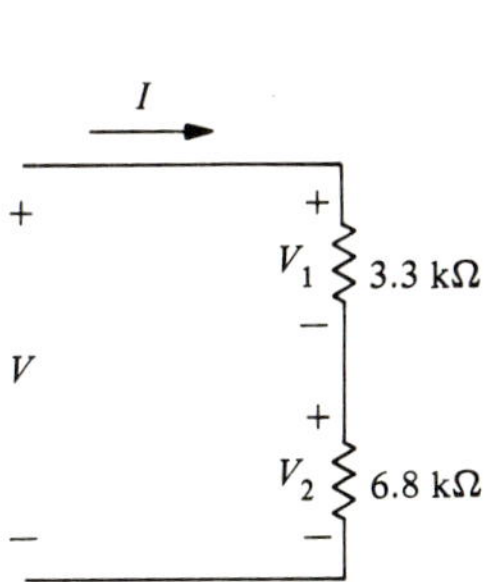

FIGURE 10.33

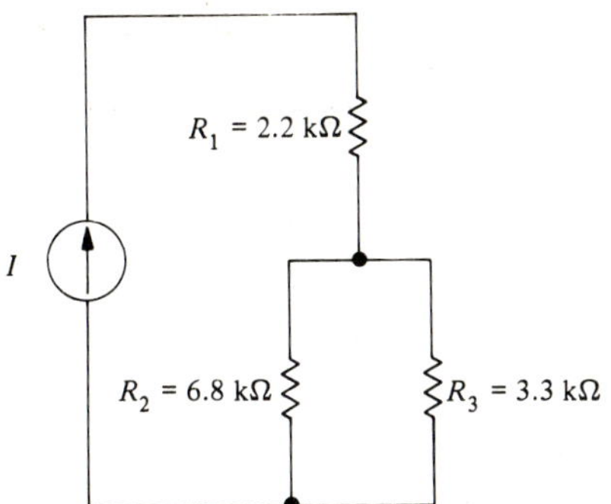

FIGURE 10.34

Sec. 10.6

44. A 10-Ω resistor is connected to the 120-V, 60-Hz supply. Plot a graph showing two cycles of $v(t)$ and $i(t)$. On the same graph plot $p(t)$.
45. Given the following peak values, find the corresponding effective values.
    a. $|V_m| = 680$ V b. $|I_m| = 62$ mA c. $|V_m| = 171$ V
46. What peak values correspond to the following effective values?
    a. $|V| = 220$ V b. $|I| = 3.2$ A c. $|V| = 58$ mV
47. A current $i(t) = 12\cos(377t + 45°)$ A passes through a 150-Ω heating element. Find the direct current that will produce the same heat, and the power that is dissipated.
48. A voltage $v(t) = 12\cos(1000\pi t + 12°)$ V is applied to an 8-Ω loudspeaker that may be considered resistive. Find the dc voltage that would dissipate the same power and the amount of power dissipated.
49. A 220-V furnace element is rated at 3 kW and can be considered resistive. Find the effective current and the resistance of the element.
50. An aircraft wing de-icer operates on 120 V, 400 Hz ac. If it is resistive and dissipates 7.3 kW, how much current does it draw?
51. A 110-V, 1/2-hp motor is used to drive a radial saw. If it can be considered resistive, how much current is required when it is operating under full load?
52. An 8-Ω loudspeaker is rated at 50-W maximum continuous power. Find the effective current and voltage at rated power.
53. A current $I = 1.6\underline{/30°}$ mA is applied to a 2.2-kΩ resistor. How much power is dissipated?

11 AC Circuit Analysis

OBJECTIVES

Upon completion of this chapter, the student should be able to

Section

11.1

1. State that the impedance of a capacitor is inversely proportional to frequency and capacitance and has a phase angle of $-90°$.
2. State that capacitive reactance represents the magnitude of the capacitive impedance and can be found from the formula $X_C = 1/\omega C$.
3. State that the current through a capacitor leads the voltage across the capacitor by 90°.
4. State that the impedance of an inductor is directly proportional to frequency and inductance and has a phase angle of 90°.
5. Find the reactance of an inductor from the formula $X_L = \omega L$.
6. State that the current through an inductor lags the voltage across the inductor by 90°.
7. Find the admittance and susceptance of inductors and capacitors.

11.2

8. Combine arbitrary impedances in series and parallel.
9. Plot complex impedance on the Z plane.
10. Combine admittances in series and parallel.
11. Find a parallel circuit equivalent to a given series circuit and vice versa.
12. Find the total impedance of a circuit consisting of a number of resistors, capacitors, and inductors. Draw a complete Z-plane diagram.

11.3

13. State that all dc circuit laws carry over to ac circuits with the exception that vector algebra must be used.

	14.	Find all currents and voltages in a simple ac series or parallel circuit.
11.4	15.	Use the superposition theorem to solve a multielement ac circuit.
	16.	Use mesh or node equations to solve an ac circuit.
	17.	Find the Thevenin or Norton equivalent for an arbitrary ac circuit.
11.5	18.	Solve an arbitrary ac circuit using the ECAP program.

INTRODUCTION

In Chap. 10 we introduced the sinusoidal waveform and studied its characteristics. We then considered the phasor method for making calculations involving sinusoids of the same frequency. This concept was applied to the analysis of networks consisting of sinusoidal sources and resistances, and the calculations were relatively simple because all currents and voltages were in phase. In this chapter we add inductance and capacitance to the circuits. We find that this complicates the circuit calculations because currents and voltages are no longer in phase.

In addition we study the use of Thevenin's and other theorems as applied to ac networks and the use of the ECAP program for ac analysis.

11.1 IMPEDANCE OF CAPACITORS AND INDUCTORS

In this section we find the sinusoidal response of L and C elements. This leads to the definition of several new terms used in connection with ac networks. Consider first that the voltage across a capacitor is (see Fig. 11.1)

$$v(t) = |V_m| \cos \omega t \tag{11.1-1}$$

The current is found from the formula [see Eq. (8.4-6)]

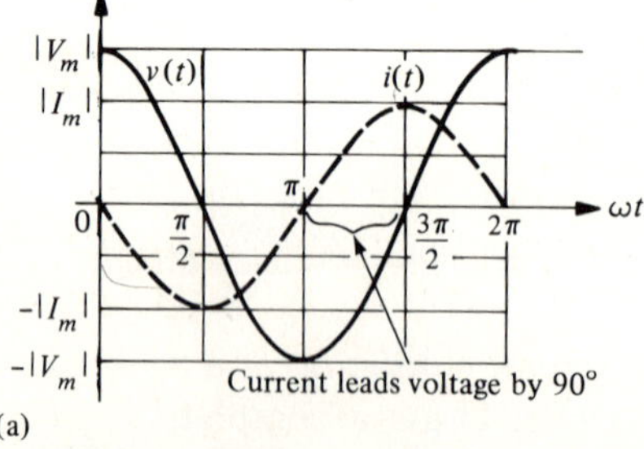

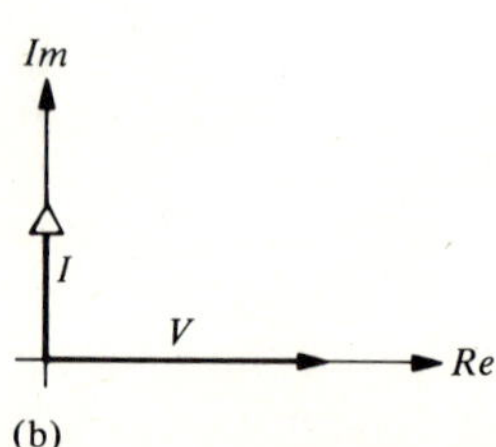

FIGURE 11.1
Sinusoidal response of a capacitor. (a) Waveforms. (b) Phasor diagram.

$$i(t) = C\frac{\Delta v(t)}{\Delta t} \tag{11.1-2}$$

where $\Delta v/\Delta t$ is the slope of the voltage waveform at time t. We can get some idea of the shape of the current curve by graphically estimating the *slope* of the voltage waveform at various points along the time scale. For the cosine curve drawn in solid lines on Fig. 11.1a we find the following:

At $\omega t = 0$

$$\frac{\Delta v}{\Delta t} = 0$$

At $\omega t = \pi/2$

$\dfrac{\Delta v}{\Delta t}$ has its maximum negative value

At $\omega t = \pi$

$$\frac{\Delta v}{\Delta t} = 0$$

At $\omega t = 3\pi/2$

$\dfrac{\Delta v}{\Delta t}$ has its maximum positive value

At $\omega t = 2\pi$

$$\frac{\Delta v}{\Delta t} = 0$$

The resulting waveform has the general shape shown in dashed lines on Fig. 11.1a. We may note that if we increase the frequency, the Δt for a given Δv will decrease. This will cause the current to become greater at higher frequencies (assuming that we do not change the voltage amplitude). Quantitatively, it can be shown using calculus that the equation for the current is

$$i(t) = \omega C|V_m|\cos(\omega t + 90°) \tag{11.1-3}$$

$$= |I_m|\cos(\omega t + 90°) \tag{11.1-4}$$

where $|I_m| = \omega C |V_m|$. The current is seen to be directly proportional to frequency.

Let us now consider the phasor version of this development. The phasor voltage across the capacitor is

$$V = |V| \underline{/0^\circ}\ V \tag{11.1-5}$$

where $|V| = |V_m|/\sqrt{2}$ is the effective value. Then, noting Eq. (11.1-3), the current is

$$I = \omega C V \underline{/90^\circ} \tag{11.1-6}$$

However, recall that $1\underline{/90^\circ} = j$. Using this, Eq. (11.1-6) can be written

$$\boxed{I = j\omega C V} \tag{11.1-7}$$

This is Ohm's law for the capacitor. The general form for Ohm's law in ac circuits is

$$\boxed{V = ZI} \tag{11.1-8}$$

where Z, called the *impedance*, is measured in ohms and plays the same role as resistance in the dc version of the law. The important difference is that, in general, Z is complex and depends on the frequency of the applied voltage or current. For the capacitor, we find from Eq. (11.1-7) that

$$\boxed{Z_C = \frac{V}{I} = \frac{1}{j\omega C} = \frac{-j}{\omega C} = \frac{1}{\omega C}\underline{/-90^\circ}} \tag{11.1-9}$$

and the impedance is *inversely* proportional to frequency.

The phasor diagram for the capacitor voltage and current is shown in Fig. 11.1b. Except for the frequency, this diagram carries all of the information in the time-function plot of Fig. 11.1a.

It is convenient to separate the magnitude of the capacitive impedance from the angle. The magnitude is then called the *capacitive reactance* symbolized by X_C. From Eq. (11.1-9) the capacitive reactance is

$$X_C = \frac{1}{\omega C} \tag{11.1-10}$$

This is measured in ohms, and is always a positive number. We can express the impedance in terms of the reactance as

$$Z_C = -jX_c = X_C \angle -90^\circ \tag{11.1-11}$$

EXAMPLE 11.1-1 Current and Voltage in a Capacitor

A 5-pF capacitor carries a current $i(t) = 2\cos(\omega t + 30^\circ)$ mA with $\omega = 10^6$ rad/s. Find the voltage across the capacitor in phasor form and as a time function.

Solution

We begin by finding the reactance as follows:

$$X_C = \frac{1}{\omega C} = \frac{1}{(10^6)(5)(10^{-12})}$$
$$= 0.2 \times 10^6 = 0.2\ \text{M}\Omega$$

The impedance is then

$$Z_C = X_C \angle -90^\circ = 0.2 \angle -90^\circ\ \text{M}\Omega$$

From the problem statement, the effective current is

$$|I| = \frac{2}{\sqrt{2}} \times 10^{-3} \angle 30^\circ\ \text{A}$$

Then the effective phasor voltage is

$$|V| = Z_C|I| = (0.2 \times 10^6 \angle -90^\circ)(1.41 \times 10^{-3} \angle 30^\circ)$$
$$= 282 \angle -60^\circ\ \text{V}$$

and the instantaneous voltage is

$$v(t) = 282\sqrt{2}\cos(10^6 t - 60^\circ)$$
$$= 400\cos(10^6 t - 60^\circ)\ \text{V}$$

The phasor diagram is shown in Fig. 11.2. Note carefully the difference between this phasor diagram and that of Fig. 11.1b. In Fig. 11.2 the whole diagram is rotated 60°

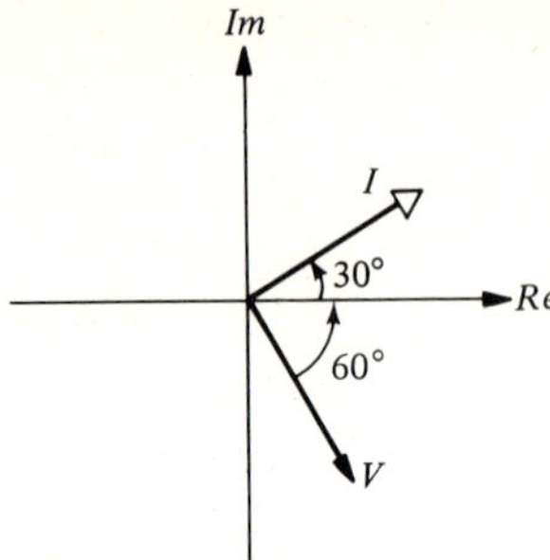

FIGURE 11.2
Phasor diagram for Example 11.1-1.

clockwise, compared to Fig. 11.1. However, in both cases, the current and voltage phasors are exactly 90° apart, with the current phasor the more counterclockwise of the two.

• • •

Phase Lead and Lag

The phase angle between two sinusoidal signals of the same frequency is the distance between corresponding points on the two signals measured in degrees or radians. The corresponding points may be the positive peaks, the negative peaks, the zero crossing preceding the positive portion, or any other convenient point.

The phasor diagram shows the phase angle even more clearly; it is the angle between the two phasors. Two phasors V_1 and V_2 are shown in Fig. 11.3a; the angle between them is ϕ. Recall from Sec. 10.4 that the phasor

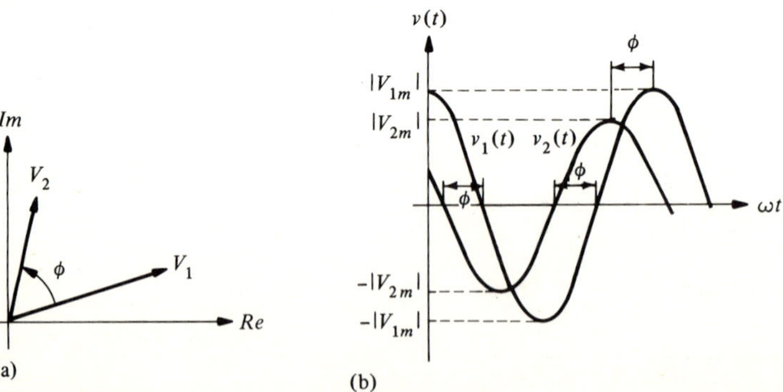

FIGURE 11.3
Phase angle between two sinusoidal signals. (a) Phasor diagram. (b) Graphs of time functions.

corresponds to the "rotating vector" representation of the sinusoidal signal at $t = 0$; as t increases the vectors rotate counterclockwise. With this in mind, we say that phasor V_2 is *leading* phasor V_1 by a phase angle of ϕ since it is further counterclockwise. An alternative statement to describe the situation is that phasor V_1 is lagging phasor V_2 by ϕ.

The situation is pictured for the time functions in Fig. 11.3b. In these graphs we see that a point on the $v_2(t)$ curve occurs $\omega t = \phi$ rad *before* the corresponding point on $v_1(t)$; thus $v_2(t)$ leads $v_1(t)$ by ϕ or $v_1(t)$ lags $v_2(t)$ by ϕ. For the capacitor, the phasor diagrams of Fig. 11.1a and 11.2 show that the current *leads* the voltage by 90°. This is an important fact so we state it as a rule.

In a capacitor driven by a sinusoid, the current leads the voltage by 90°.

Inductance

When the element is an inductance we begin by assuming an instantaneous current

$$i(t) = |I_m| \cos \omega t \tag{11.1-12}$$

Then the voltage is found from

$$v(t) = L \frac{\Delta i(t)}{\Delta t} \tag{11.1-13}$$

By analogy with the development for the capacitor

$$v(t) = \omega L |I_m| \cos(\omega t + 90°) \tag{11.1-14}$$

$$= |V_m| \cos(\omega t + 90°) \tag{11.1-15}$$

where $|V_m| = \omega L |I_m|$. For the phasor development we have

$$I = |I| \underline{/0°} \text{ A} \tag{11.1-16}$$

where $|I| = |I_m| / \sqrt{2}$ is the effective value. Then

$$V = \omega L I \underline{/90°} = j\omega L I \tag{11.1-17}$$

and the impedance of the inductor is

$$\boxed{Z_L = \frac{V}{I} = j\omega L = \omega L \underline{/90°}} \tag{11.1-18}$$

For the inductor the impedance is *directly* proportional to frequency; for the capacitor it is *inversely* proportional.

The *inductive reactance* is

$$\boxed{X_L = \omega L} \tag{11.1-19}$$

and

$$\boxed{Z_L = jX_L = X_L\underline{/90^\circ}} \tag{11.1-20}$$

The phasor diagram and time functions for the inductor current and voltage are shown in Fig. 11.4. For the inductor current and voltage we have the rule:

For an inductor driven by a sinusoid the current lags the voltage by 90°.

EXAMPLE 11.1-2 *Current and Voltage in an Inductor*

The current through the 30-μH inductance in an AM receiver input circuit when it is tuned to 1.6 MHz is

$$i(t) = 4.3 \cos(\omega t + 40^\circ)\ \mu A$$

Find the phasor and time domain voltages and plot the phasor diagram.

Solution

The reactance of the inductor is

$$\begin{aligned} X_L &= \omega L = (2\pi \times 1.6 \times 10^6)(30 \times 10^{-6}) \\ &= 302\ \Omega \end{aligned}$$

The impedance is then

$$Z_L = j302 = 302\underline{/90^\circ}\ \Omega$$

The phasor current is

$$I = \frac{4.3}{\sqrt{2}}\underline{/40^\circ} = 3.04\underline{/40^\circ}\ \mu A$$

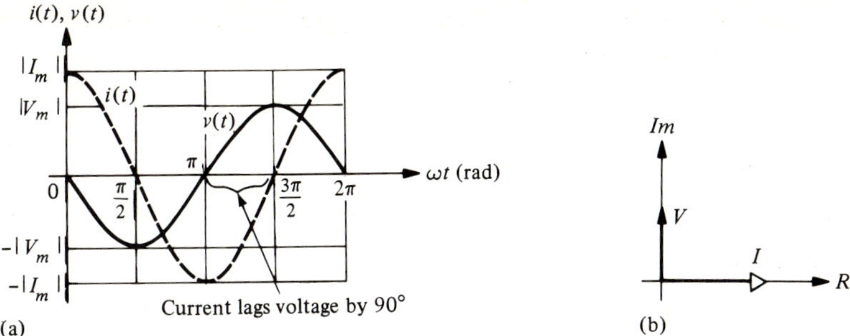

FIGURE 11.4
Sinusoidal response of an inductor. (a) Waveforms. (b) Phasor diagram.

The phasor voltage is

$$V = ZI = (302\angle 90^\circ\ \Omega)(3.04\angle 40^\circ\ \mu\text{A})$$
$$= 918\angle 130^\circ\ \mu\text{V}$$

The corresponding time function is

$$v(t) = 918\sqrt{2}\cos(\omega t + 130^\circ)\ \mu\text{V}$$
$$= 1.3\cos(\omega t + 130^\circ)\ \text{mV}$$

The phasor diagram is shown in Fig. 11.5.

• • •

Admittance

When we studied resistance, we found that we often had to use its reciprocal in equations, especially with parallel circuits. A new symbol G was defined for this and it was called conductance. The same requirement

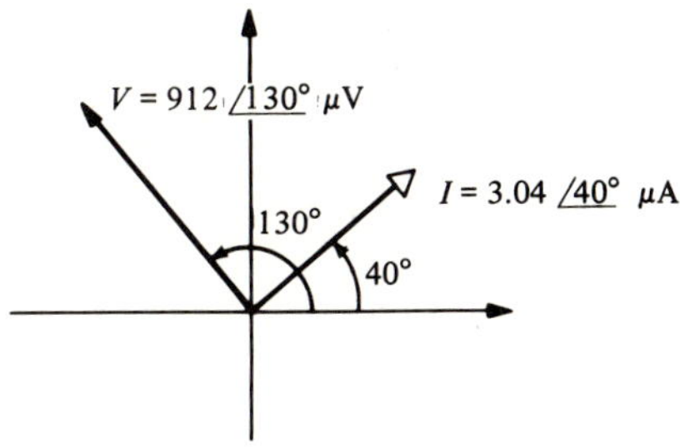

FIGURE 11.5
Phasor diagram for inductor of Example 11.1-2. Note that current and voltage are drawn to different scales.

arises with impedance: the term *admittance* is used for the reciprocal of impedance, and its symbol is Y. Thus

$$Y = \frac{1}{Z} \tag{11.1-21}$$

The units of Y are the same as for G, that is, siemens (S). As we did with impedance, it is convenient, when dealing with inductors or capacitors, to separate the magnitude from the angle. The magnitude is then called the *susceptance* B. The susceptance of a capacitor is

$$B_C = \omega C \tag{11.1-22}$$

The admittance is then

$$Y_C = jB_C = j\omega C = \omega C\angle 90^\circ \tag{11.1-23}$$

For an inductor, the susceptance is

$$B_L = \frac{1}{\omega L} \tag{11.1-24}$$

and the admittance is

$$Y_L = -jB_L = \frac{-j}{\omega L} = \omega L\angle -90^\circ \tag{11.1-25}$$

EXAMPLE 11.1-3 Admittance

Find the admittance of the capacitor of Example 11.1-1 and the inductor of Example 11.1-2.

TABLE 11.1.

	Reactance (Ω)	Impedance (Ω)	Susceptance (S)	Admittance (S)	Current-voltage phase relationship
R	—	$R\underline{/0^\circ}$	—	$\frac{1}{R}\underline{/0^\circ}$	In phase
L	$X_L = \omega L$	$\omega L\underline{/90^\circ}$	$B_L = \frac{1}{\omega L}$	$\frac{1}{\omega L}\underline{/-90^\circ}$	Current lags by 90°
C	$X_C = \frac{1}{\omega C}$	$\frac{1}{\omega C}\underline{/-90^\circ}$	$B_C = \omega C$	$\omega C\underline{/90^\circ}$	Current leads by 90°

Solution

For the 5-pF capacitor with $\omega = 10^6$ rad/s,

$$Y_C = jB_C = j\omega C = j10^6 \times 5 \times 10^{-12} = j5 \times 10^{-6}\text{ S} = 5\underline{/90^\circ}\ \mu\text{S}$$

For the 30-μH inductor with $f = 1.6$ MHz

$$Y_L = -jB_L = \frac{-j}{\omega L} = \frac{-j}{(2\pi)(1.6 \times 10^6)(30 \times 10^{-6})}$$

$$= -j0.0033\text{ S} = 3.3\underline{/-90^\circ}\text{ mS}$$

• • •

The important facts concerning individual R, L, and C elements are given in Table 11.1.

• • •

LEARNING EXERCISES FOR SEC. 11.1

1. Find the reactance of a 5-μF capacitor at a frequency of 3.5 kHz.
2. Find the impedance of the capacitor of Exercise 11.1-1.
3. Find the reactance of a 50-μH inductor at a frequency of 4.7 MHz.
4. Find the impedance of the inductor of Exercise 11.1-3.

Ans. 9.1; 1.48; $1.48\underline{/90^\circ}$; $9.1\underline{/-90^\circ}$

• • •

11.2 IMPEDANCE COMBINATIONS

Series and parallel impedances may be combined in exactly the same manner as resistances. The difference is that, in general, the values of impedances are complex numbers. Consider, for example, a resistor and capacitor in series as shown in Fig. 11.6. We wish to find the impedance looking in at terminals a-b. This is done using the same rule that we used for

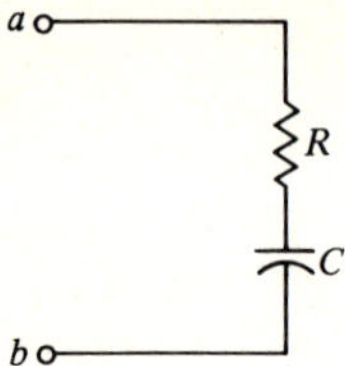

FIGURE 11.6
Series impedances.

series resistances, that is:

> *In a series circuit, the total impedance is the sum of the individual impedances.*

For this circuit we then have:

$$Z_{ab} = Z_R + Z_C \tag{11.2-1}$$

Using the results of the previous section,

$$Z_{ab} = R - jX_C = R - j\left(\frac{1}{\omega C}\right) \tag{11.2-2}$$

EXAMPLE 11.2-1 Series Impedances

In Fig. 11.6 the circuit represents a portion of a high-fidelity tone control. The resistance is 10 kΩ and the capacitance is 0.1 μF. Find the impedance of the circuit at $f = 250$ Hz.

Solution

The reactance of the capacitor is

$$X_C = \frac{1}{\omega C} = \frac{1}{2\pi f C} = \frac{1}{(2\pi)(250)(0.1 \times 10^{-6})} = 6.4 \text{ k}\Omega$$

The impedance is then

$$Z_{ab} = R - jX_C = 10 - j6.4 \text{ k}\Omega = 11.9\underline{/-32.6^\circ} \text{ k}\Omega$$

• • •

EXAMPLE 11.2-2 Impedance of an Inductive Circuit

In Fig. 11.7 we show a circuit consisting of a resistor and an inductor in series. Find the impedance looking into terminals c-d at $f = 25$ kHz.

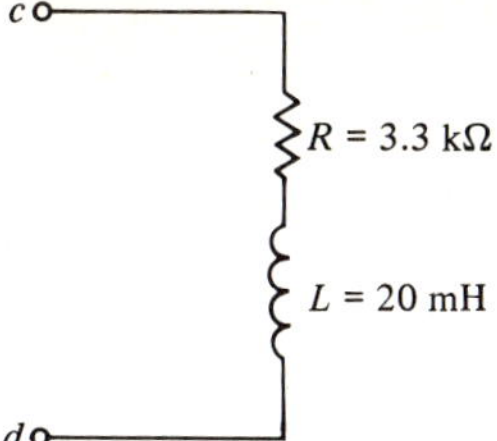

FIGURE 11.7
Circuit for Example 11.2-2.

Solution
The reactance of the inductor is

$$X_L = \omega L = 2\pi \times 25 \times 10^3 \times 0.02 = 3.14 \text{ k}\Omega$$

The impedance is then

$$Z_{cd} = R + jX_L = 3.3 + j3.14 \text{ k}\Omega = 4.56\underline{/43.6^\circ} \text{ k}\Omega$$

• • •

The Impedance Plane

It is often convenient to plot impedance in a complex plane of its own as shown in Fig. 11.8. The horizontal axis represents the real part of Z while the vertical axis represents the imaginary part of Z. For example, consider the capacitive circuit of Example 11.2-1 for which $Z_{ab} = R - jX_C = 10 - j6.4$ kΩ. This is plotted in the complex plane by adding the vectors for R and $-jX_C$ as shown. It must be noted that the complex impedance is simply a complex number (or vector) in the complex Z-plane. It is *not* a phasor which when rotated will produce a projection that represents a sinusoidal voltage or current.

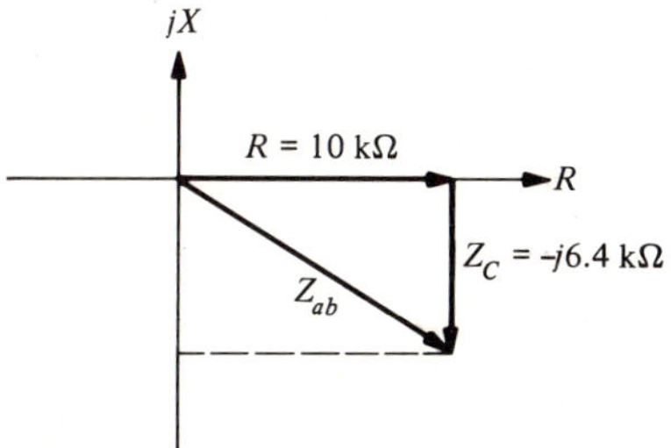

FIGURE 11.8
Complex plane for impedance (*Z*-plane).

Parallel Impedances

Consider the parallel circuit shown in Fig. 11.9a. The impedance can be found using the formula for parallel impedances

$$Z_{ab} = \frac{Z_R Z_C}{Z_R + Z_C} \tag{11.2-3}$$

or, alternately, making use of admittances:

$$Y_{ab} = Y_R + Y_C \tag{11.2-4}$$

The rule can be stated as follows:

> *In a parallel circuit, the total admittance is the sum of the individual admittances.*

The admittance of individual inductors and capacitors was found in Sec. 11.1.

EXAMPLE 11.2-3 Impedance of Parallel Circuits

In order to compare the two methods, find

a. The impedance Z_{ab} of the circuit of Fig. 11.9a using the parallel impedance formula.

b. The impedance Z_{cd} of Fig. 11.9b using admittances.

c. Plot the results in the Z-plane.

For both circuits, $R = 6.8$ kΩ, and the frequency is 5 kHz. The inductance is 0.2 H and the capacitance is 0.005 μF.

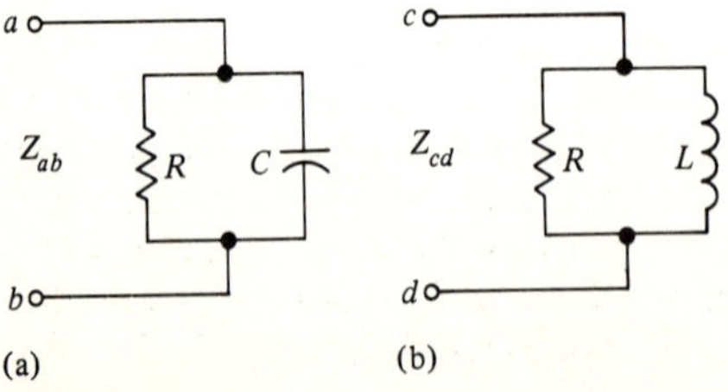

FIGURE 11.9
Parallel impedances. (a) *RC*. (b) *RL*.

Solution

a. We begin by finding the reactance of the capacitor

$$X_C = \frac{1}{2\pi fC} = \frac{1}{(2\pi)(5 \times 10^3)(0.005 \times 10^{-6})} = 6.4\ \text{k}\Omega$$

Then $Z_C = 6.4\underline{/-90^\circ}\ \text{k}\Omega$ and, keeping all impedances in units of kilohms,

$$Z_{ab} = \frac{Z_R Z_C}{Z_R + Z_C} = \frac{6.8\underline{/0^\circ} \times 6.4\underline{/-90^\circ}}{6.8 - j6.4}$$

$$= \frac{6.8\underline{/0^\circ} \times 6.4\underline{/-90^\circ}}{9.3\underline{/-43^\circ}}$$

$$= 4.7\underline{/-47^\circ}\ \text{k}\Omega$$

$$= 3.2 - j3.4\ \text{k}\Omega$$

A scientific calculator was used to do the rectangular-polar and polar-rectangular conversions.

b. The inductive susceptance is

$$B_L = \frac{1}{2\pi fL} = \frac{1}{(2\pi)(5 \times 10^3)(0.2)} = 0.16\ \text{mS}$$

Then $Y_L = -jB_L = 0.16\underline{/-90^\circ}$ mS. For the resistor, $G = 1/6.8\ \text{k}\Omega = 0.15$ mS. Next,

$$Y_{cd} = Y_R + Y_L = G - jB_L = 0.15 + 0.16\underline{/-90^\circ}\ \text{mS}$$

$$= 0.15 - j0.16\ \text{mS}$$

$$= 0.22\underline{/-47^\circ}\ \text{mS}$$

$$Z_{cd} = \frac{1}{Y_{cd}} = \frac{1}{0.22\underline{/-47^\circ}} = 4.5\underline{/47^\circ}\ \text{k}\Omega = 3.1 + j3.3\ \text{k}\Omega$$

Again, the scientific calculator was used to advantage and the two methods appear to involve about the same amount of labor.

c. The Z-plane diagrams are shown in Fig. 11.10. Note that the impedance of the circuit with the capacitor lies in the fourth quadrant. In general, any impedance that lies in this quadrant is considered to be *capacitive* because its reactive component is negative. For the circuit with the inductor, the impedance lies in the first quadrant. Any impedance in this quadrant is *inductive* because its reactive component is positive. When a circuit is composed of passive R, L, and C elements its resistive component must always be positive so that its impedance will always lie in either the first or fourth quadrants.

• • •

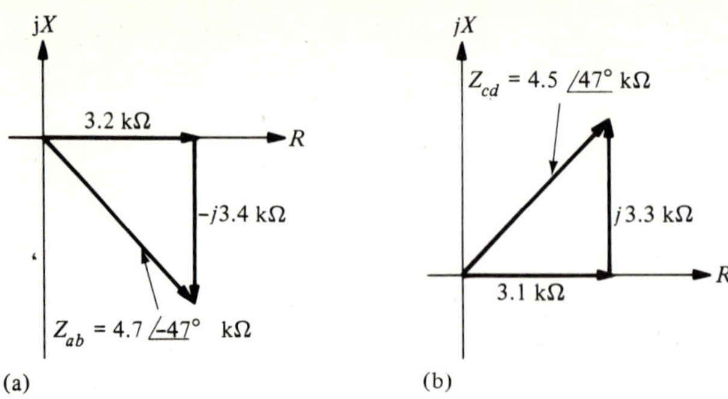

FIGURE 11.10
Example 11.2-3. (a) Z_{ab}. (b) Z_{cd}.

Equivalent Circuits

When we studied resistive networks, we found that a circuit consisting of any number of resistors could be reduced to a single equivalent resistance as far as one pair of terminals was concerned. The same is true for impedances. In general, for impedances the reduction leads to a single equivalent inpedance

$$Z = R_{eq} + jX_{eq} \tag{11.2-5}$$

where

R_{eq} = real part of impedance Z, called the *resistive* component
X_{eq} = imaginary part of Z, called the *reactive* component

A series circuit having the impedance given by Eq. (11.2-5) is shown in Fig. 11.11a.

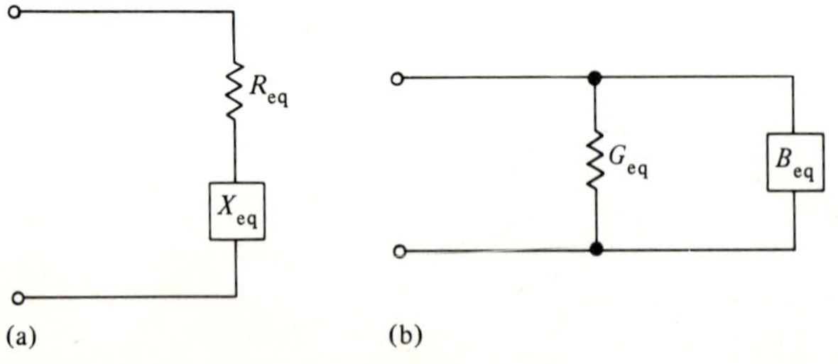

FIGURE 11.11
Equivalent circuits. (a) Series. (b) Parallel.

Like impedance, admittances can be combined into a single equivalent admittance. Thus, in general,

$$Y = G_{eq} + jB_{eq} \tag{11.2-6}$$

where

G_{eq} = real part of Y, called *conductance*, in siemens (S)
B_{eq} = imaginary part of Y, called *susceptance*, in siemens (S)

Equation (11.2-6) suggests an equivalent parallel circuit, which is shown in Fig. 11.11b.

As an example of equivalent impedance, consider the parallel R-C circuit of Fig. 11.9a. In Example 11.2-3 we found the impedance of this parallel circuit to be

$$Z_{ab} = 3.2 - j3.4 \text{ k}\Omega$$

Therefore, the series circuit shown in Fig. 11.11a with $R_{eq} = 3.2$ kΩ and $X_{eq} =$ 3.4 kΩ capacitive cannot be distinguished from the parallel circuit of Fig. 11.8a with $R = 6.8$ kΩ and $X_C = 6.4$ kΩ. The equivalence is shown with these values in Fig. 11.12. It must be emphasized that *the equivalence only holds at one frequency*, in this case 5 kHz.

Other examples of equivalent impedance and admittance are considered in Sec. 11.4 and in the problems.

Pitfalls

Consider the circuit of Fig. 11.13 and let us calculate the admittance Y_{ab}. We might be tempted to say that

$$G = \frac{1}{R} = \frac{1}{10 \text{ k}\Omega} = 0.1 \text{ mS}$$

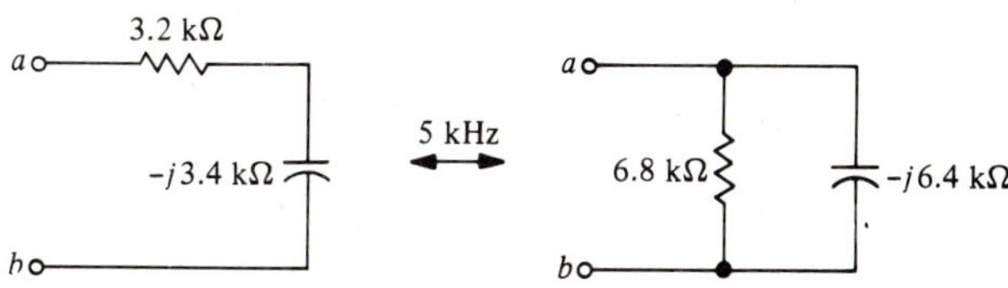

FIGURE 11.12
At a frequency of 5 kHz, the two circuits are exactly equivalent at terminals *ab*.

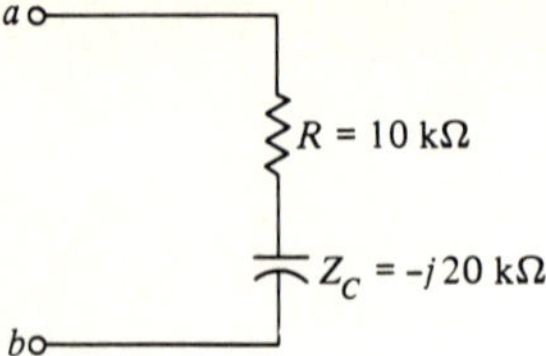

FIGURE 11.13
A series circuit.

and

$$B_C = \frac{1}{X_C} = \frac{1}{20\ \text{k}\Omega} = 0.05\ \text{mS}$$

so that

$$Y_{ab} = 0.1 + j0.05\ \text{mS}$$

However, this is *incorrect* because the rectangular coordinates 10 kΩ and $-j20$ kΩ represent the resistive and reactive components of the *series* circuit. We are looking for G_{eq} and jB_{eq}, which represent the rectangular coordinates of a *parallel* circuit. In order to find Y_{ab} correctly, we must express it in rectangular coordinates as follows:

$$Y_{ab} = \frac{1}{Z_{ab}} = \frac{1}{10 - j20\ \text{k}\Omega} = \frac{1}{22.4\underline{/-63^\circ}\ \text{k}\Omega} = 0.045\underline{/63^\circ}\ \text{mS}$$
$$= 0.02 + j0.04\ \text{mS}$$

From this result,

$$G_{eq} = 0.02\ \text{mS}$$
$$B_{eq} = 0.04\ \text{mS}$$

These answers should be compared with the wrong answers above.

This example points up the fact that care must be exercised when calculating the *admittance* of elements in *series* or the *impedance* of elements in *parallel.*

As a final example we consider a circuit with four elements in series.

EXAMPLE 11.2-9 *Series Impedances*

Find Z_{ab} in the circuit of Fig. 11.14a. The frequency is 500 Hz.

Solution

The capacitive reactance is

$$X_C = \frac{1}{2\pi fC} = \frac{1}{(2\pi)(500)(0.05 \times 10^{-6})} = 6.4\ \text{k}\Omega$$

Then

$$Z_C = 6.4\underline{/-90^\circ}\ \text{k}\Omega = -j6.4\ \text{k}\Omega$$

The inductive reactance is

$$X_L = 2\pi fL = (2\pi)(500)(0.4) = 1.3\ \text{k}\Omega$$

Then

$$Z_L = 1.3\underline{/90^\circ}\ \text{k}\Omega = j1.3\ \text{k}\Omega$$

The circuit is shown in Fig. 11.14b with the elements and their impedance values. The total impedance is (with all impedances in kilohms)

$$\begin{aligned} Z_{ab} &= 2.2 - j6.4 + 4.7 + j1.3 \\ &= 6.9 - j5.1\ \text{k}\Omega \\ &= 8.6\underline{/-36^\circ}\ \text{k}\Omega \end{aligned}$$

The complex plane diagram for this impedance is shown in Fig. 11.14c. Note how the vectors representing the individual impedances are placed tail-to-head. When

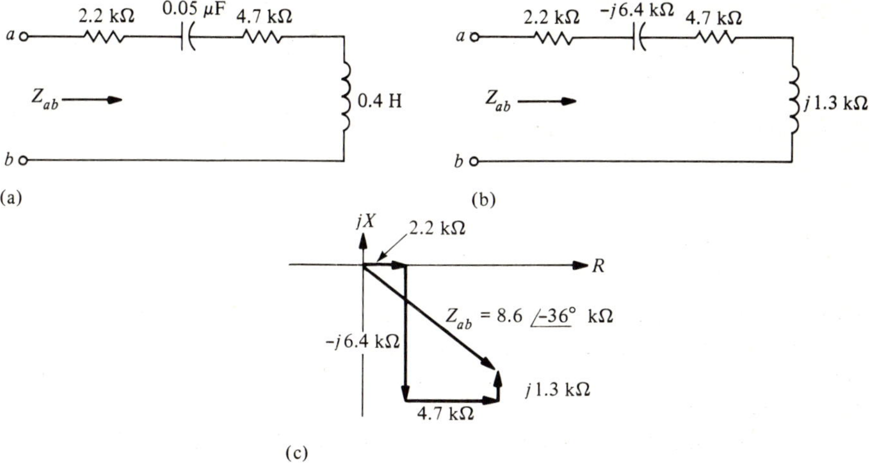

FIGURE 11.14
Example 11.2-4. (a) A series circuit. (b) Circuit with impedances. (c) Phasor diagram.

TABLE 11.2

	Impedance Series Elements, R, L, C	Admittance Parallel Elements, R, L, C
Rectangular form	$Z = R_{eq} + jX_{eq}$	$Y = G_{eq} + jB_{eq}$
Resistive component	R_{eq}	G_{eq}
Reactive component	$X_{eq} = X_L - X_C$	$B_{eq} = B_C - B_L$
Impedance/admittance diagram	jX, X_L, R_{eq}, R, X_C	jB, B_C, G_{eq}, G, B_L
Polar form	$Z = \|Z\| \underline{/\phi}$	$Y = \|Y\| \underline{/\phi}$
Magnitude	$\|Z\| = \sqrt{R_{eq}^2 + X_{eq}^2}$	$\|Y\| = \sqrt{G_{eq}^2 + B_{eq}^2}$
Phase angle	$\phi = \arctan \frac{X_{eq}}{R_{eq}}$	$\phi = \arctan \frac{B_{eq}}{G_{eq}}$

this is done, the sum of the impedances is the vector from the first tail to the last head. This is a convenient way to graphically add a large number of vectors.

• • •

Table 11.2 summarizes the important points concerning impedance and admittance for series and parallel circuits.

• • •

LEARNING EXERCISE FOR SEC. 11.2

1. A 470-Ω resistor and a 0.033-μF capacitor are first connected in series and then in parallel. Find the impedance of each connection at a frequency of 7.5 kHz.

Ans. $0.8\underline{/-54^\circ}$; $0.38\underline{/-36^\circ}$

• • •

11.3 SINUSOIDAL CIRCUIT ANALYSIS

In the last section we considered the impedance of circuits containing more than one kind of element. In this section we find the voltages and currents in these circuits when the input signals are sinusoidal. The techniques that we learned for the computation of dc circuit responses all carry over to ac circuits, with the exception that we must use vector algebra instead of ordinary algebra. However, the more complicated arithmetic is easily handled with a good scientific calculator.

We begin by considering a series R-C circuit as shown in Fig. 11.15a.

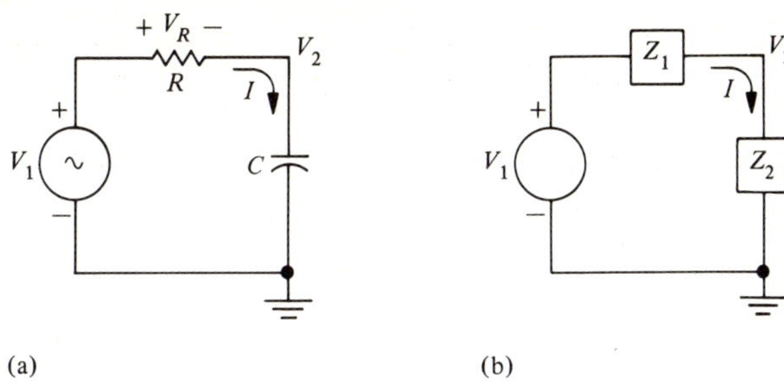

FIGURE 11.15
Sinusoidal analysis. (a) Circuit. (b) Generalized circuit.

The impedance of this circuit was found in the previous section. Now we have connected a sinusoidal voltage source to the circuit and we want to find all of the currents and voltages.

A generalized two-element series circuit is shown in Fig. 11.15b. We will set down the general theory using this circuit and then consider a numerical example.

The current is found using Ohm's law:

$$I = \frac{\mathbf{V}_1}{\mathbf{Z}_1 + \mathbf{Z}_2} \tag{11.3-1}$$

Here $\mathbf{V}_1$ and I are effective phasors and, in general, $\mathbf{Z}_1$ and $\mathbf{Z}_2$ will be complex numbers.

The voltage $\mathbf{V}_2$ can be found from Ohm's law when the current is known. However, it is more instructive to use the voltage divider formula

$$A_v = \frac{\mathbf{V}_2}{\mathbf{V}_1} = \frac{\mathbf{Z}_2}{\mathbf{Z}_1 + \mathbf{Z}_2} \tag{11.3-2}$$

For ac signals, the voltage transfer ratio A_v is, in general, a complex number (not a phasor) that represents the ratio of the output voltage phasor to the input voltage phasor. The magnitude $|A_v|$ is the ratio of the magnitudes, $|\mathbf{V}_2|/|\mathbf{V}_1|$, and the angle of A_v is the phase difference between $\mathbf{V}_2$ and $\mathbf{V}_1$. A numerical example will help to clarify these ideas.

EXAMPLE 11.3-1 *Series RC* Circuit

Find all currents and voltages in the circuit of Fig. 11.15a if $R = 4.7\ \text{k}\Omega$, $C = 0.01\ \mu\text{F}$, $\mathbf{V}_1 = 10\underline{/0^\circ}\ \text{V}$, and $f = 1\ \text{kHz}$. Plot a complete phasor diagram.

Solution

The sequence of steps to be followed in problems of this sort is:

1. Convert all input signals to phasor form, if necessary.
2. Replace all inductors and capacitors by their impedance values.
3. Calculate the desired response phasors.
4. Convert the response phasors to time functions if necessary.

We follow this sequence of steps.

1. The input signal is given in phasor form, $V_1 = 10\underline{/0°}$ V.
2. The capacitive reactance is

$$X_C = \frac{1}{2\pi fC} = \frac{1}{(2\pi)(10^3)(0.01 \times 10^{-6})} = 15.9\ \text{k}\Omega$$

The impedance is

$$Z_C = 15.9\underline{/-90°}\ \text{k}\Omega = -j15.9\ \text{k}\Omega$$

3. The unknowns in the circuit are the current I, the voltage V_R, across the resistor, and the capacitor voltage V_2. For purposes of illustration, we will use a different law or rule to find each of these.

The current is found using Ohm's law [see Eq. (11.3-1)]:

$$I = \frac{V_1}{Z_1 + Z_2} = \frac{10\underline{/0°}\ \text{V}}{4.7 - j15.9\ \text{k}\Omega} = \frac{10\underline{/0°}\ \text{V}}{16.6\underline{/-74°}\ \text{k}\Omega}$$
$$= 0.6\underline{/74°}\ \text{mA}$$

Observe that the current leads the voltage by 74° in this circuit. In a capacitive circuit, the current *always leads* the voltage.

The voltage V_2 is found using the voltage divider rule [see Eq. (11.3-2)] with all impedances in kilohms:

$$A_v = \frac{Z_2}{Z_1 + Z_2} = \frac{-j15.9}{4.7 - j15.9} = \frac{15.9\underline{/-90°}}{16.6\underline{/-74°}} = 0.96\underline{/-16°}$$

Then

$$V_2 = A_vV_1 = 0.96\underline{/-16°} \times 10\underline{/0°}\ \text{V} = 9.6\underline{/-16°}\ \text{V}$$

Note the information contained in the complex number A_v. Its magnitude, 0.96, gives us the ratio of the magnitude of V_2 to that of V_1. This information is often all that is required. The fact that the angle of A_v is $-16°$ tells us that V_2 lags V_1 by 16° as we will see in the phasor diagram.

The resistor voltage V_R will be found using KVL. Tracing the loop clockwise we

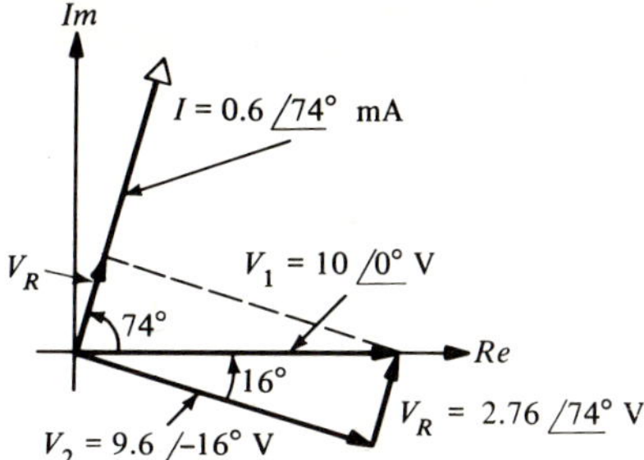

FIGURE 11.16
Phasor diagram for Example 11.3-1.

have

$$-V_1 + V_R + V_2 = 0$$

$$\begin{aligned} V_R &= V_1 - V_2 \\ &= 10\angle 0° - 9.6\angle -16° \\ &= 10 + j0 - (9.23 - j2.65) \\ &= 0.77 + j2.65 \\ &= 2.76\angle 74° \text{ V} \end{aligned}$$

Since the time functions are not required, this completes the solution.

The phasor diagram is shown in Fig. 11.16 and the following points should be carefully observed in connection with it.

1. The current through and voltage across the resistor are in phase, as they must be.
2. The current leads the capacitor voltage V_2 by 90°, as it must.
3. As a consequence of these facts the resistor voltage V_R leads the capacitor voltage V_2 by 90°.
4. The vector sum of the phasors representing V_R and V_2 is precisely the input voltage V_1, in accordance with KVL.
5. The output voltage V_2 lags the input voltage by 16°.

• • •

In the next example we illustrate KCL and the current divider rule.

EXAMPLE 11.3-2 A Parallel Circuit

Circuits like that of Fig. 11.17 are found in all radio and TV receivers. Find all currents and the voltage if $f = 5$ kHz.

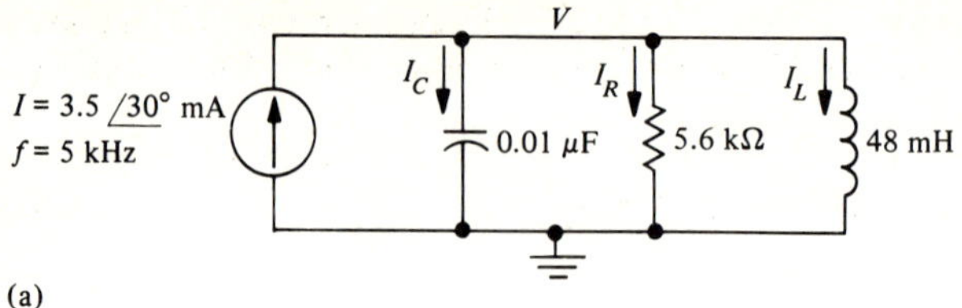

(a)

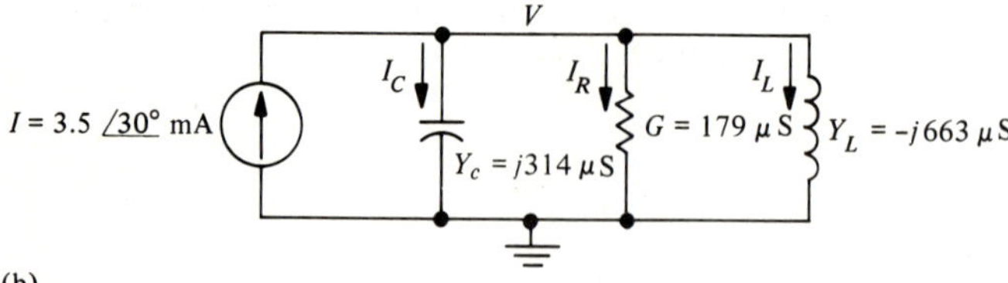

(b)

FIGURE 11.17
Example 11.3-2. (a) Circuit. (b) Admittance circuit.

Solution

We begin by finding the individual susceptances. For the capacitor

$$B_C = 2\pi fC = (2\pi)(5 \times 10^3)(0.01 \times 10^{-6}) = 314\ \mu S$$

For the inductor

$$B_L = \frac{1}{2\pi fL} = \frac{1}{(2\pi)(5 \times 10^3)(0.048)} = 663\ \mu S$$

For the resistor, the conductance is

$$G = \frac{1}{R} = \frac{1}{5.6 \times 10^3} = 179\ \mu S$$

The admittance circuit is shown in Fig. 11.17b. The total admittance is

$$\begin{aligned} Y &= G + jB_C - jB_L \\ &= 179 + j314 - j663 \\ &= 179 - j349 \\ &= 392\underline{/-63^\circ}\ \mu S \end{aligned}$$

The voltage is

$$\begin{aligned} V &= \frac{I}{Y} = \frac{3.5\underline{/30^\circ}\ \text{mA}}{0.392\underline{/-63^\circ}\ \text{mS}} \\ &= 8.93\underline{/93^\circ}\ \text{V} \end{aligned}$$

There remains to find the three element currents. For illustration we will find each of the three by a different method. First we use Ohm's law to find the capacitor current:

$$I_C = Y_C V = jB_C V = (314\underline{/90^\circ}\ \mu\text{S}) \times (8.93\underline{/93^\circ}\ \text{V})$$
$$= 2.8\underline{/183^\circ}\ \text{mA}$$

Next we find the resistor current using the current divider rule. For this circuit the current divider is written in terms of admittance as

$$A_i = \frac{I_r}{I} = \frac{Z_{LC}}{Z_R + Z_{LC}} = \frac{Y_R}{Y_R + Y_{LC}} = \frac{G}{G + j(B_C - B_L)}$$
$$= \frac{179}{392\underline{/-63^\circ}} = 0.457\underline{/63^\circ}$$

Then

$$I_R = 0.457\underline{/63^\circ} \times 3.5\underline{/30^\circ}\ \text{mA} = 1.6\underline{/93^\circ}\ \text{mA}$$

Finally, we use KCL to find the inductor current. At the upper node, we have

$$I = I_C + I_R + I_L$$

from which

$$I_L = I - I_R - I_C$$
$$= 3.5\underline{/30^\circ} - 1.6\underline{/93^\circ} - 2.8\underline{/183^\circ}$$
$$= 3.03 + j1.75 - (-0.08 + j1.6) - (-2.8 - j0.15)$$
$$= 5.9 + j0.3$$
$$= 5.91\underline{/2.9^\circ}\ \text{mA} \approx 5.9\underline{/3^\circ}\ \text{mA}$$

The phasor diagram is shown in Fig. 11.18. The following points should be noted:

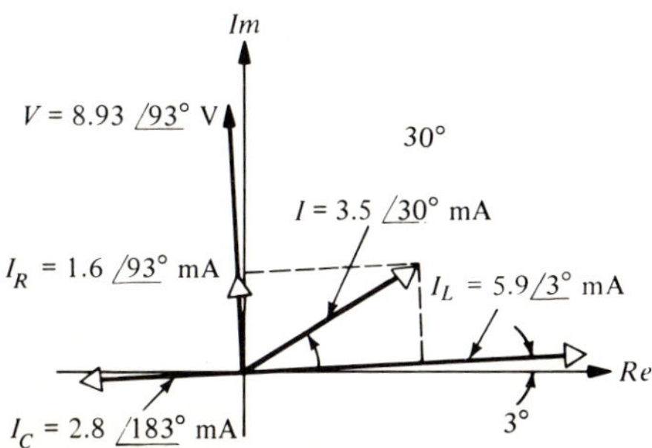

FIGURE 11.18
Phasor diagram for Example 11.3-2.

1. The resistor current and voltage are in phase.
2. The inductor current lags the voltage by 90°.
3. The capacitor current leads the voltage by 90°.
4. The three element currents add vectorially to equal the source current.
5. The source current lags the voltage by 63°. Therefore, the circuit is predominantly inductive as viewed from the source.

• • •

LEARNING EXERCISE FOR SEC. 11.3

1. A 10-kΩ resistor and a 20-mH inductor are connected in series with a 5-V, 40-kHz signal generator. Find the current through and the voltage across each element with the input voltage as a zero phase reference.

Ans. $4.5\angle -26.6°$; $0.45\angle -26.6°$; $2.3\angle 63.4°$

• • •

11.4 ADDITIONAL CIRCUIT ANALYSIS TECHNIQUES

In this section we present ac examples of the dc circuit analysis techniques studied in Chap. 5. As stated previously, all of these techniques apply to ac circuits, with vector algebra being used instead of ordinary algebra.

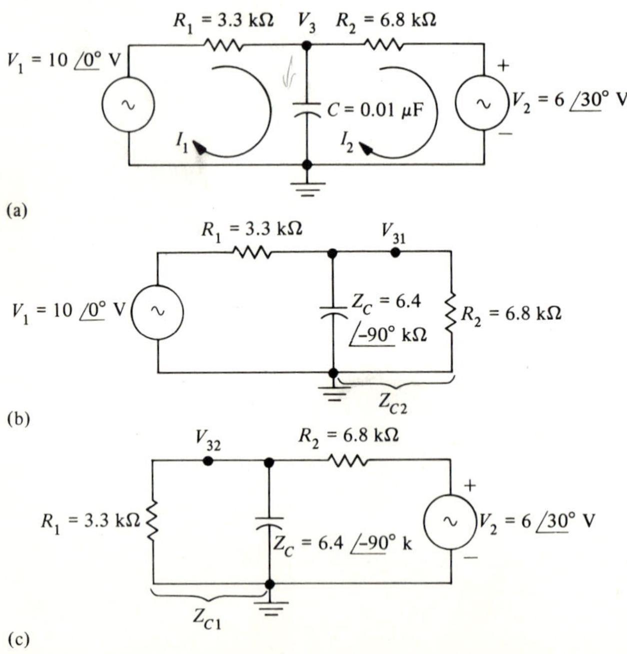

FIGURE 11.19
Example 11.4-1. (a) Original circuit. (b) Circuit with 6-V source set to zero. (c) Circuit with 10-V source set to zero.

EXAMPLE 11.4-1 Superposition in an AC Circuit

In the circuit of Fig. 11.19a, find V_3 using superposition. The frequency is 2.5 kHz.

Solution

Recall that the superposition theorem states that the response of a linear circuit with more than one source can be found by *adding* the responses to each source acting alone, all others being set to zero. Since the circuit being analyzed has two sources, we have to solve two separate circuits, one for each voltage source, with the other voltage source replaced by a short circuit. The two circuits are shown in Figs. 11.19b and c. The final superposition will then be

$$V_3 = V_{31} + V_{32}$$

The calculations proceed as follows:
For both circuits

$$Z_C = \frac{1}{2\pi f C}\angle -90^\circ = \frac{1}{(2\pi)(2.5 \times 10^3)(0.01 \times 10^{-6})}\angle -90^\circ$$
$$= 6.4\angle -90^\circ \text{ k}\Omega$$

We will use the voltage divider formula to find V_{31} and V_{32}. For these calculations we will need the impedance values for the $R_2 - C$ and $R_1 - C$ parallel combinations. These are (all in kilohms):

$$Z_{C_2} = R_2 \parallel Z_C = \frac{(6.8)(6.4\angle -90^\circ)}{6.8 - j6.4} = \frac{43.5\angle -90^\circ}{9.34\angle -43.3^\circ}$$
$$= 4.66\angle -46.7^\circ \text{ k}\Omega = 3.2 - j3.39 \text{ k}\Omega$$

$$Z_{C_1} = R_1 \parallel Z_C = \frac{(3.3)(6.4\angle -90^\circ)}{3.3 - j6.4} = \frac{21.2\angle -90^\circ}{7.2\angle -62.7^\circ}$$
$$= 2.93\angle -27.3^\circ \text{ k}\Omega = 2.6 - j1.34 \text{ k}\Omega$$

Then

$$A_{v1} = \frac{V_{31}}{V_1} = \frac{Z_{C_2}}{R_1 + Z_{C_2}} = \frac{4.66\angle -46.7^\circ}{3.3 + 3.2 - j3.39} = \frac{4.66\angle -46.7^\circ}{6.5 - j3.39} = \frac{4.66\angle -46.7^\circ}{7.34\angle -27.5^\circ}$$
$$= 0.64\angle -19^\circ$$

and

$$V_{31} = A_{v_1}V_1 = 0.64\angle -19^\circ \times 10\angle 0^\circ = 6.4\angle -19^\circ = 6.05 - j2.08 \text{ V}$$

Also

$$A_{v_2} = \frac{V_{32}}{V_1} = \frac{Z_{C_1}}{R_2 + Z_{C_1}} = \frac{2.93\angle -27.3^\circ}{6.8 + 2.6 - j1.34} = \frac{2.93\angle -27.3^\circ}{9.4 - j1.34} = \frac{2.93\angle -27.3^\circ}{9.5\angle -8.1^\circ}$$
$$= 0.31\angle -19.2^\circ$$

and

$$V_{32} = A_{v_2}V_2 = 0.31\underline{/-19.2°} \times 6\underline{/30°} = 1.86\underline{/10.8°} = 1.83 + j0.35 \text{ V}$$

Finally

$$V_3 = V_{31} + V_{32} = 6.05 - j2.08 + 1.83 + j0.35 = 7.88 - j1.73$$
$$= 8.1\underline{/-12.4°} \text{ V}$$

The time required to make these calculations is considerably reduced if a scientific calculator with direct conversion polar-rectangular and rectangular-polar capability is available.

• • •

EXAMPLE 11.4-2 Mesh Analysis

Use mesh equations to find V_3 in the circuit of Fig. 11.19a.

Solution

The clockwise mesh currents are shown on the diagram. The KVL equations are:

Mesh 1

$$-V_1 + R_1I_1 + Z_C(I_1 - I_2) = 0$$

Mesh 2

$$Z_C(I_2 - I_1) + R_2I_2 + V_2 = 0$$

Collecting terms:

$$(R_1 + Z_C)I_1 - Z_CI_2 = V_1$$
$$-Z_CI_1 + (R_2 + Z_C)I_2 = -V_2$$

The desired result is V_3. Perhaps the easiest way to find this is to use KVL in the form

$$V_3 = V_1 - R_1I_1$$

Since V_1 and R_1 are known we need to solve the mesh equations for I_1. The determinant solution is:

$$I_1 = \frac{\begin{vmatrix} V_1 & -Z_C \\ -V_2 & R_2 + Z_C \end{vmatrix}}{\begin{vmatrix} R_1 + Z_C & -Z_C \\ -Z_C & R_2 + Z_C \end{vmatrix}} = \frac{\begin{vmatrix} 10\underline{/0°} & -6.4\underline{/-90°} \\ -6\underline{/30°} & 9.3\underline{/-44°} \end{vmatrix}}{\begin{vmatrix} 7.2\underline{/-63°} & -6.4\underline{/-90°} \\ -6.4\underline{/-90°} & 9.3\underline{/-44°} \end{vmatrix}}$$

$$= \frac{93\underline{/-44°} - 38.4\underline{/-60°}}{67\underline{/-107°} - 41\underline{/-180°}} = \frac{67 - j65 - 19 + j33}{-20 - j64 + 41} = \frac{48 - j32}{21 - j64} = \frac{58\underline{/-34°}}{67\underline{/-72°}}$$

$$= 0.87\underline{/38°} \text{ mA}$$

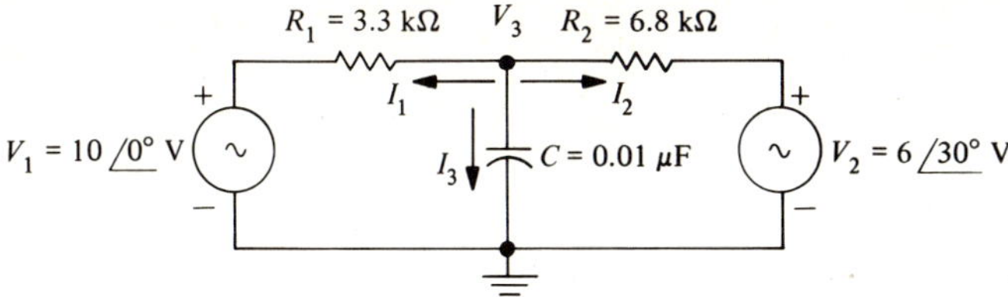

FIGURE 11.20
Circuit for Example 11.4-3.

Then

$$\begin{aligned} V_3 &= V_1 - R_1 I_1 \\ &= 10\angle 0^\circ - 3.3\text{ k}\Omega \times 0.87\angle 38^\circ\text{ mA} \\ &= 10 - (2.3 + j1.8) \\ &= 7.7 - j1.8 \\ &= 7.9\angle -13^\circ\text{ V} \end{aligned}$$

Except for a slight rounding error, this agrees with the previous example.

• • •

EXAMPLE 11.4-3 Node Analysis

Find V_3 in the circuit of Fig. 11.19a using node analysis.

Solution

Since there is only one unknown node voltage V_3, only one node equation is required. Applying KCL at the V_3 node (see Fig. 11.20) we get

$$I_1 + I_2 + I_3 = 0$$

$$G_1(V_3 - V_1) + G_2(V_3 - V_2) + Y_C V_3 = 0$$

Solving for V_3,

$$V_3 = \frac{G_1 V_1 + G_2 V_2}{G_1 + G_2 + Y_C}$$

where

$$G_1 = \frac{1}{3.3\text{ k}\Omega} = 0.3\text{ mS}$$

$$G_2 = \frac{1}{6.8\text{ k}\Omega} = 0.15\text{ mS}$$

$$Y_C = \frac{1}{Z_C} = \frac{1}{6.4\angle -90^\circ\text{ k}\Omega} = 0.16\angle 90^\circ\text{ mS} = j0.16\text{ mS}$$

Then, with admittance in millisiemens and voltage in volts,

$$V_3 = \frac{(0.3)(10) + (0.15)(5.2 + j3)}{0.3 + 0.15 + j0.16} = \frac{3.78 + j0.45}{0.45 + j0.16} = \frac{3.8\underline{/6.8^\circ}}{0.48/19.6^\circ}$$

$$= 7.9\underline{/-12.8^\circ}\text{ V}$$

Comparing with the previous two examples, the node-voltage method is clearly the easiest to apply for this particular circuit.

• • •

Thevenin's and Norton's theorems are useful for reducing a complicated resistor network to a simpler form so that calculations are more easily carried through. Both theorems carry through directly to ac circuits, as shown in the following example.

EXAMPLE 11.4-4 Thevenin's and Norton's Circuits

Find the Thevenin and Norton equivalents at terminals *ab* for the circuit shown in Fig. 11.21a. The frequency is 5 kHz.

Solution

The Thevenin impedance is found by replacing the voltage source with a short circuit and calculating the impedance looking into terminals *ab*:

$$Z_T = R_2 \parallel (R_1 + Z_L) = \frac{R_2(R_1 + j\omega L)}{R_2 + R_1 + j\omega L}$$

Substituting the given values,

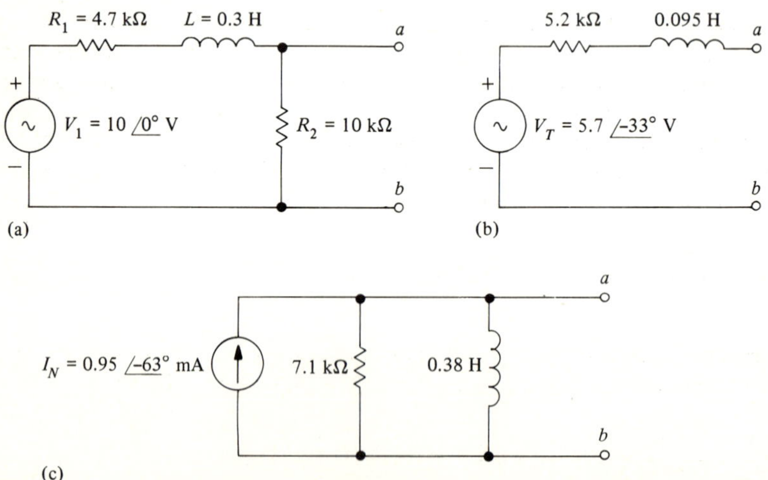

FIGURE 11.21
Example 11.4-4. (a) Original circuit. (b) Thevenin equivalent. (c) Norton equivalent.

$$\omega L = 2\pi \times 5 \text{ kHz} \times 0.3 \text{ H} = 9.4 \text{ k}\Omega$$

and, in units of kilohms,

$$Z_T = \frac{10(4.7 + j9.4)}{10 + 4.7 + j9.4} = \frac{47 + j94}{14.7 + j9.4} = \frac{105\angle 63^\circ}{17.4\angle 33^\circ} = 6\angle 30^\circ \text{ k}\Omega$$
$$= 5.2 + j3 \text{ k}\Omega$$

As expected, the Thevenin impedance is inductive. The Thevenin voltage can be found by several different methods. We use the voltage divider formula:

$$\frac{V_{ab}}{V_1} = \frac{R_2}{R_2 + R_1 + Z_L} = \frac{10}{10 + 4.7 + j9.4} = \frac{10}{17.4\angle 33^\circ} = 0.57\angle -33^\circ$$

and

$$V_T = V_{ab} = 0.57\angle -33^\circ \times V_1 = 5.7\angle -33^\circ \text{ V}.$$

The Thevenin circuit is shown in Fig. 11.21b. We can find the inductance value in the Thevenin circuit by observing that the inductive reactance is $X_T = 3$ kΩ. Then

$$X_T = \omega L_T$$

and

$$L_T = \frac{X_T}{\omega} = \frac{3000}{2\pi \times 5 \times 10^3} = 0.095 \text{ H}$$

It must be emphasized that *the Thevenin equivalent circuit is valid only at one frequency*, that is, 5 kHz. If the frequency is changed, both the resistance and inductance values in the Thevenin circuit will change. Impedance changes with frequency will be considered in Chap. 13.

The Norton current can be found by calculating the current through a short circuit connected from *a* to *b* in the circuit of Fig. 11.21a. However, we will use the circuit of Fig. 11.21b.

$$I_N = \frac{V_T}{Z_T} = \frac{5.7\angle -33^\circ \text{ V}}{6\angle 30^\circ \text{ k}\Omega} = 0.95\angle -63^\circ \text{ mA}$$

The Norton impedance is the same as the Thevenin impedance. In order to convert it to parallel circuit form, we proceed as follows: The Norton admittance is

$$Y_N = \frac{1}{Z_T} = \frac{1}{6\angle 30^\circ \text{ k}\Omega} = 0.167\angle -30^\circ \text{ mS}$$
$$= 0.14 - j0.084 \text{ mS}$$

The real part of this admittance represents the conductance of a resistor of value

$$R_N = \frac{1}{0.14 \text{ mS}} = 7.1 \text{ k}\Omega$$

This resistor is in parallel with an inductor for which

$$B_L = \frac{1}{\omega L} = 0.084 \text{ mS}$$

from which

$$L = \frac{1}{0.084 \times 10^{-3}\,\omega} = \frac{1}{0.084 \times 10^{-3} \times 2\pi \times 5 \times 10^3} = 0.38 \text{ H}$$

The Norton circuit is shown in Fig. 11.21c.

• • •

Equivalence

In the last example, we found a Thevenin impedance as two elements in series and a Norton impedance as two elements in parallel. The two circuits are shown in Fig. 11.22. They certainly appear different. For example, at $f = 0$ (dc), the series circuit is equivalent to a resistance of 5.2 kΩ while the parallel arrangement is equivalent to $R = 0$ (a short circuit). At very high frequencies, $\omega \rightarrow \infty$, the series arrangement is equivalent to an open circuit while the parallel arrangement is equivalent to a short circuit. However, at a frequency of 5 kHz, both circuits have impedance $Z_{ab} = 6\underline{/30^\circ}$ kΩ (the student should confirm this). It must be remembered that this equivalence is only valid at one frequency.

Another way to view this equivalence is as follows: if both circuits are enclosed in "black boxes" with only terminals *a* and *b* protruding from each box, then there is no way that the circuits can be distinguished from one another by measurements made *at a frequency of 5 kHz*.

In practice, equivalent circuits can, when they are appropriate, provide the designer with a variety of configurations and element values. Thus a choice can be made among the various equivalent circuits based on availability of components, cost, or other factors.

• • •

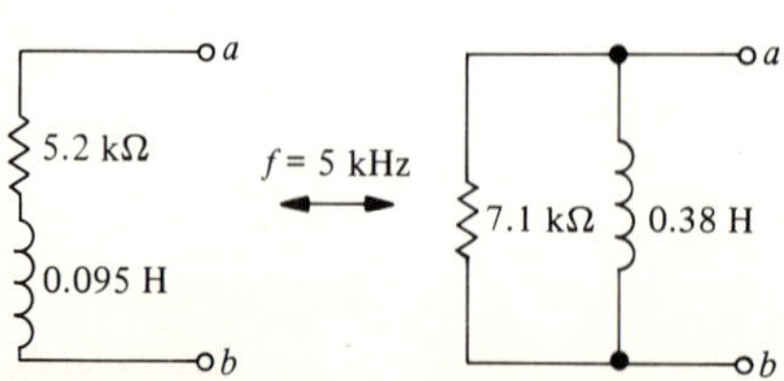

FIGURE 11.22
Equivalent circuits at 5 kHz.

LEARNING EXERCISE FOR SEC. 11.4

1. A 400-Hz control circuit consists of a 5.6-kΩ resistor in series with a 0.1-μF capacitor. What values of resistance and capacitance connected in parallel will be equivalent to this combination?

Ans. 0.033; 8.5

• • •

11.5 AC ANALYSIS USING ECAP

As noted previously, steady-state ac calculations are carried through basically like dc calculations except that all the quantities dealt with are complex rather than real, so that both magnitudes and angles must be found. The arithmetic can become very tedious, and the chance of making an error becomes high even when a scientific calculator is used.

The ECAP program can be used to perform steady-state ac analysis of circuits in essentially the same manner as it does for dc circuits. We illustrate this by means of the following example.

EXAMPLE 11.5-1 The ECAP AC Analysis Program

Use ECAP to find the node voltage and branch currents in the circuit of Fig. 11.23a.

Solution

Again an ECAP circuit with previously described branch and node numbering is prepared, as shown in Fig. 11.23b. The printout that includes the program is shown in Fig. 11.24a and differences between it and previous examples are as follows:

Line 1: Tells the computer that an ECAP ac analysis is desired.

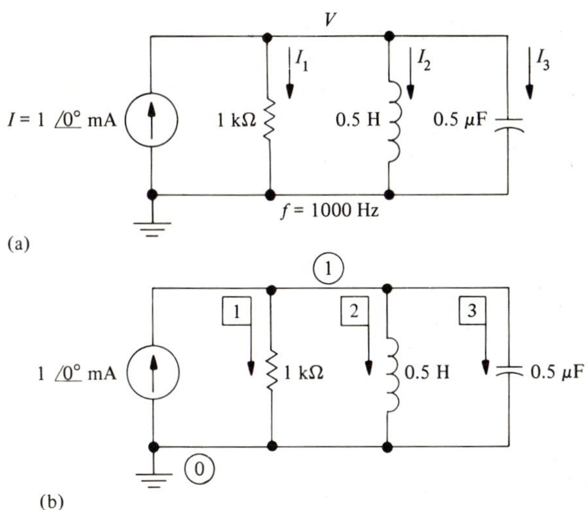

FIGURE 11.23
ECAP ac analysis. (a) Circuit for Example 11.5-1. (b) ECAP circuit.

```
      AC ANALYSIS
C       EXAMPLE 11.5-1
   B1 N(1,0), R=1E03, I=1E-03/0
   B2 N(1,0), L=0.5
   B3 N(1,0), C=0.5E-06
      FREQUENCY = 1000
      PRINT, NV, CA
      EXECUTE

EXECUTION

 FREQ =     .10000000E 04

       NODES            NODE VOLTAGES

 MAG     1- 1     .33387306E 00
 PHA              -.70495976E 02

    BRANCHES          ELEMENT CURRENTS

 MAG     1- 3     .33387306E-03    .10627509E-03    .10488931E-02
 PHA              -.70495976E 02   -.16049598E 03    .19504024E 02
       END
*EXIT*
```

1 2 3 4

$V = 0.33 \angle -70.5^\circ$ V

$I_1 = 0.33 \angle -70.5^\circ$ mA

$I_2 = 0.11 \angle -160^\circ$ mA

$I_3 = 1.05 \angle 19.5^\circ$ mA

(a)

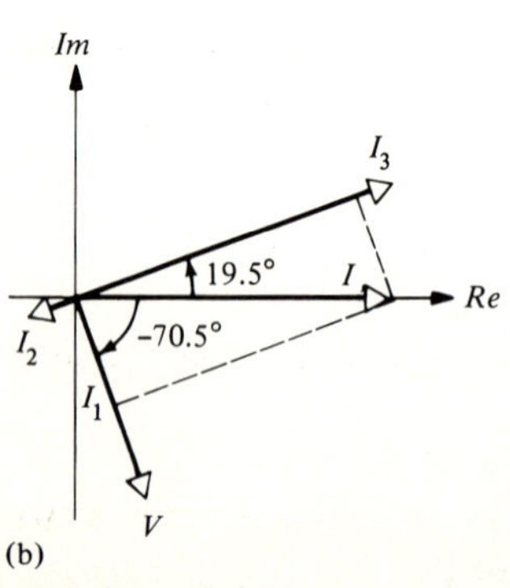

(b)

FIGURE 11.24
ECAP ac analysis. (a) Computer printout for Example 11.5-1. (b) Phasor diagram.

Line 2: The resistor and current source constitute branch 1. The resistance value is given in standard computer floating-point form. The number following the E indicates the power of 10 by which the number preceding the E is to be multiplied. Thus $R = 1E03$ means $R = 1 \times 10^3$ ($= 1$ kΩ). The current source is specified as $1E - 03 = 1 \times 10^{-3}$ A in magnitude followed by a slash (/) and a phase angle of zero degrees. The phase angles of all sources must be specified (in degrees) even if they are zero.
Line 3: Floating-point notation is used to specify the capacitance

$$C = 0.5E - 06 = 0.5 \times 10^{-6} = 0.5\ \mu F$$

Line 4: A frequency card must be included for ac analysis. The frequency must be nonzero and is specified in units of hertz.

The results printed after the program include both magnitude and phase (in degrees). A phasor diagram of the results is shown in Fig. 11.24b.

• • •

SUMMARY

In the previous chapter we learned how to handle sinusoidal waveforms of voltage and current using phasor methods. In this chapter we applied these methods to analysis of actual circuits that are typical of those encountered in all branches of electrical technology. The important points to be remembered in each section are summarized below.

Section

11.1

1. In ac circuits with sinusoidal waveforms the current through a capacitor leads the voltage across it by 90°.
2. Ohm's law for ac circuits is $V = ZI$.
3. Z is the impedance. It has the dimension of ohms and is in general a complex number that depends on the frequency of the applied voltage or current.
4. For a capacitor $Z_C = 1/\omega C \underline{/-90^\circ}$.
5. The reactance is the magnitude of the impedance. For a capacitor the reactance X_C is $1/\omega C$.
6. For an inductor the impedance is $\omega L \underline{/90^\circ}$.
7. The reactance of the inductor is $X_L = \omega L$.
8. In an inductance the current lags the voltage by 90°.
9. Admittance is the reciprocal of impedance.
10. For a capacitance the susceptance (magnitude of admittance) is $B_C = \omega C$.
11. For an inductor the susceptance is $B_L = 1/\omega L$.

11.2

12. Series and parallel impedances are combined in the same manner as series and parallel resistances.
13. Impedance, being a complex number, can conveniently be plotted in the impedance plane.
14. A circuit consisting of any number of impedances can be

reduced to a single equivalent impedance consisting of a resistor in series with a reactance.

15. Care must be exercised when finding the admittance of a number of series elements or the impedance of a number of parallel elements.
16. When a large number of impedances are to be combined in series it is convenient to draw a z-plane diagram in which the vectors representing the individual impedances are placed tail to head.

11.3 17. The circuit techniques which utilize the voltage divider formula and the current divider formula carry over directly to ac circuit analysis with complex impedance replacing resistance. The magnitude of the transfer ratio represents the ratio of the magnitudes of the voltages or currents involved. The phase angle of the transfer ratio represents the angle by which the output current or voltage leads or lags the input current or voltage.

11.4 18. The superposition theorem, mesh and node analysis, and Thevenin and Norton theorems carry over directly to ac circuits with complex impedance in place of resistance.

11.5 19. The ECAP program for ac analysis follows the same rules as the dc analysis program with the following exceptions: currents or voltages must be entered with the phase angle and a frequency must be specified.

QUESTIONS FOR REVIEW

Sec. 11.1

1. A capacitor has a sinusoidal voltage waveform. Show qualitatively how the current waveform can be found.
2. State Ohm's law for ac circuits in time domain form.
3. Express the impedance of a capacitor in as many different forms as you can.
4. What is the significance of reactance?
5. How is impedance related to reactance?
6. Why is it necessary to state an angle in addition to a magnitude when an impedance is specified in polar form?
7. If an impedance is given in rectangular coordinates, the angle is not specified. Why?
8. What is the phase relationship between voltage across and current through a capacitor?
9. What is meant by the term "phase lead" applied to two sinusoidal signals of the same frequency?
10. What is the meaning of the term "phase lag"?
11. Express the impedance of an inductor in as many forms as you can.
12. What is the reactance of an inductor? How is it related to the impedance?
13. Define the term "admittance."

14. State the formulas for capacitive susceptance and inductive susceptance.
15. How is susceptance related to admittance?

Sec. 11.2

16. How are series impedances combined?
17. What is the significance of the horizontal axis in the impedance plane?
18. What is the significance of the vertical axis in the impedance plane?
19. Why does an impedance with an angle between 0 and 90° always represent a circuit that has at least one inductor?
20. An impedance has an angle of zero degrees. What is the significance of this statement?
21. An impedance has an angle between 0 and −90°. What can we deduce about the elements that make up the impedance?
22. State the formula used to combine two series impedances.
23. State the formula for combining two parallel impedances.
24. Why are impedance diagrams not used for parallel circuits?
25. Why is it advantageous to use admittance, conductance, and susceptance when dealing with parallel circuits?
26. Explain the term "equivalent impedance."
27. Describe the significant features of an impedance diagram for the equivalent impedance of a large number of series impedances.

Sec. 11.3

28. What is Ohm's law for ac circuits in phasor form?
29. How is the angle of an impedance related to the voltage across the impedance and the current through it?
30. What is the voltage divider formula as applied to ac circuits?
31. What is the significance of the magnitude of the voltage ratio A_v?
32. What is the significance of the angle of A_v?
33. What is the sequence of steps to be followed in analyzing an ac circuit?
34. What is the current divider rule as applied to ac circuits?

Sec. 11.4

35. State the superposition theorem as applied to ac circuits.
36. Describe the use of mesh equations or node equations to solve ac circuits.
37. State Thevenin's theorem as applied to ac circuits and show the Thevenin equivalent circuit.
38. State Norton's theorem as applied to ac circuits and show the Norton equivalent circuit.
39. What is the significance of the term "equivalent circuit"?
40. Describe the ECAP ac analysis program.

PROBLEMS

Sec. 11.1

1. A voltage $v(t) = 20 \cos(2000\pi t)$ mV appears across a 200-pF capacitor. Find the current $i(t)$ through the capacitor.
2. A 5-μF motor-starting capacitor carries a current $i(t) = 2.3 \cos(377t + 90°)$ A. Find the voltage $v(t)$ across the capacitor.

3. Repeat Prob. 1 using phasors.
4. Repeat Prob. 2 using phasors.
5. Find the reactance and impedance of the following capacitors:
   a. 100 μF at $f = 330$ Hz c. 0.02 μF at $\omega = 6.3$ krad/s
   b. 22 pF at $f = 6.2$ MHz d. 1.6 nF at $f = 2.2$ GHz
6. A tone control requires a capacitive reactance of 820 Ω at a frequency of 5 kHz. Find the required capacitance.
7. What size capacitor has 6.8-kΩ reactance at 400 Hz?
8. At what frequency does a 16-pF capacitor have a reactance of 6.3 kΩ?
9. The reactance of a 10-μF capacitor is measured and found to be 130 Ω. What is the frequency?
10. A 2-μF capacitor carries a current $i(t) = 1.6\cos(377t + 30°)$ mA. Find the phasor voltage across the capacitor and draw a phasor diagram.
11. The voltage across a 300-pF tuning capacitor is $v(t) = 25\cos(3 \times 10^6\pi t - 45°)$ V. Find the phasor current through the capacitor and draw a phasor diagram.
12. List the following signals in order of leading phase, that is, the one that leads all the others first, then the one that leads all the remaining ones except itself, of course, and so on with the last one being the one that lags all the others. If any signals are in phase so indicate:

$$v_1(t) = 10\sin(100t + 30°)$$
$$v_2(t) = 50\cos 100\,t$$
$$v_3(t) = 400\cos(100t + 120°)$$
$$i(t) = -80\cos(100t - \pi/2)$$

   (You may need the identity $\sin x = \cos(x - 90°)$. Draw a phasor diagram.
13. Find the reactance and impedance of the following inductors:
   a. 16 μH at $f = 12$ MHz c. 47 mH at $f = 7.2$ kHz
   b. 3.3 H at $\omega = 377$ rad/s d. 2.2 μH at $f = 1.6$ GHz
14. A power supply inductor must have a reactance greater than 600 Ω at a frequency of 120 Hz. Find the minimum allowable inductance.
15. At 3 MHz, what inductance is required for the reactance to be 50 Ω?
16. At what frequency will a 35-μH inductor have a reactance of 2 kΩ?
17. The reactance of a 10-H telephone coil is found to be 600 Ω. What is the frequency?
18. A 30-mH inductor carries a current $i(t) = 15\cos(1000\pi t + 60°)$ mA. Find the impedance and the phasor voltage and draw a phasor diagram.
19. The voltage across a 18-μH tuning coil is $v(t) = 60\cos(10^8 t + 22°)$ μV. Find the impedance and the phasor current and draw a phasor diagram.
20. Find the susceptance and admittance of the following:
   a. A 10-μF capacitor at $f = 120$ Hz
   b. A 3-H inductor at $f = 120$ Hz
   c. A 33-pF capacitor at $f = 25$ MHz
   d. A 12-μH inductor at $f = 16$ MHz

Sec. 11.2

21. Find the impedance of the series circuits of Fig. 11.25 in rectangular and polar form and draw a complete Z-plane diagram.
22. Repeat Prob. 21 for the admittance of the parallel circuits of Fig. 11.26.

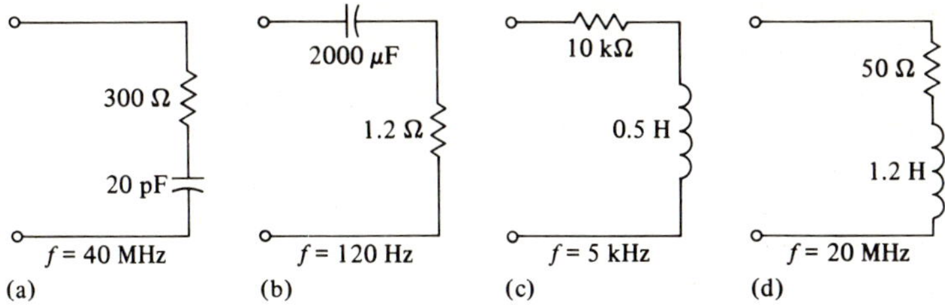

FIGURE 11.25

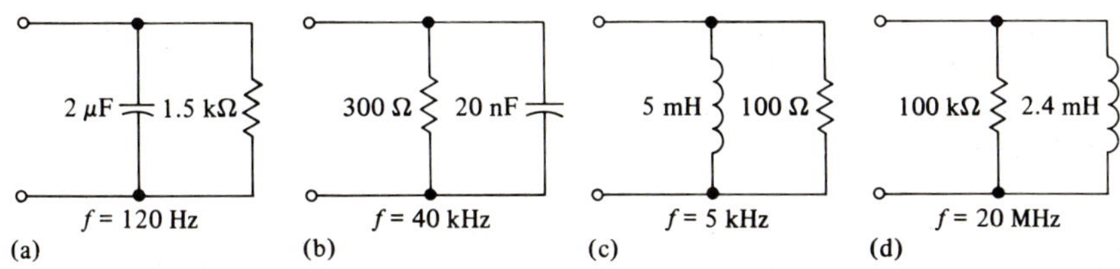

FIGURE 11.26

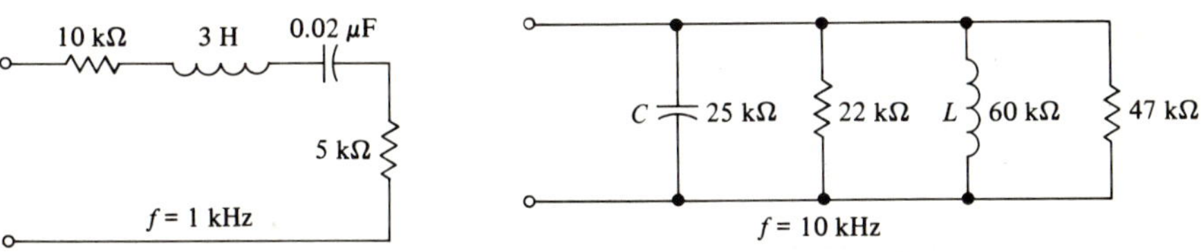

FIGURE 11.27

FIGURE 11.28

23. Find the admittance of the series circuits of Fig. 11.25 in rectangular and polar form and draw a complete Y-plane diagram. Find the corresponding two-element parallel circuit for each and observe that it is exactly equivalent to the original series circuit.
24. Find the impedance of the parallel circuits of Fig. 11.26. Find the corresponding two-element series circuit for each and observe that it is exactly equivalent to the original parallel circuit.
25. Find the impedance of the circuit of Fig. 11.27 in polar and rectangular form and draw a complete Z-plane diagram.
26. Repeat Prob. 25 at frequencies of 100 Hz and 10 kHz.
27. Find the admittance of the circuit of Fig. 11.28 in polar and rectangular form and draw a complete Y-plane diagram. Find the capacitance and inductance values.
28. Repeat Prob. 27 at frequencies of 1 kHz and 100 kHz.
29. Two elements in series have a total impedance of $10\angle 50^\circ$ kΩ at a frequency of 5 kHz. What are the elements?
30. A power supply coil has an inductance of 4.5 H. Its total impedance at 60 Hz has a magnitude of 2820 Ω. Find the resistance of the coil and the angle of its impedance.
31. A 0.01-μF capacitor and a 1.5-kΩ resistor are connected in series. At what frequency will the combination have an impedance of 50 kΩ?

32. An amplifier input circuit has an impedance of $56\angle 75°$ kΩ. Find a series equivalent circuit.
33. A high-frequency antenna has an impedance of $50\angle 60°$ Ω. Find a series equivalent circuit.
34. An impedance is measured as $5\angle 60°$ kΩ at a frequency of 1 kHz. What value of capacitance must be connected in series in order for the total impedance to be $5\angle -60°$ kΩ?
35. Impedances $Z_1 = 5\angle 30°$ kΩ, $Z_2 = 3.3\angle 80°$ kΩ, and $Z_3 = 6.8\angle -20°$ kΩ are connected in series. Find Z_T and draw a complete Z diagram.
36. The impedances of Prob. 35 are connected in parallel. Find Y_T and Z_T and draw a complete Y diagram.
37. A power supply inductor is connected to a 9-V battery and the resulting dc current is 200 mA. When the same coil is connected to the 110-V, 60-Hz power line, the ac current is 500 mA. Find the resistance and inductance of the coil.
38. A radio tuning circuit has an impedance of $50\angle 30°$ Ω at 1.6 MHz. An element is to be connected in series with it such that the total impedance is capacitive with a magnitude of 75 Ω. What kind of element is required and what is its value?
39. Find the series elements that have an equivalent admittance of $12\angle 30°$ μS.

For each problem from this point on, the solution should contain a complete phasor diagram along with an impedance or admittance diagram.

Sec. 11.3

40. In the circuit of Fig. 11.29 find all currents and voltages.
41. In the circuit of Fig. 11.30 find all currents and voltages.
42. In the circuit of Fig. 11.31 find $|V_i|$. What is the phase angle of the current with respect to V_i?
43. Find V_2 in the circuits of Fig. 11.32.
44. Find V_2 (magnitude and angle) in the circuit of Fig. 11.33.
45. A relay coil draws 1.2 A when connected to the 120-V, 60-Hz power lines. The current lags the source voltage by 65°. Find the impedance in rectangular and polar form.

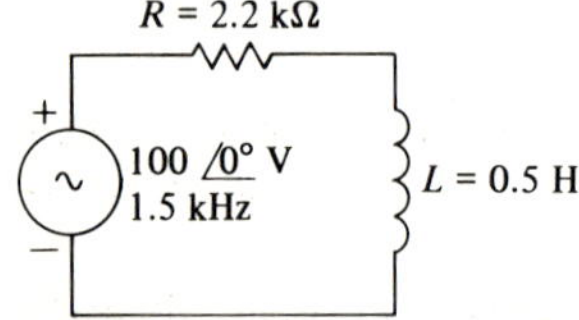

FIGURE 11.29

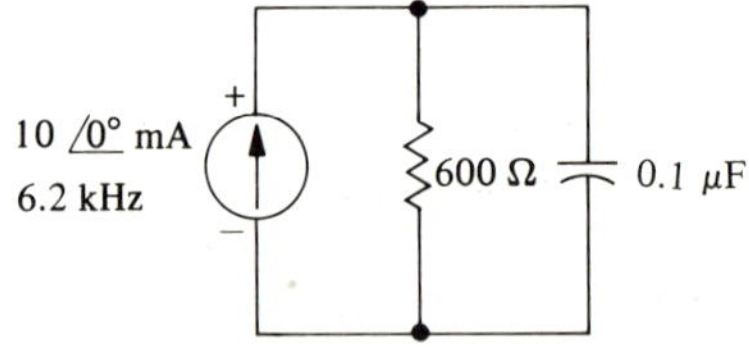

FIGURE 11.30

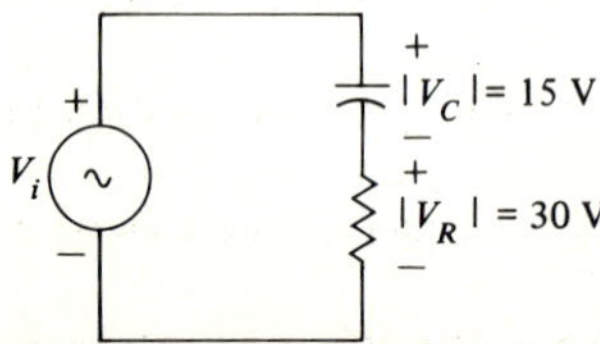

FIGURE 11.31

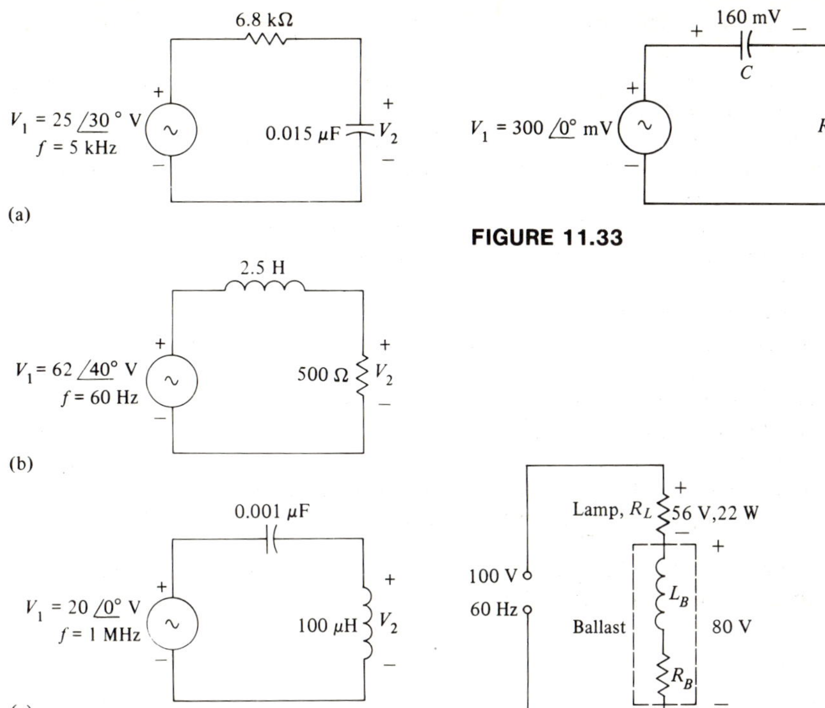

FIGURE 11.33

FIGURE 11.32

FIGURE 11.34

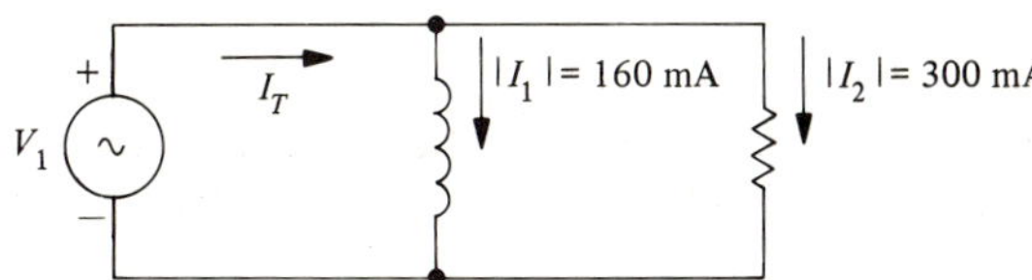

FIGURE 11.35

46. The antenna of an FM receiver supplies 1 μA at 88 MHz to the receiver input circuit. The resulting input voltage is $25\angle -30°$ μV. Find the receiver impedance.
47. A fluorescent lamp rated at 22 W and its ballast are shown in Fig. 11.34 along with measured voltages. If the lamp is purely resistive, find the resistance and inductance of the ballast.
48. In the circuit of Fig. 11.35 find the magnitude of I_T and its phase angle with respect to V_1.
49. In the circuit of Fig. 11.36 find the magnitude of I_C and its phase angle with respect to V_1.
50. A 1.5-mH inductor is paralleled by a resistor. What value of resistance is required if the impedance at 60 kHz is to be 400 Ω?
51. Find the parallel equivalent circuit of an FM amplifier circuit with an impedance of $50\angle -70°$ Ω.

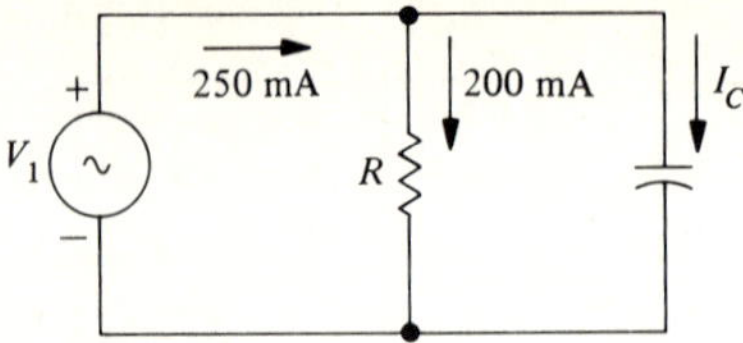

FIGURE 11.36

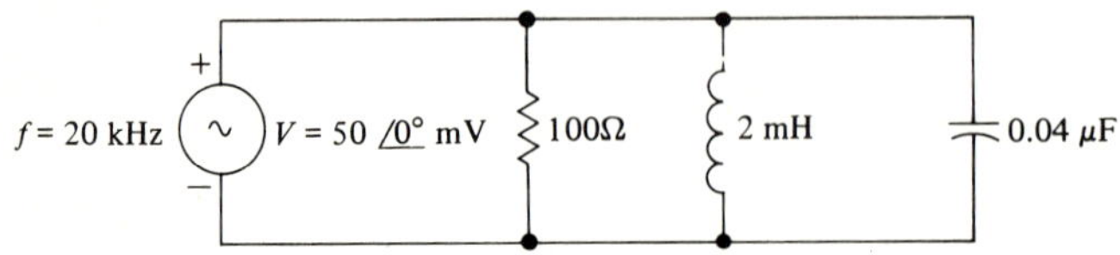

FIGURE 11.37

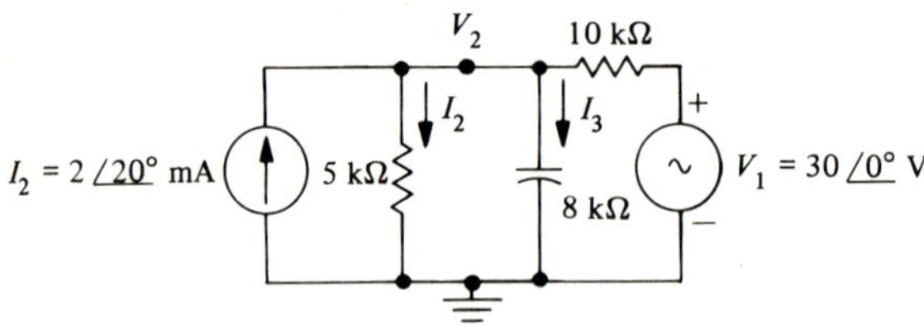

FIGURE 11.38

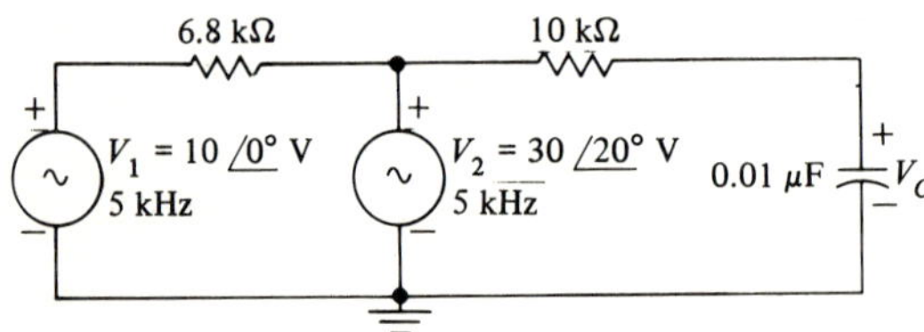

FIGURE 11.39

52. In the circuit of Fig. 11.37 find all currents and voltages.
53. In the circuit of Fig. 11.37 the voltage source is replaced by a current source of $10\angle 20^\circ$ μA at 20 kHz. Find all currents and voltages.

Sec. 11.4

54. Use superposition to find V_2 in the circuit of Fig. 11.38.
55. Repeat Prob. 54 to find I_2.
56. Repeat Prob. 54 to find I_3.
57. Use superposition to find V_C in the circuit of Fig. 11.39.
58. Use mesh equations to find V_C in the circuit of Fig. 11.39.
59. Use node equations to find V_C in the circuit of Fig. 11.39.
60. Find V_1, V_2, and V_3 in the circuit of Fig. 11.40 using node equations.
61. In the circuit of Fig. 11.41 find the current through the 1-kΩ resistor using mesh equations.
62. Find the Thevenin and Norton equivalents for the circuit of Fig. 11.42.

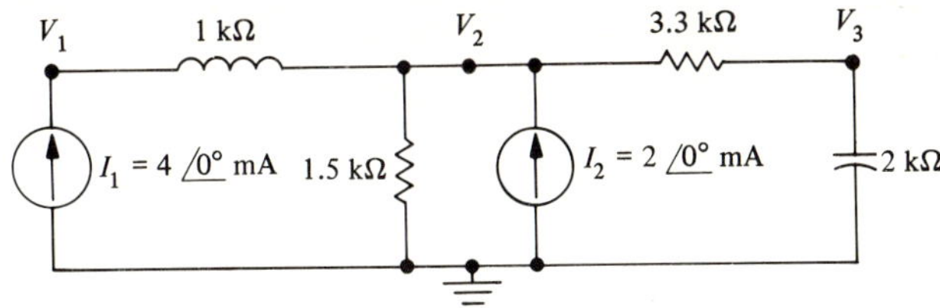

FIGURE 11.40

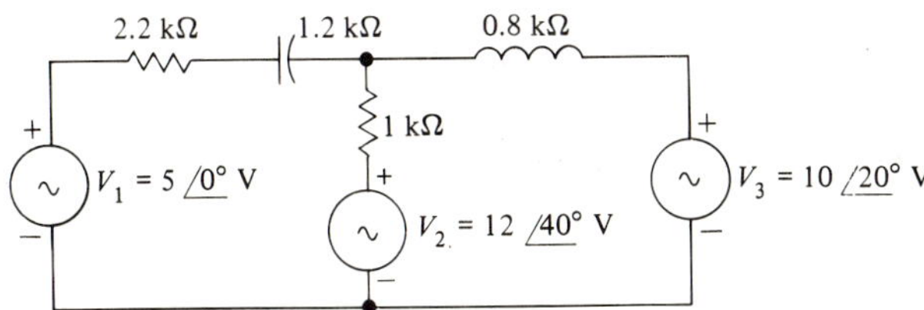

FIGURE 11.41

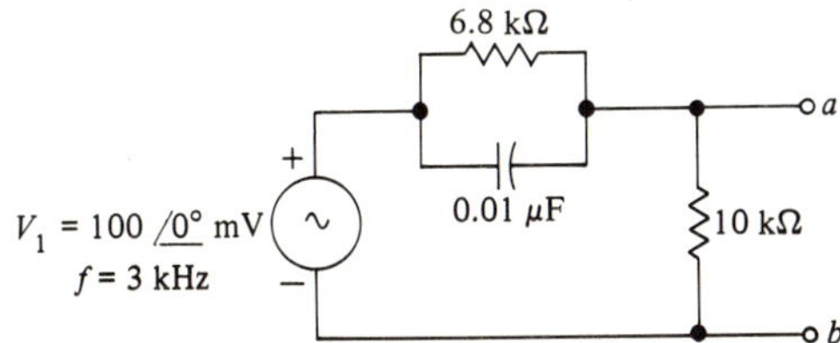

FIGURE 11.42

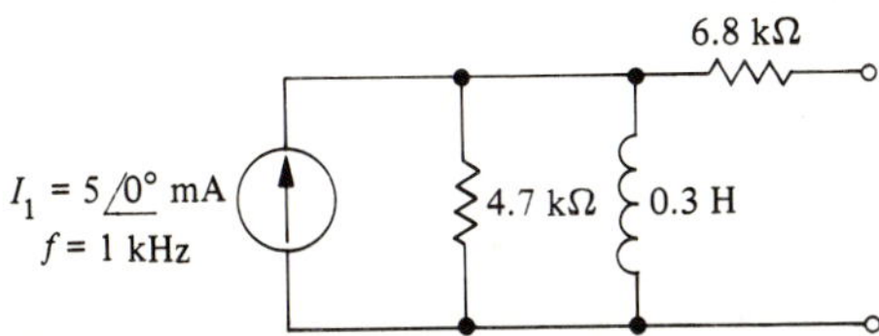

FIGURE 11.43

63. Find the Norton and Thevenin equivalents for the circuit of Fig. 11.43.
64. Find a two-element series circuit equivalent to the parallel circuit shown in Fig. 11.28.
65. Find a two-element parallel circuit equivalent to the series circuit shown in Fig. 11.27.

Sec. 11.5

66. Use ECAP or some other circuit analysis program to find all currents and voltages at $f = 1$ kHz in the circuit of Fig. 11.38.
67. Repeat Prob. 66 for the circuit of Fig. 11.39.
68. Repeat Prob. 66 at $f = 1$ kHz for the circuit of Fig. 11.40.
69. Repeat Prob. 66 at $f = 80$ kHz for the circuit of Fig. 11.41.

12 Additional Topics in AC Circuit Analysis

OBJECTIVES

Upon completion of this chapter, the student should be able to

Section

12.1
1. State that the power in an ac circuit is given by the product of the voltage, current, and power factor.
2. State that the average power in an inductance or capacitance is zero.
3. Define reactive power and apparent power.
4. Define power factor.

12.2
5. State the formula for complex power and use it to find power components in an ac circuit.
6. Explain why power factor correction is advantageous to large industrial users of electricity.
7. State the maximum power transfer theorem as applied to ac circuits.

12.3
8. Describe the three-phase generator.
9. Define phase sequence.
10. Calculate currents and voltages in a balanced wye system.
11. Calculate currents and voltages in an unbalanced wye system.
12. Find currents and voltages in a balanced delta system.

12.4
13. Use the delta-wye conversion to calculate currents and voltages in a mixed wye-delta system.
14. Find currents, voltages, and power in an Edison three-wire system.

12.5
15. State the formula for calculating power in a three-phase system using line currents and line voltage.
16. Describe the two-wattmeter method of measuring power in three-phase systems.

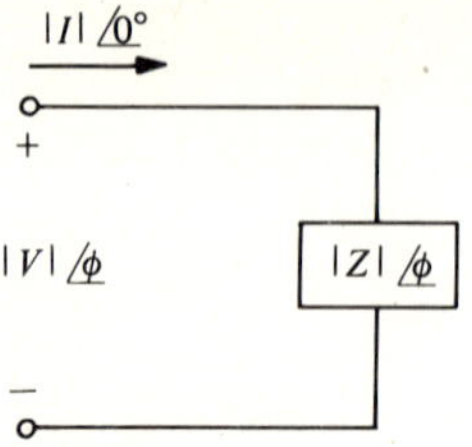

FIGURE 12.1
Reference directions for computing power delivered to an element.

12.6 17. State the basic formulas that apply to step-up or step-down of voltage and current in a transformer.
18. Determine the turns ratio required to achieve a given impedance match.
19. Use the simplified equivalent circuit to calculate current and voltage requirements in a practical iron-core transformer.

INTRODUCTION

The topics discussed in this chapter all relate to practical applications of ac networks. We consider the calculation of power in single-phase circuits, and also power factor correction, which is so important to large industrial users of electricity. We also consider three-phase systems in some detail. A section on transformers completes the chapter.

12.1 POWER IN AC CIRCUITS

In Sec. 10.6 we found the power dissipated in a resistor when its voltage and current were sinusoidal. In this section we extend this theory to cover any impedance.

In general, if we have an impedance $Z = |Z|\underline{/\phi}$ through which a current $I = |I|\underline{/0^\circ}$ flows, the voltage across the impedance will be (see Fig. 12.1)

$$V = ZI = |Z|\underline{/\phi} \times |I|\underline{/0^\circ} = |Z||I|\underline{/\phi} \qquad (12.1\text{-}1)$$

In time domain terms,

$$i(t) = |I_m| \cos \omega t \qquad (12.1\text{-}2)$$

and

$$v(t) = |V_m| \cos(\omega t + \phi) \qquad (12.1\text{-}3)$$

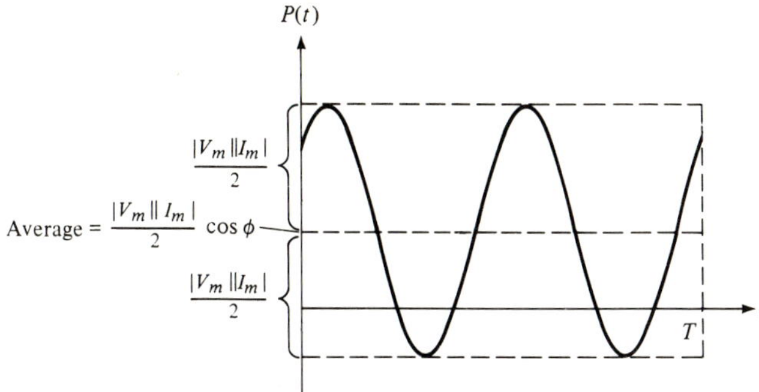

FIGURE 12.2
Instantaneous power in an impedance $|Z| < \phi$. The two cycles shown occur in the same time T as one cycle of the current and voltage.

The instantaneous power is

$$\begin{aligned} p(t) &= v(t)i(t) \\ &= |V_m|\cos(\omega t + \phi)|I_m|\cos \omega t \end{aligned} \tag{12.1-4}$$

In order to express this in a more useful form, we apply the trigonometric identity:

$$\cos x \cos y = \tfrac{1}{2}[\cos(x - y) + \cos(x + y)] \tag{12.1-5}$$

The result is

$$p(t) = \frac{|V_m||I_m|}{2}[\cos \phi + \cos(2\omega t + \phi)] \tag{12.1-6}$$

which is plotted in Fig. 12.2. The instantaneous power is seen to consist of a constant component, $(|V_m||I_m|/2)\cos \phi$, plus a sinusoidal component, $(|V_m||I_m|/2)\cos(2\omega t + \phi)$, whose frequency is twice that of the current or voltage. When we take the average over a large number of cycles, the sinusoidal component drops out since its average value is zero and we are left with the average power

$$P = \frac{|V_m||I_m|}{2}\cos \phi \tag{12.1-7}$$

It is convenient to use effective values in this equation. Recalling that $|V| = |V_m|/\sqrt{2}$ and $|I| = |I_m|/\sqrt{2}$, we have

$$\boxed{P = |V||I|\cos \phi} \tag{12.1-8}$$

where ϕ is the phase angle of the impedance, that is, the phase angle between the voltage and current.

EXAMPLE 12.1-1 Power in an Inductive Circuit

A coil is designed for the electromagnets being used with a large particle accelerator. The coil resistance is 25 Ω and its inductance is 3 mH. The voltage applied to the coil is 600 V at a frequency of 1 kHz. Find the power dissipated in the coil.

Solution

We take the voltage as our reference, so $V = 600\angle 0°$ V. The coil reactance is

$$X_L = 2\pi f L = 2\pi \times 10^3 \times 0.003 = 18.8\ \Omega$$

Then the current is

$$I = \frac{V}{Z} = \frac{600\angle 0°}{25 + j18.8} = \frac{600\angle 0°}{31.3\angle 37°} = 19.2\angle -37°\ \text{A}$$

Using $P = |V||I| \cos \phi$, with $\phi = -37°$, we have

$$P = (600)(19.2) \cos(-37°) = 9.2\ \text{kW}$$

This is the true power dissipated in the resistance of the coil.

• • •

Power in Inductance and Capacitance Elements

In a pure inductance connected to a voltage source $v(t) = |V_m| \cos \omega t$ the current will be $i(t) = |I_m| \cos(\omega t - 90°)$. The instantaneous power is

$$p(t) = v(t)i(t) = |V_m||I_m| \cos \omega t \times \cos(\omega t - 90°) \tag{12.1-9}$$

Applying the trigonometric identity in Eq. (12.1-5) this becomes

$$p(t) = \frac{|V_m||I_m|}{2}\left[\cos\frac{\pi}{2} + \cos(2\omega t - 90°)\right]$$

Since $\cos \pi/2 = 0$, this simplifies to

$$p(t) = \frac{|V_m||I_m|}{2} \cos(2\omega t - 90°)$$

$$= \frac{|V_m||I_m|}{2} \sin 2\omega t \tag{12.1-10}$$

This is plotted in Fig. 12.3 along with the current and voltage. For this case *the average power is zero*, so we conclude that the true power in an inductance is zero. The question then arises as to the significance of the

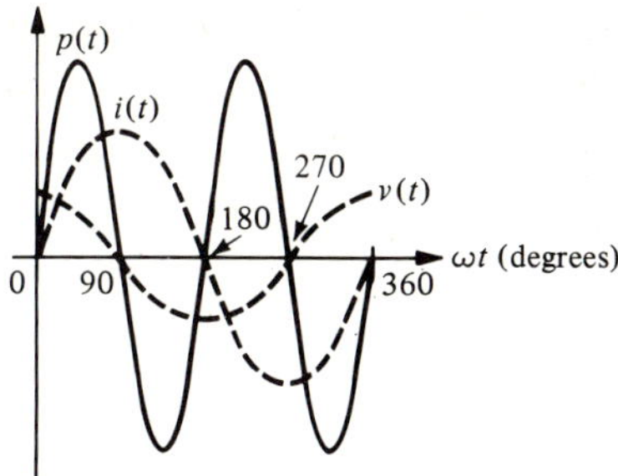

FIGURE 12.3
Waveforms in a pure inductance.

power waveform. The interpretation of this waveform is as follows: Between 0 and 90° the current is increasing. During this interval the magnetic field around the inductor is being established and the inductor is taking electric energy from the voltage source and storing it as magnetic energy in its magnetic field. The instantaneous power is positive in this interval because the inductor is receiving power. Between 90 and 180° the current is decreasing, causing the magnetic field to collapse and return the stored energy back to the voltage source. In this interval the instantaneous power is negative because the inductor is delivering power. From this discussion, we see that in a pure inductance, energy is continually being exchanged between the source and the magnetic field.

As noted before, no true power is dissipated in the inductor. Let us consider the general power formula, Eq. (12.1-8), as applied to the pure inductance. Since the phase angle $\phi = 90°$ for an inductor, $\cos \phi = 0$ and $P = 0$. This agrees with the fact that the average value of the instantaneous power curve is zero.

For a pure reactance, the $|V||I|$ product is called *reactive power*, which is symbolized by the letter Q. For a pure inductance with effective voltage $|V_L|$ and effective current $|I_L|$, the reactive power is

$$Q_L = |V_L||I_L| \text{ var} \tag{12.1-11}$$

The *dimensions* of Q are the same as those of the real power *P*. In order to avoid confusion the *unit* for reactive power is called the *reactive volt-ampere*, abbreviated *var*.

Since, in a pure inductance,

$$|V_L| = X_L|I_L| \tag{12.1-12}$$

the reactive power can be expressed in two alternate ways:

$$Q_L = X_L|I_L|^2 = \frac{|V_L|^2}{X_L} \text{ var} \tag{12.1-13}$$

Capacitance

A development identical to that for the inductance will show that similar conclusions hold for a pure capacitance. The average power is zero, and the instantaneous power oscillates at twice the supply frequency. While the voltage across the capacitor is rising, energy from the source is being used to establish the electric field between the plates. When the voltage is decreasing, the stored energy is being fed back into the source so that there is a continuous interchange of energy between the voltage source and the electric field of the capacitor.

The reactive power is

$$Q_C = |V_C||I_C| = X_C|I_C|^2 = \frac{|V_C|^2}{X_C} \qquad (12.1\text{-}14)$$

where $|V_C|$ and $|I_C|$ are effective values of voltage and current at the terminals of the capacitor and $X_C = 1/\omega C$ is the capacitive reactance.

In a circuit containing both capacitance and inductance, the net reactive power is the *difference* between the inductive reactive power and the capacitive reactive power. Thus, in general

$$\boxed{Q = Q_L - Q_C} \qquad (12.1\text{-}15)$$

EXAMPLE 12.1-2 Reactive Power

Find the reactive power in the coil of Example 12.1-1. What value of capacitance will absorb the same reactive power at the stated voltage?

Solution

In Example 12.1-1 we found the reactance of the coil to be $X_L = 18.8\ \Omega$ and the current to be 19.2 A. The reactive power is then

$$Q_L = X_L|I_L|^2 = 18.8 \times 19.2^2 = 6.9 \text{ kvar}$$

For the capacitor,

$$X_C = \frac{|V_C|^2}{Q_C}$$

The voltage is specified as 600 V and, from above, $Q_C = 6.9 \times 10^3$ var. Then

$$X_C = \frac{600^2}{6.9 \times 10^3} = 52.2\ \Omega$$

and

$$C = \frac{1}{\omega X_C} = \frac{1}{2\pi \times 10^3 \times 52.2} = 3\ \mu\text{F}$$

• • •

Apparent Power

In a dc circuit the real power is given by the product VI. In an ac circuit the $|V||I|$ product does not represent either the real power or the reactive power. It is called the *apparent* power, symbolized by S. The units of apparent power are *volt-amperes,* abbreviated VA. Then

$$\boxed{S = |V||I|} \tag{12.1-16}$$

and since $|V| = |Z||I|$,

$$S = |Z||I|^2 = \frac{|V|^2}{|Z|} \tag{12.1-17}$$

where $|V|$ and $|I|$ are the voltage and current magnitudes at the terminals of the impedance Z.

Power Factor

The ratio of the true power to the apparent power delivered to a circuit is called the *power factor,* for which there is no standard symbol. This ratio is

$$\frac{P}{S} = \frac{|V||I|\cos\phi}{|V||I|} = \cos\phi \tag{12.1-18}$$

where ϕ is the phase difference between the voltage and current. If V is the voltage across, and I the current through an arbitrary impedance $Z = R + jX$ then the phase angle ϕ is the angle of the impedance:

$$\cos\phi = \frac{\text{Re}\,Z}{|Z|} = \frac{R}{\sqrt{R^2 + X^2}} \tag{12.1-19}$$

Since $\cos\phi = \cos(-\phi)$, the power factor gives no indication as to whether the circuit is inductive or capacitive. For this reason the power factor is often specified as *leading* or *lagging.* A leading power factor occurs when the circuit is *capacitive* and the current *leads* the voltage. A lagging power factor refers to an *inductive* circuit in which the current always *lags* the voltage.

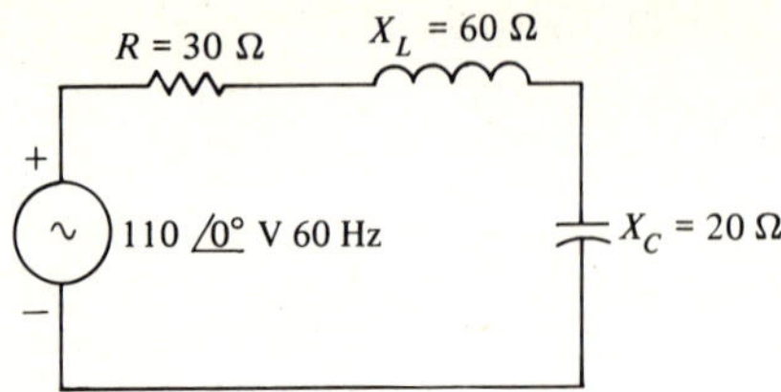

FIGURE 12.4
Circuit for Example 12.1-3.

Often the power factor is specified in *percent*. For example, a 60% power factor means that $\cos\phi = 0.6$ so that $\phi = \arccos 0.6 = 53°$. A power factor of 100% refers to a 0° phase angle.

EXAMPLE 12.1-3 Power in an Arbitrary Impedance

In the circuit of Fig. 12.4 find

a. The power factor
b. The current
c. The real power
d. The reactive power
e. The apparent power

Solution

a. We begin by finding the impedance

$$Z = R + j(X_L - X_C) = 30 + j(60 - 20) = 30 + j40 = 50\underline{/53.1°}\ \Omega$$

From Eq. (12.1-19)

$$\cos\phi = \cos 53.1° = 0.6$$

Note that Z is primarily inductive so that the phase angle is positive.

b. Using Ohm's law in phasor form

$$I = \frac{V}{Z} = \frac{110\underline{/0°}}{50\underline{/53.1°}} = 2.2\underline{/-53.1°}\ \text{A}.$$

c. The real power is found using Eq. (12.1-8):

$$P = |V||I|\cos\phi = (110)(2.2)(0.6) = 145\ \text{W}$$

This can be checked by noting that only the resistance dissipates power so that

$$P = |I|^2R = (2.2)^2(30) = 145\ \text{W}$$

d. The reactive power is found using Eqs. (12.1-13)–(12.1-15):

$$Q = I^2X = I^2(X_L - X_C) = (2.2)^2(40) = 194 \text{ var}$$

e. The apparent power is

$$S = VI = (110)(2.2) = 242 \text{ VA}$$

• • •

EXAMPLE 12.1-4 Power Factor

A device absorbs 1 kW of power at a rated voltage of 110 V. The power factor is 50% leading. Find the current and its phase relation to the voltage.

Solution

Here we use the basic power formula

$$P = |V||I|\cos\phi$$

From which

$$|I| = \frac{P}{|V|\cos\phi} = \frac{1000}{(110)(0.5)} = 18.2 \text{ A}$$

The phase angle is $\phi = \arccos 0.5 = 60°$ and since the power factor is leading, the circuit is capacitive and the current leads the voltage by 60°.

• • •

LEARNING EXERCISES FOR SEC. 12.1

1. A 220-V radial saw motor draws 5 A at a 92% power factor. How much power is dissipated?
2. Measurements on a cyclotron 60-Hz power supply yield the following data: 6300 V, 25 A, and 140 kW. Find the apparent power and the power factor.

Ans. 158; 1; 89

• • •

12.2 COMPLEX POWER AND POWER FACTOR CORRECTION

Power calculations are simplified if power is considered to be a complex quantity. We use the symbol P_c to denote complex power and define it by the equation

$$P_c = VI^* \tag{12.2-1}$$

In order to illustrate how this definition can simplify power calculations, consider the circuit of Fig. 12.5. Let us apply the complex power

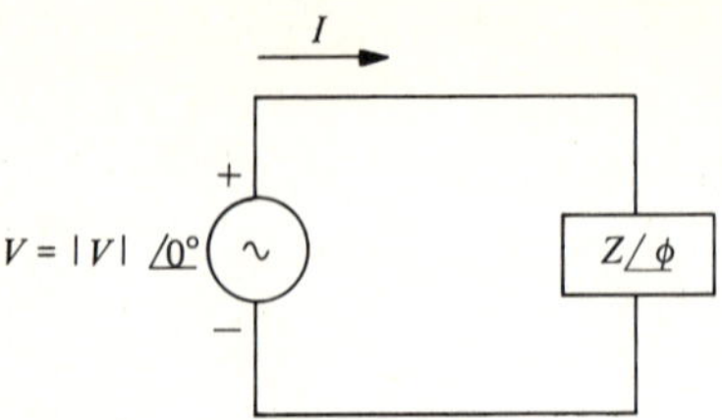

FIGURE 12.5
Complex power.

formula to this circuit. The current is

$$I = \frac{|V| \angle 0^\circ}{|Z| \angle \phi} = \frac{|V|}{|Z|} \angle -\phi = |I| \angle -\phi \tag{12.2-2}$$

The conjugate is

$$I^* = |I| \angle \phi \tag{12.2-3}$$

Applying the complex power formula

$$P_c = VI^* = |V||I| \angle \phi$$

$$= |V||I| \cos \phi + j|V||I| \sin \phi \tag{12.2-4}$$

$$= S \cos \phi + jS \sin \phi \tag{12.2-5}$$

An alternate way to proceed is to note that $Z = R + jX$ where $X = X_L - X_C$ so that

$$V = ZI = RI + jXI$$

Then

$$\begin{aligned} P_c = VI^* &= RII^* + jXII^* \\ &= R|I|^2 + jX|I|^2 \end{aligned} \tag{12.2-6}$$

Consider Eqs. (12.2-4) and (12.2-6). The real part of the complex power will be recognized as the real power P, since

$$P = |V||I| \cos \phi = R|I|^2 \tag{12.2-7}$$

Comparing the imaginary part of the complex power in Eq. (12.2-6) with Eqs. (12.1-13) and (12.1-14), we see that the imaginary part of the complex power is the reactive power

$$Q = X|I|^2 = (X_L - X_C)|I|^2 = |V||I| \sin \phi \tag{12.2-8}$$

Finally, the complex power formula becomes

$$\boxed{P_c = VI^* = S\underline{/\phi} = P + jQ} \tag{12.2-9}$$

This formula supplies values for real power, reactive power, apparent power, and power factor angle in one simple operation. The only odd point involved in remembering it is that the *conjugate* of the current is used. Other forms can be obtained using Ohm's law as follows:

$$P_c = VI^* = (ZI)I^* = Z(II^*) = Z|I|^2 \tag{12.2-10}$$

or

$$P_c = VI^* = V\left(\frac{V}{Z}\right)^* = \frac{VV^*}{Z^*} = \frac{|V|^2}{Z^*} \tag{12.2-11}$$

Since power can be expressed in complex terms, a complex power plane diagram can be drawn. The real axis is real power P and the imaginary axis is reactive power Q (inductive Q_L on the positive imaginary axis and capacitive Q_C on the negative imaginary axis). The magnitude of the complex power is the apparent power S, and the angle with the real axis is the power factor angle ϕ. These ideas will be illustrated in the example which follows.

EXAMPLE 12.2-1 Complex Power

In the circuit of Fig. 12.5 the impedance Z consists of a series combination of $R = 4\ \Omega$, $X_L = 6\ \Omega$, and $X_C = 2\ \Omega$. The source voltage is $100\underline{/0^\circ}$ V. Find P_c, S, P, Q, and the power factor. Plot a complete complex power diagram.

Solution

The first step is to find the current:

$$I = \frac{V}{Z} = \frac{100\underline{/0^\circ}}{4 + j(6-2)} = \frac{100\underline{/0^\circ}}{4 + j4} = \frac{100\underline{/0^\circ}}{4\sqrt{2}\underline{/45^\circ}} = 17.7\underline{/-45^\circ}\ \text{A}$$

Then

$$I^* = 17.7\underline{/45^\circ}\ \text{A}$$

and

$$P_c = VI^* = (100\underline{/0^\circ})(17.7\underline{/45^\circ}) = 1770\underline{/45^\circ}\text{ VA}$$
$$= 1250\text{ W} + j1250\text{ var}$$

Therefore

$$S = 1770\text{ VA}$$
$$P = 1250\text{ W}$$
$$Q = 1250\text{ var}$$
$$\cos\phi = \cos 45^\circ = 0.71$$

In order to plot a complete power diagram, we need Q_L and Q_C

$$Q_L = X_L|I|^2 = (6)\left(\frac{25}{\sqrt{2}}\right)^2 = 1875\text{ var}$$

$$Q_C = X_C|I|^2 = (2)\left(\frac{25}{\sqrt{2}}\right)^2 = 625\text{ var}$$

The net reactive power Q is the difference between the inductive and capacitive reactive power

$$Q = Q_L - Q_C = 1875 - 625 = 1250\text{ var}$$

The diagram is shown in Fig. 12.6.

• • •

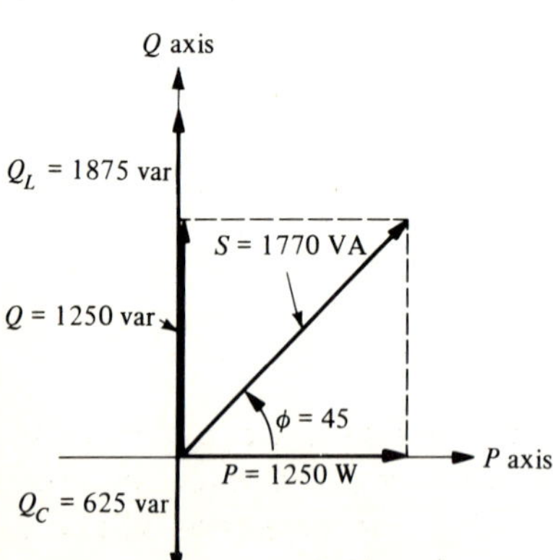

FIGURE 12.6
Complex power diagram for Example 12.2-1.

EXAMPLE 12.2-2 Nameplate Ratings

The nameplate on a large electric machine reads

440 V 60 Hz 50 kVA 0.6 lag power factor

Use this information to find

a. I b. Z c. P

Solution

a. The 50-kVA rating refers to the apparent power S. Then

$$|I| = \frac{S}{|V|} = \frac{50\,000}{440} = 114 \text{ A}$$

The power factor angle is ϕ = arccos 0.6 = 53° and if we assume the voltage as a reference, $V = 440\underline{/0^\circ}$ V, then

$$I = 114\underline{/-53^\circ} \text{ A}$$

b. The impedance is

$$Z = \frac{V}{I} = \frac{440\underline{/0^\circ}}{114\underline{/-53^\circ}} = 3.86\underline{/53^\circ}\ \Omega = 2.32 + j3.08\ \Omega$$

c. The real power is

$$P = S \cos \phi = 50 \times 0.6 = 30 \text{ kW}$$

d. The reactive power is

$$Q = S \sin \phi = 50 \sin 53^\circ = 40 \text{ kvar}$$

So that the complex power is

$$P_c = P + jQ = 30 \text{ kW} + j40 \text{ kvar}$$

• • •

Device Ratings

Much electric equipment is rated in terms of apparent power rather than real power. Ratings in volt-amperes (VA) or, more often, kilovolt-amperes (kVA) are used. In almost all cases the rated voltage for the device will be a known standard power system voltage, for example, 110 V, 220 V, or 440 V. Thus the kVA rating can be used to determine the maximum required current. This cannot be done from a power rating in watts or kilowatts unless the power factor is given. The power rating in watts is the same as the volt-ampere rating only if the power factor $\cos\phi = 1$, that is, if the load is purely resistive.

For large industrial loads, the power factor is of critical importance to the utility company supplying the power. This is illustrated in the following example.

EXAMPLE 12.2-3 The Effects of Low Power Factor

An industrial furnace must absorb 50 kW of power at 440 V, 60 Hz, in order to bring the temperature to its rated value in a reasonable time. Find the current the utility company must provide if the furnace is rated at

a. 60% power factor lagging
b. 100% power factor

Solution

a. The apparent power that must be supplied at 60% power factor is

$$S = \frac{P}{\cos\phi} = \frac{50 \text{ kW}}{0.6} = 83.3 \text{ kVA}$$

The required current is then

$$|I| = \frac{S}{|V|} = \frac{83\,300}{440} = 189 \text{ A}$$

b. If the power factor is unity, then

$$S = P = |V||I|$$

and

$$|I| = \frac{P}{|V|} = \frac{50\,000}{440} = 114 \text{ A}$$

The 60% power factor furnace requires 189 A to produce the same heat as the 100% power factor furnace when drawing only 114 A.

• • •

This example shows that low power factor loads require considerably more current. Consequently, the wire used in the generator and the transmission line between the generator and the furnace must have greater current-carrying capacity. This means that a larger diameter wire will have to be used. In addition, power losses in these wires are proportional to the square of the current. These facts all add up to increased costs for the power company. If they charge only for real power used, then feeding low-power-factor loads will reduce their profit. For this reason large customers often receive a penalty called a *demand charge* based on their average

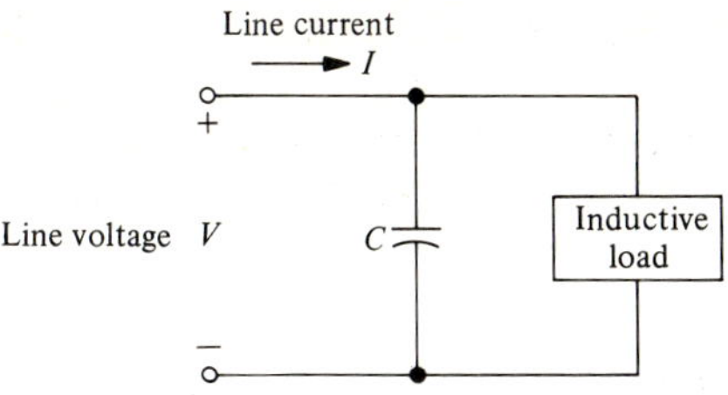

FIGURE 12.7
An external capacitor connected to improve the power factor.

power factor so that it is to their advantage to maximize the power factor in order to avoid this extra charge.

Power Factor Correction

Most industrial loads are inductive so that the load current lags the line voltage and the power factor is lagging. In order to improve the power factor, that is, bring it closer to unity, we must add something to the load which will draw a *leading* current. This is done by connecting a capacitor in parallel with the load as shown in Fig. 12.7. Since the capacitor does not dissipate any real power, the charge for real power will be the same. Calculations for determining the required capacitance will be found in the following example.

EXAMPLE 12.2-4 Power Factor Correction

In Fig. 12.7 the source is 440 V, 60 Hz, and the load dissipates 2 kW at a power factor of 80% lagging. Find the capacitance required to correct the power factor to

a. 100%
b. 95% lagging

Solution

a. This problem is most easily solved by noting that the power factor will be unity when the *net* reactive power is zero. Thus if we can find the positive inductive reactive power in the load, then the capacitor must absorb an equal negative amount so that the net reactive power is zero. To find the reactive power of the load we proceed as follows:

The power factor angle, which is the same as the angle of the load impedance, is

$$\phi = \arccos 0.8 = 36.9°$$

The power factor is

$$\cos \phi = \frac{P}{S}$$

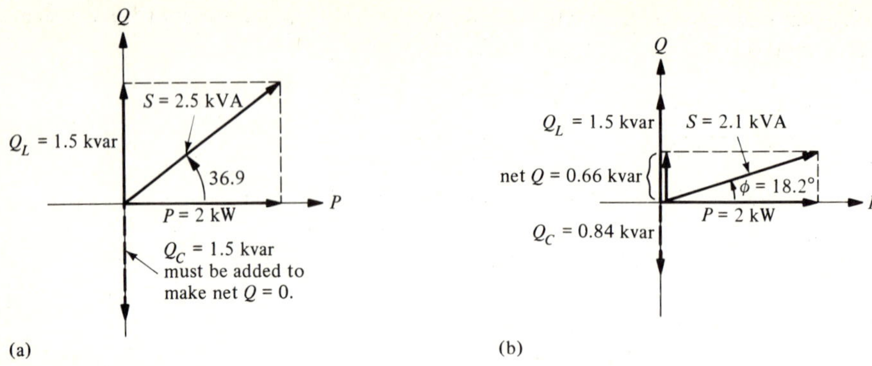

FIGURE 12.8
Complex power diagram for Example 12.2-4. (a) 100% power factor. (b) 95% power factor.

from which the apparent power is

$$S = \frac{P}{\cos\phi} = \frac{2\text{ kW}}{0.8} = 2.5\text{ kVA}$$

Since $S = |V||I|$, the reactive power of the load is

$$\begin{aligned} Q_L &= |V||I|\sin\phi = S\sin\phi = 2.5\sin 36.9^\circ \\ &= 1.5\text{ kvar} \end{aligned}$$

A complex power diagram is shown in Fig. 12.8a. In order to bring the power factor to 0°, the capacitive reactive power must be the same as the inductive reactive power, 1.5 kvar. Then, from Eq. (12.1-14),

$$Q_C = \frac{|V|^2}{X_C}$$

from which

$$X_C = \frac{|V|^2}{Q_C} = \frac{(440)^2}{1500} = 129\ \Omega$$

Finally, since $X_C = 1/\omega C$,

$$C = \frac{1}{\omega X_C} = \frac{1}{(2\pi)(60)(129)} = 0.000\,02\text{ F} = 20\ \mu\text{F}$$

b. A power factor of 95% implies a phase angle

$$\phi = \arccos 0.95 = 18.2^\circ$$

The power diagram for this angle is shown in Fig. 12.8b. The apparent power at this phase angle is

$$S = \frac{P}{\cos \phi} = \frac{2 \text{ kW}}{0.95} = 2.1 \text{ kVA}$$

The *net* reactive power at this phase angle is

$$Q = S \sin \phi = 2.1 \sin 18.2° = 0.66 \text{ kvar}$$

The inductive reactive power was found in part a to be

$$Q_L = 1.5 \text{ kvar}$$

Therefore, the capacitor must supply

$$Q_C = Q_L - Q = 1.5 - 0.66 = 0.84 \text{ kvar}$$

The required reactance is

$$X_C = \frac{|V|^2}{Q_C} = \frac{440^2}{840} = 230 \ \Omega$$

The capacitance is then

$$C = \frac{1}{\omega X_C} = \frac{1}{(2\pi)(60)(230)} = 11.5 \ \mu\text{F}$$

Note that in part a of this example, addition of the capacitor to bring the power factor up to 100% has reduced the apparent power from 2.5 to 2 kVA. The line current is then

$$|I| = \frac{S}{|V|} = \frac{2000}{440} = 4.5 \text{ A}$$

Before the capacitor was connected, the line current was

$$|I| = \frac{P}{|V| \cos \phi} = \frac{2000}{(440)(0.8)} = 5.7 \text{ A}$$

Thus, addition of the capacitor has reduced the amount of line current that the power company must provide.

In part b the line current is reduced to

$$|I| = \frac{S}{|V|} = \frac{2100}{440} = 4.8 \text{ A}$$

For both cases, the *load current* is not affected, since the load continues to draw 2 kW at 0.8 power factor.

• • •

Maximum Power Transfer

The circuit designer is sometimes faced with the problem of selecting a load impedance to connect to a Thevenin source circuit so that maximum power is delivered to the load impedance. Assume a Thevenin voltage V_T and a Thevenin impedance $Z_T = R_T + jX_T$. The problem is to determine the value of the load impedance $Z_L = R_L + jX_L$ such that the power delivered to Z_L is a maximum. The circuit is shown in Fig. 12.9a and the load current is

$$I_L = \frac{V_T}{Z_L + Z_T} = \frac{V_T}{R_L + R_T + j(X_L + X_T)} \qquad (12.2\text{-}12)$$

From Eq. (12.2-10), the complex load power is

$$P_c = Z_L |I_L|^2$$

and since $Z_L = R_L + jX_L$, the real power is

$$P = R_L |I_L|^2$$

$$= \frac{R_L |V_T|^2}{(R_L + R_T)^2 + (X_L + X_T)^2} \qquad (12.2\text{-}13)$$

This will be a maximum when the denominator is a minimum.

First consider the effect of the reactance. If $X_L = -X_T$ the reactive term in the denominator will be zero and the power P will be a maximum. Thus,

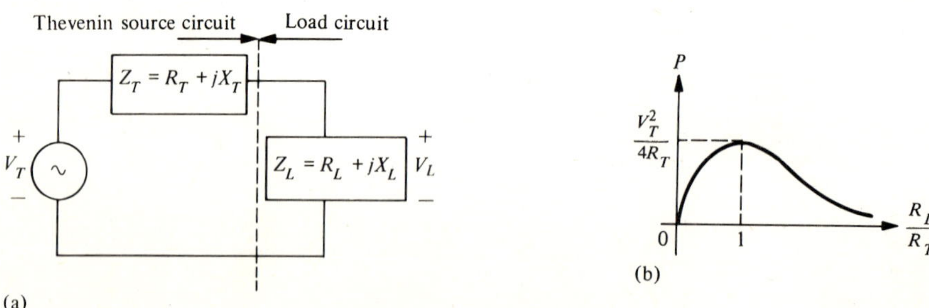

FIGURE 12.9
Maximum power transfer. (a) circuit. (b) Power as a function of R_L/R_T for $X_L = -X_T$.

if the Thevenin impedance is inductive, the load impedance should be capacitive and vice versa. With this condition, the power is

$$P = \frac{R_L|V_T|^2}{(R_L + R_T)^2} = \frac{|V_T|^2}{R_T} \times \frac{R_L/R_T}{(R_L/R_T + 1)^2} \tag{12.2-14}$$

A plot of P vs. R_L/R_T is shown in Fig. 12.9b. From this graph it is found that $R_L = R_T$ results in maximum power transfer to Z_L. When this condition is met, the power is

$$P_{L\,max} = \frac{|V_T|^2}{4R_T} \tag{12.2-15}$$

It is interesting to note that when Z_L is chosen for maximum power transfer the efficiency, that is, the ratio of power delivered to the load to that supplied by the Thevenin source, is 50%. This is easily determined by recognizing that all the power is dissipated by R_L and R_T. Since the two resistances are equal and carry the same current, they dissipate the same power and the source supplies twice the power delivered to R_L.

EXAMPLE 12.2-5 *Maximum Power Transfer*

An antenna is to be matched to a receiver for maximum power transfer. The frequency is 15 MHz and the antenna impedance at this frequency is $50 + j22\ \Omega$. The input circuit of the receiver consists of a resistor and capacitor in parallel as shown in Fig. 12.10. What values should they have in order to achieve the match?

Solution

In order to achieve maximum power transfer, the receiver should have an input impedance that is the conjugate of the antenna impedance. Therefore,

$$Z_L = 50 - j22\ \Omega$$

This result is appropriate for a series RC circuit. However, we are to find a parallel circuit so we find the load admittance

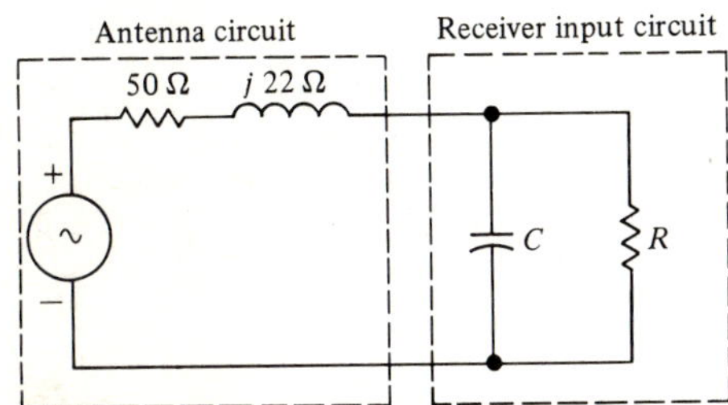

FIGURE 12.10
Circuit for Example 12.2-5.

$$Y_L = \frac{1}{Z_L} = \frac{1}{50 - j22} = \frac{1}{54.6\underline{/-23.7^\circ}} = 0.018\underline{/23.7^\circ}\ \text{S}$$
$$= 0.016 + j0.0072\ \text{S}$$

This represents a resistance $R = 1/0.016 = 62\ \Omega$ in parallel with a capacitive susceptance

$$B_C = \omega C = 0.0072\ \text{S}$$

from which

$$C = \frac{0.0072}{\omega} = \frac{0.0072}{(2\pi)(15)(10^6)} = 76 \times 10^{-12}\ \text{F} = 76\ \text{pF}$$

• • •

LEARNING EXERCISES FOR SEC. 12.2

1. A 110-V, 60-Hz transformer coil has an inductance of 2.5 H and a resistance of 250 Ω. How much real power is dissipated at the rated voltage?
2. How much reactive power is required by the coil of Learning exercise 12.2-1?
3. Find the apparent power for the coil of Learning exercise 12.2-1.

Ans. 12; 3.2; 12.4

• • •

12.3 THREE-PHASE SYSTEMS

In the introduction to Chap. 10, we showed how a sinusoidal voltage could be generated by a coil rotating in a magnetic field. This arrangement is called a *single-phase* generator because only one voltage is generated. In practice it is found that generators that provide more than one voltage, called *polyphase* generators, lead to more economical transmission of power. Therefore, *polyphase* systems are used extensively in power generation, transmission, and distribution. In this section, we consider the most popular of these, the *three-phase* system.

Fig. 12.11a shows a simplified three-phase generator in cross-section. Three identical coils are mounted on the rotor and physically spaced 120° apart and rotated in a steady magnetic field supplied by permanent magnets. The conventional representation of this arrangement schematically uses three inductance coil symbols for the generator coils spaced angularly 120° apart as shown in Fig. 12.11b. For our purposes, the three-generator representation shown in Fig. 12.11c will be most useful. The voltage induced in each coil will be sinusoidal and the three sine waves will be 120° out of phase with one another because of the physical spacing of the

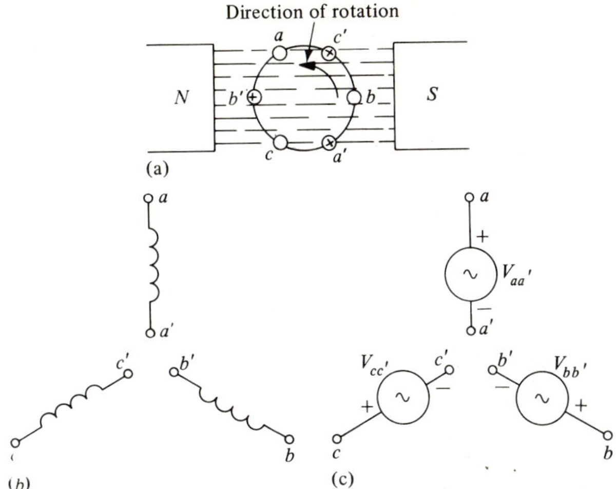

FIGURE 12.11
Three-phase generator. (a) Cross section. (b) Schematic. (c) Three-generator equivalent circuit.

coils as they rotate in the magnetic field. The waveforms are shown in Fig. 12.12a and a phasor diagram is shown in Fig. 12.12b.

Double-Subscript Notation

There are six terminals on the generators in Fig. 12.11c and they can be connected in a variety of ways. Each time we reverse the leads of one generator coil the angle of its voltage changes by 180° so that there are quite a few possible combinations. In order to minimize confusion we will use double-subscript notation throughout this section, as shown in Fig. 12.11 and 12.12. In these figures we have assumed that the coils are wound on the generator rotor so that the voltages induced in each coil have the phase

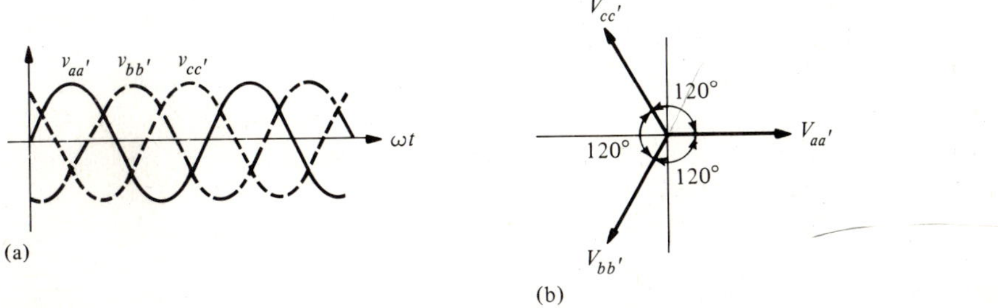

FIGURE 12.12
(a) Three-phase waveforms. (b) Phasor diagram.

relations shown. According to the time and phasor diagrams in Fig. 12.12, voltage $V_{cc'}$ lags $V_{bb'}$ by 120°; $V_{bb'}$ lags $V_{aa'}$ by 120°; and $V_{aa'}$ lags $V_{cc'}$ by 120°. The time domain equations for the voltages are, using $v_{aa'}$ as reference,

$$v_{aa'}(t) = \sqrt{2}|V_{aa'}|\cos \omega t$$

$$v_{bb'}(t) = \sqrt{2}|V_{bb'}|\cos(\omega t - 120°)$$

$$v_{cc'}(t) = \sqrt{2}|V_{cc'}|\cos(\omega t - 240°) = \sqrt{2}|V_{cc'}|\cos(\omega t + 120°) \tag{12.3-1}$$

The phasor representations are

$$V_{aa'} = |V_{aa'}|\underline{/0°}$$

$$V_{bb'} = |V_{bb'}|\underline{/-120°}$$

$$V_{cc'} = |V_{cc'}|\underline{/-240°} = |V_{cc'}|\underline{/+120°} \tag{12.3-2}$$

In one type of system, terminals a', b', and c' are connected together and called the neutral n. The generator circuit then appears as in Fig. 12.13a. The voltages V_{an}, V_{bn}, V_{cn} are called *phase voltages* and V_{ab}, V_{bc}, and V_{ca} are called *line voltages*. The phasor diagram is shown in Fig. 12.13b. Because of the appearance of the circuit and phasor diagrams, the system is said to be *wye-connected*.

The *phase sequence* is defined as the order in which the three voltages

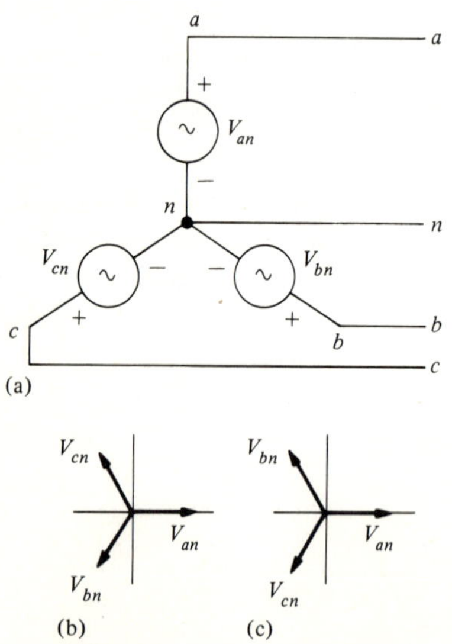

FIGURE 12.13
Wye connection. (a) Generator. (b) *abc* phase sequence. (c) *acb* phase sequence.

pass through their respective maximum values. Therefore, the phase sequence for the system in Fig. 12.13b is $V_{an}V_{bn}V_{cn}$. This is often abbreviated to *abc* and is exactly equivalent to *bca* or *cab*. Another sequence can be obtained by interchanging any two phases. The new sequence is then $V_{an}V_{cn}V_{bn}$ or *acb*, which is shown in Fig. 12.13c. The sequence can be read from the phasor diagram by reading the first subscript of each voltage while going around the diagram in a clockwise direction.

In a three-phase system the *direction* of rotation of motors depends on the phase sequence. If a large motor is accidentally connected to the wrong phase sequence, considerable damage to machinery and equipment may occur. *Phase sequence indicators* are instruments that can determine an unknown phase sequence.

The Balanced Wye-Wye System

A wye-connected generator and a wye-connected load are shown in Fig. 12.14. The load neutral is connected to the generator neutral so that we have a four-wire system. The load is shown as three resistances, R_a, R_b, and R_c and we will first consider the case where they are equal. For this case, the load is said to be *balanced*, and the calculations are much simpler than the case where the load is unbalanced. In this system the four wires are called the *lines*; the currents in lines *a*, *b*, and *c* are the *line currents*; and in the fourth line we have the *neutral* current. The generator voltages in the wye system are the *phase* voltages, that is, the voltage from each line to neutral. In addition to the phase voltages, we are interested in the *line* voltages, that is, V_{ab}, V_{bc}, and V_{ca}. The relationship between phase and line quantities will be illustrated in the following example.

EXAMPLE 12.3-1 Balanced Wye System

In Fig. 12.14, $V_{an} = 120\underline{/0°}$ V and the phase sequence is *abc*. The load consists of three identical heater coils. These may be considered purely resistive with $R = 6\ \Omega$. Find all phase and line currents and voltages and draw a phasor diagram.

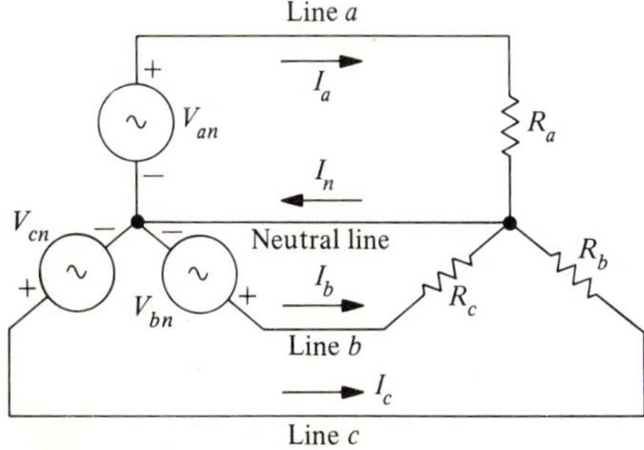

FIGURE 12.14
Wye-wye system with resistive loads.

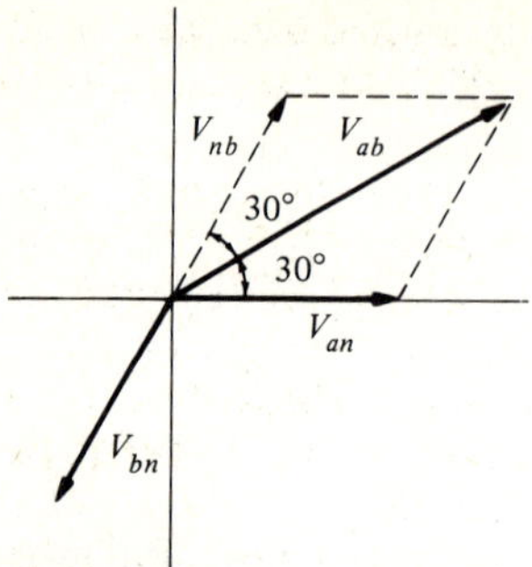

FIGURE 12.15
Finding the line voltage of a wye-connected generator.

Solution
For the indicated phase sequence

$$\mathbf{V}_{an} = 120\underline{/0°}\ \text{V}$$
$$\mathbf{V}_{bn} = 120\underline{/-120°}\ \text{V}$$
$$\mathbf{V}_{cn} = 120\underline{/+120°}\ \text{V} \tag{12.3-3}$$

These are the *phase* voltages. In order to find the line voltages, we use KVL to write

$$\begin{aligned}\mathbf{V}_{ab} &= \mathbf{V}_{an} + \mathbf{V}_{nb}\\ &= \mathbf{V}_{an} - \mathbf{V}_{bn}\\ &= 120\underline{/0°} - 120\underline{/-120°}\\ &= 120(1\underline{/0°} - 1\underline{/-120°})\\ &= 120[(1 + j0) - (-0.5 - j0.87)]\\ &= 120(1.5 + j0.87)\\ &= 120(1.73\underline{/30°})\\ &= 208\underline{/30°}\ \text{V}\end{aligned} \tag{12.3-4}$$

The phasor subtraction is shown in Fig. 12.15. It can be shown by geometrical construction on the phasor diagram (see Prob. 12.31) that this result can be written in the form

$$\boxed{\mathbf{V}_{ab} = \mathbf{V}_{an} \times \sqrt{3}\ \underline{/30°}} \tag{12.3-5}$$

Therefore, in words, the line voltage output of a wye-connected generator is $\sqrt{3}$ times the phase voltage and is displaced in phase by 30°. The complete phasor diagram is shown in Fig. 12.16a.

Observe that there are two different source voltages available from the wye generator. The phase voltages are 120 V and the line voltages are 208 V. Therefore, loads rated at either voltage can be used with this system, which is widely used for power distribution. Loads rated at 120-V single-phase are connected between any line and neutral. Loads rated at 208-V single-phase are connected between any two lines. If a three-phase load is rated at 120 V it must be wye connected; if rated at 208 V it must be delta connected. This is discussed in the next section.

In order to find the currents we note first that in the wye system the line current is the same as the phase current. Since the voltage across R_a is V_{an} we have

$$I_a = \frac{V_{an}}{R_a} = \frac{120\angle 0^\circ \text{ V}}{6\ \Omega} = 20\angle 0^\circ \text{ A}$$

Similarly,

$$I_b = \frac{V_{bn}}{R_b} = \frac{120\angle -120^\circ \text{ V}}{6\ \Omega} = 20\angle -120^\circ \text{ A}$$

$$I_c = \frac{V_{cn}}{R_c} = \frac{120\angle +120^\circ \text{ V}}{6\ \Omega} = 20\angle +120^\circ \text{ A} \qquad (12.3\text{-}6)$$

In order to find the neutral current we use KCL at node n

$$\begin{aligned} I_n &= I_a + I_b + I_c \\ &= 20(1\angle 0^\circ + 1\angle -120^\circ + 1\angle +120^\circ) \\ &= 20(1 - 0.5 - j0.866 - 0.5 + j0.866) \\ &= 0 \end{aligned}$$

We conclude from this that the neutral current in a balanced wye system is zero. Therefore, the neutral wire can be removed without affecting the system. In

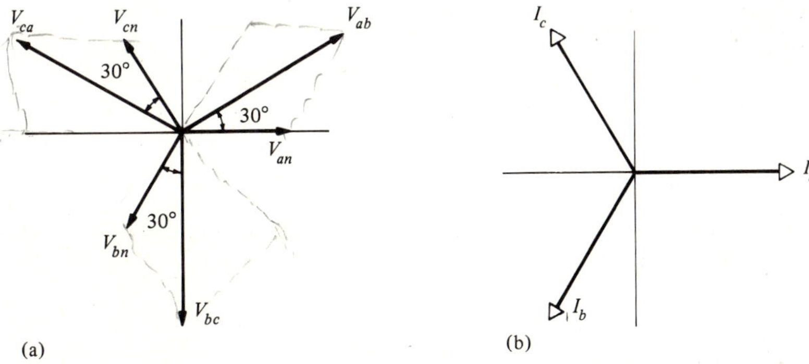

FIGURE 12.16
Example 12.3-1. (a) Phase and line voltages in a wye-connected generator. (b) The line currents.

practice, the neutral is always included and is grounded for safety. In case of load unbalance, the neutral current is not zero, as we see in a later example.

A phasor diagram of the currents is shown in Fig. 12.16b. We observe that the phase angle between the line current and phase voltage is zero because the load is a resistance. However, for example, the line current I_a lags the nearest line voltage V_{ab} by 30°.

• • •

EXAMPLE 12.3-2 Currents in an Unbalanced Wye System

Heater coils *a* and *b* in Example 12.3-1 are partially shorted so that their resistance is 4 Ω. Find the line and neutral currents and draw a phasor diagram.

Solution

The phase and line voltages are, of course, the same as in the previous example. Since phases *a* and *b* contain the partially shorted coils, then

$$I_a = \frac{V_{an}}{R_a} = \frac{120\angle 0^\circ \text{ V}}{4\ \Omega} = 30\angle 0^\circ \text{ A}$$

$$I_b = \frac{V_{bn}}{R_b} = \frac{120\angle -120^\circ \text{ V}}{4\ \Omega} = 30\angle -120^\circ \text{ A}$$

For phase *c*, the coil resistance is 6 Ω. Therefore the current is

$$I_c = \frac{V_{cn}}{R_c} = \frac{120\angle +120^\circ \text{ V}}{6\ \Omega} = 20\angle +120^\circ \text{ A}$$

The neutral current is

$$\begin{aligned} I_n &= I_a + I_b + I_c \\ &= 30\angle 0^\circ + 30\angle -120^\circ + 20\angle +120^\circ \\ &= (30 + j0) + (-15 - j26) + (-10 + j17.3) \\ &= 5 - j8.7 \\ &= 10\angle -60^\circ \text{ A} \end{aligned}$$

The phasor diagram is shown in Fig. 12.17.

• • •

The Delta-Connected System

Thus far we have considered the wye system, a common three-phase configuration. Another common configuration involves connecting the coils of the three-phase generator of Fig. 12.11 in a closed loop as shown in Fig. 12.18. In order to protect the generator from the effects of a wrong connection, a voltmeter is inserted between points *a* and *c′* *before* they are connected together. If the coils are in the proper orientation, the voltmeter will read zero and it is safe to complete the connection. If one of the coils is

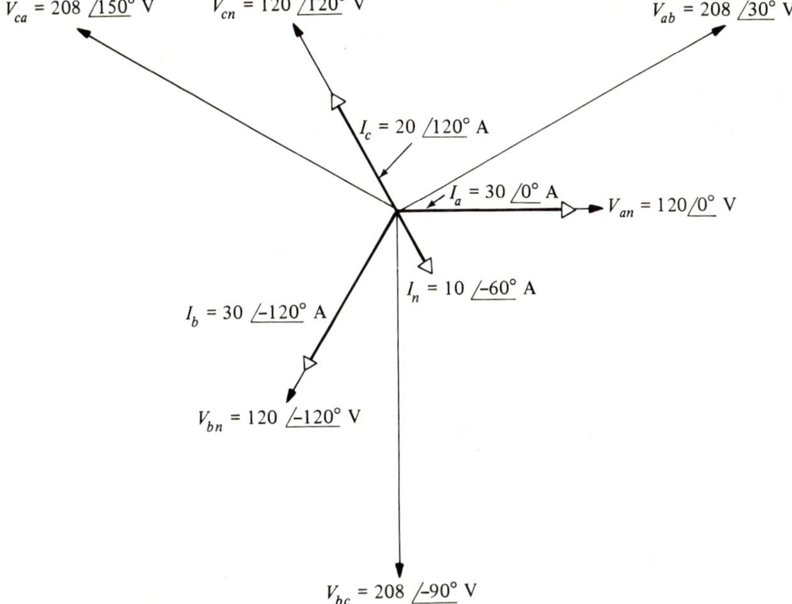

FIGURE 12.17
Phasor diagram for Example 12.3-2 showing currents when the wye load is unbalanced. Note that the neutral current is not zero.

reversed, the voltmeter will read twice the phase voltage (see Prob. 12.35) and if the connection is made, a large current will flow around the loop, limited only by the small internal impedance of the generator.

The equivalent generator circuit is shown in Fig. 12.18b. Because of the similarity to the greek letter Δ, this is called the *delta* connection. Observe that there is no neutral as in the wye connection. When discussing the delta

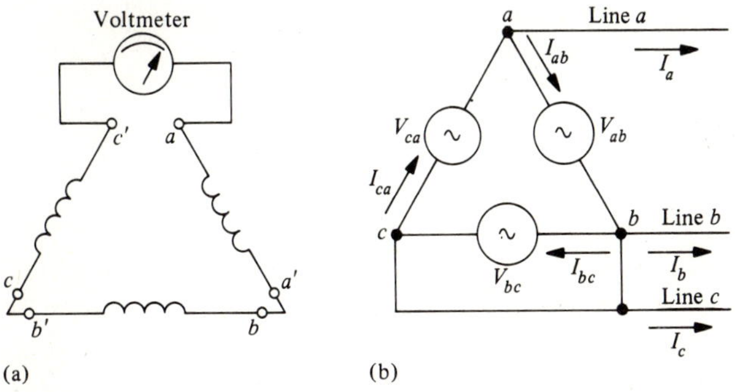

FIGURE 12.18
The delta connection. (a) Checking with a voltmeter. (b) Equivalent generator circuit.

connection we will use the notation shown in Fig. 12.18b. The significance of the subscripts is exactly the same as it has been all along, that is, V_{ab} is the equivalent generator voltage drop from *a* to *b*, and so on. Polarity signs are not required on the generator symbol because they are implicit in the definition of double-subscript notation. For the currents in the delta system, it is convenient to use double-subscript notation so that we can exploit the similarity between them and the voltages in the wye system.

In the wye system we found that line and phase *voltages* were different. In the delta system, it is line and phase *currents* that are different. This is illustrated in the next example.

EXAMPLE 12.3-3 The Balanced Delta System

The circuit of Fig. 12.19 is a balanced delta system. The impedance is $Z = 20\angle{+25°}\ \Omega$ and the sources are 220 V with phase sequence *abc*. Find the load (phase) currents I_{ab}, I_{bc}, and I_{ca} and the line currents I_a, I_b, and I_c, and draw a complete phasor diagram.

Solution

For the phase sequence specified, we have

$$V_{ab} = 220\angle{0°}\ \text{V}$$

$$V_{bc} = 220\angle{-120°}\ \text{V}$$

$$V_{ca} = 220\angle{+120°}\ \text{V}$$

Inspection of the circuit shows that the generator *phase* voltages in the delta system are the *line* voltages and each impedance is connected directly to a line voltage. This is to be compared to the wye system, where the loads are connected across the phase voltages.

In the delta system each line current is a function of the currents in two of the load impedances. In order to determine this dependence, note that the load currents

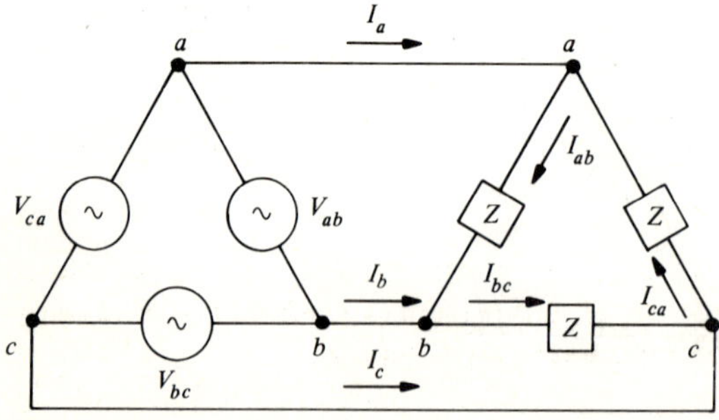

FIGURE 12.19
Balanced delta system.

are

$$I_{ab} = \frac{V_{ab}}{Z} = \frac{220\underline{/0°}\text{ V}}{20\underline{/25°}\ \Omega} = 11\underline{/-25°}\text{ A}$$

$$I_{bc} = \frac{V_{bc}}{Z} = \frac{220\underline{/-120°}\text{ V}}{20\underline{/25°}\ \Omega} = 11\underline{/-145°}\text{ A}$$

$$I_{ca} = \frac{V_{ca}}{Z} = \frac{220\underline{/+120°}\text{ V}}{20\underline{/25°}\ \Omega} = 11\underline{/95°}\text{ A}$$

The line currents are then found using KCL as follows:

$$\begin{aligned} I_a &= I_{ab} + I_{ac} \\ &= I_{ab} - I_{ca} \\ &= 11\underline{/-25°} - 11\underline{/95°} \\ &= (9.97 - j4.65) - (-0.96 + j10.96) \\ &= 10.9 - j15.6 \\ &= 19\underline{/-55°}\text{ A} \end{aligned}$$

In a similar fashion we find

$$\begin{aligned} I_b &= I_{bc} + I_{ba} \\ &= 19\underline{/-175°}\text{ A} \\ I_c &= I_{ca} + I_{cb} \\ &= 19\underline{/65°}\text{ A} \end{aligned}$$

The complete phasor diagram is shown in Fig. 12.20.

The reader will observe that the line currents lag the nearest phase current by 30° and, since $11\sqrt{3} = 19$, the phase-to-line current relationship in the delta system is the same as the phase-to-line voltage relationship in the wye system. In the delta system, the line currents are displaced from the nearest phase currents by 30° and the line current magnitude is $\sqrt{3}$ times the phase current magnitude.

• • •

LEARNING EXERCISE FOR SEC. 12.3

1. A balanced Y load as in Fig. 12.14 has a phase voltage magnitude of 120 V and a phase impedance $5\underline{/53°}\ \Omega$. Find the line current and line voltage magnitudes and the phase angle between line current I_a and line voltage V_{ab}.

Ans. 83; 208; 24

• • •

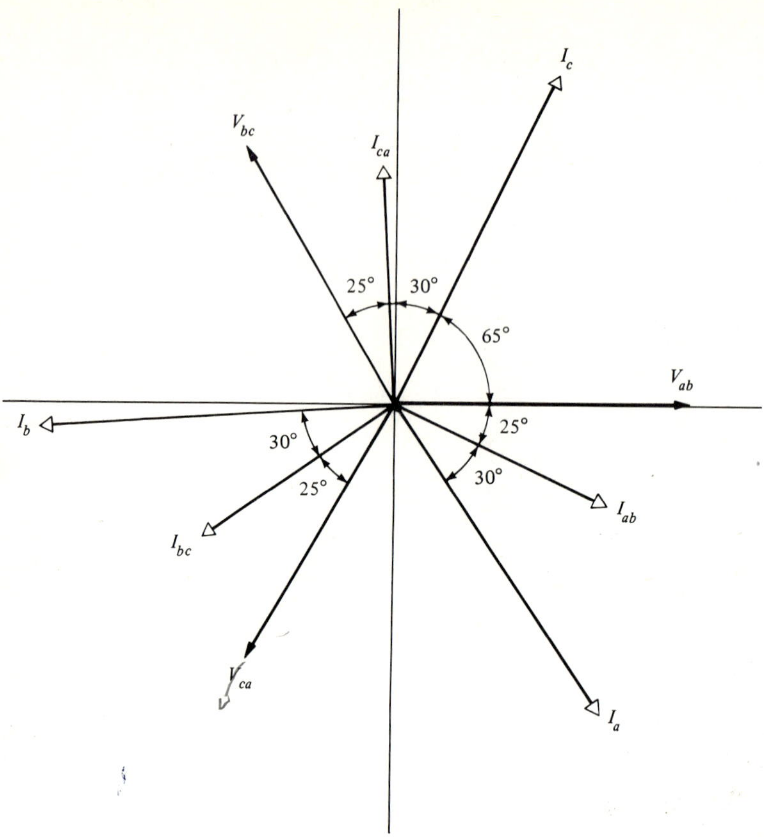

FIGURE 12.20
Phasor diagram for Example 12.3-3.

12.4 DELTA-WYE CONVERSION

Sometimes wye and delta networks are both present in a system. Analysis is simplified if the circuit contains only one type of circuit so it is useful to have a transformation by means of which we can find a wye equivalent to a given delta, and vice versa. In order to establish the conditions for equivalence consider the wye and delta circuits shown in Fig. 12.21. The two circuits will be equivalent if the resistance between terminals a and b, b and c, and c and a are identical for the two circuits. First consider terminals a–b. For the wye circuit (with terminal c open) the resistance between a and b is

$$R_{ab} = R_a + R_b \tag{12.4-1}$$

For the delta circuit, (with terminal c not connected to any other point) the resistance is

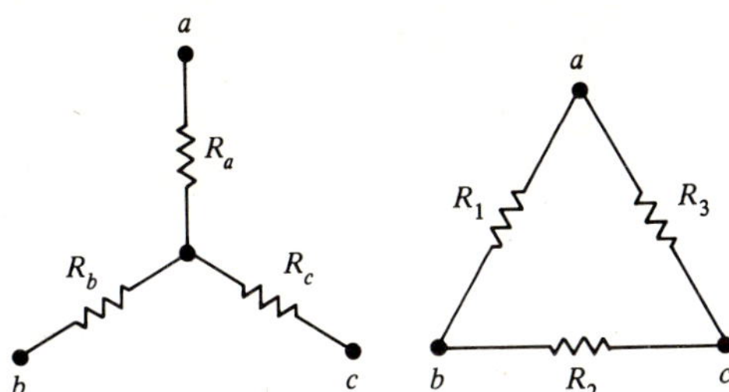

FIGURE 12.21
Wye-delta equivalence.

$$R_{ab} = R_1 \parallel (R_2 + R_3) \tag{12.4-2}$$

Equating these two for equivalence,

$$R_a + R_b = \frac{R_1(R_2 + R_3)}{R_1 + R_2 + R_3} \tag{12.4-3}$$

Similarly, for terminals b and c

$$R_b + R_c = \frac{R_2(R_1 + R_3)}{R_1 + R_2 + R_3} \tag{12.4-4}$$

and for terminals c and a

$$R_c + R_a = \frac{R_3(R_1 + R_2)}{R_1 + R_2 + R_3} \tag{12.4-5}$$

If we subtract Eq. (12.4-4) from Eq. (12.4-3) the result is

$$R_a - R_c = \frac{R_1R_3 - R_2R_3}{R_1 + R_2 + R_3} \tag{12.4-6}$$

Next we add Eqs. (12.4-5) and (12.4-6) to get

$$2R_a = \frac{2R_1R_3}{R_1 + R_2 + R_3} \tag{12.4-7}$$

and finally

$$R_a = \frac{R_1R_3}{R_1 + R_2 + R_3} \tag{12.4-8}$$

For the other resistances we find

$$R_b = \frac{R_1R_2}{R_1 + R_2 + R_3} \tag{12.4-9}$$

and

$$R_c = \frac{R_2R_3}{R_1 + R_2 + R_3} \tag{12.4-10}$$

These equations are used to find a wye circuit equivalent to a given delta. They simplify considerably if the original delta system is balanced. If this is the case $R_1 = R_2 = R_3 = R_\Delta$ and $R_a = R_b = R_c = R_y$. From any of the transformation equations

$$\boxed{R_y = \frac{R_\Delta^2}{3R_\Delta} = \frac{R_\Delta}{3}} \tag{12.4-11}$$

If we are given a wye circuit, the equivalent delta can be found by manipulating Eqs. (12.4-8), (12.4-9), and (12.4-10) (see Prob. 12.39). The results are

$$R_1 = \frac{R_aR_b + R_bR_c + R_cR_a}{R_c} \tag{12.4-12}$$

$$R_2 = \frac{R_aR_b + R_bR_c + R_cR_a}{R_a} \tag{12.4-13}$$

$$R_3 = \frac{R_aR_b + R_bR_c + R_cR_a}{R_b} \tag{12.4-14}$$

These equations are used to find a delta circuit equivalent to a given wye. If the original wye is balanced, they simplify to

$$\boxed{R_\Delta = \frac{3R_y^2}{R_y} = 3R_y} \tag{12.4-15}$$

The equations above have been derived assuming resistance elements throughout. In general, the delta or wye arms will be complex impedances and we can use all of the above results simply by replacing R by Z where appropriate. An example will illustrate the use of these transformations.

EXAMPLE 12.4-1 Balanced Wye–Delta System

In the circuit of Fig. 12.22a a wye-connected source feeds a delta load. The load is balanced with $Z = 20\angle 45° \ \Omega$. Find the line and load currents and draw a phasor diagram.

Solution

We begin by converting the delta load to a wye. From Eq. (12.4-11) the equivalent wye impedance is

$$Z_y = \frac{Z_\Delta}{3} = \frac{20\angle 45°}{3} = 6.67\angle 45° \ \Omega$$

Since the load is balanced, we can connect a neutral wire in our equivalent circuit (recall that the neutral current is zero in a balanced wye–wye system). The circuit to be analyzed in order to find the line currents is shown in Fig. 12.22b. We assume the phase sequence is *abc* so that

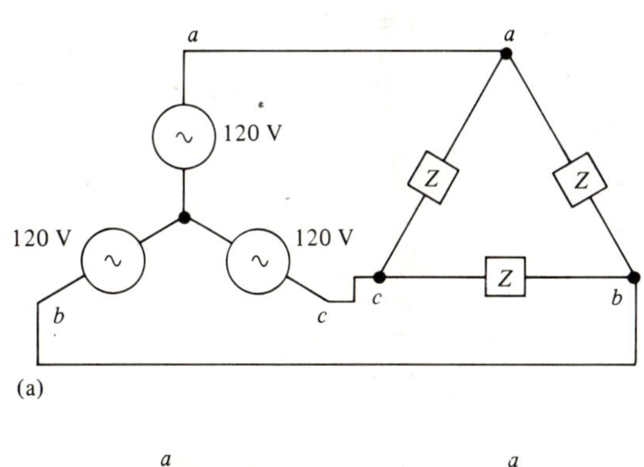

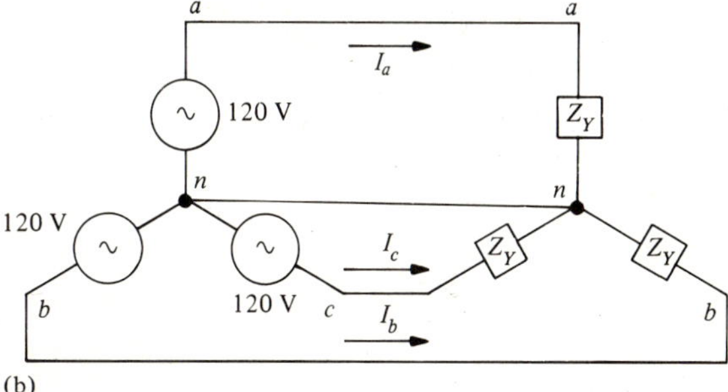

FIGURE 12.22
Example 12.4-1, a wye-delta system. (a) Original circuit. (b) Equivalent circuit after the delta load is transformed to wye.

$$\mathbf{V}_{an} = 120\underline{/0^\circ} \text{ V}$$
$$\mathbf{V}_{bn} = 120\underline{/-120^\circ} \text{ V}$$
$$\mathbf{V}_{cn} = 120\underline{/+120^\circ} \text{ V}$$

Then

$$\mathbf{I}_a = \frac{\mathbf{V}_{an}}{\mathbf{Z}_y} = \frac{120\underline{/0^\circ} \text{ V}}{6.67\underline{/45^\circ} \ \Omega} = 18\underline{/-45^\circ} \text{ A}$$

$$\mathbf{I}_b = \frac{\mathbf{V}_{bn}}{\mathbf{Z}_y} = \frac{120\underline{/-120^\circ} \text{ V}}{6.67\underline{/45^\circ} \ \Omega} = 18\underline{/-165^\circ} \text{ A}$$

$$\mathbf{I}_c = \frac{\mathbf{V}_{cn}}{\mathbf{Z}_y} = \frac{120\underline{/120^\circ} \text{ V}}{6.67\underline{/45^\circ} \ \Omega} = 18\underline{/75^\circ} \text{ A}$$

These are the line currents. As we found in Example 12.3-3, the phase currents in Δ

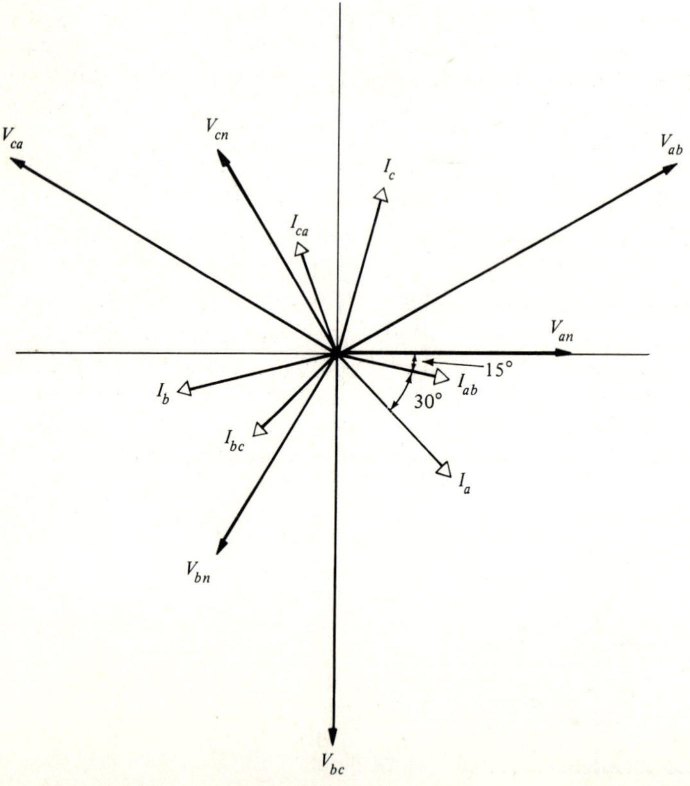

FIGURE 12.23
Phasor diagram for Example 12.4-1.

are $1/\sqrt{3}$ times the line currents and are displaced in phase by 30° from the line current whose subscript is the same as the first subscript of the phase current. Thus

$$I_{ab} = \frac{I_a}{\sqrt{3}}\underline{/+30^\circ} = \frac{18}{\sqrt{3}}\underline{/-15^\circ} = 10.4\underline{/-15^\circ}\text{ A}$$

$$I_{bc} = \frac{I_b}{\sqrt{3}}\underline{/+30^\circ} = 10.4\underline{/-135^\circ}\text{ A}$$

$$I_{ca} = \frac{I_c}{\sqrt{3}}\underline{/+30^\circ} = 10.4\underline{/105^\circ}\text{ A}$$

The phasor diagram is shown in Fig. 12.23.

• • •

The Edison Three-Wire System

A three-wire system used for distributing power to residences where the load is relatively light is shown in Fig. 12.24. The source voltages are equal in magnitude and opposite in phase with respect to the neutral n. They are usually obtained from a transformer. This is discussed later in this chapter. If the source voltages are the standard value of 115 V, then two separate 115-V, single-phase loads can be connected from lines a and b to the neutral as shown, and in addition a 230-V, single-phase load can be connected from line a to line b. An advantage of the system is that, if loads 1 and 2 are identical, the neutral current is zero. As a result the system can be constructed using a smaller wire size than if the loads were connected to a two-wire, single-phase source. This is illustrated in the following example.

EXAMPLE 12.4-2 Comparison of Three-wire and Two-wire Systems

A system is to feed two 1-kW resistive loads located 1000 ft from the source and a choice must be made between an Edison three-wire, single-phase system or a two-wire, single-phase system. For each system find the maximum line resistance such that the load voltage at full load current will not drop below 112 V when the generator voltage is 115 V. Compare the wire size required in the two systems.

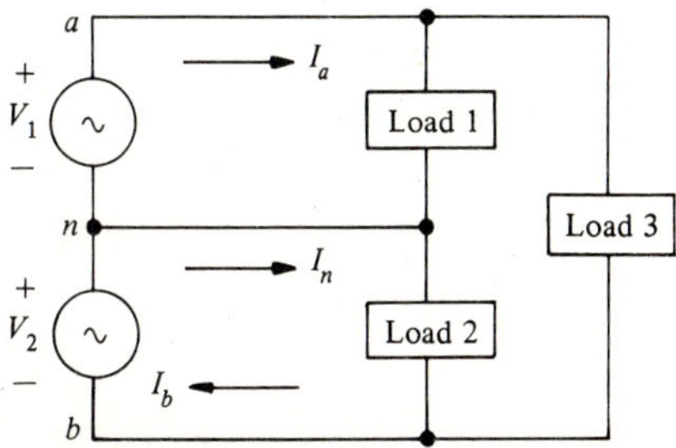

FIGURE 12.24
Edison three-wire system.

Solution

The two systems are shown in Fig. 12.25. For the Edison system the line currents are

$$I_a = I_b = \frac{P}{V} = \frac{1000 \text{ W}}{112 \text{ V}} = 8.9 \text{ A}$$

The voltage drop in the line resistance R_a must be 3 V or less. Note that the neutral current is zero, so that all of the line drop will appear across R_a. Then

$$R_a = R_b \leq \frac{3 \text{ V}}{8.9} = 0.34 \ \Omega$$

Therefore, the Edison system requires a total of 2000 ft of wire with a resistance of 0.34 Ω/1000 ft or less. According to the wire tables in Appendix B, AWG #5 wire will suffice. The 1000-ft neutral wire can be much smaller, since it is only required to carry any small unbalance current.

For the two-wire system shown in Fig. 12.25b the line current is

$$I_L = \frac{P}{V} = \frac{2 \times 1000 \text{ W}}{112 \text{ V}} = 17.9 \text{ A}$$

The 3-V allowable drop appears across $2R_L$. Thus

$$2R_L = \frac{3 \text{ V}}{17.9 \text{ A}} = 0.168 \ \Omega$$

and $R_L = 0.084 \ \Omega$.

The two-wire system requires 2000 ft of wire with a resistance of 0.084 Ω/1000 ft or less. From the wire tables, AWG #00 will be satisfactory.

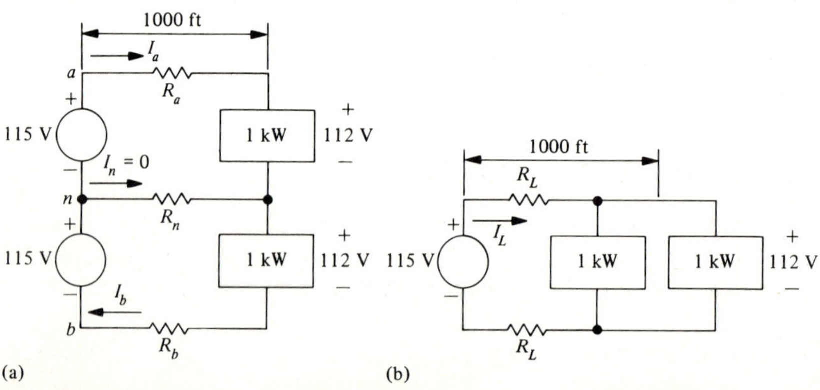

FIGURE 12.25
Example 12.4-2. (a) Edison three-wire system. (b) Two-wire system.

Since the weight of #00 wire is 400 lb/1000 ft, the total copper required for the two-wire system is 800 lb. For the Edison system, #5 wire weighs 100 lb/1000 ft, so that 200 lb of copper is required for the outer lines. Even if we use the same #5 wire for the neutral the total weight is only 300 lb, as compared to 800 lb for the two-wire system. Thus, as far as the weight of copper required to maintain a fixed voltage at the load is concerned, the Edison system is considerably better than the two-wire system.

In this example, we compared the two systems using the criterion that the line resistance should be such that the load voltage remain the same regardless which system is used. In some cases, other criteria might be more appropriate. For example, it might be more important to minimize the power lost in the transmission line.

• • •

Other Systems

The three-phase and three-wire single-phase systems are those that see the most use in power distribution applications. Other polyphase systems are used for special applications. For example, in a two-phase system, the voltages are 90° out of phase. The direction of rotation of a two-phase motor depends on whether a control voltage applied to one phase leads or lags a reference voltage applied to the second phase. Because of this ease of control, these motors are used in automatic control systems called *servomechanisms*.

Occasionally, six-phase systems are used when the ac is to be converted to dc by a process called rectification. The additional phases simplify the filters required to smooth out the pulsating unidirectional voltage at the output of the rectifier.

• • •

LEARNING EXERCISE FOR SEC. 12.4

1. A delta load has $R_1 = R_2 = 25\ \Omega$ and $Z_3 = 10\underline{/30°}\ \Omega$. Find the equivalent wye.

Ans. $4.2\underline{/25.1°}$; $10.6\underline{/-4.9°}$; $4.2\underline{/25.1°}$

• • •

12.5 POWER IN THREE-PHASE CIRCUITS

The total power to a three-phase load is simply the sum of the powers dissipated in each phase. This holds for both balanced or unbalanced loads connected in wye or delta. The power in each phase can be expressed as

$$P_P = |V_P||I_P|\cos\phi_P \tag{12.5-1}$$

where the subscript P indicates *phase* quantity. If the load is unbalanced, the power in each phase must be calculated separately and the total power is the sum of the three individual powers. If the load is balanced, the total power P is simply three times the phase power P_P

$$P = 3|V_P||I_P|\cos\phi_P \tag{12.5-2}$$

For many three-phase loads the only terminals to which meters can be connected are the line terminals. In addition, nameplate data for three-phase devices always includes the rated *line* voltage, and if a rated current is included, it is the *line* current. Therefore, we need an expression for total power in terms of line voltage and line current.

In the balanced wye system the line and phase quantities are related by

$$|V_L| = \sqrt{3}|V_P| \quad \text{and} \quad |I_L| = |I_P| \tag{12.5-3}$$

while in the balanced delta,

$$|V_L| = |V_P| \quad \text{and} \quad |I_L| = \sqrt{3}|I_P| \tag{12.5-4}$$

When either of these pairs of equations is substituted into Eq. (12.5-2) the result is

$$\boxed{P = \sqrt{3}|V_L||I_L|\cos\phi_P} \tag{12.5-5}$$

This is the desired relation involving line quantities. However, the power factor angle ϕ_P is the angle of the load (phase) impedance. Typical power calculations can be found in the following examples.

EXAMPLE 12.5-1 Power in a Balanced Load

A three-phase, 440-V heater has wye-connected coils with impedance $20\underline{/15^\circ}\ \Omega$ in each phase. Find the total power dissipated.

Solution

The line voltage is specified as 440 V. The phase voltage is then

$$|V_P| = \frac{|V_L|}{\sqrt{3}} = \frac{440}{\sqrt{3}} = 254 \text{ V}$$

The line current is found using Ohm's law:

$$|I_L| = |I_P| = \frac{|V_P|}{|Z|} = \frac{254 \text{ V}}{20\ \Omega} = 12.7 \text{ A}$$

The total power is

$$P = \sqrt{3}|V_L||I_L|\cos\phi_P$$
$$= \sqrt{3} \times 440 \times 12.7 \cos 15°$$
$$= 9350 \text{ W} = 9.35 \text{ kW}$$

• • •

EXAMPLE 12.5-2 Finding Line Current from Nameplate Data

The nameplate on an industrial power saw motor reads 5 hp, 220 V, 3 ϕ, 0.95 power factor. If the efficiency is 85%, find the line current.

Solution

The 5-hp specification refers to mechanical power output. Therefore the electric power input is

$$P = \frac{(5 \text{ hp})(0.746 \text{ kW/hp})}{0.85} = 4.4 \text{ kW}$$

Then, using Eq. (12.5-5) the line current is

$$|I_L| = \frac{P}{\sqrt{3}|V_L|\cos\phi_P} = \frac{4400}{\sqrt{3} \times 220 \times 0.95} = 12.2 \text{ A}$$

• • •

Measurement of Power in Three-phase Systems

The electrodynamometer wattmeter described in Sec. 6.6 in connection with dc power measurements is also used for ac. Its construction is such that the power factor angle is automatically taken into account. In order to measure power in a single-phase load the current coil is connected in series with the line and the voltage coil is connected across the line (see Fig. 6.19). The ± polarity indications on the terminals of the meter must be connected properly or the meter will read negative, in which case one of the coils must be reversed.

When it comes to measuring power in a three-phase load, we have a number of options. The most obvious method is to use three wattmeters, each connected to one of the load arms. A schematic showing this method applied to both wye and delta circuits is shown in Fig. 12.26. This scheme can be used only if the individual phase connections are available for the delta, Fig. 12.26b, or the neutral for the wye, Fig. 12.26a. This is not usually the case, so other arrangements must be used. The most popular method for measuring three-phase power is the *two-wattmeter method* illustrated in Fig. 12.27. In this method the unmarked end of each voltage coil is connected to one line and the current coils are placed in the remaining two lines. This arrangement has a number of advantages. One is that it uses two

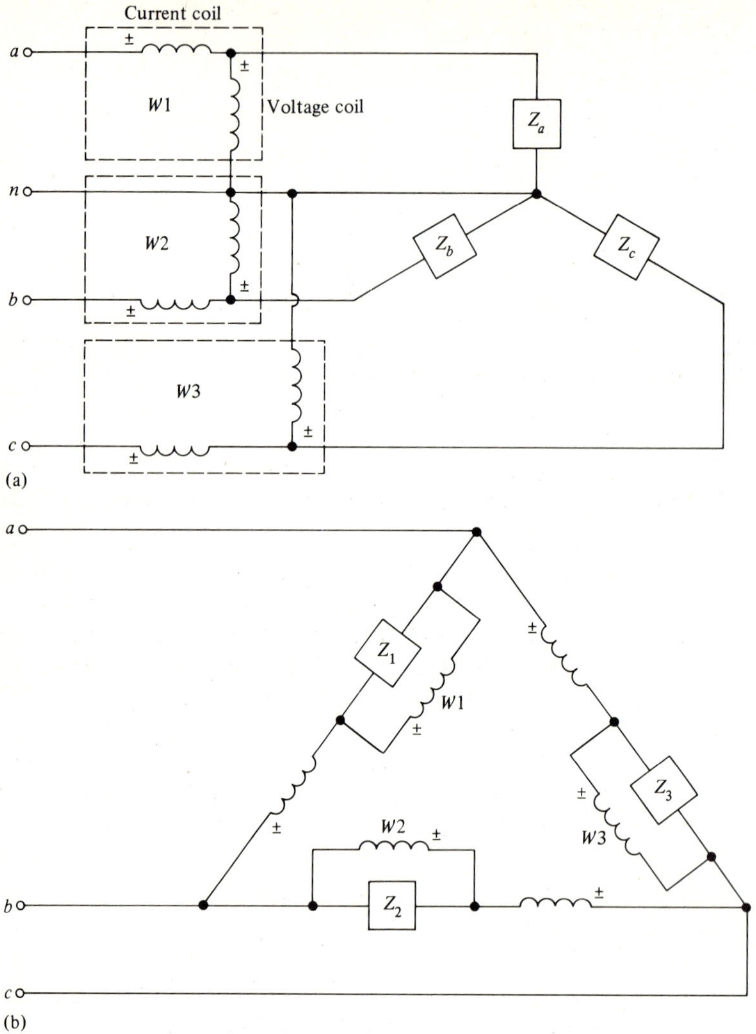

FIGURE 12.26
Three-wattmeter method for measuring three-phase power. (a) Wye load. (b) Delta load.

instead of three wattmeters and it can be used with *any* three-phase load, balanced or unbalanced; delta or wye. The total power is simply the algebraic sum of the two readings. Another advantage is that if the load is balanced we can determine the load power factor from the two wattmeter readings.

When using this scheme, the wattmeter terminals are initially connected as shown. If both wattmeters read upscale, then the total power is the sum of the readings. If one of the wattmeters reads downscale, its

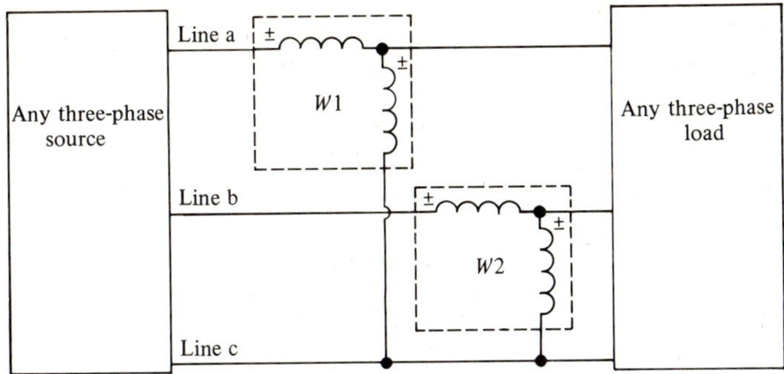

FIGURE 12.27
Two-wattmeter method for measuring three-phase power.

voltage coil is reversed and the reading is considered *negative*. In the example that follows, we calculate wattmeter readings and compare their sum with the actual total power.

EXAMPLE 12.5-3 Wattmeter Readings in the Two-wattmeter Method

A balanced three-phase wye load draws a line current of 20 A when connected to 220-V lines. The load power factor is 0.8 lagging. Find the power indicated by the two-wattmeter method and compare it with the actual total power.

Solution

The wattmeters are connected as shown in Fig. 12.27. Their readings are the product of the voltage across the voltage coil, current through the current coil, and cosine of the angle between the voltage and current. Thus

$$\begin{aligned} P_1 &= |V_{ac}||I_a|\cos\phi_a \\ P_2 &= |V_{bc}||I_b|\cos\phi_b \end{aligned} \tag{12.5-6}$$

where ϕ_a is the angle between V_{ac} and I_a, and ϕ_b is the angle between V_{bc} and I_b. The load phase angle is $\phi_P = \arccos 0.8 = 37°$ and a complete phasor diagram is shown in Fig. 12.28. From the diagram, we see that the angle ϕ_a between V_{ac} and I_a is $30° - \phi_P$ and the angle ϕ_b between V_{bc} and I_b is $30° + \phi_P$. Then, substituting into Eq. (12.5-6),

$$\begin{aligned} P_1 &= 220 \times 20 \times \cos(30° - 37°) \\ &= 4370 \text{ W} \\ P_2 &= 220 \times 20 \times \cos(30° + 37°) \\ &= 1720 \text{ W} \end{aligned}$$

and

$$\begin{aligned} P &= P_1 + P_2 \\ &= 4370 + 1720 \text{ W} \\ &= 6.1 \text{ kW} \end{aligned}$$

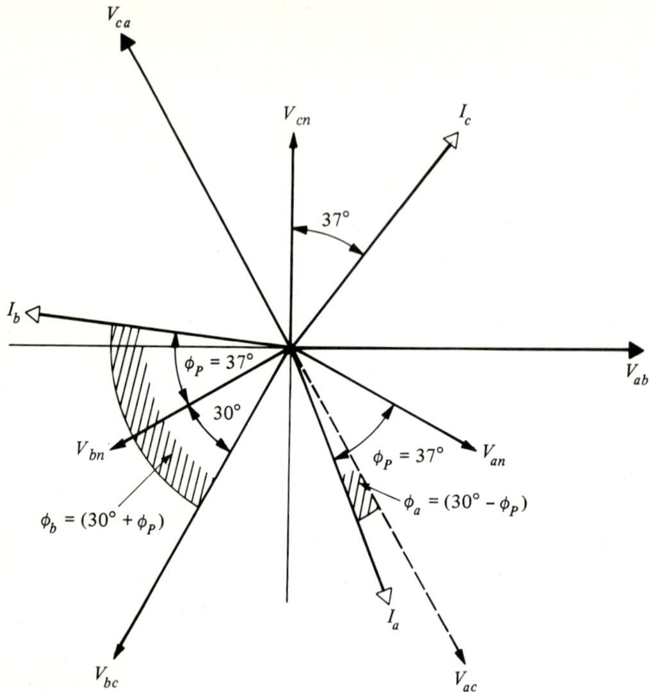

FIGURE 12.28
Phasor diagram for Example 12.5-3.

Using the formula for balanced loads, we check this result:

$$
\begin{aligned}
P &= \sqrt{3}|V_L||I_L|\cos\phi_P \\
&= \sqrt{3} \times 220 \times 20 \times 0.8 \\
&= 6.1 \text{ kW}
\end{aligned}
$$

• • •

We observe from this example that if the load phase angle ϕ_P is greater than 60°, angle ϕ_b will be greater than 90° and its cosine will be negative, leading to a negative value for P_2. This is the situation referred to previously, where one of the wattmeters reads downscale. Problem 12.53 illustrates this situation.

Finding the Power Factor when the Load is Balanced

When the load is balanced, we can find the load power factor from the ratio of the wattmeter readings in the two-wattmeter method. To show how this is done, consider Eqs. (12.5-6). If we divide one equation by the other

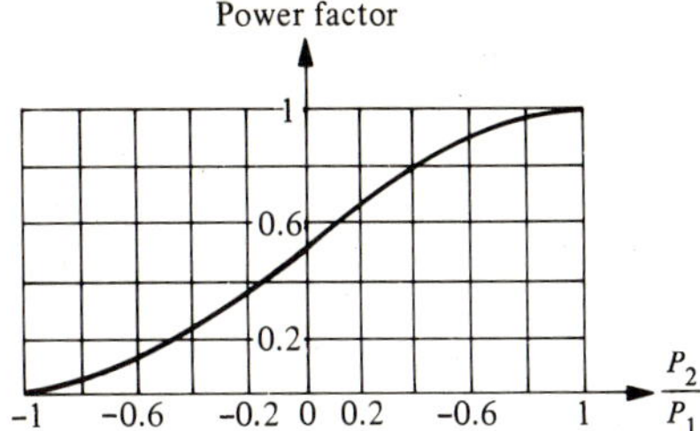

FIGURE 12.29
Power factor versus the ratio of the wattmeter readings in the two-wattmeter method.

$$\frac{P_2}{P_1} = \frac{\cos \phi_b}{\cos \phi_a} = \frac{\cos(30° + \phi_P)}{\cos(30° - \phi_P)} \tag{12.5-7}$$

This equation can be solved for the power factor angle ϕ_P (see Prob. 12.54). The result is

$$\phi_P = \arctan \sqrt{3}\left(\frac{P_1 - P_2}{P_1 + P_2}\right) \tag{12.5-8}$$

When we use this expression, P_1 is the larger of the two wattmeter readings.

In Eq. (12.5-8) we divide numerator and denominator by P_1 to put the equation in the form

$$\tan \phi_P = \sqrt{3}\left(\frac{1 - P_2/P_1}{1 + P_2/P_1}\right) \tag{12.5-9}$$

From this we can plot a curve of power factor ($\cos \phi_P$) versus P_2/P_1. This is shown in Fig. 12.29 and an example of its use follows.

EXAMPLE 12.5-4 Finding the Load Power Factor Experimentally

Find the power factor from the wattmeter readings in Example 12.5-3 by two methods:

a. Using the graph of Fig. 12.29
b. Direct substitution in Eq. (12.5-8)

Solution

From Example 12.5-3 the wattmeter readings are 4.37 and 1.72 kW, therefore,

$$\frac{P_2}{P_1} = \frac{1.72}{4.37} = 0.39$$

a. From the graph, the power factor is

$$\cos \phi_P \approx 0.81$$

b. From Eq. (12.5-8)

$$\phi_P = \arctan \sqrt{3}\left(\frac{4.37 - 1.72}{4.37 + 1.72}\right)$$

$$= \arctan\,(\sqrt{3} \times 0.435)$$

$$= 37°$$

$$\cos \phi_P = 0.799$$

Both of these answers agree closely enough with the stated power factor of 0.8

• • •

LEARNING EXERCISES FOR SEC. 12.5

1. A rolling mill motor is rated at 50 hp, 440 V, 3 ϕ, 0.9 power factor. Find the line current assuming 100% efficiency.

2. A heater draws 14 A from a three-phase 208 V source and the power dissipated is 5 kW. Find the power factor.

3. Power to a three-phase load is measured using the two-wattmeter method and the readings are 2190 W and −1050 W. Find the power factor.

Ans. 0.99; 54; 0.2

• • •

12.6 TRANSFORMERS

In Sec. 8.3 we discussed mutual induction between magnetically coupled coils. A device that makes use of this phenomenon is the *transformer*. This device has the ability to transmit power from one circuit to another without actual wire connections between the two circuits. It can step up or step down voltages in power systems so that power can be transmitted at high voltages. When this is done, less current is required so that RI^2 losses in the transmission lines are minimized. The voltage is then stepped down to safe values at the load. In addition, impedances can be made to look larger or smaller when connected to a transformer. This property sees much use in filters and electronic circuits. Finally, we can isolate one circuit from another so that they can both have different references, a property that is often useful in the laboratory.

In this section we consider only transformers in which the flux is confined to a ferromagnetic core so that the coefficient of coupling [see Eq.

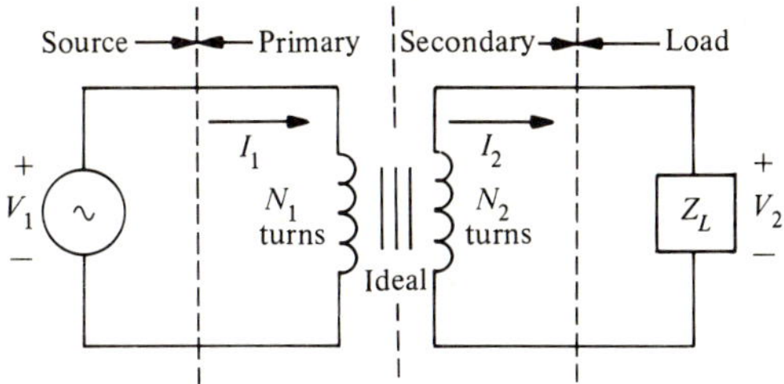

FIGURE 12.30
The ideal transformer.

(8.3-3)] is close to unity. We will introduce the terminology and basic theory by considering first the *ideal* transformer, which is shown in Fig. 12.30. The *primary* coil is connected to the power line or other voltage source. We will use the subscript 1 for the primary coil to distinguish it from the *secondary* coil, to which the load is connected and which carries the subscript 2. The primary winding has N_1 turns and the secondary winding has N_2 turns. Transformers are available with more than one secondary winding for special applications.

The fundamental relation governing transformer behavior was found in Chap. 8 by application of Faraday's law. In the ideal transformer $k = 1$ and Eq. (8.3-6) becomes

$$v_2 = \left(\frac{N_2}{N_1}\right) v_1 \tag{12.6-1}$$

where v_2 and v_1 are instantaneous values. Note that the equation is also correct in terms of rms amplitudes. If we divide through by v_1, the voltage transfer ratio becomes, in terms of rms values,

$$\boxed{\frac{V_2}{V_1} = \frac{N_2}{N_1}} \tag{12.6-2}$$

The factor N_2/N_1, called the *turns ratio*, is an important parameter which describes the transformer and it is assigned the symbol a. Thus

$$\boxed{a = \frac{N_2}{N_1}} \tag{12.6-3}$$

and, from Eq. (12.6-1)

$$\boxed{V_2 = aV_1} \tag{12.6-4}$$

Consider a transformer that has 200 primary turns and 50 secondary turns. Then $a = 50/200 = 0.25$. If the voltage applied to the primary is 100 V rms, then the voltage at the terminals of the secondary winding is $V_2 = 0.25 \times 100 = 25$ V. Therefore, when N_2 is less than N_1, a is less than one, and the secondary voltage is less than the primary voltage. In this case, the transformer is called a *step-down* transformer.

As another example, consider a transformer that has 100 primary turns and 200 secondary turns. Then $a = 200/100 = 2$ and $V_2 = 2V_1$. When the number of turns on the secondary is greater than the number of turns on the primary, a is greater than one, the secondary voltage is greater than the primary voltage, and we have a *step-up* transformer.

For the special case when the primary and secondary have the same number of turns, $a = 1$, $V_2 = V_1$, and we call this an *isolation* transformer. The word isolation is used because there is no actual wire connection between the primary and secondary circuits and the voltages across the coils are the same.

Transformer action is also obtained from a single winding that has a "tap" placed at an appropriate point, as shown in Fig. 12.31a. This arrangement is called an autotransformer. In a popular laboratory version of the autotransformer, the tap is connected to a variable wiper arm as shown in Fig. 12.31b. Thus the voltage at the tap is variable beginning at zero volts. Because of the extra turns N_3 at the top of the coil, the output voltage can be somewhat greater than the input line voltage. However, in the autotransformer there is no isolation between input and output as in the two-winding transformer.

EXAMPLE 12.6-1 Output Voltage

An audio transformer has 340 primary turns and 800 secondary turns. Find the output voltage if the input voltage is a sinusoidal signal of 200-mV rms amplitude.

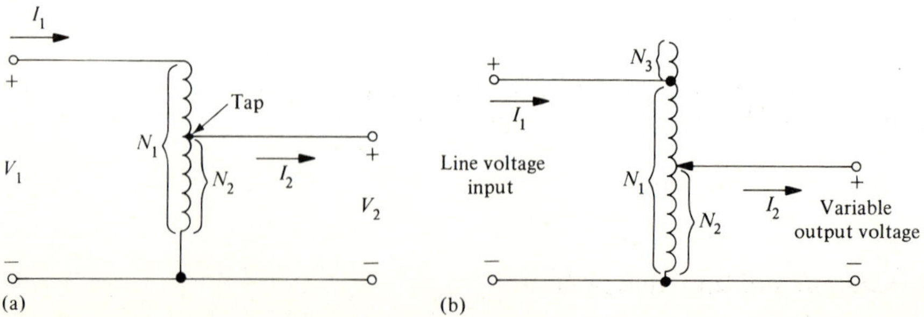

FIGURE 12.31
The autotransformer. (a) Circuit. (b) Commercial version.

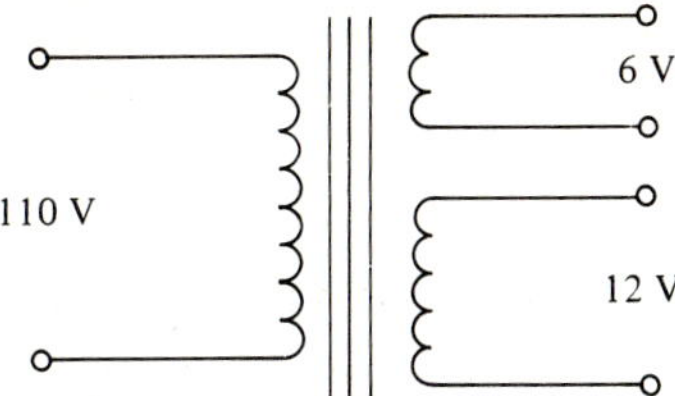

FIGURE 12.32
Three-winding transformer.

Solution
From the data we have $N_1 = 340$, $N_2 = 800$, and $V_1 = 200$ mV. Then

$$V_2 = \frac{N_2}{N_1} V_1 = \frac{800}{340} \times 200 \text{ mV} = 470 \text{ mV}$$

• • •

EXAMPLE 12.6-2 Secondary Turns

A transformer for model railroad use is to operate from the 110-V line and provide output voltages of 6 and 12 V from two separate secondary coils (see Fig. 12.32). If the primary has 360 turns, how many turns must each secondary have?

Solution
6-V output. Here the voltage ratio is

$$\frac{V_2}{V_1} = \frac{6}{110} = 0.054$$

The turns ratio is the same as the voltage ratio, that is,

$$a = \frac{N_2}{N_1} = 0.054$$

and since $N_1 = 360$ turns

$$N_2 = 0.054 \times 360 = 19 \text{ turns}$$

12-V output. For this case the output voltage is just twice as much as in the example above. Therefore, for 12-V output, we must have

$$N_2 = 2 \times 19 = 38 \text{ turns}$$

• • •

The Current and Impedance Ratios

For the ideal transformer we found that the ratio of the secondary voltage to the primary voltage is the same as the turns ratio. Now let us find the ratio of secondary current to primary current. We begin by observing that the ideal transformer is *lossless* so that all of the power delivered by the source in Fig. 12.30 must reach the load. If the load is resistive then the current and voltage are in phase and the load power is

$$P_2 = |V_2||I_2| \tag{12.6-5}$$

The input power is

$$P_1 = |V_1||I_1| \tag{12.6-6}$$

Since these two must be equal we must have

$$|V_2||I_2| = |V_1||I_1|$$

This is manipulated to yield

$$\frac{|I_2|}{|I_1|} = \frac{|V_1|}{|V_2|}$$

However, the turns ratio is $a = |V_2|/|V_1|$, therefore

$$\frac{|I_2|}{|I_1|} = \frac{1}{a} \tag{12.6-7}$$

and we see that the current ratio is equal to the *reciprocal* of the voltage ratio. This equation was derived in terms of magnitudes; however, it is equally valid for phasor currents. Therefore, in more general terms

$$\boxed{I_2 = \frac{I_1}{a}} \tag{12.6-8}$$

As we have seen, transformers can change a voltage or current level to another value by proper choice of the turns ratio. A byproduct of this capability is that impedances are transformed. We can put this on a quantitative basis by dividing the voltage and current equations as follows:

$$\frac{V_2}{I_2} = \frac{aV_1}{I_1/a} = a^2\frac{V_1}{I_1}$$

or

$$\frac{V_1}{I_1} = \frac{1}{a^2}\frac{V_2}{I_2} \tag{12.6-9}$$

Now, from the circuit of Fig. 12.30 we see that $V_2/I_2 = Z_L$, the load impedance. In addition, V_1/I_1 is the impedance Z_{in} looking into the transformer as seen by the source. Therefore,

$$Z_{in} = \frac{Z_L}{a^2} \tag{12.6-10}$$

This relation expresses the impedance-transforming property of the transformer. In words, it says that the impedance looking into an ideal transformer is the load impedance connected to the secondary divided by the square of the turns ratio. The terminology often used is "reflected" impedance. In this case we would say that the load impedance is *reflected* into the primary as Z_L/a^2.

This impedance-reflecting property of transformers is much used in electronic circuits. In audio amplifiers, it is often important that the load connected to the amplifier be a value larger than the low impedance of a typical loudspeaker. In this case a transformer is used to "match" the impedances. By proper choice of the turns ratio, a given impedance can be transformed to a larger value or a smaller value. These ideas are illustrated in the following examples.

EXAMPLE 12.6-3 Impedance Matching

A loudspeaker has an impedance of 8 Ω and a maximum power rating of 10 W. The amplifier to be used has a Thevenin resistance of 600 Ω. Therefore, for maximum power transfer a load impedance of 600 Ω is required and a transformer is to be used to reflect the 8-Ω loudspeaker impedance into the primary so that it "looks like" 600 Ω. Find the required turns ratio and the primary and secondary currents and voltages when the loudspeaker is operating at its maximum rated power.

Solution

The circuit is shown in Fig. 12.33. We find the required turns ratio by noting that the reflected impedance is to be 600 Ω while the load impedance is 8 Ω. Therefore [see Eq. (12.6-10)]

$$600 = \frac{8}{a^2}$$

$$a^2 = \frac{8}{600} = \frac{1}{75}$$

$$a = \frac{N_2}{N_1} = \frac{1}{\sqrt{75}} = \frac{1}{8.7}$$

The required step-down transformer has 8.7 primary turns for each secondary turn.

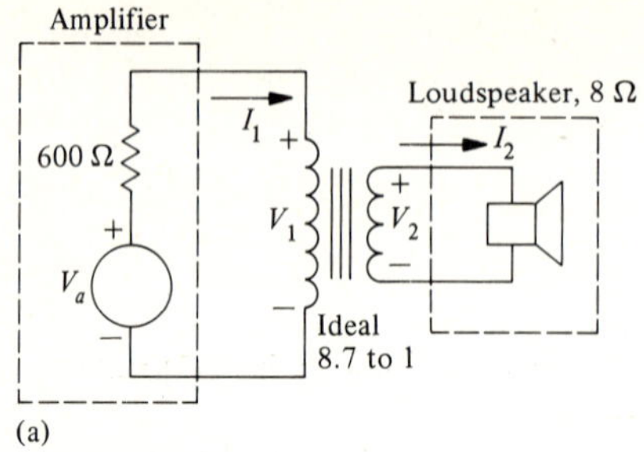

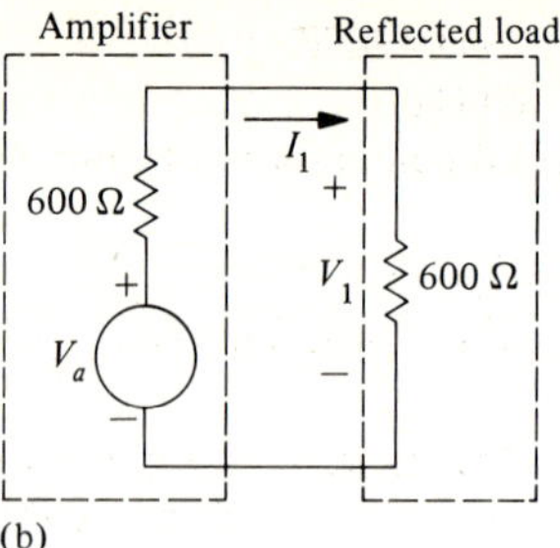

FIGURE 12.33
Audio amplifier. (a) Output circuit with amplifier. (b) Circuit with reflected impedance.

The secondary current and voltage are found from the 10-W maximum power rating as follows, assuming that the loudspeaker is essentially resistive:

$$P_2 = R_L I_2^2$$

from which

$$I_2 = \sqrt{\frac{P_2}{R_L}} = \sqrt{\frac{10}{8}} = 1.1 \text{ A}$$

Then

$$V_2 = R_L I_2 = 8 \times 1.1 = 8.8 \text{ V}$$

The primary current and voltage are found using the turns ratio as follows:

$$V_1 = \frac{V_2}{a} = 8.8 \text{ V} \times 8.7 = 77 \text{ V}$$

$$I_1 = aI_2 = \frac{1.1 \text{ A}}{8.7} = 0.13 \text{ A}$$

As a check, we calculate the power into the primary. This is

$$P_1 = V_1 I_1 = 77 \text{ V} \times 0.13 \text{ A} = 10 \text{ W}$$

Note that 10 W is also dissipated in the Thevenin resistance of the amplifier so it must be capable of supplying 20 W.

• • •

EXAMPLE 12.6-4 Nameplate Data

The nameplate on a large power transformer has the following data:

1. 100 kVA

2. 4400/220 V
3. 60 Hz

Find the maximum primary and secondary currents.

Solution

The significance of the ratings is as follows:

1. The maximum apparent power that the transformer can handle in the primary or secondary circuit without overloading is 100 kVA.
2. The transformer was designed for a primary voltage of 4400 V and a secondary voltage of 220 V. However, it can also be used for other voltages (or as a step-up instead of a step-down transformer) in which case the maximum power rating may be reduced.
3. The frequency is specified as 60 Hz. Again, the transformer can be used at other frequencies, but the maximum power rating will change.

The maximum primary current will occur when the apparent power is at its maximum value of 100 kVA. Therefore,

$$I_{1\,\max} = \frac{100\,000\ \text{VA}}{4400\ \text{V}} = 22.7\ \text{A}$$

The maximum secondary current can be found from the apparent power rating or the turns ratio. For the latter, we have

$$a = \frac{V_2}{V_1} = \frac{220}{4400} = \frac{1}{20}$$

$$I_{2\,\max} = \frac{I_{1\,\max}}{a} = 22.7 \times 20 = 454\ \text{A}$$

The secondary apparent power is 220 V × 454 A = 100 kVA.

• • •

Flexibility of Transformer Connections

Transformers can be connected in a variety of ways to provide output voltages other than those predicted purely on the basis of the voltage ratio. For example, consider the two-winding transformer of Fig. 12.34a and assume that the polarity dot markings are as shown. The turns ratio is $a = 1/10$ and the transformer is rated for 120/12 V. We can use the autotransformer connection shown in Fig. 12.34b to add the source voltage of 120 V in series with the secondary voltage of 12 V to achieve an output voltage of 132 V. When doing this we lose the isolation property of the transformer, which may make it unsuitable for some applications. If we use the subtractive connection of Fig. 12.34c, we obtain a third output voltage, 108 V. Adding a second secondary winding increases the number of available voltages still further (see Prob. 12.71).

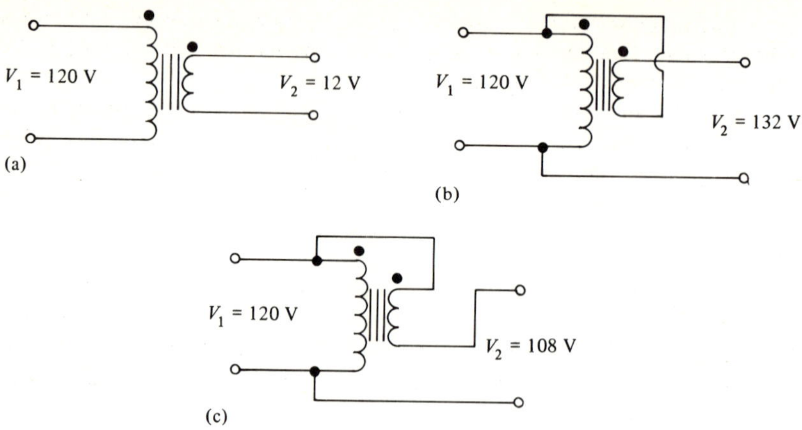

FIGURE 12.34
Transformer connections. (a) Conventional, V_2 = 12 V, isolated. (b) Autotransformer, V_2 = 132 V, nonisolated. (c) Autotransformer, V_2 = 108 V, nonisolated.

In the next example, we consider another arrangement that provides multiple output voltages.

EXAMPLE 12.6-5 Residential Service

Many residential areas are served by 12 600-V power lines. Most appliances and electrical devices we use are rated for voltages of 100/120 V, and larger units, such as electric ranges, clothes dryers, or air conditioners, use 220/240 V. The transformer arrangement shown in Fig. 12.35 is used to step down the 12 600 V from the high-voltage power line to the Edison three-wire system (see Sec. 12.4), which provides the required voltages. Observe that the residential service consists of two separate 110-V circuits with a common neutral. These are used for lighting and small appliances and the load should be distributed as evenly as possible between line 1 and line 2. The 220-V service for heavy appliances is obtained by connecting between lines 1 and 2 as shown. In practice the neutral wire is carefully grounded so

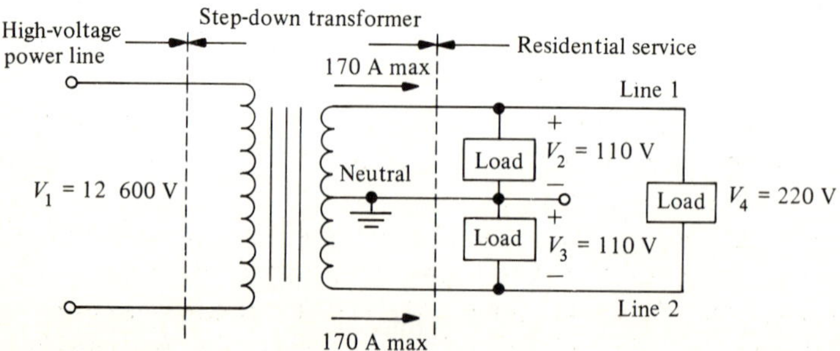

FIGURE 12.35
Transformer for residential three-wire system.

that the maximum voltage between any wire in the home and ground is 110 V. Thus the maximum shock that can be experienced by anyone touching ground and a conducting wire is due to 110 V rather than 220 V.

In a typical home, there are lighting and other 110-V circuits protected by individual 15- and 20-A fuses, and collectively by a 60-A fuse. In addition, 220-V service is provided for a central air conditioner (40 A), a clothes dryer (30 A), and a range (40 A). Find the maximum currents in the transformer.

Solution

With all major appliances on, the 220-V lines will carry 40 + 40 + 30 = 110 A. For the 110-V service the maximum current in lines 1 and 2 is 60 A so that, considering both 110- and 220-V loads, the maximum current in lines 1 and 2 is 110 + 60 = 170 A. The maximum neutral current is 60 A, which occurs when the 110-V load is all being drawn by one of the 110-V services.

The turns ratio is

$$a = \frac{220}{12\,600} = \frac{1}{57.3}$$

Therefore, the maximum primary current is

$$I_1 = \frac{170}{57.3} = 3 \text{ A}$$

• • •

Practical Iron-Core Transformers

Up to this point, we have considered ideal transformers. Real transformers depart from the ideal for a variety of reasons. In this section we discuss equivalent circuits which take into account some of these non-ideal characteristics.

To begin, we note that ideal transformer theory [Eq. (12.6-8)] predicts that when a transformer secondary is open-circuited so that the secondary current is zero, then the primary current will be zero. However, in a practical transformer, if the primary current is zero, no flux will be produced to induce a voltage across the open secondary. Therefore, some primary current must always flow in order to establish flux so that voltage is induced across the windings. This current is called the *magnetizing current* of the transformer and it is usually less than 5% of the primary current which would flow when the secondary is fully loaded.

Another departure from the ideal occurs because it is impossible for all of the flux produced by current in either the primary or the secondary to be confined to the iron core. There will always be some magnetic lines of force that do not link the opposite winding. This is called the *leakage flux*. Another departure from the ideal occurs because of the actual resistance of the wire used to wind the coils.

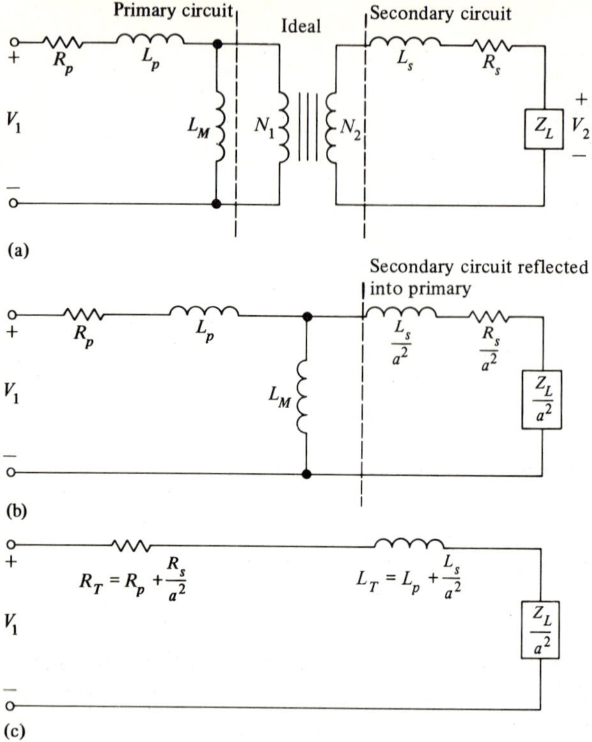

FIGURE 12.36
Nonideal transformers. (a) Equivalent circuit. (b) Circuit with secondary reflected into primary. (c) Simplified equivalent circuit when the transformer is under full load.

These effects can be accounted for by using the equivalent circuit shown in Fig. 12.36a. The elements of the circuit which represent departures from the ideal are as follows:

R_p accounts for primary circuit winding resistance
L_p represents the primary winding leakage inductance
L_m represents the primary magnetizing inductance, which accounts for the magnetizing current
R_s accounts for the secondary circuit winding resistance
L_s represents the secondary winding leakage inductance

The ideal transformer has the same turns ratio as the practical transformer.

More detailed investigation would lead to elements in the equivalent circuit that represent second-order effects. We will confine our attention to the first-order circuit shown.

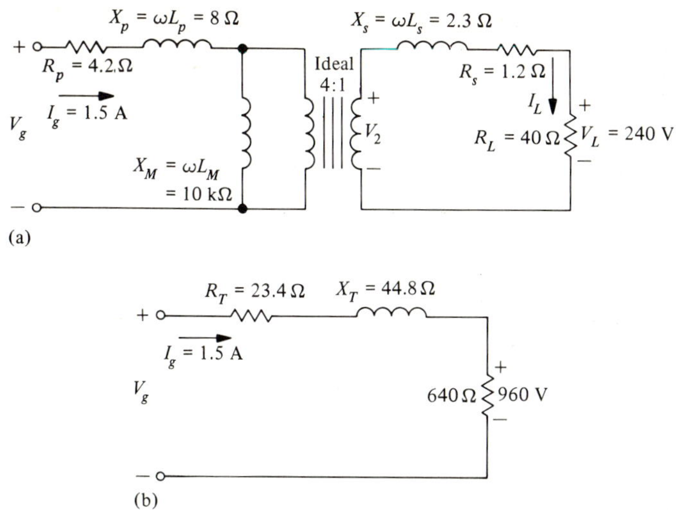

FIGURE 12.37
Example 12.6-6. (a) Equivalent circuit. (b) Simplified equivalent circuit.

For purposes of analysis, it is often convenient to reflect the secondary circuit into the primary. This leads to the circuit shown in Fig. 12.36b. A further simplification can be made when the secondary current is at or near its full-load value. For this case, the magnetizing current through L_m is less than 5% of the load current so it can be ignored in comparison with the load current and L_m can be considered an open circuit. The simplified equivalent circuit that results is shown in Fig. 12.36c.

An example will illustrate the effects of the elements in the equivalent circuit.

EXAMPLE 12.6-6 Nonideal Transformer

An iron-core step-down transformer has the equivalent circuit shown in Fig. 12.37a. Find the generator voltage V_g if the generator current is 1.5 A. Compare with the voltage required if the transformer is assumed to be ideal.

Solution

The turns ratio is

$$a = \frac{N_2}{N_1} = \frac{1}{4} = 0.25$$

The reflected secondary circuit elements are

$$\frac{X_s}{a^2} = \frac{2.3}{0.25^2} = 36.8\ \Omega$$

$$\frac{R_s}{a^2} = \frac{1.2}{0.25^2} = 19.2\ \Omega$$

$$\frac{R_L}{a^2} = \frac{40}{0.25^2} = 640\ \Omega$$

The 10-kΩ magnetizing reactance is large enough so that we may neglect it and use the simplified equivalent circuit of Fig. 12.36c. The element values are

$$R_T = R_p + \frac{R_s}{a^2} = 4.2 + 19.2 = 23.4\ \Omega$$

$$X_T = X_p + \frac{X_s}{a^2} = 8 + 36.8 = 44.8\ \Omega$$

The circuit is shown in Fig. 12.37b.

If we assume the generator current as our zero phase reference, the load current is

$$I_L = \frac{I_g}{a} = \frac{1.5\underline{/0^\circ}}{0.25} = 6\underline{/0^\circ}\ \text{A}$$

and the load voltage is

$$V_L = R_L I_L = 40 \times 6\underline{/0^\circ} = 240\underline{/0^\circ}\ \text{V}$$

Using KVL in the circuit of Fig. 12.37b, the generator voltage is

$$\begin{aligned} V_g &= (R_T + jX_T)I_g + \frac{R_L}{a^2} I_g \\ &= (23.4 + j44.8 + 640) \times 1.5\underline{/0^\circ} \\ &= 665\underline{/3.9^\circ} \times 1.5\underline{/0^\circ} \\ &= 998\underline{/3.9^\circ}\ \text{V} \end{aligned}$$

If the transformer is ideal, $R_T = X_T = 0$ and

$$V_g = \frac{R_L}{a^2} I_g = 960\underline{/0^\circ}\ \text{V}$$

As we would expect, the generator voltage must be larger in the practical case because we must provide enough voltage to compensate for the voltage drops in the winding resistances and leakage reactances.

• • •

Transformer Applications

As we have seen in this section, transformers are used for many applications in ac circuits. These include:

1. Transforming voltages to higher or lower levels.
2. Transforming currents to higher or lower values.
3. Transforming impedances.
4. Supplying two or more loads with different voltage requirements simultaneously from one source through the use of more than one secondary.
5. By using more than one secondary, several independent replicas of the input signal are made available.
6. Electrical isolation of one circuit from another for safety or other reasons.

• • •

LEARNING EXERCISES FOR SEC. 12.6

1. A residential doorbell transformer is to step 120 V down to 8 V. What turns ratio *a* is required? If the primary has 420 turns, how many turns should the secondary have? If the doorbell requires 2 A to operate efficiently, what will the primary current be?

2. An audio transformer is to match a special vacuum-tube amplifier to an 8-Ω loudspeaker. The amplifier must "see" 5 kΩ for optimum performance. Find the required turns ratio.

Ans. 0.13; 0.067; 28; 0.04

• • •

SUMMARY

In this chapter we considered a number of topics with important practical applications. At this point the student should be able to perform an analysis of a single-phase or three-phase circuit in order to determine current, voltage, and power. The important points covered in each section are summarized below.

Section

12.1
1. The power dissipated in a single-phase ac circuit is $P = |V||I|\cos\phi$
2. The average power in a pure capacitor or pure inductor is zero.
3. The VI product for a pure reactance is called reactive power, Q.
4. The VI product for an arbitrary impedance is called apparent power, S.
5. The power factor, $\cos\phi$, can be found as the ratio of the real power to the apparent power, P/S.
6. The power factor of an inductive circuit is said to be lagging and the power factor of a capacitive circuit is said to be leading.

12.2
7. Complex power is given by the formula $P_c = VI^*$

8. In order to bring the power factor of a large industrial load close to unity, a capacitor is connected in parallel with the load. If the reactive power drawn by the capacitor exactly equals the inductive reactive power of the load, then the overall power factor will be unity.
9. Maximum power transfer is achieved in an ac circuit when the load impedance is the conjugate of the Thevenin impedance of the source.

12.3 10. A balanced three-phase generator supplies three voltages, equal in magnitude, and spaced 120° in phase angle.
11. Phase sequence is important because it determines the direction of rotation of three-phase motors.
12. In a wye-connected generator, the line voltage is $\sqrt{3}$ times the phase voltage and is displaced in phase by 30°. In a balanced delta system, the line current is $\sqrt{3}$ times the phase current and is displaced from it by 30°.

12.4 13. By using the appropriate conversion formulas, wye and delta circuits can be converted from one form to the other. If the phase impedances are balanced then

$$R_\Delta = 3R_y$$

14. The Edison three-wire system provides two 115-V circuits and one 230-V circuit. It is used primarily for residential loads.

12.5 15. In terms of line quantities, the power in a three-phase load is given by the formula

$$P = \sqrt{3}|V_L||I_L|\cos\phi_P$$

16. Power in a three-phase system can be found using two wattmeters.

12.6 17. In a transformer the step-up or step-down ratio of voltage or current is determined by the turns ratio $a = N_2/N_1$.
18. The secondary voltage is $V_2 = aV_1$.
19. The secondary current in an ideal transformer is $I_2 = I_1/a$.
20. The impedance seen looking into an ideal transformer with load Z_L is

$$Z_{in} = \frac{Z_L}{a^2}$$

21. Practical iron-core transformers differ from ideal transformers because of magnetizing inductance, leakage reactance, and winding resistance.

QUESTIONS FOR REVIEW

Sec. 12.1

1. What is the formula for power in an arbitrary impedance Z?
2. Why is the average power in a pure inductance equal to zero?
3. What is the significance of the term "reactive power"?
4. What is the unit for reactive power?
5. What is the significance of the term "apparent power"?
6. What are the units of apparent power?
7. What is the significance of the term "power factor"?
8. Express the power factor in as many different forms as you can.
9. What is the significance of a leading power factor? A lagging power factor?

Sec. 12.2

10. What is the formula for complex power?
11. Express the complex power in as many different forms as you can.
12. Why are many electrical machines rated in terms of kVA rather than kW?
13. Why are low power factor loads undesirable from the point of view of the power company?
14. How is power factor correction implemented?
15. An ac source has Thevenin impedance $R + jX$. What value of load impedance will result in maximum power transfer?

Sec. 12.3

16. What is the phase angle between the phase voltages in a three-phase system?
17. How are three-phase voltages generated?
18. What is the significance of the term "phase sequence?"
19. What is the sum of the three voltages of a balanced three-phase system at any instant of time?
20. How is line voltage related to phase voltage in a wye system?
21. How is phase current related to line current in a wye system?
22. What is the neutral current in a balanced wye system?
23. In an unbalanced wye, how does one find the neutral current?
24. How is phase voltage related to line voltage in a balanced delta system?
25. How is phase current related to line current in a balanced delta system?

Sec. 12.4

26. How is the phase impedance of an equivalent wye circuit related to the phase impedance of the given delta circuit?
27. Name several advantages of the Edison three-wire system.
28. What is the neutral current in an Edison three-wire system if the low-voltage loads are balanced?

Sec. 12.5

29. What is the formula for power in a three-phase system in terms of line voltage, line current, and phase power factor?
30. Describe the connections used in the two-wattmeter method for measuring three-phase power.

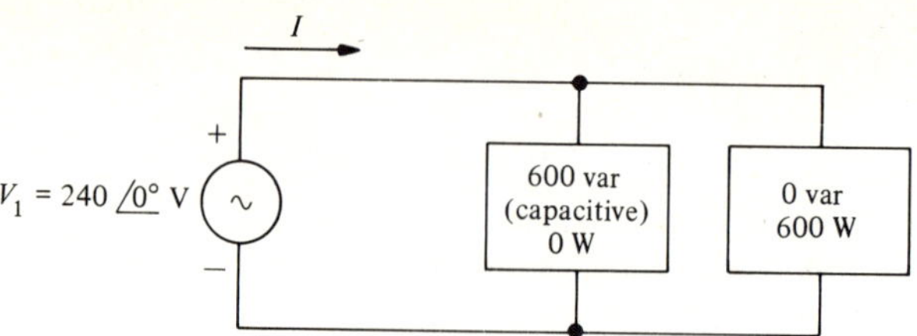

FIGURE 12.38

Sec. 12.6

31. Why does a transformer have a steel core?
32. Name some advantages of the transformer.
33. What is the turns ratio of a transformer?
34. Express the voltage ratio in terms of the turns ratio.
35. What is the significance of the terms *step-up transformer? Step-down transformer?*
36. What is the significance of the term *primary winding? Secondary winding?*
37. Describe the *autotransformer.*
38. How is the current ratio related to the turns ratio?
39. What is meant by the term *reflected impedance?*
40. How is the impedance ratio related to the turns ratio?
41. Why is it necessary to match impedances?
42. Describe how transformer connections can be made so as to provide different output voltages.
43. Why can the primary current not be zero in a practical transformer?
44. What is the significance of the term *magnetizing current?*
45. What is the significance of the term *leakage flux?*
46. Describe a number of transformer applications.

PROBLEMS

Sec. 12.1

1. The voltage across an impedance of $1.5\underline{/30^\circ}$ kΩ is 30 cos 377*t* V. Plot the instantaneous voltage, current, and power for $0 < t < 35$ ms.
2. A relay coil has a resistance of 24 Ω, an inductance of 0.45 H, and it is connected to a 120-V, 60-Hz supply. Find

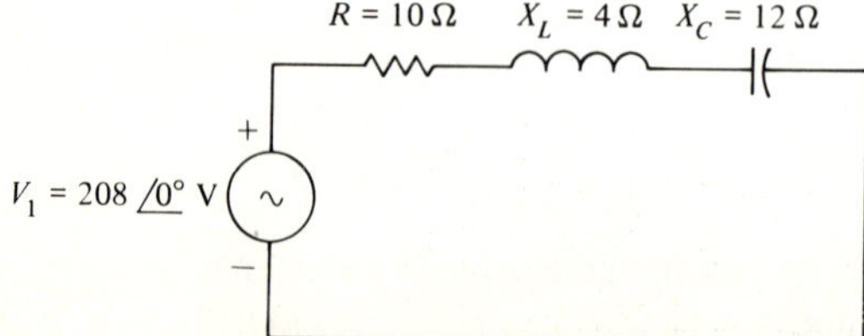

FIGURE 12.39

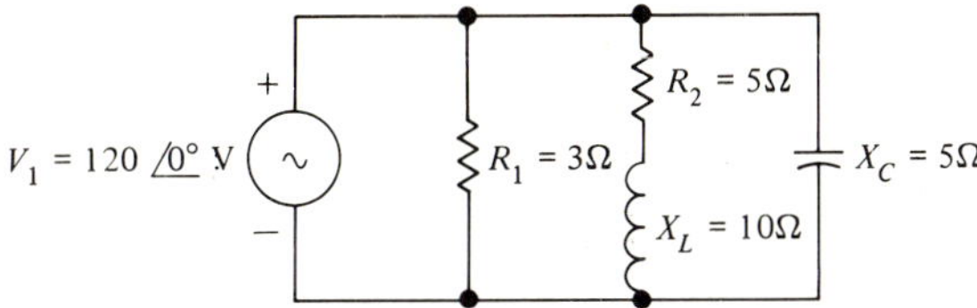

FIGURE 12.40

a. The apparent power c. The reactive power
b. The real power d. The power factor

3. Repeat Prob. 2 for a 10-μF capacitor in series with a 330-Ω resistor connected to the 120-V, 60-Hz supply.
4. A 110-V emergency generator must supply 2.5 kW to an 80% lagging power factor load. How much current must the generator supply?
5. A device is rated at 5 kVA, 120 V, 0.6 lag power factor. What is its impedance?
6. A 3-H inductance is connected to a 20-V, 1-kHz oscillator. Find
   a. The true power b. The reactive power
7. Repeat Prob. 6 for a 3300-pF capacitor connected to a 1.5-V, 1-MHz oscillator. What value of inductance will absorb the same reactive power at the stated voltage and frequency?
8. In the circuit of Fig. 12.38, find the total watts, vars, current I, power factor, and volt-amperes.
9. A new lamp bank being tested absorbs 0.8 kW of power from the 110-V line at a power factor of 0.75 lagging. Find the current and its phase angle relative to the voltage.
10. A resistor and inductor in series are connected to a 24-V, 400-Hz supply. What values of R and L will cause the circuit to draw 100 W at 60% power factor from the supply?

Sec. 12.2

11. In the circuit of Fig. 12.39, find P_c, S, P, Q, and the power factor. Plot a complete complex power diagram.
12. Repeat Prob. 11 for the circuit of Fig. 12.40.
13. The nameplate on a large industrial heater reads

    208 V 60 HZ 10 kVA 0.85 lag power factor

    Find the current, impedance, true power, and complex power.
14. In the system shown in Fig. 12.41 find

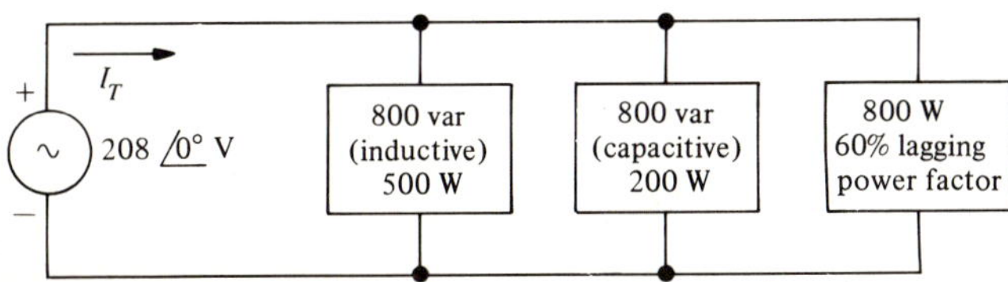

FIGURE 12.41

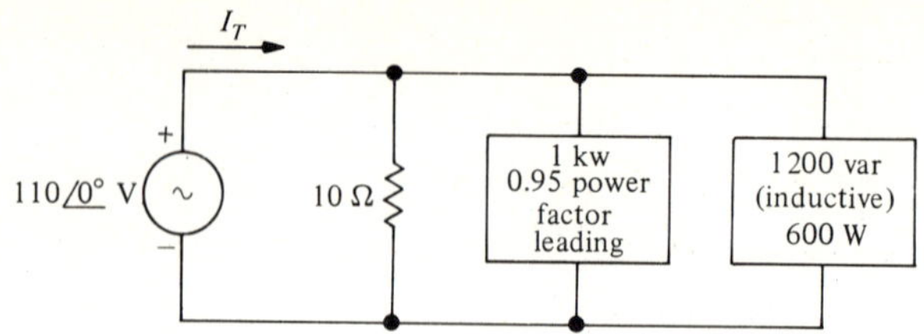

FIGURE 12.42

a. The complex power for each load
b. The total real, reactive, and apparent power
c. The current I_T
d. The overall power factor

15. Repeat Prob. 14 for the system shown in Fig. 12.42.
16. In the system of Fig. 12.43 the 2-Ω resistor represents the resistance of a transmission line. Find
    a. The generator voltage V_g required to maintain 208 V at the load
    b. The current
    c. Draw a complete phasor diagram and complex power diagram.
17. Repeat Prob. 16 for the 440-V system shown in Fig. 12.44.
18. A compressor motor draws 6.5 A at 0.8 lagging power factor from a 208-V, 60-Hz source.
    a. What capacitance value must be connected in parallel with the motor to raise the power factor to 100%?
    b. With the capacitor connected, find the motor current, capacitor current, and line current.
    c. Draw a complete phasor diagram and a complex power diagram.
19. Find the capacitance value required to bring the overall power factor to 90% in Prob. 18.
20. A steel mill requires 1.2 MW of power at 4400 V, 60 Hz. Find the current which is required if the overall power factor of the mill is
    a. 80% lagging b. 100%
21. In Prob. 20 find the capacitance required to correct the power factor from 80% to
    a. 100% b. 95%
22. A 220-pF capacitor is connected in series with a 50-μH inductor and a 10-Ω resistor. Find the frequency at which this circuit has unity power factor.
23. A loudspeaker has a nominal impedance of 8 Ω resistive. The amplifier driving it

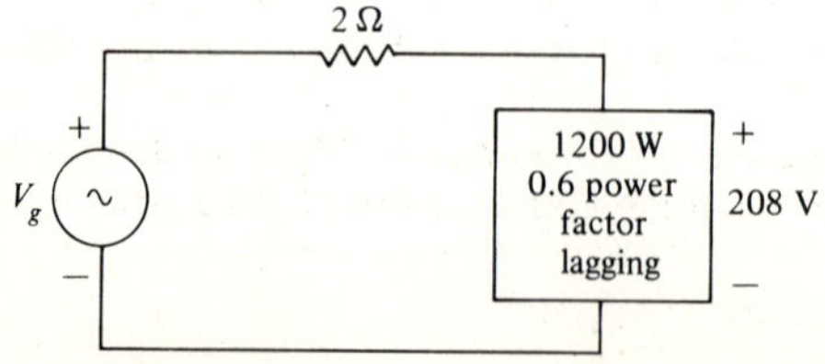

FIGURE 12.43

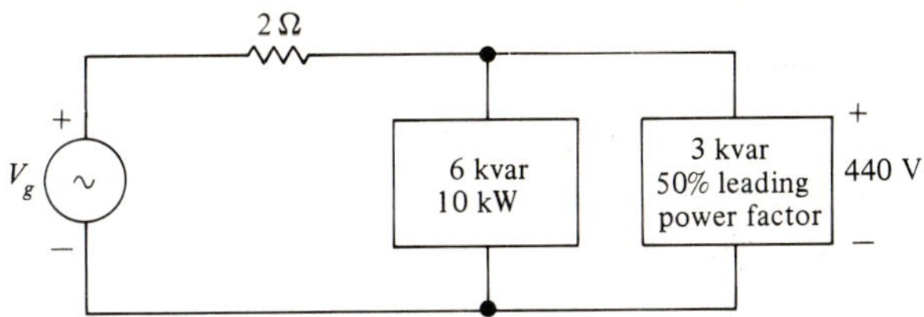

FIGURE 12.44

has a Thevenin impedance of $8 + j2\ \Omega$. How may the speaker circuit be modified to match it for maximum power transfer? Draw the final circuit.

24. Find the parallel *R-C* circuit matching the amplifier of Prob. 23.

Sec. 12.3

25. Show that the three-phase voltages in Eq. (12.3-2) add up to zero if the magnitudes are all the same.
26. The line voltage in a balanced three-phase four-wire system is 240 V. What voltage would be measured from each line to neutral?
27. A wye-connected generator has a voltage of 120 V from line to neutral. What is the magnitude of the line-to-line voltage?
28. A balanced four-wire wye generator has a phase voltage of 120 V and the phase sequence is *acb*. The load is a balanced wye consisting of 12-Ω resistors.
    a. Draw a neatly labeled schematic.
    b. Find all line voltages and currents.
    c. Draw a complete phasor diagram.
    d. Find the neutral current.
29. Repeat Prob. 28 if the load has three-phase impedances $Z_p = 12 + j4\ \Omega$.
30. Repeat Prob. 28 if the load consists of three-phase impedances drawing 600 W at 90% lagging power factor.
31. Prove Eq. (12.3-5) by geometrical construction on the phasor diagram.
32. Repeat Prob. 28 if the wye impedances are unbalanced with $Z_{an} = 30\underline{/20^\circ}\ \Omega$; $Z_{bn} = 45\underline{/30^\circ}\ \Omega$; and $Z_{cn} = 50\underline{/-45^\circ}\ \Omega$.
33. Repeat Prob. 28 if the wye impedances are unbalanced with Z_{an} = 600 W, 80% lag power factor; Z_{bn} = 1 hp motor, 85% power factor, 80% efficiency; Z_{cn} = 1 kW, 100% power factor heater.
34. In the circuit of Fig. 12.45, the voltage across each 10-Ω load resistor is to be 120 V. What must the generator line voltage be? Draw a complete phasor diagram.
35. Before three coils are connected in delta as shown in Fig. 12.18a, it is possible for the voltage across the open terminals to be twice the phase voltage. Prove this statement.
36. A balanced delta-connected source has line voltages of 550 V and a phase sequence *abc*. To this source is connected a balanced delta load consisting of three 22-Ω resistors.
    a. Find all phase and line currents and voltages.
    b. Draw a complete phasor diagram.
37. Repeat Prob. 36 if the load impedances are each $20\underline{/-45^\circ}\ \Omega$.

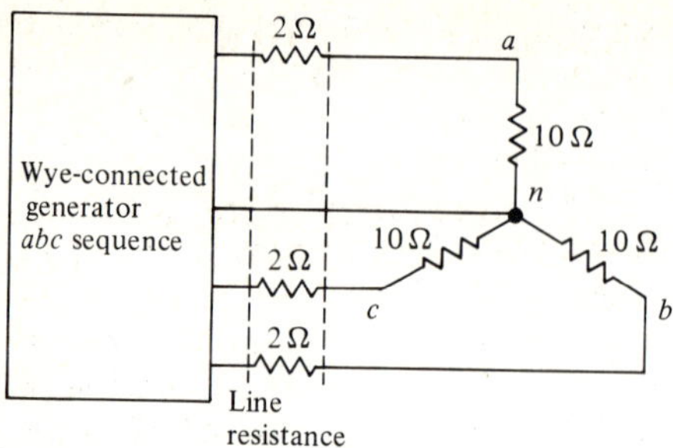

FIGURE 12.45

38. Repeat Prob. 36 if the load is unbalanced with $R_{ab} = 12\ \Omega$, $R_{bc} = 20\ \Omega$, and $R_{ca} = 8\ \Omega$.

Sec. 12.4

39. Prove Eqs. (12.4-12), (12.4-13), and (12.4-14).
40. In the wye-delta system shown in Fig. 12.22a the load impedances are each $40\underline{/-20^\circ}\ \Omega$. Transform the delta to a wye and find all line currents. Find the phase currents for the original delta and draw a complete phasor diagram.
41. A delta-connected load has $Z_{ab} = 10\ \Omega$, $Z_{bc} = 5 + j15\ \Omega$, and $Z_{ca} = -j10\ \Omega$. Transform this circuit to a wye.
42. Reduce the load in Fig. 12.46 to a single wye and find all currents and voltages. Use these results to find all currents and voltages in the original circuit.
43. Repeat Prob. 42 by reducing the load to a single delta.

Sec. 12.5

44. Find the total power required by the load in Prob. 28.
45. Find the total power required by the load in Prob. 29.
46. How much power is required by the load in Prob. 30?
47. How much power is required by the unbalanced load in Prob. 32?
48. In the circuit of Fig. 12.45, how much power is dissipated in the load? In the lines?
49. How much power is required by the delta load in Prob. 36?
50. A 15-hp motor operates at full load from a 230-V three-phase source. The motor

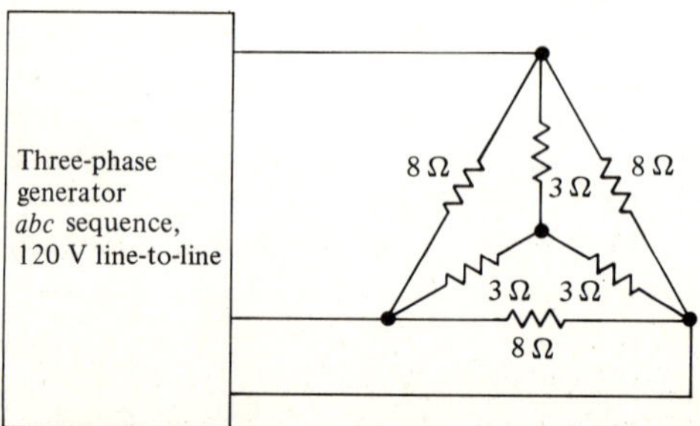

FIGURE 12.46

has a power factor of 92% lagging and an efficiency of 86% and is connected in delta. Find the line and phase currents and the power required.

51. The nameplate on an industrial motor reads 60 hp, 660 V, 3 ϕ, 0.94 power factor, full-load efficiency = 96%. Find the line current.
52. In the circuit of Fig. 12.47, show how a second wattmeter is connected so that the two will properly measure the total load power. Find the readings of the two meters and compare with the calculated power.
53. A balanced wye load connected to 440-V three-phase lines with phase sequence *acb* has a phase impedance $Z_p = 20\underline{/75°}\ \Omega$. The two-wattmeter method is used to measure the total power. What will each of the meters read if the current coils are connected in lines *a* and *b*?
54. Prove Eq. (12.5-8) by solving Eq. (12.5-7) for ϕ_P. You may need the identity $\cos(x \pm y) = \cos x \cos y \mp \sin x \sin y$.
55. Find the load power factor from the ratio of the wattmeter readings in Prob. 52.
56. Find the load power factor from the ratio of the wattmeter readings in Prob. 53.

Sec. 12.6

57. A transformer has 20 turns on the primary and 360 turns on the secondary. What is the turns ratio?
58. A bell transformer operates on 120 V, 60 Hz and has a secondary voltage of 8 V. What is the turns ratio?
59. A transformer primary has 80 turns and is connected to a 60-V, 400-Hz supply. The secondary has 240 turns. What is the secondary voltage?
60. A transformer is rated as 110 V to 6.3 V. If the secondary has 30 turns, how many turns must the primary have?
61. A 120-V transformer is designed to have a 100-V secondary. In addition, the secondary is to have taps at 30 and 60 V. If the primary has 380 turns, how many turns must be on the secondary and at what points should the taps be connected?
62. In the autotransformer shown in Fig. 12.31b, $N_1 = 200$ turns and $V_1 = 110$ V. Find N_3 if V_2 is to be variable from 0 to 140 V.
63. A 1.2-Ω load is connected to the 36-V secondary of a transformer. The primary current is measured and found to be 6 A. Is this a step-up or step-down transformer? What is the primary voltage?
64. In the bell transformer of Prob. 58 the bell draws 1.2 A at 8 V. What is the primary current?

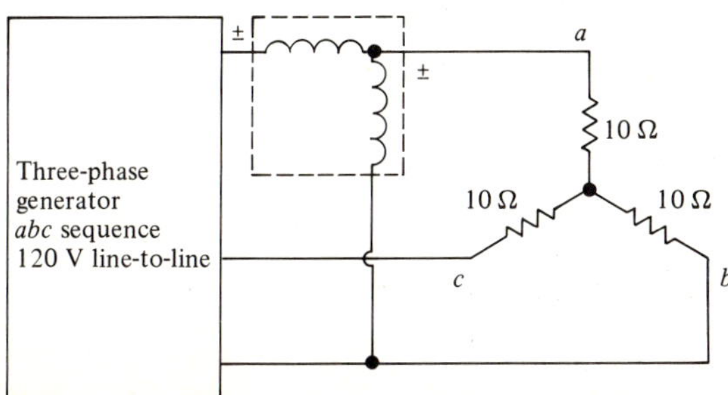

FIGURE 12.47

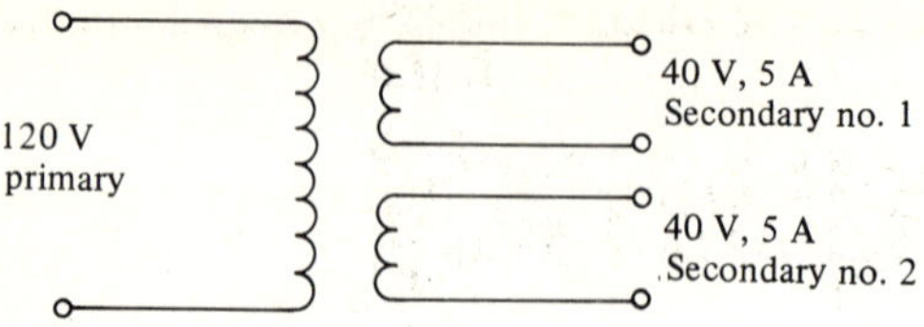

FIGURE 12.48

65. In the transformer of Prob. 59 the primary current is 360 mA. What is the secondary current?
66. An audio output transformer has 1500 primary turns and 40 secondary turns. A 4-Ω load is connected to the secondary. What does this load look like as viewed from the primary?
67. An audio interstage transformer is designed to match a 50-Ω load to a 600-Ω source. If an 8-Ω load is connected in place of the rated 50-Ω load, what impedance will be reflected into the primary?
68. An audio transformer has 1000 primary turns and two secondaries, each with 40 turns. The amplifier driving the transformer requires a 5-kΩ load for optimum performance. What load must be connected to the secondary when the two coils are connected in series aiding? In parallel?
69. A 100-Ω source is to be matched to a 60-Ω, 2-W load. Find the required turns ratio and the primary and secondary voltages and currents.
70. A single-phase power transformer nameplate contains the following data:

    13 600/440 V 100 kVA 60 Hz

    Find the maximum primary and secondary current.
71. How many different output voltages are possible for the transformer of Fig. 12.48? Use both isolation connections and autotransformer connections and draw the circuit for each possibility.
72. In the transformer equivalent circuit of Fig. 12.37 $R_p = 3\ \Omega$, $X_p = 6\ \Omega$, $R_s = 2\ \Omega$, $X_s = 1.1\ \Omega$, $R_L = 10\ \Omega$, the turns ratio is $a = 0.2$, and X_m can be neglected. Find R_T and X_T and draw the simplified equivalent circuit. Find the generator voltage V_g if the generator current is 1 A. Compare with the voltage required if the transformer is assumed to be ideal.

13 Frequency Response

OBJECTIVES

Upon completion of this chapter, the student should be able to

Section

13.1 1. State that a nonsinusoidal-periodic signal is composed of a fundamental and a series of harmonics.
     2. Describe the frequency spectrum of a signal.
13.2 3. Find the ac transfer function of an arbitrary network.
13.3 4. Find and plot the frequency response of a network from its transfer function.
13.4 5. Design a parallel resonant circuit for specified center frequency and bandwidth.
13.5 6. Use a computer analysis program to find the frequency response of a network.
13.6 7. State that there is a correlation between frequency response and transient response and that, in general, short pulse rise times require wide bandwidth circuits.

INTRODUCTION

In previous chapters, we considered the response of typical circuits to single-frequency sine waves. This is the signal wave-form used in power distribution, radio transmission, and as a test signal in electronics laboratories; however, other signals encountered in practice are seldom single-frequency sine waves. Some examples of typical signals were shown in Fig. 2.16.

In this chapter we consider the response of typical circuits to *periodic* nonsinusoidal signals. We find that such signals can be represented as a *sum* of sine waves of different frequencies called *harmonics.* We then consider the way in which a circuit affects the various frequency components of its input signal in order to find the frequency components of the output signal. With this information we will be able to determine the waveform of the output signal.

13.1 HARMONICS

In this section we show how a periodic signal, such as a square wave, can be constructed by adding together a number of sine waves of different frequencies. The lowest frequency is called the *fundamental* and the others are the *harmonics.* If the sine waves are to add up to yield a periodic signal, they must be *harmonically* related to the fundamental. The harmonic relationship is illustrated in Fig. 13.1, which shows a fundamental at a frequency $f_0 = 500$ Hz. Harmonically related frequencies are found by multiplying the fundamental frequency by integers. For example, the second harmonic is at $2 \times f_0 = 2 \times 500 = 1000$ Hz; the third harmonic at $3 \times 500 = 1500$ Hz, and so on. The harmonically related sine waves in Fig. 13.1 can be written in a general form as follows with $\omega_0 = 2\pi f_0$:

Fundamental: (first harmonic)

$$v_1(t) = |V_{1m}|\cos(\omega_0 t + \phi_1) \tag{13.1-1}$$

Second harmonic:

$$v_2(t) = |V_{2m}|\cos(2\omega_0 t + \phi_2) \tag{13.1-2}$$

Third harmonic:

$$v_3(t) = |V_{3m}|\cos(3\omega_0 t + \phi_3) \tag{13.1-3}$$

and so on.

The final waveform is the sum of the fundamental and all of the harmonics:

$$\begin{aligned} v(t) &= V_0 + v_1(t) + v_2(t) + v_3(t) + \cdots \\ &= V_0 + |V_{1m}|\cos(\omega_0 t + \phi_1) + |V_{2m}|\cos(2\omega_0 t + \phi_2) \\ &\quad + |V_{3m}|\cos(3\omega_0 t + \phi_3) + \cdots \end{aligned} \tag{13.1-4}$$

V_0 represents the dc, or average value of the waveform. In general, the amplitudes $|V_{1m}|$, $|V_{2m}|$, $|V_{3m}|$, etc., and the individual phase angles ϕ_1, ϕ_2, ϕ_3, etc., are different for each periodic waveform. In order to determine

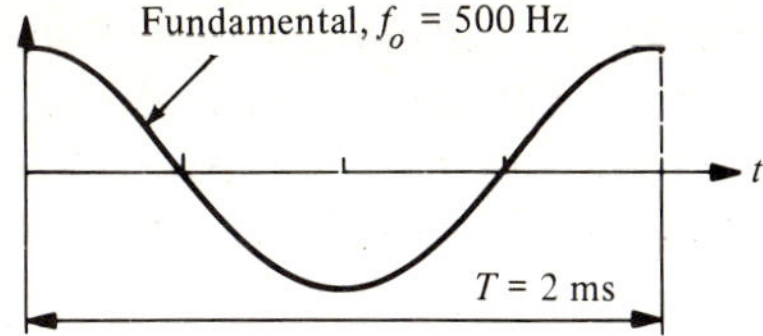

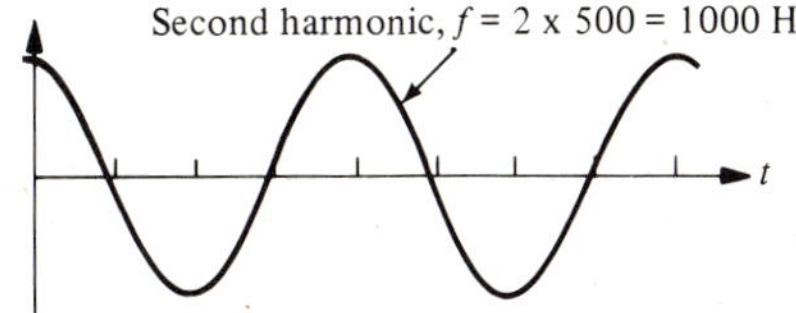

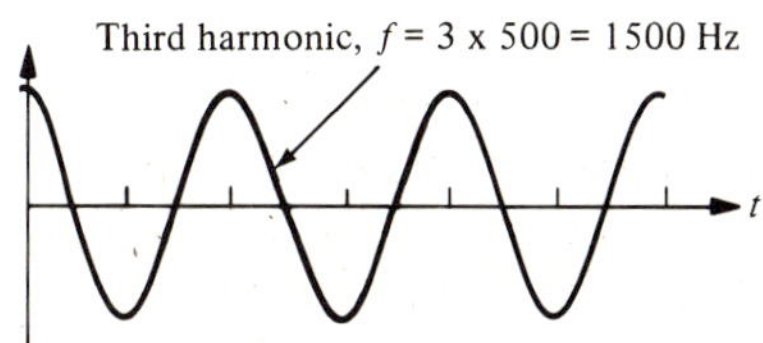

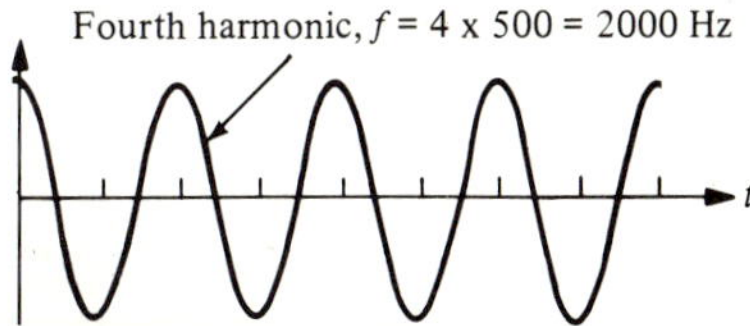

FIGURE 13.1
Fundamental and first three harmonics.

numerical values for these quantities, a mathematical technique called Fourier analysis must be used. The mathematics involved makes use of calculus, which is beyond the scope of this course. For our purposes, it will be sufficient to present the equation (called a *Fourier series*) for a given periodic waveform and discuss its properties.

The Square Wave

The Fourier series for the normalized square wave shown in Fig. 13.2a can be shown to be

$$\frac{v(t)}{\frac{4}{\pi}V_m} = \cos \omega_0 t - \frac{1}{3}\cos 3\omega_0 t + \frac{1}{5}\cos 5\omega_0 t - \frac{1}{7}\cos 7\omega_0 t + \cdots \qquad (13.1\text{-}5)$$

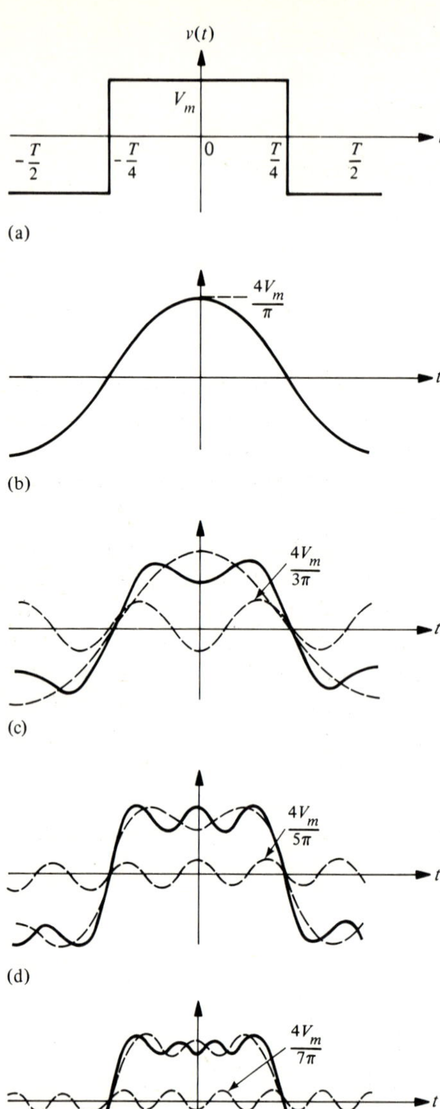

FIGURE 13.2
Fourier series for a square wave. (a) Original waveform. (b) Fundamental. (c) Sum of fundamental and third harmonic. (d) Sum of fundamental, third, and fifth harmonics. (e) Sum of fundamental, third, fifth, and seventh harmonics.

where

$$V_m = \text{peak value of square wave voltage}$$

$$\omega_0 = 2\pi f_0 = \frac{2\pi}{T} = \text{fundamental angular frequency}$$

$$3\omega_0 = \text{third harmonic frequency}$$

$$5\omega_0 = \text{fifth harmonic frequency}$$

$$7\omega_0 = \text{seventh harmonic frequency}$$

and so on. The fundamental is shown in Fig. 13.2b. Observe that its shape is not unlike the square wave since it has one positive portion and one negative portion, each lasting one-half period, or $T/2$ seconds. In Fig. 13.2c the solid line shows the fundamental and third harmonic added together, that is, $\cos \omega_0 t - 1/3 \cos 3\,\omega_0 t$. This is beginning to look more like the square wave. In Fig. 13.2d, e the fifth and seventh harmonics are added. The waveform is seen to approach the square shape with the sides getting steeper and the ripple along the top decreasing in amplitude as more harmonics are added. As we add more harmonics, the sum looks more and more like the original square wave. From Eq. (13.1-5) we see that the amplitude of each harmonic is inversely proportional to its frequency. Also, for this particular Fourier series, all harmonics are of odd-order (3, 5, 7, 9, . . .) and all are in phase with the fundamental.

The amplitudes of the fundamental and harmonics are often plotted in the form shown in Fig. 13.3a. This is called the *amplitude-frequency*

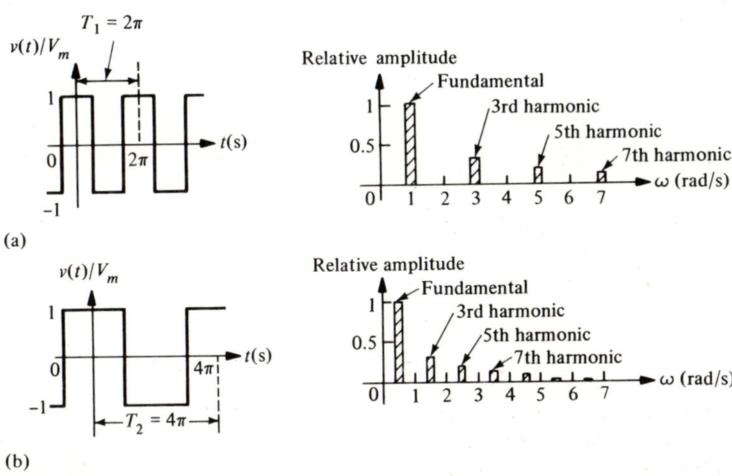

FIGURE 13.3
Frequency spectra. (a) Square wave with period of 2π seconds. (b) Square wave with period of 4π seconds.

spectrum and it clearly shows how the amplitudes of the harmonics decrease as the order of the harmonics increases. The spectrum is shown for a square wave with a period $T_1 = 2\pi$ s. Thus the fundamental angular frequency is $\omega = 2\pi/T_1 = 1$ rad/s. In Fig. 13.3b we show a square wave with a period $T_2 = 4\pi$ s (fundamental at 1/2 rad/s), along with its spectrum plotted to the same scale as the square wave with a period half as long from Fig. 13.3a. When we compare the two spectra we see that the frequency interval between harmonics becomes smaller as the period of the square wave increases. The harmonics therefore decrease to a much smaller amplitude in a given frequency range for the square wave with the longer period. Another way to state this is that the longer period square wave has less high-frequency content than the shorter period square wave.

Typically, humans can hear sine waves of sound in the range from about 30 Hz to 15 kHz. Consider listening to square waves of sound. If the fundamental frequency of the square wave is, for example, 50 Hz, most of the harmonics will be within the range of the ear, so the listener will get a relatively good idea of what a square wave sounds like. However, if the fundamental frequency is 10 kHz, the lowest harmonic (the third) will be at 30 kHz, which is outside the range of human hearing. For this signal, the person listening would effectively hear only the 10-kHz sine wave (the fundamental) because all of the harmonics would be out of the hearing range and would therefore be rejected.

When a square wave is the input to a circuit containing inductors or capacitors, the output of the circuit will contain the same fundamental and harmonics as the original square wave. However, in passing through the circuit, each frequency component will, in general, be changed in amplitude and phase so that the waveform at the output will no longer be a square wave. Methods for calculating the output waveform will be considered in the next section.

• • •

LEARNING EXERCISE FOR SEC. 13.1

1. A certain waveform consists of a fundamental at 5 krad/s and a second and fifth harmonic. Find the three frequencies in units of hertz.

Ans. 1590; 796; 3980

• • •

13.2 AC TRANSFER FUNCTIONS

In Chaps. 11 and 12 we considered the response of circuits containing resistors and inductors or capacitors to single-frequency sine waves. In this section we extend the concept of the transfer function so that we will be able to find the response of these circuits to sine waves of any frequency.

Consider the voltage divider network shown in Fig. 13.4a. In Sec. 4.5

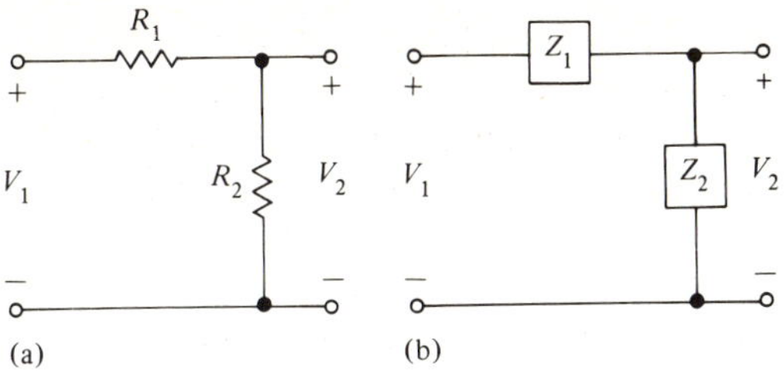

FIGURE 13.4
Transfer functions. (a) dc. (b) ac.

we found that the output voltage V_2 was related to the input voltage V_1 by the voltage ratio transfer function

$$A_v = \frac{V_2}{V_1} = \frac{R_2}{R_1 + R_2} \tag{13.2-1}$$

Now we are led to consider the network shown in Fig. 13.4b, in which V_1 is the phasor input voltage, and Z_1 and Z_2 may contain all three types of elements, R, L, and C. In order to find the output phasor voltage V_2 we use the same voltage divider formula except that resistances are replaced by impedances. Thus

$$\boxed{A_v(\omega) = \frac{V_2}{V_1} = \frac{Z_2}{Z_1 + Z_2}} \tag{13.2-2}$$

If either impedance contains inductance or capacitance, the voltage ratio will usually be complex and will depend on the frequency of V_1. We have explicitly indicated this by writing $A_v(\omega)$. These ideas are illustrated in the following example.

EXAMPLE 13.2-1 Finding Transfer Functions

The circuits of Fig. 13.5 are typical of those used to couple transistors in a multistage amplifier. Use the transfer function concept to find expressions for the output voltage in the circuit of Fig. 13.5a and the output current in the circuit of Fig. 13.5b.

Solution

Circuit a:

$$Z_1 = R \qquad Z_2 = \frac{1}{j\omega C}$$

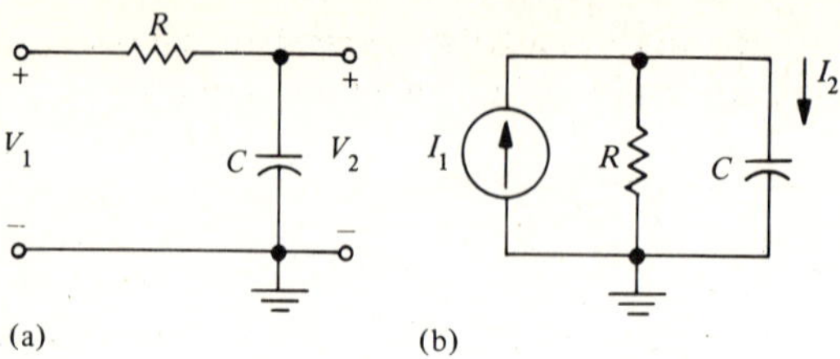

FIGURE 13.5
Circuits for Example 13.2-1.

Therefore:

$$A_v(\omega) = \frac{1/j\omega C}{R + 1/j\omega C}$$

Multiplying numerator and denominator by $j\omega C$, this becomes

$$A_v(\omega) = \frac{1}{1 + j\omega RC} \tag{13.2-3}$$

Observe that the voltage transfer ratio depends on the frequency of V_1. The only unknown on the right-hand side of this expression is the angular frequency ω since R and C will be known. Thus the voltage transfer ratio is easily evaluated for any desired frequency. It is interesting to note that the values of R and C appear only as the product RC. This is the *time constant*, which we found in Chap. 9, and which determined the speed of the exponential response of the circuit. As we will see shortly, this value also determines the frequency dependence of the circuit. The output voltage can then be found from

$$\boxed{V_2 = A_v(\omega)V_1} \tag{13.2-4}$$

Circuit b: For this circuit the current divider is appropriate.

$$A_i(\omega) = \frac{I_2}{I_1} = \frac{R}{R + 1/j\omega C} = \frac{j\omega RC}{1 + j\omega RC} \tag{13.2-5}$$

The output current can be found from

$$\boxed{I_2 = A_i(\omega)I_1}$$

The reader will note that both transfer functions in the example are complex. In the next section we consider the dependence of these functions on frequency.

• • •

LEARNING EXERCISE FOR SEC. 13.2

1. In the circuit of Fig. 13.5a, $R = 1\ \text{k}\Omega$ and $C = 0.1\ \mu\text{F}$. Find A_v at $\omega = 100$, 10^4, and 10^6 rad/s.

Ans. $0.01\angle{-90°}$; 1; $0.707\angle{-45°}$

• • •

13.3 FREQUENCY RESPONSE OF RC AND RL CIRCUITS

The term *frequency response* refers to information concerning the response of a system to sinusoids of all frequencies that lie in an appropriate range between dc (zero frequency) and very high frequencies. It is often necessary to design circuits having a frequency response curve designed to achieve a specific objective. Such circuits are called "filters." An example is the tone control circuitry included in high-fidelity sound reproduction systems. A treble boost circuit is a filter designed to emphasize the high frequencies of the audio range while a bass boost circuit emphasizes the lower frequencies.

Frequency response can be determined experimentally using the circuit of Fig. 13.6. The frequency of the oscillator is varied over the range of interest and at each frequency setting the output amplitude and phase angle are measured. The input voltage is maintained at a convenient amplitude and phase, for example, $1\angle{0°}$ V during the measurements. The frequency response is then displayed in the form of graphs of the output vs. frequency. The plot of output amplitude vs. frequency is called the amplitude response, while the plot of phase angle vs. frequency is called the phase response. Sometimes the amplitude response conveys sufficient information and the phase response is omitted.

A typical amplitude response curve for a high-fidelity amplifier is shown in Fig. 13.7. The frequency scale is logarithmic and the transfer

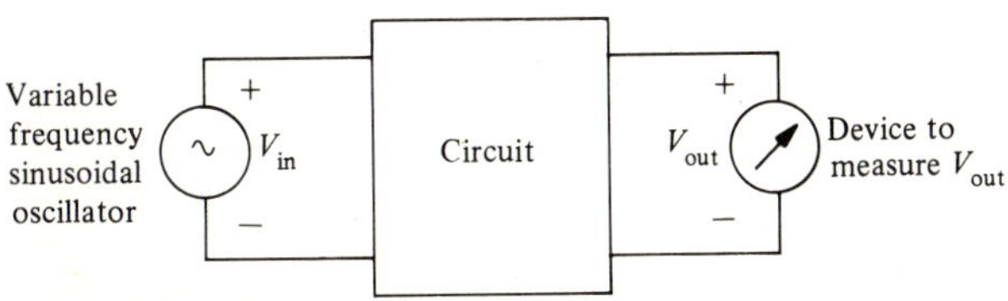

FIGURE 13.6
Test circuit to determine frequency response.

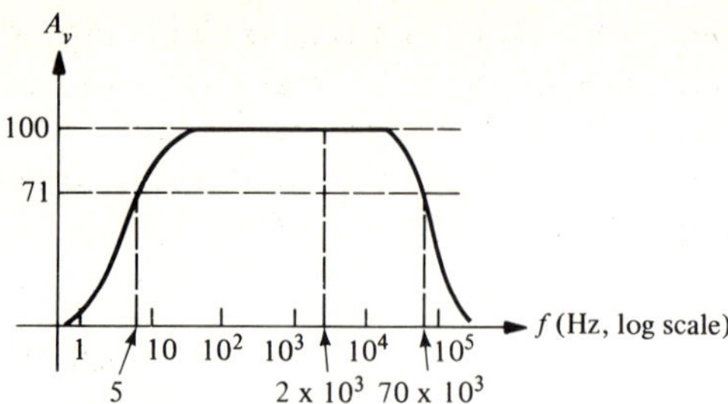

FIGURE 13.7
High-fidelity amplifier response,
$A_v = V_{speaker}/V_{input}$.

function which is plotted represents the voltage amplitude ratio

$$A_v = \frac{V_{speaker}}{V_{input}} = \frac{V_s}{V_i}$$

The graph should be interpreted in the following way:

At $f = 5$ Hz the transfer function is $A_v = 71$. Thus if the input voltage amplitude is 0.1 V at this frequency, then the output voltage amplitude will be $A_v V_i = (71)(0.1) = 7.1$ V.

At $f = 2$ kHz the transfer function is $A_v = 100$. If the input is 0.1 V at this frequency, the output will be $(100)(0.1) = 10$ V.

At $f = 70$ kHz the transfer function is again 71, so the output amplitude for 0.1 V would be 7.1 V.

The reader will note that for frequencies below about 1 Hz and above about 100 kHz the response is very small compared with the response in the frequency band from 5 Hz to 70 kHz. Thus this amplifier is classified as a *band-pass* amplifier, with a pass-band that extends from about 5 Hz to 70 kHz. In a later section we describe the specification of frequency pass-bands such as these in more detail.

Using the Transfer Function to Find Frequency Response

In the previous paragraph, an experimentally determined plot of frequency response was described. Frequency response information can also be obtained directly from a transfer function in the following way.

Consider the transfer function found in Example 13.2-1. It is repeated for convenience:

$$A_v(\omega) = \frac{V_2}{V_1} = \frac{1}{1 + j\omega RC} \qquad (13.2\text{-}3)$$

For a given frequency, $A_v(\omega)$ is a complex value that relates the complex output phasor to the complex input phasor. Most often we are interested in the magnitude and phase of $A_v(\omega)$ as ω varies. Since the magnitude of the quotient of two complex numbers is equal to the quotient of the individual magnitudes, we can write (recalling that $|x + jy| = \sqrt{x^2 + y^2}$)

$$|A_v(\omega)| = \left|\frac{V_2}{V_1}\right| = \frac{1}{\sqrt{1 + (\omega RC)^2}} \tag{13.3-1a}$$

The phase angle is [recall that $\arg(x + jy)^{-1} = -\arctan(y/x)$]

$$\phi = \arg A_v(\omega) = -\arctan(\omega RC) \tag{13.3-1b}$$

A graph of Eq. (13.3-1a) is the amplitude response and a graph of Eq. (13.3-1b) is the phase response. An example follows.

EXAMPLE 13.3-1 Response of a Low-pass Filter

a. Plot the amplitude and phase response for the transistor coupling circuit of Fig. 13.8a.

b. Find the output voltage when the input voltage is $2.5\angle 30^\circ$ V at a frequency of 250 Hz.

Solution

a. For the amplitude response we use Eq. (13.3-1a) with

$$RC = (10 \times 10^3)(0.1 \times 10^{-6}) = 10^{-3}\text{ s}$$

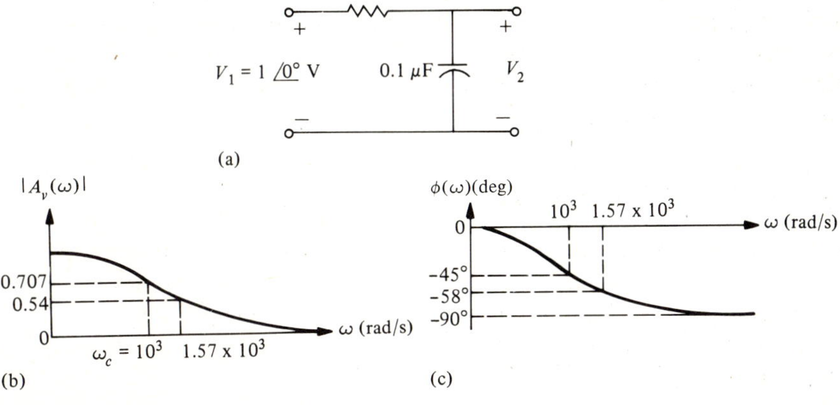

FIGURE 13.8
RC circuit for Example 13.3-1. (a) Circuit. (b) Amplitude-frequency response. (c) Phase angle-frequency response.

Then

$$A_v(\omega) = \frac{1}{\sqrt{1 + (\omega RC)^2}} = \frac{1}{\sqrt{1 + (\omega/1000)^2}}$$

For the phase response we use Eq. (13.3-1b):

$$\phi = -\arctan(\omega RC) = -\arctan\left(\frac{\omega}{1000}\right)$$

The graphs are shown in Fig. 13.8b and c. Some key points are calculated as follows:
At $\omega = 0$,

$$A_v = 1$$
$$\phi = -\arctan 0 = 0°$$

At $\omega = 1000$,

$$A_v = \frac{1}{\sqrt{1 + 1}} = 0.707$$

$$\phi = -\arctan 1 = -45°$$

As $\omega \rightarrow \infty$,

$$A_v \rightarrow 0$$
$$\phi \rightarrow -\arctan \infty = -90°$$

b. At $f = 250$ Hz, $\omega = 2\pi f = 1570$ rad/s. Then, using the transfer function,

$$|A_v(1570)| = \frac{1}{\sqrt{1 + (1570/1000)^2}} = \frac{1}{\sqrt{1 + (1.57)^2}} = 0.54$$

and

$$\phi = -\arctan(1570/1000) = -\arctan(1.57) = -58°$$

These points are shown on the graphs.

Now

$$A_v(\omega) = \frac{V_2}{V_1}$$

so that

$$V_2 = A_v(\omega)V_1 = (0.54\underline{/-57°})(2.5\underline{/30°}) = 1.35\underline{/-27°} \text{ V}$$

• • •

Cutoff Frequency

The amplitude response shown in Fig. 13.8b is typical of a large class of system responses that exhibit the same behavior. They all allow low-frequency signals to pass much more readily than high-frequency signals. The higher frequencies are said to be *attenuated*. Such systems are classified as *low-pass* and the circuit of Fig. 13.8a is called a *low-pass filter*. In order to distinguish between those frequencies that are passed and those that are stopped, a *cutoff* frequency, $f_c = \omega_c/2\pi$, is defined as the frequency at which the amplitude response is $1/\sqrt{2} = 0.707$ times its maximum value. For the low-pass filter of Fig. 13.8a the maximum value of $|A_v(\omega)|$ is 1. Therefore, the amplitude response will be 0.707 times its maximum value at the frequency for which $|A_v(\omega)| = 0.707$. We find this frequency from Eq. (13.3-1a):

$$|A_v(\omega_c)| = \frac{1}{\sqrt{1 + (\omega_c RC)^2}} = 0.707 = \frac{1}{\sqrt{2}}$$

Squaring both sides, we get

$$\frac{1}{1 + (\omega_c RC)^2} = \frac{1}{2}$$

from which $\omega_c RC = 2\pi f_c RC = 1$

It is interesting to observe that if we solve this equation for R, the result is

$$R = \frac{1}{2\pi f_c C} = X_c \qquad \text{at } \omega_c$$

Thus, at the cutoff frequency the reactance of the capacitor is exactly equal to the resistance. For our example the cutoff frequency is

$$f_c = \frac{1}{2\pi RC} = \frac{1}{2\pi \times 10^{-3}} = 159 \text{ Hz} \tag{13.3-2}$$

Since power is proportional to the square of the voltage, the power at f_c is $(1/\sqrt{2})^2 = 1/2$ of the value at low frequencies. For this reason, f_c is sometimes called the *half-power frequency*. An example of low-pass response to a square wave follows.

EXAMPLE 13.3-2 Square Wave Response

The input voltage to the circuit of Fig. 13.9a is the square-wave signal shown in Fig. 13.9b. Find the output waveform, assuming that the square wave is adequately represented by the first four terms of its Fourier series.

Solution

The first step is to represent the square wave by its Fourier series. Its fundamental angular frequency is $\omega_0 = 2\pi/T = 2\pi/0.02 = 314$ rad/s and its amplitude is 10 V.

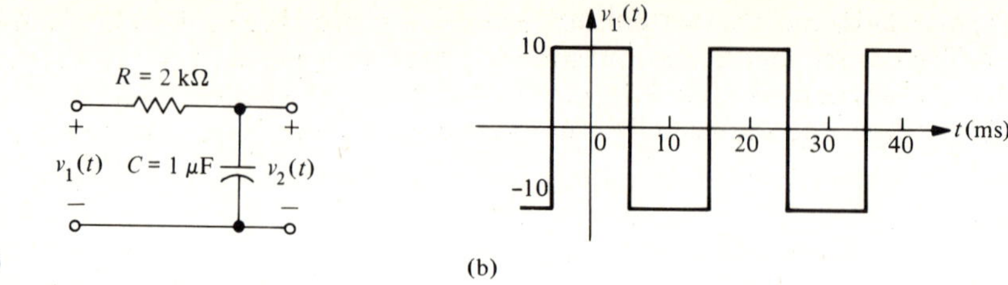

FIGURE 13.9
Low-pass filter for Example 13.3-2. (a) Circuit. (b) Square-wave input voltage.

Using Eq. (13.1-5), the Fourier series is

$$v_1(t) \approx 12.7 \cos \omega_0 T - 4.24 \cos 3\omega_0 t + 2.55 \cos 5\omega_0 t - 1.82 \cos 7\omega_0 t \qquad (13.3\text{-}3)$$

This equation can be represented in circuit form using four voltage sources in series as shown in Fig. 13.10a. In order to find the output voltage $v_2(t)$, we are going to apply the superposition theorem. This requires that we consider the response to each of these sources separately, with the other three replaced by short circuits.

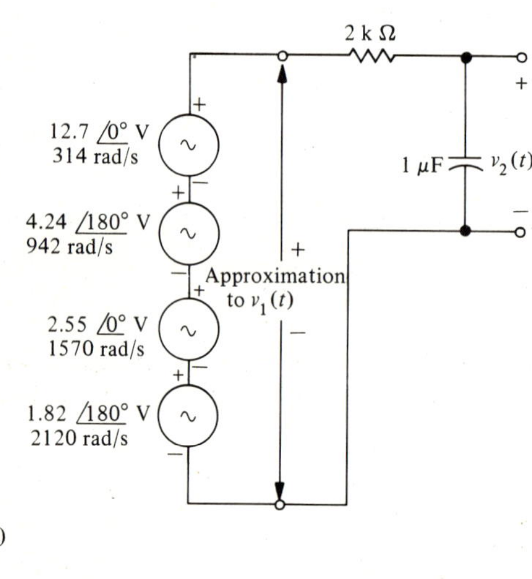

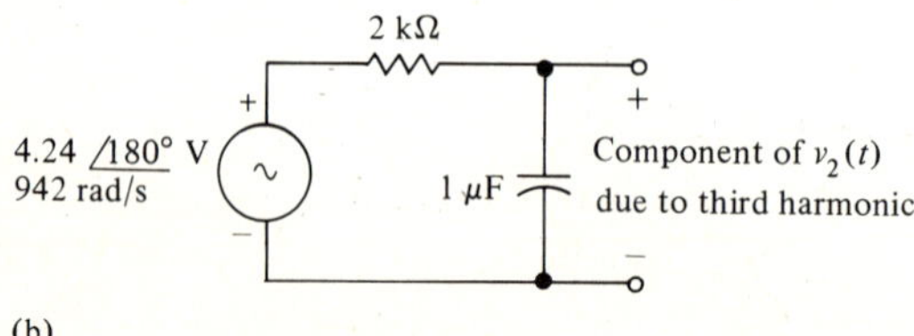

FIGURE 13.10
Applying superposition to find the square-wave response. (a) Circuit with sources for first four terms of Fourier series; all voltages are peak values. (b) Circuit for third harmonic.

We begin by calculating the transfer function at each harmonic in the Fourier series. The transfer function magnitude and phase are given by Eqs. (13.3-1a) and (13.3-1b). The time constant is $RC = (2 \times 10^3\ \Omega)(1 \times 10^{-6}\ \text{F}) = 2$ ms.

Using this value, the general formulas for the amplitude and phase are

$$|A_v(\omega)| = \frac{1}{\sqrt{1 + (\omega/500)^2}}$$

and

$$\phi = -\arctan\left(\frac{\omega}{500}\right)$$

A tabulation of the values at the fundamental and first three harmonics of the square wave input is shown in Table 13.1.

The next step is to find the output due to each source. This is done by multiplying the amplitude of each harmonic by the value of the transfer function at the harmonic frequency and adding the phase angle. For example, Fig. 13.10b shows the circuit for the third harmonic at $\omega = 942$ rad/s. At this frequency we find from the table that $|A_v(\omega)| = 0.47$ and $\phi = -62°$. From the Fourier series [Eq. (13.3-3)] the peak value of the input voltage at this frequency is $V_{1m} = 4.24\underline{/180°}$ V. Then the peak value of the output voltage at $\omega = 942$ rad/s is

$$\begin{aligned} V_{2m} &= A_v(\omega) \times V_{1m} \\ &= 0.47\underline{/-62°} \times 4.24\underline{/180°} \\ &= 2\underline{/118°}\ \text{V} \end{aligned}$$

Then, for the third harmonic component of the output,

$$v_2(t) = 2\cos(3\omega_0 t + 118°)$$

The other components of the output voltage are calculated in the same way (Prob. 13.18). Finally, applying superposition, we add the outputs due to each of the four sources acting alone. The result is

$$\begin{aligned} v_2(t) = {} & 11\cos(\omega_0 t - 32°) \\ & + 2\cos(3\omega_0 t + 118°) \\ & + 0.77\cos(5\omega_0 t - 72°) \\ & + 0.4\cos(7\omega_0 t + 103°) \end{aligned} \tag{13.3-4}$$

A plot of this voltage is shown in Fig. 13.11.

• • •

Table 13.1

ω	314	942	1570	2120
$\lvert A_v(\omega)\rvert$	0.85	0.47	0.3	0.23
$\phi(\omega)$	−32°	−62°	−72°	−77°

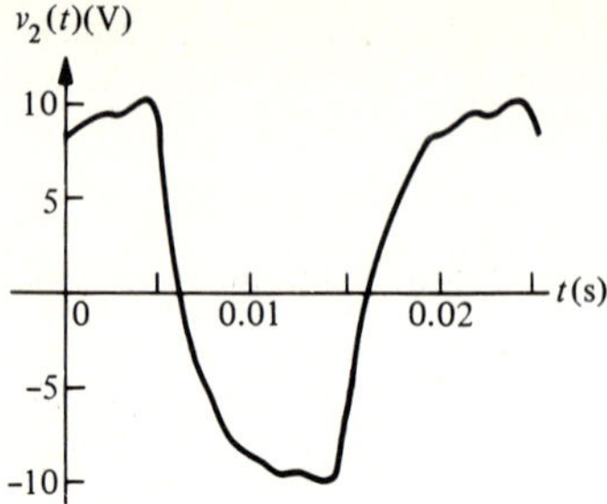

FIGURE 13.11
Square wave response of low-pass filter based on four-term Fourier series approximation.

EXAMPLE 13.3-3 High-pass Filter

The circuit of Fig. 13.12a is a *high-pass* filter. Draw the frequency response plots for the circuit if $R = 10\ \text{k}\Omega$ and $L = 30$ mH.

Solution

The transfer function is

$$A_v(\omega) = \frac{V_2}{V_1} = \frac{j\omega L}{R + j\omega L} = \frac{j\omega(L/R)}{1 + j\omega(L/R)} \tag{13.3-5}$$

For the indicated component values

$$\frac{L}{R} = \frac{30 \times 10^{-3}\ \text{H}}{10 \times 10^{3}\ \Omega} = 3 \times 10^{-6}\ \text{s} = 3\ \mu\text{s}$$

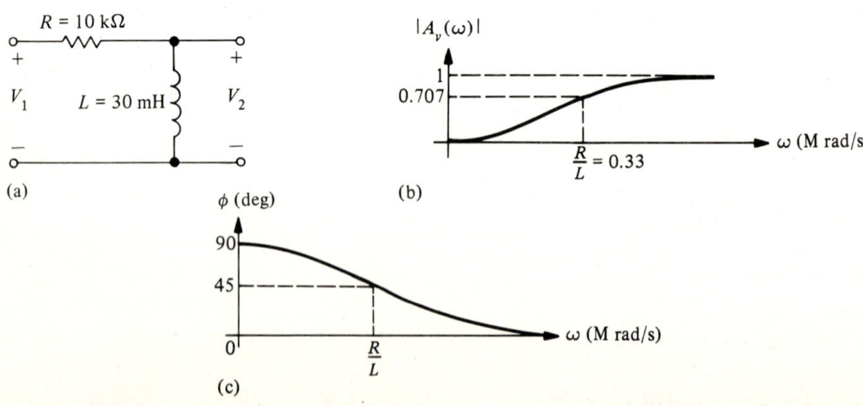

FIGURE 13.12
High-pass filter frequency response. (a) Circuit. (b) Amplitude response. (c) Phase response.

and

$$A_v(\omega) = \frac{j3 \times 10^{-6}\,\omega}{1 + j3 \times 10^{-6}\,\omega}$$

If we specify the units of ω as Mrad/s (= 10^6 rad/s), the formula becomes

$$A_v(\omega) = \frac{j3\omega}{1 + j3\omega} \tag{13.3-6}$$

The amplitude response is

$$|A_v(\omega)| = \frac{\omega(L/R)}{\sqrt{1 + [\omega(L/R)]^2}} = \frac{3\omega}{\sqrt{1 + (3\omega)^2}} \tag{13.3-7}$$

and the phase response is

$$\phi(\omega) = 90^\circ - \arctan\left(\omega \frac{L}{R}\right) = 90^\circ - \arctan(3\omega) \tag{13.3-8}$$

Some key points are:
At $\omega = 0$,

$$|A_v(\omega)| = 0$$

$$\phi = 90^\circ$$

As $\omega \rightarrow \infty$,

$$|A_v(\omega)| \rightarrow 1$$

$$\phi \rightarrow 90^\circ - 90^\circ = 0^\circ$$

The response is seen to have the same characteristics as the *RC* low-pass filter with the values at $\omega = 0$ and $\omega = \infty$ interchanged. As in the low-pass case we define a cutoff angular frequency. Frequencies above the cutoff are passed while those below cutoff are attenuated (rejected). For the *R-L* circuit the cutoff frequency occurs where the inductive reactance equals the resistance, that is, where $X_L = \omega_c L = R$. Thus

$$\omega_c = \frac{R}{L} = \frac{10\ \text{k}\Omega}{30\ \text{mH}} = \frac{1}{3}\ \text{Mrad/s}$$

and when $\omega = \omega_c$

$$|A_v(\omega_c)| = \frac{1}{\sqrt{2}} = 0.707$$

$$\phi = 90^\circ - \arctan 1 = 45^\circ$$

The response curves for this high-pass filter are shown in Figs. 13.12b and c.

• • •

Logarithmic Frequency Response Plots

In practice, frequency response graphs often are drawn with logarithmic scales on both axes. When this is the case, the graphs for low-pass and high-pass RC or RL filters take on a particularly simple form. They can be approximated quite well by two straight line segments. The filter whose response we will consider was analyzed in Example 13.3-1 and its response sketched on linear scales in Fig. 13.8.

The approximation is shown in Fig. 13.13. Observe that the horizontal line segment covers the pass band. The sloped line segment intersects the horizontal segment at the cutoff frequency and its slope is such that the amplitude response decreases by a factor of 10 over each decade of frequency (a decade represents a 10 to 1 range). The exact value of $|A_v|$ at the cutoff frequency is 0.707 (see Fig. 13.8) and it is at this point that the exact curve in dashed lines has its maximum departure from the straight-line approximation. For frequencies below cutoff the exact curve is asymptotic to the horizontal segment and for frequencies above cutoff it is asymptotic to the sloped segment. The two-segment straight-line approximation is called a "Bode plot" after the engineer who first described it. Only two pieces of information are required to construct a Bode plot when the slope is known. These are the pass-band value of $|A_v|$ and the cutoff frequency, as shown in the following example.

EXAMPLE 13.3-4 Bode Plot for a High-pass Filter

Sketch the Bode plot for the RL high-pass filter of Example 13.3-3.

Solution

In Example 13.3-3 we found the pass-band value $|A_v| = 1$. The cutoff frequency is

$$f_c = \frac{\omega c}{2\pi} = \frac{1}{2\pi} \times \frac{1}{3} \text{ Mrad/s} = 53 \text{ kHz}$$

The plot is shown in Fig. 13.14. To draw it, we first locate the cutoff frequency, 53 kHz, on the logarithmic frequency scale. This value and the pass-band amplitude

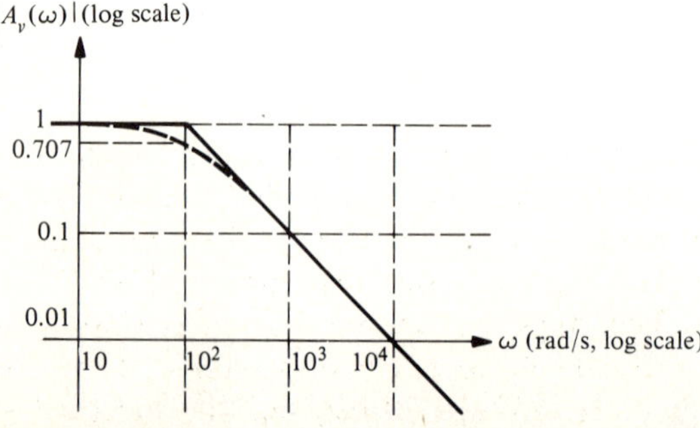

FIGURE 13.13
Logarithmic low-pass filter frequency response. The dashed line is the exact curve.

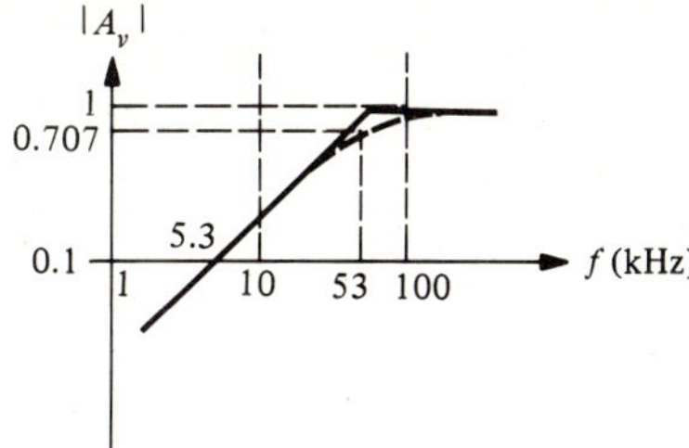

FIGURE 13.14
Bode plot for Example 13.3-4. The dashed line is the exact curve.

response value $|A_v| = 1$ give the intersection point for the two line segments. Since this is a high-pass filter, the horizontal segment covers the pass band from f_c to ∞. The sloped segment decreases to 0.1 at $f = 53/10 = 5.3$ kHz.

• • •

In this section we have considered both low-pass and high-pass filters. In the next section we will consider a circuit that passes only a small band of frequencies.

• • •

LEARNING EXERCISE FOR SEC. 13.3

1. A circuit has the voltage transfer ratio

$$A_v(\omega) = \frac{1}{1 + j10^{-2}\omega}$$

Find the output when the input is $0.1\underline{/-30^\circ}$ V, $\omega = 1$ rad/s; $0.36\underline{/20^\circ}$ V, $\omega = 100$ rad/s; and $2.7\underline{/70^\circ}$ V, $f = 1$ kHz.

Ans. $0.1\underline{/-30^\circ}$; $0.25\underline{/-25^\circ}$; $0.043\underline{/-20^\circ}$

• • •

13.4 RESONANT CIRCUITS

In many applications, especially in the field of radio and television, all of the information being transmitted or received is contained in a relatively narrow band of frequencies around a center frequency f_0. In conventional AM radio, the center frequencies of the radio stations range from about 0.6 to 1.6 MHz in discrete steps. The steps are chosen far enough apart in frequency so that the stations will not interfere with one another. A typical station has a center frequency at 0.66 MHz (660 kHz) and the information being transmitted is contained in a band 5 kHz on either side of the center frequency. Receivers and transmitters designed for such signals often require filters that will pass only the narrow information band around the

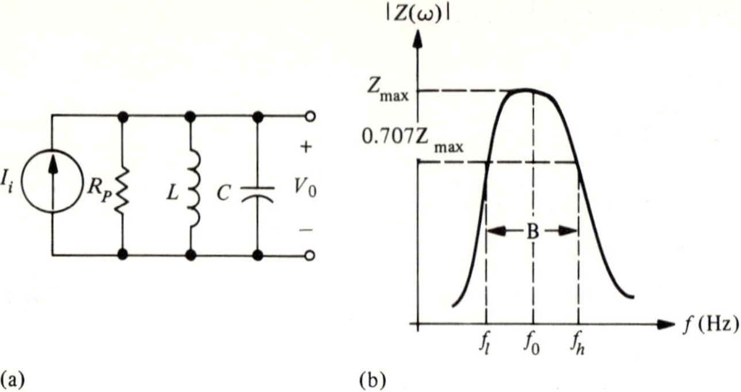

FIGURE 13.15
Bandpass filter. (a) Circuit. (b) Response.

center frequency and reject all other frequencies. This is necessary in order to ensure that there will be no interference between stations.

The parallel "tuned" circuit of Fig. 13.15a is often used as a band-pass filter when the signal is narrow-band as described above. For this case there are two cutoff frequencies f_h and f_l (above and below the center frequency f_0) at which the response is $1/\sqrt{2}$ times the center-frequency value, as shown in Fig. 13.15b. The object of our analysis is to find these frequencies, and hence the bandwidth, in terms of R_p, L, and C.

We begin by writing the transfer function

$$\frac{V_o}{I_i} = Z(\omega) = \frac{1}{1/R_p + j\omega C + 1/j\omega L} \tag{13.4-1a}$$

Note that the transfer function of interest is the impedance. To put this in a form suitable for interpretation, we first multiply numerator and denominator by R_p to get

$$Z(\omega) = \frac{R_p}{1 + j\omega R_p C + R_p/j\omega L}$$

Factoring the denominator,

$$Z(\omega) = \frac{R_p}{1 + jR_pC(\omega - 1/\omega LC)} \tag{13.4-1b}$$

Noting that LC must have the dimensions of seconds squared, we define

$$\boxed{\omega_0^2 = \frac{1}{LC}} \tag{13.4-2}$$

Substituting Eq. (13.4-2) into Eq. (13.4-1b)

$$Z(\omega) = \frac{R_p}{1 + jR_pC(\omega - \omega_0^2/\omega)}$$

The denominator is rearranged as follows:

$$= \frac{R_p}{1 + j\omega_0 R_p C(\omega/\omega_0 - \omega_0/\omega)} \qquad (13.4\text{-}3)$$

It is convenient to define *normalized* frequency

$$x = \frac{\omega}{\omega_0} \qquad (13.4\text{-}4)$$

and a *quality factor* (to be explained shortly)

$$Q = \omega_0 R_p C = \omega_0^2 \frac{R_p C}{\omega_0} = \frac{1}{LC}\frac{R_p C}{\omega_0} = \frac{R_p}{\omega_0 L} \qquad (13.4\text{-}5)$$

The impedance is converted to a normalized transfer function $H(x)$ by dividing through by R_p. Making use of the definitions of Q and x above, the transfer function becomes

$$H(x) = \frac{Z(x)}{R_p} = \frac{1}{1 + jQ(x - 1/x)} \qquad (13.4\text{-}6)$$

Inspection of the equation indicates that $H(x)$ will be a maximum at that value of x for which the denominator is a minimum. Observation of the denominator indicates that its magnitude will be a minimum when $x = 1$, at which value the imaginary part $(x - 1/x)$ is zero. Therefore the maximum amplitude of $H(x)$ occurs at $x = 1$, that is, $\omega = \omega_0$. Thus $\omega_0/2\pi$ is the *center frequency*, and at this frequency the magnitude of H is unity and the impedance is simply R_p, a pure resistance. At $\omega = \omega_0$, the inductive reactance X_L is equal to the capacitive reactance X_C so that the equivalent impedance of the inductor in parallel with the capacitor is infinite. However, there is current in the capacitor so that energy is being stored in its electric field. Also there is current in the inductor, so that energy is being stored in its magnetic field. At the center frequency, this energy is continually exchanged without attenuation between the inductance and capacitance and it can be shown that the net reactive power is zero. This is further borne out by the fact that the circuit is purely resistive at the center frequency, and a resistive circuit has zero reactive power.

To find the cutoff frequencies, we look for those values of x for which the magnitude of $H(x) = 1/\sqrt{2}$, that is,

$$|H(x)| = \frac{1}{|1 + jQ(x - 1/x)|} = \frac{1}{\sqrt{2}} \tag{13.4-7}$$

Since $|1 \pm j1| = \sqrt{2}$, we see that the solution to this equation occurs where

$$\pm Q\left(x - \frac{1}{x}\right) = 1 \tag{13.4-8a}$$

Multiplying through by x and collecting terms yields

$$x^2 \mp \frac{x}{Q} - 1 = 0 \tag{13.4-8b}$$

Solving for x,

$$x = \pm\frac{1}{2Q} \pm \sqrt{\frac{1}{(2Q)^2} + 1} \tag{13.4-8c}$$

Thus there are four distinct values of x that satisfy Eq. (13.4-8c) because there are four possible combinations of the $\pm$ signs.

We can eliminate two of these by observing that the cutoff frequency x must always be positive. The value $-1/2Q - \sqrt{1/(2Q)^2 + 1}$ can be discarded because it is always negative. Also, since Q is always positive,

$$\sqrt{\frac{1}{(2Q)^2} + 1} > \frac{1}{2Q}$$

so that the value $1/2Q - \sqrt{1/(2Q)^2 + 1}$ will be negative and can be discarded. The remaining values are positive. They are:

$$x_h = \frac{\omega_h}{\omega_0} = \sqrt{1 + \frac{1}{(2Q)^2}} + \frac{1}{2Q} \tag{13.4-9a}$$

$$x_1 = \frac{\omega_1}{\omega_0} = \sqrt{1 + \frac{1}{(2Q)^2}} - \frac{1}{2Q} \tag{13.4-9b}$$

These yield the required upper and lower cutoff frequencies. The *bandwidth B* of the circuit is defined as the *difference* between the upper and lower cutoff frequencies. Thus, subtracting Eq. (13.4-9b) from (13.4-9a), we have

$$\omega_h - \omega_l = \frac{\omega_0}{Q} \text{ rad/s} \tag{13.4-10a}$$

The bandwidth in hertz is then

$$\boxed{B = \frac{\omega_0}{2\pi Q} = \frac{f_0}{Q} \text{ Hz}} \tag{13.4-10b}$$

The meaning of *quality factor* can now be explained. It is a measure of the *sharpness* of the filter. Higher values of Q lead to smaller relative bandwidths, B/f_0, or what is the same thing, sharper filters. Systems which exhibit the type of frequency response shown in Fig. 13.15b are often called *resonant* systems, and the response curve is called a *resonance* curve. The phenomenon of resonance is not significant unless the Q is at least 5 or more. For conventional *RLC* networks, values of Q in the range 10 to 100 are typical, while much higher Qs can be obtained by using transistors. The effect of changes in Q is shown in Fig. 13.16.

Specifications for the design of this type of network usually include f_0 and B. From Eq. (13.4-2)

$$\boxed{f_0 = \frac{1}{2\pi\sqrt{LC}} \text{ Hz}} \tag{13.4-11}$$

and using Eqs. (13.4-2) and (13.4-5),

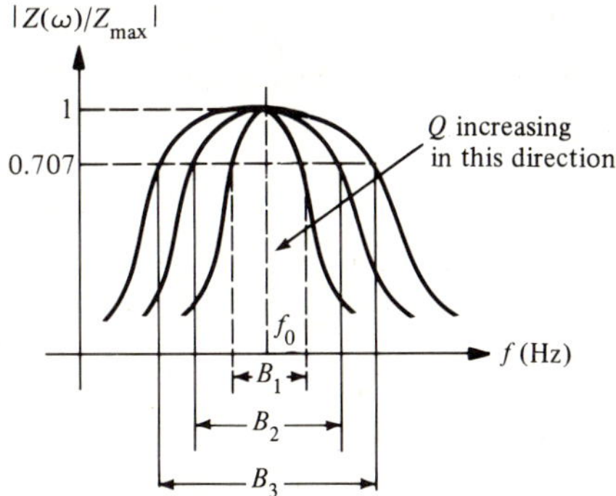

FIGURE 13.16
The effect of changes in *Q*. Note how the bandwidth decreases as *Q* increases.

$$B = \frac{1}{2\pi R_p C} = \frac{R_p}{2\pi L} \text{ Hz} \tag{13.4-12}$$

These may be used as design equations, as shown in the following example.

EXAMPLE 13.4-1 Filter Design

Design an RLC band-pass filter as in Fig. 13.15a with a bandwidth of 3.5 kHz which is centered at 100 kHz. At the center frequency, the impedance is to be 10 kΩ.

Solution

At the center frequency, Eq. (13.4-3) indicates that $Z(\omega) = R_p$. Thus we choose R_p = 10 kΩ to meet the specification on impedance. Next, the bandwidth requirement of 3.5 kHz when substituted into Eq. (13.4-12) yields an RC product of

$$R_pC = \frac{1}{2\pi B} = \frac{1}{2\pi(3.5)(10^3)\text{Hz}} \approx 0.045 \times 10^{-3} \text{ s}$$

Thus, since $R = 10$ kΩ,

$$C = \frac{(0.045)(10^{-3}) \text{ s}}{10^4 \ \Omega} = 0.0045 \ \mu\text{F} = 4.5 \text{ nF}$$

Finally, from Eq. (13.4-2)

$$L = \frac{1}{\omega_0{}^2C} = \frac{1}{(2\pi \times 10^5)^2(4.5)(10^{-9})}$$

$$= 560 \ \mu\text{H}$$

The circuit Q is, from Eq. (13.4-10b),

$$Q = \frac{f_0}{B} = \frac{10^5 \text{ Hz}}{(3.5)(10^3)\text{Hz}} = 29$$

This value is readily achieved with standard components.

• • •

Series Resonant Circuit

The series RLC circuit shown in Fig. 13.17a exhibits resonant behavior similar to that of the parallel RLC tuned circuit. In the parallel circuit we found that the *voltage* across the circuit reached a maximum when the input current was at the resonant frequency. For the series tuned circuit at resonance, the inductive and capacitive reactances are equal, so that the net

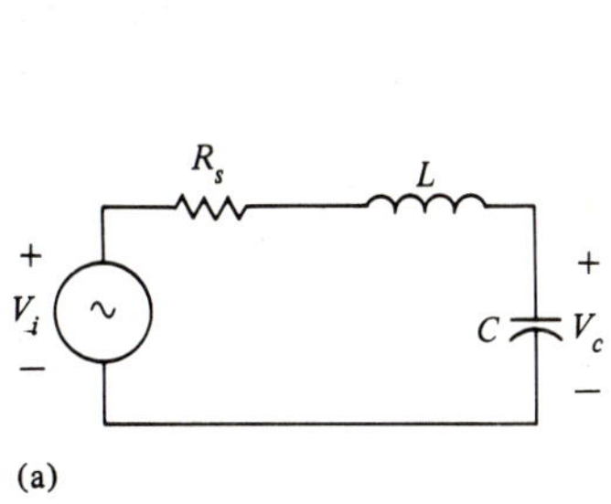

(a)

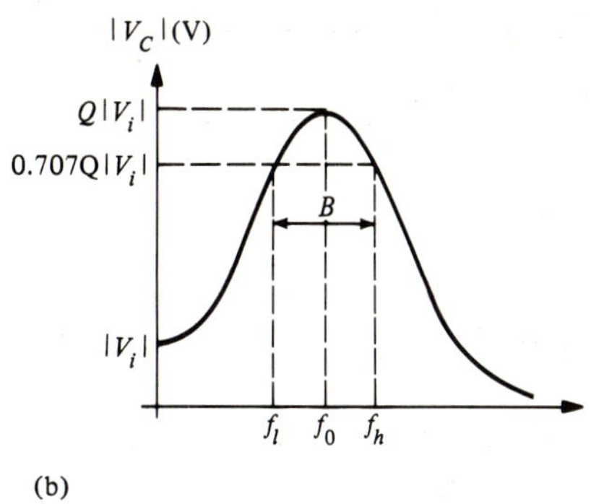

(b)

FIGURE 13.17
Series resonance. (a) Circuit. (b) Resonance curve.

reactance $X_L - X_C$ is zero and the impedance of the circuit is simply the resistance R_s which includes the coil resistance and the internal resistance of the voltage source. This is typically a relatively small value in contrast to the parallel tuned circuit, which has a very high impedance at resonance. Therefore we find that the *current* through the series circuit is a maximum when the input *voltage* is at the resonant frequency.

A graph showing how the capacitor voltage varies with frequency is shown in Fig. 13.17b. The conditions at resonance are established by observing that

$$Z = R_s + j(X_L - X_C) \tag{13.4-13}$$

However, at resonance $X_L = X_c$ so that

$$Z(\omega_0) = R_s \tag{13.4-14}$$

In general, the capacitor voltage is found using the voltage divider:

$$\left|\frac{V_C}{V_i}\right| = \left|\frac{Z_C}{Z}\right| \tag{13.4-15}$$

At resonance, this becomes

$$\left|\frac{V_C}{V_i}\right| = \frac{X_C}{R_s} \tag{13.4-16}$$

For the series circuit, the quality factor Q is

$$\boxed{Q = \frac{X_L}{R_s} = \frac{\omega_0 L}{R_s} = \frac{X_C}{R_s} = \frac{1}{\omega_0 R_s C}} \tag{13.4-17}$$

Therefore, at resonance, Eq. (13.4-16) becomes

$$V_C = QV_i \tag{13.4-18}$$

and the output voltage across the capacitor is Q times the input voltage. Note that this amplification takes place only in a relatively narrow band of frequencies.

Like the parallel circuit, the bandwidth of the series tuned circuit is

$$\boxed{B = \frac{\omega_0}{2\pi Q} = \frac{f_0}{Q} \text{ Hz}} \tag{13.4-19}$$

EXAMPLE 13.4-2 Series Tuned Circuit

A series *RLC* as shown in Fig. 13.17a has $R_s = 2\ \Omega$, $L = 0.1$ mH, and $C = 33$ nF. The input voltage magnitude is 10 μV. Find:

a. The center frequency and bandwidth.
b. The voltage across the capacitor at the resonant frequency.

Solution

a. The center frequency is

$$f_0 = \frac{1}{2\pi\sqrt{LC}} = \frac{1}{2\pi\sqrt{0.1 \times 10^{-3} \times 33 \times 10^{-9}}} = 87.6 \text{ kHz}$$

To find the bandwidth we first find

$$Q = \frac{\omega_0 L}{R_s} = \frac{2\pi \times 87.6 \times 10^3 \times 0.1 \times 10^{-3}}{2} = 28$$

Then

$$B = \frac{f_0}{Q} = \frac{87.6 \text{ kHz}}{28} = 3.13 \text{ kHz}$$

b. At ω_0, the capacitor voltage is [see Eq. (13.4-18)]

$$V_C = QV_i = 28 \times 10\ \mu\text{V} = 280\ \mu\text{V}.$$

• • •

Comparison of series and parallel resonance

	Series	Parallel
Center frequency	$f_0 = \dfrac{1}{2\pi\sqrt{LC}}$	$f_0 = \dfrac{1}{2\pi\sqrt{LC}}$
Impedance at f_0	R_s	R_p
Impedance below f_0	Capacitive	Inductive
Impedance above f_0	Inductive	Capacitive
Q	$\dfrac{\omega_0 L}{R_s} = \dfrac{1}{\omega_0 R_s C} = \dfrac{f_0}{B}$	$\omega_0 R_p C = \dfrac{R_p}{\omega_0 L} = \dfrac{f_0}{B}$
Output voltage	$V_C = QV_i$	$V_o = R_p I_i$
Requirement for high-Q	Source with small internal resistance	Source with high internal resistance

• • •

LEARNING EXERCISE FOR SEC. 13.4

1. A parallel resonant circuit has $L = 3\ \mu$H, $C = 12$ pF, and $R_p = 12$ kΩ. Find the center frequency, quality factor, and bandwidth.

Ans. 1.1; 26.5; 24

• • •

13.5 FREQUENCY RESPONSE USING ECAP

Calculations of frequency response can become quite tedious. If ECAP (or a similar program) is available, they can be carried through with ease. The frequency is varied by using a MODIFY routine (Sec. 5.8) as illustrated in the following example.

EXAMPLE 13.5-1 ECAP

Use the ECAP program to determine the frequency response (amplitude and phase) of the circuit of Fig. 13.18a with $R = 1$ kΩ and $C = 1\ \mu$F. Cover the frequency range 0 to 300 Hz and plot the results on a logarithmic frequency scale.

Solution

The ECAP circuit is shown in Fig. 13.18b and the program in Fig. 13.19a. The only signifiant difference between this and previous ECAP examples is the MODIFY routine, which requests a frequency iteration, that is, a calculation repeated at different frequencies. Two types of iteration are available:

Multiplicative This is used to obtain equally spaced points on a logarithmic scale. An example of the required frequency modification statement is

FREQ = 40(1.1)300

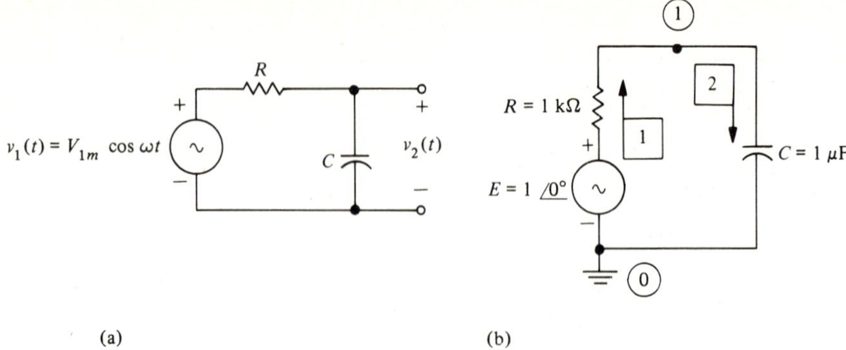

FIGURE 13.18
ECAP frequency response. (a) Circuit for Example 13.5-1. (b) ECAP circuit.

Here 40 Hz is the lower frequency limit, and 300 Hz the upper limit. The number 1.1 in parentheses is a *multiplication* factor. This statement requests calculation at the frequencies 40 Hz, (1.1)(40) Hz, $(1.1)^2(40)$ Hz, . . . , up to the nearest step greater than 300 Hz.

Additive This is used to obtain equally spaced points on a linear scale. The frequency statement is shown in Fig. 13.19b after the MODIFY statement. The lower frequency limit is 40 Hz, and the upper limit is 300 Hz. The +13 in parentheses requests calculations in 13 equal frequency steps between 40 and 300 Hz, that is, at intervals of (300 Hz − 40 Hz)/13 = 20 Hz.

Note that the presence or absence of the plus sign in the parentheses tells the computer which type of iteration is desired.

A partial printout is shown in Fig. 13.19b and a plot of the results is shown in Fig. 13.20 using semilog coordinates.

• • •

EXAMPLE 13.5-2 Resonant Circuit Response

Use the ECAP program to determine the frequency response of the circuit designed in Example 13.4-1. Cover the frequency range 95 to 105 kHz on a linear scale.

Solution

The ECAP circuit is shown in Fig. 13.21. The program is essentially the same as that used in Example 13.5-1. Since a *linear* frequency response calculation is required in this example, the additive iteration frequency statement follows the MODIFY statement in the program shown in Fig. 13.22 along with a partial printout. The lower frequency limit is 0.96 E 05(96 kHz)—95 kHz was done in the original analysis—and the upper limit is 0.105 E 06(105 kHz). The +9 in parentheses requests calculations in nine equal frequency steps between 96 and 105 kHz, that is, at intervals of 1 kHz.

A plot of the results is shown in Fig. 13.23. The center frequency is very close to 100 kHz, and the half-power frequencies occur at approximately 98.5 and 102 kHz.

```
C       FREQUENCY RESPONSE
C       NEW YORK INSTITUTE OF TECHNOLOGY
C
        AC ANALYSIS
    B1  N(0,1),R=1E 03,E=1E 00/0
    B2  N(1,0),C=1E-06
C
        FREQUENCY=40
C
        PRINT,NV
        EXECUTE


FREQ =   0.40000000E 02



        NODES                 NODE VOLTAGES


MAG       1-  1   0.96983886E 00
PHA              -0.14107789E 02
```

(a)

```
C
C
        MODIFY
        FREQUENCY=40(+13)300
        EXECUTE


FREQ =   0.40000000E 02



        NODES                 NODE VOLTAGES


MAG       1-  1   0.96983886E 00
PHA              -0.14107789E 02

FREQ =   0.59999985E 02



        NODES                 NODE VOLTAGES


MAG       1-  1   0.93571538E 00
PHA              -0.20655960E 02
```

(b)

FIGURE 13.19
Example 13.5-1. (a) Program. (b) Partial printout.

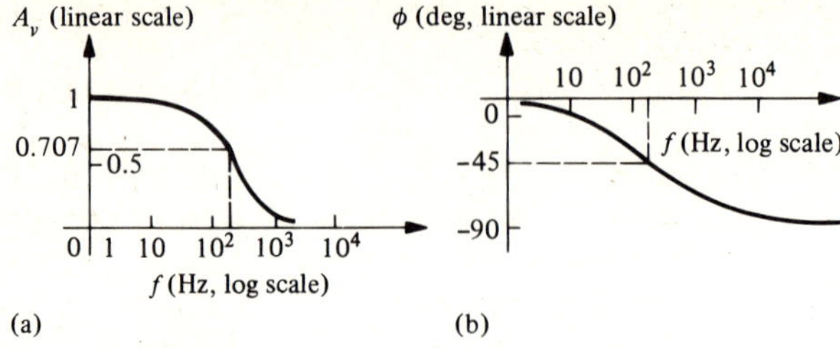

FIGURE 13.20
Low-pass filter response. (a) Amplitude characteristic. (b) Phase characteristic.

This agrees well with the original design values of Example 19.4-1, that is, $f_0 = 100$ kHz and $B = 3.5$ kHz.

• • •

13.6 CORRELATION OF TIME AND FREQUENCY RESPONSE

In Sec. 9.8 we studied the pulse response of the RC low-pass filter and found its 10 to 90% rise time to be

$$t_r = 2.2RC \tag{9.8-6}$$

In Sec. 13.3 the cutoff frequency of the same filter was found to be

$$f_c = \frac{1}{2\pi RC} \tag{13.3-2}$$

Solving for RC in Eq. (13.3-2) and substituting the result into Eq. (9.8-6)

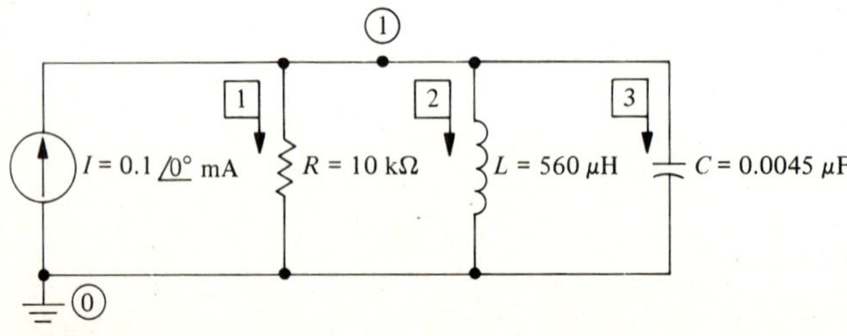

FIGURE 13.21
ECAP circuit for Example 13.5-2.

```
C         ECAP SOLUTION FREQUENCY RESPONSE


C
C
          AC ANALYSIS
     B1   N (1,0), R=1E 04,I=IE-04/0
     B2   N (1,0), L=0.56E-03
     B3   N (1,0), C=0.45E-08
          FREQUENCY=0.95E 05
          PRINT, NV
          EXECUTE

FREQ =  0.94999937E 05

          NODES                  NODE VOLTAGES


MAG       1-  1      0.31102085E 00
PHA                  0.71879211E 02

          MODIFY
          FREQ=0.96E 05 (+9) 0.105E 06
          EXECUTE


FREQ =  0.95999937E 05

          NODES                  NODE VOLTAGES


MAG       1-  1      0.37639797E 00
PHA                  0.67889236E 02


FREQ =  0.96999937E 05

          NODES                  NODE VOLTAGES


MAG       1-  1      0.47089422E 00
PHA                  0.61907623E 02


FREQ =  0.97999937E 05

          NODES                  NODE VOLTAGES


MAG       1-  1      0.61215973E 00
PHA                  0.52254150E 02


FREQ =  0.98999937E 05

          NODES                  NODE VOLTAGES


MAG       1-  1      0.81306171E 00
PHA                  0.35603836E 02
```

FIGURE 13.22
Example 13.5-2; program and partial printout of results.

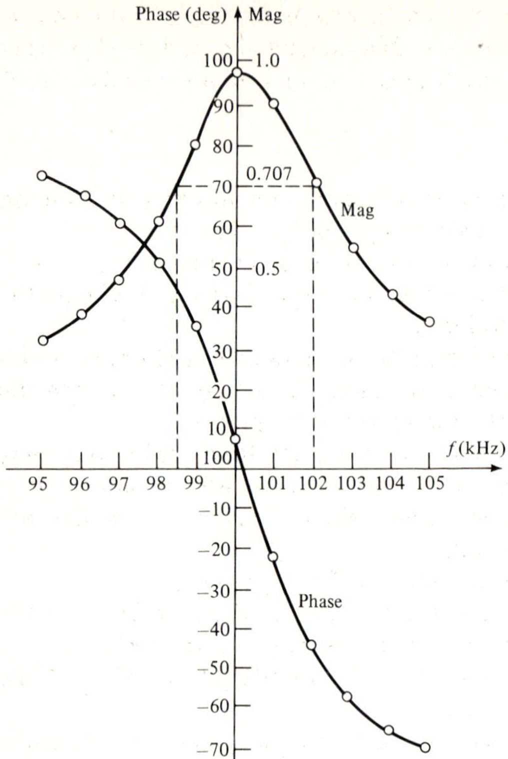

FIGURE 13.23
Example 13.5-2; plot of results.

we obtain

$$t_r = \frac{2.2}{2\pi f_c} = \frac{0.35}{f_c} \tag{13.6-1}$$

Thus if we require a filter with a specific rise time, its cutoff frequency is fixed by this relation.

In the design of digital systems it is often required to design for the shortest possible rise time. Analysis of different systems leads to relations similar to Eq. (13.6-1), indicating that rise time is inversely proportional to cutoff frequency. Thus in order to achieve short rise times the bandwidth must be as wide as possible.

SUMMARY

In this chapter we have considered the important concept of the ac transfer function and the frequency response of typical filters. In addition we studied the harmonic relationships existing among the terms of the Fourier series. The student should now be able to analyze and design basic low-pass, high-pass, and bandpass filters. The important concepts in each section are listed below:

Section

13.1 1. The *Fourier series* is an infinite sum of sinusoidal components that can be used to represent a periodic waveform.
a. There is a dc component equal to the average value.
b. The *fundamental*, or first, *harmonic* has a period equal to the period of the waveform.
c. The frequencies of the other components are integral multiples of the fundamental frequency. The frequency of the nth harmonic is n times the fundamental frequency.

13.2 2. Transfer functions for ac networks are found using the same techniques we used for dc networks. These include voltage divider, current divider, superposition, Thevenin's theorem, and mesh and node equations.

13.3 3. The *frequency response* consists of two graphs:
a. The *amplitude response* is a plot of the magnitude of the complex transfer function versus the angular frequency, ω.
b. The *phase response* is a plot of the phase angle of the transfer function versus ω.

4. A *low-pass filter* attenuates high-frequency components while passing low-frequency components. *A high-pass filter* does the opposite.

5. The *cutoff* frequency ω_c is that frequency at which the filter's amplitude response is equal to its maximum response divided by $\sqrt{2}$.

6. A *half-power frequency* is a frequency at which a circuit's amplitude response is $1/\sqrt{2}$ times its maximum value.

7. In general, the shape of a nonsinusoidal signal will change when it is passed through a circuit containing inductance and/or capacitance. The output waveform is found by applying the transfer function to each harmonic and adding the results.

8. The Bode plot is a straight-line approximation to the amplitude response of a simple filter. It is plotted using the pass-band response and the cutoff frequency.

13.4 9. A *resonant* circuit has an amplitude response that peaks at a center frequency determined by its inductive and capacitive components. The *bandwidth* B is the frequency interval between the upper and lower cutoff frequencies. The *quality factor* Q measures the "sharpness" of the resonant peak, higher values of Q leading to sharper peaks.

13.5 10. The MODIFY routine is used with ECAP to iterate the frequency and obtain the frequency response of a circuit.

13.6 11. For simple filters the rise time of the pulse response is inversely proportional to the cutoff frequency.

QUESTIONS FOR REVIEW

Sec. 13.1

1. Define the fundamental frequency of a nonsinusoidal waveform.
2. What are *harmonic* frequencies?
3. Describe the Fourier series for a square wave.
4. Describe the amplitude-frequency spectrum of a square wave.

Sec. 13.2

5. Compare ac transfer functions with dc transfer functions.

Sec. 13.3

6. What is meant by the term *frequency response*?
7. Describe a circuit that can be used to measure frequency response.
8. Describe several situations in which it might be necessary to design a circuit which has a specific frequency response.
9. How is frequency response found using the ac transfer function?
10. Explain the difference between amplitude response and phase response.
11. What is the difference between low-pass filters, high-pass filters, and band-pass filters?
12. Define *cutoff frequency.*
13. A square wave with a fundamental frequency of 1 kHz is the input to a low-pass filter with a cutoff frequency of 2 kHz. Describe the output of the filter qualitatively.
14. A square wave with a fundamental frequency of 1 kHz is applied to the input of a high-pass filter with a cutoff frequency of 3 kHz. Describe the output of the filter.
15. What is a *Bode* plot?
16. What information is required for the construction of a Bode plot?
17. For a low-pass filter, what is the slope of the Bode plot above the cutoff frequency?

Sec. 13.4

18. Describe as many situations as you can which might require a sharply tuned band-pass filter.
19. Define the center frequencies and the cutoff frequencies of a parallel *RLC* tuned circuit.
20. What is the meaning of *bandwidth*?
21. What is the meaning of *quality factor*?
22. Two parallel *RLC* circuits are tuned to the same center frequency. One has $Q = 10$, and the other $Q = 100$. Describe the difference between these two in terms of bandwidth.

Sec. 13.5

23. When using ECAP, how is a multiplicative frequency iteration obtained? Additive frequency iteration?

Sec. 13.6

24. In an RC low-pass filter, how is the pulse rise time related to the cutoff frequency?

PROBLEMS

Sec. 13.1

1. The musical note A has a fundamental frequency of 440 Hz. What are the frequencies of the second, fourth, and seventh harmonics?
2. The waveform of a tone from a bass guitar can be represented by the Fourier series

$$v(t) = 2 \cos 200t + 0.3 \cos(400t + 20°) + 0.02 \cos(800t - 32°)$$

   a. What is the fundamental frequency in hertz?
   b. What harmonics are present?
   c. What are the amplitude and phase of the fundamental and each harmonic?
3. What is the fundamental frequency of each of the waveforms of Fig. 13.24?

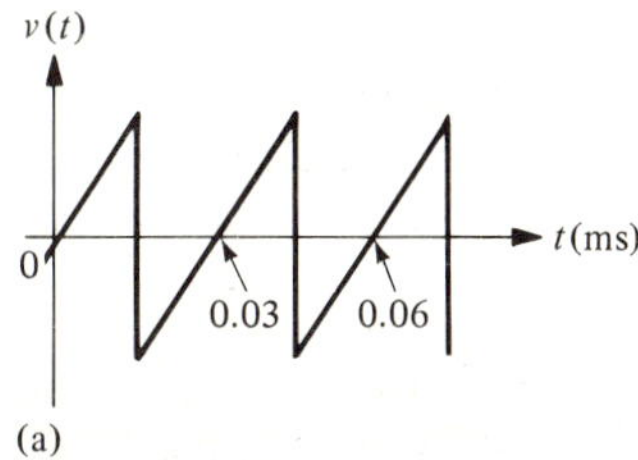

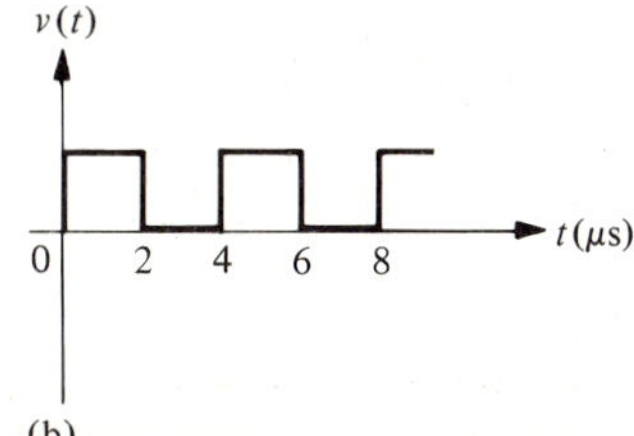

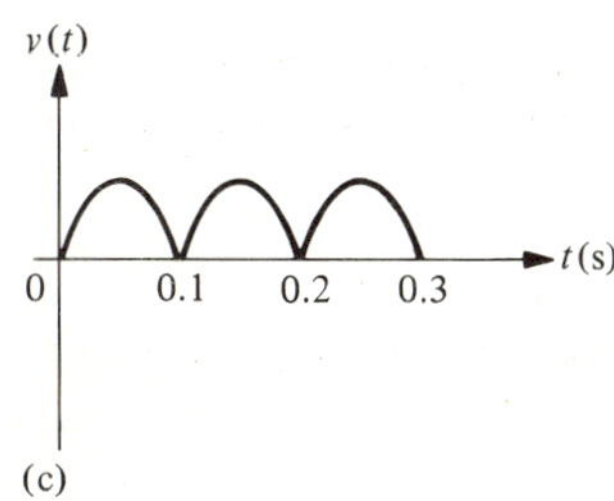

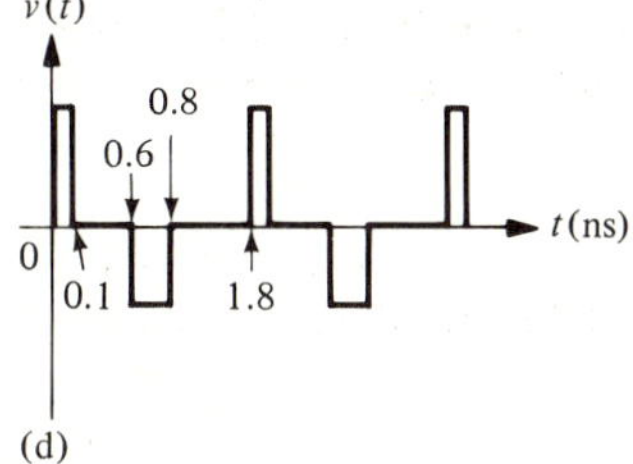

FIGURE 13.24

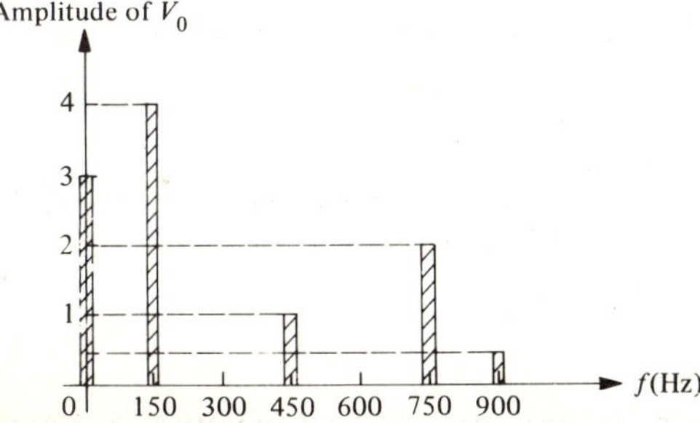

FIGURE 13.25

4. Draw the amplitude-frequency spectrum for the Fourier series of Prob. 2.
5. Write the Fourier series for the amplitude-frequency spectrum shown in Fig. 13.25. Assume that the phase angle for each term is zero.
6. The Fourier series for the triangular wave shown in Fig. 13.26 is

$$v(t) = 1 + \frac{8}{\pi^2}\sin \omega t + \frac{8}{9\pi^2}\sin 3\,\omega t + \frac{8}{25\pi^2}\sin 5\,\omega t + \cdots$$

   a. If $T = 1s$, find the average value, period, and fundamental frequency.
   b. Which harmonics are present?
   c. Draw the amplitude-frequency spectrum.
   d. What is the amplitude of the seventh harmonic?
   e. Compare the amplitudes of the harmonics with those of the square wave of Eq. (13.1-5).

Sec. 13.2

7. Find the complex transfer functions $A_v(\omega) = V_2/V_1$ for each of the circuits of Fig. 13.27.
8. Repeat Prob. 7 for the circuits of Fig. 13.28.
9. Find the complex transfer function $A_i(\omega) = I_2/I_1$ for each of the circuits of Fig. 13.29.
10. Find the complex transfer function I_2/V_1 for each of the circuits of Fig. 13.30.

Sec. 13.3

11. For the circuit of Fig. 13.27a
   a. Find expressions for $|A_v(\omega)|$ and $\phi(\omega)$.
   b. Determine $|A_v(0)|$ and $|A_v(\infty)|$.
   c. Find $\phi(0)$ and $\phi(\infty)$.
   d. Find the half-power frequency and the phase angle at this frequency.
   e. Sketch $|A_v(\omega)|$ and $\phi(\omega)$ as functions of frequency.

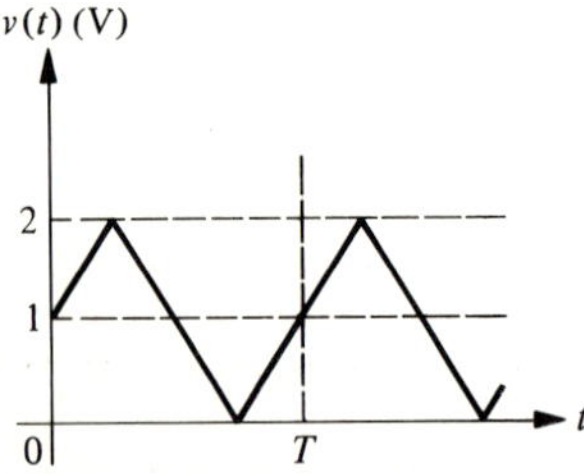

FIGURE 13.26

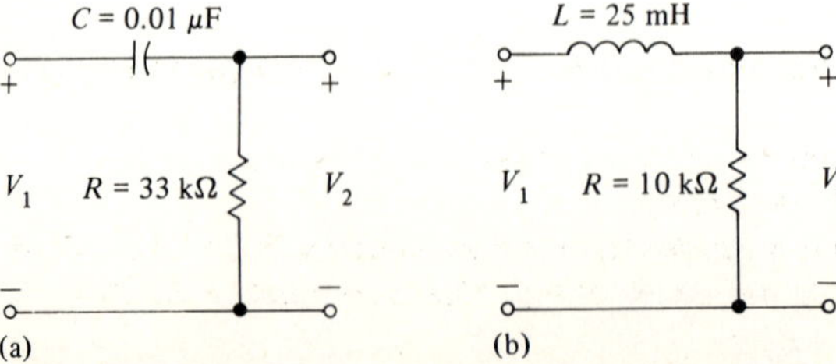

FIGURE 13.27

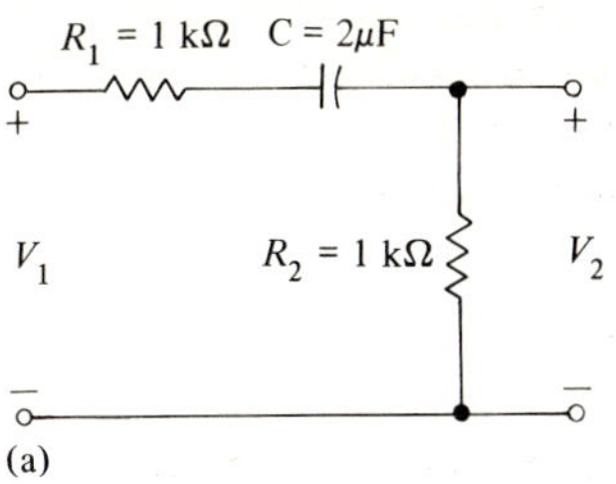

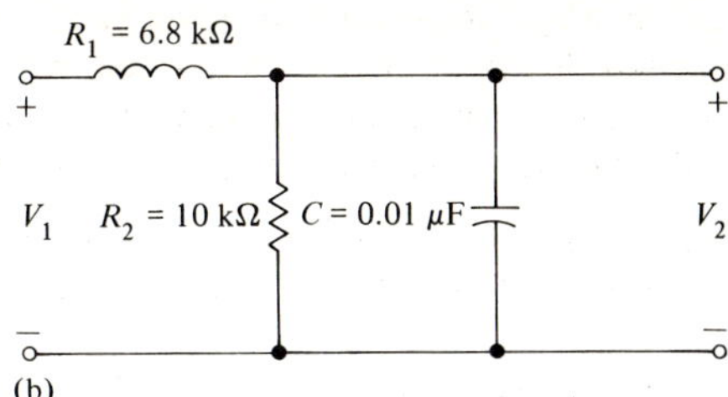

FIGURE 13.28

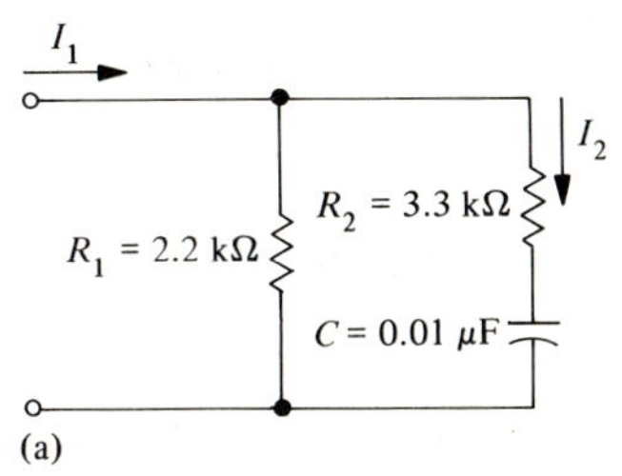

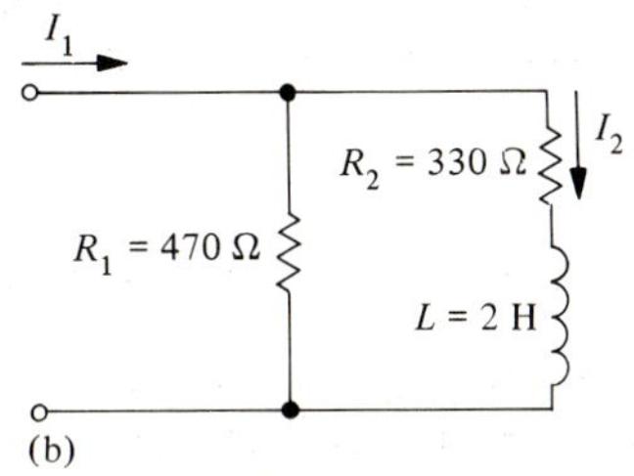

FIGURE 13.29

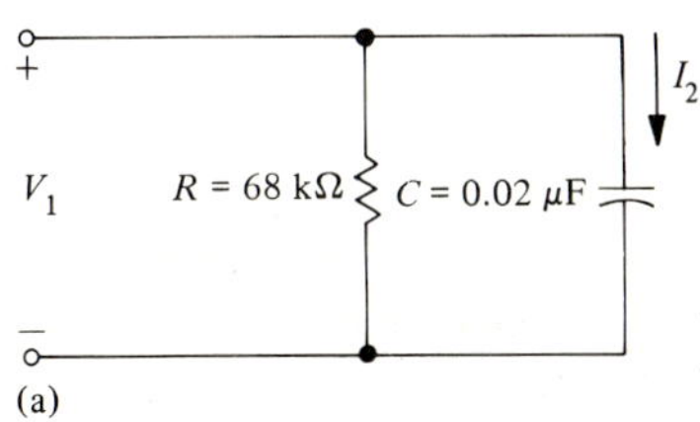

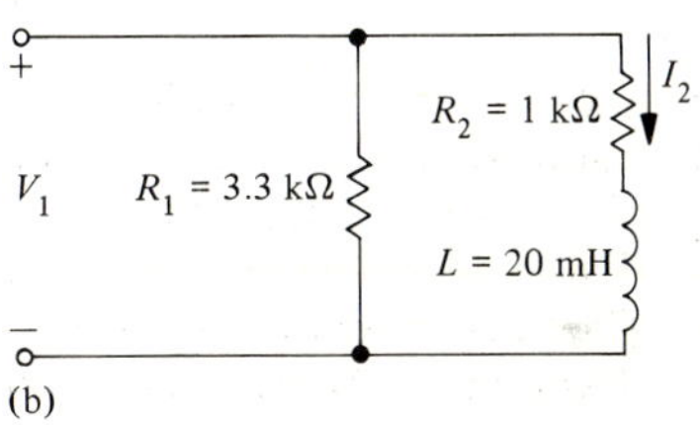

FIGURE 13.30

12. Repeat Prob. 11 for the circuit of Fig. 13.27b.
13. Repeat Prob. 11 for the circuit of Fig. 13.28a.
14. Repeat Prob. 11 for the circuit of Fig. 13.28b.
15. Repeat Prob. 11 for the circuit of Fig. 13.30a.
16. In the circuit of Fig. 13.27a find $|A_v(\omega)|$ and $\phi(\omega)$ at frequencies of 200, 500, and 2000 Hz.
17. In the circuit of Fig. 13.28b find $|A_v(\omega)|$ and $\phi(\omega)$ at frequencies of 0.5, 1.6, 4, and 10 kHz.
18. In Example 13.3-2 find the fundamental component and the fifth and seventh harmonics.
19. The waveform whose Fourier series is given in Prob. 2 is the input to the circuit of Fig. 13.31.
    a. Find the output for each term of the series.
    b. Neatly plot the output waveform.
20. The square wave which is represented by the Fourier series in Eq. (13.3-3) with ω_0 = 3 krad/s is the input to the circuit of Fig. 13.27a. Repeat Prob. 19 for this situation.
21. Sketch the Bode plot for V_2/V_1 in both circuits of Fig. 13.27a.
22. Repeat Prob. 21 for I_2/V_1 in the circuit of Fig. 13.30a.

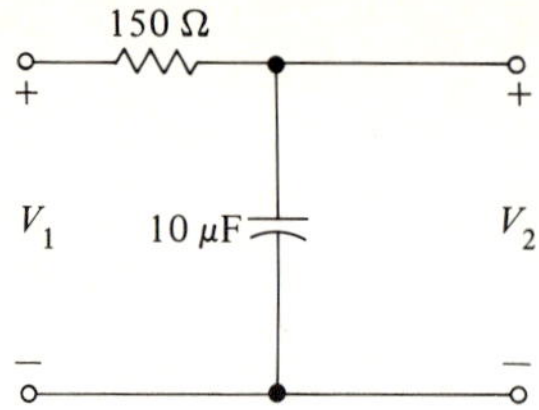

FIGURE 13.31

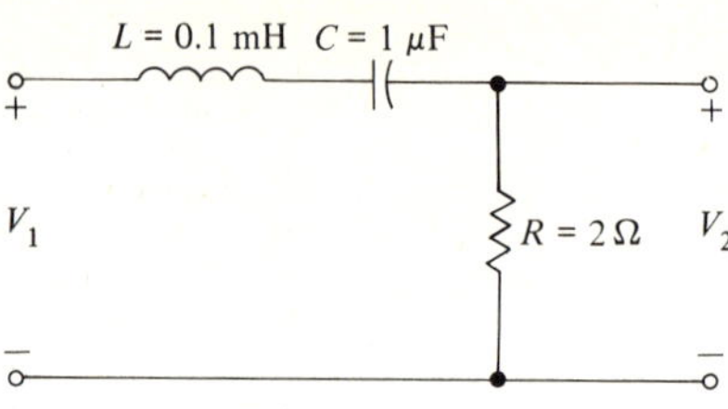

FIGURE 13.32

23. The triangular wave of Prob. 6 is applied to the low-pass *RC* filter of Fig. 13.5a with $R = 1$ MΩ and $C = 1$ μF. Assuming that four terms of the series are adequate, find the Fourier series for the output.
24. Program a computer to find the amplitude and phase response of the circuit of Example 13.3-1. If possible, have the computer plot the results. Otherwise, plot amplitude and phase vs. frequency using log-log graph paper.
25. Program a computer to find the output waveform in Example 13.3-2. If possible, have the computer plot the waveform.

Sec. 13.4

26. The circuit of Fig. 13.32 is a *series* tuned circuit.
    a. Find the expression for the voltage transfer ratio

$$A_v(\omega) = V_2 / V_1$$

    b. Find the center frequency, ω_0.
    c. What is the input impedance at ω_0?
    d. What is the Q of the circuit?
    e. Find the bandwidth.
    f. Sketch $|A_v(\omega)|$ vs. frequency.
27. Repeat Prob. 26 if the resistance in the circuit is decreased to 0.5 Ω. Compare with the results of Prob. 26.
28. Design a series *RLC* circuit such that the resonant frequency is 100 kHz. The bandwidth is to be 10 kHz, and the impedance at resonance is to be 100 Ω.
29. Design a parallel resonant circuit to be used with an FM receiver. The center frequency is 10.7 MHz and the cutoff frequencies are to be ±15 kHz from the center frequency. The impedance is to be 200 kΩ at resonance.
30. A section of an AM radio requires a parallel *RLC* circuit tuned to 455 kHz with a bandwidth of 10 kHz. Design a circuit for these frequencies to have an impedance at resonance of 10 kΩ.
31. In practice, the coil in a tuned circuit always has appreciable resistance so the circuit could be represented as shown in Fig. 13.33. For this circuit
    a. Find the transfer function V_o / I_i.
    b. Assuming the Q of the coil, $\omega_0 L / R_s$, to be high, find the center frequency of the circuit.
32. The triangular wave of Prob. 6 represents a current with a fundamental frequency of 10 kHz and amplitude of 2 mA peak-to-peak. It is applied to a parallel *RLC* circuit with $f_0 = 50$ mHz, $Q = 100$, and $R_p = 100$ Ω. Find the output voltage.

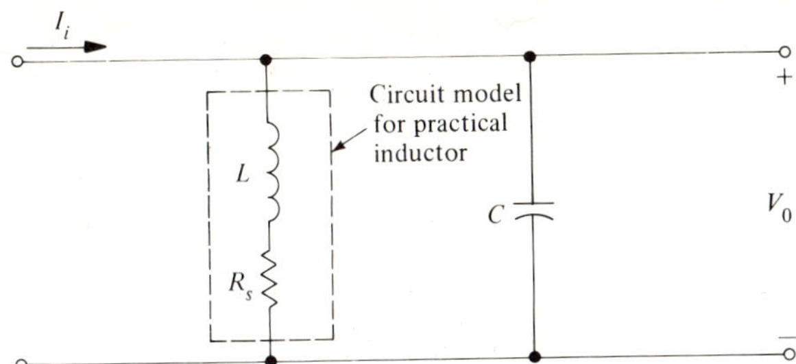

FIGURE 13.33

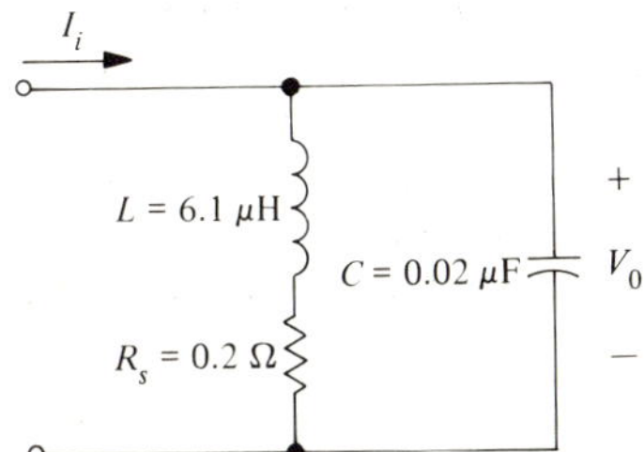

FIGURE 13.34

Sec. 13.5

33. Use ECAP or a similar circuit analysis program to find the frequency and phase response of the circuit of Fig. 13.28b. Cover the frequency range of 500 Hz to 10 kHz.
34. Repeat Prob. 33 for the parallel tuned circuit shown in Fig. 13.34. Cover the frequency range 445 to 465 kHz in 1-kHz steps.

Sec. 13.6

35. The input to a pulse amplifier in a home computer is effectively a low-pass filter as shown in Fig. 13.5a. If the rise time must be less than 50 ns and the resistance is 10 kΩ, what is the maximum capacitance that can be allowed? With this much capacitance in the circuit, what is the cutoff frequency of the filter?

14 Two-Port Networks

OBJECTIVES

Upon completion of this chapter, the student should be able to

Section		
14.1	1.	Draw the block diagram of a two-port and indicate the conventional reference directions.
14.2	2.	Define power amplification as a ratio and in decibels.
	3.	State the formula for voltage gain in decibels.
14.3	4.	Enumerate the four different kinds of controlled sources.
14.4	5.	Describe ideal voltage and current amplifiers and calculate output voltages and currents.
14.5	6.	Describe the effect of input and output resistance in amplifiers.
14.6	7.	Define the hybrid parameters for a two-port.
	8.	Draw the hybrid equivalent circuit.
	9.	State the physical meaning of each parameter in terms of the equivalent circuit.
	10.	Calculate the hybrid parameters for a given network.
14.7	11.	State the chain rule as applied to cascaded two-ports.
	12.	State that the overall decibel gain of cascaded two-ports is equal to the sum of the individual decibel gains.
14.8	13.	Describe a typical op-amp and draw its equivalent circuit.
	14.	Define the term *differential input.*
	15.	Describe the difference between the inverting input and the noninverting input.
	16.	Describe linear negative and positive scalers and state the formulas for overall gain.
	17.	Describe the op-amp summing circuit and state the formula for overall gain.

18. Describe the input and output impedances for the scaler circuits.
19. Define the term *feedback* as used in the op-amp scalers.

INTRODUCTION

In previous chapters we discussed the laws governing current and voltage distribution in networks containing R, L, and C elements and independent sources. Such networks cannot *amplify* voltages or currents. In order to provide amplification, transistors or other active devices are required. These devices are modeled by circuits containing *controlled* sources, the new element being introduced in this chapter.

Most of the networks are classified as *two-ports*. The general representation for a two-port is shown in Fig. 14.1 where the two ports are the input and output terminal pairs. In this chapter, we consider methods used for the description and analysis of such networks.

14.1 TRANSFER FUNCTIONS

In the two-port network shown in Fig. 14.1 the conventional reference directions for current and voltage are shown. Currents are taken as positive into the port, so a negative answer indicates that the actual current is flowing out of the port. The network in the box, if it is linear, can be described mathematically by a transfer function, which represents the ratio of the output signal to the input signal. This is a natural way to characterize such a system, since we can easily find the output in response to any input, given the transfer function.

We first introduced the idea of the transfer function in Sec. 4.5 in connection with the voltage divider network. At that point, our interest centered on the transfer voltage ratio

$$A_v = \frac{v_2}{v_1} \tag{14.1-1}$$

where A_v is the transfer function for the voltage divider network. If we solve for the output voltage v_2, the result is

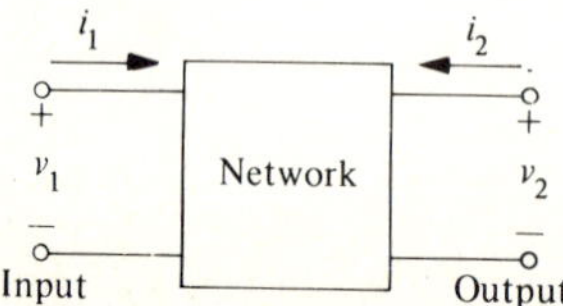

FIGURE 14.1
Two-port network.

$$V_2 = A_v V_1 \tag{14.1-2}$$

and we see that the output is simply the product of the transfer function and the input. In this case, as shown in Sec. 4.5, the transfer function is a real number that lies between 0 and 1 so that the output has the same waveshape as the input, except that it may be reduced in amplitude.

In Sec. 13.2 we extended the transfer function concept to ac circuits containing inductance and capacitance in addition to resistance. In this chapter we make use of this technique to analyze typical amplifier configurations. Before proceeding, we present some review examples of transfer function calculations.

EXAMPLE 14.1-1 Voltage Divider Transfer Function

Consider the voltage divider network shown in Fig. 14.2a. Assume that $R_1 = R_2 = 10\ \text{k}\Omega$. Find $v_2(t)$ if $v_1(t) = 10e^{-2t}$ V.

Solution

If $i_2 = 0$, we can use the voltage divider formula

$$A_v = \frac{V_2}{V_1} = \frac{R_2}{R_1 + R_2} = \frac{1}{2} \tag{14.1-3}$$

Then

$$v_2(t) = A_v v_1(t) = \frac{1}{2} \times 10e^{-2t} = 5e^{-2t}\ \text{V} \tag{14.1-4}$$

and we see that the output has the same waveshape as the input, but one-half the amplitude, as shown in Fig. 14.2b.

• • •

This example is deceptively simple because we assumed in the course of the solution that the output current i_2 was zero. In fact, this assumption

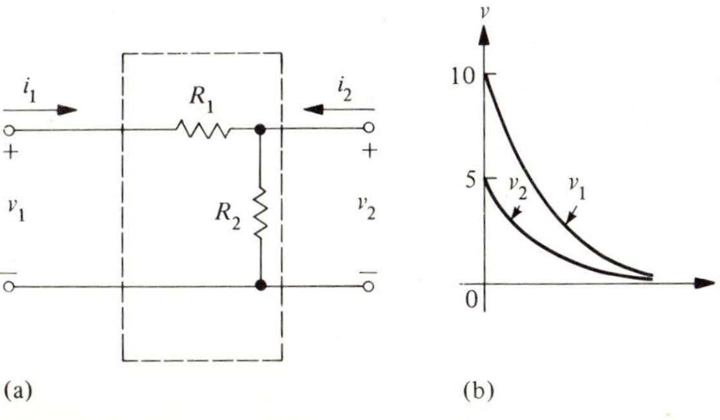

FIGURE 14.2
(a) Network. (b) Waveforms.

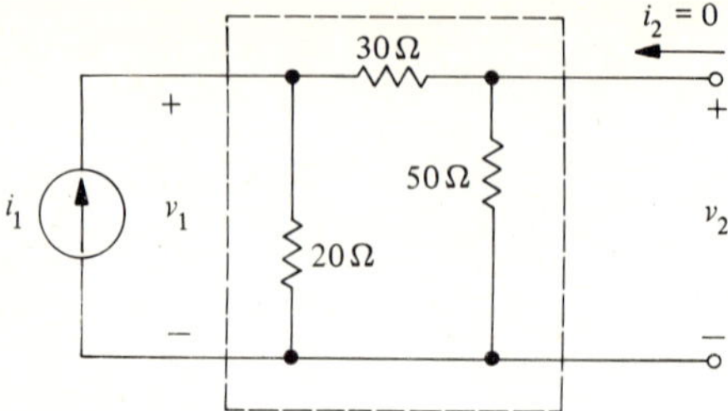

FIGURE 14.3
Two-port for Example 14.1-2.

was made in the original derivation of the voltage-divider relation. In general, when dealing with a two-port, the output current will not be zero and the specification of the transfer function becomes much more complicated. In a later section we find that it takes four transfer functions to completely specify a two-port. However, in many cases, a single transfer function will suffice, as shown in the following example.

EXAMPLE 14.1-2 Chain Calculation

In the circuit of Fig. 14.3, find

a. The transfer resistance $R_T = v_2/i_1$, with $i_2 = 0$.
b. The output voltage $v_2(t)$ when the input current is $i_1(t) = 5 \cos \omega t$ mA.

Solution

a. Up to now we have considered transfer functions that were voltage ratios or current ratios. Here we are interested in a transfer function that relates an output voltage to an input current and thus has dimensions of volts per ampere, or ohms.

We can find this transfer function, denoted R_T, by writing it in the following way:

$$R_T = \left(\frac{v_1}{i_1}\right)\left(\frac{v_2}{v_1}\right) = \frac{v_2}{i_1} \tag{14.1-5}$$

This "chain" type of calculation reduces a relatively complicated problem to two simpler problems since each of the terms in the chain is calculated separately as follows:

Using Ohm's law, with all resistances in ohms,

$$\frac{v_1}{i_1} = 20 \parallel (30 + 50) = \frac{(20)(80)}{(100)}$$

Using the voltage divider relation

$$\frac{v_2}{v_1} = \frac{50}{30 + 50} = \frac{(50)}{(80)}$$

Finally, carrying out the indicated multiplication

$$R_T = \frac{v_2}{i_1} = \frac{(20)(80)}{(100)}\frac{(50)}{(80)} = 10 \text{ V/A}$$

In this case, it is best to think of R_T as 10 volts per ampere rather than 10 ohms. The units immediately suggest that R_T is 10 V at the output terminals per ampere into the input terminals. This is the proper way to view a *transfer* resistance.

b. $$\begin{aligned} v_2(t) &= R_T i_1(t) \\ &= (10)(5 \times 10^{-3} \cos \omega t) \\ &= 50 \cos \omega t \text{ mV} \end{aligned}$$

Once again, the output voltage has exactly the same waveshape as the input current.

• • •

When we deal with networks that contain only resistances, the transfer functions are always real numbers and the output waveshape is always a scaled replica of the input waveshape. If inductors and capacitors are added to the network as in Chap. 13, we find more complicated transfer functions, which modify the input waveshape so that the output waveshape is different from the input waveshape.

14.2 AMPLIFIERS AND APPLICATIONS

In most systems involving signal transmission, amplification is required either because input signals are too small to actuate the load directly or because losses in signal strength occur due to long transmission distances.

A familiar example is the radio receiver. In this system, the signal picked up by the antenna has a power level measured in microwatts and a voltage level measured in microvolts. In order to actuate the load, which consists of one or more loudspeakers requiring perhaps 0.05 to 50 or more watts for proper operation, the power level must be amplified by large factors, often 1×10^6 or more. This amplification, or *gain*, is provided by electronic amplifiers.

Another example is the long-distance telephone circuit. The power level at the output of the telephone instrument is only a few milliwatts. As this signal is transmitted along the telephone cable it is subject to power loss because of the resistance of the cable. In order to maintain sufficient signal strength for satisfactory operation it is necessary to place amplifiers at suitable points along the cable.

Numerous other examples exist in all fields of technology. Some of these will be considered in later sections.

Amplifier Gain

The basic function of an amplifier is to provide amplification of power, voltage, or current. A block diagram of such an amplifier is shown in Fig. 14.4. In a power amplifier the output power from the amplifier is greater than the input power. The power amplification is expressed as

$$\boxed{A_p = \frac{\text{output signal power}}{\text{input signal power}}} \tag{14.2-1}$$

Since the power out is greater than the power in, it might appear that the principle of conservation of energy is being violated. This is not so because the additional power is supplied by the dc power source shown in the figure. The power from this source is converted to signal power in the amplifier.

In the sections that follow, we begin to look at electronic amplifiers from the two-port point of view.

Gain and the Decibel

In this section we introduce the decibel (dB), a logarithmic unit often used in the specification and measurement of amplifier performance. This unit was originally introduced because it was found that the ear perceives as approximately equal changes in sound intensity that are in the same *ratio*, so that a change from 5 to 10 mW would seem the same as a change from 50 to 100 mW (because 5/10 = 50/100). Since these changes in the ear's response are approximately equally spaced on a logarithmic scale, it is appropriate that a logarithmic unit be used for expressing power levels and ratios in communication circuits. A change of 0.5 to 1 dB is about the smallest discernible change for the average person.

A further reason for the use of the decibel is that the use of a logarithmic unit converts the operation of multiplication into simple addi-

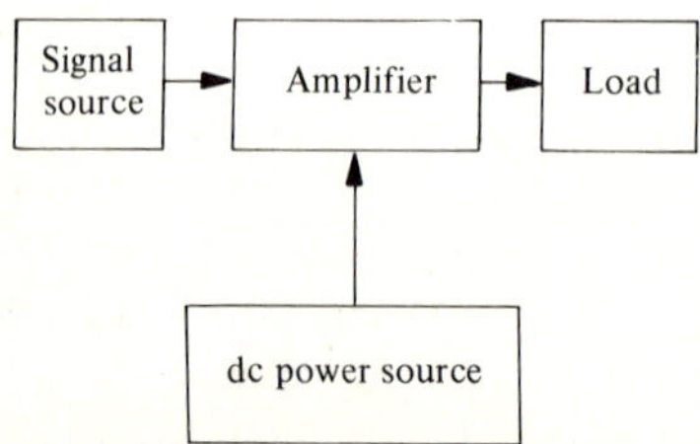

FIGURE 14.4
Amplifier block diagram.

tion. This arises frequently in amplifier work. In addition, the logarithmic unit compresses a very large range of numbers into a much smaller range, a decided convenience in many instances.

The *power gain* of an amplifier measured in decibels is*

$$G_p = 10 \log \frac{P_2}{P_1} \text{ dB} \tag{14.2-2}$$

where log means $\log_{10}$ and P_2 and P_1 are output and input power, respectively, expressed in the same units.

This definition is used in two ways: first, to express a power *ratio* in logarithmic units, and second, to express a power *level* with respect to a fixed reference power level. One often-used reference level is 1 mW, for which the units are called dBm. Thus for a power level in decibels with respect to 1 mW, we have, with P_2 in watts,

$$\begin{aligned} G_p &= 10 \log \frac{P_2}{10^{-3}} \\ &= 10 \log (P_2 \times 10^3) \text{ dBm} \end{aligned} \tag{14.2-3}$$

Decibel Voltage Gain

The decibel was originally defined in terms of power gain. Because of its usefulness as a logarithmic measure, it is generally applied directly to voltage gain and current gain based on the following: If P_2 and P_1 in Eq. (14.2-2) are dissipated in identical resistances R, then

$$P_2 = \frac{V_2^2}{R} = RI_2^2 \quad \text{and} \quad P_1 = \frac{V_1^2}{R} = RI_1^2 \tag{14.2-4}$$

where I_1, V_1, I_2, and V_2 are rms amplitudes in appropriate units.

Substituting Eq. (14.2-4) into Eq. (14.2-2)

$$G_v = 10 \log \frac{V_2^2}{V_1^2} = 20 \log \frac{V_2}{V_1} \text{ dB} \tag{14.2-5}$$

*From here on we will use the symbol G for decibel gain and A for amplification ratio.

and

$$G_i = 10 \log \frac{I_2^2}{I_1^2} = 20 \log \frac{I_2}{I_1} \text{ dB} \qquad (14.2\text{-}6)$$

where we have used $\log X^2 = 2 \log X$.

From the foregoing, we see that voltage and current gains in decibels are equal to power gain only if the resistances in which the powers are dissipated are equal. However, due to long usage, Eqs. (14.2-5) and (14.2-6) are usually used for voltage and current gain regardless of the resistances.

EXAMPLE 14.2-1 Decibel Calculations

a. Find G_v in decibels for $V_2/V_1 = 0.1, 1, 2, 5, 10, 20, 100, 1000$.
b. Find G_p for $P_2/P_1 = 1, 2, 10$.
c. Find G_p in dBm for $P_2 = 10^{-6}$ W and 10 W.

Solution

Some of the calculations are shown below. All of the results are tabulated in Table 14.1.

a. $V_2/V_1 = 0.1$

$$G_v = 20 \log \frac{V_2}{V_1} = 20 \log 0.1 = 20 \log 10^{-1} = -20 \log 10 = -20 \text{ dB}$$

When the voltage ratio is less than one, the decibel value is negative.

$V_2/V_1 = 1$

$$G_v = 20 \log 1 = 20(0) = 0 \text{ dB}$$

$V_2/V_1 = 2$

$$G_v = 20 \log 2 = 20(0.3) = 6 \text{ dB}$$

$V_2/V_1 = 1000$

$$G_v = 20 \log 1000 = 20 \log 10^3 = 60 \log 10 = 60 \text{ dB}$$

We have carried through all of these calculations in detail for illustration. They are all more easily completed using the log X key on a pocket scientific calculator. Simply enter the V_2/V_1 ratio, depress the log X key, and multiply the result by 20.

Note that when the output voltage is greater than the input voltage the gain in decibels is a positive number; when V_2 is less than V_1, it is a negative number; and when $V_2 = V_1$ the gain in decibels is zero (see Table 14.1).

TABLE 14.1

V_2/V_1	0.1	1*	2*	5*	10*	20	100	1000
G_v (dB)	−20	0	6	14	20	26	40	60

Those entries marked with an asterisk (*) are worth memorizing. Decibel values for many round numbers can easily be found from these in the following way: consider $V_2/V_1 = 4000$. Then

$$4000 = \underset{6\text{ dB}}{(2)} \times \underset{+\ 6\text{ dB}}{(2)} \times \underset{+\ 60\text{ dB}}{(1000)}$$

Therefore,

$$G_v = 72 \text{ dB}$$

b. $\dfrac{P_2}{P_1} = 1 \qquad G_p = 10 \log \dfrac{P_2}{P_1} = 10 \log 1 = 0 \text{ dB}$

$\dfrac{P_2}{P_1} = 2 \qquad G_p = 10 \log 2 = 3 \text{ dB}$

$\dfrac{P_2}{P_1} = 10 \qquad G_p = 10 \log 10 = 10 \text{ dB}$

c. $P_2 = 10^{-6}$ W

$$G_p = 10 \log \frac{P_2}{1 \text{ mw}} = 10 \log(P_2 \times 10^3) = 10 \log(10^{-6} \times 10^3) = 10 \log 10^{-3}$$
$$= -30 \log 10$$
$$= -30 \text{ dBm}$$

$P_2 = 10$ W

$$G_p = 10 \log(10 \times 10^3) = 10 \log 10^4 = 40 \log 10$$
$$= 40 \text{ dBm}$$

• • •

LEARNING EXERCISE FOR SEC. 14.2

1. Find the decibel gain for each of the following gain ratios:

$$\frac{V_2}{V_1} = 246 \qquad \frac{P_2}{P_1} = 6.3 \times 10^6 \qquad A_i = 8.3 \times 10^3$$

Ans. 78; 48; 68

• • •

14.3 CONTROLLED SOURCES

In order to model a real amplifier, we must add the *controlled source* to our list of elements. The sources introduced in Chap. 4 were *independent* sources, whose current or voltage was independent of any of the currents or voltages in the network, that is, the current or voltage was determined by the source itself. The voltage or current from a controlled source, on the other hand, *depends on* a voltage or current at some point in the network. In this sense it is dependent on the form of the network and also the independent sources; it is sometimes called a *dependent* source.

We have occasion to use both current and voltage sources, which can in turn be controlled by a current or voltage. Thus there are four possibilities:

1. *Voltage-controlled voltage source.* This is shown in Fig. 14.5a. In the diagram v_x is the *controlling voltage* and K_1 is the factor by which the controlled source multiplies v_x. The current i will be determined by the circuit connected to the controlled source.
2. *Current-controlled voltage source.* This is shown in Fig. 14.5b. Here i_x is the controlling current and K_2 is the factor by which this current must be multiplied in order to obtain the voltage of the source. K_2 must be in ohms, so that when multiplied by a current, the result will be volts.
3. *Voltage-controlled current source.* This is shown in Fig. 14.5c. Here v_x is the controlling voltage and K_3 is the factor by which v_x is multiplied in order to obtain the source current. K_3 must have units of conductance.

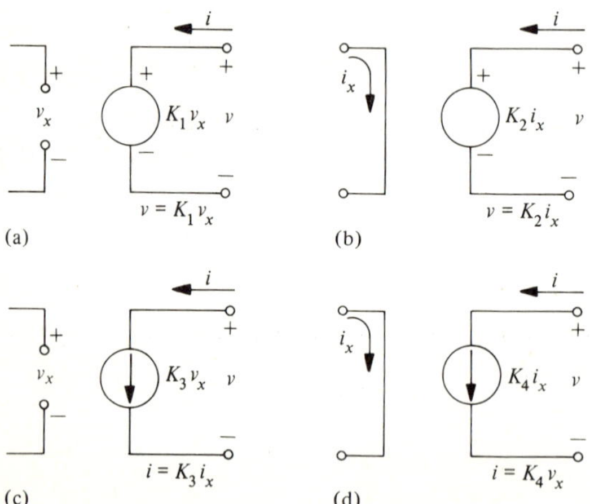

FIGURE 14.5
Controlled sources. (a) Voltage-controlled voltage source. (b) Current-controlled voltage source. (c) Voltage-controlled current source. (d) Current-controlled current source.

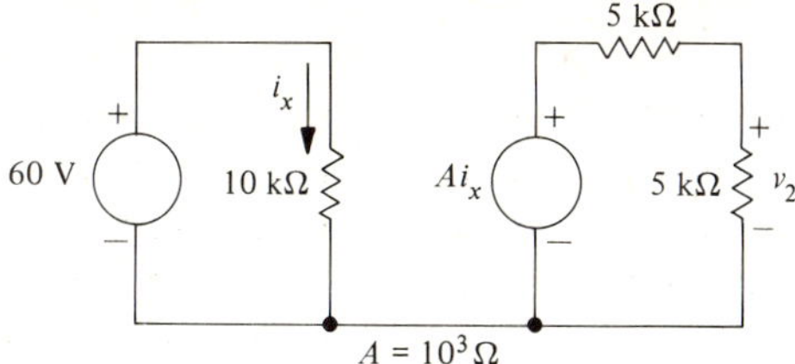

FIGURE 14.6
Circuit for Example 14.3-1.

Observe that the voltage v across the source will depend on the external circuit.

4. *Current-controlled current source.* This is shown in Fig. 14.5d. K_4 must be dimensionless.

In the following sections we show how these sources are used to model actual amplifiers.

EXAMPLE 14.3-1 Controlled Source

The circuit of Fig. 14.6 contains an independent voltage source of 60 V and a current-controlled voltage source. Find the output voltage v_2.

Solution

We must first find the controlling current i_x. Using Ohm's law

$$i_x = \frac{60 \text{ V}}{10 \text{ k}\Omega} = 6 \text{ mA}$$

Thus the voltage output of the controlled source is 10^3 V/A $\times 6 \times 10^{-3}$ A $= 6$ V. Next we use the voltage divider:

$$v_2 = \frac{5}{5+5} \times 6 \text{ V} = 3 \text{ V}$$

Note that it was necessary to find the numerical value of the controlling current before the solution could be completed.

• • •

LEARNING EXERCISE FOR SEC. 14.3

1. The circuit of Fig. 14.7 contains a current-controlled current source and represents the equivalent circuit of a transistor amplifier. Find

   a. i_2

   b. $A_i = \dfrac{i_2}{i_i}$

c. $G_p = 10 \log \frac{P_L}{P_i}$

where P_L is the power dissipated in R_L and P_i the power from the signal source.

Ans. 41; −2.7; −270

• • •

14.4 IDEAL AMPLIFIERS

When we study certain topics, it is convenient to begin by considering an ideal model of the device being considered. The behavior of practical devices is then accounted for by adding additional elements to the ideal model in order to account for departures from the ideal. This is the procedure we follow as we consider voltage and current amplifiers.

The Ideal Voltage Amplifier

The ideal voltage amplifier takes the form shown in Fig. 14.8a. The voltage-controlled voltage source provides an output voltage v_2 directly proportional to v_1, that is,

$$v_2 = A_v v_1 \tag{14.4-1}$$

where the constant A_v is the *voltage* amplification factor. In most cases of interest A_v will be a real number greater than unity. The *vi* characteristic of the output circuit is shown in Fig. 14.8b. This differs from previous *vi* characteristics in that it consists of a family of vertical straight lines with the input voltage v_1 as a parameter. Note that the output voltage is independent of the output current and the output current is determined by the external circuit, that is, the load. As we stated in Sec. 14.1, the reference direction for the output current is positive *into* the network. For the amplifier in the figure, taking note of the indicated reference directions

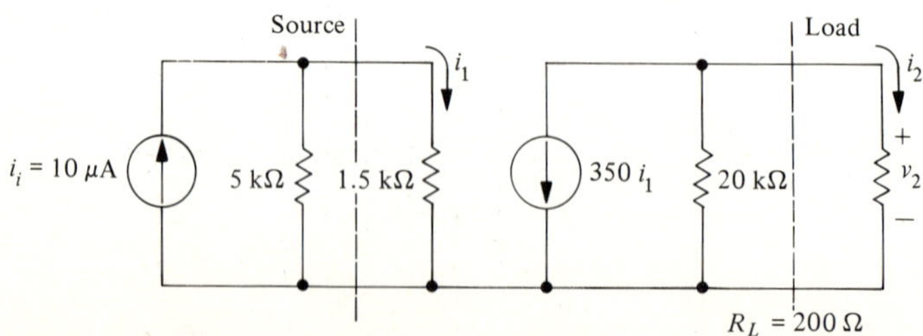

FIGURE 14.7
Circuit for Learning exercise 14.3-1.

$$i_2 = \frac{-v_2}{R_L} = \frac{-A_v v_1}{R_L} \qquad (14.4\text{-}2)$$

EXAMPLE 14.4-1 Ideal Voltage Amplifier

In a certain ideal amplifier, the amplification factor is $A_v = 150$, the load is $R_L = 600\ \Omega$, and the input signal is 10 mV. Find

a. The output voltage v_2, current i_2, and power P_2

b. Power gain

Solution

a. From Eq. (14.4-1)

$$v_2 = A_v v_1 = (150)(0.01\text{ V}) = 1.5\text{ V}$$

From Eq. (14.4-2)

$$i_2 = -\frac{v_2}{R_L} = -\frac{1.5\text{ V}}{600\ \Omega} = -2.5\text{ mA}$$

The negative sign simply indicates that the actual current is flowing out of the upper output terminal. The output power is

$$P_2 = \frac{v_2^2}{R_L} = \frac{(1.5\text{ V})^2}{600\ \Omega} = 3.75\text{ mW}$$

b. The input power is zero since $i_1 = 0$. Thus the power gain is infinite for the ideal voltage amplifier.

• • •

Current Amplifiers

Amplifiers that utilize junction transistors as amplifying elements are best modeled by current amplifiers. The ideal current amplifier and its *vi*

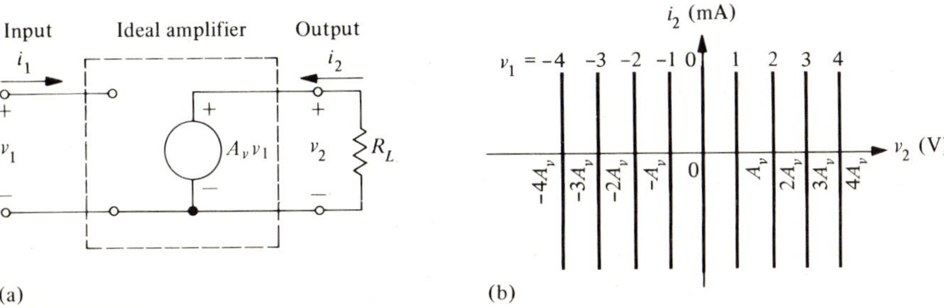

FIGURE 14.8
Ideal voltage amplifier. (a) Circuit. (b) *vi* characteristic.

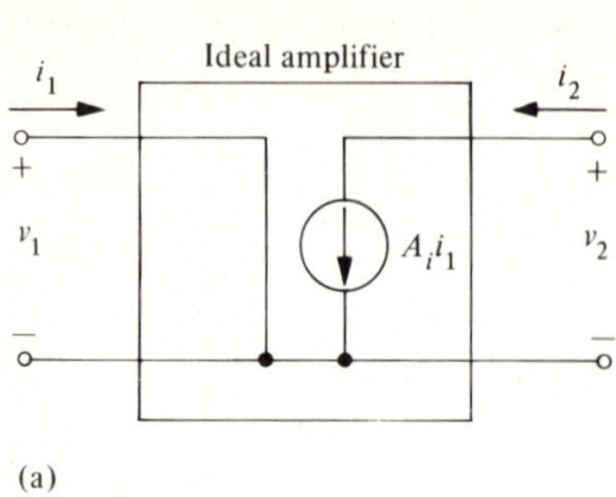

(a)

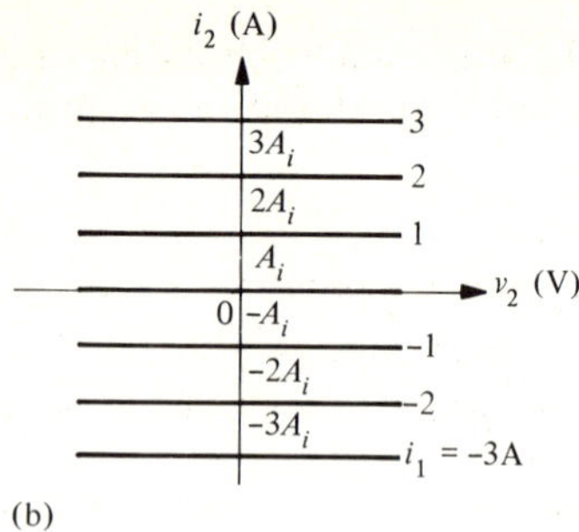

(b)

FIGURE 14.9
Ideal current amplifier. (a) Circuit. (b) *vi* characteristic.

characteristic are shown in Fig. 14.9. Note that the ideal current amplifier has a short circuit ($R_i = 0$) across its input terminals, in contrast to the open-circuited input ($R_i = \infty$) of the ideal voltage amplifier.

From the *vi* characteristic we observe that the current from the controlled source is A_i amperes for each ampere into the input terminals so that

$$A_i = \frac{i_2}{i_1} \tag{14.4-3}$$

where A_i is the current amplification factor. The output current is independent of the output voltage; it depends only on input current according to Eq. (14.4-3). The output voltage will depend on the circuit connected to the output terminals.

EXAMPLE 14.4-2 Ideal Current Amplifier

An ideal current amplifier is characterized by a current-amplification factor $A_i = 350$ and a load $R_L = 50\ \Omega$. The input current is $i_1 = 1\ \mu A$. Find

a. Output current i_2, voltage v_2, and power P_2
b. Power gain

Solution

a. From Eq. (14.4-3)

$$i_2 = A_i i_1 = (350)(10^{-6}\ A) = 0.35\ mA$$

Then

$$v_2 = -R_L i_2 = -(0.05\ k\Omega)(0.35\ mA) = -0.0175\ V = -17.5\ mV$$
$$P_2 = R_L i_2^2 = (50\ \Omega)(0.35\ mA)^2 = 6.31\ \mu W$$

b. Since the input power is zero because $R_i = 0$, the power gain is infinite.

• • •

In the next section we add elements to account for some of the nonideal characteristics of actual amplifiers.

• • •

LEARNING EXERCISES FOR SEC. 14.4

1. An ideal voltage amplifier has an amplification factor $A_v = -375$, a load resistance of 2.2 kΩ, and the input voltage is 5 mV. Find v_2 and i_2.

2. An ideal current amplifier has an amplification factor $A_i = -25$ and a load resistance $R_L = 2.3\ \Omega$. The input current is 2.8 A. Find i_2 and P_2.

Ans. −70; 0.9; −1.9; 11.3

• • •

14.5 THE EFFECT OF INPUT AND OUTPUT RESISTANCE

Voltage Amplifier Input Resistance

An important characteristic of any voltage amplifier is its input resistance, that is, the resistance seen by the signal source. For the ideal amplifier the input resistance is infinite, that is, an open circuit. Practical voltage amplifiers always have finite input resistance, typically ranging from kilohms to several tens of megohms. Figure 14.10 shows the circuit diagram of a voltage amplifier with finite input resistance R_i. Included in the diagram is a practical signal source v_i with Thevenin resistance R_s driving the amplifier. Note carefully that the controlled source amplifies the input voltage v_1. This will always be less than the signal source voltage v_i because of the voltage divider network formed by R_s and R_i. This effect is considered in the next example.

EXAMPLE 14.5-1 Effect of Input Resistance

The amplifier of Example 14.4-1 has $A_v = 150$ and $R_L = 600\ \Omega$. Assume that its input resistance is 2 kΩ and that the 10-mV signal source has 600-Ω internal resistance. Find

a. The overall gain* $G_{vT} = 20 \log (v_2/v_i)$ in decibels.
b. v_2 (compare with Example 14.4-1)

Solution

a. From the circuit of Fig. 14.10

$$\frac{v_2}{v_1} = A_v \qquad \text{and} \qquad \frac{v_1}{v_i} = \frac{R_i}{R_s + R_i}$$

*From here on we will use the subscript T to indicate total, or overall, gain.

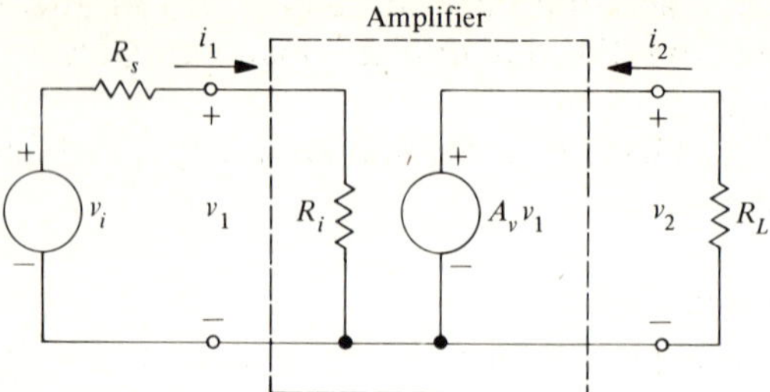

FIGURE 14.10
Voltage amplifier with input resistance.

Thus, using the chain rule

$$A_{vT} = \frac{V_2}{V_i} = \left(\frac{V_2}{V_1}\right)\left(\frac{V_1}{V_i}\right) = A_v\left(\frac{R_i}{R_s + R_i}\right) = \frac{(150)(2000)}{(600 + 2000)} = 115$$

To convert to decibels, if a scientific calculator is not available,

$$\begin{aligned} G_{vT} &= 20 \log (115) = 20 \log (1.15 \times 10^2) \\ &= 20 \log 1.15 + 20 \log 10^2 \\ &= 1.2 + 40 = 41.2 \text{ dB} \end{aligned}$$

b. $V_2 = A_{vT}V_i = (115)(0.01 \text{ V}) = 1.15 \text{ V}$

This is smaller than the 1.5-V output voltage in Example 14.4-1 where the input resistance is infinite. Otherwise, the circuits are the same.

• • •

Voltage Amplifier Output Resistance

Another important characteristic of real voltage amplifiers is that they do not have zero output resistance. Figure 14.11 shows the circuit diagram of a

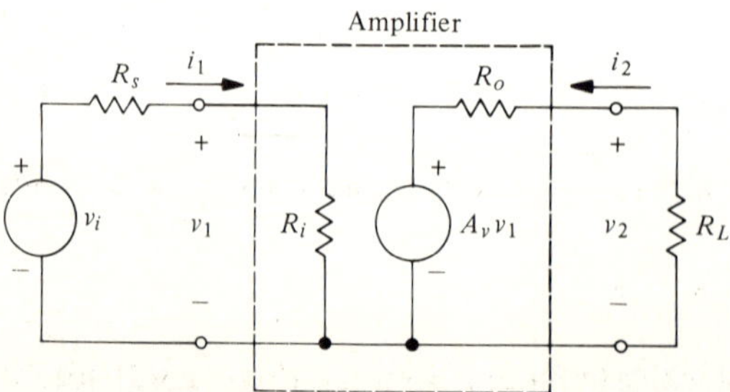

FIGURE 14.11
Voltage amplifier with input and output resistance.

voltage amplifier with both input and output resistance. Both of these elements tend to reduce the gain from the theoretical maximum value A_v, which is the gain when both are not present.

The amplifier model in Fig. 14.11 is adequate for many practical calculations of amplifier performance.

EXAMPLE 14.5-2 Effect of Output Resistance

In the amplifier of Example 14.5-1 the output resistance is $R_o = 200\ \Omega$. Find

a. $G_{vT} = 20 \log (v_2/v_i)$ in decibels.

b. v_2 (compare with Examples 14.4-1 and 14.5-1).

Solution

a. From Fig. 14.11 we have for the output circuit

$$\frac{v_2}{v_1} = \frac{A_v R_L}{R_o + R_L} \tag{14.5-1}$$

For the input circuit

$$\frac{v_1}{v_i} = \frac{R_i}{R_s + R_i} \tag{14.5-2}$$

The overall gain A_{vT} is

$$\begin{aligned} A_{vT} &= \frac{v_2}{v_i} = \left(\frac{v_2}{v_1}\right)\left(\frac{v_1}{v_i}\right) = A_v \left(\frac{R_L}{R_o + R_L}\right)\left(\frac{R_i}{R_s + R_i}\right) \\ &= 150\left(\frac{600}{200 + 600}\right)\left(\frac{2000}{600 + 2000}\right) \doteq 85.6 \end{aligned} \tag{14.5-3}$$

In decibels

$$\begin{aligned} G_{vT} &= 20 \log(86.5) = 20 \log(8.65 \times 10) \\ &= 20 \log 8.65 + 20 \log 10 \\ &= 18.7 + 20 = 38.7 \text{ dB} \end{aligned}$$

b. The output voltage is

$$v_2 = A_{vT} v_i = (86.5)(0.01) = 0.865 \text{ V} \tag{14.5-4}$$

As expected, the output voltage is further reduced compared to the previous examples due to the output resistance.

• • •

Note that the "chain" type of calculation done in Eq. (14.5-3) serves several purposes. First it breaks the complicated transfer function v_2/v_i into two separate simpler transfer functions v_2/v_1 and v_1/v_i, which are multiplied together. These simple transfer functions usually result when the input and output circuits are isolated as they are in this model. Second, it places clearly in evidence those factors that exert most influence on the overall gain. This is important in design, where tradeoffs often have to be made in order to meet specifications.

Current Amplifier Input and Output Resistance

Practical current amplifiers can often be modeled with sufficient accuracy by an ideal current amplifier with input and output resistance as shown in Fig. 14.12a. The graphical characteristics of this circuit are shown in Fig. 14.12b. These should be compared with Fig. 14.9b. The effect of the output resistance is to cause each of the family of straight lines to have a slope equal to the reciprocal of the output resistance. Typical gain calculations are illustrated in the following example.

EXAMPLE 14.5-3 Current Amplifier

In the circuit of Fig. 14.12, $R_s = 50\ \Omega$, $R_i = 10\ \Omega$, $A_i = 350$, $R_o = 10\ \text{k}\Omega$, $R_L = 1\ \text{k}\Omega$, and $i_i = 5\ \mu\text{A}$. Find

a. Overall current gain $G_{iT} = 20 \log(i_2/i_i)$ in decibels
b. i_2
c. Power gain in decibels.

Solution

a. For overall gain A_{iT} we use a "chain" equation as follows:

$$A_{iT} = \frac{i_2}{i_i} = \left(\frac{i_2}{i_1}\right)\left(\frac{i_1}{i_i}\right)$$

where, using the current divider rule

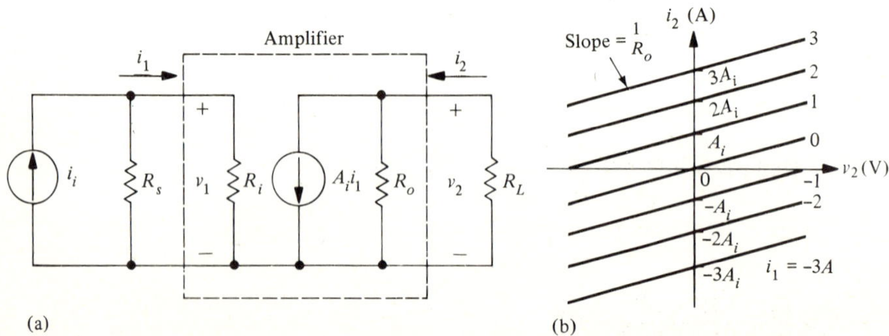

FIGURE 14.12
Current amplifier with input and output resistance. (a) Circuit. (b) Output characteristic.

$$\frac{i_2}{i_1} = A_i\left(\frac{R_o}{R_o + R_L}\right)$$

and

$$\frac{i_1}{i_i} = \left(\frac{R_s}{R_s + R_i}\right)$$

Thus

$$A_{iT} = A_i\left(\frac{R_o}{R_o + R_L}\right)\left(\frac{R_s}{R_s + R_i}\right)$$

$$= 350\left(\frac{10}{10 + 1}\right)\left(\frac{50}{50 + 10}\right)$$

$$= 265$$

In decibels,

$$G_{iT} = 20 \log 265 = 20 \log(2.65 \times 10^2) = 20 \log 2.65 + 20 \log 10^2$$
$$= 8.5 + 40 = 48.5 \text{ dB}$$

b. $i_2 = A_{iT} i_i = (265)(5) = 1325\ \mu\text{A} \approx 1.32$ mA.

c. The power dissipated in the load is

$$P_2 = R_L i_2^2 = (10^3)(1.33 \times 10^{-3})^2 = 1.77 \times 10^{-3} \text{ W} = 1.77 \text{ mW}$$

The power delivered to the amplifier is

$$P_1 = R_i i_1^2$$

where

$$i_1 = i_i \frac{R_s}{R_s + R_i} = (5)\left(\frac{50}{50 + 10}\right) = 4.17\ \mu\text{A}$$

Thus

$$P_1 = (10)(4.17 \times 10^{-6})^2 = 1.74 \times 10^{-10} \text{ W} = 0.174 \text{ nW}$$

and

$$A_p = \frac{P_2}{P_1} = \frac{1.77 \times 10^{-3}}{1.74 \times 10^{-10}} \approx 10^7$$

$$G_p = 10 \log 10^7 = 70 \text{ dB}$$

• • •

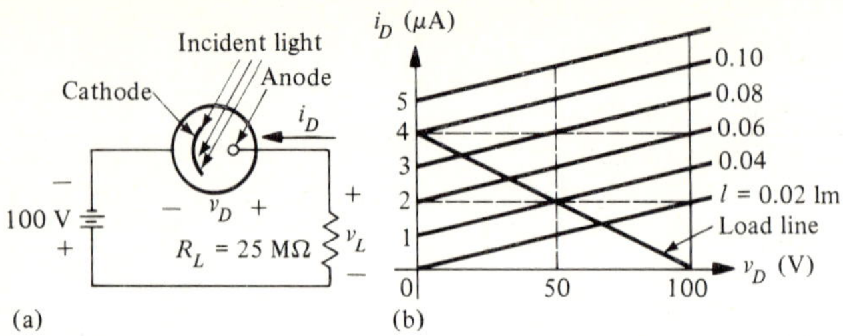

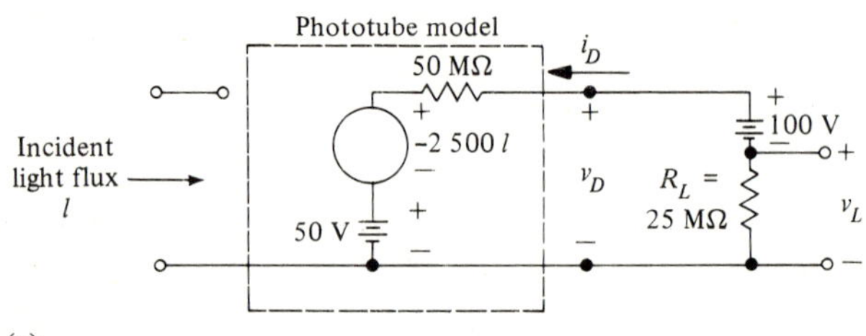

FIGURE 14.13
Modeling a phototube. (a) Circuit. (b) *vi* characteristic. (c) Derived circuit model.

*EXAMPLE 14.5-4 Controlled Source Circuit Model for a Phototube**

This example illustrates the utility of the controlled source when it is desired to find a circuit model for a device that is not completely electrical. We are going to find a model for a *phototube*, a light-sensitive electronic vacuum tube in which current flow depends not only on the voltage, but also on the intensity of the light incident on the cathode, or light-gathering surface of the internal structure of the tube. This light causes electrons to be emitted from the cathode, which are attracted to the anode and thus cause a current to flow. A schematic diagram of a phototube connected in series with a battery and load resistor is shown in Fig. 14.13a.

A set of linearized characteristics for a typical phototube is shown in Fig. 14.13b and we see that they are very similar to the *vi* characteristics of the amplifier shown in Fig. 14.12. The slope can be accounted for by including an output resistance in series with the controlled source. The parameter in the curve family is the light flux, measured in lumens (abbreviated lm).

Our aim is to find a model for the phototube, and also to predict the effects of changes in light intensity on the load voltage v_L. We begin by noting that in the first quadrant, the phototube voltage is a linear function of both the current and light intensity. Thus we can write

$$v_D = Al + Bi_D + C \tag{14.5-5}$$

where l is the incident light intensity in lumens. The coefficients A, B, and C are found from the *vi* characteristics by noting that if we keep i_D constant at a value of

*This example may be omitted without loss of continuity.

2 μA, then

$$A = \frac{\Delta v_D}{\Delta l} = \frac{100 \text{ V} - 0 \text{ V}}{0.02 \text{ lm} - 0.06 \text{ lm}} = \frac{100 \text{ V}}{-0.04 \text{ lm}} = -2500 \text{ V/lm} \tag{14.5-6}$$

If we keep l constant at 0.04 lm, then

$$B = \frac{\Delta v_D}{\Delta i_D} = \frac{100 \text{ V} - 0 \text{ V}}{3\ \mu\text{A} - 1\ \mu\text{A}} = \frac{100 \text{ V}}{2\ \mu\text{A}} = 50 \text{ M}\Omega \tag{14.5-7}$$

Now we have

$$v_D = -2500l + 50 \times 10^6 i_D + C$$

In order to find the value of C, we can evaluate this equation at any point on the vi characteristic. For convenience, we use the point $i_D = 1\ \mu$A, $v_D = 50$ V, $l = 0.02$ lm. This yields

$$50 = (-2500)(0.02) + (50 \times 10^6)(1 \times 10^{-6}) + C$$

Solving for C, we find

$$C = 50 \text{ V}$$

Finally, then, Eq. (14.5-5) becomes

$$v_D = -2500l + 50i_D + 50 \tag{14.5-8}$$

With l in lumens and i_D in microamperes, v_D is in volts. The student should check several points on the vi characteristic to make sure this is correct.

Equation (14.5-8) suggests that the phototube voltage is made up of the *sum* of three component voltages and can thus be modeled by a three-element series circuit. The first voltage, $-2500l$, is represented by a controlled source whose controlling variable is the light intensity l. The second, $50i_D$, represents the voltage drop across a 50-MΩ resistor, and a third, a 50-V battery. The series circuit that satisfies Eq. (14.5-8) and is our desired circuit model for the phototube is shown in Fig. 14.13c along with the load resistance R_L and 100-V battery of the original circuit.

In order to predict the effect of changes in light level on the load voltage, we construct the load line (see Chap. 5). This represents the Thevenin characteristic of the circuit connected to the phototube, that is, the load resistance R_L and the 100-V battery. The load line intersects the v_D axis at 100 V and the i_D axis at 100 V/25 MΩ = 4 μA. This is shown in Fig. 14.13b. If the ambient light level is 0.04 lm, then the operating point (Sec. 5.4) is at $i_D = 2\ \mu$A, $v_D = 50$ V. To find the load voltage under ambient light conditions we note that when $i_D = 2\ \mu$A, $v_L = -i_D R_L = -(2\ \mu\text{A})(25\ \text{M}\Omega) = -50$ V.

In order to find how V_L changes when the light level changes from its ambient value, we make use of the circuit model of Fig. 14.13c. Using the voltage divider formula, and being very careful with reference polarities, we find

$$v_L = \frac{25\ \text{M}\Omega}{50\ \text{M}\Omega + 25\ \text{M}\Omega}(-50 - 2500l) = -\frac{1}{3}(50 + 2500l) \tag{14.5-9}$$

The change in v_L in terms of the change in l is found from this equation as

$$\frac{\Delta v_L}{\Delta l} = -\frac{2500}{3} = -833\ \text{V/lm}$$

This gives the *change* in v_L from its ambient value per lumen of incident light. For example, if $\Delta l = 0.01$ lm, that is, if l increases from the ambient value of 0.04 lm to 0.05 lm then $\Delta v_L = -8.33$ V.

The technique presented in this section can be used to model and predict behavior for any device whose characteristics can be linearized without departing too much from the actual characteristics.

• • •

LEARNING EXERCISES FOR SEC. 14.5

1. In the voltage amplifier of Fig. 14.11, $v_i = 6\ \mu\text{V}$, $R_s = 50\ \Omega$, $R_i = 1\ \text{k}\Omega$, $A_v = 10^4$, $R_o = 600\ \Omega = R_L$. Find v_2 and G_{vT}.

2. In the current amplifier of Fig. 14.12, $i_i = 2\ \mu\text{A}$, $R_s = 50\ \text{k}\Omega$, $R_i = 5\ \text{k}\Omega$, $A_i = 10^3$, $R_o = 2\ \text{k}\Omega$, $R_L = 1\ \text{k}\Omega$. Find i_2 and G_{iT}.

Ans. 74; 55.6; 29; 1.21

• • •

14.6 HYBRID PARAMETERS

When we considered two-terminal networks, we found that their terminal behavior could be completely characterized by a single linear equation involving the voltage and current at the terminals as variables. When we consider a two-port network, there are four variables to consider; voltage and current at both input and output terminals. Therefore, we would expect that the mathematical description of the behavior of the two-port would involve more than one simple equation. In fact, two equations are required to completely specify the terminal behavior of the two-port. These two equations can be written in a number of different ways. For example, we can write

$$i_1 = f_1(v_1, v_2)$$
$$i_2 = f_2(v_1, v_2)$$

or

$$v_1 = f_3(i_1, i_2)$$
$$v_2 = f_4(i_1, i_2)$$

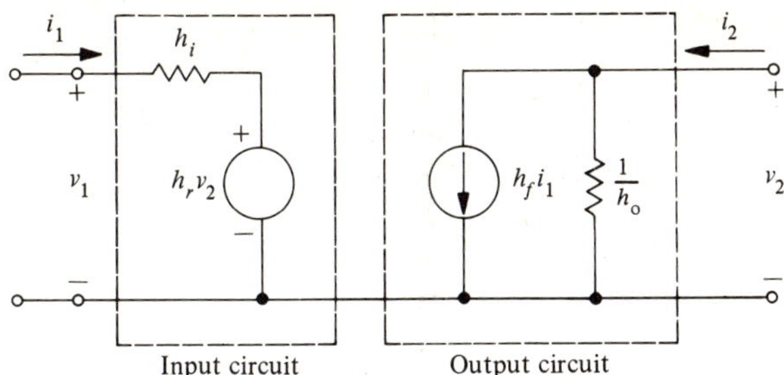

FIGURE 14.14
Hybrid equivalent circuit.

or the variables can be mixed. One popular version of the two equations uses output current i_2 and input voltage v_1 as the dependent variables with input current and output voltage the independent variables. The resulting set of equations, called "hybrid" equations, turns out to be particularly appropriate for transistor models.* These equations for a linear two-port with no independent sources are

$$v_1 = h_i i_1 + h_r v_2 \tag{14.6-1a}$$
$$i_2 = h_f i_1 + h_o v_2 \tag{14.6-1b}$$

The subscripts identify the physical nature of the parameter, and can best be explained by reference to the hybrid equivalent circuit shown in Fig. 14.14. This circuit is derived from the hybrid equations in the following way: Consider Eq. (14.6-1a), which leads to the input circuit. Each term is a voltage and the equation is an expression of KVL. The first term on the right-hand side indicates that h_i is dimensionally a resistance and the term represents the voltage drop across the resistance due to current i_1 flowing through it. The second term represents a voltage-controlled voltage source, with h_r the factor by which v_2 is multiplied. Each of these terms corresponds to one of the elements in the input circuit.

The output circuit is derived from Eq. (14.6-1b), which is an expression of KCL. The first term on the right represents a current-controlled current source where h_f is the factor by which i_1 is multiplied. In the second term h_o must have the dimensions of conductance since the term represents a current.

Physical Interpretation of Parameters

The physical meaning of the h parameters can be explained by considering both the equations and the equivalent circuit. For example, consider h_i.

*In most transistor amplifier circuits the signals are a combination of ac and dc. For clarity, we will assume that the signals are ac only. The actual mixture of ac and dc will be studied in electronics courses.

From Eq. (14.6-1a) we can write

$$h_i = \left.\frac{v_1}{i_1}\right|_{v_2=0} \tag{14.6-2}$$

This indicates that h_i is the input resistance with the output short-circuited ($v_2 = 0$). The subscript i stands for input. Similarly, h_r is dimensionless and represents the *reverse* voltage ratio with the input open-circuited ($i_1 = 0$). Definitions for all four parameters are as follows:

$$h_i = \left.\frac{v_1}{i_1}\right|_{v_2=0} = \text{short-circuit } \textit{input} \text{ resistance} \tag{14.6-3}$$

$$h_r = \left.\frac{v_1}{v_2}\right|_{i_1=0} = \text{open-circuit } \textit{reverse} \text{ voltage gain} \tag{14.6-4}$$

$$h_f = \left.\frac{i_2}{i_1}\right|_{v_2=0} = \text{short-circuit } \textit{forward} \text{ current gain} \tag{14.6-5}$$

$$h_o = \left.\frac{i_2}{v_2}\right|_{i_1=0} = \text{open-circuit } \textit{output} \text{ conductance} \tag{14.6-6}$$

These definitions, when applied to the current amplifier of Fig. 14.12, yield the following equivalences:

$$h_i = R_i$$

$$h_r = 0$$

$$h_f = A_i$$

$$\frac{1}{h_o} = R_o$$

Thus the hybrid model fits the current amplifier perfectly.

The equivalent circuit of Fig. 14.14 is useful for several reasons: First, it isolates the input and output circuits, their interaction being accounted for by the two controlled sources; second, the two parts of the circuit are in a form that makes it easy to take into account source and load circuits.

The definitions also suggest methods of measurement for the parameters. For example, Eq. (14.6-5) indicates that h_f can be measured by placing a short circuit* across the output (so that $v_2 = 0$), applying a small known input

*This cannot always be done in practice since it may cause physical damage. For ac signals a large capacitance can often be used as a short circuit since it has no effect on any dc which may be present and it is chosen to have a small reactance at frequencies of interest.

current i_1 and measuring the resulting output current i_2. The parameter h_f is then the ratio of these currents.

The hybrid equivalent circuit is used in a number of ways. It can be used like the Thevenin and Norton equivalent circuits, that is, to reduce a complicated circuit to an equivalent, easier to handle, form. For the hybrid circuit the equivalence is at both input and output terminals. Another use is for modeling of new or unknown devices. For example, if we are given an amplifier or other two-port in a "black box" where we have no idea of the circuitry inside the black box, we can perform the measurements indicated previously in order to determine the four h parameters, and thus have a circuit model for the black box. This model can then be used to determine the effects of different sources and loads or for any other purpose involving the behavior of the black box at the two pairs of terminals. Clearly, we cannot use our hybrid model to predict the behavior of the individual circuit elements inside the box.

EXAMPLE 14.6-1 Finding the h Parameters for a Given Circuit

a. Find the hybrid parameters for the voltage divider circuit of Fig. 14.15a.
b. Verify the voltage divider formula for the case when $i_2 = 0$.

Solution

a. The simplest way to find the parameters is to apply the definitions in Eq. (14.6-3) through (14.6-6).

In order to find h_i we short-circuit the output as shown in Fig. 14.15b. This short-circuits resistor R_2 so that

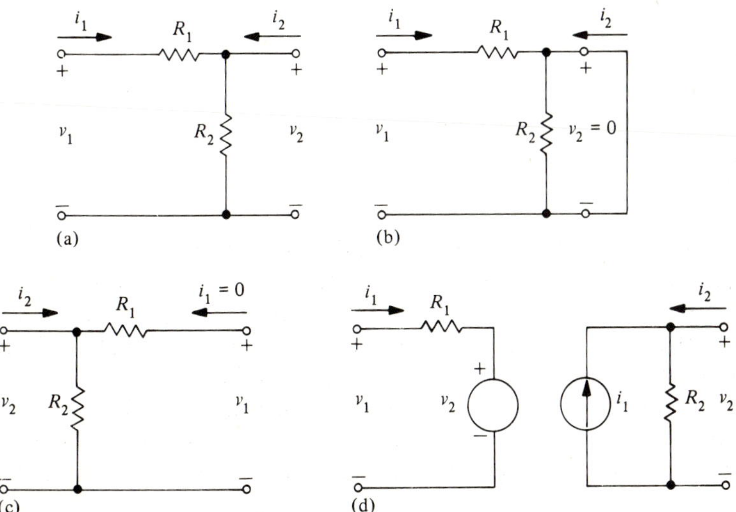

FIGURE 14.15
Circuits for Example 14.6-1 (a) Voltage divider. (b) Circuit for $v_2 = 0$. (c) Circuit for $i_1 = 0$. (d) Hybrid circuit.

$$h_i = \left.\frac{v_1}{i_1}\right|_{v_2=0} = R_1 \tag{14.6-7}$$

To find h_r we turn the circuit around as shown in Fig. 14.15c. Since $i_1 = 0$ there is no voltage drop across R_1 and we must have $v_1 = v_2$. Then

$$h_r = \left.\frac{v_1}{v_2}\right|_{i_1=0} = 1 \tag{14.6-8}$$

For h_f we again consider Fig. 14.15b. Since no current flows through R_2 we must have $i_2 = -i_1$ so that

$$h_f = \left.\frac{i_2}{i_1}\right|_{v_2=0} = -1 \tag{14.6-9}$$

Finally, for h_o we again consider Fig. 14.15c. With the input terminals open we have simply

$$h_o = \left.\frac{i_2}{v_2}\right|_{i_1=0} = \frac{1}{R_2} \tag{14.6-10}$$

The hybrid equations (14.6-1) become

$$v_1 = R_1 i_1 + v_2 \tag{14.6-11}$$

$$i_2 = -i_1 + \frac{v_2}{R_2} \tag{14.6-12}$$

The equivalent circuit is shown in Fig. 14.15d. The student will note that it is more complicated than the original circuit but has the advantage that input and output are isolated.

b. The voltage divider relation can be verified by setting $i_2 = 0$ in Eq. (14.6-12). Then

$$i_1 = \frac{v_2}{R_2}$$

This relation is substituted into Eq. (14.6-11):

$$v_1 = \frac{R_1 v_2}{R_2} + v_2 = v_2\left(\frac{R_1 + R_2}{R_2}\right)$$

Finally

$$\frac{v_2}{v_1} = \frac{R_2}{R_1 + R_2} \tag{14.6-13}$$

• • •

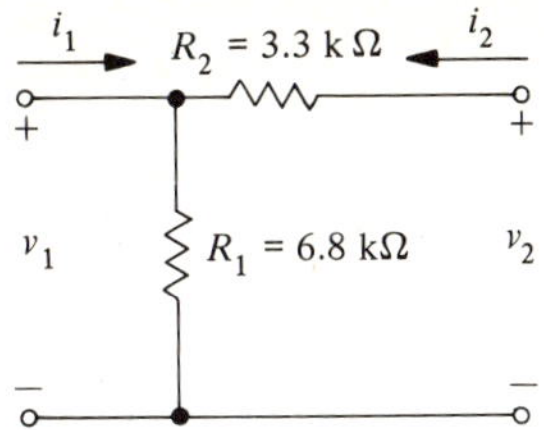

FIGURE 14.16
Circuit for Learning exercise 14.6-1.

LEARNING EXERCISE FOR SEC. 14.6

1. Find the hybrid parameters for the circuit of Fig. 14.16.

Ans. 0.67; 2.2; −0.67; 99

• • •

14.7 CASCADED TWO-PORTS

It often becomes necessary to connect the output of one two-port to the input of another as shown in Fig. 14.17a. They are then said to be connected in *cascade*. If each individual two-port is the model for a practical current amplifier as discussed in Sec. 14.5, then the cascade takes the form shown in Fig. 14.17b, where a source and load have been added to the cascade. Gain calculations are carried out using a chain equation as illustrated in the following example.

EXAMPLE 14.7-1 Cascaded Amplifiers

In the circuit of Fig. 14.17b both amplifiers have $h_i = 1\ \text{k}\Omega$, $1/h_o = 10\ \text{k}\Omega$, and $h_f = 150$. The source has $v_i = 10\ \mu\text{V}$ and $R_s = 500\ \Omega$. The load is $R_L = 500\ \Omega$. Find the output voltage.

Solution

We first find the voltage gain $A_{vT} = v_4/v_i$. This is done using a chain equation:

$$A_{vT} = \frac{v_4}{v_i} = \left(\frac{v_4}{i_3}\right)\left(\frac{i_3}{i_2}\right)\left(\frac{i_2}{i_1}\right)\left(\frac{i_1}{v_i}\right) \tag{14.7-1}$$

The four individual terms in the chain are:

1. Using Ohm's law with all resistors in kilohms

$$\frac{v_4}{i_3} = -h_{f2}\left(\frac{1}{h_{o2}} \parallel R_L\right) = -150(10 \parallel 0.5) = \frac{-(150)(5)}{(10.5)}\ \text{k}\Omega$$

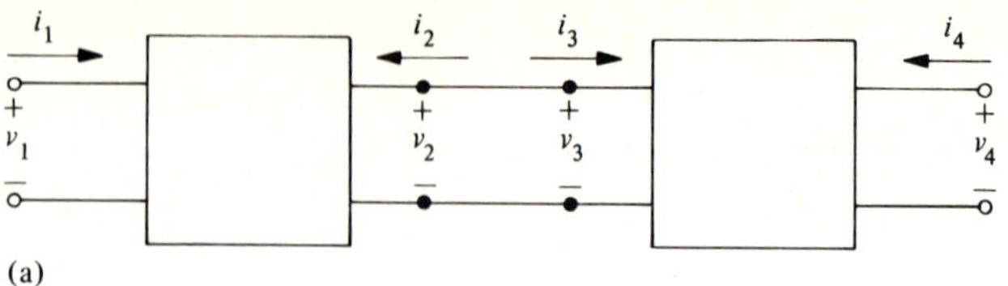

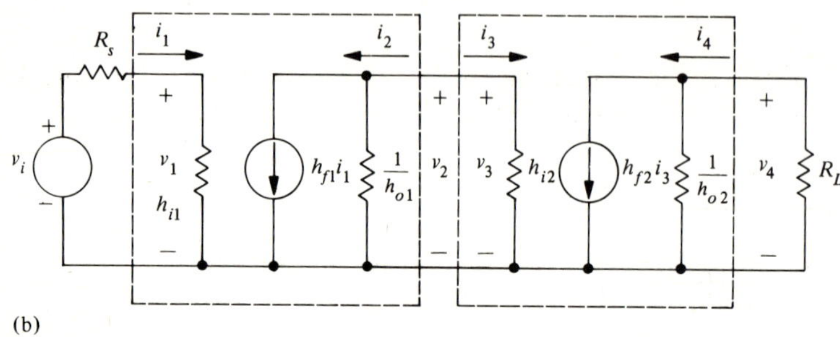

FIGURE 14.17
Two-port networks. (a) Cascade connection. (b) Cascade of current amplifiers.

2. By inspection, $i_3 = -i_2$ so that $i_3/i_2 = -1$.
3. Using the current divider formula, and taking careful note of current directions,

$$\frac{i_2}{i_1} = h_{f1}\left(\frac{1/h_{o1}}{h_{i2} + 1/h_{o1}}\right) = \frac{(150)(10)}{(11)}$$

4. Again using Ohm's law,

$$\frac{i_1}{v_i} = \frac{1}{R_s + h_{i1}} = \frac{1}{1.5\ \text{k}\Omega}$$

Substituting into Eq. (14.7-1) we obtain

$$A_{vT} = \frac{v_4}{v_i} = \frac{(-150)(5)(-1)(150)(10)}{(10.5)(11)(1.5)} = 6500$$

Finally, since $v_i = 10\ \mu\text{V}$,

$$\begin{aligned} v_4 &= A_{vT}v_i \\ &= (6.5)(10^3)(10)(10^{-6}) = 65 \times 10^{-3}\ \text{V} = 65\ \text{mV} \end{aligned}$$

• • •

Gain of Cascaded Amplifiers

When amplifiers are connected in cascade, the overall gain can often be found simply by adding the individual gains expressed in decibels. To

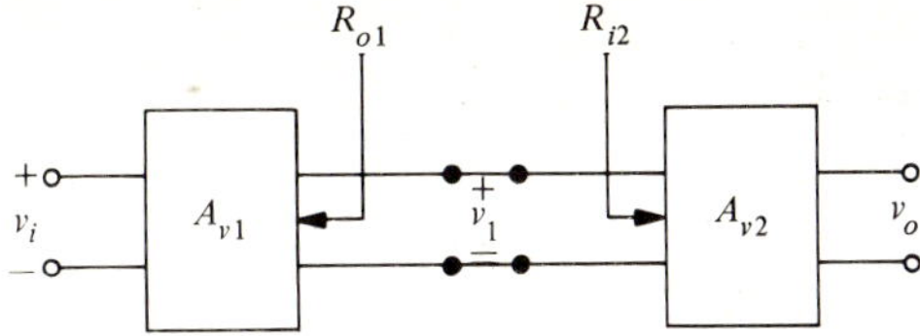

FIGURE 14.18
Cascade voltage amplifiers.

illustrate this, consider the system shown in Fig. 14.18 where two voltage amplifiers are connected in cascade. If $R_{i2} \gg R_{o1}$, we can write

$$A_{vT} = \frac{V_o}{V_i} \approx \left(\frac{V_o}{V_1}\right)\left(\frac{V_1}{V_i}\right) = A_{v1}A_{v2} \tag{14.7-2}$$

where A_{v1} and A_{v2} are the open-circuit individual gains.

Then, converting to decibels

$$\begin{aligned} G_{vT} &= 20 \log A_{vT} = 20 \log (A_{v1}A_{v2}) = 20 \log A_{v1} + 20 \log A_{v2} \\ &= G_{v1} + G_{v2} \end{aligned} \tag{14.7-3}$$

If the input resistance of the second amplifier were not much greater than the output resistance of the first, there would be a loss in gain due to the loading effect and the overall gain calculation would be more complicated.

EXAMPLE 14.7-2 Decibel Gain

Three identical voltage amplifiers are cascaded. Each has an open-circuit gain of 10, an input resistance of 10 kΩ, and an output resistance of 100 Ω. Find the overall gain.

Solution

Since $R_i \gg R_o$ between stages, we can use Eq. (14.7-3). Thus

$$G_{v1} = G_{v2} = G_{v3} = 20 \log 10 = 20 \text{ dB}$$

and

$$G_{vT} = G_{v1} + G_{v2} + G_{v3} = 3 \times 20 = 60 \text{ dB}$$

• • •

LEARNING EXERCISE FOR SEC. 14.7

1. Three stages in a phonograph preamplifier have voltage gains of 22, 168, and 3.2×10^3, respectively. Loading between states is negligible and the input voltage is 1.2 μV. Find A_{vT}, the output voltage, and G_{vT}.

Ans. 11.8×10^6; 14.2; 141

• • •

14.8 THE OP-AMP AND ITS APPLICATIONS

The term op-amp (short for *operational amplifier*) is applied to very high gain amplifiers used to implement a wide variety of linear and nonlinear *operations*. The different operations are obtained by simply changing a few external elements, such as resistors, capacitors, and diodes. A large variety of op-amps is available in integrated circuit form, and as many as four op-amps can be obtained in a single integrated circuit package. An example is shown in Fig. 14.19. Prices of these units range down to as little as $1. As a result, op-amps have become an important element in analog system design. They are used for such applications as signal processing, interfacing, instrumentation, and waveshaping. In this section we discuss the basic op-amp and some of its linear applications.

The Amplifier

The typical op-amp is a high-gain voltage amplifier designed to approximate an ideal voltage amplifier. This means that its input resistance is very high, often 50 MΩ or more, and its output resistance is very low, in the range of 50 to 100 Ω.

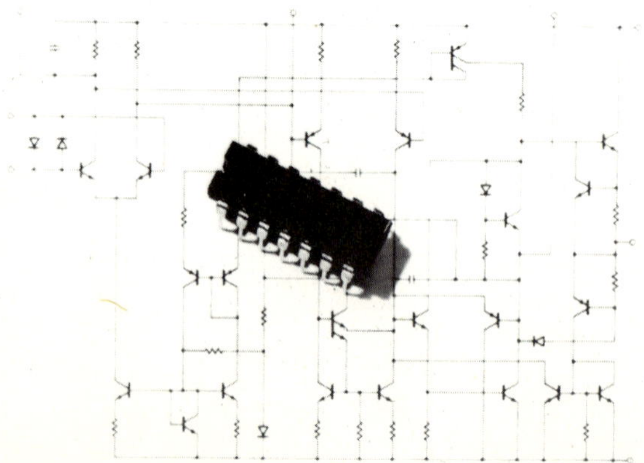

FIGURE 14.19
Integrated circuit op-amp (courtesy of Signetics Corp.).

In many systems the input voltage is connected between an input terminal and ground. The ground terminal is the reference for all of the voltages, both ac and dc, in the amplifier or system of which the amplifier is a part. This configuration is called a single-ended input, and it places constraints on the input to the amplifier because one side is grounded. An alternate configuration is called a differential input and consists of two input terminals, neither of which is grounded, to which the input voltage is applied. The circuit symbol for a differential input op-amp is shown in Fig. 14.20a. This is often called a double-ended input. The (−) input terminal is called the inverting input because the output produced by a signal applied to this terminal alone will be 180° out of phase with the input. The (+) terminal is called the noninverting input because the output due to this input alone will be in phase with the input.

The amplifier output is the amplification factor multiplied by the voltage difference between these two terminals. The differential input voltage is

$$v_d = v_2 - v_1 \tag{14.8-1}$$

and the output voltage is

$$v_o = A_d v_d = A_d(v_2 - v_1) \tag{14.8-2}$$

A_d is the differential voltage gain, an important characteristic of the op-amp. Typical values range from 10^5 to 10^6 (100 to 120 db).

The op-amp output can be single-ended if one output terminal is grounded, or doubled-ended if not. Most op-amps have single-ended

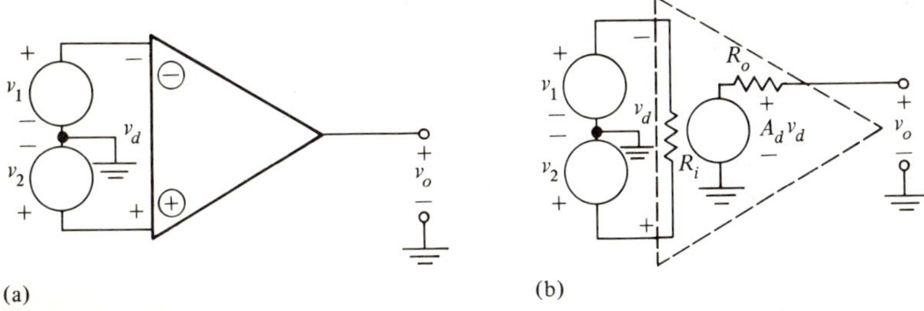

FIGURE 14.20
Operational amplifier. (a) Circuit symbol. (b) Equivalent circuit.

output and the equivalent circuit is shown in Fig. 14.20b. Typically, A_d is 10^5 to 10^6, R_i is 10 to 100 MΩ, and R_o is 50 to 100 Ω.

There are other features of the op-amp that have not been discussed and that are not indicated in the equivalent circuit. One important characteristic in this category is the amplifier's frequency response. The operational amplifier is basically a low-frequency device; its amplification is relatively constant from dc to a rather low cutoff frequency, typically 1 to 10 Hz. Above the cutoff frequency, the amplification decreases with increasing frequency. The frequency response does not have a sharp break; rather it is a gradual decrease in amplitude that becomes more rapid after the cutoff frequency.

Another important general characteristic of operational amplifiers is their *stability*. This refers to the constancy of their amplification. Early operational amplifiers for analog computers used a very expensive technique, called *chopper stabilization*, to achieve good stability. Modern solid-state techniques have made it possible to achieve relatively good stability without the need for chopper stabilization. This is one of the reasons for the great reduction in cost of these amplifiers. Chopper stabilization is still used where very high precision is needed.

Other amplifier characteristics of importance include drift, common-mode rejection ratio, and noise, but a discussion of all of these is beyond the scope of this text. Nor is it necessary for the following discussion.

The Linear Inverting Amplifier

The configuration shown in Fig. 14.21a and often called a *scaling* circuit functions as a linear amplifier. For this application resistors R_1 and R_2 are connected as shown and the noninverting terminal is connected to ground. The circuit is sometimes called a *scaler* because the output voltage is a *scaled* replica of the input voltage, that is, the output voltage is equal to the input voltage multiplied by a constant.

Overall Gain

In Fig. 14.21b we have drawn the equivalent circuit for the scaler. The transfer function of interest is the overall voltage gain, that is, the ratio of the output voltage to the input voltage,

$$A_{vT} = \frac{v_o}{v_i} \tag{14.8-3}$$

In order to find A_{vT} we begin by observing that the input current is

$$i_1 = \frac{v_i + v_d}{R_1} = \frac{v_i}{R_1} + \frac{v_d}{R_1} \tag{14.8-4}$$

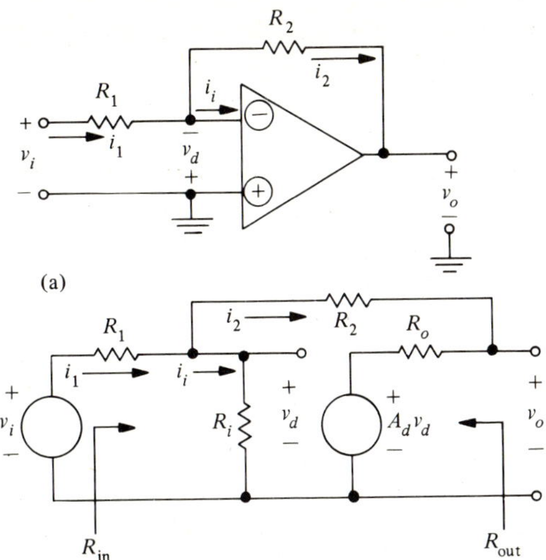

FIGURE 14.21
The negative scaler. (a) Circuit. (b) Equivalent circuit.

Consider the second term of this equation. The differential input voltage v_d is related to the output voltage by Eq. (14.8-2), which can be rearranged as

$$v_d = \frac{v_o}{A_d}$$

Thus if the differential gain A_d is 10^5 or more, v_d will be very small compared to the output voltage. In addition, since the overall gain A_{vT} is seldom more than 100, v_d will be very small compared to v_i. This condition is a key concept in op-amp analysis and it is called a *virtual* short circuit. Stated in different words, the amplifier input voltage v_d is forced to such a small value that it may be assumed to be zero for all practical purposes. With this assumption, the negative input terminal (the inverting input) is essentially at zero voltage and Eq. (14.8-4) becomes

$$i_1 \approx \frac{v_i}{R_1} \qquad (14.8\text{-}5)$$

The current through R_2 is

$$i_2 = \frac{-v_o - v_d}{R_2} \qquad (14.8\text{-}6)$$

Because $v_d \approx 0$ this simplifies to

$$i_2 \approx \frac{-v_o}{R_2} \tag{14.8-7}$$

At the junction of R_1 and R_2, KCL yields

$$i_1 - i_2 = i_i = \frac{-v_d}{R_i} \tag{14.8-8}$$

Since R_i is so large and v_d is essentially zero, $i_i \approx 0$ and this equation becomes

$$i_1 - i_2 \approx 0 \tag{14.8-9}$$

Substituting Eqs. (14.8-5) and (14.8-7) into Eq. (14.8-9),

$$\frac{v_i}{R_1} + \frac{v_o}{R_2} = 0$$

Rearranging, this becomes

$$\boxed{A_{vT} = \frac{v_o}{v_i} = -\frac{R_2}{R_1}} \tag{14.8-10}$$

This is the final result for the scaler. In words, it states that the overall gain is simply the negative of the ratio of the two external resistors. The negative sign leads to the name *inverting* amplifier. Observe that the overall gain is *independent* of the gain A_d of the internal amplifier as long as A_d is large. For most applications, if $A_d \geq 100\ R_2/R_1$, the assumption that $v_d \approx 0$ will be valid.

Input Resistance

The input resistance seen by the signal source v_i can be found from the equivalent circuit of Fig. 14.21b. It is

$$R_{in} = \frac{v_i}{i_1} = \frac{R_1 i_1 - v_d}{i_1} = R_1 - \frac{v_d}{i_1}$$

In a practical op-amp, $v_d \approx 0$ (virtual short circuit) so that the end of R_1 connected to the amplifier input may be considered grounded and as a result

$$R_{in} \approx R_1 \tag{14.8-11}$$

This result can be obtained by inspection if we note that the virtual short circuit at the amplifier input terminal causes the corresponding end of R_1 to be grounded so that the v_i source sees resistor R_1 connected to ground.

Output Resistance

Consider the problem of finding the output resistance of the circuit of Fig. 14.22a. One way to do this is to set $v_i = 0$, connect a test voltage source v_o at the output terminal, and measure the current drawn from the test source as shown in Fig. 14.22b. The output (Thevenin) resistance is then $R_{out} = v_o/i_o$. When this technique is applied to the circuit of Fig. 14.21b with $R_o \ll R_2$ and $R_i \gg R_1$ or R_2, the output resistance is found to be (Prob. 14.49)

$$R_{out} = \frac{R_o}{1 + R_1 A_d/(R_1 + R_2)} \tag{14.8-12}$$

This is usually much less than the R_o of the internal amplifier as shown by the following example.

EXAMPLE 14.8-1 Characteristics of a Negative Scaler

The configuration of Fig. 14.21a is to be designed to obtain a gain $A_{vT} = -50$ for use in a high-fidelity preamplifier. The op-amp has $R_i = 100$ kΩ, $A_d = 100\,000$, and $R_o = 100$ Ω. The input resistance R_{in} is to be 1 kΩ.

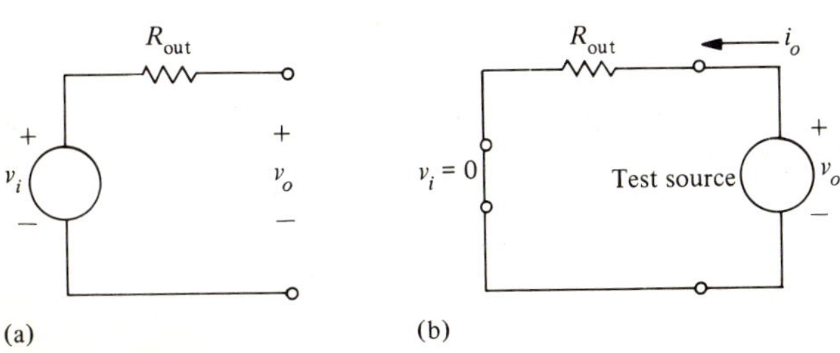

FIGURE 14.22
Output resistance. (a) Circuit. (b) Circuit with test source.

Find:

a. R_1
b. R_2
c. R_{out}

Solution

a. Since $R_{in} \approx R_1$ we choose $R_1 = 1\text{ k}\Omega$ to meet the input resistance specification.
b. The overall gain is to be $A_{vT} = -50$. Therefore, from Eq. (14.8-10),

$$A_{vT} = -50 = -\frac{R_2}{R_1} = -\frac{R_2}{1\text{ k}\Omega}$$

and

$$R_2 = 50\text{ k}\Omega$$

c. Using Eq. (14.8-12), the output resistance is

$$R_{out} = \frac{R_o}{1 + R_1A_d/(R_1 + R_2)} = \frac{100}{1 + [10^3)(10^5)/(10^3 + 50 \times 10^3)}$$
$$= 0.05\ \Omega$$

• • •

The Positive Scaler

The circuit of Fig. 14.23 scales the input by a *positive* constant, in contrast to the negative scaler just considered. Observe that the junction of R_1 and R_2 is still at the inverting terminal but the signal source v_i is connected to the noninverting terminal. If we assume an ideal op-amp with $R_o = 0$, R_i and $A_d \rightarrow \infty$, so that $v_d \approx 0$, then the equivalent circuit appears as shown in Fig.

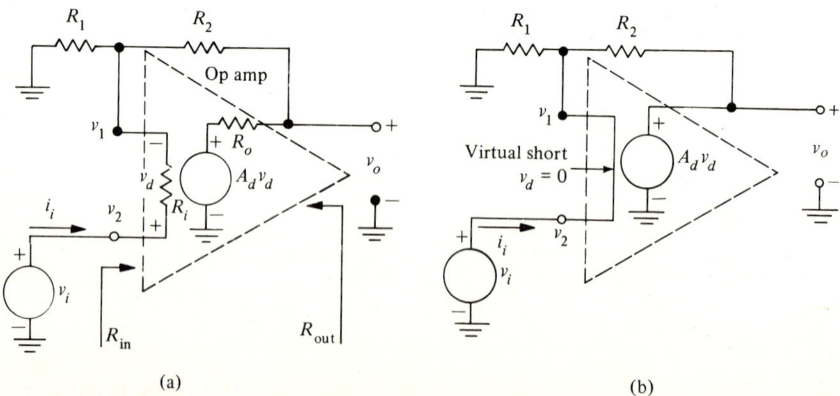

FIGURE 14.23
Positive scaler. (a) Circuit. (b) Equivalent circuit assuming ideal op-amp.

14.23b. From the circuit, with the virtual short in place, we see that

$$v_i = v_2 = v_1 \tag{14.8-13}$$

and since R_1 and R_2 form a voltage divider across the output voltage v_o,

$$v_1 = \frac{R_1}{R_1 + R_2} v_o \tag{14.8-14}$$

Substituting Eq. (14.8-14) into Eq. (14.8-13),

$$v_i = v_1 = \frac{R_1}{R_1 + R_2} v_o$$

and rearranging

$$\boxed{A_{vT} = \frac{v_o}{v_i} = \frac{R_1 + R_2}{R_1} = 1 + \frac{R_2}{R_1}} \tag{14.8-15}$$

This is the desired formula for the overall gain of this configuration. It is always positive and greater than unity. As with the inverting configuration, the gain depends only on the external resistors R_1 and R_2 and is approximately independent of A_d.

Voltage Follower

The gain can be made equal to unity by replacing R_2 by a short circuit and R_1 by an open circuit ($R_1 \rightarrow \infty$). This configuration, shown in Fig. 14.24, is called a *voltage follower* because $v_o = v_i$ and the output voltage "follows" the input voltage.

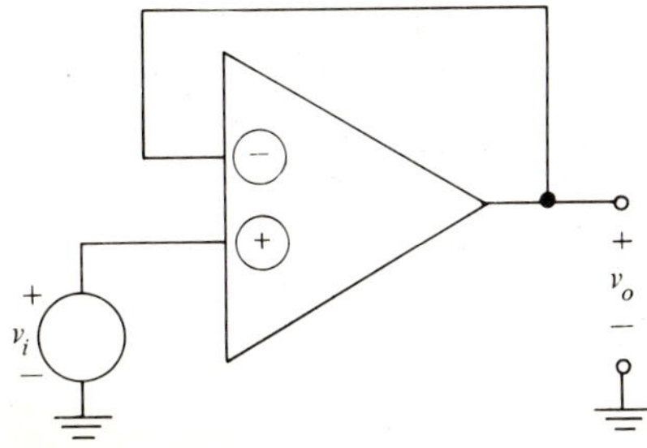

FIGURE 14.24
Voltage follower.

Input Impedance

From Fig. 14.23a the input impedance of the positive scaler is

$$R_{in} = \frac{v_i}{i_i} \tag{14.8-16}$$

To determine R_{in} we proceed as follows:

$$i_i = \frac{v_d}{R_i}$$

but $v_d = v_o/A_d$ so that

$$i_i = \frac{v_o}{A_d R_i} \tag{14.8-17}$$

From Eq. (14.8-15)

$$v_o = \left(1 + \frac{R_2}{R_1}\right) v_i$$

Substituting into Eq. (14.8-17),

$$i_i = \frac{1 + R_2/R_1}{A_d R_i} v_i$$

from which the input resistance is

$$\boxed{R_{in} = \frac{v_i}{i_i} = \frac{A_d R_i}{1 + R_2/R_1}} \tag{14.8-18}$$

In a good op-amp, A_d is large so that R_{in} is large.

Output Impedance

In order to find the output impedance we replace the signal source v_i by a short circuit. The resulting equivalent circuit is then identical to the circuit used to calculate R_{out} for the negative scaler. Therefore, R_{out} can be found from Eq. (14.8-12).

EXAMPLE 14.8-2 Positive Scaler

A positive scaler is to be designed for use in an automobile cassette deck. The overall gain is to be $A_{vT} = 50$. The op-amp is characterized by the values $A_d = 10^5$, $R_i = 100\ \text{k}\Omega$, and $R_o = 100\ \Omega$. If R_1 is 1 kΩ, find

a. R_2
b. R_{in}
c. R_{out}

Solution

a. The overall gain is

$$A_{vT} = 1 + \frac{R_2}{R_1}$$

Substituting the given values

$$50 = 1 + \frac{R_2}{1\ \text{k}\Omega}$$

from which

$$R_2 = 49\ \text{k}\Omega$$

b. The input resistance is

$$R_{in} = \frac{A_d R_i}{1 + R_2/R_1} = \frac{(10^5)(10^5\ \Omega)}{50} = 0.02 \times 10^9\ \Omega = 0.02\ \text{G}\Omega$$

c. The output impedance is the same as that of the negative scaler of Example 14.8-1, $R_{out} = 0.05\ \Omega$.

• • •

From this example we see that the positive scaler has extremely high input resistance and very low output resistance. These characteristics make it useful as a "buffer" amplifier, which is used between a source and a load as shown in Fig. 14.25. With the buffer in the circuit, changes in the load cannot be reflected back to the source.

The Summing Amplifier

The amplifier shown in Fig. 14.26 accepts any number of inputs and provides an output voltage proportional to the linear sum of the input voltages. The equation for the output voltage can be derived by observing that a virtual short circuit exists across the op-amp input terminals so that $v_d = i_i = 0$. As a consequence, KCL applied to the inverting input terminal

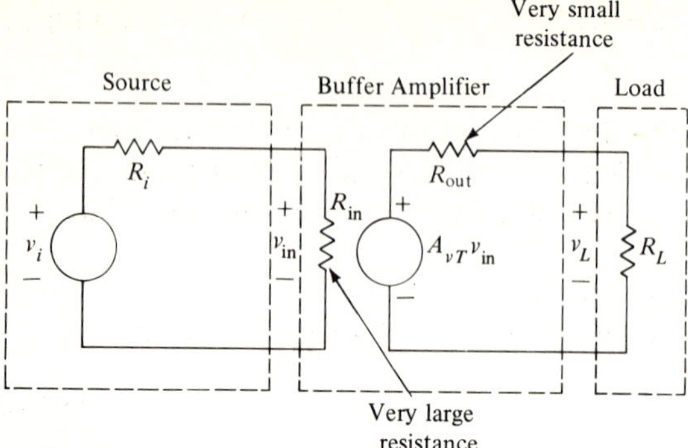

FIGURE 14.25
Buffer amplifier connected between a source and a load.

yields

$$i_{11} + i_{12} + i_{13} = i_2 \tag{14.8-19}$$

Since $v_d = 0$, we have

$$i_{11} = \frac{v_1}{R_{11}} \qquad i_{12} = \frac{v_2}{R_{12}} \qquad i_{13} = \frac{v_3}{R_{13}} \tag{14.8-20}$$

and

$$i_2 = -\frac{v_o}{R_2}$$

Substituting Eqs. (14.8-20) into Eq. (14.8-19) we find

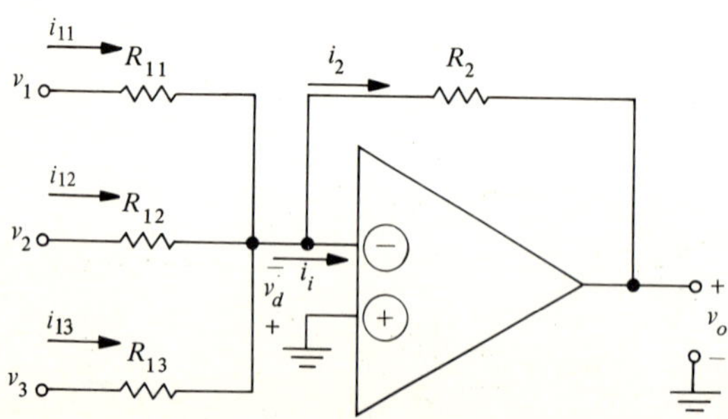

FIGURE 14.26
Three-input summing amplifier.

$$v_o = -\left(\frac{R_2}{R_{11}} v_1 + \frac{R_2}{R_{12}} v_2 + \frac{R_2}{R_{13}} v_3\right) \tag{14.8-21}$$

Thus each input can be multiplied by a different scale factor and then added. The inversion (negative sign) cannot be avoided in a summer, unless an additional amplifier is used, as illustrated in the following example.

EXAMPLE 14.8-3 Design of a Summing Amplifier

The analog computer in an automotive ignition control requires an output

$$v_o = 2v_1 + 5v_2 \tag{14.8-22}$$

Design an op-amp circuit to provide this output.

Solution

In order to achieve the positive sign, we use two op-amps connected in cascade. One will scale and add v_1 and v_2 and the other will change the sign from negative to positive. The proposed circuit is shown in Fig. 14.27. The resistors in the first circuit will be chosen to do the required scaling while the second circuit will provide a gain of -1 to change the sign. Since the output resistance of the scaling circuit will typically be less than 1 Ω, the loading between the two circuits can be neglected as long as R_3 is at least 1 kΩ.

The output voltage is found as follows:
For the first circuit

$$v_3 = -\left(\frac{R_2}{R_{11}} v_1 + \frac{R_2}{R_{12}} v_2\right)$$

For the second circuit

$$v_o = -\frac{R_4}{R_3} v_3$$

Combining these two equations:

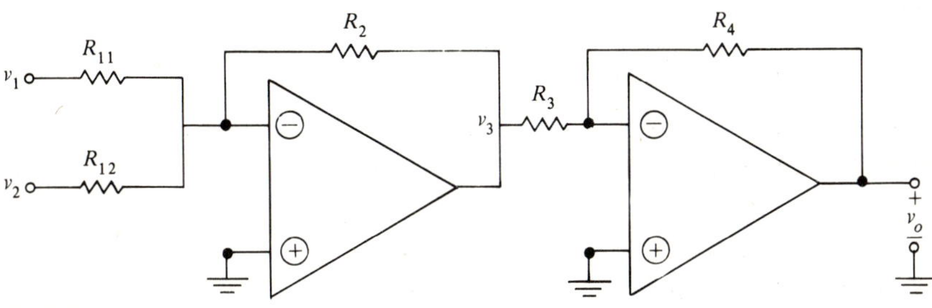

FIGURE 14.27
Circuit for Example 14.8-3.

$$v_o = \frac{R_4}{R_3}\left(\frac{R_2}{R_{11}} v_1 + \frac{R_2}{R_{12}} v_2\right)$$

In the absence of other specifications we choose $R_2 = R_3 = R_4 = 10\ \text{k}\Omega$. Then

$$v_o = \frac{10}{R_{11}} v_1 + \frac{10}{R_{12}} v_2 \tag{14.8-23}$$

Comparing this with Eq. (14.8-22), if $R_{11} = 5\ \text{k}\Omega$ and $R_{12} = 2\ \text{k}\Omega$, the desired output will be obtained.

• • •

Feedback

The op-amp circuits that we have discussed all yield transfer functions which are independent of the gain of the internal op-amp as long as that gain is high. This characteristic is a consequence of the fact that these amplifiers are basically *feedback* systems. *A feedback system is one in which a portion of the output is fed back and combined with the input.* In the negative scaler circuit shown in Fig. 14.21a, for example, a portion of the output voltage v_o is fed back through R_2 and combined with a portion of the input signal being fed through R_1. The combination takes place at the inverting input terminal of the op-amp. As we have seen, this arrangement improves the performance of the circuit in the sense that the overall transfer function is independent of the gain of the internal amplifier. Since the gain of typical op-amps without feedback may vary by as much as $\pm 50\%$ this is a considerable improvement. The price we pay for this is that the gain is very much lower with the feedback. However, gain is relatively cheap and we can usually well afford to give some away in order to achieve increased stability.

Feedback is used in most engineering systems, from home heating controls to missile guidance systems. Many of the functions of the human body involve feedback. For example, consider reaching for an object on a table. The position of the hand as it moves toward the object is continually fed back and compared to the position of the object via the eye and the brain.

Feedback offers other advantages in addition to stabilization against component changes. It also reduces the effects of external disturbances over which the designer has no control and can be used to speed up the response of systems which are inherently slow. Students of electrical technology will study feedback in more detail in courses on electronics and automatic control.

• • •

LEARNING EXERCISE FOR SEC. 14.8

1. An op-amp negative scaler has $R_1 = 10\ \text{k}\Omega$, and $R_2 = 50\ \text{k}\Omega$. The integrated circuit op-amp is specified as having $R_i = 500\ \text{k}\Omega$, $A_d = 10^6$ and $R_o = 200\ \Omega$. Find A_{vT}, R_{in}, and R_{out}.

Ans. 1.2; 10; −5

• • •

SUMMARY

In this chapter we have made use of the circuit theory learned in previous chapters to calculate the important characteristics of voltage and current amplifiers, along with several important op-amp configurations. This information will provide a firm foundation for later, more detailed study of amplifiers in electronics courses. The important points of each section are detailed below:

Section

14.1 1. The two-port is described by transfer functions that relate output quantities to input quantities.

14.2 2. Power gain is the ratio of output signal power to input signal power.

3. In decibels, the power gain is $G_p = 10 \log(P_2/P_1)$.

4. In decibels, the voltage gain is $G_v = 20 \log(V_2/V_1)$.

14.3 5. In a controlled source, the voltage or current from the source *depends on* the voltage or current at some point in the network.

14.4 6. The ideal voltage amplifier provides voltage gain and has infinite input resistance and zero output resistance.

7. The ideal current amplifier provides current gain and has zero input resistance and infinite output resistance.

14.5 8. Nonideal amplifiers have finite input and output resistances, reducing the gains compared to the ideal values.

14.6 9. The hybrid parameters relate input voltage and output current to input current and output voltage. They reduce complicated circuits to simpler form and are particularly appropriate when applied to current amplifiers.

14.7 10. The gain of a cascade of two-ports is equal to the product of the individual gains if the interstage loading can be neglected.

11. The decibel gain of a cascade of two-ports is the sum of the individual decibel gains when interstage loading can be neglected.

14.8 12. The op-amp is a high-gain voltage amplifier with relatively high input impedance and low output impedance. Most op-amps have double-ended input and single-ended output.

13. In the inverting configuration the gain is $A_{vT} = -R_2/R_1$, the input

resistance is approximately R_1 and the output resistance is very low.

14. In the noninverting configuration the gain is $A_{vT} = (1 + R_2/R_1)$, the input resistance is very high, and the output resistance is very low.
15. In the summing amplifier any number of input voltages can be multiplied by different constants and added together.
16. In a feedback system, a portion of the output is fed back and compared to the input. This leads to many improvements in performance.

QUESTIONS FOR REVIEW

Sec. 14.1

1. What is a transfer function?
2. Sketch a general two-port network showing reference directions of currents and voltages at each port.

Sec. 14.2

3. Define the word *amplification*, or *gain*.
4. How is amplification expressed?
5. What is the formula for power gain in decibels?
6. What are dBm?
7. What is the formula for decibel voltage gain?
8. What is the decibel voltage gain for voltage ratios of 1, 2, and 10?

Sec. 14.3

9. Describe the controlled source.
10. What are the four different kinds of controlled sources?

Sec. 14.4

11. Draw a diagram of an ideal voltage amplifier.
12. What is the significance of the voltage amplification factor?
13. Draw the circuit of an ideal current amplifier.
14. What is the significance of the current amplification factor?

Sec. 14.5

15. Draw a voltage amplifier circuit with input and output resistance.
16. What effect do these resistances have on the gain as compared to the ideal voltage amplifier?
17. Draw the circuit diagram of a current amplifier with input and output resistance.
18. What effect do these resistances have on the gain as compared to the ideal current amplifier?

Sec. 14.6

19. Why are two equations required to completely specify the terminal behavior of a two-port?
20. Draw the hybrid equivalent circuit for a general two-port.

21. Write the hybrid equations from the circuit diagram.
22. What is the physical interpretation of each of the h parameters?

Sec. 14.7

23. Describe the chain equation that is useful for analyzing cascaded amplifiers.
24. How is the overall voltage gain of a number of cascaded amplifiers related to the individual amplifier gains? Your answer should be in terms of ratios and decibels.

Sec. 14.8

25. Describe the typical integrated circuit op-amp.
26. Draw the equivalent circuit for a differential-input op-amp with single-ended output.
27. What is the significance of the term *differential input* voltage?
28. How is the output voltage of an op-amp related to the differential input voltage?
29. Explain the term *virtual short circuit.*
30. Draw the circuit for a negative scaler.
31. What is the overall gain of the negative scaler?
32. Repeat the previous two questions for the positive scaler.
33. Draw the op-amp circuit that can be used to find the weighted sum of three different input voltages.
34. What is the equation for the output voltage in the previous question?
35. What is a feedback system?
36. Name some of the advantages of feedback.

PROBLEMS

Sec. 14.1

1. Find the transfer function $A_v = v_2/v_1$ for the two-port shown in Fig. 14.28 if $i_2 = 0$. (Hint: Use the chain rule but be careful of loading.)
2. In the circuit of Fig. 14.28, $v_1 = 12\cos(320t - 42°)$ mV. If $i_2 = 0$, what is v_2?
3. Find the transfer function $A_i = i_2/i_1$ for the two-port shown in Fig. 14.28 if $v_2 = 0$.
4. In the circuit of Fig. 14.28, $i_1 = 12 + 42e^{-3t}$ mA. If $v_2 = 0$ what is i_2?

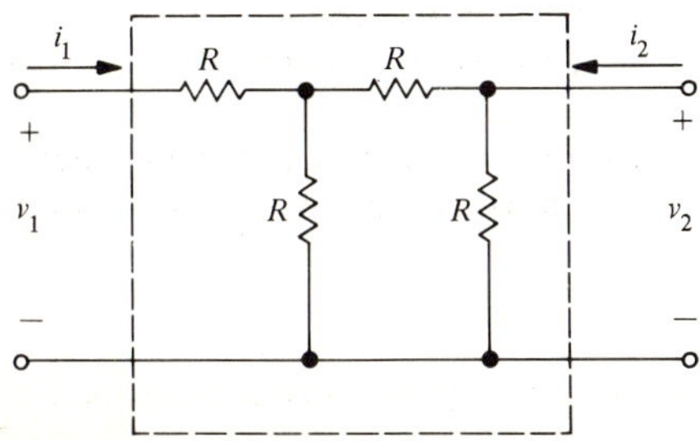

FIGURE 14.28

Sec. 14.2

5. Find the power gain in decibels corresponding to power ratios of 0.027, 0.13, 1.3, 2.7, 130, and 27 000.
6. How many dBm correspond to $P = 0.001$, 0.01, 1, 20, and 100 W?
7. Find the gain in decibels for
   a. $V_2/V_1 = 10^{-4}, 10^{-2}, 10^{0}$
   b. $V_2/V_1 = 12, 263, 4\,630$
   c. $I_2/I_1 = 3.2 \times 10^{-4}, 6.8 \times 10^{-3}, 32 \times 10^{6}$
8. In a certain system, the output power is 2 hp and the input power is 1 W. Find the power gain.
9. The output voltage of an amplifier is 30 V when the input is 2 mV. The input resistance is 10 kΩ and the load resistance is 600 Ω. Find
   a. Power gain in decibels
   b. Input and output power in dBm
   c. Voltage gain in decibels
10. A current amplifier has an input resistance of 10 Ω. When the input current is 2 μA and the load resistance is 5 kΩ, the output current is 10 mA. Find
    a. Power gain in decibels
    b. Input and output power in dBm
    c. Current gain in decibels
11. The input power to an audio transformer is 27 W and the output power is 25 W. What is the power loss in decibels?
12. What is the voltage across 600 Ω at 0 dBm? 10 dBm?
13. An amplifier has a power gain of 120 dB. If the input power is 5 mW, what is the output power?
14. In a 100-W broadcast studio amplifier, the hum level is 80 decibels below full output. How much hum power does this represent?
15. An amplifier has a power output of 32 dBm and a power gain of 50 decibels. The input resistance of the amplifier is 25 kΩ. Find the input voltage.
16. An amplifier for a public address system has an input resistance of 10 kΩ and load resistance of 600 Ω. The output power is 60 W when the input voltage is 2.3 V. Find
    a. Power input
    b. Power gain in decibels
    c. Voltage gain in decibels

Sec. 14.3

17. Find v_2 in the circuit of Fig. 14.29.
18. Find v_2 in the circuit of Fig. 14.30.
19. In the circuit of Fig. 14.31 find the output current i_2 and the current gain in decibels.
20. In the circuit of Fig. 14.32, $v(t) = 30 \cos 377t$ V. Find $v_2(t)$.
21. Find the cutoff frequency and plot the frequency response of $A_v = v_2/v_i$ in the circuit of Fig. 14.33.
22. Repeat Prob. 21 for the circuit of Fig. 14.34.

Sec. 14.4

23. An ideal voltage amplifier is connected to a load of 2.2 kΩ. The input voltage is 10 mV and the output power is 200 mW. Find the output voltage and the voltage amplication factor.

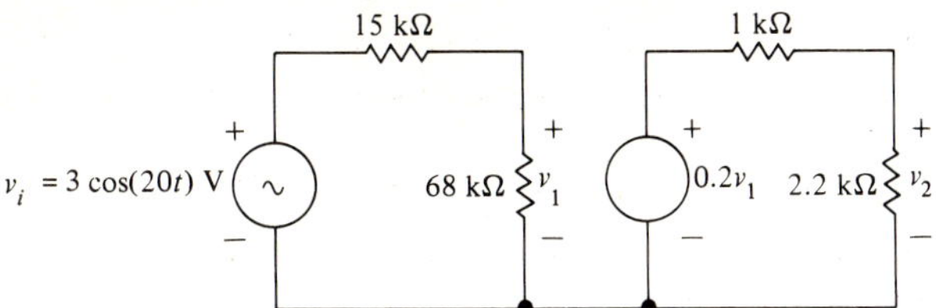

FIGURE 14.29

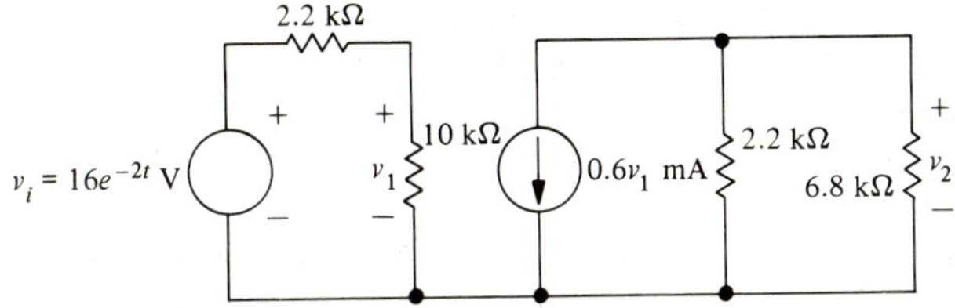

FIGURE 14.30

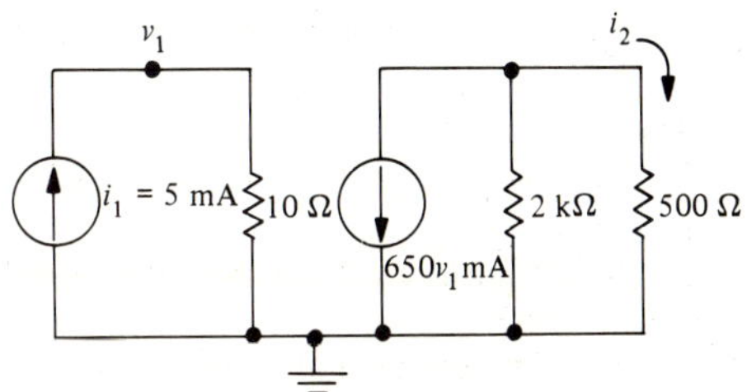

FIGURE 14.31

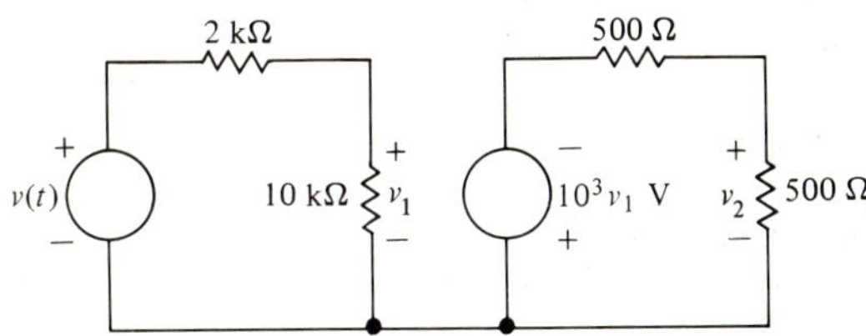

FIGURE 14.32

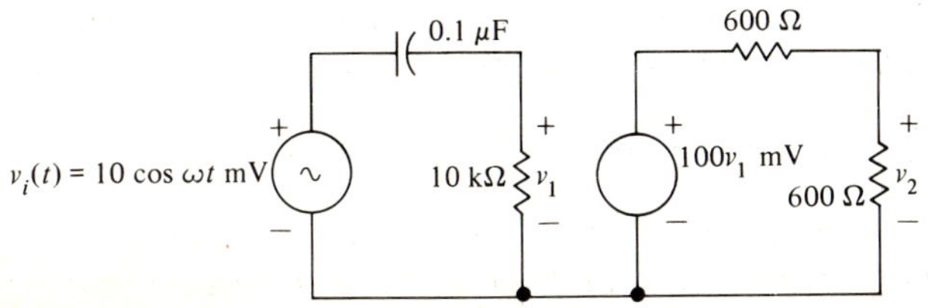

FIGURE 14.33

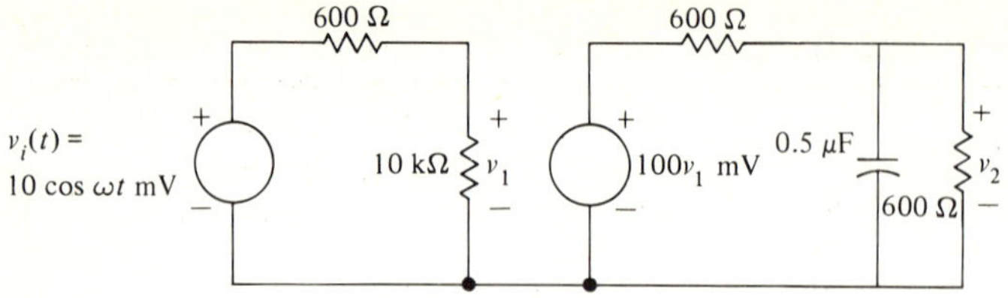

FIGURE 14.34

24. An ideal current amplifier is connected to a load of 25 Ω. The input current is 12 μA and the output power is 3.6 W. Find the output current and the current amplification factor.

Sec. 14.5

25. A voltage amplifier has an input resistance of 20 kΩ, amplification factor of 500, and output resistance of 1000 Ω. The signal source has an internal resistance of 1 kΩ and the load resistance is 470 Ω. Find the overall voltage gain.
26. In the amplifier of Prob. 25 the output power is 500 mW. Find the input voltage and the power gain.
27. A voltage amplifier has $R_i = 1$ kΩ, and $R_o = 500$ Ω. The source has $R_s = 50$ Ω and the load is a loudspeaker rated at 10 Ω. If the power to the loudspeaker is to be 10 W and the source voltage is 10 mV, find the required voltage amplification factor.
28. In the current amplifier of Fig. 14.12, $i_i = 10 \sin \omega t$ mA, $R_s = 10$ kΩ, $R_i = 100$ Ω, $A_i = 10^4$, $R_o = 10$ kΩ, and $R_L = 10$ Ω. Find i_2 and the current gain.
29. In the amplifier of Prob. 28 the average power into R_L is to be 10 W. Find the required value of i_i.
30. A certain current amplifier has $R_s = 5$ kΩ, $R_i = 50$ Ω, $R_o = 5$ kΩ, and $R_L = 10$ Ω. The output power is to be 5 W and the input current is 2 mA. Find the required value of A_i.
31. In the voltage amplifier of Prob. 25, a 1-μF capacitor is connected in series with R_L to block any dc present. Find the cutoff frequency and plot the frequency response of $A_v = v_2 / v_i$.

Sec. 14.6

32. A black box has two pairs of terminals, with reference directions as shown in Fig. 14.1. When the output terminals are short-circuited and a 1-mA current is applied to the input terminals, the short-circuit current is 5 mA and the voltage across the input terminals is 25 mV. When the input terminals are left open and a 1-mV source is applied at the output, the voltage across the input terminals is 0.1 mV and the current drawn from the 1-mV source is 250 μA. Find the hybrid parameters for the black box.
33. Find the hybrid parameters for the circuit of Fig. 14.35.
34. Find the hybrid parameters for the circuit of Fig. 14.36.
35. The equations

$$v_1 = R_{11}i_1 + R_{12}i_2$$
$$v_2 = R_{21}i_1 + R_{22}i_2$$

define four *open-circuit resistance* parameters similar to the hybrid parameters.

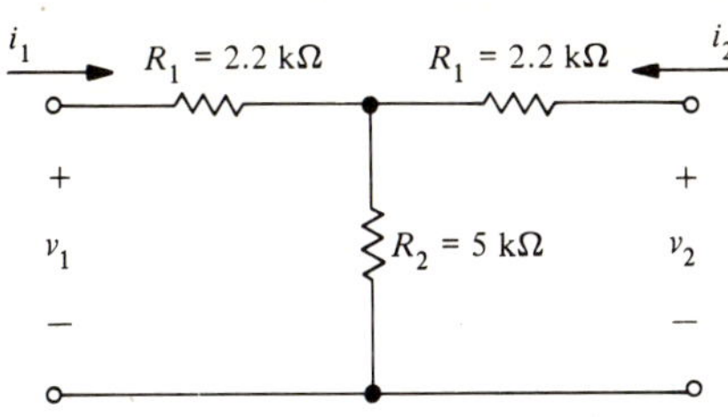

FIGURE 14.35

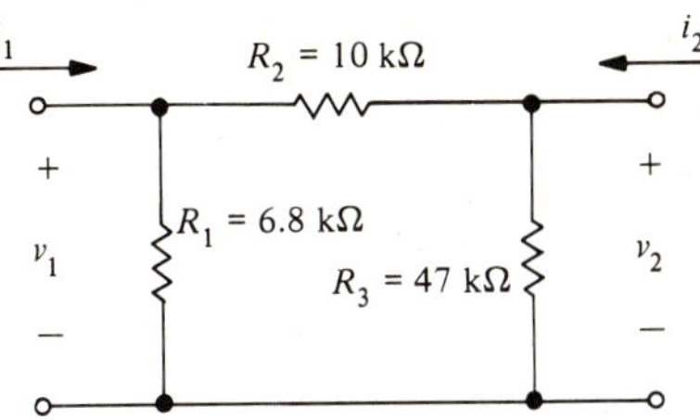

FIGURE 14.36

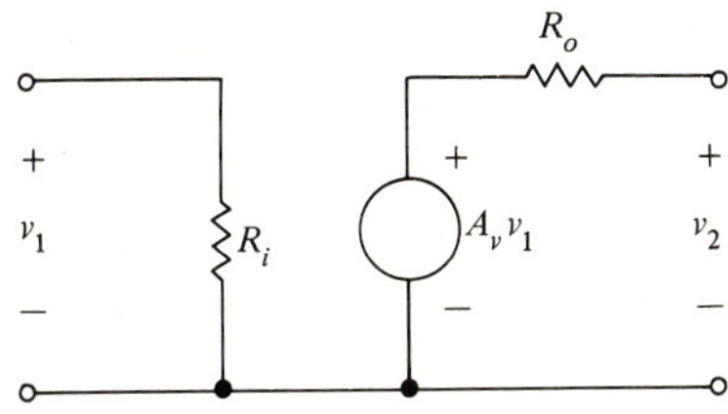

FIGURE 14.37

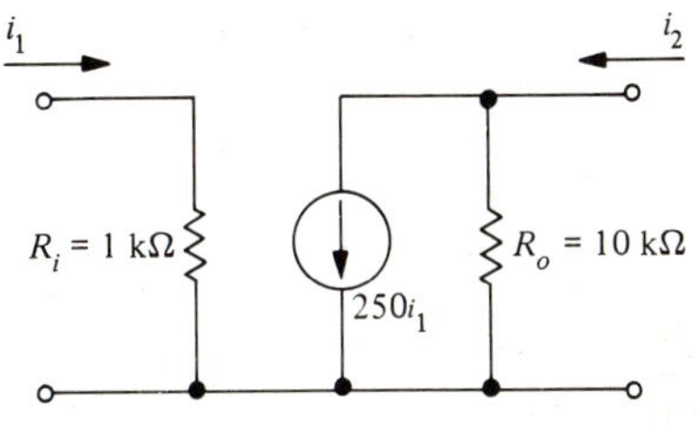

FIGURE 14.38

Discuss the physical significance of these parameters. [Hint: See Eqs. (14.6-3) through (14.6-6).] Draw an equivalent circuit similar to Fig. 14.14.

36. A transistor is specified with the following hybrid parameters: $h_i = 1\ k\Omega$, $h_r = 0$, $h_f = 350$, $1/h_o = 20\ k\Omega$. It is used in a current amplifier with $R_s = 10\ k\Omega$ and $R_L = 600\ \Omega$. Find the overall current gain.

Sec. 14.7

37. In the circuit of Fig. 14.17 the amplifiers are identical with $R_L = R_s = 50\ \Omega$, $h_i = 250\ \Omega$, $1/h_o = 100\ k\Omega$, and $h_f = 350$. Find the overall gain $A_v = v_4/v_i$.
38. A voltage amplifier circuit model is shown in Fig. 14.37. In the circuit $R_i = 1\ M\Omega$, $R_o = 600\ \Omega$, and $A_v = 250$. Find the overall gain if two of these amplifiers are cascaded to drive a load resistance of 10 kΩ.
39. Repeat Prob. 38 for a load resistance of 600 Ω.
40. Repeat Prob. 38 with R_i changed to 600 Ω.
41. In the voltage amplifier circuit of Fig. 14.37, $R_i = R_o = 600\ \Omega$ and $A_v = 250$. Find the overall gain if two of these amplifiers are cascaded to drive a load resistance of 600 Ω. Compare your results with the solution to Prob. 38 and comment on the effect of the loading.
42. A current amplifier model is shown in Fig. 14.38. Find the overall gain if three of these amplifiers are cascaded to drive a load resistance of 1 kΩ.
43. Repeat Prob. 42 if $R_o = R_i = R_L = 1\ k\Omega$. Comment on the loading effect.
44. Show that Eq. (14.7-2) is valid if $R_{i2} \gg R_{o1}$. (Hint: Use a hybrid equivalent circuit with $h_r = 0$ for each amplifier.)

Sec. 14.8

45. In the difference amplifier of Fig. 14.20, $v_2 = 0$, $v_1 = 10 \sin \omega t$ mV, $A_d = 5000$, $R_i = 100\ k\Omega$, $R_o = 200\ \Omega$, and a load resistance $R_L = 470\ \Omega$ is connected to the amplifier. Find the voltage gain and power gain in decibels.
46. The waveforms shown in Fig. 14.39 are applied to a difference amplifier. Neatly sketch the waveform of v_d.

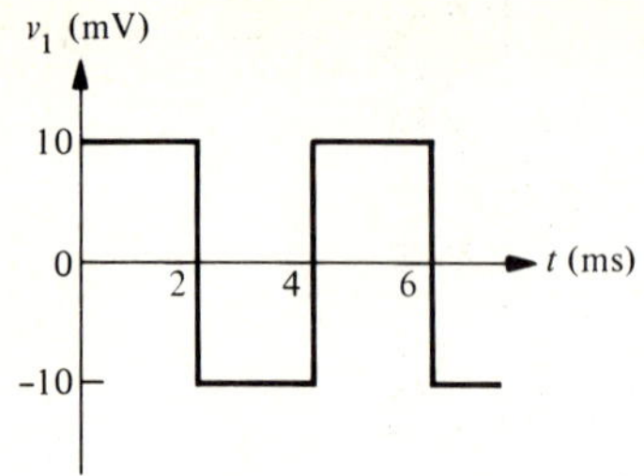

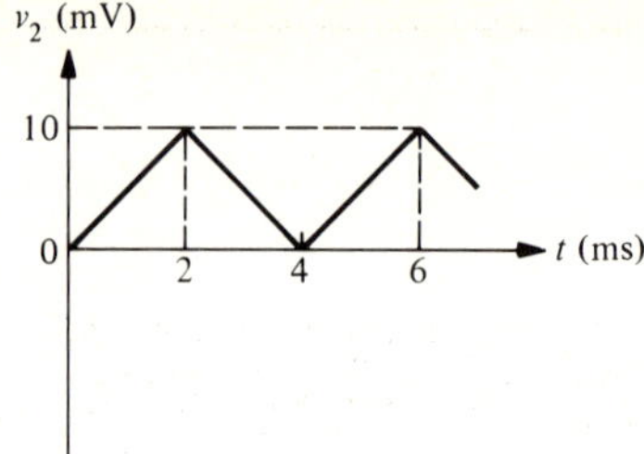

FIGURE 14.39

47. Draw the circuit model for a differential input operational amplifier whose voltage amplification is 5×10^5, input resistance is 20 MΩ, and output resistance is 100 Ω.
48. Using as many 1-kΩ, 10-kΩ, 100-kΩ, and 1-MΩ resistances as required, but no more than required, draw circuits to implement the following functions using one operational amplifier per circuit.
    a. $v_o = -10v_1$ b. $v_o = 21v_1$
49. Derive Eq. (14.8-12) using the technique discussed in the paragraph preceding the equation.
50. An inverting scaler for use in a desktop analog computer is to have a transfer function $A_{vT} = v_o/v_1 = -52.6$ and the input resistance is to be at least 5 kΩ. Find suitable values for R_1 and R_2.
51. Design a positive scaler to have $A_{vT} = 12$. The op-amp specifications include the values $A_d = 10^6$, $R_i = 250$ kΩ, and $R_o = 50$ Ω. If $R_2 = 25$ kΩ, find R_1, R_{in}, and R_{out}.
52. The op-amp specified in Prob. 51 is used in a voltage follower circuit ($R_2 = 0$, $R_1 = \infty$). Find R_{in} and R_{out} and draw the equivalent circuit.
53. In the voltage follower of Prob. 52, what value of load resistance must be connected in order to have the output voltage drop to 90% of its open-circuit value?
54. Design an op-amp system to realize the following:

$$v_o = 6v_1 + 2v_2 + v_3.$$

    Assume op-amps with very high gain and input resistance, and low output resistance. Use resistances in the range 10 to 100 kΩ.
55. Repeat Prob. 54 for the output

$$v_o = 6v_1 - 2v_2 + v_3$$

15 AC Measurements

OBJECTIVES

Upon completion of this chapter, the student should be able to

Section

15.1 1. Explain the operation of the half-wave and full-wave rectifier-type ac meter.
2. Find the conversion factor to be applied when using a rectifier-type meter for waveforms other than sinusoids.

15.2 3. Describe the operation of a typical digital frequency meter.

15.3 4. Find the two balance conditions for an ac bridge.

15.4 5. Describe the basic operation of the cathode ray tube.
6. Explain how the time variation of a signal is displayed on an oscilloscope using a linear sweep.
7. Find voltage and time from an oscilloscope display by application of appropriate voltage and time calibration factors.

INTRODUCTION

In Chap. 6 we discussed the theory of dc instruments and the important concepts of instrument loading and measurement error. In this chapter we consider instruments used for ac measurements. The theory of loading and error calculation given previously for dc circuits carries over directly to ac circuits.

As in dc measurements, two types of instruments are available: digital and analog. We will describe the most important of these. In addition, the oscilloscope, on which waveforms are displayed, will be described in some detail because of its importance. It may well be the most often used instrument in the typical electronics laboratory.

15.1 RECTIFIER-TYPE METERS

As noted in Sec. 6.6 the electrodynamometer meter movement will respond directly to ac waveforms, but its effective frequency range is limited to a few hundred hertz or less. Most ac voltmeters operate by converting some property of the ac waveform into a dc current, and passing this current through a D'Arsonval movement. The meter scale is calibrated in terms of the desired property of the ac waveform, even though the meter is actually responding to the average value of the current. A block diagram of the typical system is shown in Fig. 15.1.

Most often it is the rms value of the sine wave that is required, so most ac meters are calibrated to read rms. However, most types of ac meter do not actually respond to the rms value of the applied waveform, so that incorrect values will be obtained unless the input is a sine wave.

Different types of meters respond to various waveform properties. These include the actual rms value, the average value of half-wave or full-wave rectified waveforms, and the peak or peak-to-peak value. These are shown for sine waves in Fig. 15.2. The full-wave rectified waveform shown in Fig. 15.2c sees the most use in analog meters. It is obtained by applying a sinusoidal signal to a rectifier circuit containing diodes or other active elements. A description of these circuits is beyond the scope of this course; however, they will be studied in a later course in electronics. For each waveform, a different amount of dc current will flow through the D'Arsonval movement. In all but the true rms meter, the following problem arises: If, for example, a meter that responds to the full-wave rectified average is calibrated in terms of the rms value of a sine wave, it will not correctly indicate the rms value of any other waveform because the relationship between the rms value and the full-wave rectified average is different for sine waves than for other waveforms.

In order to use such a meter for waveforms other than sine waves, the waveform to be measured must be known, and a correction factor must be applied. For example, consider that a meter with a full-wave rectifier is calibrated for sine waves but is to be used to measure the amplitude of a square wave. The rms value of the sine wave is $V_m/\sqrt{2}$ and the full-wave rectified average is $2V_m/\pi$. Thus the scale reading is $\pi/2\sqrt{2} = 1.11$ times the

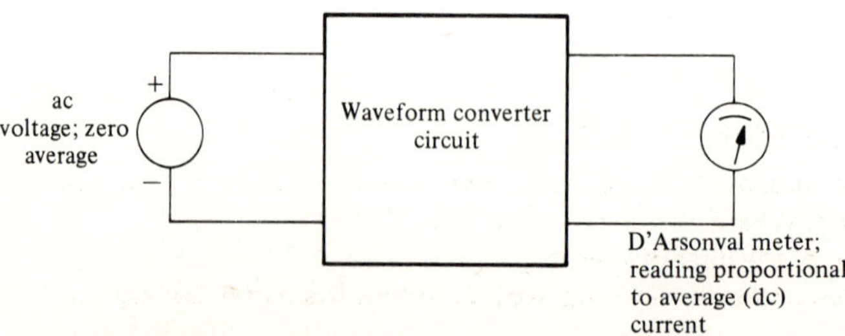

FIGURE 15.1
Typical ac meter.

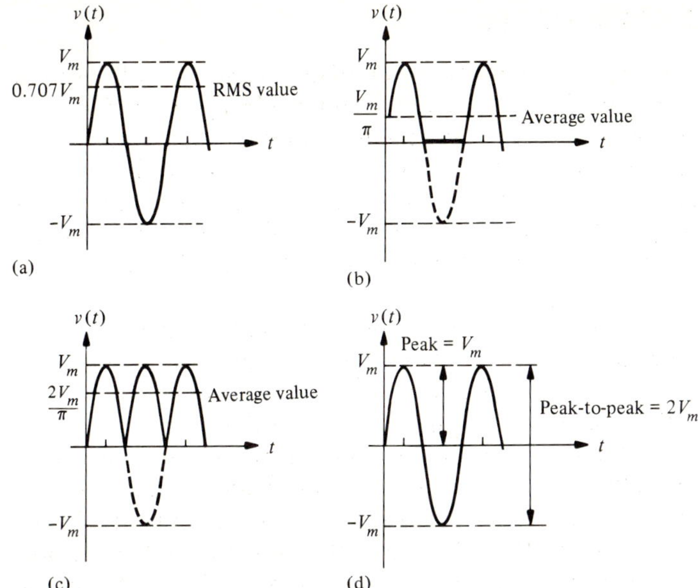

FIGURE 15.2
Waveform properties used in converter circuits. (a) True RMS value. (b) Half-wave rectified waveform average value. (c) Full-wave rectified waveform average value. (d) Peak or peak-to-peak value.

actual full-wave rectified average value of the input waveform. If the square wave to be measured has a peak value of $\pm V_0$, then its full-wave rectified average is also V_0. The scale reading on the meter would be 1.11 times this or $1.11V_0$. Thus the scale reading must be multiplied by 1/1.11 = 0.9 in order to obtain the amplitude of the square wave.

Sometimes the original ac waveform has a dc component. If the waveform converter has a capacitor in series to block the dc, the conversion factor may be found as in the preceding paragraph, otherwise the dc level must be taken into account.

The Simpson 260 VOM (schematic shown in Fig. 6.22) uses the full-wave rectified average and has a series capacitor present only when the OUTPUT terminal is used.

• • •

LEARNING EXERCISE FOR SEC. 15.1

1. A square wave is applied to a voltmeter that uses a full-wave rectifier and is calibrated for sine waves. The voltmeter reads 42 V. Find
   a. The amplitude of the square wave
   b. The peak value of a sine wave that would produce the same reading

Ans. 59; 38

• • •

15.2 DIGITAL FREQUENCY METERS

The accurate measurement of frequency is often required in the laboratory. There are many methods for making this measurement, all of which fall into one of two categories: analog and digital.

Some common types of analog frequency meters include the vibrating-reed type, the resonant-circuit type, and the active filter type. The accuracy of these meters is about $\pm 1\%$ and they are usually limited to frequencies below a few megahertz. The oscilloscope can be used for frequency measurement, but its accuracy is also limited to about $\pm 1\%$.

In general, digital meters provide much higher accuracy and excellent stability along with their digital readout. They tend to be more expensive than their analog counterparts at present, but their cost is dropping due to the use of integrated circuits.

The *digital frequency meter,* also called a *digital counter,* is a very accurate, easy-to-use instrument. Measurement ranges are from dc to over 40 GHz (1 GHz = 10^9 Hz), and typical accuracy is 1 part in 10^8 ($=10^{-6}\%$).

A photograph of a commercial frequency meter is shown in Fig. 15.3 and a block diagram with waveforms is shown in Fig. 15.4. The input pulse generating circuit converts the input signal, waveform *A*, in Fig. 15.4b into a pulse train of one pulse per cycle, as in waveform *B*. The pulses are controlled by some property of the input waveform. In our case, each pulse occurs at the instant that the input waveform is going through zero with negative slope. This pulse train is applied to the gate circuit. The gating pulse (signal *C*) is derived from the internal time base oscillator by the control circuit and is 0.1 to 10 s in duration. This is called the counting time or sampling time. The action of the gate circuit is such that only those pulses from the input waveform that occur during the counting time are counted by the decade counter. Thus if the counting time is exactly 1 s the count displayed on the counter will be exactly equal to the number of cycles per second of the input waveform, that is, the frequency, because we have exactly one pulse for each cycle. In Fig. 15.4b, four pulses will be counted during the counting time, so the number 4 appears on the counter display, indicating that the frequency of the input waveform is 4 Hz.

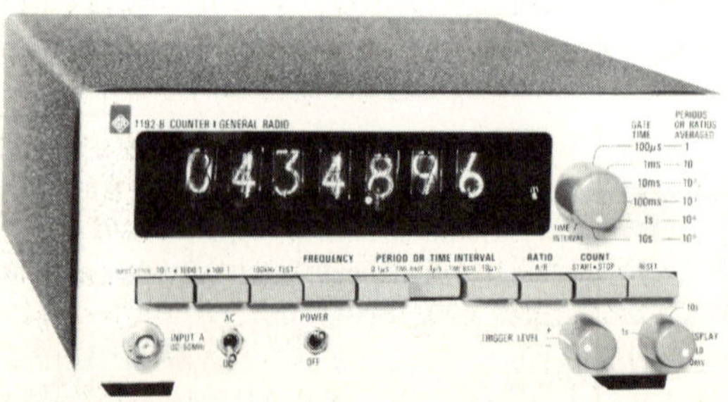

FIGURE 15.3
Frequency meter (courtesy of GenRad, Inc.).

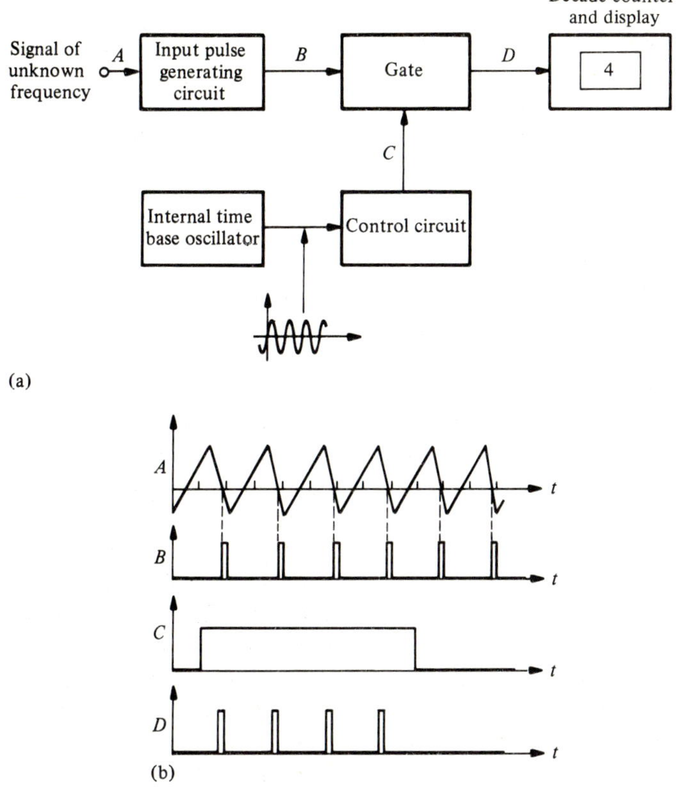

FIGURE 15.4
Digital frequency meter. (a) Block diagram. (b) Waveforms.

The internal time-base oscillator is usually crystal-controlled and operates at a frequency of 100 kHz or 1 MHz. The control circuit is a frequency divider, which counts a predetermined number of cycles of the time base oscillator output and from this generates gating signal C. If, for example, a gating signal of 1 s duration is desired and the time base frequency is 1 MHz, then the control circuit must count 10^6 cycles of the time-base output signal.

After each gating signal, there is a short time interval during which no measurements are made. The system is then reset, a gating pulse is generated, and a new measurement is made. While the new measurement is being made the previously measured value is displayed. The reset frequency is called the *sampling rate* or *display time* and is set by a control on the instrument.

The accuracy of the system is controlled by the accuracy of the counting time, which in turn depends on the stability of the time base oscillator. Typical accuracy is one part in 10^7 or 10^8. For higher accuracy, external time base signals may be fed into the instrument in place of the internal time base. The external signal would be obtained from a highly stable frequency standard.

Digital frequency meters may also be used in other ways. Some of these include direct display of the *period* of a waveform, display of the *ratio* of the frequencies of two different signals, or counting of events, either random or periodic.

• • •

LEARNING EXERCISES FOR SEC. 15.2

1. A digital frequency meter as in Fig. 15.4 is used to check the frequency of a Citizens Band transmitter. If the gating signal is 0.2 s and the number of pulses counted during this interval is 3.2×10^4, what is the transmitter frequency?

2. The frequency meter in Learning Exercise 15.2-1 has an internal time base oscillator operating at 1 MHz. How many cycles must be counted to generate a gating signal of 0.25 s?

Ans. 160 000; 250 000

• • •

15.3 IMPEDANCE BRIDGES

In Sec. 6.5 we showed how an unknown resistor could be measured by a *bridge* circuit. This idea can be extended to include the measurement of complex ac impedance as well. A commercial impedance bridge used in many laboratories is shown in Fig. 15.5 and an automatic digital-output

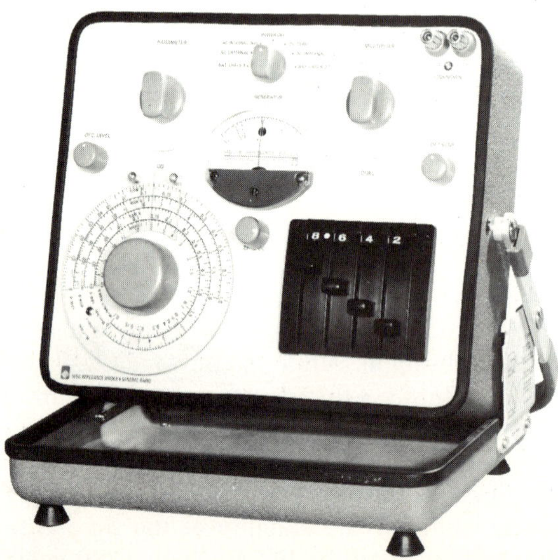

FIGURE 15.5
Laboratory impedance bridge (courtesy of GenRad, Inc.).

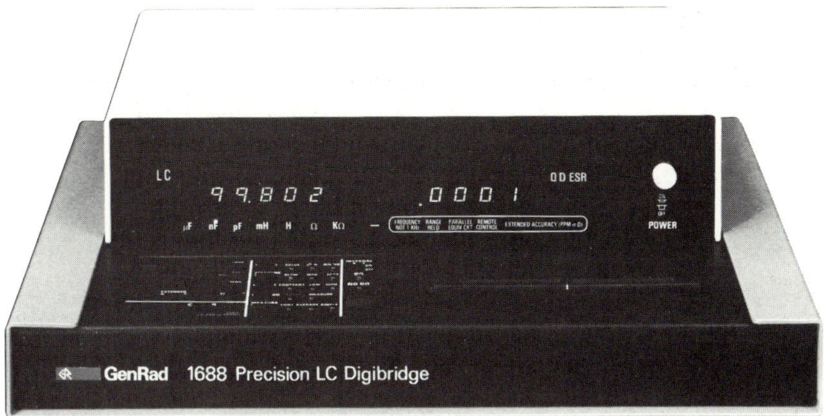

FIGURE 15.6
Automatic impedance bridge with digital readout (courtesy of GenRad, Inc.).

bridge is shown in Fig. 15.6. The circuit is shown in Fig. 15.7. It is similar to the Wheatstone bridge, but the arms may be complex impedances, the source is usually a sinusoidal oscillator, and the null detector may be a sensitive ac voltmeter, or a set of earphones if the signal is in the audio range. As in the Wheatstone bridge, no current flows through the detector at balance, so the voltage across it is zero, and a development paralleling Eqs. (6.5-1) through (6.5-4) leads to the balance equation

$$Z_x = \frac{Z_3 Z_2}{Z_1} \tag{15.3-1}$$

The unknown impedance Z_x will in general be complex, thus

$$\boxed{Z_x = R_x + jX_x = \frac{Z_3 Z_2}{Z_1}} \tag{15.3-2}$$

This is a complex equation, which will be satisfied only when the real parts of both sides are equal, and also the imaginary parts of both sides. It is sometimes convenient to choose Z_1 and Z_2 to be resistors such that their ratio is some appropriate power of 10. Then Z_3 must be complex and both its resistive and reactive parts must be adjustable. The balance equation becomes

$$R_x + jX_x = (R_3 + jX_3)\frac{R_2}{R_1} \tag{15.3-3}$$

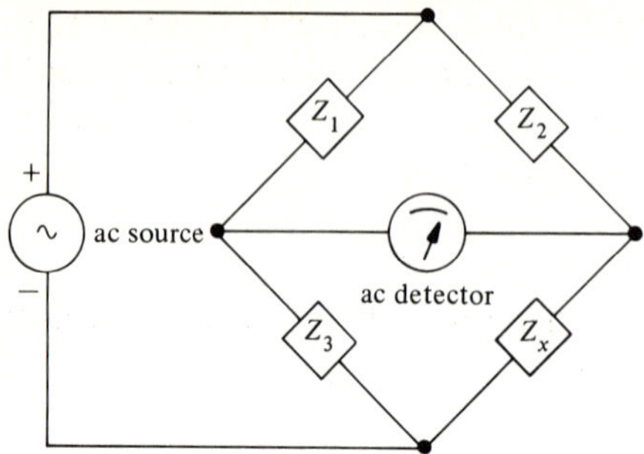

FIGURE 15.7
ac impedance bridge.

Equating real and imaginary parts separately, we have *two* balance equations

$$\boxed{R_x = R_3 \frac{R_2}{R_1}} \tag{15.3-4}$$

$$\boxed{X_x = X_3 \frac{R_2}{R_1}} \tag{15.3-5}$$

where R_2/R_1 is usually made equal to some convenient power of 10.

From the balance equation we see that two adjustments must be made in order to achieve balance, one to satisfy Eq. (15.3-4) and one to satisfy Eq. (15.3-5). Ideally, the two adjustments should be independent, that is, when the real-part adjustment is being made to satisfy Eq. (15.3-4) there should be

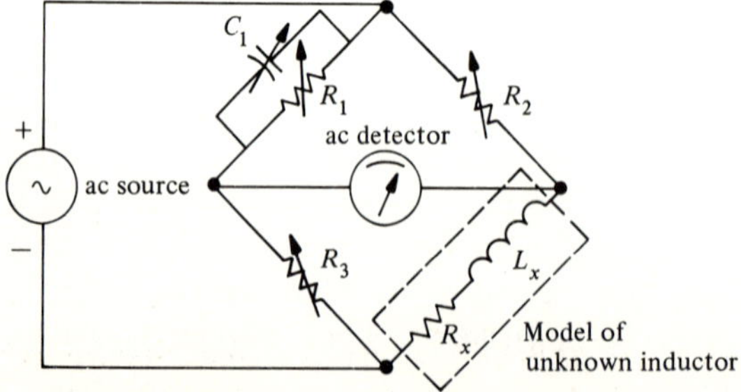

FIGURE 15.8
Maxwell bridge.

no effect on the balance in Eq. (15.3-5). If this is not so, then a number of successive adjustments of each component will have to be made in order to arrive at balance. Sometimes this process converges very slowly and special techniques must be used to speed up the procedure.

EXAMPLE 15.3-1 The Maxwell Bridge

The bridge circuit shown in Fig. 15.8 is called a Maxwell bridge and is used to measure inductance. Find the balance equations.

Solution

Because of the configuration of this bridge is it convenient to use Eq. (15.3-2) as the balance equation. The various impedances are

$$Z_x = R_x + j\omega L_x$$

$$Z_2 = R_2 \qquad Z_3 = R_3$$

$$Z_1 = R_1 \left|\right| \frac{1}{j\omega C_1} = \frac{R_1/j\omega C_1}{R_1 + 1/j\omega C_1} = \frac{R_1}{1 + j\omega R_1 C_1}$$

Substituting into Eq. (15.3-2)

$$R_x + j\omega L_x = \frac{R_2 R_3}{R_1}(1 + j\omega R_1 C_1)$$

Equating real parts on both sides

$$\boxed{R_x = \frac{R_2 R_3}{R_1}} \tag{15.3-6}$$

Equating imaginary parts on both sides

$$\omega L_x = \frac{R_2 R_3}{R_1}(\omega R_1 C_1)$$

This reduces to

$$\boxed{L_x = C_1 R_2 R_3} \tag{15.3-7}$$

When using this bridge R_2 and R_3 would first be set to give a convenient power of 10 multiplier. Next, R_1 would be adjusted for the best possible null. Then C_1 would be adjusted to improve this null. Then back to R_1 and then C_1 until the best possible null is obtained.

Additional bridge configurations will be considered in the homework problems.

• • •

LEARNING EXERCISE FOR SEC. 15.3

1. A radio-frequency inductance coil is to be measured using the Maxwell bridge of Example 15.3-1. The standard bridge resistors are $R_2 = R_3 = 10$ kΩ. At balance, $R_1 = 163$ kΩ and $C = 250$ pF. Find R_x and L_x.

Ans. 25; 613

• • •

15.4 THE OSCILLOSCOPE

In the study of physical devices and systems, we are usually concerned with input and output signals that vary, sometimes periodically, with time. It is very often desirable to view the actual time variations, and the cathode-ray oscilloscope ("scope" for short) is the instrument which provides the means for doing this if the signal can be transformed into a current or voltage. Modern oscilloscopes (see Fig. 15.9) can operate over a voltage amplitude range from microvolts to kilovolts and a time range of minutes to nanoseconds.

In the oscilloscope the voltage is viewed on the face of the cathode-ray tube (CRT), which is the heart of the system (see Fig. 15.10a). Within the CRT there are three main components. An electron gun and focusing system provide a sharply focused beam of electrons, which is directed at the fluorescent material coated on the inside of the CRT. This produces the spot that traces out the display. The deflecting system usually consists of two pairs of parallel plates, referred to as the vertical and horizontal plates, through which the electron beam passes. The system is usually arranged so that a positive voltage applied to the vertical input terminals deflects the beam upward in direct proportion to the applied voltage and a positive voltage on the horizontal input deflects the beam to the right in direct proportion to the applied voltage. Since the deflections are independent of each other, we can consider the tube face as an xy plane and the position of the spot can be expressed in terms of its x and y coordinates as follows:

$$x = K_x v_x \qquad y = K_y v_y \qquad \text{cm} \tag{15.4-1}$$

where v_x and v_y are the voltages applied to the input terminals of the scope and K_x and K_y are gain constants (the units are centimeters per volt) relating spot position to input voltage. These gains include the effects of the amplifiers connected between the input terminals and the plates as shown in Fig. 15.10b.

We note at this point that the voltages v_x and v_y can be chosen so that the display is a TV picture, a graphical plot of a voltage at a particular point

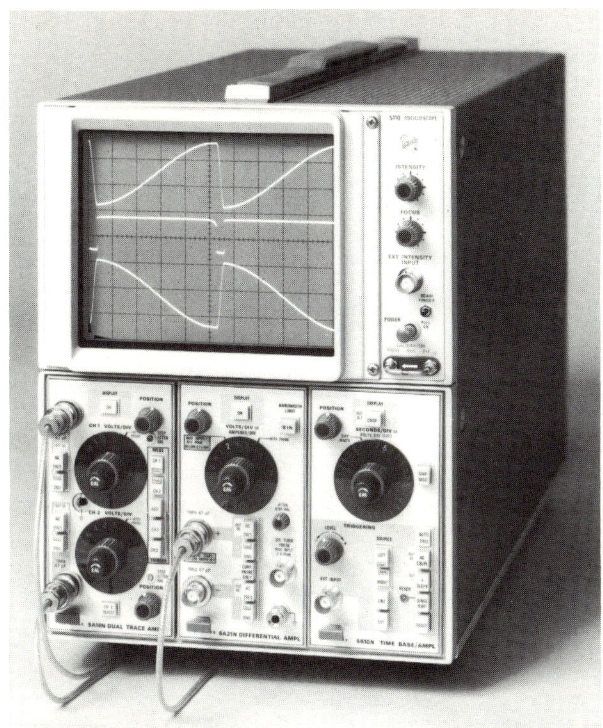

FIGURE 15.9
Tektronix type 5110 oscilloscope (courtesy of Tektronix, Inc.).

in a circuit, or even an artistic abstraction. In our work we shall most often be concerned with displaying voltage as a function of time. We next discuss means for accomplishing this.

Display of Variations with Respect to Time

In order to display the variation of a voltage with time we start by applying a linearly increasing voltage to the x axis. This voltage can be expressed in the form

$$v_x = K_1 t \qquad (K_1 \text{ in volts per second}) \tag{15.4-2}$$

Thus the horizontal position of the spot is

$$x = K_x K_1 t$$

and if we combine constants so that $K_h = K_x K_1 =$ spot velocity in centimeters

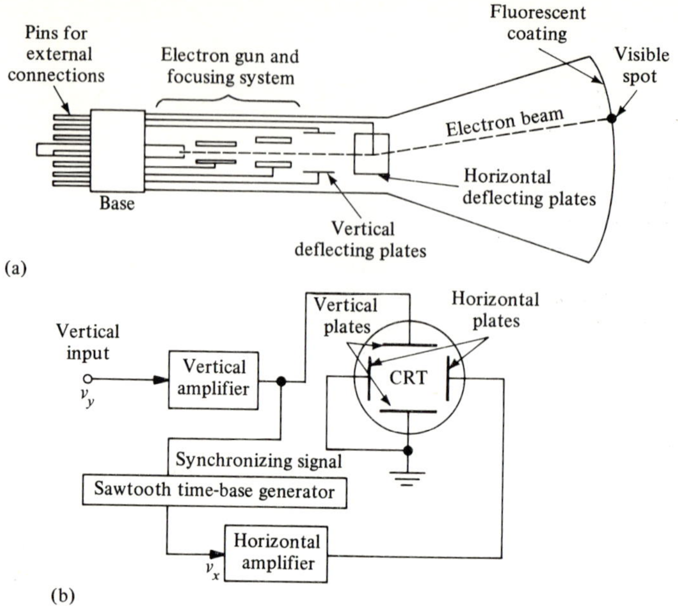

FIGURE 15.10
The oscillocope. (a) The cathode-ray tube. (b) Simplified block diagram.

per second, then

$$x = K_h t \tag{15.4-3}$$

and the beam sweeps across the face of the tube with constant velocity K_h. Clearly this is impractical because as soon as x becomes greater than the distance to the outer edge of the tube, our spot will disappear. This difficulty is overcome by using a *sawtooth* (called a linear sweep or time base), for v_x as shown in Fig. 15.11. With this waveform applied to the horizontal plates the beam *sweeps* back and forth across the tube face, with the return time interval very much less than the forward time interval. If v_y is a constant and the spot velocity is high enough, then persistence of vision will make the display appear to be a straight line.

Suppose now that v_y is a function of time $f(t)$. Let us determine the shape of the resulting display, first analytically and then graphically.

We have, assuming for convenience that $K_y = 1$ cm/V,

$$v_y = y = f(t) \tag{15.4-4}$$

Also, in the time interval from 0 to t_1 shown in Fig. 15.11, assuming $K_h =$

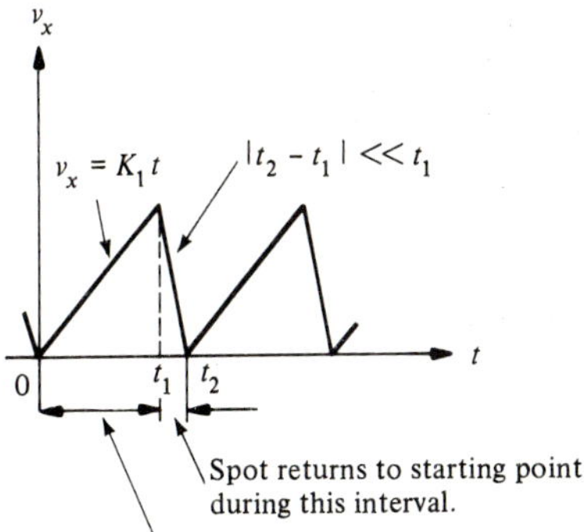

FIGURE 15.11
Linear sweep waveform.

1 cm/s,

$$v_x = x = t \tag{15.4-5}$$

Substituting Eq. (15.4-5) into (15.4-4),

$$y = f(x) = f(t) \tag{15.4-6}$$

and the display is a normalized graph of the functional relation between v_y and t.

During the second and succeeding cycles of the sawtooth, the display may be completely different, depending on the function $f(t)$. This will usually produce an indecipherable jumble on the screen. If, on the other hand, $f(t)$ *repeats* itself one or any integral number of times during the 0 to t_1 interval, then the display will be the *same* during all sweep cycles. Such a function is called *periodic* and the display, due again to persistence of vision, will appear as a normalized graph of the number of *cycles* of $f(t)$ contained in the t_1 interval. Almost all oscilloscopes contain a sawtooth time-base generator and synchronizing circuits that can be set to automatically adjust K_h, the spot velocity, so that an integral number of cycles of the periodic waveform will appear on the screen.

Graphical Determination of Display

The graphical method for determining the shape of the display is illustrated in Fig. 15.12 for the case where $f(t)$ is an exponential function.

The interval t_1 is divided into equal subintervals Δt. At $t = 0$, the spot will be at $x = y = 0$. At $t = \Delta t$ the spot will move to the point P_1, at the intersection of the projections from $v_y(\Delta t)$ and $v_x(\Delta t)$. At $t = 2\Delta t$ the spot will

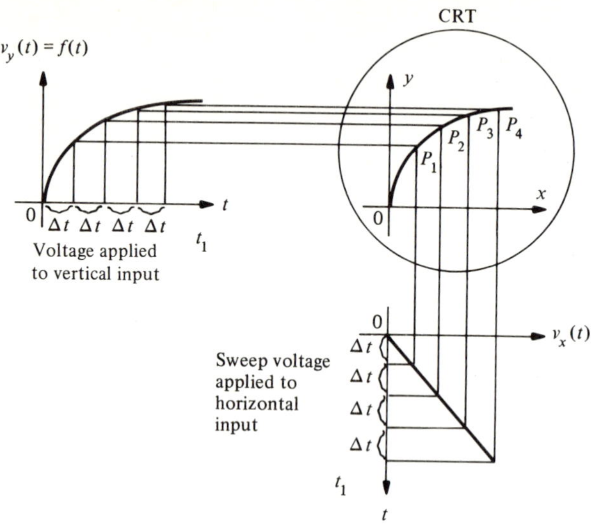

FIGURE 15.12
Graphical construction for CRT display.

be at P_2, the intersection of the projections from $v_y(2\Delta t)$, and $v_x(2\Delta t)$, and so on. Clearly, the curve generated by the sequence of points 0, P_1, P_2 is the same as $v_y(t)$, because of the linear relation between v_x and t. At $t = t_1$ the spot rapidly returns to $x = y = 0$. Note carefully that this graphical procedure was applied only for the time interval 0 to t_1. Unless $v_y(t)$ repeats itself from t_1 to $2t_1$, the display will change as succeeding sweep cycles come into view. Many oscilloscopes have very slow sweep speeds available (1 s to traverse 2 cm and slower) so that slowly varying nonperiodic waveforms can be observed. In addition, long-persistence cathode-ray tubes are available so that a single sweep may be retained on the face of the tube for long periods of time.

Oscilloscope Measurement of Sinusoids

A typical display of voltage vs. time is shown in Fig. 15.13 for a sinusoidal voltage applied to the vertical input terminals and a sawtooth sweep to the horizontal.

Most modern scopes have calibrated voltage and time axes. Thus it is a relatively simple matter to read voltage amplitude and time information directly from the display. The time information can be used to calculate frequency and phase.

The actual voltage is obtained by applying the calibration factor indicated on the *vertical sensitivity* dial. For example, if the vertical sensitivity is set at 100 mV/cm, then the waveform of Fig. 15.13 has a

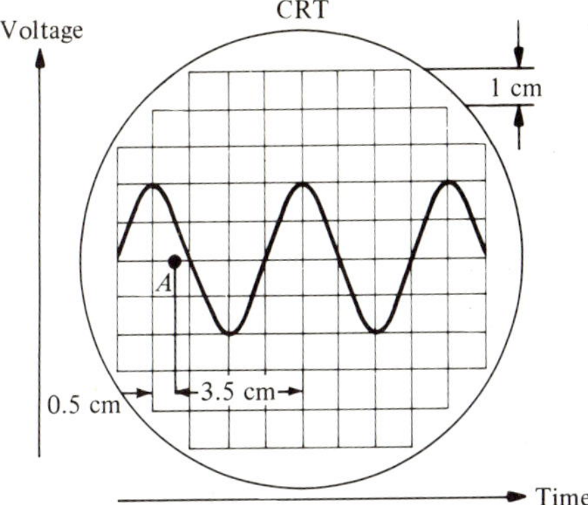

FIGURE 15.13
Typical display.

peak-to-peak voltage of

$$V_{p\text{-}p} = 4 \text{ cm} \times 100 \text{ mV/cm} = 400 \text{ mV}$$

Thus

$$V_m = \frac{1}{2} V_{p\text{-}p} = 200 \text{ mV}$$

The calibration of the time scale is given by the sweep speed on the *horizontal display* dial. If in Fig. 15.13 the sweep speed is 5 ms/cm, then one complete cycle occurs in

$$T = 4 \text{ cm} \times 5 \text{ ms/cm} = 20 \text{ ms} = 0.02 \text{ s}$$

The frequency is then

$$f = \frac{1}{t} = \frac{1}{0.02} = 50 \text{ Hz}$$

and the angular frequency is $2\pi f \approx 314$ rad/s.

In order to determine the phase we must know the location of the origin of the time scale. For present purposes, let us assume that the origin of the time scale is at point A in Fig. 15.13. Then, noting that 2 cm on the time scale is equivalent to 180° of phase shift (one-half of a complete cycle) we have

$$\theta = \frac{3.5 \text{ cm}}{2 \text{ cm}} \times 180° = 315°$$

We now have sufficient information to write the equation for the voltage being displayed. It is

$$v(t) = 200 \cos(2\pi \times 50t - 315°)\ \text{mV} \qquad (t \text{ in seconds})$$

In the example above, the accuracy of the result depends on the accuracy with which we can read the dimensions of the display and on the accuracy of the calibration factors as indicated on the dials. Typically, the overall accuracy of this type of measurement ranges from ±3 to ±10%.

Modern-day oscilloscopes are extremely versatile and can be used to measure many properties of signals other than those described above. The basic principles, however, remain the same, the signal processing taking place in the amplifier circuits. Because of the differences in control designations, etc., among scopes of different manufacturers, additional details will be left for laboratory courses.

• • •

LEARNING EXERCISE FOR SEC. 15.4

1. The waveform of a typical electrocardiogram signal displayed on an oscilloscope is shown in Fig. 15.14. If the vertical sensitivity is set at 100 mV/cm and the sweep speed is set at 2 ms/cm find the peak-to-peak amplitude of the signal and the frequency that represents the heart rate of the person to whom the electrodes are attached.

Ans. 250; 71

• • •

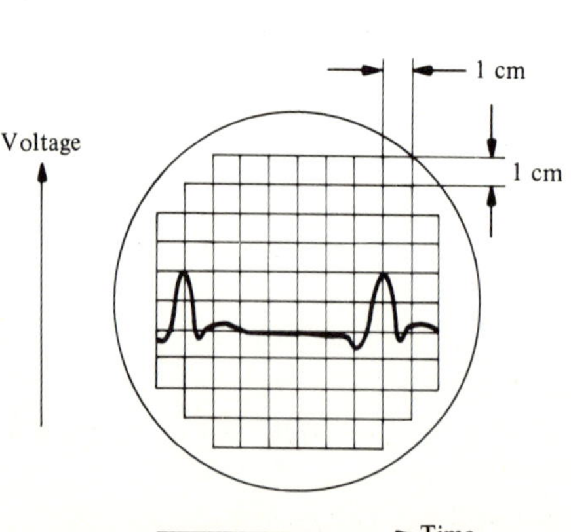

FIGURE 15.14
Oscilloscope display of electrocardiogram signal.

SUMMARY

In this chapter, we have described some of the ways in which quantities of interest in ac circuits are measured. These include voltage, current, impedance, and frequency. The student should now have a better understanding of the instruments he or she may already have encountered in the laboratory. The important points covered in each section are summarized below.

Section

15.1 1. In the usual rectifier-type ac meter, the rectifier converts the ac to pulsating dc and the D'Arsonval movement responds to the average value of the rectified waveform. The scale is usually calibrated in terms of the rms value of a sine wave, so corrections must be applied for other waveforms.

15.2 2. In a digital frequency meter, the number of cycles in a precisely known interval of time, for example, 1 s, are counted and displayed. Such meters are characterized by excellent accuracy and negligible loading.

15.3 3. Impedance bridges for ac require two separate balancing operations to account for the real and imaginary parts of the unknown impedance.

15.4 4. An oscilloscope is used to view the actual time variation of the signal applied to its terminals. Calibrated controls allow for easy amplitude and time measurement.

QUESTIONS FOR REVIEW

Sec. 15.1

1. How is a D'Arsonval meter movement used to measure ac waveform properties?
2. What properties of ac waveforms are used in analog meters?
3. Describe a full-wave rectified sine wave.
4. If a rectifier-type meter is calibrated in terms of the rms value of a sine wave, will it read the rms value of a triangular wave?

Sec. 15.2

5. Describe qualitatively the operation of a typical digital frequency meter.

Sec. 15.3

6. What is a *bridge* circuit?
7. What is a *null detector?*
8. What is the formula for the balance condition of an ac bridge in terms of the four arm impedances?
9. Why are two balancing operations required for an ac bridge?

Sec. 15.4

10. Why is the oscilloscope such a useful instrument?
11. Describe the operation of the cathode-ray tube.

12. Describe how time variations are displayed on the oscilloscope.
13. Why is a *sawtooth* waveform required on the horizontal axis of the oscilloscope?
14. What is the significance of the *vertical sensitivity* reading on an oscilloscope dial?
15. How is the calibration of the time scale on the horizontal display dial used?

PROBLEMS

Sec. 15.1

1. A D'Arsonval meter with a half-wave rectifier, calibrated to read rms for sine waves, is to be used to measure the peak value of a square wave. There is no dc component. Find the factor by which the scale reading must be multiplied to obtain the correct peak value of the square wave.
2. The full-wave rectifier meter described in Sec. 15.1 is connected to a 30-V peak-to-peak square wave having no dc component. What will the meter reading be? The half-wave rectifier meter of Prob. 1 is connected to the same square wave. What will it read?

Sec. 15.2

3. The frequency meter shown in Fig. 15.4 is set for a gating signal of 2×10^4 pulses from the internal oscillator which operates at 100 kHz. If 230 pulses derived from the input waveform are counted during the gating pulse, what is the frequency of the input signal?

Sec. 15.3

4. Derive the balance equation for the inductance bridge shown in Fig. 15.15.
5. The bridge circuit shown in Fig. 15.16 is called a Hay bridge. Show that the balance equations are

$$L_x = \frac{C_1 R_2 R_3}{1 + \omega^2 C_1^2 R_1^2}$$

and

$$R_x = \frac{\omega^2 C_1^2 R_1 R_2 R_3}{1 + \omega^2 C_1^2 R_1^2}$$

Find L_x and R_x for the values indicated.

6. In the Maxwell bridge of Fig. 15.8 $R_1 = 2$ kΩ, $C_1 = 3$ μF, $R_2 = 0.5$ kΩ, and $R_3 = 4$ kΩ. Find L_x and R_x.
7. The circuit shown in Fig. 15.17 is a *capacitance comparison* bridge in which the unknown C_x and its parasitic series resistance R_x are balanced against a precision standard capacitor C_s. Find the equations for C_x and R_x at balance.
8. In the general bridge circuit of Fig. 15.7, $Z_1 = 680\underline{/0°}$ Ω, Z_2 is a 0.1-μF capacitor in series with a 680-Ω resistor, $Z_3 = 5\underline{/0°}$ kΩ, and $f = 1.5$ kHz. Find Z_x as a series combination of two elements. What are the element values?

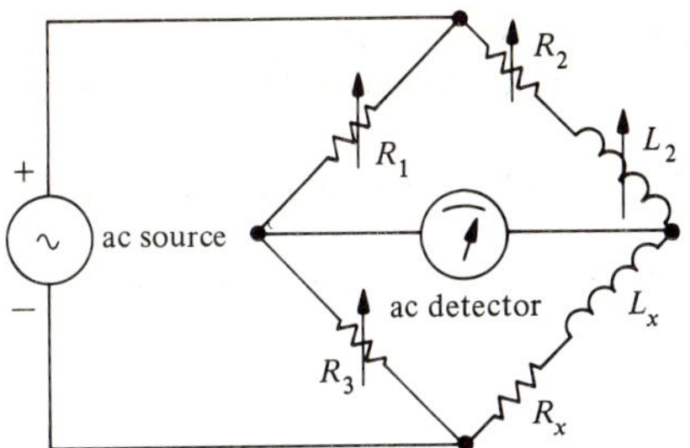

FIGURE 15.15

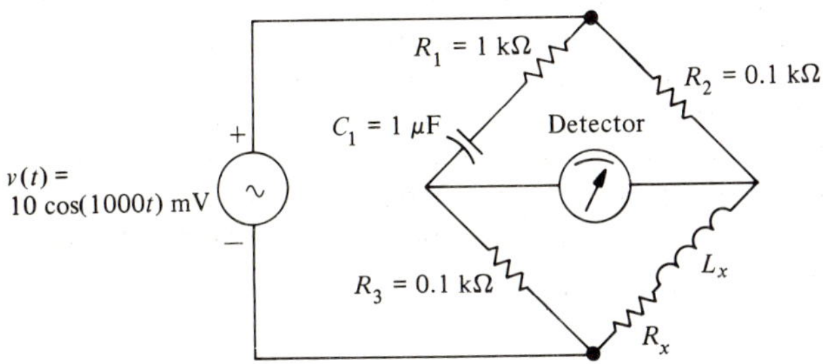

FIGURE 15.16

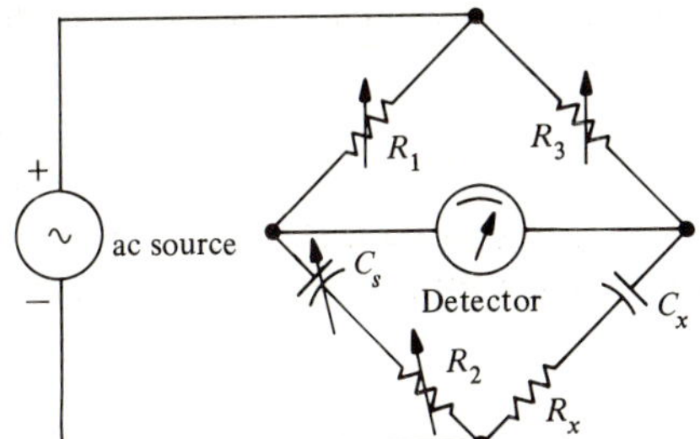

FIGURE 15.17

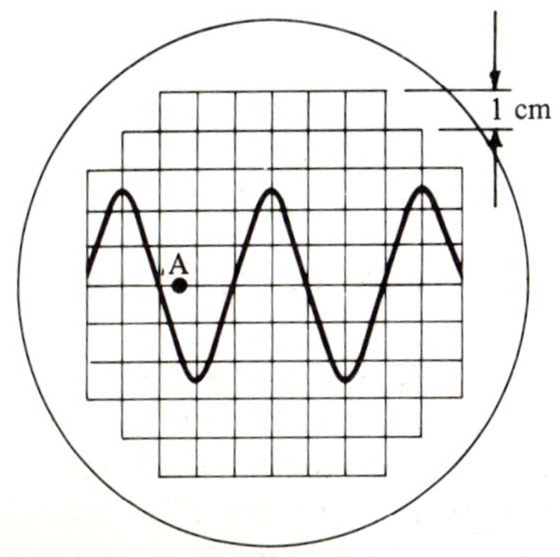

FIGURE 15.18

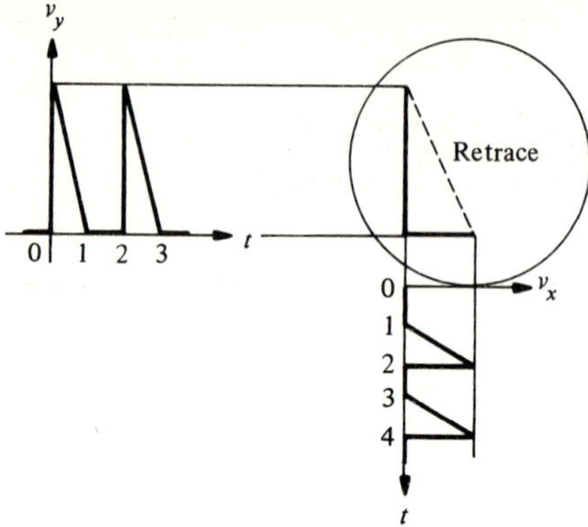

FIGURE 15.19

Sec. 15.4

9. For the display shown in Fig. 15.18, the vertical amplifier control is set at 0.05 V/cm and the sweep speed is 0.02 ms/cm. Find the amplitude and frequency of the sine wave.
10. In Prob. 9 it is known that the $t = 0$ reference occurs at point A on the oscilloscope display. Find the phase angle and write the equation for the sinusoid.
11. In Fig. 15.19 we have shown one of many pairs of waveforms that generate the letter L on the scope face. Find a pair of waveforms that will generate a capital Greek letter ⅂ (gamma). Assume that no discernible indication will appear during any required retracing.
12. Repeat Prob. 11 for a capital V.

Appendixes

APPENDIX A Conversion Factors for the International System of Units (SI)

	Given	Multiply by	To obtain
Length	inches (in)	25.4	millimeters (mm)
	feet (ft)	30.5	centimeters (cm)
	yards (yd)	0.914	meters (m)
	miles (mi)	1.61	kilometers (km)
Area	circular mils (cmil)	0.51×10^{-3}	square millimeters (mm^2)
	square inches (in^2)	6.45	square centimeters (cm^2)
	square feet (ft^2)	0.093	square meters (m^2)
Volume	fluid ounces, U.S. (fl oz)	29.6	cubic centimeters (cm^3) or milliliters (mL)
	gallons, U.S. (gal)	3.78	liters (L)
	cubic inches (in^3)	16.4	cubic centimeters (cm^3)
	cubic feet (ft^3)	0.0283	cubic meters (m^3)
Speed	feet per minute (ft/min)	5.08	millimeters per second (mm/s)
	miles per hour (mi/hr)	0.447	meters per second (m/s)
	kilometers per hour (km/h)	0.278	meters per second (m/s)
Mass	ounces, avdp (oz)	28.4	grams (g)
	pounds, avdp	0.454	kilograms (kg)
	slug	14.6	kilograms (kg)
Density	pounds per cubic foot (lb/ft^3)	16	kilograms per cubic meter (kg/m^3)
Force	ounces-force (oz-f)	0.278	newtons (N)
	pounds-force (lb-f)	4.45	newtons (N)
Work/energy	foot-pounds-force (ft·lb-f)	1.36	joules (J)
	calorie (cal)	4.18	joules (J)
	British Thermal Units (BTU)	1055	joules (J)
Power	horsepower (hp)	746	watts (W)
	foot pounds-force per second (ft·lb-f/s)	1.36	watts (W)
	To obtain	**Divide by**	**Given**

APPENDIX B Wire Table

Gage Number	Diameter (mils)	Area (cir mils)	Resistance (Ω/1000 ft) 25°C (77°F)	Weight (lb/1000 ft)	Allowable Current Capacity, A: Rubber Insulation	Varnished Cambric Insulation	Other Insulations
0000	460.0	211 600.0	0.0500	641.0	225	270	325
000	410.0	167 800.0	0.0630	508.0	175	210	275
00	365.0	133 100.0	0.0795	403.0	150	180	225
0	325.0	105 500.0	0.100	319.0	125	150	200
1	289.0	83 690.0	0.126	253.0	100	120	150
2	258.0	66 370.0	0.159	201.0	90	110	125
3	229.0	52 640.0	0.201	159.0	80	95	100
4	204.0	41 740.0	0.253	126.0	70	85	90
5	182.0	33 100.0	0.319	100.0	55	65	80
6	162.0	26 250.0	0.403	79.5	50	60	70
7	144.0	20 820.0	0.508	63.0			
8	128.0	16 510.0	0.641	50.0	35	40	50
9	114.0	13 090.0	0.808	39.6			
10	102.0	10 380.0	1.02	31.4	25	30	30
11	91.0	8 234.0	1.28	24.9			
12	81.0	6 530.0	1.62	19.8	20	25	25
13	72.0	5 178.0	2.04	15.7			
14	64.0	4 107.0	2.58	12.4	15	18	20
15	57.0	3 257.0	3.25	9.86			
16	51.0	2 583.0	4.09	7.82	6		
17	45.0	2 048.0	5.16	6.20			
18	40.0	1 624.0	6.51	4.92	3		
19	36.0	1 288.0	8.21	3.90			
20	32.0	1 022.0	10.4	3.09			
21	28.5	810.0	13.1	2.45			
22	25.3	642.0	16.5	1.95			
23	22.6	509.0	20.8	1.54			
24	20.1	404.0	26.2	1.22			
25	17.9	320.0	33.0	0.970			
26	15.9	254.0	41.6	0.769			
27	14.2	202.0	52.5	0.610			
28	12.6	160.0	66.2	0.484			
29	11.3	127.0	83.4	0.384			
30	10.0	100.0	105.0	0.304			
31	8.9	79.7	133.0	0.241			
32	8.0	63.2	167.0	0.191			
33	7.1	50.1	211.0	0.152			
34	6.3	39.8	266.0	0.120			
35	5.6	31.5	335.0	0.0954			
36	5.0	25.0	423.0	0.0757			
37	4.5	19.8	533.0	0.0600			
38	4.0	15.7	673.0	0.0476			
39	3.5	12.5	848.0	0.0377			
40	3.1	9.9	1070.0	0.0299			

Standard annealed copper wire solid. American wire gage.

APPENDIX C DETERMINANT SOLUTION OF SIMULTANEOUS EQUATIONS

The method of determinants is an organized "mechanical" method for solving sets of simultaneous equations. In this section we present the method in sufficient detail so that the reader will be able to apply it to sets of two or three equations. Extension to sets of four or more equations requires additional theory, which will be covered in mathematics courses.

Second-order Determinants

Consider the mesh equations

$$3I_1 - 2I_2 = 10$$
$$-2I_1 + 5I_2 = 20$$

The determinant of the coefficients of these equations is an array consisting of two rows and two columns containing the coefficients in the following form:

$$\begin{vmatrix} 3 & -2 \\ -2 & 5 \end{vmatrix}$$

This is called a *second*-order determinant because it has two rows and two columns. Determinants have a numerical value which is determined by a process of *diagonal multiplication*, shown diagrammatically below:

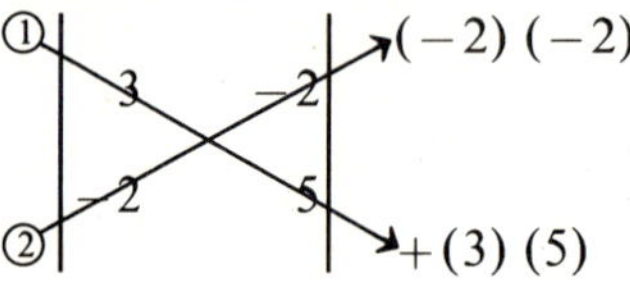

The product of the terms along each diagonal is computed. The value of the determinant is the difference of the two products, the product for diagonal ① minus the product for diagonal ②:

$$+(3)(5) - (-2)(-2) = 15 - 4 = 11$$

The solutions for I_1 and I_2 are expressed as ratios of second-order determinants, the denominator being the determinant of the coefficients. The rule for formulating the numerator determinant is quite simple: It is the same as the denominator except for the column corresponding to the variable for which we are solving (I_1 in the example below). The coefficients in this column are replaced by the column of constants, that is, the right-hand sides of the equations

$$I_1 = \frac{\begin{vmatrix} 10 & -2 \\ 20 & 5 \end{vmatrix}}{\begin{vmatrix} 3 & -2 \\ -2 & 5 \end{vmatrix}} = \frac{(10)(5) - (20)(-2)}{(3)(5) - (-2)(-2)} = \frac{50 + 40}{15 - 4} = \frac{90}{11}\ \text{A}$$

The current I_2 is given by

$$I_2 = \frac{\begin{vmatrix} 3 & 10 \\ -2 & 20 \end{vmatrix}}{\begin{vmatrix} 3 & -2 \\ -2 & 5 \end{vmatrix}} = \frac{(3)(20) - (-2)(10)}{11} = \frac{60 + 20}{11} = \frac{80}{11}\ \text{A}$$

As a further example, we solve the set of equations

$$3x + 2y = 8$$
$$5x + y = 11$$

$$x = \frac{\begin{vmatrix} 8 & 2 \\ 11 & 1 \end{vmatrix}}{\begin{vmatrix} 3 & 2 \\ 5 & 1 \end{vmatrix}} = \frac{(8)(1) - (11)(2)}{(3)(1) - (5)(2)} = \frac{8 - 22}{3 - 10} = \frac{-14}{-7} = 2$$

$$y = \frac{\begin{vmatrix} 3 & 8 \\ 5 & 11 \end{vmatrix}}{-7} = \frac{(3)(11) - (5)(8)}{-7} = \frac{33 - 40}{-7} = 1$$

Third-order Determinants

When we have three simultaneous equations to solve, we arrive at third-order determinants, each of which has three rows and three columns. For example, consider the set of equations

$$x + y + z = 15$$
$$2x - y - z = 10$$
$$3x + 2y - 5x = 14$$

The determinant of the coefficients is

$$\begin{vmatrix} 1 & 1 & 1 \\ 2 & -1 & -1 \\ 3 & 2 & -5 \end{vmatrix}$$

In order to evaluate this determinant using diagonal multiplication we rewrite it with the first two columns repeated to the right as follows:

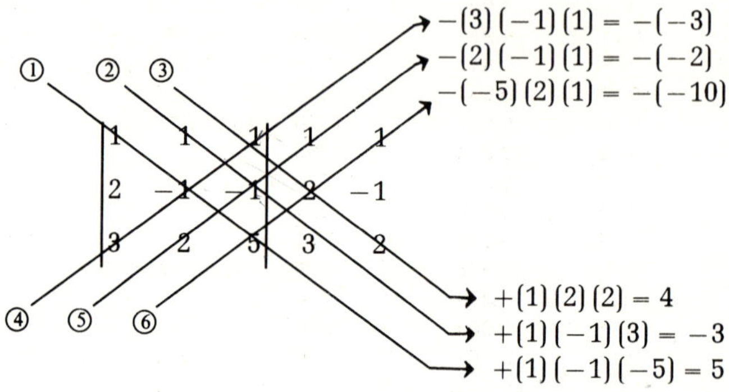

The diagonal multiplication is then performed as shown. For this case diagonals ①, ②, and ③ take a positive sign while diagonals ④, ⑤, and ⑥ take a negative sign. The result is

$$4 - 3 + 5 - (-3) - (-2) - (-10) = 21$$

The next step in the solution is to formulate the numerator determinants according to the rule previously stated. The complete expression for x is then

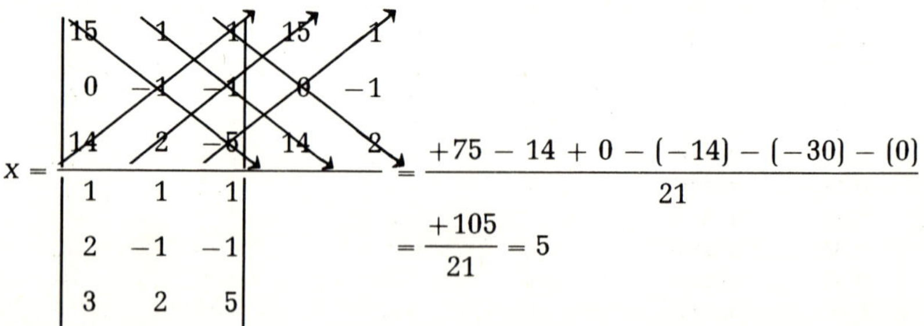

The proof that $y = 7$ and $z = 3$ is left for the reader.

APPENDIX D TABLES

TABLE D-1
The exponential function e^{-x}

X	Exp $(-X)$	X	Exp $(-X)$	X	Exp $(-X)$
0.00	1.000	0.50	0.607	1.00	0.368
0.01	0.990	0.51	0.600	1.01	0.364
0.02	0.980	0.52	0.595	1.02	0.361
0.03	0.970	0.53	0.589	1.03	0.357
0.04	0.961	0.54	0.583	1.04	0.353
0.05	0.951	0.55	0.577	1.05	0.350
0.06	0.942	0.56	0.571	1.06	0.346
0.07	0.932	0.57	0.566	1.07	0.343
0.08	0.923	0.58	0.560	1.08	0.340
0.09	0.914	0.59	0.554	1.09	0.336
0.10	0.905	0.60	0.549	1.10	0.333
0.11	0.896	0.61	0.543	1.11	0.330
0.12	0.887	0.62	0.538	1.12	0.326
0.13	0.878	0.63	0.533	1.13	0.323
0.14	0.869	0.64	0.527	1.14	0.320
0.15	0.861	0.65	0.522	1.15	0.317
0.16	0.852	0.66	0.517	1.16	0.313
0.17	0.844	0.67	0.512	1.17	0.310
0.18	0.835	0.68	0.507	1.18	0.307
0.19	0.827	0.69	0.502	1.19	0.304
0.20	0.819	0.70	0.497	1.20	0.301
0.21	0.811	0.71	0.492	1.21	0.298
0.22	0.803	0.72	0.487	1.22	0.295
0.23	0.795	0.73	0.482	1.23	0.292
0.24	0.787	0.74	0.477	1.24	0.289
0.25	0.779	0.75	0.472	1.25	0.287
0.26	0.771	0.76	0.468	1.26	0.284
0.27	0.763	0.77	0.463	1.27	0.281
0.28	0.756	0.78	0.458	1.28	0.278
0.29	0.748	0.79	0.454	1.29	0.275
0.30	0.741	0.80	0.449	1.30	0.273
0.31	0.733	0.81	0.445	1.31	0.270
0.32	0.726	0.82	0.440	1.32	0.267
0.33	0.719	0.83	0.436	1.33	0.264
0.34	0.712	0.84	0.432	1.34	0.262
0.35	0.705	0.85	0.427	1.35	0.259
0.36	0.698	0.86	0.423	1.36	0.257
0.37	0.691	0.87	0.419	1.37	0.254
0.38	0.684	0.88	0.415	1.38	0.252
0.39	0.677	0.89	0.411	1.39	0.249
0.40	0.670	0.90	0.407	1.40	0.247
0.41	0.664	0.91	0.403	1.41	0.244
0.42	0.657	0.92	0.399	1.42	0.242
0.43	0.651	0.93	0.395	1.43	0.239
0.44	0.644	0.94	0.391	1.44	0.237
0.45	0.638	0.95	0.387	1.45	0.235
0.46	0.631	0.96	0.383	1.46	0.232
0.47	0.625	0.97	0.379	1.47	0.230
0.48	0.619	0.98	0.375	1.48	0.228
0.49	0.613	0.99	0.372	1.49	0.225

TABLE D-1
continued

X	Exp (−*X*)	*X*	Exp (−*X*)	*X*	Exp (−*X*)
1.50	0.223	2.00	0.135	2.5	0.082
1.51	0.221	2.01	0.134	2.6	0.074
1.52	0.219	2.02	0.133	2.7	0.067
1.53	0.217	2.03	0.131	2.8	0.061
1.54	0.214	2.04	0.130	2.9	0.055
1.55	0.212	2.05	0.129	3.0	0.050
1.56	0.210	2.06	0.127	3.1	0.045
1.57	0.208	2.07	0.126	3.2	0.041
1.58	0.206	2.08	0.125	3.3	0.037
1.59	0.204	2.09	0.124	3.4	0.033
1.60	0.202	2.10	0.122	3.5	0.030
1.61	0.200	2.11	0.121	3.6	0.027
1.62	0.198	2.12	0.120	3.7	0.025
1.63	0.196	2.13	0.119	3.8	0.022
1.64	0.194	2.14	0.118	3.9	0.020
1.65	0.192	2.15	0.116	4.0	0.018
1.66	0.190	2.16	0.115	4.1	0.017
1.67	0.188	2.17	0.114	4.2	0.015
1.68	0.186	2.18	0.113	4.3	0.014
1.69	0.185	2.19	0.112	4.4	0.012
1.70	0.183	2.20	0.111	4.5	0.011
1.71	0.181	2.21	0.110	4.6	0.010
1.72	0.179	2.22	0.109	4.7	0.009
1.73	0.177	2.23	0.108	4.8	0.008
1.74	0.176	2.24	0.106	4.9	0.007
1.75	0.174	2.25	0.105	5.0	0.007
1.76	0.172	2.26	0.104		
1.77	0.170	2.27	0.103		
1.78	0.169	2.28	0.102		
1.79	0.167	2.29	0.101		
1.80	0.165	2.30	0.100		
1.81	0.164	2.31	0.099		
1.82	0.162	2.32	0.098		
1.83	0.160	2.33	0.097		
1.84	0.159	2.34	0.096		
1.85	0.157	2.35	0.095		
1.86	0.156	2.36	0.094		
1.87	0.154	2.37	0.093		
1.88	0.153	2.38	0.093		
1.89	0.151	2.39	0.092		
1.90	0.150	2.40	0.091		
1.91	0.148	2.41	0.090		
1.92	0.147	2.42	0.089		
1.93	0.145	2.43	0.088		
1.94	0.144	2.44	0.087		
1.95	0.142	2.45	0.086		
1.96	0.141	2.46	0.085		
1.97	0.139	2.47	0.085		
1.98	0.138	2.48	0.084		
1.99	0.137	2.49	0.083		

TABLE D-2

Natural trigonometric functions for decimal fractions of a degree. From H. D. Larsen, "Rinehart Mathematical tables, Formulas, and Curves," Holt, Rinehart and Winston, 1953.

Deg.	Sin	Tan	Ctn	Cos	Deg.
0–	.00000	.00000	—	1.00000	**90–**
.1	.00175	.00175	572.96	1.00000	.9
.2	.00349	.00349	286.48	0.99999	.8
.3	.00524	.00524	190.98	.99999	.7
.4	.00698	.00698	143.24	.99998	.6
.5	.00873	.00873	114.59	.99996	.5
.6	.01047	.01047	95.489	.99995	.4
.7	.01222	.01222	81.847	.99993	.3
.8	.01396	.01396	71.615	.99990	.2
.9	.01571	.01571	63.657	.99988	.1
1–	.01745	.01746	57.290	.99985	**89–**
.1	.01920	.01920	52.081	.99982	.9
.2	.02094	.02095	47.740	.99978	.8
.3	.02269	.02269	44.066	.99974	.7
.4	.02443	.02444	40.917	.99970	.6
.5	.02618	.02619	38.188	.99966	.5
.6	.02792	.02793	35.801	.99961	.4
.7	.02967	.02968	33.694	.99956	.3
.8	.03141	.03143	31.821	.99951	.2
.9	.03316	.03317	30.145	.99945	.1
2–	.03490	.03492	28.636	.99939	**88–**
.1	.03664	.03667	27.271	.99933	.9
.2	.03839	.03842	26.031	.99926	.8
.3	.04013	.04016	24.898	.99919	.7
.4	.04188	.04191	23.859	.99912	.6
.5	.04362	.04366	22.904	.99905	.5
.6	.04536	.04541	22.022	.99897	.4
.7	.04711	.04716	21.205	.99889	.3
.8	.04885	.04891	20.446	.99881	.2
.9	.05059	.05066	19.740	.99872	.1
3–	.05234	.05241	19.081	.99863	**87–**
.1	.05408	.05416	18.464	.99854	.9
.2	.05582	.05591	17.886	.99844	.8
.3	.05756	.05766	17.343	.99834	.7
.4	.05931	.05941	16.832	.99824	.6
.5	.06105	.06116	16.350	.99813	.5
.6	.06279	.06291	15.895	.99803	.4
.7	.06453	.06467	15.464	.99792	.3
.8	.06627	.06642	15.056	.99780	.2
.9	.06802	.06817	14.669	.99768	.1
4–	.06976	.06993	14.301	.99756	**86–**
.1	.07150	.07168	13.951	.99744	.9
.2	.07324	.07344	13.617	.99731	.8
.3	.07498	.07519	13.300	.99719	.7
.4	.07672	.07695	12.996	.99705	.6
.5	.07846	.07870	12.706	.99692	.5
.6	.08020	.08046	12.429	.99678	.4
.7	.08194	.08221	12.163	.99664	.3
.8	.08368	.08397	11.909	.99649	.2
.9	.08542	.08573	11.664	.99635	.1
5–	.08716	.08749	11.430	.99619	**85–**
Deg.	Cos	Ctn	Tan	Sin	Deg.

Deg.	Sin	Tan	Ctn	Cos	Deg.
5–	.08716	.08749	11.430	.99619	**85–**
.1	.08889	.08925	11.205	.99604	.9
.2	.09063	.09101	10.988	.99588	.8
.3	.09237	.09277	10.780	.99572	.7
.4	.09411	.09453	10.579	.99556	.6
.5	.09585	.09629	10.385	.99540	.5
.6	.09758	.09805	10.199	.99523	.4
.7	.09932	.09981	10.019	.99506	.3
.8	.10106	.10158	9.8448	.99488	.2
.9	.10279	.10334	9.6768	.99470	.1
6–	.10453	.10510	9.5144	.99452	**84–**
.1	.10626	.10687	9.3572	.99434	.9
.2	.10800	.10863	9.2052	.99415	.8
.3	.10973	.11040	9.0579	.99396	.7
.4	.11147	.11217	8.9152	.99377	.6
.5	.11320	.11394	8.7769	.99357	.5
.6	.11494	.11570	8.6427	.99337	.4
.7	.11667	.11747	8.5126	.99317	.3
.8	.11840	.11924	8.3863	.99297	.2
.9	.12014	.12101	8.2636	.99276	.1
7–	.12187	.12278	8.1443	.99255	**83–**
.1	.12360	.12456	8.0285	.99233	.9
.2	.12533	.12633	7.9158	.99211	.8
.3	.12706	.12810	7.8062	.99189	.7
.4	.12880	.12988	7.6996	.99167	.6
.5	.13053	.13165	7.5958	.99144	.5
.6	.13226	.13343	7.4947	.99122	.4
.7	.13399	.13521	7.3962	.99098	.3
.8	.13572	.13698	7.3002	.99075	.2
.9	.13744	.13876	7.2066	.99051	.1
8–	.13917	.14054	7.1154	.99027	**82–**
.1	.14090	.14232	7.0264	.99002	.9
.2	.14263	.14410	6.9395	.98978	.8
.3	.14436	.14588	6.8548	.98953	.7
.4	.14608	.14767	6.7720	.98927	.6
.5	.14781	.14945	6.6912	.98902	.5
.6	.14954	.15124	6.6122	.98876	.4
.7	.15126	.15302	6.5350	.98849	.3
.8	.15299	.15481	6.4596	.98823	.2
.9	.15471	.15660	6.3859	.98796	.1
9–	.15643	.15838	6.3138	.98769	**81–**
.1	.15816	.16017	6.2432	.98741	.9
.2	.15988	.16196	6.1742	.98714	.8
.3	.16160	.16376	6.1066	.98686	.7
.4	.16333	.16555	6.0405	.98657	.6
.5	.16505	.16734	5.9758	.98629	.5
.6	.16677	.16914	5.9124	.98600	.4
.7	.16849	.17093	5.8502	.98570	.3
.8	.17021	.17273	5.7894	.98541	.2
.9	.17193	.17453	5.7297	.98511	.1
10–	.17365	.17633	5.6713	.98481	**80–**
Deg.	Cos	Ctn	Tan	Sin	Deg.

TABLE D-2
(continued)

Deg.	Sin	Tan	Ctn	Cos	Deg.
10–	.17365	.17633	5.6713	.98481	**80–**
.1	.17537	.17813	5.6140	.98450	.9
.2	.17708	.17993	5.5578	.98420	.8
.3	.17880	.18173	5.5026	.98389	.7
.4	.18052	.18353	5.4486	.98357	.6
.5	.18224	.18534	5.3955	.98325	.5
.6	.18395	.18714	5.3435	.98294	.4
.7	.18567	.18895	5.2924	.98261	.3
.8	.18738	.19076	5.2422	.98229	.2
.9	.18910	.19257	5.1929	.98196	.1
11–	.19081	.19438	5.1446	.98163	**79–**
.1	.19252	.19619	5.0970	.98129	.9
.2	.19423	.19801	5.0504	.98096	.8
.3	.19595	.19982	5.0045	.98061	.7
.4	.19766	.20164	4.9594	.98027	.6
.5	.19937	.20345	4.9152	.97992	.5
.6	.20108	.20527	4.8716	.97958	.4
.7	.20279	.20709	4.8288	.97922	.3
.8	.20450	.20891	4.7867	.97887	.2
.9	.20620	.21073	4.7453	.97851	.1
12–	.20791	.21256	4.7046	.97815	**78–**
.1	.20962	.21438	4.6646	.97778	.9
.2	.21132	.21621	4.6252	.97742	.8
.3	.21303	.21804	4.5864	.97705	.7
.4	.21474	.21986	4.5483	.97667	.6
.5	.21644	.22169	4.5107	.97630	.5
.6	.21814	.22353	4.4737	.97592	.4
.7	.21985	.22536	4.4373	.97553	.3
.8	.22155	.22719	4.4015	.97515	.2
.9	.22325	.22903	4.3662	.97476	.1
13–	.22495	.23087	4.3315	.97437	**77–**
.1	.22665	.23271	4.2972	.97398	.9
.2	.22835	.23455	4.2635	.97358	.8
.3	.23005	.23639	4.2303	.97318	.7
.4	.23175	.23823	4.1976	.97278	.6
.5	.23345	.24008	4.1653	.97237	.5
.6	.23514	.24193	4.1335	.97196	.4
.7	.23684	.24377	4.1022	.97155	.3
.8	.23853	.24562	4.0713	.97113	.2
.9	.24023	.24747	4.0408	.97072	.1
14–	.24192	.24933	4.0108	.97030	**76–**
.1	.24362	.25118	3.9812	.96987	.9
.2	.24531	.25304	3.9520	.96945	.8
.3	.24700	.25490	3.9232	.96902	.7
.4	.24869	.25676	3.8947	.96858	.6
.5	.25038	.25862	3.8667	.96815	.5
.6	.25207	.26048	3.8391	.96771	.4
.7	.25376	.26235	3.8118	.96727	.3
.8	.25545	.26421	3.7848	.96682	.2
.9	.25713	.26608	3.7583	.96638	.1
15–	.25882	.26795	3.7321	.96593	**75–**
Deg.	Cos	Ctn	Tan	Sin	Deg.

Deg.	Sin	Tan	Ctn	Cos	Deg.
15–	.25882	.26795	3.7321	.96593	**75–**
.1	.26050	.26982	3.7062	.96547	.9
.2	.26219	.27169	3.6806	.96502	.8
.3	.26387	.27357	3.6554	.96456	.7
.4	.26556	.27545	3.6305	.96410	.6
.5	.26724	.27732	3.6059	.96363	.5
.6	.26892	.27921	3.5816	.96316	.4
.7	.27060	.28109	3.5576	.96269	.3
.8	.27228	.28297	3.5339	.96222	.2
.9	.27396	.28486	3.5105	.96174	.1
16–	.27564	.28675	3.4874	.96126	**74–**
.1	.27731	.28864	3.4646	.96078	.9
.2	.27899	.29053	3.4420	.96029	.8
.3	.28067	.29242	3.4197	.95981	.7
.4	.28234	.29432	3.3977	.95931	.6
.5	.28402	.29621	3.3759	.95882	.5
.6	.28569	.29811	3.3544	.95832	.4
.7	.28736	.30001	3.3332	.95782	.3
.8	.28903	.30192	3.3122	.95732	.2
.9	.29070	.30382	3.2914	.95681	.1
17–	.29237	.30573	3.2709	.95630	**73–**
.1	.29404	.30764	3.2506	.95579	.9
.2	.29571	.30955	3.2305	.95528	.8
.3	.29737	.31147	3.2106	.95476	.7
.4	.29904	.31338	3.1910	.95424	.6
.5	.30071	.31530	3.1716	.95372	.5
.6	.30237	.31722	3.1524	.95319	.4
.7	.30403	.31914	3.1334	.95266	.3
.8	.30570	.32106	3.1146	.95213	.2
.9	.30736	.32299	3.0961	.95159	.1
18–	.30902	.32492	3.0777	.95106	**72–**
.1	.31068	.32685	3.0595	.95052	.9
.2	.31233	.32878	3.0415	.94997	.8
.3	.31399	.33072	3.0237	.94943	.7
.4	.31565	.33266	3.0061	.94888	.6
.5	.31730	.33460	2.9887	.94832	.5
.6	.31896	.33654	2.9714	.94777	.4
.7	.32061	.33848	2.9544	.94721	.3
.8	.32227	.34043	2.9375	.94665	.2
.9	.32392	.34238	2.9208	.94609	.1
19–	.32557	.34433	2.9042	.94552	**71–**
.1	.32722	.34628	2.8878	.94495	.9
.2	.32887	.34824	2.8716	.94438	.8
.3	.33051	.35020	2.8556	.94380	.7
.4	.33216	.35216	2.8397	.94322	.6
.5	.33381	.35412	2.8239	.94264	.5
.6	.33545	.35608	2.8083	.94206	.4
.7	.33710	.35805	2.7929	.94147	.3
.8	.33874	.36002	2.7776	.94088	.2
.9	.34038	.36199	2.7625	.94029	.1
20–	.34202	.36397	2.7475	.93969	**70–**
Deg.	Cos	Ctn	Tan	Sin	Deg.

TABLE D-2
(continued)

Deg.	Sin	Tan	Ctn	Cos	Deg.
20–	.34202	.36397	2.7475	.93969	**70–**
.1	.34366	.36595	2.7326	.93909	.9
.2	.34530	.36793	2.7179	.93849	.8
.3	.34694	.36991	2.7034	.93789	.7
.4	.34857	.37190	2.6889	.93728	.6
.5	.35021	.37388	2.6746	.93667	.5
.6	.35184	.37588	2.6605	.93606	.4
.7	.35347	.37787	2.6464	.93544	.3
.8	.35511	.37986	2.6325	.93483	.2
.9	.35674	.38186	2.6187	.93420	.1
21–	.35837	.38386	2.6051	.93358	**69–**
.1	.36000	.38587	2.5916	.93295	.9
.2	.36162	.38787	2.5782	.93232	.8
.3	.36325	.38988	2.5649	.93169	.7
.4	.36488	.39190	2.5517	.93106	.6
.5	.36650	.39391	2.5386	.93042	.5
.6	.36812	.39593	2.5257	.92978	.4
.7	.36975	.39795	2.5129	.92913	.3
.8	.37137	.39997	2.5002	.92849	.2
.9	.37299	.40200	2.4876	.92784	.1
22–	.37461	.40403	2.4751	.92718	**68–**
.1	.37622	.40606	2.4627	.92653	.9
.2	.37784	.40809	2.4504	.92587	.8
.3	.37946	.41013	2.4383	.92521	.7
.4	.38107	.41217	2.4262	.92455	.6
.5	.38268	.41421	2.4142	.92388	.5
.6	.38430	.41626	2.4023	.92321	.4
.7	.38591	.41831	2.3906	.92254	.3
.8	.38752	.42036	2.3789	.92186	.2
.9	.38912	.42242	2.3673	.92119	.1
23–	.39073	.42447	2.3559	.92050	**67–**
.1	.39234	.42654	2.3445	.91982	.9
.2	.39394	.42860	2.3332	.91914	.8
.3	.39555	.43067	2.3220	.91845	.7
.4	.39715	.43274	2.3109	.91775	.6
.5	.39875	.43481	2.2998	.91706	.5
.6	.40035	.43689	2.2889	.91636	.4
.7	.40195	.43897	2.2781	.91566	.3
.8	.40355	.44105	2.2673	.91496	.2
.9	.40514	.44314	2.2566	.91425	.1
24–	.40674	.44523	2.2460	.91355	**66–**
.1	.40833	.44732	2.2355	.91283	.9
.2	.40992	.44942	2.2251	.91212	.8
.3	.41151	.45152	2.2148	.91140	.7
.4	.41310	.45362	2.2045	.91068	.6
.5	.41469	.45573	2.1943	.90996	.5
.6	.41628	.45784	2.1842	.90924	.4
.7	.41787	.45995	2.1742	.90851	.3
.8	.41945	.46206	2.1642	.90778	.2
.9	.42104	.46418	2.1543	.90704	.1
25–	.42262	.46631	2.1445	.90631	**65–**
Deg.	Cos	Ctn	Tan	Sin	Deg.

Deg.	Sin	Tan	Ctn	Cos	Deg.
25–	.42262	.46631	2.1445	.90631	**65–**
.1	.42420	.46843	2.1348	.90557	.9
.2	.42578	.47056	2.1251	.90483	.8
.3	.42736	.47270	2.1155	.90408	.7
.4	.42894	.47483	2.1060	.90334	.6
.5	.43051	.47698	2.0965	.90259	.5
.6	.43209	.47912	2.0872	.90183	.4
.7	.43366	.48127	2.0778	.90108	.3
.8	.43523	.48342	2.0686	.90032	.2
.9	.43680	.48557	2.0594	.89956	.1
26–	.43837	.48773	2.0503	.89879	**64–**
.1	.43994	.48989	2.0413	.89803	.9
.2	.44151	.49206	2.0323	.89726	.8
.3	.44307	.49423	2.0233	.89649	.7
.4	.44464	.49640	2.0145	.89571	.6
.5	.44620	.49858	2.0057	.89493	.5
.6	.44776	.50076	1.9970	.89415	.4
.7	.44932	.50295	1.9883	.89337	.3
.8	.45088	.50514	1.9797	.89259	.2
.9	.45243	.50733	1.9711	.89180	.1
27–	.45399	.50953	1.9626	.89101	**63–**
.1	.45554	.51173	1.9542	.89021	.9
.2	.45710	.51393	1.9458	.88942	.8
.3	.45865	.51614	1.9375	.88862	.7
.4	.46020	.51835	1.9292	.88782	.6
.5	.46175	.52057	1.9210	.88701	.5
.6	.46330	.52279	1.9128	.88620	.4
.7	.46484	.52501	1.9047	.88539	.3
.8	.46639	.52724	1.8967	.88458	.2
.9	.46793	.52947	1.8887	.88377	.1
28–	.46947	.53171	1.8807	.88295	**62–**
.1	.47101	.53395	1.8728	.88213	.9
.2	.47255	.53620	1.8650	.88130	.8
.3	.47409	.53844	1.8572	.88048	.7
.4	.47562	.54070	1.8495	.87965	.6
.5	.47716	.54296	1.8418	.87882	.5
.6	.47869	.54522	1.8341	.87798	.4
.7	.48022	.54748	1.8265	.87715	.3
.8	.48175	.54975	1.8190	.87631	.2
.9	.48328	.55203	1.8115	.87546	.1
29–	.48481	.55431	1.8040	.87462	**61–**
.1	.48634	.55659	1.7966	.87377	9
.2	.48786	.55888	1.7893	.87292	.8
.3	.48938	.56117	1.7820	.87207	.7
.4	.49090	.56347	1.7747	.87121	.6
.5	.49242	.56577	1.7675	.87036	.5
.6	.49394	.56808	1.7603	.86949	.4
.7	.49546	.57039	1.7532	.86863	.3
.8	.49697	.57271	1.7461	.86777	.2
.9	.49849	.57503	1.7391	.86690	.1
30–	.50000	.57735	1.7321	.86603	**60–**
Deg.	Cos	Ctn	Tan	Sin	Deg.

TABLE D-2
(continued)

Deg.	Sin	Tan	Ctn	Cos	Deg.
30–	.50000	.57735	1.7321	.86603	**60–**
.1	.50151	.57968	1.7251	.86515	.9
.2	.50302	.58201	1.7182	.86427	.8
.3	.50453	.58435	1.7113	.86340	.7
.4	.50603	.58670	1.7045	.86251	.6
.5	.50754	.58905	1.6977	.86163	.5
.6	.50904	.59140	1.6909	.86074	.4
.7	.51054	.59376	1.6842	.85985	.3
.8	.51204	.59612	1.6775	.85896	.2
.9	.51354	.59849	1.6709	.85806	.1
31–	.51504	.60086	1.6643	.85717	**59–**
.1	.51653	.60324	1.6577	.85627	.9
.2	.51803	.60562	1.6512	.85536	.8
.3	.51952	.60801	1.6447	.85446	.7
.4	.52101	.61040	1.6383	.85355	.6
.5	.52250	.61280	1.6319	.85264	.5
.6	.52399	.61520	1.6255	.85173	.4
.7	.52547	.61761	1.6191	.85081	.3
.8	.52696	.62003	1.6128	.84989	.2
.9	.52844	.62245	1.6066	.84897	.1
32–	.52992	.62487	1.6003	.84805	**58–**
.1	.53140	.62730	1.5941	.84712	.9
.2	.53288	.62973	1.5880	.84619	.8
.3	.53435	.63217	1.5818	.84526	.7
.4	.53583	.63462	1.5757	.84433	.6
.5	.53730	.63707	1.5697	.84339	.5
.6	.53877	.63953	1.5637	.84245	.4
.7	.54024	.64199	1.5577	.84151	.3
.8	.54171	.64446	1.5517	.84057	.2
.9	.54317	.64693	1.5458	.83962	.1
33–	.54464	.64941	1.5399	.83867	**57–**
.1	.54610	.65189	1.5340	.83772	.9
.2	.54756	.65438	1.5282	.83676	.8
.3	.54902	.65688	1.5224	.83581	.7
.4	.55048	.65938	1.5166	.83485	.6
.5	.55194	.66189	1.5108	.83389	.5
.6	.55339	.66440	1.5051	.83292	.4
.7	.55484	.66692	1.4994	.83195	.3
.8	.55630	.66944	1.4938	.83098	.2
.9	.55775	.67197	1.4882	.83001	.1
34–	.55919	.67451	1.4826	.82904	**56–**
.1	.56064	.67705	1.4770	.82806	.9
.2	.56208	.67960	1.4715	.82708	.8
.3	.56353	.68215	1.4659	.82610	.7
.4	.56497	.68471	1.4605	.82511	.6
.5	.56641	.68728	1.4550	.82413	.5
.6	.56784	.68985	1.4496	.82314	.4
.7	.56928	.69243	1.4442	.82214	.3
.8	.57071	.69502	1.4388	.82115	.2
.9	.57215	.69761	1.4335	.82015	.1
35–	.57358	.70021	1.4281	.81915	**55–**
Deg.	Cos	Ctn	Tan	Sin	Deg.

Deg.	Sin	Tan	Ctn	Cos	Deg.
35–	.57358	.70021	1.4281	.81915	**55–**
.1	.57501	.70281	1.4229	.81815	.9
.2	.57643	.70542	1.4176	.81714	.8
.3	.57786	.70804	1.4124	.81614	.7
.4	.57928	.71066	1.4071	.81513	.6
.5	.58070	.71329	1.4019	.81412	.5
.6	.58212	.71593	1.3968	.81310	.4
.7	.58354	.71857	1.3916	.81208	.3
.8	.58496	.72122	1.3865	.81106	.2
.9	.58637	.72388	1.3814	.81004	.1
36–	.58779	.72654	1.3764	.80902	**54–**
.1	.58920	.72921	1.3713	.80799	.9
.2	.59061	.73189	1.3663	.80696	.8
.3	.59201	.73457	1.3613	.80593	.7
.4	.59342	.73726	1.3564	.80489	.6
.5	.59482	.73996	1.3514	.80386	.5
.6	.59622	.74267	1.3465	.80282	.4
.7	.59763	.74538	1.3416	.80178	.3
.8	.59902	.74810	1.3367	.80073	.2
.9	.60042	.75082	1.3319	.79968	.1
37–	.60182	.75355	1.3270	.79864	**53–**
.1	.60321	.75629	1.3222	.79758	.9
.2	.60460	.75904	1.3175	.79653	.8
.3	.60599	.76180	1.3127	.79547	.7
.4	.60738	.76456	1.3079	.79441	.6
.5	.60876	.76733	1.3032	.79335	.5
.6	.61015	.77010	1.2985	.79229	.4
.7	.61153	.77289	1.2938	.79122	.3
.8	.61291	.77568	1.2892	.79016	.2
.9	.61429	.77848	1.2846	.78908	.1
38–	.61566	.78129	1.2799	.78801	**52–**
.1	.61704	.78410	1.2753	.78694	.9
.2	.61841	.78692	1.2708	.78586	.8
.3	.61978	.78975	1.2662	.78478	.7
.4	.62115	.79259	1.2617	.78369	.6
.5	.62251	.79544	1.2572	.78261	.5
.6	.62388	.79829	1.2527	.78152	.4
.7	.62524	.80115	1.2482	.78043	.3
.8	.62660	.80402	1.2437	.77934	.2
.9	.62796	.80690	1.2393	.77824	.1
39–	.62932	.80978	1.2349	.77715	**51–**
.1	.63068	.81268	1.2305	.77605	.9
.2	.63203	.81558	1.2261	.77494	.8
.3	.63338	.81849	1.2218	.77384	.7
.4	.63473	.82141	1.2174	.77273	.6
.5	.63608	.82434	1.2131	.77162	.5
.6	.63742	.82727	1.2088	.77051	.4
.7	.63877	.83022	1.2045	.76940	.3
.8	.64011	.83317	1.2002	.76828	.2
.9	.64145	.83613	1.1960	.76717	.1
40–	.64279	.83910	1.1918	.76604	**50–**
Deg.	Cos	Ctn	Tan	Sin	Deg.

TABLE D-2
(continued)

Deg.	Cos	Ctn	Tan	Sin	Deg.
40–	.64279	.83910	1.1918	.76604	**50–**
.1	.64412	.84208	1.1875	.76492	.9
.2	.64546	.84507	1.1833	.76380	.8
.3	.64679	.84806	1.1792	.76267	.7
.4	.64812	.85107	1.1750	.76154	.6
.5	.64945	.85408	1.1708	.76041	.5
.6	.65077	.85710	1.1667	.75927	.4
.7	.65210	.86014	1.1626	.75813	.3
.8	.65342	.86318	1.1585	.75700	.2
.9	.65474	.86623	1.1544	.75585	.1
41–	.65606	.86929	1.1504	.75471	**49–**
.1	.65738	.87236	1.1463	.75356	.9
.2	.65869	.87543	1.1423	.75241	.8
.3	.66000	.87852	1.1383	.75126	.7
.4	.66131	.88162	1.1343	.75011	.6
.5	.66262	.88473	1.1305	.74896	.5
.6	66393	.88784	1.1263	.74780	.4
.7	.66523	.89097	1.1224	.74664	.3
.8	.66653	.89410	1.1184	.74548	.2
.9	.66783	.89725	1.1145	.74431	.1
42–	.66913	.90040	1.1106	.74314	**48–**
.1	.67043	.90357	1.1067	.74198	.9
.2	.67172	.90674	1.1028	.74080	.8
.3	.67301	.90993	1.0990	.73963	.7
.4	.67430	.91313	1.0951	.73846	.6
.5	.67559	.91633	1.0913	.73728	.5
Deg.	Cos	Ctn	Tan	Sin	Deg.

Deg.	Cos	Ctn	Tan	Sin	Deg.
42–	.66913	.90040	1.1106	.74314	**48–**
.6	.67688	.91955	1.0875	.73610	.4
.7	.67816	.92277	1.0837	.73491	.3
.8	.67944	.92601	1.0799	.73373	.2
.9	.68072	.92926	1.0761	.73254	.1
43–	.68200	.93252	1.0724	.73135	**47–**
.1	.68327	.93578	1.0686	.73016	.9
.2	.68455	.93906	1.0649	.72897	.8
.3	.68582	.94235	1.0612	.72777	.7
.4	.68709	.94565	1.0575	.72657	.6
.5	.68835	.94896	1.0538	.72537	.5
.6	.68962	.95229	1.0501	.72417	.4
.7	.69088	.95562	1.0464	.72297	.3
.8	.69214	.95897	1.0428	.72176	.2
.9	.69340	.96232	1.0392	.72055	.1
44–	.69466	.96569	1.0355	.71934	**46–**
.1	.69591	.96907	1.0319	.71813	.9
.2	.69717	.97246	1.0283	.71691	.8
.3	.69842	.97586	1.0247	.71569	.7
.4	.69966	.97927	1.0212	.71447	.6
.5	.70091	.98270	1.0176	.71325	.5
.6	.70215	.98613	1.0141	.71203	.4
.7	.70339	.98958	1.0105	.71080	.3
.8	.70463	.99304	1.0070	.70957	.2
.9	.70587	.99652	1.0035	.70834	.1
45–	.70711	1.00000	1.0000	.70711	**45–**
Deg.	Cos	Ctn	Tan	Sin	Deg.

Answers to Selected Odd-Numbered Problems

CHAPTER 1

SEC. 1.1

1. **a.** 1.0×10^6 **b.** 1.0×10^{-5} **c.** 1.0×10^4 **d.** 1.0×10^{-7} **e.** 1.0×10^7 **f.** 1.0×10^{-2}
3. **a.** 10^5 **b.** 10^{-6} **c.** 10^1 **d.** 10^3 **e.** 10^{-2} **f.** 10^6 **g.** 10^{-8} **h.** 10^{-6}
5. **a.** 1.1×10^2 **b.** 1.1×10^{-1} **c.** 9×10^3 **d.** 9.9×10^{-5} **e.** 1.001×10^6 **f.** -9×10^{-2}
7. Prob. 5: **a.** 110 A **b.** 0.11 A **c.** 9kA **d.** 0.99 μA **e.** 1.001 MA **f.** −90 mA

SEC. 1.2

9. 0.779 **b.** 1.05

SEC. 1.4

11. **a.** 3×10^6 μs **b.** 10 000 μF **c.** 102 s **d.** 11 520 s **e.** 6300 ns **f.** 6.3 μs
13. **a.** 48.3 km/h **b.** 6.1 m **c.** 1.61 km **d.** 32.5 N **e.** 15.2 cm **f.** 16.3 J **g.** 263 kg
15. **a.** yes **b.** yes **c.** no **d.** no

SEC. 1.5

17. **a.** 3.8×10^5 **b.** 4.5×10^{-3} **c.** 1.24×10^3 **d.** 1.25×10^3 **e.** 8.0×10^{-3} **f.** 0.2

SEC. 1.6

19. **a.** 144 mi
23. $y = \frac{1}{2}x + 2.5$
25. $V = -0.26I + 0.7$

CHAPTER 2

SEC. 2.2

3. 2.3×10^{-8} N repulsion, 0.575×10^{-8} N
5. -2×10^{-9} C

SEC. 2.3

7. 2000 s **9.** 2.5 A **11.** 24 C
13. 53 A left to right

SEC. 2.5

17. 6 J **19.** 3.6 V
21. 40 mJ delivered to
23. 0.333 V **25.** 101 J

SEC. 2.6

27. 4.4 A **29.** 6 h

SEC. 2.7

31. 1.89×10^9 J
35. 36 mW delivered to
37. ⅓ A plus to minus
41. 10 V

CHAPTER 3

SEC. 3.1

1. **a.** 6.67×10^{-3} S, 6.67 mS, 6.67×10^{3} μS
b. 45.5×10^{-6} S, 45.5×10^{-3} mS, 45.5 μS
c. 0.303×10^{-6} S, 0.303×10^{-3} mS, 0.303 μS

SEC. 3.2

3. 375 Ω **5.** 0.2 m **7.** 3.27 Ω
9. 0.558 m **11.** 8×10^{-4} cm², 1.24×10^{-4} in²
13. 55.6 m

SEC. 3.3

15. 1 cmil = 0.785×10^{-6} in²
17. **a.** 64 cmil **b.** #32 **c.** 200 Ω
19. 21.2 Ω **21.** #3 **23.** #14

SEC. 3.4

25. 250 Ω **27.** L = 20 mm

SEC. 3.5

29. 0.0134 Ω **31.** 27.1 Ω
33. −236.4°C
35. **a.** ΔT = 16.7°C **b.** ΔT = 166°C

SEC. 3.7

37. brown, green, blue, silver
39. 1 kΩ: brown, black, red, silver
2.2 kΩ: red, red, red, silver
4.7 kΩ: yellow, violet, red, silver

SEC. 3.8

41. $N = 10$ $9 < N < 11$
.
.
$N = 22$ $19.8 < N < 24.2$
.
.
$N = 47$ $42.3 < N < 51.7$
.
.
$N = 68$ $61.2 < N < 74.8$

43. 39 kΩ, max departure = 4.9 kΩ

CHAPTER 4

SEC. 4.1

1. 720 V **3.** 0.16 A **5.** 60 V
7. 360 Ω **9.** 0.02 Ω **11.** 83 V

SEC 4.2

13. 10.1 kΩ **15.** 1.3 mA, 2.86 V, 6.11 V
17. 0.596 mA, 3.2 V, 5.37 kΩ

SEC. 4.3

19. −9 V **21.** $V_2 = 45$ V, $V_4 = 25$ V
23. $V_{ab} = 70$ V, $V_4 = 40$ V
25. $V_{eq} = 13$ V, $R_{eq} = 32$ kΩ

SEC. 4.4

27. **a.** $V_{ac} = 80$ V, $V_{ab} = 16.8$ V, $V_{bc} = 63$ V
b. same as (a)

SEC. 4.5

29. **a.** 3.97 V, 18 V **b.** 11.6 V, 25.6 V, 52.8 V **31.** **a.** 0.788 **b.** 0.429
33. $R_1 = 33$ kΩ, $R_2 = 19.8$ kΩ is one possible solution
35. $A_v = x$

SEC. 4.6

37. **a.** $G_T = 1.01$ S, $R_T = 0.99$ Ω
b. $G_T = 0.517$ mS, $R_T = 1.94$ kΩ
c. $G_T = 0.1$ mS, $R_T = 10$ kΩ
d. $G_T = 21.2$ μS, $R_T = 47$ kΩ
39. 31 kΩ **41.** 3.43 MΩ
43. $V_{6.8} = V_{22} = 80$ V, $I_{6.8} = 11.8$ mA, $I_{22} = 3.64$ mA, $I_T = 15.4$ mA

SEC. 4.7

45. **a.** 20 mA **b.** 27 mA
47. $I_T = 28.6$ mA, $V_1 = 28.6$ V, $V_2 = 26.6$ V, $V_3 = 18.6$ V, $V_4 = 15.6$ V, $V_5 = 8.6$ V
49. **a.** $I_{eq} = 17$ A, $R_{eq} = 4.44$ Ω
b. $I_{eq} = -1$ mA, $R_{eq} = 2.22$ kΩ

SEC. 4.8

51. **a.** $A_{i1} = 0.409$, $A_{i2} = 0.591$
b. $A_{i1} = 0.571$, $A_{i2} = 0.429$
55. 55 kΩ

57. **a.** $V_1 = 45$ V, $I_2 = 1.36$ mA,
$I_i = 4.36$ mA, $R_T = 10.3$ kΩ
b. $V_1 = 81.6$ V, $I_2 = 12$ mA, $I_3 = 5.44$ mA,
$I_i = 29.44$ mA, $R_T = 2.7$ kΩ

SEC. 4.9

59. 1.1 A (use standard 1.5 A)
61. 20 A

SEC. 4.10

63. **a.** 0.0934 **b.** 3.27 V
65. 2.88 kΩ
67. Bleeder: use $R_2 = 220$ Ω
R_1 use 1.5 kΩ
8.8 V $< V_L <$ 10.8 V
69. **a.** $I_6 = 9.1$ A, $I_3 = 10.9$ A,
$I_4 = 5.45$ A, $V_3 = 21.8$ V, $V_6 = 54.1$ V

SEC. 4.11

71. 0.59 W **73.** 225 μW
75. 60.5 W
77. $P_1 = 267$ W, $P_2 = 534$ W
79. **a.** 7.07 mA **b.** 6.52 mA
c. 0.389 A **d.** 70.7 mA

SEC. 4.12

81. \$2.88 **83.** \$0.09
85. 81.4% **87.** 77.7%
89. $\eta_G = 85.7\%$, $\eta_{TL} = 91.7\%$,
$\eta_{OV} = 78.6\%$

CHAPTER 5

SEC. 5.1

1. 3.56 mA **3.** 1.33 mA
5. −5.88 mA

SEC. 5.2

7. $I_1 = 1.36$ mA, $I_2 = 0.033$ mA
9. $V_3 = 33$ V, $V_5 = -3$ V

SEC. 5.3

13. $V_1 = -2.12$ V, $V_2 = -12.5$ V
15. $V_1 = 18$ V, $V_2 = 26.2$ V,
$V_x = -8.2$ V
19. Fig. 5.55: 3 loop, 4 node
Fig. 5.51: 4 loop, 3 node
Fig. 5.49: 2 loop, 1 node

SEC. 5.4

21. 33.3% **23.** 0.088 Ω
25. **a.** (80 V, 160 A), (98 V, 100 A),
(113 V, 60 A)
b. 10 V, 115 V **c.** 20 A **d.** 34.7%

SEC. 5.6

27. $V_T = 16.5$ V, $R_T = 3.3$ kΩ
29. $V_T = 22$ V, $R_T = 2.2$ kΩ
31. $V_T = 5$ V, $R_T = 2$ Ω; $I_N = 2.5$ A,
$R_N = 2$ Ω
33. $V_0 = -1.42$ V
35. $V_T = -1$ V, $R_T = 1$ kΩ

SEC 5.7

39. 9.3 Ω
41. 5.78 Ω, 0.02 Ω; 94.3%; 5.7%

CHAPTER 6

SEC. 6.2

1. 2.22% **3.** 50 mΩ, 0.505 Ω
5. $R_1 = 0.5$ Ω, $R_2 = 1$ Ω, $R_3 = 2$ Ω,
$R_4 = 5$ Ω
7. 2 mΩ

SEC. 6.3

9. $R_{se} = 5.98$ kΩ,
200 Ω/V **11.** 2.99 kΩ
13. 28 V to 30 V
15. 50 μA, 50 kΩ, 200 kΩ, 1 MΩ,
5 MΩ
17. **a.** 133 V **b.** 129 V **c.** 80 V

SEC. 6.4

21. **a.** $R_s + R_z = 29$ kΩ
$R_{fs} = 0$ Ω, $R_{3/4} = 10$ kΩ, $R_{1/2} = 30$ kΩ,
$R_{1/4} = 90$ kΩ
23. **a.** $R_s = 28$ kΩ
b. $R_{fs} = \infty$, $R_{3/4} = 2.9$ kΩ, $R_{1/2} = 484$ Ω,
$R_{1/4} = 322$ Ω
25. no **29.** 100 V

SEC. 6.5

31. 163.2 kΩ
33. 0 to 1 kΩ $R_3/R_1 = -0.133$
0 to 10 kΩ $R_3/R_1 = 1.33$
0 to 100 kΩ $R_3/R_1 = 13.3$
35. $l_1 = 56.1$ cm, $l_3 = 43.9$ cm

CHAPTER 7

SEC. 7.2

1. 0.533 T **3.** 25.2 μWb
5. 13.6 mm^2 **7.** 1.13 T

SEC. 7.3

13. 4.17 A **15.** 7.67 A
17. 1.1 mWb **19.** 230 MAt/Wb
21. +0.342 mWb/Atm
23. 3.14×10^{-3} Wb/Atm
25. 175 kAt/Wb **27.** 1.02 mWb
29. **a.** 0.55 T **b.** 1.43 T **c.** 1.58 T
31. 1 m
33. **a.** 48.6 kAT/Wb **b.** 1.54 mWb **c.** 105.6 At/m **d.** 0.331 T

SEC. 7.4

35. **a.** 0.22 mWb **b.** 0.209 μWb **c.** 64 μWb
37. **a.** 11.9 At **b.** 238 **c.** 0.952 MAt/Wb

SEC. 7.5

39. 82.2 At; μ_{ss} = 2.6 mWb/Atm; μ_{CI} = 0.24 mWb/Atm
41. 8.46 A **43.** 0.45 mWb
45. 0.238 A

SEC. 7.6

47. 0.75 A **49.** 4330 turns

SEC. 7.8

51. 1.56 turns **55.** 2.25 V
57. 49 mV

SEC. 7.9

59. 0.187 T **61.** 9 N
63. 35 N in negative Z direction

CHAPTER 8

SEC. 8.1

1. 12.5 H **3.** 26 **5.** 170 μH
7. 24

SEC. 8.2

15. 2, 3, 5, 7, 8, 10, 0.968, 1.2, 1.43, 1.88, 3.88, 4.43, 6.2 mH

Sec. 8.3

17. **a.** M = 1.21 H **b.** v_1 = 85 V, v_2 = 295 V
19. M = 0.0632 H **b.** v_1 = 10 V, v_2 = 0.5 V
21. k = 0.327, L_{eq} = 340 mH
23. **a.** 0.34 mH **b.** 1.72 mA **c.** 3.5 mH
25. L_{eq} = 5.4 mH

SEC. 8.4

27. −0.24 nN attraction
29. 108 GV/m **31.** 45 V
33. 0.22 μC **35.** −4 mA
37. 0.075 C

SEC. 8.5

39. **a.** 3.54 pF **b.** 17.7 pF **c.** 2.66 nF
41. 133 nF **43.** ϵ_r = 2.7
45. **a.** 3 kV **b.** 200 kV **c.** 3 kV

SEC. 8.6

47. 420 V **49.** $v(t) = -10 + 3000t$

SEC 8.7

53. **a.** 3.9 μF **b.** 31.5 pF
55. 0.1, 0.22, 0.5, 0.32, 0.6, 0.72, 0.82, 0.0604, 0.0688, 0.0833, 0.153, 0.567, 0.303, 0.253 μF **57.** 33.3 kV

SEC. 8.8

59. 0.144 J
61. W_{10} = 8 mJ, W_5 = 16 mJ
63. 11.3 kJ **65.** 12.2 kW
67. $W_{0.05}$ = 10 μJ, $W_{0.01}$ = 50 μJ
69. 56.3 mJ

CHAPTER 9

SEC. 9.1

1. $RC\dfrac{\Delta v_c}{\Delta t} + v_c = RI_0\mu(t)$

3. $v_c(0^+) = v_c(0^-) = 0$ $i_c(\infty) = 0$
$i_R(0^+) = 0$ $i_R(\infty) = 2$ mA
$i_c(0^+) = 2$ mA $v_c(\infty) = 9.4$ V

SEC. 9.3

7. 3.58 ms
9. $i_1 = 0.147e^{-t/\tau}$mA, $v_R = -7.33e^{-t/\tau}$ V, τ = 4.95 ms

SEC. 9.4

11. Ampl = 30 V, τ = 20 μs
13. $v(t) = 21.7e^{-t/5\text{ ms}}$ V, A = 21.7 V
15. $v(0^+) = 0$, $v(\infty) = 10$ V, τ = 3 ms
17. decay

SEC. 9.5

19. $i_L(t) = 6.67(1 - e^{-t/\tau})$ mA,
τ = 0.152 μs
21. $i_L(t) = 10(1 - e^{-t/\tau})$ mA,
τ = 0.0638 μs
23. $i_L(t) = 6.67e^{-t/\tau}$ mA, τ = 0.152 μs

SEC. 9.6

25. $v_c(t) = 4.5 - 2.5e^{-t/\tau}$ V, τ = 0.33 ms
27. $i_L(t) = 6 - 4e^{-t/\tau}$ mA, τ = 2.9 ms
29. 0.686 s
31. $v_c(t) = 8.77 + 1.3e^{-t/\tau}$ mA,
τ = 82.5 μs

SEC. 9.8

33. $v_c(t) = 22e^{-t/\tau}$ V, $t < 5$ ms
$v_c(t) = 22 - 14e^{-(t-5)/\tau}$ V, $t > 5$ ms;
τ = 4.95 ms
35. $i_L(t) = 5e^{-t/\tau_1}$ mA, $t < 0.3$ μs,
τ_1 = 0.6 μs
$i_L(t) = 5 - 1.97e^{-(t-0.3\mu s)/\tau_2}$ mA,
0.3 μs $< t <$ 0.6 μs;
τ_2 = 0.3 μs
$i_L(t) = 4.28e^{-(t-0.6\mu s)/\tau_1}$ mA, $t > 0.6$ μs

SEC. 9.9

37. $v_0(t) = 10 - 5e^{-t/\tau}$ V,
$i(t) = 50e^{-t/\tau}$ μA, $t < 0.75$ ms;
τ = 0.5 ms
$v_0(t) = 8.88e^{-(t-0.75\text{ms})/\tau}$,
$i(t) = -88.8e^{-(t-0.75\text{ms})/\tau}$ μA;
$t > 0.75$ ms
39. τ = 9.66 μs

CHAPTER 10

SEC. 10.1

1. Ampl = 50 mV, ω = 6280 rad/s,
f = 1000 Hz, T = 1 ms, ϕ = −32°
5. **a.** −19.7 mA **b.** 19.7 mA
7. **a.** 50 cos(370t + 55°)
b. 50 cos(2π × 10^6t − 235°)
9. 3 V

SEC. 10.2

11. **a.** $4 + j2 = 4.47\angle 27°$
b. $5 + j0 = 5\angle 0°$
c. $5\angle 45° = 3.54 + j3.54$
d. $0 - j2 = 2\angle -90°$
13. $-3.2 + j12$
15. Im z = 2.31, |z| = 4.62

SEC. 10.3

17. **a.** −10. **b.** $17.3 - j10$
c. $-70.7 - j70.7$
d. $-0.0129 + j0.0606$
19. **a.** $5\angle 27°$ **b.** $14.1\angle -135°$
c. $53.9\angle 68.2°$ **d.** $16\angle 90°$
21. **a.** $10.8\angle 21.8°$ **b.** $11\angle -123°$
c. $7.7 + j9.2$ **d.** $11.3 + j11.3$
23. **a.** $7.07\angle -8.1°$ **b.** $117\angle 133°$
c. $10.3\angle -61°$ **d.** $105\angle 56.8°$
25. **a.** $13.5\angle 31.3°$ **b.** $81\angle 80°$
c. $0.125\angle 235°$ **d.** $2.85\angle 37.9°$
27. $3.83\angle 45.8°$
29. **a.** $4.47\angle 153°$ **b.** $1.73 + j1$

SEC. 10.4

31. **a.** $3.54\angle 30°$ mA
b. $121\angle 80°$ mV **c.** $8.84\angle -70°$ μA
33. **a.** $v(t) = 14.1\cos(2\pi \times 10^6 t + 30°)$ V
b. $i(t) = 3.68\cos(2\pi \times 50t + 110°)$ mA
c. $v(t) = 141\cos(600t + 30°)$ mV
35. $i(t) = 32.2\cos(\omega t + 44.4°)$ mA

SEC. 10.5

37. $I = 0.136\angle -21°$ mA,
$i(t) = 0.192\cos(\omega t - 21°)$ mA
39. $I_1 = 6.48\angle 30°$ mA,
I_2 = 9.52
$\angle 30°$ mA, $V = 64.8\angle 30°$ V
41. $I = 5.94\angle 10°$ mA, $V_1 = 19.6\angle 10°$ V,
$V_2 = 40.4\angle 10°$ V
43. $V_T = 26.4\angle 20°$ V,
$V_{R1} = V_{R2} = V_{R3} = 13.2\angle 20°$ V,
$I_{R2} = 1.94\angle 20°$ mA
$I_{R3} = 4.06\angle 20°$ mA, $I_{R1} = I = 6\angle 20°$ mA,

SEC. 10.6

45. **a.** 481 V **b.** 43.8 mA **c.** 121 V
47. 8.5 A **49.** 13.6 A, 16.2 Ω
51. 3.38 A **53.** 5.6 mW

CHAPTER 11

SEC. 11.1

1. $0.025 \cos(2000\pi t + 90°)\ \mu A$

3. $0.0177\angle 90°\ \mu A$

5. **a.** $Z = 4.8\angle -90°\ \Omega$ **b.** $X = 1.17\ k\Omega$ **c.** $X = 7.94\ k\Omega$ **d.** $X = 45.2\ m\Omega$

7. 58.5 nF **9.** 122 Hz

11. $56\angle 45°$ mA

13. **a.** $Z = 1.2\angle 90°\ k\Omega$ **b.** $X = 1.24\ k\Omega$ **c.** $X = 2.1\ k\Omega$ **d.** $X = 22\ k\Omega$

15. $2.65\ \mu H$ **17.** 9.55 Hz

19. $Z = 1800\angle 90°\ \Omega$, $I = 0.024\angle -68°\ \mu A$

SEC. 11.2

21. **a.** $Z = 300 - j199 = 360\angle -33.6°\ \Omega$ **c.** $Z = 10 + j15.7 = 18.6\angle 57.5°\ k\Omega$

23. **a.** $Y = 2.78\angle 33.6° = 2.32 + j1.54$ mS, 431-Ω resistor in parallel with 6.13-pF capacitor
c. $Y = 53.8\angle -57.5° = 28.9 - j45.4\ \mu S$, 34.6 kΩ resistor in parallel with 0.711-H inductor

25. $Z = 15 - j10.8 = 18.5\angle -35.7°\ k\Omega$

27. $Y = 70.7\angle 19.3° = 66.7 + j23.3\ \mu S$

29. $R = 6.43\ \Omega$, $L = 0.244$ mH

31. $f = 318$ Hz

33. $R = 25\ \Omega$ in series with $X_L = 43.3\ \Omega$

35. $Z_{T1} = 16\angle 12.4°\ k\Omega$

37. $R = 45\ \Omega$, $L = 0.57$ H

39. $R = 72.2\ k\Omega$, $X_c = 41.7\ k\Omega$

SEC. 11.3

41. $V = 2.36\angle -66.8°$ V, $I_R = 3.93\angle -66.8°$ mA, $I_C = 9.18\angle 23.2°$ mA

43. **a.** $7.44\angle -42.7°$ V **b.** $29\angle -22°$ V, **c.** $26.8\angle 0°$ V

45. $10\angle 65° = 4.2 + j9.1\ \Omega$

47. $R_B = 58.7\ \Omega$, $X_B = 197\ \Omega$

49. $149\angle +90°$ mA

51. $Y = 6.84 + j18.8$ mS

53. $V = 1\angle 12.4°$ mV, $I_R = 10\angle 12.4°\ \mu A$
$I_L = 3.7\angle -77.6°\ \mu A$,
$I_C = 5\angle 102.4°\ \mu A$

Sec. 11.4

55. $3.04\angle -14.6°$ mA

57. $V_C = 9.09\angle -52.4°$ V

59. $V_C = 9.09\angle -52.4°$ V

61. $0.646\angle 85.7°$ mA

63. $V_T = 8.75\angle 68.2°$ V, $Z_T = 7.62\angle 12.1°\ k\Omega$, $I_N = 1.15\angle 56.1°$ mA

65. $R = 22.8\ k\Omega$, $C = 5.03$ nF

SEC. 11.5

67.

```
200        AC
300   B1   N(0, 1), R = 6.8E03,
           E = 10/0
400   B2   N(0, 1), R = 0.001,
           E = 30/20
500   B3   N(1, 2), R = 10E03
600   B4   N(2, 0), C = 1E-08
800        FREQUENCY = 5000
900        PRINT, NV, CA
```

CHAPTER 12

1. $i(t) = 20 \cos(377t - \pi/6)$mA, $p(t) = 0.26 + 0.3 \cos(754t - \pi/6)$W

3. **a.** $S = 3.41$ VA **b.** 26.5 W **c.** -21.4 var **d.** 0.78

5. $2.88\angle -53°\ \Omega$

7. **a.** $P = 0$ **b.** -8.3 var; $7.7\ \mu H$

9. $9.7\angle -41.4°$ A

SEC. 12.2

11. $P_C = 3.39\angle 38.7°$ kVA, $S = 3.39$ kVA, $P = 2.65$ kW, $Q = -2.12$ kvar, $\cos\phi = 0.78$

13. $I = 48\angle -31.8°$ A, $Z = 4.3\angle 31.8°\ \Omega$, $P = 8.5$ kW, $P_C = 10\angle 31.8°$ kVA

15. **a.** $P_{C1} = 1.21 + j0$ kVA, $P_{C2} = 1 - j0.329$ kVA, $P_{C3} = 0.6 + j1.2$ kVA
b. $P_{CT} = 2.81 + j0.871$ kVA, $= 2.94\angle 17.2°$ kVA
$P_T = 2.81$ kW, $S_T = 2.94$ kVA
$Q_T = 0.871$ kvar
c. $I_T = 26.6\angle -17.2°$ A
d. $\cos\phi = 0.96$

17. **a.** $v_g = 495\angle -1.3°$ V
b. $I = 27.5\angle -14.3°$ A

19. $17.7\ \mu F$

21. **a.** $123\ \mu F$ **b.** $69.1\ \mu F$

23. $-j2\ \Omega$ in series

SEC. 12.3

27. 208 V

29. $I_a = 6.52\angle -18.4°$ A, $I_b = 6.52\angle 101.6°$ A, $I_c = 6.52\angle -138.4°$ A, $I_n = 0$

33. $I_a = 6.25\angle -36.9°$ A, $I_b = 7.3\angle 88.2°$ A, $I_c = 8.3\angle -120°$ A, $I_n = 10.1\angle -21.2°$ A

37. $I_{ab} = 27.5\angle 45°$ A, $I_{bc} = 27.5\angle -75°$ A, $I_{ca} = 27.5\angle -165°$ A, $I_a = 47.6\angle 15°$ A, $I_b = 47.6\angle -105°$ A, $I_c = 47.6\angle 135°$ A

SEC. 12.4

41. $Z_a = 6.33\angle -108.4°\ \Omega$, $Z_b = 10\angle 53.2°\ \Omega$, $Z_c = 10\angle -36.8°\ \Omega$,

43. $I_{8\Omega} = 15\angle 0°$ A, $I_{3\Omega} = 23.1\angle -30°$ A

SEC. 12.5

45. 2.23 kW **47.** 932 W

49. 41.2 kW **51.** 43.4 A

53. $P_1 = -1.45$ kW, $P_2 = 3.95$ kW

55. 1.0

SEC. 12.6

57. 18 **59.** 180 V

61. $N_2 = 317$ turns, $N_{60} = 190$ turns, $N_{30} = 95$ turns

63. step-down, 180 V

65. 120 mA **67.** 96 Ω

69. $a = 0.775$, $V_2 = 11$ V, $V_1 = 14.2$ V, $I_2 = 0.183$ A, $I_1 = 0.142$ A

71. 40 V, 80 V, 160 V, 200 V

CHAPTER 13

SEC. 13.1

1. 880 Hz, 1760 Hz, 3080 Hz

3. **a.** 33.3 kHz **b.** 10 Hz **c.** 556 MHz

5. $v(t) = 3 + 4\cos 2\pi \times 150t + \cos 3\pi \times 150t + 2\cos 5\pi \times 150t + 0.5\cos 6\pi \times 150t$

Sec. 13.2

7. **a.**

$$\frac{V_2}{V_1} = \frac{\dfrac{j\omega}{3 \times 10^3}}{1 + \dfrac{j\omega}{3 \times 10^3}}$$

b.

$$\frac{V_2}{V_1} = \frac{1}{1 + \dfrac{j\omega}{400 \times 10^3}}$$

9. **a.**

$$\frac{I_2}{I_1} = \frac{0.4\left(\dfrac{j\omega}{18.2 \times 10^3}\right)}{1 + \dfrac{j\omega}{18.2 \times 10^3}}$$

b.

$$\frac{I_2}{I_1} = \frac{0.588}{1 + \dfrac{j\omega}{400}}$$

SEC. 13.3

11. **a.**

$$|Av| = \frac{\dfrac{\omega}{3 \times 10^3}}{\sqrt{1 + \left(\dfrac{\omega}{3 \times 10^3}\right)^2}}$$

$$\phi = 90° - \arctan\left(\frac{\omega}{3 \times 10^3}\right)$$

b. $Av(0) = 0$, $Av(\infty) = 1$

c. $\phi(0) = 90°$, $\phi)\infty) = 0$

d. 300 rad/s, 45°

13. **a.**

$$|Av| = \frac{0.5\left(\dfrac{\omega}{250}\right)}{\sqrt{1 + \left(\dfrac{\omega}{250}\right)^2}},$$

$$\phi = 90° - \arctan\frac{\omega}{250}$$

b. $Av(0) = 0$, $Av(\infty) = 0.5$

c. $\phi(0) = 90°$, $\phi(\infty) = 0$

d. 250 rad/s, 45°

15. **a.** $\left|\dfrac{I_2}{V_1}\right| = \dfrac{50 \times 10^6}{\omega}$, $\phi(\omega) = -90°$

b. $\left|\dfrac{I_2}{V_1}\right|_0 = \infty$, $\left|\dfrac{I_2}{V_1}\right|_\infty = 0$

c. $\phi(0) = \phi(\infty) = -90°$

d. no half-power frequency

17.

0.5 kHz: $A_v = 0.59\angle -6.6°$

1.6 kHz: $A_v = 0.559\angle -20.2°$

4 kHz: $A_v = 0.439\angle -42.5°$

10 kHz: $A_v = 0.238\angle -66.4°$

23. $v_0(t) = 1 + 0.127\sin(2\pi t - 81°) + 0.0048\sin(3\pi t - 87°) + 0.00103\sin(5\pi t - 88°)$

SEC. 13.4

27. $Q = 20$, $B = 795$ Hz, $f_0 = 15.9$ kHz
29. $C = 26.6$ pF, $L = 8.32$ μH, $R = 200$ kΩ
31. **a.** $\dfrac{V_0}{I_i} = \dfrac{R_s + j\omega L}{1 - \omega^2 LC + j\omega R_s C}$
b. $\omega_0 = \dfrac{1}{\sqrt{LC}}$
35. 2.27 pF, 7 MHz

CHAPTER 14

SEC. 14.1

1. $A_v = 0.2$ **3.** $A_i = -0.5$

SEC. 14.2

5. −15.7, −8.9, 1.1, 4.3, 21.1, 44.3 dB
7. **a.** −80, −40, 0 dB
b. 21.6, 48.4, 73.3 dB
c. −79.9, −43.3, 150 dB
9. **a.** $G_p = 95.7$ dB **b.** $G_{p_{in}} = -64$ dBm, $G_{p_{out}} = 31.7$ dBm
c. $G_v = 83.5$ dB
11. 0.334 dB
13. 5 GW
15. 0.63 V

SEC. 14.3

17. $v_2 = 0.338 \cos 20t$ V
19. 26 mA, 14.3 dB
21. $f_c = 159$ Hz

SEC. 14.4

23. 21 V, $A_v = 2100$

SEC. 14.5

25. $A_{vT} = 152$
27. $A_v = 53 \times 10^3$
29. 0.101 mA
31. 108 Hz

SEC. 14.6

33. $h_f = -0.694$, $h_i = 3.77$ kΩ, $h_r = 0.694$, $h_o = 0.139$ mS

SEC. 14.7

37. $A_v = 20.4 \times 10^3$
39. $A_{vT} = 31.3 \times 10^3$
41. $A_{vT} = 15.6 \times 10^3$
43. $A_{iT} = -1.9 \times 10^6$

SEC. 14.8

45. $G_v = 71$ dB, $G_p = 94.3$ dB
51. $R_1 = 2.3$ kΩ, $R_{in} = 20.8$ GΩ, $R_{out} = 0.6$ mΩ
53. 450 μΩ

CHAPTER 15

SEC. 15.1

1. 0.9

SEC. 15.2

3. 1150 Hz

SEC. 15.3

7. $R_x = \dfrac{R_3 R_2}{R_1}$ $C_x = C_s \dfrac{R_1}{R_3}$

SEC. 15.4

9. 0.125 V, 12.5 kHz

Index